Thermochemistry and Its Applications to Chemical and Biochemical Systems

The Thermochemistry of Molecules, Ionic Species and
Free Radicals in Relation to the Understanding of
Chemical and Biochemical Systems

NATO ASI Series
Advanced Science Institutes Series

A series presenting the results of activities sponsored by the NATO Science Committee, which aims at the dissemination of advanced scientific and technological knowledge, with a view to strengthening links between scientific communities.

The series is published by an international board of publishers in conjunction with the NATO Scientific Affairs Division

A	Life Sciences	Plenum Publishing Corporation
B	Physics	London and New York
C	Mathematical and Physical Sciences	D. Reidel Publishing Company Dordrecht, Boston and Lancaster
D	Behavioural and Social Sciences	Martinus Nijhoff Publishers
E	Engineering and Materials Sciences	The Hague, Boston and Lancaster
F	Computer and Systems Sciences	Springer-Verlag
G	Ecological Sciences	Berlin, Heidelberg, New York and Tokyo

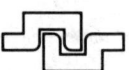

Series C: Mathematical and Physical Sciences Vol. 119

Thermochemistry and Its Applications to Chemical and Biochemical Systems

The Thermochemistry of Molecules, Ionic Species and
Free Radicals in Relation to the Understanding of
Chemical and Biochemical Systems

edited by

Manuel A.V. Ribeiro da Silva

Department of Chemistry, Faculty of Sciences, University of Oporto, Portugal

D. Reidel Publishing Company

Dordrecht / Boston / Lancaster

Published in cooperation with NATO Scientific Affairs Division

Proceedings of the NATO Advanced Study Institute on
Thermochemistry Today and Its Role in The Immediate Future
Viano do Castelo, Portugal
July 5-15, 1982

Library of Congress Cataloging in Publication Data

NATO Advanced Study Institute on Thermochemistry Today and Its Role in the
    Immediate Future (1982 : Viano do Castelo, Portugal)
    Thermochemistry and its applications to chemical and biochemical systems.

    (NATO ASI series. Series C, Mathematical and physical sciences ; v. 119)
    "Published in cooperation with NATO Scientific Affairs Division."
    "Proceedings of the NATO Advanced Study Institute on Thermochemistry
Today and Its Role in the Immediate Future, Viano do Castelo, Portugal, July 5–15,
1982."—Verso t.p.
    Includes index.
    1. Thermochemistry—Congresses. 2. Calorimeters and calorimetry—
Congresses. I. Silva, Manuel A. V. Ribeiro da, 1940- . II. Title.
III. Series.
QD510.N38   1982   541.3'6   83-24632
ISBN 90-277-1698-6

Published by D. Reidel Publishing Company
P.O. Box 17, 3300 AA Dordrecht, Holland

Sold and distributed in the U.S.A. and Canada
by Kluwer Academic Publishers,
190 Old Derby Street, Hingham, MA 02043, U.S.A.

In all other countries, sold and distributed
by Kluwer Academic Publishers Group,
P.O. Box 322, 3300 AH Dordrecht, Holland

D. Reidel Publishing Company is a member of the Kluwer Academic Publishers Group

All Rights Reserved
© 1984 by D. Reidel Publishing Company, Dordrecht, Holland.
No part of the material protected by this copyright notice may be reproduced or utilized
in any form or by any means, electronic or mechanical, including photocopying, recording
or by any information storage and retrieval system, without written permission from the
copyright owner.

Printed in The Netherlands.

TABLE OF CONTENTS

PREFACE ix

TECHNIQUES AND THEORY OF TITRATION CALORIMETRY 1
James J. Christensen

APPLICATIONS OF TITRATION CALORIMETRY
James J. Christensen 17

MICROCALORIMETRY
Ingemar Wadsö 31

DIFFERENTIAL SCANNING CALORIMETRY
C. T. Mortimer 47

AN OXYGEN FLOW CALORIMETER FOR DETERMINING THE HEATING VALUE
OF KILOGRAM-SIZE SAMPLES OF MUNICIPAL SOLID WASTE
S. Abramowitz, E. S. Domalski, K. L. Churney, A. E. Ledford,
R. V. Ryan and M. L. Reilly 61

THERMOGENESIS BY HEAT CONDUCTION CALORIMETRY
Henri Tachoire, J. L. Macqueron and V. Torra 77

THE MEASUREMENT OF VAPOUR PRESSURE
A. S. Carson 127

THE USE OF A SIMULTANEOUS TORSION AND MASS-LOSS EFFUSION
APPARATUS FOR DETERMINING VAPOUR PRESSURES AND ENTHALPIES
OF SUBLIMATION
C. G. de Kruif 143

VAPORIZATION STUDIES BY KNUDSEN CELL-MASS SPECTROMETRY AND
THERMODYNAMIC STABILITY OF HIGH TEMPERATURE GASEOUS SPECIES
G. De Maria 157

TRANSPIRATION MASS SPECTROMETRY - A NEW THERMOCHEMICAL TOOL
J. W. Hastie and D. W. Bonnell 183

THERMODYNAMIC ACTIVITY AND VAPOR PRESSURE MODELS FOR
SILICATE SYSTEMS INCLUDING COAL SLAGS
J. W. Hastie, D. W. Bonnell, E. R. Plante and W. S. Horton   235

METAL-LIGAND HEATS AND RELATED THERMODYNAMIC QUANTITIES BY
TITRATION CALORIMETRY
James J. Christensen   253

MEASUREMENT OF COMPLEXING CONSTANTS AND ENTHALPIES OF
COMPLEXING BY BATCH CALORIMETRY
Michael H. Abraham   275

METAL LIGAND BOND STRENGTHS AND BONDING THEORY
C. T. Mortimer   289

THERMOCHEMISTRY OF β-DIKETONES AND METAL-β-DIKETONATES.
METAL-OXYGEN BOND ENTHALPIES
Manuel A. V. Ribeiro da Silva   317

THERMOCHEMISTRY OF METAL-POLYAMINE COMPLEXES
P. Paoletti   339

THERMOCHEMISTRY OF MAIN-GROUP ORGANOMETALLIC COMPOUNDS
Geoffrey Pilcher   353

EXPERIMENTAL THERMOCHEMISTRY OF TRANSITION-METAL
ORGANOMETALLIC COMPOUNDS
Geoffrey Pilcher   367

THERMOCHEMISTRY OF MOLECULAR TRANSITION METAL COMPOUNDS
J. A. Connor   383

MEASUREMENT OF ENTHALPIES OF SOLUTION OF ELECTROLYTES
Michael H. Abraham   393

HYDROPHOBIC HYDRATION OF ALKYLAMMONIUM BROMIDES IN
AQUEOUS MIXED SOLVENTS
G. Somsen   411

SOLVATION OF ALCOHOLS, AMINES, UREAS AND AMIDES IN MIXED
SOLVENTS
G. Somsen   425

MICROCALORIMETRY AS AN ANALYTICAL TOOL IN BIOLOGY
Ingemar Wadsö   439

APPLICATION OF CALORIMETRY TO THE LIFE SCIENCES
H. Klump   461

TABLE OF CONTENTS

SUPERHELICITY AND ENERGETICS OF STRUCTURAL TRANSITIONS
OF NUCLEIC ACIDS
H. Klump — 489

CALORIMETRIC STUDIES IN BINDING REACTIONS IN BIOCHEMISTRY
G. Rialdi and S. Raffanti — 511

CELLULAR THERMOCHEMISTRY: RESTING STATE AND STIMULATION
G. Rialdi and S. Raffanti — 531

DIFFERENTIAL SCANNING CALORIMETRY APPLICATIONS TO PROTEIN
SOLUTIONS
Pedro L. Mateo — 541

THERMOCHEMISTRY OF MOLECULES, MOLECULE-IONS, AND FREE
RADICALS. FACTORS INFLUENCING BOND ENERGIES IN THESE SPECIES
Henry A. Skinner — 569

EXPERIMENTAL METHODS TO STUDY THE THERMOCHEMISTRY OF GAS
PHASE IONS AND ION-MOLECULE REACTIONS
Rod S. Mason — 601

THE THERMOCHEMISTRY OF GAS PHASE ION-MOLECULE REACTIONS
Rod S. Mason — 627

THE APPLICATION OF LOW-TEMPERATURE CALORIMETRY TO SOME
CONTEMPORARY PROBLEMS IN SOLID STATE CHEMISTRY
L. A. K. Staveley — 653

THERMODYNAMICS OF CRYSTALS - 1982
Edgar F. Westrum, Jr. — 671

CALORIMETRY OF PHASE, SCHOTTKY AND OTHER TRANSITIONS
Edgar F. Westrum, Jr. — 695

THERMOPHYSICS OF MINERAL AND ROCK SYSTEMS
Edgar F. Westrum, Jr. — 719

COMPUTERIZED ADIABATIC THERMOPHYSICAL CALORIMETRY
Edgar F. Westrum, Jr. — 745

LECTURES ON THERMOCHEMISTRY AND KINETICS
Sidney W. Benson — 769

BARRIERS TO INTERNAL ROTATION IN INORGANIC SPECIES
Stanley Abramowitz — 789

CRITICAL EVALUATION OF THERMODYNAMIC DATA - A RESEARCH
ACTIVITY
S. Abramowitz, D. D. Wagman, V. B. Parker, and D. Garvin — 803

THERMOCHEMISTRY TODAY AND ITS ROLE IN THE IMMEDIATE FUTURE
H. A. Skinner   815

LIST OF PARTICIPANTS   823

SUBJECT INDEX   827

PREFACE

The progress that has recently been made in the field and applications of thermochemistry, having as a consequence an increase in the interest of the subject, undoubtedly can be considered highly remarkable.

Traditionally, the thermochemist has provided accurate thermal data on *chemical compounds of practical importance*, mainly by calorimetric and by equilibrium studies. The scope has been considerably extended in recent years, following the development of microcalorimetric techniques, of flow calorimetry, of titration calorimetry, and of high temperature calorimetry. The impact has been most noticeable in biochemical studies, in metallurgical studies, and in organometallic and inorganic thermochemistry.

A parallel development has led to increasing output of significant thermal data on *gas-phase transient species* (e.g. free radicals, radical ions) by kineticists, and by use of photoionization spectroscopy, mass spectroscopy and ion-cyclotron resonance spectroscopy. These species are outside the scope of traditional calorimetric study, but as more data on them become available they vastly add to the value of traditional thermochemical data and enable bond energies to be evaluated, and the inter-relation of molecular structure and bonding energy to be more closely examined.

It happens that the application of newer calorimetric techniques has been mainly pursued in Europe, whereas the newer techniques involving kinetics, mass spectroscopic (including ICR) methods, PI spectra, etc. have been mainly carried out in North America. Time is ripe for closer liaison between the more "classical" thermochemists and the "chemical physicists" (now providing data of special value to thermochemistry in the widest sense), and the ASI held in Portugal intended to achieve this objective.

The chapters in this book were written by the review lecturers at the NATO Advanced Study Institute on *"Thermochemistry Today and Its Role in the Immediate Future"*, held in Viana do Castelo, Portugal, from 5th to 15th July 1982. I should like to thank all the lecturers for their care in preparing not only their lectures but the review chapters presented here.

Sincere thanks are also due to the scientific advisers of this ASI, Prof. H. A. Skinner (Manchester, U.K.) and Prof. I. Wadsö (Lund, Sweden), for the unfailing and invaluable advice, help and encouragement given throughout this Institute, from the very days of preparation to the end of the lectures. Thanks are also extensive to the organisers and other participants who made the 1982 ASI such a success.

I am greatly indebted to NATO Scientific Affairs Division and Calouste Gulbenkian Foundation, Lisbon, for the generous sponsoring of this ASI; I also thank Instituto Nacional de Investigação Científica, (Lisboa) and Junta Nacional de Investigação Científica e Tecnológica (Lisboa) for their financial support.

I think this is the appropriate place to show my gratitude to Prof. Roger Irving (Surrey, U.K.), my former supervisor, for initiating me in the field of thermochemistry and since then for his constant encouragement, help and invaluable advice. The facilities given to work in the Manchester Thermochemistry Laboratory by Prof. H. Skinner and Dr. G. Pilcher, as well as their permanent help, are also deeply appreciated.

Finally, that so many authors should co-operate to produce the collection of chapters in this volume, is a tribute to the enthusiasm and good-will of all those who participated in the Institute, to whom I wish all the success in their scientific careers. It is our hope that the readers of this book will find it both interesting and useful for their scientific purposes.

University of Oporto, Portugal
January, 1983

Manuel A. V. Ribeiro da Silva

TECHNIQUES AND THEORY OF TITRATION CALORIMETRY

James J. Christensen

Department of Chemical Engineering
and the Thermochemical Institute
Brigham Young University
Provo, Utah 84602, USA

The purpose of this lecture is to outline the techniques used in titration calorimetry and to describe instruments and data evaluation procedures which have been developed to obtain thermodynamic properties for interacting chemical systems.

A common property of all chemical reactions is enthalpy change (ΔH) and it is the ability to quantitatively measure this property that makes calorimetry useful for the study of many chemical systems. Indeed, the general nature of the measured parameter, temperature or heat change, is the major asset in applying titration calorimetry to a wide range of problems.

Calorimeters applicable for the measurement of thermodynamic parameters by the titration technique have existed for about fifteen years. During this time, problems associated with the components of the calorimeter (i.e. constant temperature bath, constant rate buret, reaction vessel, temperature sensing circuit, and data analysis procedure) have gradually been solved so that the titration method now gives results comparable in accuracy to those obtained in conventional solution calorimeters. Inexpensive research quality titration calorimeters have become available commercially in the last few years. Initial analytical applications of titration calorimetry were based primarily on the analysis of the thermogram for end points. The development of titration calorimeters capable of measuring heat changes accurately has made possible the determination of additional information from the thermogram.

The additional information which can be obtained includes reaction enthalpies, stoichiometries and sometimes equilibrium constants for the reaction(s) of interest (1-6). In this talk techniques, instruments and data evaluation procedures used in titration calorimetry will be discussed. A more thorough coverage of much of the material contained in this talk can be found in the laboratory manual "Experiments in Thermometric Titrimetry and Titration Calorimetry (2).

1. TITRATION CALORIMETRY

Titration calorimetry has proven to be an especially useful technique for the determination of ΔH where consecutive or simultaneous reactions exist (7), for the determination of ΔG for reactions involving weakly interacting species (8-10), for the identification of species present in complex reaction mixtures (5-7,11) and for analytical determinations. (5,6,11) The methods, techniques and data analysis involved in titration calorimetry have been extensively described in the literature. (1,2,5,7-10,12) Titration calorimetry is the calorimetric technique where one reactant is titrated into another reactant and the temperature of the system is measured as a function of the titrant added. The temperature change may be produced by a chemical reaction (for example, proton ionization from protein or a metal ion complexing with a macrocyclic molecule) or by physical interaction between the titrate and the titrant (for example the adsorption of an organic molecule on a solid surface such as Zeolite or rock). There are two types of titrant addition -- incremental and continuous. In the first type the titrant is added incrementally and the temperature is usually readjusted to the initial temperature before each additional increment of titrant is added. This procedure has the advantage that reactions which are kinetically hindered may be accurately studied. In continuous titration, the titrant is introduced at a constant rate during a run. This continuous addition of the titrant has the advantage that a complete record is obtained of the heat effects during a reaction. Apparatus based on the continuous addition of titrant must have quick response to temperature changes and can be used only with systems in which the reactions take place rapidly.

As there are two ways to add the titrant there are two types of titration calorimeters -- isoperibol and isothermal. Isoperibol calorimetry is based on the continuous monitoring of the temperature of the contents of an adiabatic reaction vessel. The reaction vessel is usually located in a constant temperature environment. Isothermal calorimetry is based on the continuous monitoring of heat flux between the reaction vessel and its surroundings with the maintenance of the reaction vessel and its contents at a constant temperature equal to the temperature of the surroundings. Isothermal calorimetry has the advantage over isoperibol calorimetry that no heat capacity measurements are required and no corrections are necessary for the heat exchanged between the reaction vessel and its environment.

Regardless of the mode of titrant delivery and/or the type of calorimeter used the object in titration calorimetry is to produce a thermogram of the heat of reaction as a function of moles of titrant added. A typical thermogram is shown in Figure 1 where p correspond to the total heat produced for the total moles of titrant added to point p.

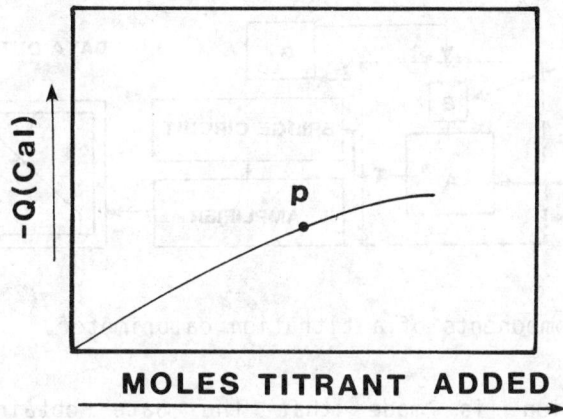

Figure 1. Typical thermogram

The thermogram from a single continuous titration is equivalent to that constructed from data obtained from an incremental titration where a large number of measurements were taken. For the sake of simplicity most of the material in this lecture refers to the continuous titration technique unless the incremental method is specifically mentioned. However, whatever applies to one method usually applies equally to the other, with the exception that more work must be done in the case of the incremental method to produce the same amount of data. It should also be noted that thermograms similar to the one shown in Figure 1 can be produced using either conventional non-titrational batch calorimeters or flow calorimeters, again with the expenditure of much more work and time than if a continuous titration calorimeter were used.

The main components of a titration calorimeter are indicated in the block diagram shown in Figure 2. The titrant

containing one of the reactants is introduced from the buret B
into the reaction vessel A. The resulting temperature change of
the reaction is sensed by the temperature sensor T and converted
to either a corresponding voltage or a heat production rate in
the bridge circuit. This voltage is amplified in the amplifying
circuit and recorded on either a strip-chart recorder (shown) or
another data collection system. The temperature of the bath
containing the reaction vessel A and the buret B is measured by
the sensor S and controlled by the temperature controller G.
The tempertures of the titrant and reaction vessel either must
be equal, or their difference must be known very precisely.

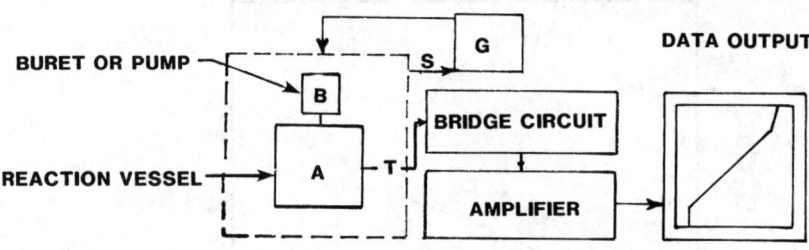

Figure 2. Main components of a titration calorimeter.

The assumption is made that the data obtained are
representative of the calorimeter system in both thermal and
chemical equilibrium. In practice, exact thermal equilibrium is
not obtained. With an isoperibol titration calorimeter the
reaction cell can be so designed that the response for
attainment of equilibrium is about two seconds. With an
isothermal titration calorimeter the response time will be two
or three orders of magnitude larger, suggesting some form of
data manipulation must be used to simulate thermal
equilibrium. Obviously, if chemical kinetics are slower or of
the same order of magnitude as the thermal response time of the
calorimeter, an additional complication is added to the data
analysis.

2. ISOPERIBOL TITRATION CALORIMETRY

Isoperibol titration calorimetry is a technique where the
temperature of a reaction vessel, immersed in a constant
temperature environment, is monitored as a function of time.
The reaction vessel used in an isoperibol titration calorimeter
should be designed to minimize the rate of heat exchange between

the vessel and its surroundings. A typical reaction vessel for precision isoperibol continuous titration calorimetry (13,14) is shown in Figure 3. The reaction vessel is placed in a constant-temperature bath and the run initiated when the temperature of the reaction vessel and its contents is the same as the bath temperature. Temperature changes are monitored while titrant (at the bath temperature) is continuously added to the reaction vessel. The thin inside wall of the Dewar vessel and low mass of the various elements inside the vessel (stirrer, thermistor, and heater) are specially designed to allow rapid thermal equilibration. The total heat capacity of a reaction vessel containing 100 cm^3 of water can be as low as 427 J°C^{-1} with a corresponding heat leak modulus of 1 x 10^{-3} min^{-1} (13). Constant-temperature environment titration calorimeters of this type have a precision of better than 0.2% and are capable of measuring temperature changes of 0.01 °C to an accuracy of 0.2%.

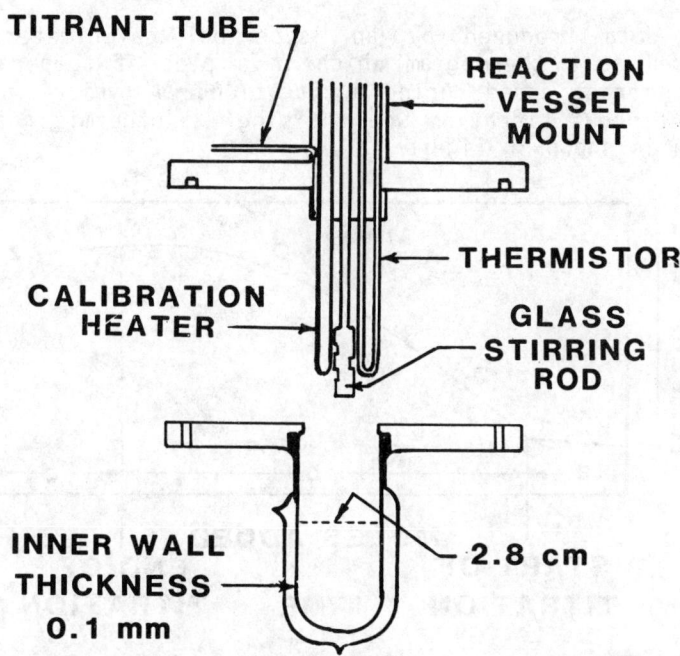

Figure 3. Schematic representation of a small volume isoperibol titration calorimeter reaction vessel.

Miniaturized reaction vessels have been developed with volumes as small as 1.5 cm^3. These vessels have a total heat capacity, when filled with 1.5 cm^3 of water, of only 7.0 J°C^{-1} (14,15). A reaction vessel containing 2.0 cm^3 of water will have a heat capacity of 9.12 J°C^{-1} and a heat leak modulus of 5 x 10^{-3} min^{-1}. Two problems complicate the development of small volume isoperibol titration calorimeters. First, the heat leak modulus increases as the volume of the reaction vessel decreases. As a result, the magnitude of the heat leak correction relative to the heat measured may be large. Uncertainties in the measured heat changes, resulting from uncertainties in the heat leak correction, are directly dependent on both the total measured heat, and the total titration time. Second, the fraction of the total heat capacity of the calorimeter due to unstirred material (i.e. glass dewar walls, stirrer, thermistor, heater) increases as the volume of the reaction vessel decreases, leading to a lowered response time by the instrument. However, a 2.0 cm^3 vessel has been reported which is capable of measuring temperature changes of 0.01°C and heats of 90 mJ to an accuracy and precision of 0.5% (15).

The data produced by an isoperibol calorimeter can be represented as a thermogram which is a plot of temperature vs. moles of titrant added during a titration. A typical thermogram for a continuous titration where a single exothermic reaction is occurring is shown in Figure 4.

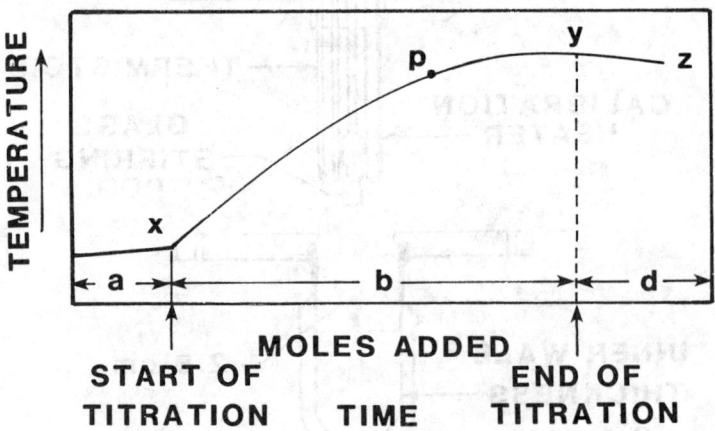

Figure 4. Thermogram for a continuous isoperibol titration for a single exothermic reaction.

The thermogram from a single continuous titration is equivalent to that constructed from data obtained from an incremental titration where a large number of measurements are taken. However, whatever applies to one method usually applies equally to the other, with the exception that more work must be done in the case of the incremental method to produce the same amount of data.

Region a (initial or lead period) in Figure 4 indicates the net heat loss or gain of the reaction vessel and contents before the titration begins. The slope is a function of the composite of effects of heating by stirring, IR drop across the thermistor, and heat effects due to conduction, radiation, convection and evaporation. Region b indicates the heat rise due to the reactions taking place in the reaction vessel, plus the effects of dilution of titrant and titrate, temperature differential of titrant and titrate, and those mentioned for region a. Region c is not represented in Figure 4. This region would consist of that portion of the curve in which the titration continues but the reaction is complete. This portion of the curve exists only for those systems in which the equilibrium constant, K, for the reaction is large enough so that the products are quantitatively formed as titrant is added. Region d (final or trial period) is generated after the titration is completed, and the slope is a function of the same effects as were mentioned for region a. Regions a and d are used to calculate the corrections for the heat loss from the reaction vessel during the titration. Other corrections must be made to allow for the effects of dilution, temperature difference of titrant and titrate solutions, and heat capacity changes due to the addition of titrant.

In order for the data represented by Figure 4 to be used for the calculation of the thermodynamic quantities associated with the reaction, the equipment must be calibrated and the data analyzed to produce a data array of heat produced vs. the quantity of titrant added. (2) Data analysis requires calibration of the titrant delivery system and the temperature-sensing device as well as the determination of the heat capacity of the reaction vessel. Analysis of a thermogram involves taking temperature data at strategic points along the curve, correcting the data for thermal effects due to non-chemical energy terms, and expressing the corrected data in heat units. The greater the complexity of the system, the greater the number of data points required. A large number of points are represented on a strip-chart recording, and the investigator may choose as many points as the complexity of the system dictates.

3. ISOTHERMAL TITRATION CALORIMETRY

Isothermal calorimetry is a technique in which the temperature of the system is kept constant and heat flux through the system is measured as a function of time or titrant added (16). The temperature of the calorimeter reaction vessel, its contents and its environmental surroundings is maintained at the same precise point so that radiation heat losses are eliminated and all changes in heat flux out of the system are due only to chemical or physical changes occurring in the calorimeter reaction vessel. Since heat is measured directly, the amount of required auxillary data is reduced; thus, it is not necessary to determine the heat capacity of either the reaction vessel or its contents. Many calorimetric measurements made with an isoperibol calorimeter can be made using an isothermal calorimeter. In addition, the close control of the environment and accurate monitoring of the heat flux make this instrument particularly useful in the following situations. (a) Processes involving slow reaction rates such as microbial growth and metabolism. (b) Processes in which large amounts of heat are produced. These processes include reactions occurring in concentrated aqueous solution and heats of mixing of organic liquids. (c) Systems involving large changes in heat capacities during the titration. Examples include heats of mixing, reactions in concentrated solutions, and reactions involving two liquid phases. The major drawback to isothermal calorimetry is that the time response of the instrument is on the order of minutes, rather than seconds as is the case with isoperibol calorimetry. Thus, rapid changes in heat production are difficult to follow.

An isothermal titration calorimeter is represented schematically in Figure 5. Heat is removed from the calorimeter reaction vessel at a constant rate q_c using a Peltier thermoelectric cooler. This heat effect is balanced against the heat due to stirring and self heating of the thermistors by means of a variable heater q_H to maintain a constant temperature in the reaction vessel. The temperatures of the reaction vessel and the surrounding constant-temperature water bath are the same thereby eliminating thermal radiation between the bath and reaction vessel. As titrant is added, q_H is adjusted to maintain the isothermal condition. The heat loss due to addition of titrant is then given by this change in the q_H value. An isothermal titration calorimeter reaction vessel using this principle is shown in Figure 6. Calorimeters of this type are capable of maintaining the temperature in the reaction vessel to $\pm 5 \times 10^{-6}$ °C during a run (6).

Data obtained using an isothermal titration calorimeter for the titration of a strong base with a strong acid are illustrated in Figure 7. During period $\underline{a}$ no titrant is added to the reaction vessel and the data represent the heat input from the control heater required to balance the heat effects from the thermoelectric cooler and heat effects in the reaction vessel arising from stirring, thermistor self-heating, etc. At time $\underline{x}$,

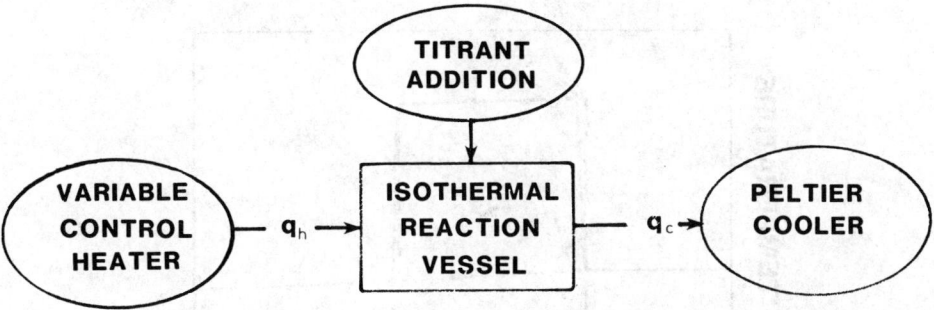

Figure 5. Schematic representation of an isothermal titration calorimeter.

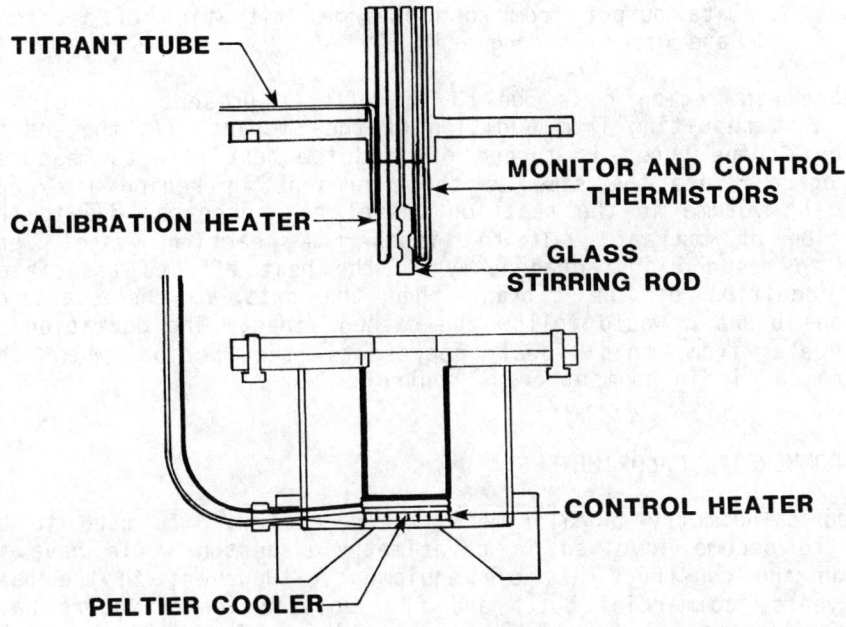

Figure 6. Schematic represenation of a small volume isothermal titration calorimeter reaction vessel.

titrant is added at a constant rate and an increased heat effect is seen due to the combined effects of addition of the titrant and neutralization of the strong acid with the base. At the end of region b, the neutralization process is complete and the heat

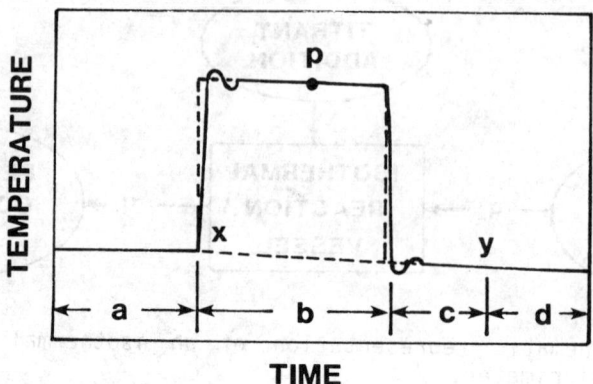

7. Data output from an isothermal titration of a strong base with a strong acid.

produced in region c is due to the effects present in region a plus that resulting from addition of the titrant. At the end of region c, the buret is turned off and the heat effects measured in region d are the same as those present in region a except that the volume in the reaction vessel has increased due to the addition of titrant. If the isothermal reaction vessel were able to respond instantaneously to the heat effects associated with addition of the titrant then the data at the start of regions b and c would follow the dashed lines. The deviation of the data from this ideal represents the period when the instrument is in nonisothermal control.

4. COMMERCIAL CALORIMETERS

To do calorimetry a calorimeter is necessary. It used to be that to become involved in calorimetry a person would have to design and construct his own equipment. However, in the past ten years, commercial batch and flow solution calorimeters have become increasingly available and popular. Today in the field of calorimetry, as in many other fields such as spectroscopy, the beginning investigator can select and purchase his experimental equipment from a wide variety of ready-to-use units. Table 1 lists most of the commercially available solution calorimeter presently on the market together with the names and addresses of manufacturers and representatives. These units are either sold as solution calorimeters or can be adapted to use with liquid solutions. The operating characteristics of these commercial solution calorimeters are listed in Table 2.

Table 1. Commercially available solution calorimeters

Designation	Manufacturer or Representative	
Arnett Solution Calorimeter	SKC Inc. P.O. Box 8538 Pittsburgh, PA 15220	Luminon Incorp. 120 Coit Street Irvington, NJ 07111
Calvet Microcalorimeters C.R.M.T. Calorim.	Setaram 101-103 Rue de Seze 69 Lyon 6° France	
Enthalpimeter	American Instrument Co. 8030 Georgia Avenue Silver Spring, MD 20910	
High Pressure, Moderate temperature Flow Calorimeter	Hart Scientific P.O. Box 934 Provo, Utah 84603	
LKB Precision Calorimetry and Microcalorimetry Systems	LKB Instruments AB Fack, 161 25 Bromma 1 Stockholm, Sweden LKB Strumenti S.p.A. Via Morgagni 30/e piano 3°, int. 7, 00161 Roma	LKB Instruments 12221 Parklawn Dr. Rockville, MD 20852
Parr Solution Calorimeter	Parr Instrument Co. 211 Fifty-third Street Moline, IL 61265	
Picker Microcalorimeter	Techneurop Inc. CIL Building Suite 2475 630 Dorchester Blvd. West Montreal, Quebec Canada H3B 1S6	
Sanda Thermo--Titrators	Sanda Inc. 4343 East River Drive Philadelphia, PA 19129	
Tronac Isoperibol and Isothermal Calorimetry Systems	Tronac Inc. 1804 So. Columbia Lane Orem, UT 84057	

Table 2. Operating characteristics of selected commercial solution calorimeters

Calorimeter	Measurement Range (Watts, W or Joules, J)	Temperature Range C°	Detectibility Limit	Precision %	Volume cm³	Price Range $ (US)
Arnett Calorimeter		25	1.26 J	1	300	1,900-2,100
Calvet Microcalorimeter	<0.5 W, <50 J <0.5 W, <50 J	-206-1500	0.1 μW, 50 μJ 0.2 μW, 200 μJ	0.2 0.2	15 100	30,000-50,000
CRMT Calorimeter	<2 W, <200 J <2 W, <500 J	0-100	200 μW, 50 mJ 400 μW, 150 mJ	0.2 0.2	15 100	20,000
Enthalpimeter		-5-40	0.063 J(15 cm³)	1-5	15 (0.5)	1,500
High Pressure; Moderate Temp., Flow Calorimeter (Pressure capability up to 130 atm)	0.01-1 J	-20-200	25 μW	0.5	2.5	13,000-55,000
LKB 8700 LKB 10700 (Batch) LKB 10700 (Flow)		5-60 0-50 0.50	0.021 J(100 cm³) 1 μW, 200 μJ 1 μW	0.1 0.1-1 0.1-1	100 (25) 2 and 4 1	17,000-19,000

Table 2. (cont.) Operating characteristics of selected commercial solution calorimeters

Calorimeter	Measurement Range (Watts, W or Joules, J)	Temperature Range C°	Detectibility Limit	Precision %	Volume cm³	Price Range $ (US)
Parr Solution Calorimeter	8.4-4186 J	20-30	0.84 J	1	90-120	1,200
Picker Flow Calorimeter		10-60	0.3 µW	2		13,000-17,000
Sanda Thermo-Titrators		25	1.25 J	0.1-1	50	3,000-5,500
Tronac Isoperibol	<52 J	-10-60	0.01 J (25 cm³)	0.4	2.5-200	4,000-12,000
Tronac Isothermal	0.005-2 J	-10-60	25 µW (25 cm³)	0.3	2.5-100	18,000-26,000
Tronac Series 300	0.02 W	0-60	0.5 µW	0.1	40	15,000-25,000
Tronac Flow Calorimeter (Pressure capability up to 400 atm)	0.005-2 J	-10-75	25 µW	0.5	2.5	18,000-30,000

As can be seen, the commercial units cover a wide range of operating variables (temperatures range from -206 to 1500°C, volume of sample from 0.5 to 3000 cm^3) and measurement variables (measurement range 0.005 to 500 joules, precision 0.1 to 5%). The numbers in Table 2 should be considered as only indicating the normal range of operation since many of the instruments can be modified in one way or another to extend the range given.

5. CALCULATION OF ΔH VALUES

The addition of titrant to the titrate solution produces either one or more reactions where the extent of the reaction(s) and the energy produced are related to the corresponding equilibrium constant(s) and enthalpy change(s) for the reaction(s). The equations relating the heat produced, the equilibrium constant(s), and enthalpy change(s) for the reaction(s) are generally complex. It is convenient to express the relationship among these quantities for the general case of n reactions occurring in the reaction vessel in the form given by equation 1

$$Q_{c,p} = \sum_{i=1}^{n} \Delta H_i \Delta n_{i,p} \qquad (1)$$

where $\Delta n_{i,p}$ is the change in the moles of product i formed from point x to point p and is a function of the equilibrium constant for reaction i. In general, the best values of ΔH are calculated by a least squares analysis of equation 1. (2)

6. CALCULATION OF K VALUES

The equilibrium constant for a given reaction can be determined by titration calorimetry if the magnitudes of K and ΔH for the overall reaction taking place in the reaction vessel are within certain limits. The thermograms for systems with K values greater than approximately 10^4 differ only slightly from one another; hence it is difficult to make accurate calculations of K in these cases. For reactions with K values less than 10, very little reaction takes place and hence very little heat is evolved. In addition, the shape of the thermogram is little affected by incremental decreases in K for K values either greater than 10^4 or less than approximately 10. In order to obtain data which can be used to calculate reliable K and ΔH values, the lower the K value is, the higher the ΔH must be. It is emphasized that, while the calculated value of ΔH is directly related to the magnitude of the measured Q values, the calculated value of K depends only on the curvature of the thermogram. For this reason, one can obtain accurate K values even though ΔH is erroneous due to calibration errors, incorrectly handled heat of dilution data, etc. More detailed information on error analysis and factors which evaluate possible errors in the calorimetric determination of K and ΔH is available (17-21).

The successful application of the calorimetric method of determining equilibrium constants to a given system, therefore, depends on (a) the equilibrium constant and concentrations of reacting species being within the bounds required to yield a sufficiently curved thermogram and (b) the ΔH value being large enough that the total measured heat is known with a reproducibility of better than 5%. It has been shown that by use of selective titrants the method can be extended to the determination of equlibrium constants for proton ionization (20) and metal-ligand interaction (19) of almost any magnitude. Specific examples illustrating the use of this method in the analysis of calorimetric data for a variety of chemical systems is given in the second lecture of this series.

REFERENCES
(1) Bark, L.S., Bark, S.M.: 1969, "Thermometric Titrimetry," Pergamon Press, New York.
(2) Eatough, D.J., Christensen, J.J., Izatt, R.M.: 1974, "Experiments in Thermometric Titrimetry and Titration Calorimetry," Brigham Young University Press, Provo.
(3) Eatough, D.J., Izatt, R.M., Christensen, J.J.: 1982, "Titration and Flow Calorimetry: Instrumentation and Data Calculation" chapter in "Comprehensive Analytical Chemistry," Ed. N.D. Jespersen, Part B, "Thermal Analysis," Vol. XII, Elsevier Scientific Pub. Co., New York, pp. 3-37.
(4) Jordan, J.: 1968, "Thermometric and Enthalpy," chapter in "Treatise on Analytical Chemistry," Eds. I.M. Kolthoff and P. Elving, Part 1, "Theory and Practice" Vol. 8, Interscience, New York, pp. 5175-5242.
(5) Tyrrell, H.J.V., Beezer, A.E.: 1968, "Thermometric Titrimetry," Chapman and Hall, Ltd., London.
(6) Vaughan, G.A.: 1973, "Thermometric and Enthalpimetric Titrimetry," Van Nostrand Reinhold Company, London.
(7) Christensen, J.J. and Izatt, R.M.: 1968, "Thermochemistry in Inorganic Solution Chemistry," chapter in "Techniques in Advanced Inorganic Chemistry," Ed. P. Day and A. Hill, John Wiley and Sons, New York.
(8) Christensen, J.J., Eatough, D.J., Ruckman, J., and Izatt, R.M.: 1972, Thermochimica Acta 3, p. 203.
(9) Eatough, D.J., Christensen, J.J., and Izatt, R.M.: 1972, Thermochimica Acta 3, p. 219.
(10) Eatough, D.J., Izatt, R.M., and Christensen, J.J.: 1972, Thermochimica Acta 3, p. 233.
(11) Hansen, L.D., Izatt, R.M., and Christensen, J.J.: 1974, "Applications of Thermometric Titrimetry to Analytical

Chemistry" chapter in "New Developments in Titrimetry," Ed. J. Jordon, Marcel Dekker Inc., New York.
(12) Carr, P.W.: 1972, in CRC "Critical Reviews in Analytical Chemistry," Ed. L. Meites, Chemical Rubber Co., Cleveland, Ohio, pp. 491-557.
(13) Christensen, J.J., Izatt, R.M., and Hansen, L.D.: 1965 Rev. Sci. Instrum., 36, pp. 779-83.
(14) Hansen, L.D., Izatt, R.M., Eatough, D.J., Jensen, T.E. and Christensen, J.J.: 1974, "Analytical Calorimetry", Eds. R.S. Porter and J.F. Johnson, Vol. 3, Plenum Press, pp. 7-16.
(15) Hansen, L.D., Jensen, T.E., Mayne, S, Eatough, D.J., Izatt, R.M. and Christensen, J.J.: 1975, J. Chem. Thermodynamics, 7, pp. 919-926.
(16) Christensen, J.J., Gardner, J.W., Eatough, D.J., Izatt, R.M., Watts, P.J., and Hart, R.M.: 1973, Rev. Sci. Instrum., 44, pp. 481-484.
(17) Christensen, J.J., Wrathall, D.P., Oscarson, J.L. and Izatt, R.M.: 1968, Anal. Chem., 40, pp. 1713-1717.
(18) Izatt, R.M., Eatough, D.J., Snow, R.L., and Christensen, J.J.: 1968, J. Phys. Chem., 72, pp. 1208-1213.
(19) Eatough, D.J.: 1970, Anal. Chem., 42, pp. 635-639.
(20) Christensen, J.J., Wrathall, D.P., and Izatt, R.M.: 1968, Anal. Chem., 40, pp. 175-181.
(21) Christensen, J.J., Rytting, J.H. and Izatt, R.M.: 1969, J. Chem. Soc. A, pp. 47-53.

APPLICATIONS OF TITRATION CALORIMETRY

James J. Christensen

Department of Chemical Engineering
and the Thermochemical Institute
Brigham Young University
Provo, Utah 84602 USA

In this lecture the concepts and calculation procedures developed in the previous lecture for titration calorimetry are applied to several chemical systems. The examples considered are chosen because of their relevance to understanding the techniques and potential uses of titration calorimetry both as a thermodynamic and analytical tool.

The availability of equipment and a growing acceptance of titration calorimetry as a convenient and reliable analytical technique suggest that this method may provide unique solutions to calorimetric and analytical problems in specific areas. The justifications for this are that the method is rapid, sensitive and general, and that the equipment is relatively inexpensive and lends itself well to automation.

Application of the technique to several varied chemical systems will be illustrated by examples taken from the laboratory manual to be handed out in class (1) and from the current literature (2-6). Many other applications are described in several review articles, books and chapters dealing with the analytical and thermodynamic application of titration calorimetry (1-12). A good general review of the applications of titration calorimetry to analytical chemistry has been published (3). This review covers the analytical applications including aqueous acid-base, precipitation, and metal complexation reactions; the determination of the stoichiometry of a reaction; and the determination of equilibrium constants by titration calorimetry. The reader is refered to this and the other publication given above for a comprehensive discussion and presentation of the many application of titration calorimetry as a thermodynamic and analytical tool.

1. AQUEOUS REDOX REACTIONS

Potentiometric titration methods generally become inaccurate at concentrations below 0.1 M because the curve of voltage (or log [concentration]) versus milliliters of titrant becomes too flat for accurate determination of the end point. Calorimetric titration methods, on the other hand, do not show this loss of end-point sharpness even in solutions more dilute than 0.1 M as the following example will illustrate.

We used potassium dichromate solution to determine Fe(II) in acid solution at concentrations of 0.05 M and 0.007 M (3). The results are shown in Figure 1 and Table 1. As can be seen

Table 1. Standardization of ferrous sulfate solution by a thermometric titration of Fe^{2+} with $K_2Cr_2O_7$[a]

$K_2Cr_2O_7$ titrant (ml)	$[Fe^{2+}]$ found, M	$[Fe^{2+}]$ added, M
0.9876 N $K_2Cr_2O_7$		
5.432	0.05332	
5.430	0.05330	
5.434	0.05334	0.05333
5.431	0.05331	
5.435	0.05335	
0.1640 N $K_2Cr_2O_7$		
4.724	0.007751	
4.722	0.007748	
4.721	0.007746	0.007746
4.719	0.007743	
4.719	0.007743	
4.722	0.007748	

[a] 99.95 ml solution of Fe^{2+}.

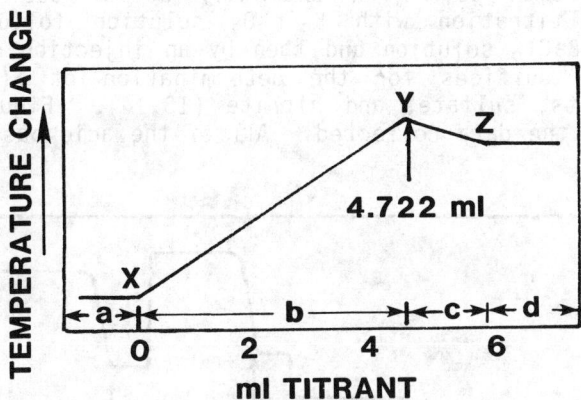

Figure 1. Thermogram for the titration of 99.95 ml of 0.007748 N $FeSO_4$ solution with 0.1640 N $K_2Cr_2O_7$: (a) initial stirring slope, (x) point at which titrant addition begins: (b) reaction: $Cr_2O_7^{2-} + 6Fe^{2+} + 14H^+ = 2Cr^{3+} + 6Fe^{3+} + 7H_2O$, (y) equivalence point: (c) continued addition of titrant, (z) point at which titrant addition stops: and (d) final stirring slope.

in Table 1 the method gives results comparable to those obtained by conventional procedures and is not affected by a tenfold dilution of the Fe(II) solution.

2. DETERMINING CHEMICAL SPECIES IN AIRBORNE PARTICULATE MATTER

While extensive determinations of the elemental composition of airborne particulate matter have been made and can now be done routinely with automated equipment, little information is available concerning the exact chemical species and compounds present. This is especially true of the minor inorganic components. Interestingly enough, it is these same minor components that cause concern when released to the atmosphere. Specific information on the compounds present is extremely important in understanding the environmental impact of fossil fuel combustion because the behavior of an element in the environment can differ markedly with oxidation state and with the other species with which it is combined. Specifically we have been interested in identifying acids, bases, and the S, N, and As species present in airborne particulate matter produced by the smelting of metals and the combustion of fossil fuels.

The Tronac model 450 micro isoperibol titration calorimeter (4,12) has proven to be an extremely useful tool for these studies. A titration with $K_2Cr_2O_7$ solution followed by an injection of $BaCl_2$ solution and then by an injection of sulfamic acid solution suffices for the determination of S(IV), other reducing agents, sulfate, and nitrite (13,14). Figure 2 shows schematically the data collected. All of the acid-base active

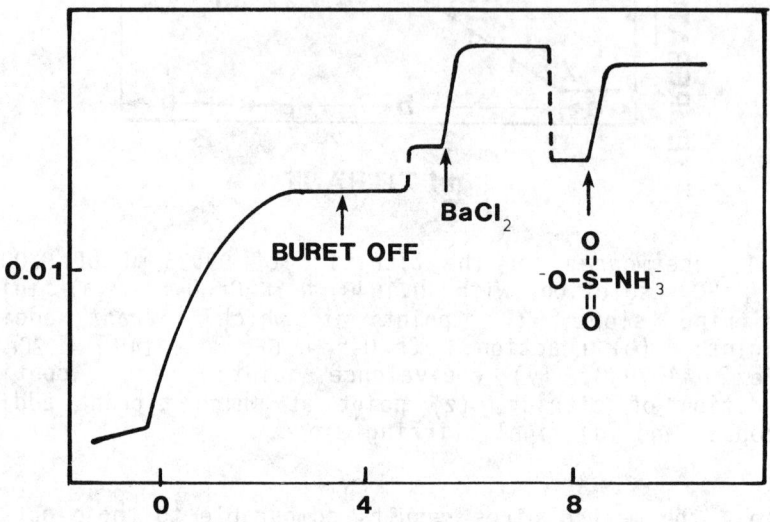

Figure 2. Typical thermogram obtained in the determination of species present in a water extract of particulate matter.

species present in a water extract of the particulate matter can often be determined in a separate experiment using the same model of calorimeter with the addition of a set of micro pH electrodes to the reaction vessel (15). The titrants in this case are $HClO_4$ and NaOH solutions.

The advantage of the titration calorimetry approach are (a) freedom from interferences, (b) rapidity, (c) ability to handle sub milligram samples, and (d) the equipment is portable and relatively inexpensive. Because the thermogram produced by titration calorimetry provides data on both the amount (end points) and the identity (ΔH) of the substance undergoing reaction it is essentially free from interferences. The direct injection enthalpimetric determinations of sulfate and nitrite are interference free because of the specificity of the reagents used. Since a titration and an injection can be run in about twenty minutes, the method is fast enough for routine work. The

APPLICATIONS OF TITRATION CALORIMETRY

detection limit of the method depends mainly on the ΔH value for the analytical reaction. Typically it ranges from about thirty nanomoles (ΔH~40 kj/mole) to about three nanomoles (ΔH~400 kj/mole) in two ml of solution. Development of a micro isoperibol titration calorimeter has made this analytical application of calorimetry possible (4).

3. ADSORPTION OF AROMATIC COMPOUNDS BY ZEOLITE (LMX 13X)

The adsorption of aniline, nitrobenzene, and toluene by Zeolite (ground rock) is an interesting example of the use of isothermal calorimetry to determine a concentration when no end point can be observed directly(16). Results from several runs for the addition of aniline, nitrobenzene, or toluene to a hexane suspension of Zeolite are given in Figure 3.

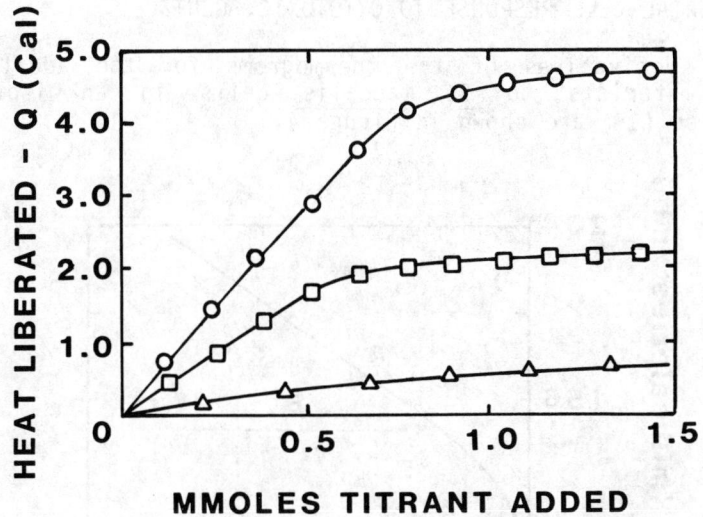

Figure 3. Typical calorimetric titration curves for the interaction of toluene (△) nitrobenzene (□) or aniline (O) with 0.5 grams of Zeolite in hexane.

An equation has been derived from which the number of binding sites and log K and ΔH values resulting from the interactions are obtained by fitting the equation to the thermogram (Figure 3) (19). Representative of the results obtained is the log K values for the adsorption of aniline by Zeolite in hexane, 3.08 ± 0.14. This value compares favorably with the value, 3.05 ± 0.03, obtained independently by a direct analytical method.

The adsorption capacity of Zeolite for aniline in hexane has also been determined by calorimetric and analytical methods to be 1.64 ± 0.10 and 1.67 ± 0.01, respectively. It is interesting that the accuracy of the calorimetrically determined values is good even though the thermogram has no definite end point. These data indicate that reliable thermodynamic and stoichiometry values may be obtained from the calorimetric data without the requirement for quantitative reactions or visible end points if the reaction is known. Although the method outlined above is applicable to any system of chemical reactions, the particulate nature of the system studied imposes certain restrictions. The rate of reactions generally decreases as particle size increases above small colloid dimensions. Thus, incremental isothermal, rather than continuous isoperibol titration, calorimetry must often be used in studying these systems.

4. BACTERIAL CELL RESPONSE TO CYTOTOXIC AGENTS

The time derivatives of the thermograms for the addition of cytoxic materials to S. faecalis cells in an isothermal calorimeter (18) are shown in Figure 4.

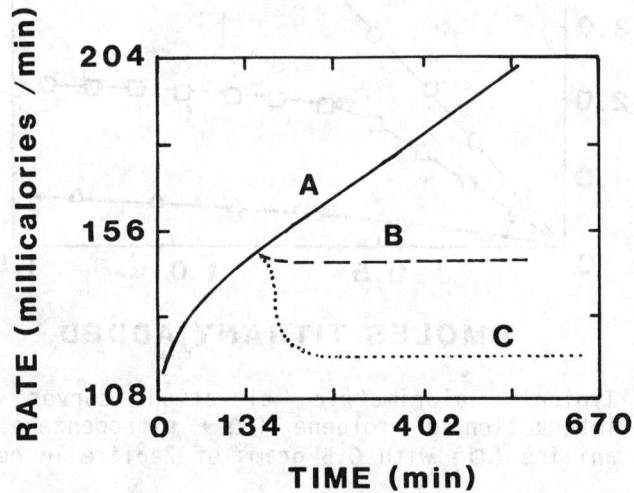

Figure 4. The time derivative of the thermogram for the addition of cytotoxic material to S. faecalis (4% DMSO broth), A) standard curve. The arrow indicates injection of: B) 0.1 ml of 12.6 mg penicillin 'G'/ml (4% DMSO broth), and C) 0.1 ml 11.8 mg tetracycline HCl/ml (4% DMSO broth).

The thermogram provides information about the cell metabolism and thus becomes an indicator of the extent of cytotoxicity of the added agent. The degree of cell response and the character of the added agent. The degree of cell response and the character of that response appear to be a property of the concentration and type of cytotoxic agent used.

One complete determination of cell response can be accomplished in less than one hour. The range of possible uses of this technique include: 1) the study of the types of response of growing cells to cytotoxic agents, 2) the determination of the concentrations of cytotoxic agents, and 3) the study of other factors that influence cell metabolism, i.e. serum factors such as antibody, complement, and β-lysin.

5. HEAT OF IONIZATION OF WATER

To illustrate the evaluation of a heat of reaction in solution and to compare different technique of titration calorimetry several determinations of the heat of ionization of water were made using the reaction of $HClO_4$ with NaOH (4). In all cases the acid (titrant) was titrated into the base (titrate). The results obtained with the isothermal and isoperibol calorimeters are given in Tables 2 and 3 respectively. The uncertainty given in each case is the standard deviation of the mean of each series of runs.

The results from the isothermal calorimeter (Table 2), ΔH_w = -13.39±0.03, -13.36±0.02, and -13.35±0.05 kcal/mole at ionic strengths of 0.0097, 0.0050, and 0.00097, respectively, are in excellent agreement with previously determined values of -13.38, -13.36 and -13.35 kcal/mole at these respective ionic strengths (19). The total heats measured in these runs were 0.55, 0.28, and 0.055 cal, respectively.

The results obtained using the isoperibol calorimeter (Table 3) ΔH_w = -13.44±0.04 at μ = 0.010 and -13.33±0.10 at μ = 0.0029, are in good agreement with previously determined values of -13.38 and -13.36 at the respective ionic strengths (19). Total heats measured in these runs were 0.28 and 0.077 cal, respectively.

Not unexpectedly, the results from the isothermal calorimeter are somewhat more precise than those from the isoperibol instrument. There do not appear to be any significant systematic errors present in the operation of either instrument.

Table 2. Heat of ionization of H_2O at 25°C as determined in a 4 ml isothermal calorimeter[a]

$-\Delta H/\text{kcal mol}^{-1}$

$\mu = 0.0097$[b]	$\mu = 0.0050$[c]	$\mu = 0.00097$[d]
13.47	13.37	13.30
13.29	13.32	13.18
13.47	13.42	13.43
13.35	13.32	13.36
13.37		13.47
13.42		
Av. = 13.39±0.03	Av. = 13.36±0.02	Av. = 13.35±0.05

[a] Neutralization of NaOH with $HClO_4$.
[b] In each run 0.5 ml of 0.2059 M $HClO_4$ was titrated into 4.0 ml of 0.01030 M NaOH.
[c] In each run 0.5 ml of 0.2059 M $HClO_4$ was titrated into 4.0 ml of 0.005150 M NaOH.
[d] In each run 0.5 ml of 0.02026 M $HClO_4$ was titrated into 4.0 ml of 0.001031 M NaOH.

Table 3. The heat of ionization of H_2O at 25° as determined in a 3 ml isoperibol titration calorimeter[a]

$-\Delta H/\text{kcal mol}^{-1}$

$\mu = 0.010$[b]	$\mu = 0.0029$[c]
13.40	13.10
13.35	13.06
13.34	13.27
13.36	13.12
13.61	13.68
13.50	13.55
13.52	13.56
Ave. = 13.44±0.0	Ave. = 13.33±0.10

[a] Neutralization of NaOH with HCl.
[b] In each run 0.25 ml of 0.2581 M HCl was titrated into 2.7 ml of 0.01030 M NaOH.
[c] In each run 0.25 ml of 0.06727 M HCl was titrated into 2.7 ml of 0.00287 M NaOH.

6. BINDING OF ADENOSINE DIPHOSPHATE TO BOVINE LIVER GLUTAMATE DEHYDROGENASE

This system involves the reaction of the inhibitor adenosinediphosphate (ADP) with the enzyme bovine liver glutamate dehydrogenase (GDH) to form a complex. The calorimetric data of Beaudette and Langerman (20) are illustrated in Figure 5 and given in Table 4. The results

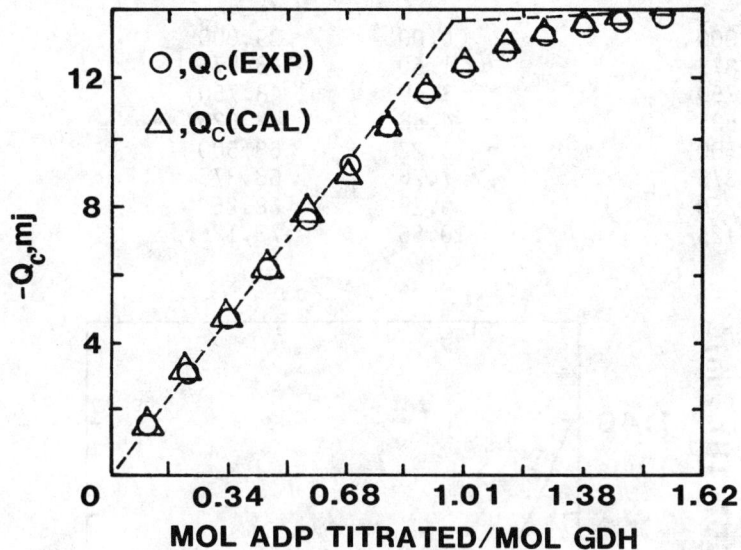

Figure 5. Experimental and calculated Q_C values vs. the mol ratio of ADP/GDH for data given in Table 4. The calculated Q_C values were obtained by using $K = 2.70 \times 10^5$ and $\Delta H = -54.35$ kJ mol^{-1}.

suggest formation of a weak 1:1 complex with an end point at about 1 mol ADP to 1 mol GDH. While the data may be fitted by least squares method to give both K and ΔH for the binding process, Beaudette and Langerman chose to use the value determined by Subramanian et al. (21) of $K = 2.6 \times 10^5$ M^{-1} and calculated a ΔH value of -54.4 ± 2.9 kJ mol^{-1}. We have taken their data (Table 1) and calculated by least squares analysis the values $K = 2.70 \times 10^5$ M^{-1} and $\Delta H = -54.35$ kJ mol^{-1} in excellent agreement with the values reported by the authors illustrating that both K and ΔH values may be reliably obtained by least squares analysis of the calorimetric data for the inhibition reaction (2). The Q_C values calculated from these K and ΔH values are shown in Figure 5. The plot of the error square sum vs. log K for the analysis is shown in Figure 6.

Table 4. Calorimetric titration data (20) for the addition of 6.17 mM ADP to 2.00 cm^3 of 0.1340 mM GDH at pH 7.6, 25°C.

Titrant Volume, μ dm^3	$-Q_{C,p}$, mJ	Titrant Volume, μ dm^3	$-Q_{C,p}$, mJ
0.000	0.00	39.000	11.57
4.875	1.50	43.875	12.33
9.750	3.10	48.750	12.87
14.625	4.68	53.625	13.31
19.500	6.22	58.500	13.61
24.375	7.76	63.375	13.72
29.250	9.28	68.250	13.74
34.125	10.56	73.125	13.92

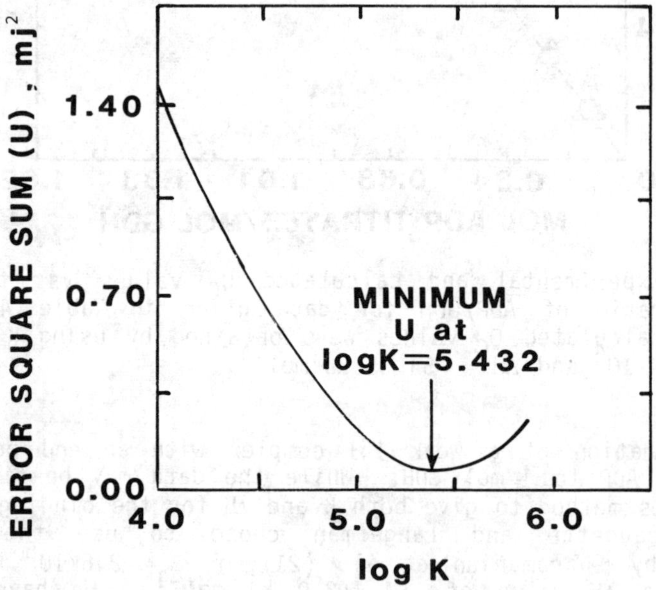

Figure 6. Error square sum, U, vs. log K for the data given in Table 4.

At the minimum the data are fitted to ± 0.071 mJ, in agreement with the expected precision of the data.

7. BINDING OF Zn^{2+} AND 1,10-PHENANTHROLINE TO THERMOLYSIN

A least squares analysis of calorimetric data has been used to deduce the chemistry occurring for the interactions between Zn^{2+}, 1,10-phenanthroline (P) and thermolysin ($ZnTh^{n+}$) (22). The ligand P is an inhibitor of $ZnTh^{n+}$, the Zn^{2+} ion being removed from the protein by the ligand. The calorimetric data obtained from the addition of P to a solution of the enzyme in 0.10 M KCl, 0.012 M $CaCl_2$ at pH 5.5 are shown in Figure 7a. The reactions occurring are represented by equations 1-5.

$$ZnTh^{n+} + P = PZnTh^{n+} \qquad (1)$$

$$PZnTh^{n+} = ZnP^{2+} + Th^{n-2} \qquad (2)$$

$$ZnP^{2+} + P = ZnP_2^{2+} \qquad (3)$$

$$ZnP_2^{2+} + P = ZnP_3^{2+} \qquad (4)$$

$$Ca^{2+} + P = CaP^{2+} \qquad (5)$$

The K and ΔH values for the reactions 3-5 as well as the formation of ZnP^{2+} from Zn^{2+} and P may be determined in the absence of protein, simplifying the least squares treatment of the data presented in Figure 7. The data were analyzed by an iterative least squares procedure to determine K and ΔH values associated with reactions 1 and 2. These analyses were performed (a) assuming that both reactions 1 and 2 occurred during the titration (i.e. that the inhibitor complex $PZnTh^{n+}$ did not form). The data were adequately represented only by condition (a). The resulting thermodynamic values for reactions 1 and 2 are given in Table 5. The removal of Zn^{2+} from the enzyme by P depends on formation of the 3:1 ZnP_3^{2+} complex, the concentrations of ZnP^{2+} and ZnP_2^{2+} being negligible throughout the titration. The calculated changes

Table 5. Thermodynamic results obtained from the least squares an analysis of the data given in figure 7

REACTION	log K	ΔH (kJ mol^{-1})
$ZnTh^{n+} + P = PZnTh^{n+}$	3.8 ± 0.1	-56.1 ± 0.8
$PZnTh^{n+} = ZnP^{2+} + Th^{n-2}$	-6.5 ± 0.3	-43.1 ± 7.5

in concentration of the species involved in the principal reactions are summarized in Figure 7.

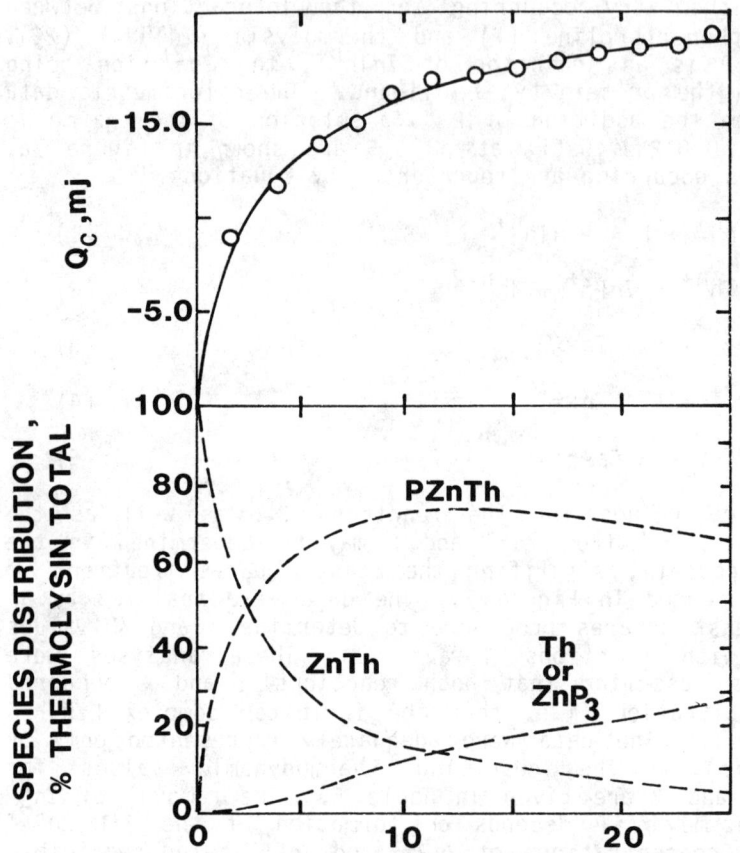

Figure 7. Plot of Q_C vs. the mol ratio of phenanthroline, P, to thermolysin (ZnTh) for the addition of 12.7 mM P to 2.98 cm^3 of 0.106 mM ZnTh in 0.012 M CaCl$_2$ at 25°C, pH = 5.5(1) and corresponding species calculated to be present in the solution(b).

REFERENCES

(1) Eatough D.J., Christensen, J.J., Izatt, R.M.: 1974, Experiments in "Thermometric Titrimetry and Titration Calorimetry," Brigham Young University Press, Provo.

(2) Eatough, D.J., Rehfeld, S.J., Izatt, R.M., Christensen, J.J.: 1982, "Titration and Flow Calorimetry: Application to Proteins and Lipids" chapter in "Comprehensive Analytical Chemistry," Ed. N.D. Jespersen, Part B, "Thermal Analysis," Vol. XII, Elsevier Scientific Pub. Co., New York, pp. 112-134.

(3) Hansen, L.D., Izatt, R.M., Christensen, J.J.: 1974, "Applications of Thermometric Titrimetry to Analytical Chemistry" chapter in "New Developments in Titrimetry," Ed. J. Jordan, "Treatise on Titrimetry", Vol. 2, Marcell Dekker, Inc., New York.

(4) Hansen, L.D., Izatt, R.M., Eatough, D.J., Jensen, T.E., Christensen, J.J.: 1974, "Recent Advances in Titration Calorimetry," chapter in "Analytical Calorimetry" Eds. R.S. Porter and J.F. Johnson, Vol. 3, Plenum Press, New York, pp. 7-16.

(5) Hansen, L.D., Christensen, J.J., Eatough, D.J., Izatt, R.M.: 1979, "Recent Developments in Calorimetric Instrumentation: Applications to Problems of Energy Storage and Recovery," Bureau of Mines Information Circular/1981, U.S. Dept. of Interior. "Proceedings of the Workshop of Techniques for Measurements of Thermodynamic Properties," Albany, Oregon, August 21-23, pp. 81-99.

(6) Izatt, R.M., Hansen L.D., Eatough, D.J., Jensen, T.E., Christensen, J.J.: 1974, "Recent Analytical Application of Solution Calorimetry" chapter in "Analytical Calorimetry" Eds. R.S. Porter and J.F. Johnson, Vol. 3, Plenum Press, New York, pp. 237-248.

(7) Bark, L.S., Bark, S.M.: 1969, "Thermometric Titrimetry," Pergamon Press, New York.

(8) Jordan, J.: 1968, "Thermometric and Enthalpy," chapter in "Treatise on Analytical Chemistry," Eds. I.M. Kolthoff and P. Elving, Part 1, "Theory and Practice" Vol. 8, Interscience, New York, pp. 5175-5242.

(9) Tyrrell, H.J.V., Beezer, A.E.: 1968, "Thermometric Titrimetry," Chapman and Hall, Ltd., London.

(10) Vaughan, G.A.: 1973, "Thermometric and Enthalpimetric Titrimetry," Van Nostrand Reinhold Company, London.

(11) Carr, P.W.: 1972, in CRC Critical Reviews in Analytical Chemistry, Ed. L. Meites, Chemical Rubber Co., Cleveland, Ohio, pp. 491-557.

(12) Hansen, L.D., T.E. Jensen, S. Mayne, D.J., Eatough, R.M. Izatt, and J.J. Christensen: 1975, J. Chem. Thermodynamics, 7, pp. 919-926.

(13) Hansen, L.D., B.E. Richter, and D.J. Eatough: 1977, Analyt. Chem., 49, pp. 1779-1781.

(14) Hansen, L.D., L. Whiting, D.J. Eatough, T.E. Jensen, and R.M. Izatt: 1976, Analyt. Chem., 48, pp. 634-638.

(15) Eatough, D.J., L.D. Hansen, R.M. Izatt, and N.F. Mangelson: 1977, "Determination of Acidic and Basic Species in Particulates by Thermometric Titration Calorimetry" in "Methods and Standards for Environmental Measurements." Proceedings of the 8th Materials Research Symposium, NBS Special Publication 464, U.S. Government Printing Office, Washington, DC, 1977, pp. 643-649.
(16) Eatough, D.J., Salim, S., Izatt, R.M., Christensen, J.J., and Hansen, L.D.: 1974, Anal. Chem., 46, p. 126.
(17) Eatough, D.J., Izatt, R.M., and Christensen, J.J.: 1972, Thermochimica Acta, 3, p. 233.
(18) Jensen, T.E., Hansen, L.D., Eatough, D.J., Sagers, R.D., Izatt, R.M., and Christensen, J.J.: 1976, Thermochimica Acta 17, p. 65.
(19) Christensen, J.J., Hansen, L.D., and Izatt, R.M.: 1976, "Handbook of Proton Ionization Heats and Related Thermodynamic Quantities," John Wiley & Sons, New York.
(20) Beaudette, N.V. and Langerman, N.: 1978, Analytical Biochemistry, 90, p. 693-704.
(21) Subramanian, S., Stickel, D.C., and Fisher, H.F.: 1975, J. Biol. Chem., 250 pp. 5885-5889.
(22) Lewis, E.A., Gardner, J.W., Hansen, L.D., Izatt, R.M., Christensen, J.J., and Eatough, D.J.: 1982, submitted to J. Inorganic Biochemistry.

MICROCALORIMETRY

Ingemar Wadsö

Thermochemistry Laboratory, University of Lund,
Chemical Center, P O Box 740, S-220 07 Lund 7, Sweden

A condensed status report is given for some areas of thermochemistry where sensitive calorimeters using small sample volumes are employed. Particular emphasis is placed on developments in instrumentation. Areas for which progress recently has been significant include DSC-instrumentation suitable for studies on dilute solutions and suspensions, calorimeters for studies of dissolution of slightly soluble compounds and instrumentation for thermodynamic as well as analytical work in biochemistry and cell biology.

INTRODUCTION

There are many types of experiments in thermochemistry which are best performed on the "macro" level involving e.g. mmole quantities of material. However, modern microcalorimeters have dramatically expanded the range accessible for high quality thermochemical work on expensive or hazardous material, on compounds which are slightly soluble or have low volatility, and on reactions involving high molecular compounds. Other important areas for microcalorimetry involve monitoring of many types of processes, e.g. life processes on the cellular level and for processes taking place in technically important materials.

There exists no well-defined difference between "ordinary" calorimeters or "macrocalorimeters" and "microcalorimeters". However, currently the micro-prefix normally indicates a sensitive instrument which only requires small sample quantities. For instance, for "micro-reaction calorimeters" the power sensitivity is typically of the order 1 μW or better and sample volumes are a few cm^3 or less. The amount of materials reacting is typically on the μmolar level corresponding to a heat evolution on the order of mJ.

COMBUSTION CALORIMETRY

One should probably talk about "miniaturized combustion calorimeters" rather than "microcalorimeters", keeping in mind the comparatively large quantities of heat typically evolved in a combustion process involving only a few mg of organic material. With a modern "micro" combustion calorimeter it is possible to approach a precision of 0.01% using only 10 mg of sample. (Combustion of 10 mg of benzoic acid yields about 260 J.) The development of miniaturized instruments in this classical field of calorimetry now seems to have reached a state where further improvement of precision and sensitivity hardly is called for (1). Rather, a continued attention has to be given to improvements in the control of the chemistry of the process.

VAPORIZATION CALORIMETRY

There are today very few laboratories where calorimetric measurements of enthalpy of vaporization are performed. However, during the last couple of years progress has been made in precise vaporization measurements on mg quantities of samples over the temperature range 300-420 K (2). In the work by Adedeji et al., a high temperature Calvet microcalorimeter was used for determination of enthalpies of sublimation through a simple drop technique (3).

DECOMPOSITION STUDIES

Skinner and co-workers have shown that well-defined decomposition processes can be studied in a high-temperature Calvet microcalorimeter by use of a drop technique, see eq. (3, 4). A sample (ca. 5 mg) is enclosed in a glass capillary tube and is transferred by free fall from the outside of the calorimeter to the hot reaction zone.

HEAT CAPACITY MEASUREMENTS

There has as yet been made no real miniaturization of the classical low temperature adiabatic heat capacity calorimeter.

Temperature scanning calorimeters, most commonly with a differential arrangement (DSC, differential scanning calorimetry), represent a well-established type of heat capacity calorimeters normally considered as "microcalorimeters". The most widely used instrument of this type is the DSC-instrument marketed by Perkin-Elmer (Norwalk, Conn.). Its main application area, which also applies to other commercial DSC-instruments, is in different kinds of analytical experiments, but they have also found use in thermodynamic work, e.g. in studies of transitions of biopolymers in concentrated solutions. DSC-measurement on dilute biopolymer solutions and on suspensions of biological membranes and membrane models is one of the most active areas of current biothermochemistry, see e.g. (5). For such experiments a very high sensitivity is essential. Instruments designed by Privalov et al. (6) and by Jackson and Brandts (7) are examples of adiabatic shield calorimeters which have been used successfully for this kind of work. Figure 1 shows schematically the design of the Privalov instrument. (See following page.)

The following characteristics were reported for this instrument:

Operational interval of temperatures	0-100 $^{\circ}$C
Rate of heating	0.1-2.0 $^{\circ}$C/min
Operational volume	1.0 ml
Sensitivity expressed as heat capacity	4×10^{-6} cal/$^{\circ}$C ml
Precision of heat capacity determination of the sample relative to that of the reference sample	2×10^{-5} cal/$^{\circ}$C ml
Precision of the temperature recording	0.1°

Suurkuusk et al. (8) have reported a sensitive DSC-instrument, based on the heat conduction principle, which has been used for a large number of studies of thermal transitions of lipid material in aqueous suspensions. Setaram (Lyon, France) is marketing a DSC instrument which also utilizes the heat conduction principle and which seems to be sensitive enough for work in this area. Arntz (9) has recently described a DSC instrument for use at high pressure over a wide temperature range.

Several novel methods for heat capacity determinations have recently been developed: the short heat pulse method, the diffusion method, the a.c. method and the relaxation method (for references see (10)). With all these techniques, very small samples are used but the methods have not yet found any significant application in thermochemical work. The "a.c." technique, measuring amplitudes of periodic heat pulses propagated through the sample, have a very high resolution of heat capacity differences (0.01%) and temperature resolution of transitions (10^{-3}K), see e.g. (11). The method does not as yet seem to be able to be suitable for accurate heat capacity determinations. However, it is judged that a.c. calorimetry will develop rapidly

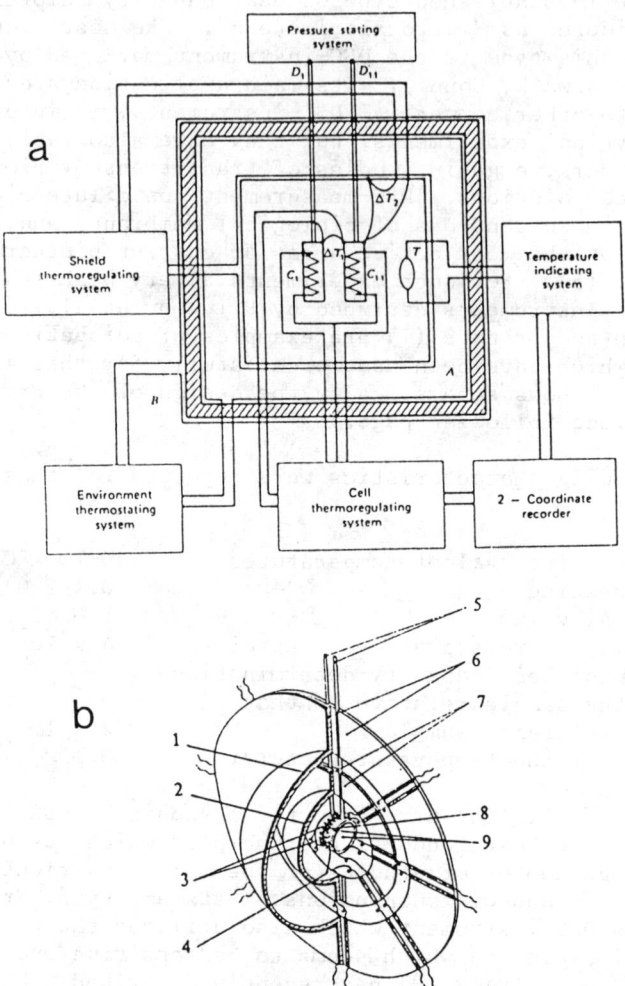

Fig. 1. Schematic diagram of the Privalov DSC-instrument (6).

a: A, adiabatic shield; B, thermostat; C, calorimetric cells; D, capillary inputs; ΔT, thermal sensors.

b: Calorimetric block with adiabatic enclosure: 1, internal shield heaters; 2, external shield heaters; 3, calorimetric chambers; 4, shield thermal sensor; 5, capillary inlets; 6, external shield rim; 7, internal shield rim; 8, thermopile; 9, calorimetric chamber heaters.

in the near future.

The Picker flow heat capacity calorimeter (12), commercially available from Setaram, has been used in many Cp-studies on dilute solutions. The principle of this instrument is very elegant. The instrument is shown schematically in Figure 2.

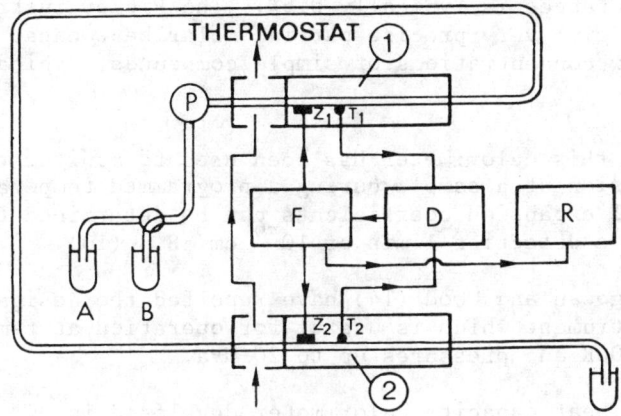

Fig. 2. Principles of the Picker flow heat capacity calorimeter (12). (A) solvent, (B) solution, (z_1, z_2) heaters, (T_1, T_2) thermistors, (P) pump, (F) feed back, (D) null detector, (R) recorder.

It consists of two flow cells, each equipped with a heater and a thermistor. The reference liquid A (e.g. pure water) is thermostated, passed through the flow cell (1) where it is heated, z_1, and the temperature T_1 is measured. The liquid is again thermostated and is passed through the second closely identical flow cell (2) where heat is supplied, z_2, and the temperature T_2 is measured as in cell (1). Once the steady-state condition has been reached, liquid B (e.g. a dilute aqueous solution) is circulated, but because of the length of the flow circuit, there is a time interval during which different liquids are passing through the two cells. A thermal feedback procedure maintains the same temperature gradient in both flow cells regardless of differences in heat capacities of the two liquids. The instrument thus measures the change in power input ΔP that is necessary to maintain the final temperature of the liquid in the "working flow cell" equal to that of the reference liquid in the other cell. Knowing the densities ρ^o and ρ for liquids A and B respectively, and the heat capacity C_p^o of liquid A, it is possible to calculate the heat capacity C_p for liquid B:

$$\frac{C_p}{C_p^o} = (1 + \frac{\Delta P}{P})\frac{\rho^o}{\rho} \tag{1}$$

It is thus not necessary to know the flow rate exactly nor the temperature rise in the flow cells, ΔT. The derived C_p value is a mean value referring to the temperature interval ΔT. This is usually a few K. The precision of the measured heat capacity differences is typically 0.5%. The Picker instrument has proved to give very precise partial molar heat capacity values for low concentrations of simple compounds, typically $\sim 1 J K^{-1} mol^{-1}$.

Recently this calorimeter has been used to monitor the thermal expansion of a sample during a programmed temperature scan. Thermal expansion coefficients can be determined to within 1% and a detection limit of 10^{-6} $cm^3 \cdot S^{-1}$ (13).

Smith-Magowan and Wood (14) have reported the design of a similar instrument which is useful for operation at temperatures up to 600K and pressures up to 20 MPa.

The drop heat capacity calorimeter developed in our laboratory (15) can be used for precise (0.01%) measurements of small solid or liquid samples (0.7 cm^3) at ambient temperatures. The instrument uses a twin thermopile heat conduction calorimeter as receiver (16). Absolute accuracies of heat capacity values are better than 0.1%.

MIXING AND DILUTION

In studies of various types of association processes, or of deviations from ideal behaviour of solutions and mixtures, mixing or dilution experiments are employed. To an increasing extent commercial microcalorimeters (batch, titration or flow, cf. e.g. (17)) are used in such work. It may be noted that flow calorimeters work without a gas phase and are thus very well suited for experiments involving volatile liquids.

Development of flow-mixing calorimetry at high pressure has recently attracted significant interest. Christensen et al. (18) have described an isothermal flow-mixing calorimeter, utilizing Peltier effect cooling, which can be operated at temperatures and pressures in the range of 273-343 K and 1-400 bar, respectively. Liquids are delivered at programable rates. Precision at a power of 3 mW is reported to be 0.4% (this power is significantly higher than normally used in "microcalorimetric" measurements). Another Peltier effect flow-mixing calorimeter working at pressures up to 600 bar has been described by Heintz and Lichtenthaler (19) and Wood and Smith-Magowan have recently reported a flow-mixing calorimeter used for aqueous solutions up to 600 K (20).

SOLUTION OF SLIGHTLY SOLUBLE COMPOUNDS

In solution experiments involving slightly soluble compounds, only small quantities of material are dealt with and sensitive instruments suitable for measurements of slow processes have to be used. Recently significant progress has been made in this field involving gaseous (21-24), liquid (25, 26) and solid (27, 28) solutes. To a significant extent the development work has been connected with studies of hydrophobic solutes in aqueous solution. In the work conducted in our laboratory (24, 25, 26, 28), a twin heat conduction calorimeter (16) employing insert flow solution vessels has been used. Fig. 3 (see next page) shows schematically our gas solution calorimeter (24). A known amount of gas (ca. 0.4 ml) is injected into the dissolution vessel, where it is dissolved by gas-free solvent which is pumped continuously through the vessel at a flow rate of about 40 ml/h. The instrument is currently used for studies of ΔH and ΔCp of solution in water of hydrocarbons and the rare gases.

BIOCHEMICAL REACTIONS

Measurements of reaction enthalpies in biochemical systems typically require microcalorimetric methods. Biopolymers have high molecular weights giving very small enthalpy changes per unit mass, e.g. in ligand-binding reactions. It is often necessary to work with dilute solutions of biopolymers (< 1%) in order to avoid unwanted aggregation. In addition, biochemical substances are often available only in µmole quantities. By far most experiments are today performed with well-established commercial instruments (see e.g. (17)). In some cases interesting modifications of these instruments have been employed. Cooper and Converse (29) have incorporated fiber optics light guides into an LKB batch microcalorimeter, enabling studies of photochemical reactions. Several workers have attached motor-driven syringes to the same instrument in order to speed up the determination of ligand-binding isotherms, and very recently a commercial titration unit has become available (LKB 2107-350), cf. (30).

Mountcastle et al (31) have described a continuous exponential dilution technique which is suited for use with flow microcalorimeters. It is claimed that the method can be used for precise generation of ligand-binding isotherms.

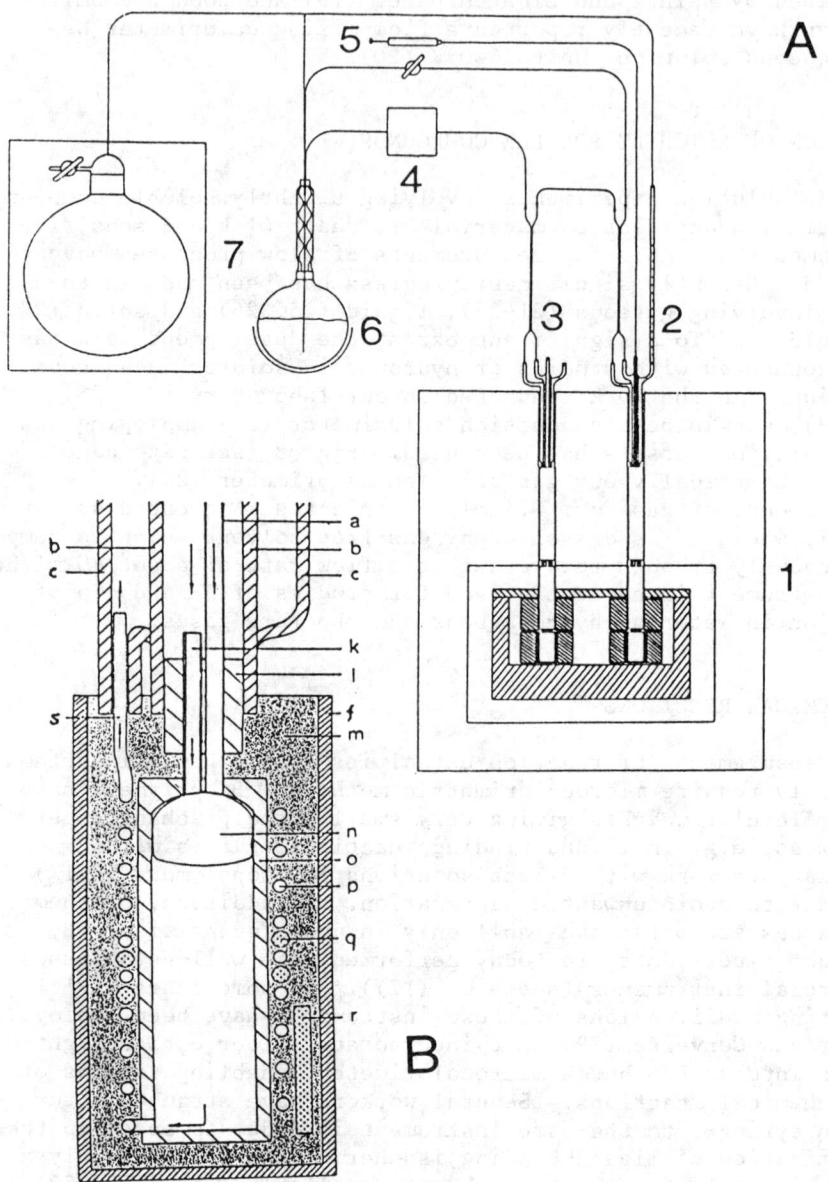

Fig. 3. Calorimeter for solution of slightly soluble gases (Gill and Wadsö, 24). A: instrument assembly. 1, twin calorimeter of the thermopile heat conduction type (16); 2, reaction cell; 3. reference cell; 4, peristaltic pump; 5, gas syringe; 6, still producing gas-free solvent; 7, buffer volume.

B: section through dissolution vessel (lowest part of the reaction cell). k, stainless steel tube; l, steel rod; m, Wood's metal filling; n, gas bubble; o, steel dissolution vessel; p, equilibration tube; q, r, electrical calibration heaters; solvent outlet.

A very sensitive adiabatic shield titration calorimeter has been reported by Spokane and Gill (32). Typical heat measurements range from 0.1-1 mJ with a reproducibility of about 0.01 mJ. Typical volume of injected liquid is 10 µl. The reaction cell is a thin-walled 1 ml glass bulb. Copper-plated constantan thermopiles are used for registration of the temperature change. A cross section of the instrument is shown in Fig. 4 (see next page).

As part of a new 4-channel microcalorimeter (cf. below), an insertion titration vessel was designed (33) which also can be used in connection with perfusion experiments on pieces of biological tissue. This calorimetry system also includes a new flow-mixing calorimeter, primarily intended for continuous flow experiments in biochemistry and biology. At low flow rates (ca. 10 + 10 ml), the useful sensitivity of this instrument is 0.1 µW.

Several simple thermistor-operated flow calorimeters, primarily used for substrate determinations and other analytical experiments, have been developed, see e.g. (34, 35). In the most simple case, a thermistor positioned in a small column containing immobilized enzyme will sense the difference in temperature between a solution containing substrate and a reference solution. Alternatively, thermistors may be positioned before and after the reaction zone of the column. The sensitivity of these instruments is significantly lower than "regular" micro-flow calorimeters.

MEASUREMENTS OF REACTION RATES

Microcalorimetry has repeatedly been suggested as a tool for obtaining reaction kinetic data. In the methodological work by Johnson and Biltonen (36), it was demonstrated that flow calorimeters with a rather high time constant, ca. 75 s, can be used for accurate determination of reaction rates for reactions whose half-lives exceed only a few s.

Nakamura (37) has designed a thermistor-operated stopped flow calorimeter for kinetic and enthalpy measurements of biochemical reactions.

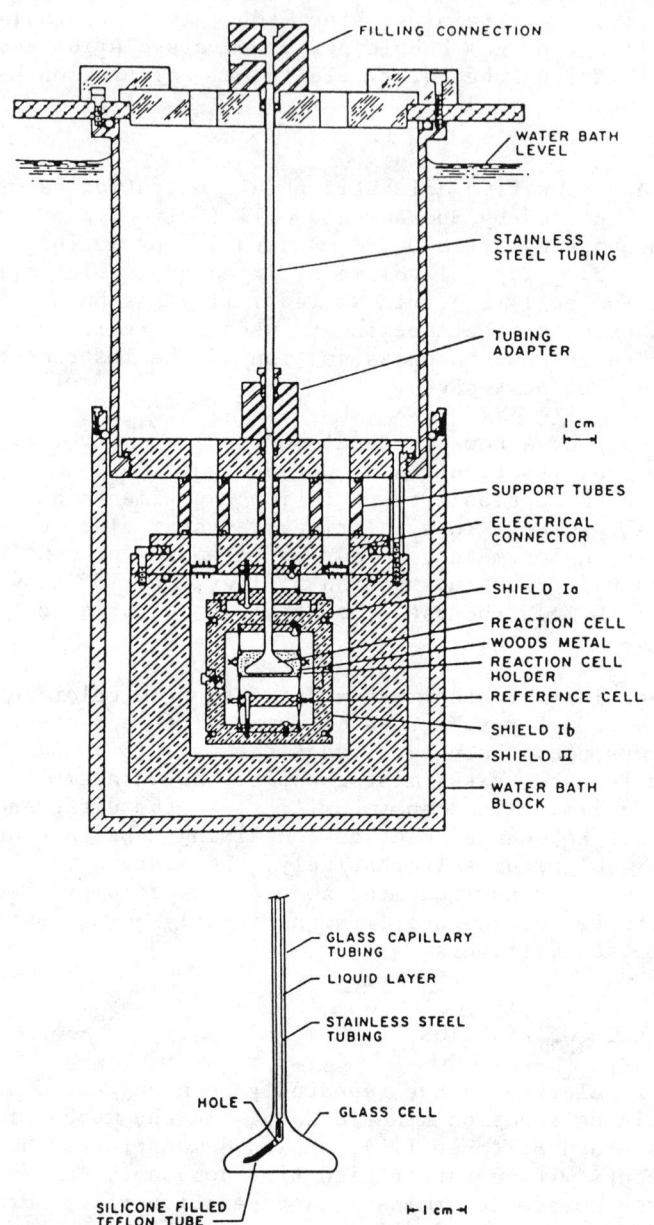

Fig. 4. Adiabatic shield titration calorimeter for very small sample volumes. Spokane and Gill (32).

Recently Berger and co-workers (38) described a thermistor-operated stopped-flow microcalorimeter for studies of very fast reactions. By use of an "uptake syringe" mechanically coupled to a set of driving syringes, the pressure-induced thermal disturbance normally encountered could be largely eliminated. The instrument can be used for studies of reactions with total reaction times close to the response time of the thermistor, 7 ms.

POWER MEASUREMENTS IN LIVING SYSTEMS

The thermal power produced by living systems provides a useful measure for the "activity" of the system. As such, power values are of immediate practical value in different areas of biological analysis, e.g. in biotechnology, ecology and in clinical work. A large amount of calorimetric methodological work of this nature has been conducted during the last few years, see e.g. (5). To a small but significant extent such studies are also concerned with thermochemical problems on a molecular level, i.e. the connection between identified chemical reactions and the observed power-time curves.

Frequently the power evolved in a biological sample (as well as the volume of the sample) is large and there is no need for microcalorimetric techniques. This is e.g. the case in studies on whole animals and on rapidly growing microbiological systems. In other cases, e.g. for purified fractions of blood cells, samples may only be available in small quantities (about 1 cm^3 of suspension, producing a minute power (a few $\mu W \cdot cm^{-3}$). This latter field currently presents a large number of challenging microcalorimetric problems, not the least in connection with the proper control of factors like pH, oxygen pressure, sedimentation and adhesion, see chapter by Wadsö in (5). By far most microcalorimeters employed in this field are of the thermopile heat conduction type and are mainly of commercial origin.

A new microcalorimetric system, mainly intended for work on living cells, was recently developed in our laboratory (3). It is a modular system where up to 4 different, or identical, microcalorimeters can be operated simultaneously using the same water thermostat. The system incorporates several instruments with different functions: flow-mixing and flow-through calorimeters, static ampoule instruments and perfusion and titration calorimeters. Figure 5A shows schematically the basic instrument assembly. It consists of a very precise water thermostat (d), (long term stability 10^{-4}K), an electric console (a), and calorimeters enclosed in steel cans (e). Normally the calorimeters are twin heat conduction calorimeters using semi-conduction thermopiles as heat flow sensors. Fig. 5B shows a

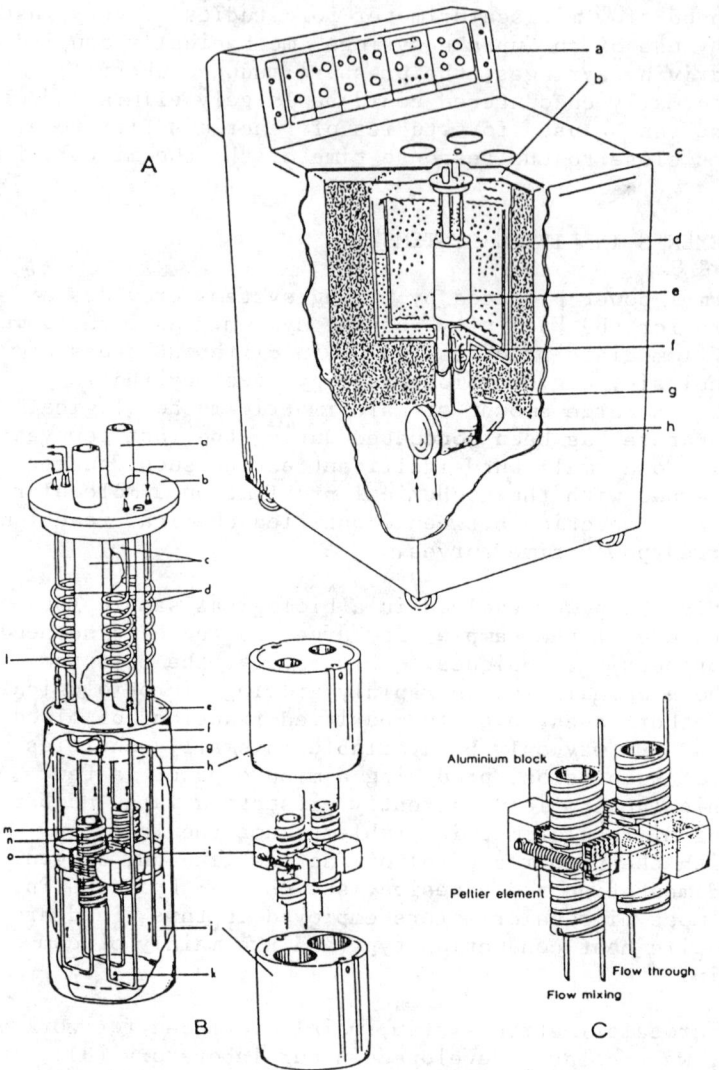

Fig. 5. A 4-channel microcalorimetric system (Suurkuusk and Wadsö, (31)).

"combination channel" with flow-through and flow-mixing vessels. The flow tubes are wound around thin-walled aluminium tubes, which can be used as holders for ampoules. Fig. 5C shows details of the twin calorimetric unit. The sensitivities and the baseline stabilities are improved significantly compared to our earlier flow (36) and ampoule (26) microcalorimeters.

OTHER POWER MEASUREMENTS

Sensitive calorimeters have long been employed as analytical tools for power measurements on unstable compounds like explosives and other organic substances and on radioactive materials. For characterization of thermally decomposable materials, DSC methods are currently used routinely in many laboratories. However, it would be of great practical as well as basic importance if decomposition power measurements could be conducted at ambient, or slightly increased, temperatures. The characterization of slow processes in metals, ceramics and plastic materials are also of significant interest. A special area with similar problems and prospects is the one dealing with calorimetric characterization of self-discharge of batteries, see e.g. (41).

The requirements on the calorimetric instrumentation for these types of power measurements are in many respects identical to those asked for in work on cellular systems, although the complex physiological aspects can be neglected.

SOME TRENDS AND NEEDS IN MICROCALORIMETRY

It is interesting to note that microcalorimeters in current use are to a large extent produced commercially. "Macro"-instruments appear to be laboratory products to a much larger degree. It is believed that this trend will be even more pronounced in the near future, in particular for calorimeters used in biochemistry and cell biology and for very sensitive instruments needed for studies of processes under isothermal conditions in technically important materials. For DSC instruments commercially available equipment is already completely dominating.

Future developments of microcalorimeters will probably concentrate to a large extent on flow and injection techniques which are well-suited for microprocessor-controlled operation. It is believed that the current significant development in microcalorimeters suited for high-pressure work will continue. Possibly pressure-scanning calorimeters will become available which should open up a new dimension in thermochemical-thermodynamic work. Calorimetric investigations on living cells represent a vast experimental area, where many specialized instrumental properties and working procedures are needed. Requirements for adequate sensitivity, stability and precision of the instruments are often satisfied today but the important characterization of the physiological conditions of the cells during the calorimetric measurements is frequently neglected. It is felt that further attention must be given to the design of specialized calorimetric vessels, where the conditions of the cells (e.g. supply of oxygen, sedimentation, adhesion, cell concentration in a flow vessel, pH, etc.) can be conveniently confirmed. More attention should also be given to the development of processes suitable for tests and calibrations of microcalorimeters used for these studies, in particular for flow vessels where electrical calibration procedures may be very uncertain. Recently a new test and calibration process was developed for this field (42).

REFERENCES

1. Månsson, M., in "Experimental Chemical Thermodynamics, Vol. 1, Combustion Calorimetry" (S. Sunner and M. Månsson, Eds.), pp. 388-394. Pergamon Press, Oxford, 1979.

2. Sunner, S. and Svensson, C., J.C.S. Faraday I (1979), 75, pp 2359-2365.

3. Adedeji, F. A., Brown, D. S., Connor, J. A., Leung, M. L., Paz-Andrade, J. M., and Skinner, H. A., J. Organometal. Chem. (1975), 97, pp 221-228.

4. Connor, J. A., Skinner, H. A., and Virmani, Y., J. Chem. Soc. Faraday Trans. 1 (1972, 68, pp 1754-1763.

5. Beezer, A. E. (ed.), "Biological Microcalorimetry", Academic Press, London, 1980.

6. Privalov, P. L.,. Plotnikov, V. V., Filimonov, V. V., J. Chem. Thermodynamics (1975),7. pp 41-47.

7. Jackson, W. M. and Brandts, J. F., Biochemistry (1970) 9, p. 2294.

8. Suurkuusk, J., Lentz, B. R., Barenhols, Y., Biltonen, R. L. and Thompson, T. E., Biochemistry (1976) 15, pp 1393-1401.

9. Arntz, H., Rev. Sci. Instr. (1980), 51, pp. 965-967.

10. Forgan, E. M. and Nedjat, S., Rev. Sci. Instr. (1980), 51, pp. 411-417.

11. Tanasijczuk, O. S. and Oja, T., Rev. Sci. Instr. (1978), 49, pp. 1545-1548.

12. Picker, P.,Leduc, P.-A., Philip, P. R. and Desnoyers, J., J. Chem. Thermodynamics (1971), 3, pp 631-642.

13. Fortier, J.-L., Simard, M.-A., Picker, P. and Jolicoeur, C. Rev. Sci. Instr. (1979), 50, pp. 1474-1480.

14. Smith-Magowan, D. and Wood, R. H., J. Chem. Thermodynamics (1981), 13, pp. 1047-1073.

15. Suurkuusk, J. and Wadsö, I., J. Chem. Thermodynamics (1974), 6, pp. 667-679.

16. Wadsö, I., Science Tools (1974), 21, pp. 18-21.

17. Spink, C. and Wadsö, I. in "Methods of Biochemical Analysis" (D. Glick, Ed.), Vol. 23, pp. 1-151. John Wiley and Sons, inc., New York, 1976.

18. Christensen, J. J., Hansen, L. D., Eatough, D. J. and Izatt, R. M., Rev. Sci. Instr. (1976), 47, pp. 730-734.

19. Heintz, A. and Lichtenthaler, R. N., Ber. Bunsenges. Phys. Chem. (1979), 83, pp. 853-856.

20. Wood, R. H. and Smith-Magowan, D., ACS Symp. Ser. 133, Thermodyn. Aqueous Systems. Industrial Applications (1979), pp. 569-580.

21. Cone, J., Smith, L. E. S. and van Hook, W. A., J. Chem. Thermodynamics (1979), 11, pp. 277-285.

22. Battino, R., and March, K. N., Aust. J. Chem. (1980), 33, pp. 1977-2003.

23. Krestov, G. A., Prorokov, V. N., and Dolotov, V. V., Zh. Fiz. Khim. (1982), 56. pp. 238-239.

24. Gill, S. J. and Wadsö, I., J. Chem. Thermodynamics (1982). In press.

25. Gill, S. J., Nichols, N. and Wadsö, I., J. Chem. Thermodynamics (1975), 7, pp. 175-183.

26. Nilsson, S.-O. and Wadsö. I. To be published.

27. Gill, S. J. and Seibold, M. L., Rev. Sci. Instr. (1976), 47, pp. 1399-1401.

28. Wadsö, I., to be published.

29. Cooper, A. and Converse, C. A., Biochemistry (1976), 15. pp. 2970-2978.

30. Chen, A.-t. and Wadsö, I., Biochemical and Biophysical Methods (1982), in press.

31. Mountcastle, D. B., Freire, E. and Biltonen, R. L., Biopolymers (1976), 15, pp. 355-371.

32. Spokane, R. B. and Gill, S. J., Rev. Sci. Instr. (1981). 52. pp. 1728-1733.

33. Suurkuusk, J. and Wadsö, I., Chemica Scripta (1982), in press.

34. Danielsson, B., Gadd, K., Mattiasson, B. and Mosbach, K., Clinica Chimica Acta (1977), 81. pp. 163-175.

35. Bowers, L. D. and Carr, P. W., Clin. Chem. (1976), 22, pp. 1427-1433.

36. Johnson, R. E. and Biltonen, R. L., J. Am. Chem. Soc. (1975), 97, pp. 2349-2355.

37. Nakamura, T., J. Biochem. (1978), 83, pp. 1077-1083.

38. Bowen, P., Balko, B., Blevins, K., Berger. R. L. and Hopkins, H. P., Anal. Biochemistry (1980), 102, pp. 434-440.

39. Nakamura, T. and Matsuoko, I., J. Biochem. (1978), 84, pp. 34-46.

40. Monk, P. and Wadsö. I., Acta Chem. Scand (1968), 22, pp. 1842-1852.

41. Prosen, E. J. and Colbert, J. C., J. Res. Natl. Bur. Stand. (1980), 85, pp. 193-203.

42. Chen, A.-t. and Wadsö, I., Biochem. Biophys. Methods (1982), in press.

DIFFERENTIAL SCANNING CALORIMETRY

C.T. Mortimer

Chemistry Department, University of Keele, Keele, Staffs., ST5 5BG, U.K.

ABSTRACT

The purpose of this paper is to review briefly the technique of differential scanning calorimetry (1), and to indicate how it may be used to determine enthalpies of phase change and of certain chemical reactions.

TECHNIQUE

When the temperature of a substance is raised or lowered, physical or chemical changes may occur; for example, a phase change from solid to gas, the transition from one crystalline form to another, or a chemical change. The differential scanning calorimeter has been designed to determine the enthalpies of these changes by measuring the differential heat flow required to maintain a sample of the substance and an inert reference at the same temperature. The temperature is usually programmed to scan a temperature range by increasing linearly at a predetermined rate, but the apparatus can also be used isothermally.

Historically, the scanning calorimeter was developed from the differential thermal analysis apparatus, Figure 1(a). Here, the

Figure 1

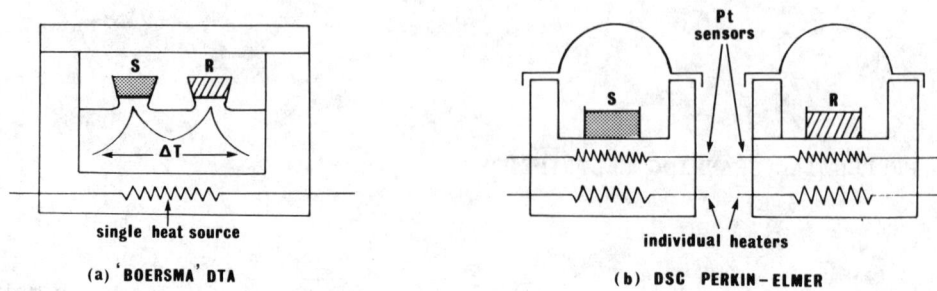

(a) 'BOERSMA' DTA (b) DSC PERKIN-ELMER

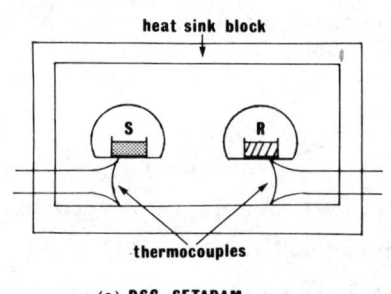

(c) DSC SETARAM

sample and reference are heated by a single source and temperatures are measured by sensors attached to pans which contain the sample and reference materials. A plot is made of the temperature difference, $\Delta T = T_s - T_r$, between the sample and reference, against time. The magnitude of ΔT, at a given time, is proportional to (a) the enthalpy change, (b) the heat capacities and (c) the thermal resistance to heat flow. With this apparatus it is not easy to make a simple conversion of peak height from the plot of ΔT against time, into energy units. This difficulty of conversion is largely overcome in the scanning calorimeter. Two of the available commercial instruments are (i) the Perkin Elmer DSC-2 and (ii) the Setaram DSC 111.

The Perkin Elmer DSC-2 apparatus is schematically represented in Figure 1(b), which shows the sample and reference each provided with individual heaters. This makes it possible to

use a 'null-balance' principle. It is convenient to think of
the system as divided into two control loops, shown schematically
in Figure 2. One is for average temperature control, so that
the temperature, T, of the sample and reference may be increased

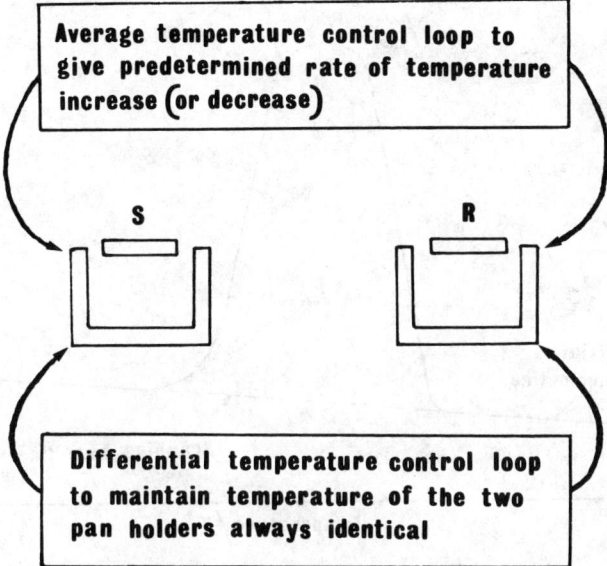

Figure 2. Schematic representation of the DSC control loops

at a predetermined rate, which is recorded. The second loop
ensures that if a temperature difference develops between the
sample and reference (because of exothermic or endothermic
reaction in the sample), the power input is adjusted to remove
this difference. This is the null-balance principle. Thus,
the temperature of the sample holder is always kept the same as
that of the reference holder by continuous and automatic
adjustment of the heater power. A signal, proportional to the
difference between the heat input to the sample and that to the
reference, dH/dt, is fed into a recorder. In practice, this
recorder is also used to register the average temperature of the
sample and reference.

Figure 3 shows an idealised thermogram or record of the differential heat input dH/dt, against temperature T (or time, t, on the same axis).

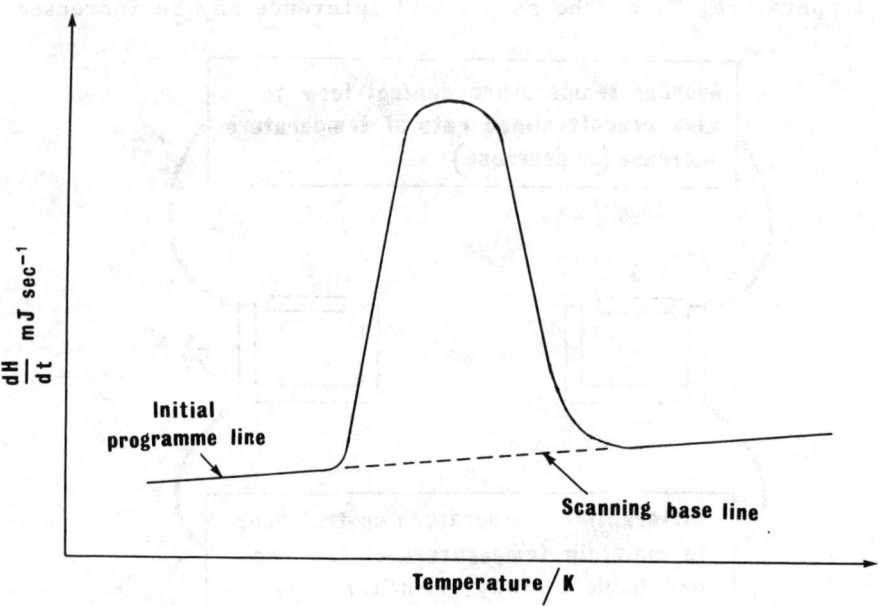

Figure 3. Idealised DSC thermogram.

The Setaram DSC 111 apparatus reverts to the use of a single heater for both sample and reference, although it takes the form of a thermostatically programmed block, through which run two thin refractory tubes, shown in section in Figure 1(c). The tubes are connected thermally to the block by thermocouples, which can be used to monitor the difference between the heat flow to the sample and reference. The principle is the same as that used in the Tian-Calvet microcalorimeters (2). Both the Perkin Elmer and Setaram calorimeters have similar maximum sensitivity of about 1.0 mJ sec^{-1} for a full scale deflection. They use milligram quantities of sample and can be operated in the temperature range from about -150 to +800°C. The apparatus can be calibrated by dissipating electrically generated heat from a small wire coil placed in the sample pan, or by use of high purity metals, of accurately known enthalpies of fusion, as calibration standards. The most commonly used calibrant is indium, $\Delta H(\text{fusion}) = 28.5 \text{ J g}^{-1}$, m.p. $156.4°C$. Between 5 and 10 mg are weighed in an aluminium sample pan on a microbalance and a thermogram of the melting peak is run at a selected heating rate, (dT/dt), sensitivity (range) and chart speed of the recorder. The scanning base line (Figure 3) is drawn in from the point at which the trace begins to depart from the initial programme line to the point at which the trace returns to the programme line. The area between the interpolated base line and peak is integrated by some device, usually a planimeter. The information is then used to calculate the calibration constant, k, in mJ (unit area)$^{-1}$ from the relationship

$$k = \frac{\Delta H(\text{fusion}) \times m_c}{A_c} \quad / [\text{mJ (unit area)}^{-1}]$$

where $\Delta H(\text{fusion})$ is the enthalpy of fusion of the calibrant in mJ mg^{-1}, m_c is the mass of calibrant in milligram and A_c is the peak area of the calibration thermogram. Table 1 shows the enthalpies of fusion of other metals which can be used to calibrate at higher temperatures.

Table 1.

Enthalpies of fusion (Jg^{-1}) of metals

	Tin	Lead	Zinc	Aluminium
M.p./°C	231.9	327.4	419.5	660.4
$\Delta H(\text{fusion})$	59.6 ± 1.0	23.2 ± 0.6	111.3 ± 2.1	387.9 ± 4.6

ENTHALPIES OF FUSION AND VAPORIZATION

In an open system the solid→liquid change is sharp and independent of pressure, but the liquid→vapour change is controlled by the pressure and will only occur at a constant temperature if the vapour pressure over the liquid is kept constant. Melting is also considered as a more or less instantaneous process, whereas boiling requires mass transfer from the liquid surface to the vapour phase. Consequently, the peak shape will depend on the rate of vaporisation, which is a function of temperature and of the rate at which vapour is removed. The surface area of the liquid and the size of the orifice through which the vapour escapes to the surroundings are therefore important factors.

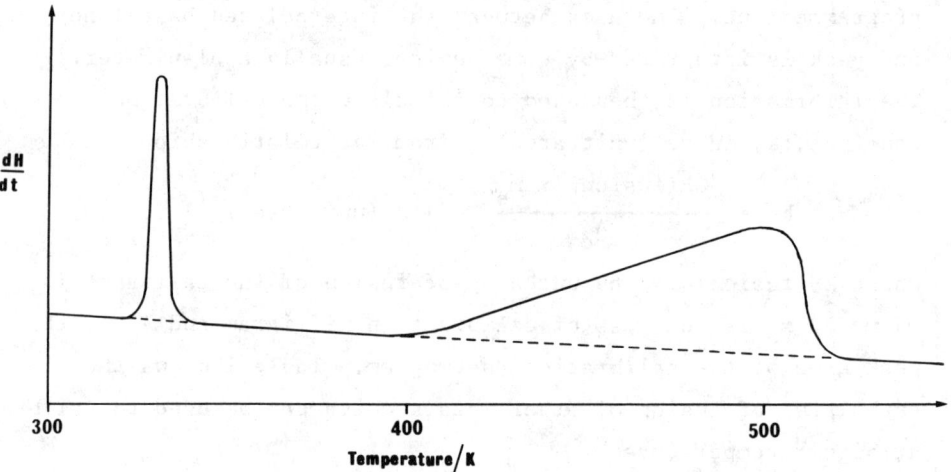

Figure 4. Fusion and vaporization of Salicylaldoxime

Figure 4 shows a thermogram of the fusion and vaporization of a sample of salicylaldoxime. Melting is sharp at 336 K, but vaporization occurs over a wide temperature range 400-530 K. Values obtained were $\Delta H(\text{fusion}) = 15.4 \pm 0.1$ kJ mol^{-1} and $\Delta H(\text{vaporization}) = 81.3 \pm 9$ kJ mol^{-1}.

SPECIFIC HEATS

When a sample material is subjected to a linear temperature

increase, the rate of heat flow into the sample is proportional to its instantaneous specific heat. By regarding this rate of heat flow as a function of temperature and comparing it with a standard material under the same conditions, we can obtain the specific heat, C_p, as a function of temperature. The procedure is briefly described as follows. Empty aluminium pans are placed in the sample and reference holders and a base line is recorded as shown in Figure 5. The procedure is repeated with a known mass of sample in the sample pan and a trace of dH/dt against

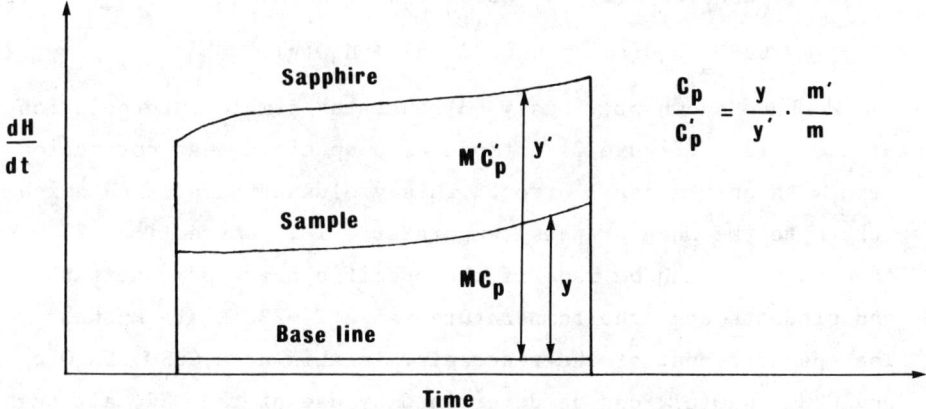

Figure 5. Specific heat determination by ratio method.

time is recorded. There is a pen displacement due to the absorption of heat by the sample, and we may write

$$\frac{dH}{dt} = mC_p \cdot \frac{dT}{dt}$$

where m is the mass of the sample in grams, C_p is the specific heat in cal g^{-1} and dT/dt is the programmed rate of temperature increase.

This equation could be used to obtain values of C_p directly, but any errors in ordinate read-out, dH/dt, and in programming rate, dT/dt, would reduce the accuracy. To minimise these errors the procedure is repeated with a known mass of sapphire, the specific heat of which is well established, and a new trace is recorded. Thus, only two ordinate deflections at the same temperature (y and y' of Figure 5 are required to yield a ratio of the C_p values of sample and sapphire.

Correction of $\Delta H T_p$ to ΔH^{298}. The enthalpy of a reaction, derived from the DSC thermogram, must be corrected for the difference between the specific heats of the reactant and products. This is particularly important for a reaction in which a solid compound dissociates, on heating, with the loss of a gaseous product, e.g.

$$CuSO_4 \cdot 5H_2O(c) \rightarrow CuSO_4 \cdot H_2O(c) + 4H_2O(g) \qquad (1)$$

$$CuSO_4 \cdot H_2O(c) \rightarrow CuSO_4(c) + H_2O(g) \qquad (2)$$

We have shown previously (3) that the simple interpolation of a base line (Figure 3) introduces a specific heat correction term with only a small error. This yields a value of ΔH which refers to the mean or peak temperature, T_p. Correction of ΔH_{T_p} to ΔH^{298} can be made if the specific heats of reactants and products over the temperature range $T_p - 298$ K are known. The specific heat of the reactants, in this case $CuSO_4 \cdot 5H_2O(c)$ and $CuSO_4 \cdot H_2O(c)$, can be determined by use of the DSC, although some extrapolation of experimental data is required to obtain values close to T_p, where the onset of decomposition has occurred.

Although separate measurements of these specific heats are desirable, a simpler experimental approach can be adopted. The thermogram of thermal decomposition is recorded, line x of Figure 6. The thermogram of the solid product is recorded, line y. Line Z is that for an empty sample pan. The area a is a measure of ΔH at temperature T_p, whilst the area (a+b) represents this value of ΔH corrected to 298 K for the specific heats of reactant (e.g. $CuSO_4 \cdot 5H_2O$) and product (e.g. $CuSO_4 \cdot H_2O$), though not that for the gaseous product ($4H_2O$).

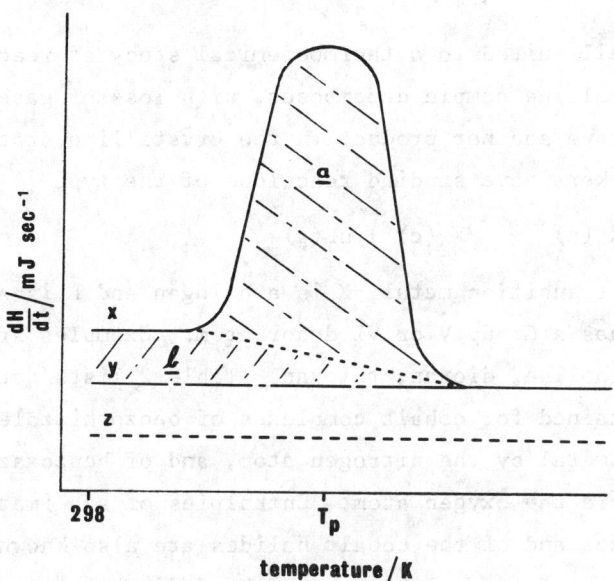

Figure 6. Correction of ΔH^{T_p} to ΔH^{298}

For reactions (1) and (2) the enthalpies are ΔH^{370} = +213 ± 2.5, ΔH^{510} = +72.4 ± 0.8 which yield ΔH^{298} = +215 ± 2.5 and ΔH^{298} +73.2 ± 0.8 kJ mol^{-1} respectively. These latter values can be compared with the 'selected' values of +226.6 and +72.6 kJ mol^{-1}, derived from solution calorimetry. There is clearly some discrepancy between the ΔH^{298} values for reaction (1). Barnard et al. (4) have studied reactions of the following type.

$$\text{Ni(glycinate)}_2 \cdot 2H_2O(c) \rightarrow \text{Ni(glycinate)}_2(c) + 2H_2O(g) \quad (3)$$
$$\text{Cu(alaninate)}_2 \cdot 2NH_3(c) \rightarrow \text{Cu(alaninate)}_2(c) + 2NH_3(g) \quad (4)$$

Values of ΔH^{298} have been calculated from enthalpies of formation derived from solution calorimetry and from scanning calorimetry data at 400 K, corrected to 298 K. Results agree to within about ±5 kJ mol^{-1}. Incidentally, this is the difference between the enthalpies of formation of the metal amino-acid complexes obtained in different thermochemical laboratories (5).

TRANSITION METAL COMPLEXES

DSC is well suited to a thermochemical study of reactions in which a crystalline sample decomposes, with loss of gaseous ligand, to leave another product in the crystalline state. A number of workers have studied reactions of the type

$$ML_nX_2(c) \rightarrow MX_2(c) + nL(g)$$

where M is a transition metal, X is a halogen and L is a ligand which often has a Group V or VI donor atom. Examples are pyridine, quinoline, dioxan, thioxan. Table 2 lists data which have been obtained for cobalt complexes of benzothiazole, which bonds to the metal by the nitrogen atom, and of benzoxazole, which bonds via the oxygen atom. Enthalpies of sublimation of these complexes and of the cobalt halides are also known, so that it is possible to calculate mean bond dissociation energies $\overline{D}(Co-N)$ and $\overline{D}(Co-O)$. The values are very similar.

Table 2
Enthalpies of decomposition of benzothiazole and benzoxazole complexes (6), values in kJ mol^{-1}

Complex[a]	ΔH_d[b]	ΔH_{sub}[c]	$\overline{D}(M-N)$
$Co(bt)_2Br_2$	160.2 ± 1.7	124.2 ± 4.2	129.1 ± 3
$Co(2Mebt)_2Cl_2$	167.8 ± 2.5	122.6 ± 1.3	139.6 ± 2
$Co(2Mebt)_2Br_2$	161.1 ± 1.7	113.8 ± 4.2	134.7 ± 3
			$\overline{D}(M-O)$
$Co(2Mebo)_2Cl_2$	146.9 ± 1.2	92.4 ± 2.5	144.3 ± 2
$Co(2Mebo)_2Br_2$	160.2 ± 3.3	111.3 ± 4.2	135.5 ± 4
$Co(2,5diMebo)_2Cl_2$	156.6 ± 2.1	95.4 ± 4.6	147.5 ± 3
$Co(2,5diMebo)_2Br_2$	138.9 ± 2.9	104.6 ± 5.8	128.2 ± 4

a (bt) benzothiazole   (bo) benzoxazole

b $ML_2X_2(c) \rightarrow ML_2(c) + 2L(g)$, at 520-550 K
c at 345 - 390 K.

The effect on the strength of the cobalt-nitrogen bond, caused by substitution in the pyridine ring, can be seen from the data shown in Table 3, which have also been obtained from scanning calorimetry.

Table 3

Enthalpies of decomposition of methyl- and halogen-substituted pyridine complexes (3,7), values in kJ mol^{-1}

Complex	ΔH_d [b]	ΔH_{sub} [c]	$\bar{D}$(M-N)
Co(py)$_2$Cl$_2$ [a]	119.7 ± 2.1	(100 ± 10) [d]	127 ± 6
Co(2Mepy)$_2$Cl$_2$	109.6 ± 2.1	86.6 ± 3.8	128.5 ± 3
Co(2Clpy)$_2$Cl$_2$	115.1 ± 1.3	101.2 ± 6.7	124.0 ± 4
Co(3Clpy)$_2$Cl$_2$	134.7 ± 2.5	77.0 ± 4.2	145.7 ± 3

[a] blue tetrahedral form
[b] ML$_2$X$_2$(c) → MX$_2$(c) + 2L(g), at 480 - 560 K
[c] ML$_2$X$_2$(c) → ML$_2$X$_2$(g), at 345 - 365 K
[d] estimated.

Bleijerveld and Vrieze (8) have used scanning calorimetry in the isothermal mode to measure the enthalpy of replacement of CH$_3$CN in W(CO)$_5$(CH$_3$CN), by CO. Incorporation of the enthalpies of sublimation of reactant and product leads to a value for the

$$W(CO)_5(CH_3CN)(c) + CO(g) \xrightarrow[kJ\ mol^{-1}]{\Delta H^{298}=70.7} W(CO)_6(c) + CH_3CN(g)$$

$$\Delta H_{sub} = 48.1 \text{ kJ mol}^{-1} \qquad \Delta_{sub} = 17.7 \text{ kJ mol}^{-1}$$

$$W(CO)_5(CH_3CN)(g) + CO(g) \xrightarrow[kJ\ mol^{-1}]{\Delta H^{298}=96.7} W(CO)_6(g) + CH_3CN(g)$$

enthalpy of the gas-phase process at 298 K. This enthalpy is a direct measure of the difference between the bond dissociation energies

$$D(W-CO) - D(W-CH_3CN) = 96.7 \text{ kJ mol}^{-1}$$

Both Pt(II) and Ir(I) form complexes of the type PtX(CH$_3$)A$_2$L and IrX(CO)A$_2$L, in which X is a halogen, A is a phosphine or arsine and L is a substituted olefin or acetylene. On heating, the olefin or acetylene is lost to yield the 4-coordinate, trans isomer, e.g.

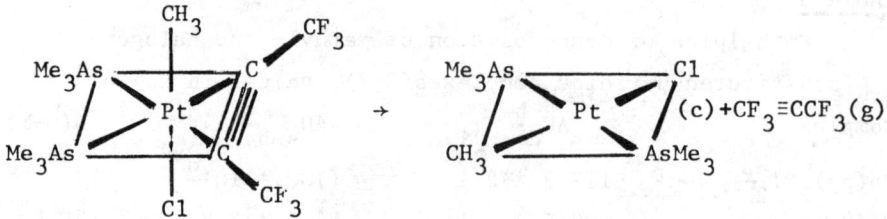

Scanning calorimetry has been used (9) to good effect to show how the enthalpies of such reactions vary with (a) different halogens bonded to the metal, (b) different Group V ligands and (c) either tetrafluoroethylene or hexafluorobut-2-yne as the unsaturated organic ligand. Details of some results are shown in Table 4. Because of the lack of enthalpies of sublimation of the complexes, it has not been possible to correct the measured enthalpies to refer to the gas phase. The assumption has been made that differences between enthalpies of sublimation of solid reactant and product, e.g. PtCl(CH$_3$)(AsMe$_3$)$_2$C$_2$F$_4$ and PtCl(CH$_3$)(AsMe$_3$)$_2$, will be reasonably constant and small for differing halogen, Group V ligand and organic ligand. If this assumption is justified, then the measured enthalpies of decomposition will be close to the gas-phase dissociation energies, D(metal-olefin) and D(metal-acetylene).

Considering first the data for the platinum compounds, we note the following points.
(i) The ΔH values are all in the range 40-80 kJ except for the compound PtCl(CH$_3$)(PPh$_3$)$_2$(C$_2$CN$_4$) where the value is much higher at 142 kJ mol^{-1}.
(ii) Where AsMe$_3$ is the Group V ligand, ΔH values are larger for X=Cl than X=Br. With AsMe$_2$Ph, this trend is reversed.

For the compounds IrX(CO)(PPh$_3$)$_2$L, we have ΔH values (11)

Table 4

Enthalpies of reactions $PtX(CH_3)A_2L(c) \rightarrow \underline{t}\text{-}PtX(CH_3)A_2(c) + L(g)$

X	L	A	T/K	$\Delta H/\text{kJ mol}^{-1}$
Cl	C_2F_4	$AsMe_3$	355	51.4 ± 0.4
Br	C_2F_4	$AsMe_3$	365	48.1 ± 0.8
Cl	C_4F_6	$AsMe_3$	350	68.6 ± 1.6
Br	C_4F_6	$AsMe_3$	345	61.1 ± 1.6
Cl	C_4F_6	$AsMe_2Ph$	310	73.2 ± 1.2
Br	C_4F_6	$AsMe_2Ph$	320	80.3 ± 0.8
Cl	$C_2(CN)_4$	PMe_2Ph	298	142.3 ± 0.4[a]

[a] from solution calorimetry

Enthalpies of reactions $IrX(CO)(PPh_3)_2L(c) \rightarrow \underline{t}\text{-}IrX(CO)(PPh_3)_2(c) + l(g)$

X	L	T/K	$\Delta H/\text{kJ mol}^{-1}$
F	C_2F_4	480	79.5 ± 1.6
Cl	C_2F_4	450	67.4 ± 1.6
Br	C_2F_4	450	41.0 ± 0.8
I	C_2F_4	460	57.3 ± 2.0
F	C_4F_6	450	99.2 ± 0.4
Cl	C_4F_6	430	95.8 ± 1.6
Br	C_4F_6	450	78.7 ± 0.4
I	C_4F_6	480	82.4 ± 2.8

for a wider range of halogens bonded to the metal. A plot of ΔH (for the removal of either C_2F_4 or C_4F_6) against the electronegativity of the halogen, X, shows a minimum value at X = Br. These trends in ΔH values can be explained in terms of the molecular orbital description of metal-olefin and metal-acetylene bonds in these compounds.

REFERENCES

1. For a more comprehensive review see, for example, McNaughton, J.L. and Mortimer, C.T., 1975, "Differential Scanning Calorimetry", International Review of Science, Physical Chemistry Series Two, Vol. 10, Ed. H.A. Skinner, Butterworths, p.1.
2. Calvet, E. and Prat, H., 1963, "Recent Progress in Microcalorimetry" translated by H.A. Skinner, Pergamon, Oxford.
3. Beech, G., Mortimer, C.T. and Tyler, E.G., 1967, J. Chem. Soc. A, p.925.
4. Barnard, M.A., Bois, N. and Daireaux, M., 1967, Thermochim. Acta, 16, p.283.
5. Burkinshaw, M.P. and Mortimer, C.T., 1982, Coord. Chem., in press.
6. Mortimer, C.T. and McNaughton, J.L., 1973, Thermochim. Acta, 6, p.269.
7. Mortimer, C.T. and McNaughton, J.L., 1974, Thermochim. Acta, 10, p.125.
8. Bleijerveld, R.H.T. and Vrieze, K., 1976, Inorg. Chim. Acta, 19, p.195.
9. Mortimer, C.T., McNaughton, J.L. and Puddephatt, R., 1972, J. Chem. Soc. Dalton, p.1265.
10. Mortimer, C.T. and Wilkinson, M.P., unpublished results.
11. Mortimer, C.T., McNaughton, J.L., Burgess, J., Hacker, M.J., Kemmitt, R.D.W., Bruce, M.I., Shaw, G. and Stone, F.G.A., 1973, J. Organomet. Chem., 47, p.439; 1974, J. Organomet. Chem., 71, p.287.

AN OXYGEN FLOW CALORIMETER FOR DETERMINING THE HEATING VALUE OF
KILOGRAM-SIZE SAMPLES OF MUNICIPAL SOLID WASTE

S. Abramowitz, E. S. Domalski, K. L. Churney, A. E.
Ledford, R. V. Ryan and M. L. Reilly

National Bureau of Standards, Washington, D.C. 20234,
U.S.A.

ABSTRACT

A new calorimeter is being developed at the National Bureau of Standards to determine the enthalpies of combustion of kilogram-size samples of minimally processed municipal solid waste (MSW) in flowing oxygen near atmospheric pressure. The organic fraction of 25 gram pellets of highly processed MSW has been burned in pure oxygen to CO_2 and H_2O in a small prototype calorimeter. The carbon content of the ash and the uncertainty in the amount of CO in the combustion products contribute calorimetric errors of 0.1 percent or less to the enthalpy of combustion. Large pellets of relatively unprocessed MSW have been successfully burned in a prototype kilogram-size combustor at a rate of 15 minutes per kilogram with CO/CO_2 ratios not greater than 0.1 percent. The design and construction of the kilogram-size calorimeter has been completed. Preliminary trial combustions are being performed.

INTRODUCTION

NBS is developing a calorimeter to determine the calorific value of kilogram-size samples of minimally processed municipal solid waste (MSW). A large scale calorimeter affords greater credibility because kilogram-size samples of MSW should represent the properties of the heterogeneous bulk material more reliably than the highly processed gram-size samples currently used in bomb calorimetric determinations. This calorimeter will also be useful to determine the calorific values of other heterogeneous fuels.

Combustion of samples in flowing oxygen near atmospheric pressure rather than in the high pressure oxygen of a combustion bomb was adopted for safety considerations. The flow technique has not been used in any substantial way to determine the enthalpy of combustion of solids since the 1880's. Its development was discontinued because of the simpler techniques and more quantitative results obtained with the bomb calorimeter. Attaining complete combustion was particularly difficult as were the problems in quantitativly determining the states of the initial and final products. As a consequence, the first task of the NBS project was to demonstrate that the oxygen flow technique could be used to obtain the calorific values.

To calculate the calorific value of the combustible fraction of MSW, the ash produced in each experiment must be determined quantitatively. Ash collection is simplified by burning pellets of the sample in such a way as to prevent ash dispersal. Therefore the combustors were designed to burn pellets.

The first step in the program was to build a calorimeter for combustion of 25 gram pellets of highly processed MSW. The purpose was to establish the equivalence of the flow results with those obtained by bomb calorimetry with the new NBS 25 gram capacity bomb calorimeter (1,2). The preliminary results obtained with the 25 gram flow calorimetric system are described in the following section.

Experiments with kilogram-size samples of minimally processed MSW were undertaken as the second step of the NBS program. The primary aim was to develop a method of burning a large heterogeneous sample completely with a minimum scattering of ash. Calorimetric run-time considerations required the combustion time be 15 min/kg or less. Additional experiments were carried out to test the suitability of cellulose as a combustion calibrant material for the kilogram-size calorimeter. These experiments are described in the middle section.

The design of the new kilogram-size calorimeter which is based on the results of the initial work is discussed in the final section. Some initial results of trial combustions have been recently obtained.

25 GRAM FLOW EXPERIMENTS

The combustor used in the 25 gram oxygen flow experiments is shown in Fig. 1. The sample pellet (D of Fig. 1) was placed on a quartz plate (F) which had eight radial slots and a small central hole which permitted circulation of oxygen beneath the sample. The plate sat in a quartz crucible (E) which was supported by a

AN OXYGEN FLOW CALORIMETER

nichrome stand (J). Primary oxygen (H) was supplied locally to the sample through a three-port tubular quartz ring which was located immediately above the crucible. The lower edge of the Pyrex thermal shield (C) enclosed the ring and upper edge of the crucible. The outer boundary of the combustor consisted of the stainless steel top (N) and base (I) plus the Pyrex wall (L) which were sealed by rubber gaskets (B). A secondary flow of oxygen (G) swept down between the wall of the combustor and the thermal shield and confined the flow of the products of combustion to the interior of the thermal shield. Gases left the combustor through a stainless steel exit line (A) in the top. The temperatures of the product gases near the exit port of the thermal shield

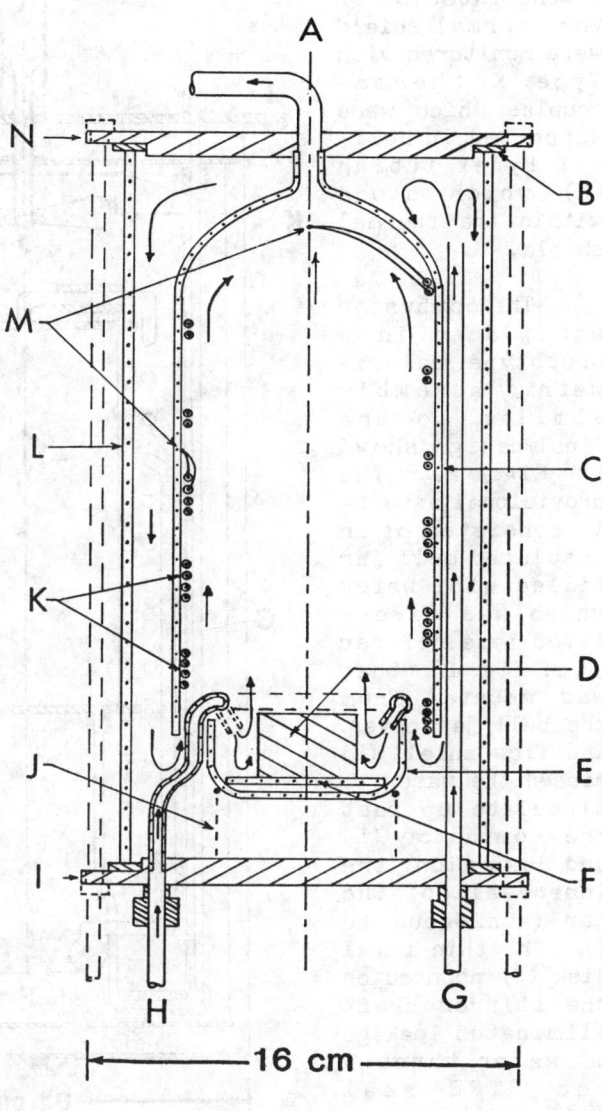

Fig. 1. The 25 Gram Combustor. A denotes the product gas exit line, B the gasket, C the thermal shield, D the RDF-4 sample, E the quartz crucible, F the quartz plate, G the secondary oxygen, H the primary oxygen, I the combustor base, J the crucible support, K the thermocouple helixes, L the combustor walls, M the thermocouple junctions and N the combustor top.

and of the midpoint of the interior of the thermal shield were monitored with Type K thermocouples which were supported in helical Pyrex tubing (K) which stood within the thermal shield.

The combustor was placed in a prototype calorimetric assembly similar to the final design shown in Fig. 2. The provisional assembly consisted of an insulated bell jar filled with water which was circulated by a stirrer (J of Fig. 2) that was mounted from the bell jar cover. The flow shield (E) caused the water to circulate up past the combustor (L) and down along the inner wall of the jar (analogous to B). The thin metal disk (J) mounted on the stirrer shaft eliminated leakage of water through the PTFE seal between the stirrer shaft and the bell

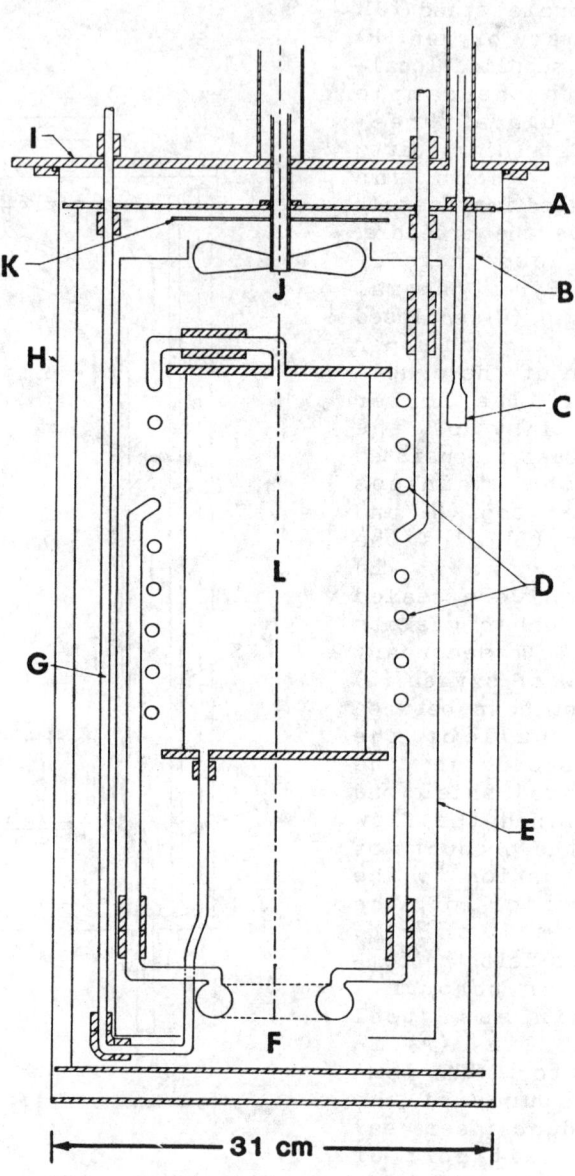

Fig. 2. The 25 Gram Flow Calorimeter. A denotes the vessel lid, B the vessel can, C the thermometer, D the exit gas coils, E the flow shield, F the water trap, G an oxygen supply line, H the submarine can, I the submarine lid, J the stirrer, K the metal disk, and L the combustor.

jar cover (cf. between the shaft and A of Fig. 2). The temperature of the water was measured with a long stem platinum resistance thermometer (C). The product gases flowed through the helical exit gas coil (D) into the water trap (F) and then through a second coil before leaving the calorimeter. The coils and trap were made from Pyrex.

After leaving the calorimeter the product gases were scrubbed by bubbling through water and then the entire gas stream was passed successively through a variable orifice flowmeter and filter-type infrared monochrometers for measuring the concentrations of CO and CO_2 in real time. Preliminary analysis of the product gases by both mass spectrometry and conventional infrared spectroscopy showed CO to be the only species due to incomplete combustion.

In addition to temperatures within the combustor, Type K thermocouples were used to measure the temperature of the oxygen as it entered the calorimeter and the temperature of the product gases as they left the calorimeter as well as at the CO and CO_2 detectors. The pressure of the product gases at the CO and CO_2 detectors was monitored with a digital capacitance manometer.

Sample pellets for the 25 gram experiments were prepared from a blended powder of minus 0.5 mm particle size. The powder was made by milling large batches of minus 15 cm MSW from which most of the metals, glass and entrained inorganics had been first removed. This powder is referred to as RDF-4. The cylindrical pellets, which were 3.5 cm in diameter and 2.5 cm high, were prepared by pressing the powder in a die using a force ranging from 45 to 160 kN. The residual moisture content of the powder was determined to be about 5% using the ASTM Standard Test E 790-81 (Residual Moisture in a Refuse-Derived Analysis Fuel Sample).

In a typical experiment, the calorimeter was assembled and the combustor was flushed with pure oxygen. The temperature of the calorimeter water was recorded as a function of time for approximately twenty minutes after a steady drift rate was attained. Just before ignition, the primary and secondary oxygen flow rates were set at approximately 8 and 5 L/min, respectively. Except for two experiments, the inlet oxygen was saturated with water vapor. The sample was ignited by passing electrical current through an iron fuse wire (not shown in Fig. 1) which was in contact with the top of the pellet. A 25 gram pellet typically burned within fourteen minutes.

Upon ignition the pellet burned with a diffusion flame which spread over its entire surface within the first 30 seconds. The flame temperature was determined to be greater than 1500°C. A small but detectable peak in the CO concentration occurred at about 3 minutes. The surface flame disappeared at about the eighth minute. The glowing sample continued to burn internally leaving a porous ash structure that had almost the same shape as the original pellet. The product gases contained increasing amounts of CO, which reached a peak near the eleventh minute. However, the ratio of CO to CO_2 remained low. A typical $CO-CO_2$ composition profile of the product gases is shown in Fig. 3.

The temperature of the product gas stream at the top of the combustor rose quickly to a peak near 600°C some 30 seconds after ignition and then decayed exponentially as the combustion proceeded. The product gases were cooled in the exit gas coils and approximately 85% of the water formed was collected in the trap. The gas was further cooled in the second exit coil. Throughout the combustion, the temperature of the product gases leaving the calorimeter was less than 0.1°C above the temperature of the calorimeter water.

During the first seven minutes the temperature of the calorimeter water rose linearly at a rate near 0.25°C/min. Subsequently, the temperature of the calorimeter water continued to rise at a progessively diminishing rate until a final steady-state drift rate was attained about forty-five minutes after ignition. To establish the final drift rate, temperatures were recorded for at least thirty additional minutes. The temperature rise (from the end of the initial drift period to the beginning of the final drift period) was typically 2.3°C.

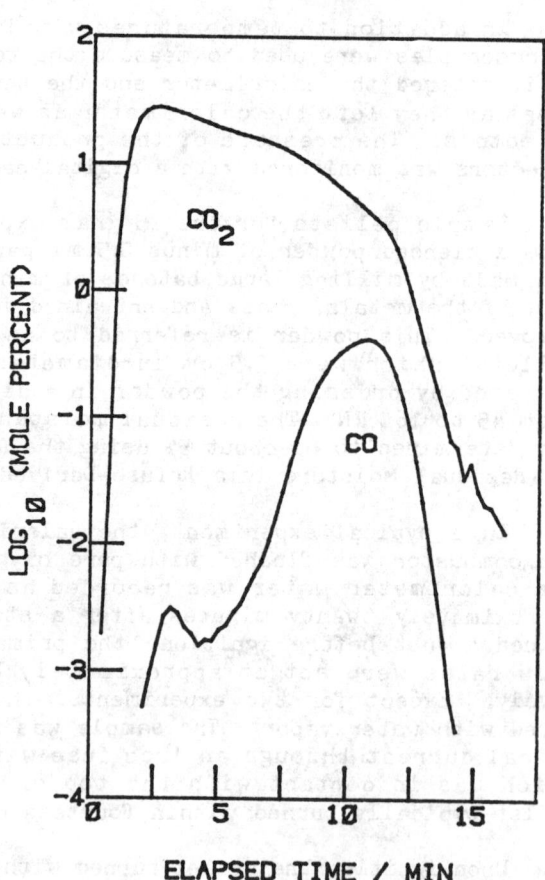

Fig.3. The $CO-CO_2$ Composition of the Product Gases vs the Elapsed Time from Ignition for a Typical Experiment.

After disassembly of the calorimeter, the ash and the quartz crucible were weighed. The carbon content of the ash was determined from the sum of the mass decrease produced by heating the ash and crucible with an oxygen-gas flame plus subsequent analysis for total occluded carbon. The carbon in the ash ranged from 0.015 to 0.1% of the initial sample mass.

A small amount (<1mg) of white residue was deposited on the thermal shield during each run. In about one-half of the experiments, the water condensed in the trap (see Fig. 2) was greenish blue rather than clear. The trap also contained a black residue. Apparently this was trapped fly ash which had partially dissolved. Analysis showed that no carbon was present in the residue and that the colored solution contained Fe, Cr, Ni, Na and K (i.e. in the range of 100 μg/ml) in addition to the usual acidity (0.1 to 0.2 milliequivalents per milliliter).

The prototype flow calorimeter was calibrated by burning five pellets of RDF-4 whose enthalpy of combustion had been determined previously at NBS with a gram-size bomb calorimeter. The heat of combustion of a different lot (i.e. different source) was measured in six experiments. For the flow experiments, the imprecision at the 95% confidence level was 0.9% for the five calibration measurements using Lot 1 and 1.2% for the six measurements of Lot 2. The corresponding imprecision obtained with the bomb calorimeter for four samples of Lot 1 was 0.8% and for four samples of Lot 2 was 0.5%. The ratio of the enthalpies for the two samples measured by the flow technique was 1.009 with an imprecision of 1.5% and an overall uncertainty (which includes estimated systematic error) of 2.7%. The corresponding ratio for the bomb calorimeter data was 1.027 with an imprecision and overall uncertainty of 1.0%. The ratios agree within their combined uncertainties. The uncertainty of the flow ratio is greater than its imprecision because of the errors associated with the residual moisture determination for each lot.

We believe the major error in the flow measurements is in the correction for heat leak applied to the observed temperature rise; corrections ranged from -0.7 to +4% of the observed rise. The corrections assume that the temperature of the calorimeter environment is held constant and that steady state heat transfer between the calorimeter water and the enviornment is achieved rapidly. These conditions were not met with the prototype calorimeter.

Other significant corrections and their approximate magnitudes are: for incomplete combustion of CO to CO_2 (~0.6%), for heat transport by gas flow (~0.4%), for vaporization of water (~1.4% for saturated and ~ 4.3% for dry inlet gas). No corrections for carbon content of the ash or the residue on the thermal shield were made. The error incurred is less than 0.1%.

In addition to measurements on MSW samples, several candidate substances were also burned to test their suitability as calibrant materials. Benzoic acid (the standard calibrant used for bomb calorimetry) was found to be unsatisfactory for flow work. Pellets melted and burned with a fuel rich flame that produced large amounts of soot. Ultra-pure carbon powder also proved to be unsatisfactory; it was very difficult to ignite. Flame calorimetry calibrants (gaseous H_2 or CH_4) were not tested because apparatus necessary for the quantitative (i.e. 0.01%) determination of the amounts of calibrants combusted was not available.

After the RDF-4 experiments, one gram pellets of pure cellulose were test burned in a similar combustor. The pellets were found to ignite easily and burn completely. Subsequently, bomb calorimetric measurements have established the enthalpy of combustion of this material so that it can be used as a standard solid calibrant for the flow calorimetry measurements (3).

TRIAL KILOGRAM COMBUSTIONS

A large combustor was designed to burn kilogram-size pellets of dried RDF-2. RDF-2 is municipal solid waste which has been processed to reduce the particle size so that 95 mass-percent passes through a 15 cm square mesh screen. The burning characteristics of the kilogram-size pellets in oxygen were unknown and were expected to differ from those of the 25 gram pellets. The kilogram-size pellet composition was more variable and the physical heterogeneity measured against the sample size was greater than the 25 gram pellet.

The goal of the trial experiments was to develop a method for completely burning 2.5 kg samples in approximately 40 minutes or less. This time limit was estimated to be the maximum time which would guarantee that the imprecision contributed by the calorimetric measurements would be less than one percent.

Trial combustions of samples up to 2.2 kg of RDF-2 were carried out in a burner cooled by convective and radiant heat losses to ambient temperature. The burner was mounted in a large exhaust hood. The burner design was similar to that used in the 25 gram experiments in that a thermal shield was employed to keep the hot, reacting, combustion product gases from being cooled by contact with the outer burner walls. The combustor differed from that used in the 25 gram experiments in that all of the oxygen was supplied locally to the sample; no oxygen flowed between the thermal shield and the cool combustor walls. Oxygen was supplied in the form of high velocity jets which were directed either at the top of or the side of the sample or both depending on the experiment. A diffuse, slower, flow of oxygen was directed upward at the bottom of the sample.

The combustor, as it appeared in the first of the trial burns, is shown in cross section in Fig. 4. The unit consists of two chambers. The lower chamber, which was the cylindrical region enclosed by the lower burner jacket (E of Fig. 4), contained the sample and two oxygen supply inlets. The sample was supported on vertical stainless steel pins which fit into the holes of a perforated stainless steel plate (H). This support system was analogous to the quartz plate of the 25 gram combustor. Primary oxygen was supplied to the bottom of the sample from the inlet I. Secondary oxygen was directed at the top of the sample by three jets aimed radially and horizontally by the inlets F. The intent of the sample-oxygen supply

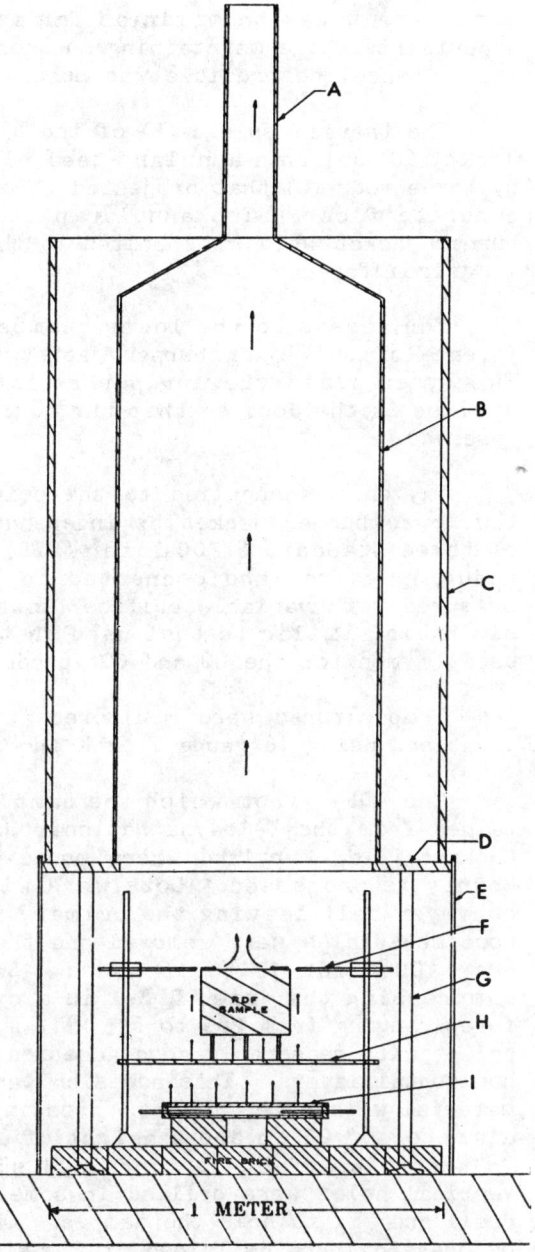

Fig. 4. The 2.5 Kilogram Combustor. A denotes the stack, B the upper thermal shield, C the upper burner jacket, D the annular plate, E the lower burner jacket, F the secondary oxygen inlets, G the support rod, H the sample support and I the primary oxygen intlet.

arrangement was to mimic as far as possible the 25 gram flow experiments while maintaining an unobstructed view of the combustion. Hence, no crucible was used.

The thermal shield (B) of the upper chamber and upper burner jacket (C) sat on a annular steel plate (D) which was supported by three rods (G) that projected through the lower chamber to the concrete floor. The annular plate also supported the lower burner jacket which was bolted to the plate around two-thirds of its circumference.

For access to the lower chamber, one-third of the circumference of the lower burner jacket was a semi-cylindrical door. Two Pyrex glass viewing ports (not shown in Fig. 4) were installed in the door so that the course of a combustion could be observed.

Oxygen was supplied to the primary and secondary inlets in the lower burner jacket by independent sources, each consisting of three standard 6,200 liter (STP) oxygen tanks equipped with reducing valves and connected in parallel. Flow rates were measured with variable orifice flowmeters. A product gas analysis train similar to that used in the 25 gram experiments was used to monitor the CO and CO_2 production.

Temperatures were monitored at as many as twelve different locations using 18 gauge Type K thermocouples.

The RDF-2 from which the sample pellets were made was obtained from the Teledyne National Resource Recovery Facility in Cockeysville, Maryland where municipal solid waste of Baltimore County is processed. Lots were withdrawn at random from the conveyer belt leaving the primary shredder. At NBS, the large noncombustibles were removed and the remainder was dried in air near 100°C for 12 to 16 hours. Sample pellets were made by compressing the dried RDF-2 in a cylindrical die piece with a force ranging from 265 to 625 kN. A single compression yielded a pellet with reasonably good adhesion of the various heterogeneous horizontal layers. This adhesion was not improved by wetting the material with water prior to pressing. The finished pellet had a diameter of 22 cm and a height of about 5.9 cm/kg sample mass. To test the effect of increased surface area, three or seven vertical holes were drilled in some of the pellets using a metal drill and jig to hold the sample. One cellulose pellet was made by pressing pure cellulose fluff using the same technique.

In all, eighteen experiments were carried out to test the effectiveness of: (1) various arrangements of primary and secondary oxygen inlets, (2) preheating the oxygen, and (3) reducing heat losses from the sample by the use of a crucible and a radiation shield. Thirteen experiments were run with RDF-2 pellets. Five experiments were run with pure cellulose or its substitute (a stack of unglazed paper plates).

Changes in the apparatus made as a result of these tests are illustrated by Fig. 5 which shows the configuration of the lower chamber of the combustor used in the final two experiments. An RDF-2 sample with seven vertical 2.5 cm diameter holes sat on a horizontal lattice of alumina rods which rested on a stainless steel support (see D of Fig. 5). Two tiers of secondary oxygen inlets (B and C) were aimed radially inward and horizontally at the side of the pellet. The lower tier of six inlets was supplied with oxygen that was preheated by passing it through coiled tubing (F) which was wound on the outside of the crucible. The upper tier of three inlets was supplied with oxygen that was preheated by passing it through the coil (I) which was inside of the crucible. Oxygen was supplied to the bottom of the sample from the primary oxygen inlet (E). Most of the ash fell through the center of the inlet to the bottom of the crucible. A radiation shield which just fit the inside diameter of the annular steel plate (D of Fig. 4) was placed around the crucible to reduce heat losses.

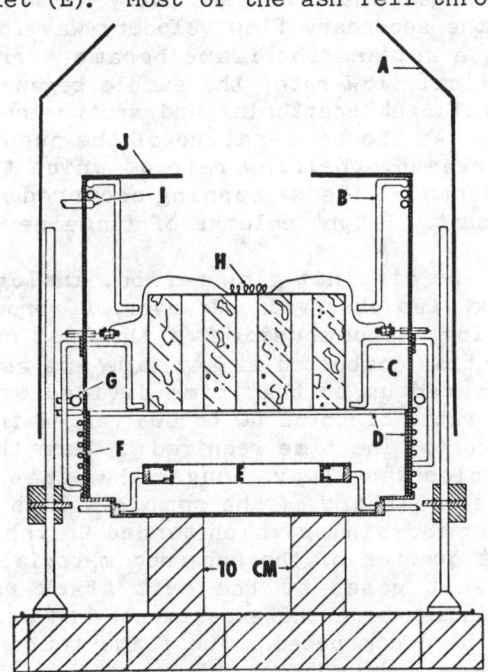

Back flow of the product gases between the crucible side and the lower thermal shield was restricted by a low upward flow of diffuse oxygen from the multiport ring inlet (G of Fig. 5). Rectangular openings were cut in the crucible and thermal shield in order to observe the combustion.

Fig.5. The Configuration of the Lower Chamber of the Combustor for Experiments 17 and 18. A denotes the lower thermal shield, B the upper tier of secondary oxygen inlets, C the lower tier of secondary oxygen inlets, D the sample support, E the primary oxygen inlet, F the lower tier preheat coil, G the multiport ring oxygen inlet, H the iron fuse, I the upper tier preheat coil and J the crucible lid.

A Pyrex window covered the opening in the thermal shield.

In a typical experiment, the combustor was flushed with oxygen for ten minutes. The sample was then ignited by passing electrical current through an iron fuse wire (H of Fig. 5) touching the top of the sample. The flow rates were adjusted to preselected initial values for the experiment. Primary and secondary oxygen flows were varied to study the effectiveness of different rates and inlet arrangements. The ratio of diffuse to directed flow was of the order of 3:1 or greater and the total flow rate ranged between 3.5 and 11 moles per minute. At the end of the burn, the combustor was flushed with oxygen and allowed to cool. The ash was collected and weighed. The ash contents of RDF-2 pellets ranged from 15 to 32% of the initial mass and no unburned organic material was identified in the ash.

In general the most rapid burning of the sample occurred on the areas where the secondary jets of oxygen struck the sample. As the secondary flow velocity was increased, these areas became white hot and the flame became ever more turbulent. Above a critical flow rate, the sample began to fragment vigorously with significant scattering and sputtering of the burning matter and ash. As the temperature of the preheated secondary oxygen was increased, the flow rate at which these changes occurred was lowered. Intense burning occurred in the vertical holes, when present. Bright columns of flame were observed above the holes.

No distinct glow period, analogous to that of the 25 gram work, was observed. However, CO production tended to be larger during the combustion of the last quarter of the sample. The fraction combusted at any time was assumed to be the rato of CO_2 produced up to that time divided by the total CO_2 production. The ratio of total CO to CO_2 ranged from 2 to less than 0.1 mole percent. The time required to burn the last quarter of all RDF-2 samples was always longer than that required to burn the first three quarters of the sample. This was due to the presence of noncombustibles which tended to inhibit the combustion of the last quarter of the unburned material. Peak temperatures of the exhaust gases at the exit stack ranged from 365 to 500°C, depending upon the experiment, and occurred before half of sample had been combusted. The flame temperature above the sample was determined to be in excess of 1400°C. The peak temperature of the combustor components nearest the sample ranged from 500 to 1200°C for the top of the primary oxygen supply and 440 to 500°C for various parts of the crucible and lower thermal shield. For the most part, only minor surface corrosion of the these components occurred. Actual ignition of the combustor components only took place when the burning sample fell from its support and came into contact with the oxygen inlets.

The total burn time of RDF-2 pellets was reduced from 77 minutes per kilogram initial mass in the early runs with the experimental arrangement illustrated in Fig. 4 to an acceptable 15 min/kg using the arrangement of Fig. 5. As more than one change in sample and/or combustor configuration was made in each trial experiment, interpretation of the effects due to the individual changes tends to be somewhat ambiguous. However, we draw the following conclusions: (1) The introduction of a crucible (having a side wall one half the height of that shown in Fig. 5) caused a reduction in burn times for RDF-2 pellets of 33%. (2) The introduction of vertical holes in the sample reduced burn times between 10 and 50%. (3) Burn times for a single tier of lower secondary oxygen inlets (equivalent to C of Fig. 5) were about 50% less than those obtained using a single tier of upper secondary inlets (which were directed downward at a 45 degree angle toward the top edge of the pellet).

TABLE I.

Specifications for the 2.5 kg Capacity Flow Calorimeter

Dimensions of Main Components

Component	Height (cm)	Outside Diameter (cm)	Wall Thickness (mm)	Total Mass (kg)
Combustor	155	41	3.2	74
Combustor Enclosure	163	61	4.8	196
Flow Shield	213	79	1.6	73
Calorimeter Vessel	236	91	4.8	390
Water-Wall Jacket	244	102	25.	454

Additional Specifications

Calorimeter Water	983 liters	
Heat Capacity	322 kJ/K	(Assembly)
	4111	(Water)
	4433 kJ/K	(Total)

We found that cellulose pellets were easily ignited and burned smoothly, leaving negligible ash. Burn times for cellulose samples were up to 50% shorter than those for RDF-2. Cellulose appears to be a satisfactory calibrant.

2.5 KILOGRAM CAPACITY FLOW CALORIMETER

The final design adopted for the large scale calorimeter is illustrated in Fig. 6 with dimensions of interest presented in Table I. Stainless steel (type 304) will be used in the fabrication of all components except as noted.

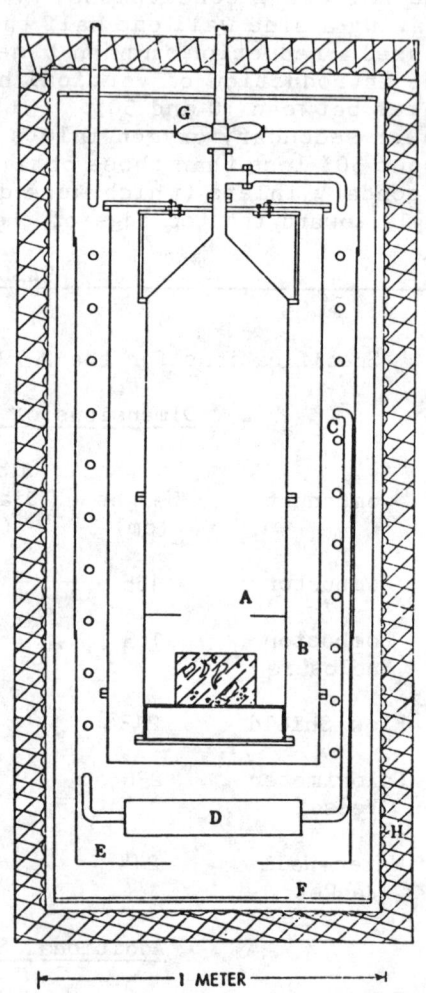

The combustor (A of Fig. 6) will incorporate all of the features which were evolved during the trial burns plus additional ideas which we were unable to test completely due to the constraints of time. The sample pellet is to be supported by a parallel horizontal lattice of alumina rods held by notches in the upper edge of the pan in which the ash is to be collected. The bottom of the ash pan will be covered by a layer of sand to prevent its ignition if large pieces of burning sample fall on it. Three independent tiers of up to six secondary inlets each (not shown in drawing) will penetrate the wall of the combustor in the region surrounding the sample. A similar tier of six inlets will also penetrate the ash pan to provide oxygen flow beneath the sample. This arrangement replaces the primary inlet of the

Fig. 6. The 2.5 Kilogram Capacity Flow Calorimeter. A denotes the combustor, B the combustor enclosure, C the exhaust cooling coils, D the exhaust condenser, E the flow shield, F the calorimetric vessel, G the stirrer and H the water wall enclosure.

trial combustor and overcomes the awkward ash recovery procedure used during the trial burns. An annular disk divides the combustor into two zones. The diameter of the combustor (41 cm) and the space between the sample pellet and the annular disk (15 cm) are comparable to the dimensions of the crucible and lid arrangement shown in Fig. 5. An additional tier of six oxygen inlets is to be placed in the upper zone. All oxygen can be preheated in coils welded to the outer wall of the combustor. Provision is made to introduce one or more interior preheat coils in the lower zone of the combustor if desired. Initially the entire combustor is to be made from 316 stainless steel with welded construction and gas-tight seals at the flanges. The various oxygen inlets will be sealed at the combustor wall.

The combustor enclosure (B) is to be filled with argon which will provide a clean, inert environment in which all tubing and connections as well as electrical and thermocouple wiring will be protected.

The combustor enclosure and exhaust cooling coils (C) will be immersed in the calorimeter water held by the calorimeter vessel (F). There will be three complete turns of the cooling coils preceding the exhaust condenser (D) and seven turns before the exit from the calorimeter vessel. The exhaust cooling coils and the exhaust condenser will be made from Incoloy 825.

The flow shield (E) and stirrer (G) are designed to circulate the calorimeter water completely about once in five minutes. For a 2.5 kg pellet of pure cellulose, we expect a temperature rise of about $10°$ C.

To reduce the total mass and volume of the assembly, we plan to enclose the calorimeter vessel with a water-wall jacket (H). Water from a thermostatted reservoir will be circulated through its walls to provide the necessary constant temperature environment for the calorimeter. The jacket replaces the submarine vessel and lid (cf. H and I of Fig. 2) and the thermostatted, stirred-water bath into which the conventional isoperibol calorimeter is immersed. The water bath and submarine for a calorimeter of this size would weigh 2800 kg.

The assembly of the calorimeter has been completed. Initial trial combustions have been made.

ACKNOWLEDGEMENT

This work is jointly sponsored by the NBS Office of Recycled Materials and the U. S. Department of Energy, Office of Energy from Municipal Waste.

REFERENCES

1. Kirklin, D. R., Domalski, E. S., and Mitchell, D. J.: 1980, NBSIR 80-1968.

2. Kirklin, D. R., Colbert, J., Decker, P., Abramowitz, S., and Domalski, E. S.: 1981, NBSIR 81-2278.

3. Colbert, J. C., Xiheng, H., and Kirklin, D. R.: 1981, J. Res. NBS. 86, pg. 655.

THERMOGENESIS BY HEAT CONDUCTION CALORIMETRY

Henri TACHOIRE

Laboratoire de thermochimie, Université de
Provence, 13331 MARSEILLE cedex 3, France

Jean-Luc MACQUERON

Laboratoire de traitement du signal, INSA,
69621 VILLEURBANNE cedex, France

Vincent TORRA

Departamento de termolgia, Universidad de
PALMA DE MALLORCA, España

In the course of the lecture, we will describe digital
and electronic technics which enable us to balance the
first poles and zeros of the instrumental transfer
function. We'll also present a comparative application
of the harmonic analysis and filtering to the same calorimetric output.

In a following part, we'll show, by several examples,
the interest of these deconvolution technics. These
examples concern the kinetic study of a solid-solide
transformation and the thermodynamic study of molecular mixtures. They will prove the very important increase of the application field of the heat-flux calorimetry by using these technics.

(A)

THE DATA·PROCESSING

I - INTRODUCTION

Heat conduction calorimetry is now a well established technique (1), in widespread use and a certain number of instruments can be easily found in the market (2).

Such devices are extremely sensitive (the sensor is a thermal fluxmeter) and the differential mounting of two identical elements ensures a great stability in time. They allow an accurate determination of the total energy of a given phenomenon when properly calibrated (i.e. in the absence of systematic error).

When the thermogenesis of the phenomenon under study slowly changes in time, experimental thermograms give themselves a good representation of the thermogenesis except for a time delay (velocity error).

However, in view of their structure, heat-conduction-calorimeters possess an inertia which may be considerable.

Thus, recorded thermograms no longer represent the thermogeneses when they show a quick variation in time. This fact is a serious handicap if we are willing to using these instruments to measure an instantaneous thermal power (determination of kinetic parameters, description of equilibria in solution by microcalorimetric titration, application of this titration in chemical analysis,...).

Then, paradoxically, in spite of the fact that they are thermal fluxmeters, these instruments are not much used in this field.
We will present here two techniques for the deconvolution of calorimetric responses (harmonic analysis and inverse filtering) together with a critical comparison of their efficiency and an illustration, by means of a few examples, of the areas where they may be applied and, thus, of their significance.

First of all, though, it should be emphasized that a deconvolution of the instrumental response is not absolutly essential when the thermogenesis of the phenomenon under study shows a slow variation in time (3).

II - BEHAVIOUR OF HEAT CONDUCTION CALORIMETERS AT LOW FREQUENCIES

The kinetic properties of a heat conduction calorimeter are perfectly defined by its frequency response or transfer function (amplitude characteristic $A/dB = f(\nu)$ and phase characteristic $\Phi/rad = f(\nu)$). Figure 1 shows the transfer function of a specific calorimeter.

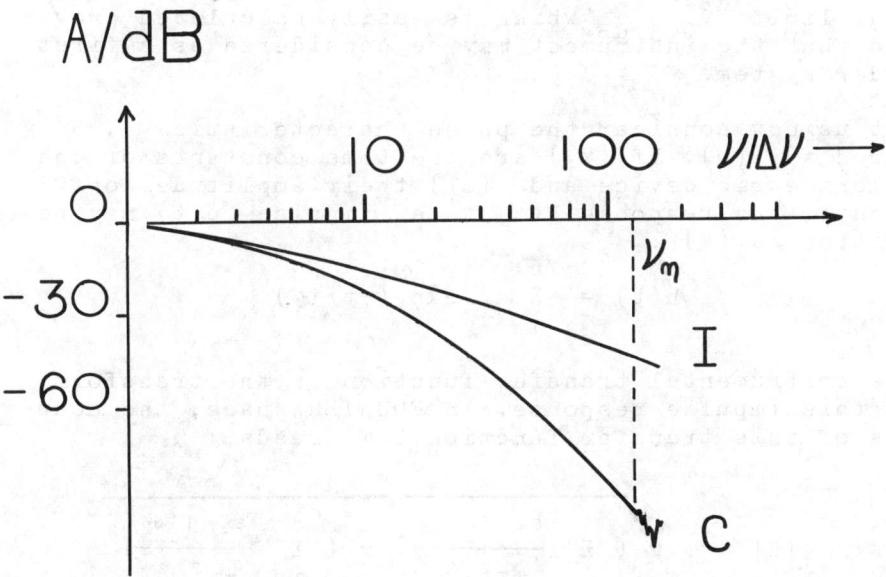

Figure 1 — Amplitude characteristic ($A/dB = f(\nu)$ in ν log scale) of a typical heat conduction calorimeter (graph C)
Curve I represents the amplitude characteristic of a first order system whose time constant is equal to the first time constant τ_1 of the calorimeter (frequency sample $\Delta\nu = 1/2048$ hertz)

Experimental noise affecting the recorded thermograms hinders the determination of the transfer function beyond a certain frequency ν_n (figure 1, graph (C)).

A study of the amplitude characteristic $A/dB = f(\nu)$ shows that, at low frequencies, any heat conduction calorimeter may be considered as a first order system characterized by a time constant τ_1 (graph I). A study of ten or so devices showed that a weakening of the signal of 2% (0.3 dB) happens at a frequency $\nu_{0.3\,dB}$ in such a way that the product $\nu_{0.3\,dB} \cdot \tau_1$ is close to 0.028 in all cases (for a group of 3 calorimeters whose time constant τ_1 is between 40 and 1000 seconds).

When the amplitude of the signal is concerned, at very low frequencies, the oscillograph behaviour of the devices is governed by the previous value of the product $\nu_{0.3\,dB} \cdot \tau_1$. This value gives an upper frequential limit $\nu_{0.3\,dB}$ which is easily calculated provided that the instrument may be considered as a first order system.

Let us now consider the phase characteristic $\Phi/rad = f(\nu)$. If $\{\tau_i\}$ are the time constants of the calorimetric device and $\{a_i\}$ their amplitude coefficients, the response $h(t)$ to an impulse $\delta(t)$ may be written as (4):

$$h(t) = \sum_{1}^{n} a_i \exp(-t/\tau_i)$$

The instrumental transfer function is the transform of this impulse response. In FOURIER space, the modulus of this transfer function $H(\omega)$ reads:

$$|H| = \sqrt{\left(\sum_i \frac{a_i \tau_i}{1+\tau_i^2 \omega^2}\right)^2 + \left(\sum_i \frac{a_i \tau_i^2 \omega}{1+\tau_i^2 \omega^2}\right)^2}$$

where $\omega = 2\pi\nu$

The phase of $H(\omega)$ is given by

$$\Phi = \arctan \frac{-\sum_i a_i \tau_i^2}{\sum_i a_i \tau_i} \omega$$

at very low frequencies where $\omega^2 \ll (1/\tau_i)^2$

If the phase Φ does not exceed 15 or so degrees, we

may approximate the tangent by the corresponding arc

$$\Phi \sim \tan \Phi$$

So, we have

$$\Phi = - \frac{\sum_i a_i \tau_i^2}{\sum_i a_i \tau_i} \omega = -2\pi \frac{\sum_i a_i \tau_i^2}{\sum_i a_i \tau_i} \nu = -2\pi k \nu$$

Thus, the phase Φ is proportional to the frequency ν.

Therefore, at very low frequencies (this is the case for thermogenesis showing very slow variation), provided that the weakening in the signal does not exceed 0.3 decibel and that the phase shift is less than 15 degrees, the thermograms obtained give a good representation of the thermogenesis except for a constant delay in time k (or velocity error). This delay depends on the time constants $\{\tau_i\}$ and on the amplitude coefficients $\{a_i\}$ that is to say on the contents of the laboratory cell and on the location of the heat source.

It can be shown that, if the heat source is not located near the detector thermocouples (central source for example), the proportionality factor k is simply equal to the sum of time constants τ_i (3).

$$k = \sum_i \tau_i$$

Up to now, we have been concerned only with the behaviour of heat conduction calorimeters at low frequencies (thermogenesis showing very slow variation). We have studied the upper limits governing the use of these devices in thermokinetic studies without using any deconvolutive procedure. But, if the signal spectrum appreciably extends beyond $\nu_{0.3dB}$, a deconvolution of the thermograms turns out to be essential.

The deconvolution techniques, electronic or digital, are based on the fact that calorimetric devices may be considered as linear systems. It is commonly accepted that there exists a linear relationship between "cause" and "effect" amplitudes. This functional relationship does not change under a given set of experimental conditions (contents of the calorimetric cell and location of the source of thermogenesis in that cell).

When these conditions change to any marked degree, though a linear relationship still exists, its functional expression will be obviously different. Very many experiments, confirmed by theoretical analysis, have shown, for example, the effects that a simple manipulation of the laboratory cell (when it is removable) or a change in the location of the heat sources inside that cell may bring about on the transfer function of the device (5), figure 2a and 2b.

The linearity of the instrument must be tested by analysing its behaviour under addition of different input signals and multiplication a given input signal times an arbitrary scalar factor. Here, we can use either power steps $u(t)$ or $\delta(t)$ impulses (intervals not exceeding about 1/300 of the main time constant τ_1 of the device).

III - DECONVOLUTION BY HARMONIC ANALYSIS

The calorimetric system is now considered as a black box (6). The first step is to identify the system, in other words, to obtain its transfer function (amplitude and phase characteristics). This identification must be performed under conditions resembling as closely as possible those of the thermogenesis to be reproduced. In view of the fact that the deconvolution technique is based on the invariability of the transfer function, it is essential that geometrical and thermal properties of the calorimetric device and the location of the heat source remain unchanged (both during identification of the system and during the process under study). This is sometimes difficult to achieve.

The functional relation between the input $e(t)$ and the output $s(t)$ may be written

$$s(t) = M \{e(t)\} = h(t) * e(t)$$

In linear systems, if $S(\omega)$ and $E(\omega)$ represent the FOURIER transforms of the signals $s(t)$ and $e(t)$, $H(\omega)$ the transfer function of the experimental system, we have

$$S(\omega) = H(\omega) \; E(\omega)$$

In order to determine $H(\omega)$, one must know the input

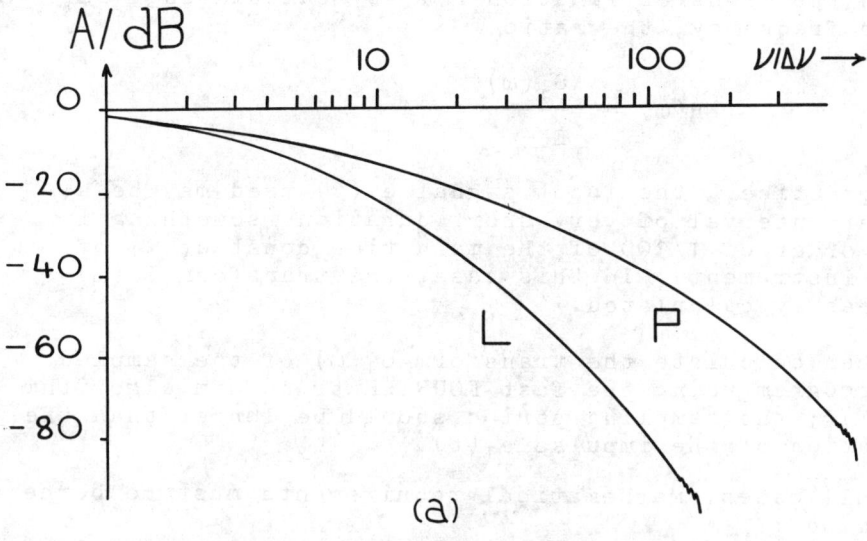

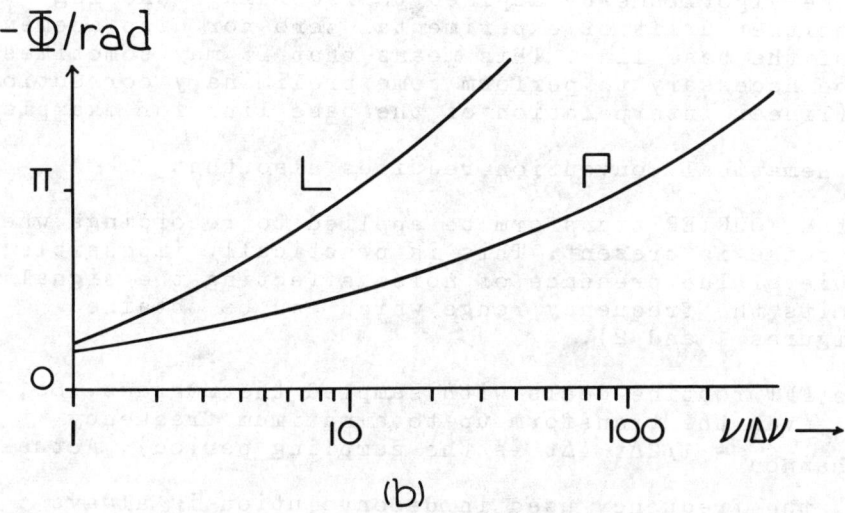

Figure 2 - Transfer function of a typical heat conduction calorimeter, $\Delta\nu = 1/2048$ Hz
(a) amplitude characteristic when the heater is located near the thermocouples (curve P) and far from the thermocouples (curve L)
(b) corresponding phase characteristic

$e_1(t)$ corresponding to an output $s_1(t)$ and, then, calculate the FOURIER transforms $S_1(\omega)$ and $E_1(\omega)$. We obtain the transfer function $H(\omega)$ by calculating, for each frequency, the ratio

$$H(\omega) = \frac{S_1(\omega)}{E_1(\omega)}$$

In practice, the input signal $e_1(t)$ used may be a power interval of very short duration (something in the order of 1/300 of the main time constant τ_1 of the instrument). In this case, the transform $E_1(\omega)$ is easily calculated.

We can calculate the transform $S_1(\omega)$ of the sampled thermogram using the fast FOURIER transform algorithm (FFT) ; the sampling period should be longer than the duration of the impulse $e_1(t)$.

In all cases, mathematical requirements must be borne in mind :
- the integrals be extended to the whole time scale concerned (i.e. the total duration of the phenomenon)
- the algorithms be applied to recordings showing neither drift of experimental zero nor displacement of the base line. This means that it may sometimes be necessary to perform some preliminary corrections (linear interpolation of the base line for example).

Mathematical convention requires also that

- the FOURIER transform be applied to recordings where no noise is present. This is practically impossible to achieve. The presence of noise affecting the signal limits the frequency range which can be obtained (figures 1 and 2).

The FFT routine deals with sampled thermograms. So, it gives the transform up to a maximum frequency $\nu_{Shannon} = 1/2\Delta t$ (Δt is the sampling period). Actually, the frequency used in deconvolution is always lower than $\nu_{Shannon}$.

Let us look now at the amplitude characteristic $A/dB = f(\nu)$ (v.s. a log frequency scale) of any calorimetric system (figure 3). We have already indicated that, beyond the frequency ν_n ($\nu_n < \nu_{Shannon}$), it is

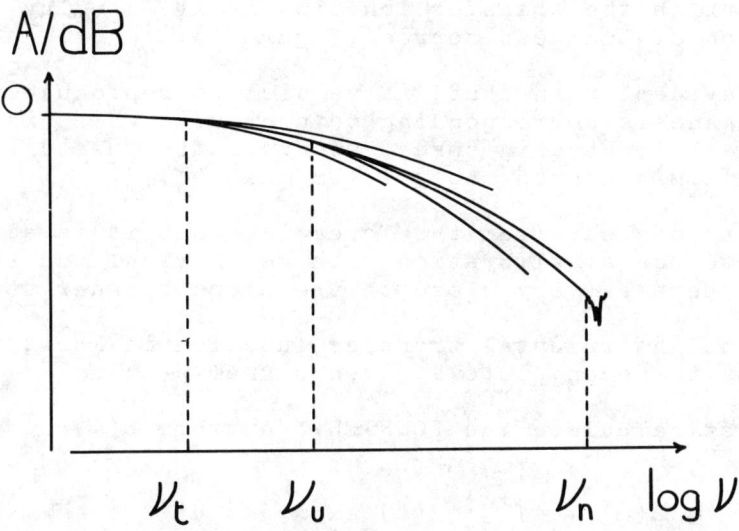

Figure 3 - Amplitude characteristic of a heat conduction calorimeter
For $\nu > \nu_n$, the calorimetric response is affected by noise
For $\nu > \nu_u$, the assembly and dismantling of the laboratory-cell introduce uncertainties in the determination of the transfer function
For $\nu > \nu_t$, the transfer function is affected by the location of the heat source

impossible to determine the transfer function accurately since the calorimetric response is affected by noise (oscillations of the amplitude and phase characteristics).

If, during the deconvolutive calculus, we include frequencies higher than ν_n, we obtain a calculated thermogenesis showing very pronounced oscillations.

Moreover, experience shows, and theoretical analysis confirms, that minor changes in contact thermal resistance values have an appreciable effect on the transfer function at high frequencies. What is more, the assembly and dismantling of the laboratory cell introduce an element of uncertainty in the determination of the transfer function above ν_u.

Finally, there exists another frequency limit ν_t

above which the transfer function is affected by the location of the heat source (figure 3).

It is evident then that, if we wish to reproduce the thermogenesis corresponding to a process which is not very well located in space, we must often use a frequency limit considerably lower than ν_n.

It must be clear from the foregoing that it is advisable to use a calibration with geometrical and thermal properties very close to the process under study.

Once the instrumental transfer function is know, it is easy to deconvolute a given signal s(t).

We first calculate the FOURIER transform $S(\omega)$

$$S(\omega) = \int_{-\infty}^{+\infty} s(t) \exp(-j\omega t) dt$$

Subsequently, the relation

$$E(\omega) = S(\omega)/H(\omega)$$

allows us to calculate the transform $E(\omega)$ of the thermogenesis to be reproduced. The division is performed up to an appropriate cut-off frequency ν_c ($\nu_c < \nu_n$). Then, the inverse transformation gives the thermogenesis e(t).

$$e(t) = T^{-1}\{E(\omega)\} = \frac{1}{2\pi} \int_{-\infty}^{+\infty} E(\omega) \exp(j\omega t) d\omega$$

The introduction of a cut-off frequency ν_c causes an extra oscillation whose period is equal to $1/\nu_c$. A final smoothing is therefore necessary (figure 4a and 4b).

IV - DECONVOLUTION BY INVERSE FILTERING

The use of electronic or digital filtering is restricted to a certain range of the instrumental transfer function. We obtain then, a priori, a less reliable reproduction of the thermogenesis than we would get from harmonic analysis.

In the LAPLACE space, the linear relation between the input and the output may be written now

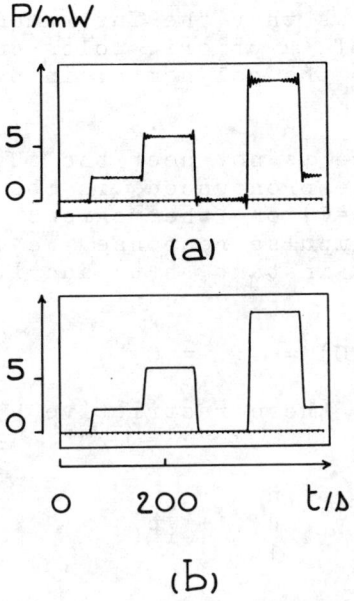

Figure 4 - When using harmonic analysis, the cut-off frequency ν_c causes an oscillation whose period is equal to $1/\nu_c$ (a). A final smoothing is then necessary (b)

$$S(p) = H(p) E(p)$$

The transfer function $H(p)$ is the transform of the impulse response $h(t)$

$$h(t) = \sum_{1}^{n} a_i \exp(-t/\tau_i)$$

Namely,

$$H(p) = \sum_{1}^{n} \{a_i \tau_i (1 + \tau_i p)^{-1}\}$$

Generally speaking, then, $H(p)$ may be expressed as

$$H(p) = \text{polynomial in } p \text{ (degree } (n-2) \text{ at most)} / \prod_{1}^{n}(1+\tau_i p)$$

n represents the number of exponential terms used to describe the impulse response $h(t)$; theoretically, n tends to infinity.

This expression shows that the instrumental transfer function consists of, a priori, poles and zeros (the zeros correspond to the polynomial in p in the numerator).

When the heat source is not near the thermocouples, a lull sometimes quite pronounced is observed right at the beginning (t=0) on thermograms which correspond to h(t), the impulse response. We can assume, then, that at that instant, h(t) and its first derivatives are zero.

$$h(0) = h'(0) = \ldots = 0$$

Taking into account these restrictive hypotheses, we may express H(p) as

$$H(p) = \sum_1^n a_i \tau_i \Big/ \prod_1^n (1+\tau_i p)$$

If we assume that the static gain is equal to one, we have

$$H(p) = 1 \Big/ \prod_1^n (1+\tau_i p)$$

However, we usually need a few terms in the product because the sequence of time constants τ_1, $\tau_2, \ldots, \tau_i$, ... decreases very quickly (7).

Figure 5 (graph C) shows the transfer function (amplitude characteristic) of a heat conduction calorimeter where the heat source is located far from the thermocouples.

At very low frequencies (ν < 0.0015 Hz), the device can be considered as a first order system of time constant τ_1 (=198 seconds), graph I. When ν lies between 0.0015 and 0.0075 Hz, it behaves as a second order system of time constants τ_1 (=198 seconds) and τ_2 (=49 seconds), graph II. With increasing frequency, the apparatus must correspondingly behave as a higher order system.

If this kind of device is associated with an inverse (digital or electronic) filtering whose transfer function is $(1+\tau_1 p)(1+\tau_2 p)$ for example, the signal $s_2(t)$ obtained no longer contains the first two poles

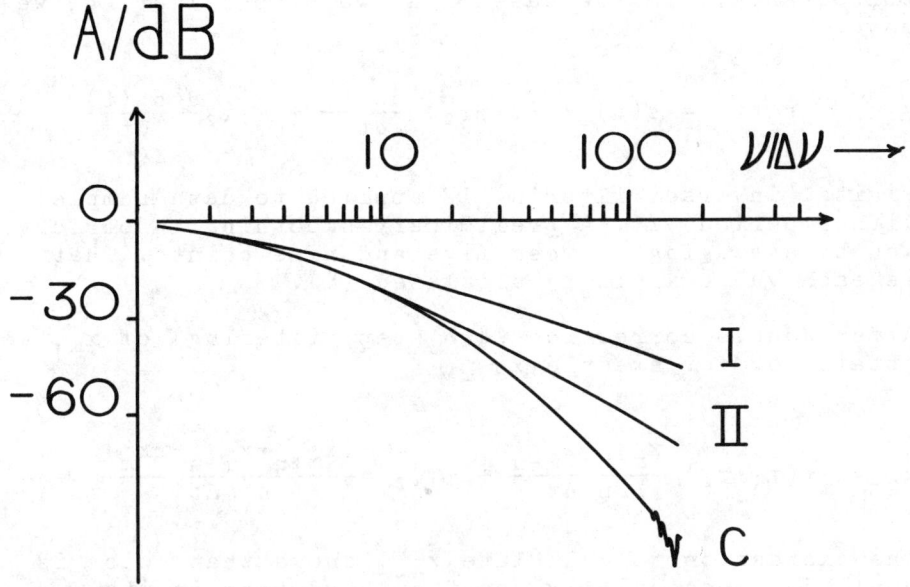

Figure 5 - Amplitude characteristic of a typical heat conduction calorimeter. As the frequency ν rises, the apparatus behaves as a correspondingly higher-order system (first-order : curve I, second-order : curve II,...). $\Delta\nu$ = 1/2048 Hz

of the function $H(p)$ since the transfer function of the whole assembly is now

$$\frac{1}{\prod_{1}^{n}(1+\tau_i p)} (1+\tau_1 p)(1+\tau_2 p) = 1/\prod_{3}^{n}(1+\tau_i p)$$

The calorimetric response is speeded up to a remarkable extent since τ_3 becomes the main time constant. One step filtering (either electronic or digital) gives as response $s_1(t)$

$$s_1(t) = s(t) + \tau_1 \frac{ds(t)}{dt}$$

since its transfer function equals $(1 + \tau_1 p)$

For a second step,

$$s_2(t) = s_1(t) + \tau_2 \frac{ds_1(t)}{dt}$$

Consequently, in the case of a two step filtering, we have

$$s_2(t) = s(t) + (\tau_1+\tau_2)\frac{ds(t)}{dt} + \tau_1\tau_2\frac{d^2s(t)}{dt^2}$$

Digital inverse filtering is applied to data sampled with a period Δt. A preliminary smoothing is performed by averaging between five and nine points. That is each value x_m is recalculated (8).

After double correction (two step filtering) of x_m, we obtain now for example $x_{2,m}$

$$x_{2,m} = x_m + (\tau_1+\tau_2)\frac{x_{m+n}-x_{m-n}}{2n\,\Delta t} + \tau_1\tau_2\frac{(x_{m+n}+x_{m-n}-2x_m)}{n^2\,\Delta t^2}$$

The expression to calculate $x_{2,m}$ shows that noise is amplified, essentially, by the third term in its second member (particularly when a low value is taken for n). Figure 6 shows how the value of n affects the

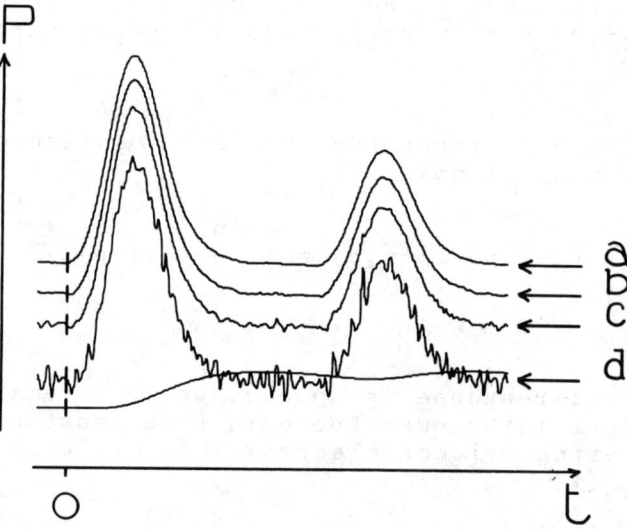

Figure 6 - Deconvolution by digital filtering. The figure shows how the value of the time increment $n\,\Delta t$ affects the quality of the signal after filtering (see text). $n\Delta t$ = 20 s (curve a), 15 s (curve b), 10 s (curve c), 5 s (curve d). The experimental thermogram is also given

quality of the signal obtained after digital filtering.
In the example shown, provided that the increment
$n\Delta t$ does not exceed ten seconds or so, its value does
not result in an appreciable loss of information. It
simply causes a marked decrease in the noise affecting
the reproduced thermogenesis.

This can be justified. Adopting an increment equal to
$n\Delta t$ is equivalent to introducing a cut-off frequency
$1/2n\Delta t$. If its magnitude is greater than ν_2 (the
frequency above which the calorimeter no longer behaves as a second order system), the quality of the
reproduced thermogenesis is not affected by the selection of n except as regards noise intensity.

In the opposite case, the two step inverse filtering
would lose some of its effectiveness.

Concerning electronic inverse filtering (9), the operations defined above are performed with the aid of
analogical circuits.

Let us consider the circuit shown in figure 7 (A is
an operational amplifier having very high gain and

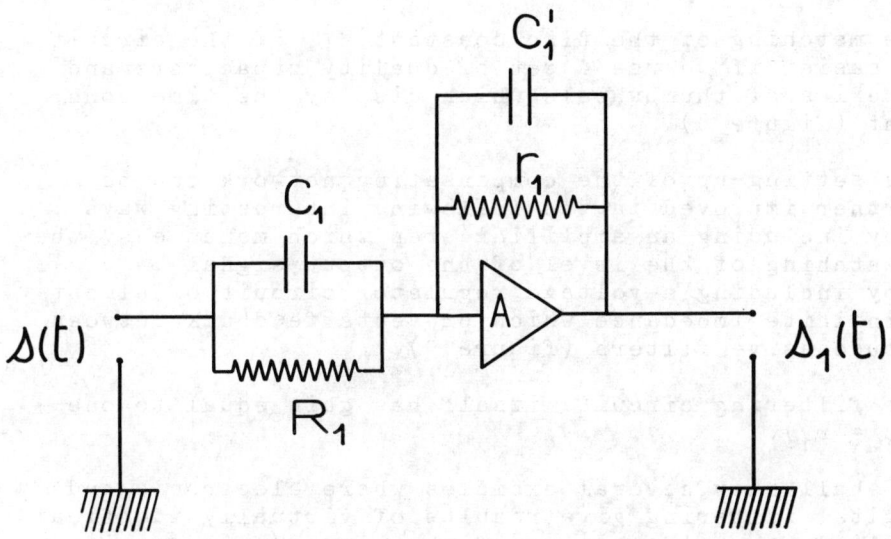

Figure 7 - One step electronic inverse filtering

input impedance). Its transfer function is $(\frac{1}{R_1}+C_1 p)r_1$.
Let us define

$$\tau_1 = R_1 C_1$$

Then, it becomes $(1 + \tau_1 p)$ with

$$R_1 = r_1 \quad \text{and} \quad C_1 = \tau_1/R_1$$

If an input signal $s(t)$ is received by this circuit, it gives a response $s_1(t)$ so that

$$s_1(t) = s(t) + \tau_1 \frac{ds(t)}{dt}$$

In practice, in order to avoid considerable amplification of electric noise and interference, a capacitor C_1' must be placed in parallel to the resistor r_1. Its presence must not affect the transfer function of the circuit (we select the ratio C_1'/C_1 to be close to 0.1). If the influence of the capacitor C_1' is significant, the transfer function of the filtering circuit is no longer $((1/R_1)+C_1 p)/(1/r_1)$ but $((1/R_1)+C_1 p)/((1/r_1)+C_1' p)$ instead.

The matching of the time constant τ_1 of the circuit is easier if we use a set of quality capacitors and a series of thumbwheels which display the time constant (figure 8).

The setting-up of the compensating network can be further improved in the following interesting ways :
- by including an amplifier step which makes easy the matching of the level of the output signal ,
- by including a voltage regulator circuit of almost infinite impedance which prevents feedback between successive filters (figure 9).

The filtering circuit itself has gain equal to one ($R_1 = r_1$).

We shall show several examples where electronic and digital filtering give results of virtually identical quality (the final value of the signal/noise ratio being the same), figures 10, 11 and 12.

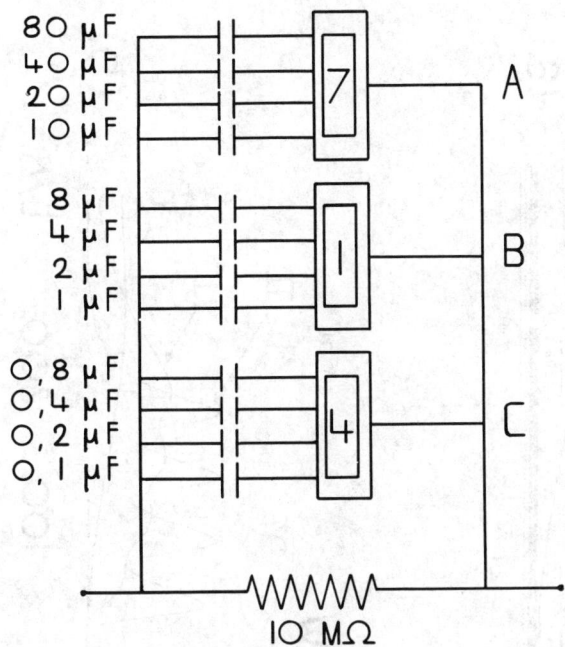

Figure 8 - Electronic filtering
Matching of a time constant (714 s) by using a series of thumbwheels

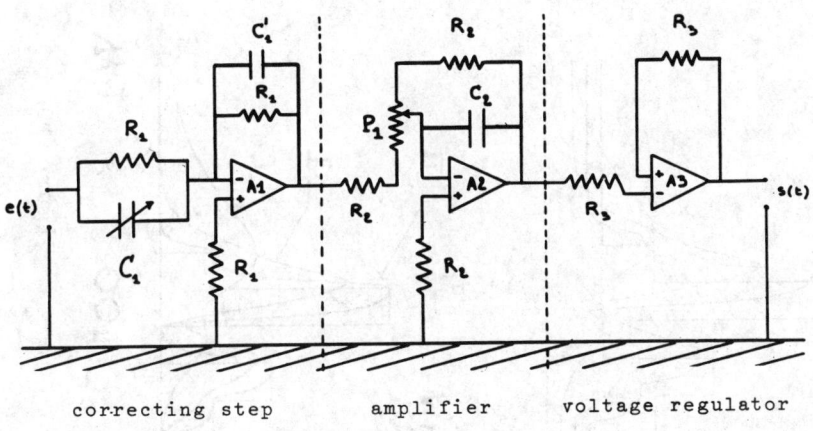

Figure 9 - One step electronic compensating network

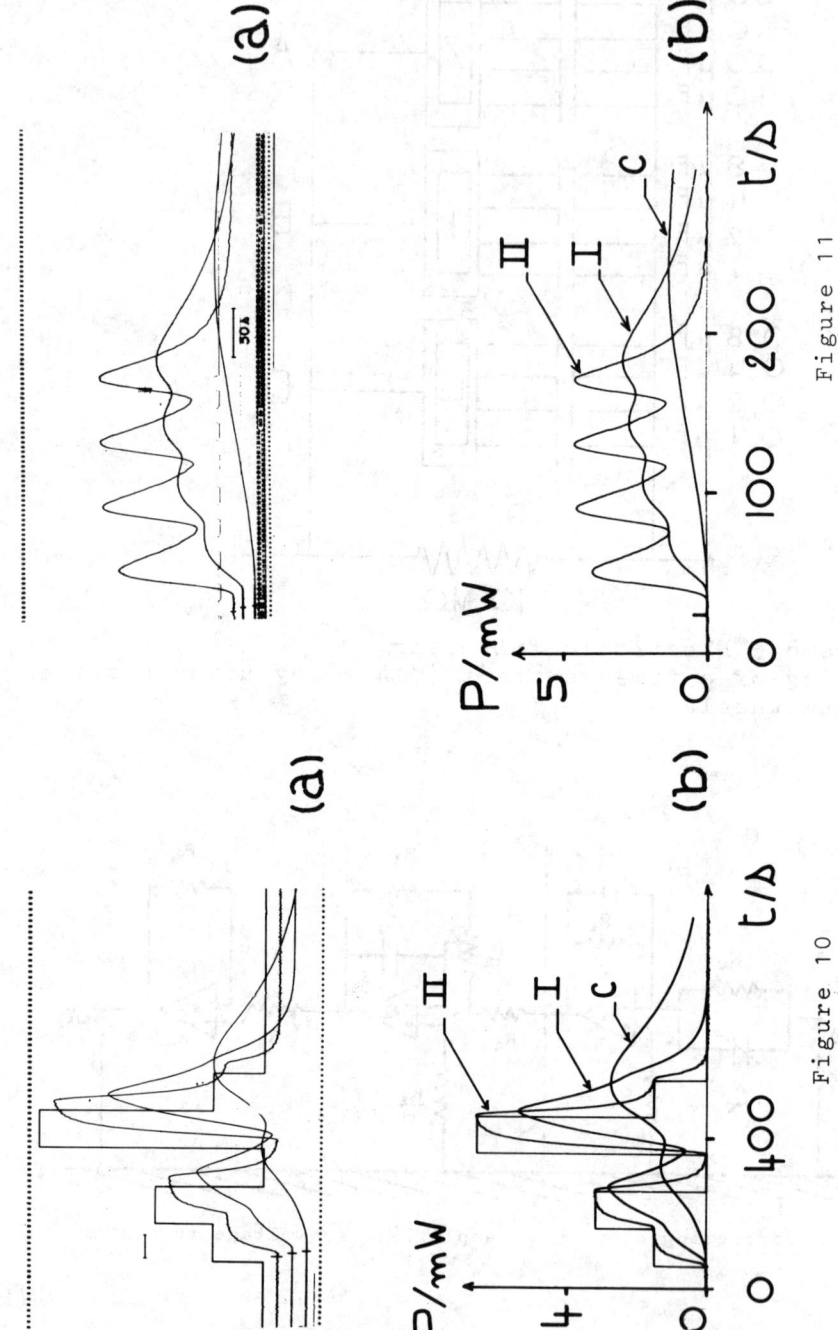

Figure 10

Figure 11

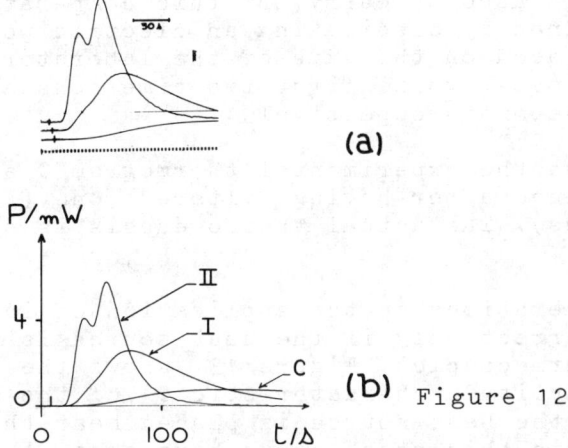

Figures 10,11,12 - Comparative study of electronic (a) and digital (b) filtering for a typical heat conduction calorimeter
C experimental thermogram, I response given by one step filtering, II response given by two step filtering. In the case of figure 10, the thermogenesis is represented

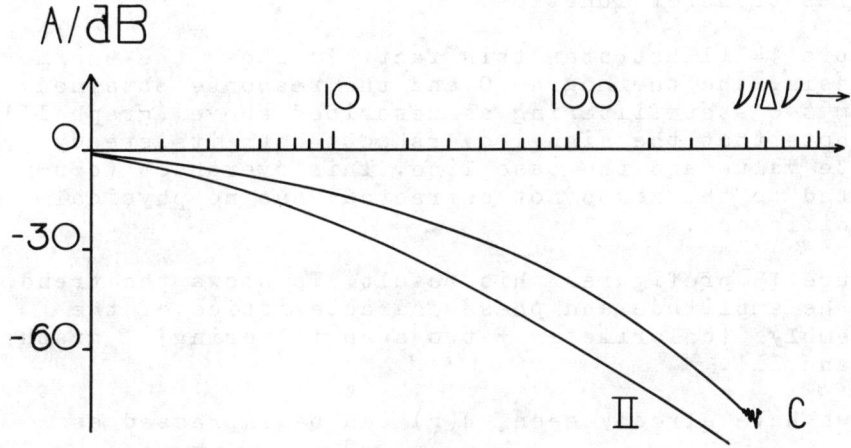

Figure 13 - Amplitude characteristic C of a typical heat conduction calorimeter when the source of the thermogenesis is located near the thermocouples
II represents the amplitude characteristic of a second order system whose time constants (τ_1 and τ_2) are equal to the first two time constants of the calorimeter

The thermogenesis used to carry out this comparative study were obtained by dissipating an electric power in a resistor placed on the axis of the laboratory cell of a calorimeter whose first two time constants are 198 and 49 seconds respectively.

Each figure shows the experimental thermogram C and the signal obtained after having filtered (one (I) or two (II) steps). The actual thermogenesis is not plotted there.

The above considerations on the application of inverse filtering are correct only if the heat source is not close to the thermocouples. Figure 13 shows the transfer function (amplitude characteristic C) of the same instrument when the heat source is placed near the thermocouples : in this case, as we have seen, the function $H(p)$ contains one or more relevant zeros.

We can see now the amplitude characteristic C bears no resemblance to that of the second order system of time constants τ_1 and τ_2 even at very low frequencies (graph II). The use of a two step filtering does, then, suppress the first two time constants of the calorimetric system but does not the zero (or zeros) of its transfer function.

Figure 14 illustrates this fact. It shows the thermogenesis, the thermogram C and the response obtained from two step filtering as described above (graph II). We note that the signal overshoots both its steady state value and the base line. This overshoot (connected to the zeros not corrected) has no physical significance.

Figure 15 prefigures this result. It shows the trend of the amplitude and phase characteristics of the assembly {calorimeter + two step filtering} , graph II and II'.

As we have already seen, $H(p)$ can be expressed as

$$H(p) = \frac{\text{polynomial in } p}{\prod_{1}^{n}(1+\tau_i p)} \quad \text{or, equivalently,} \quad \frac{\prod_{1}^{m}(1+\tau_j^* p)}{\prod_{1}^{n}(1+\tau_i p)}$$

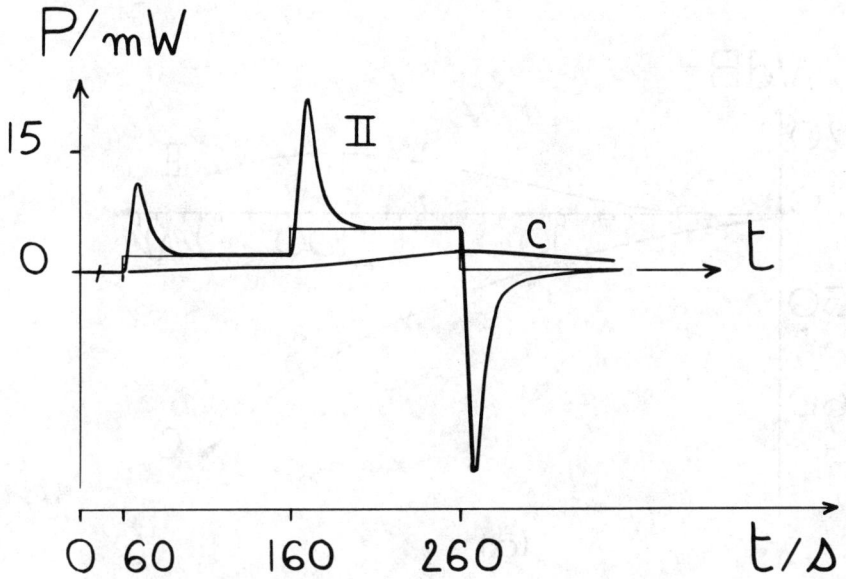

Figure 14 - When the source of the thermogenesis is located near the thermocouples, the response II given by the two step filtering overshoots the actual thermogenesis. C represents the experimental thermogram

We have shown that by applying to the response s(t) the operation defined by the expression s(t) + τ_1 ds(t)/dt we obtained a signal in which the time constant τ_1 no longer has any effect.

On the other hand, if we now apply to s(t) the operation defined by the relation

$$s(t) = s_1^*(t) + \tau_1^* \frac{ds_1^*(t)}{dt}$$

we obtain a signal $s_1^*(t)$ in which the effect of the first zero of the transfer function is lost (10).

So, the suppression of one pole of the transfer function is achieved by "derivation" whereas a zero is suppressed by "integration" (integrating the differential equation given above).

We can show for example that this approach allows us to succeed in suppressing the overshoot as figure 16 shows.

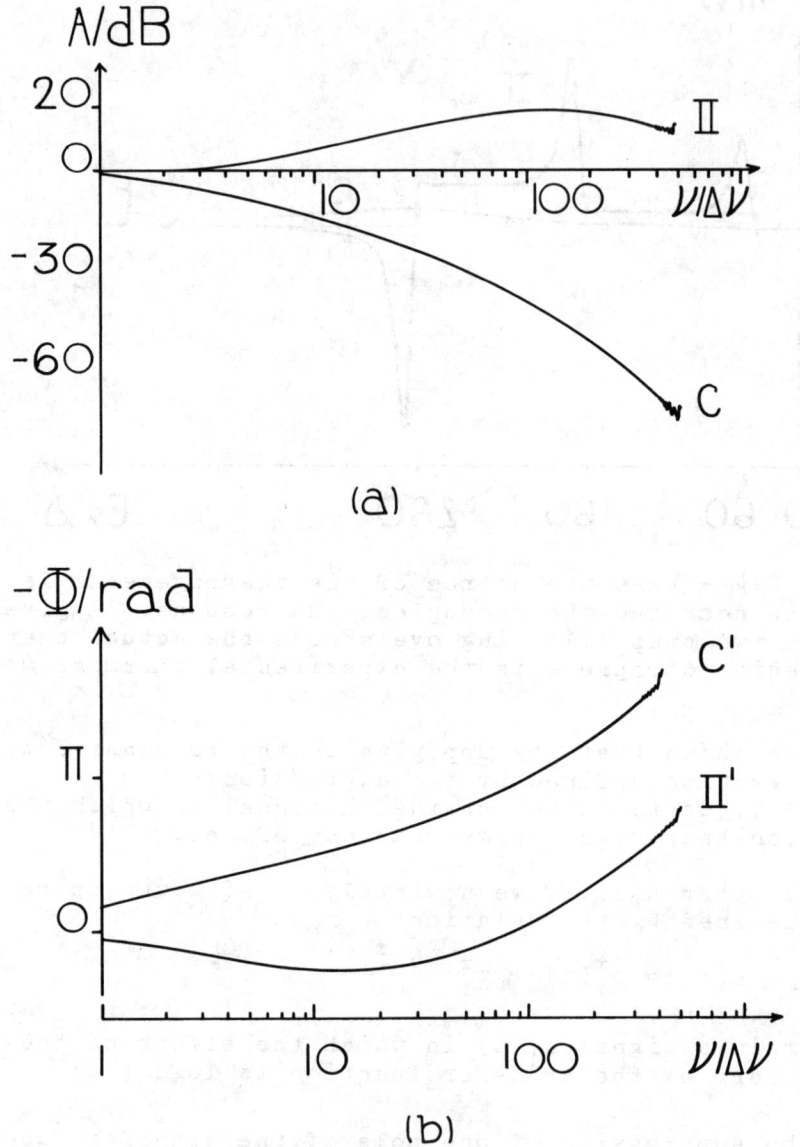

Figure 15 - Transfer function of a typical heat-conduction calorimeter when the source of the thermogenesis is located near the thermocouples.
(a) curve II represents the amplitude characteristic after compensating the two first time constants τ_1 and τ_2
(b) II' represents the corresponding phase characteristics.

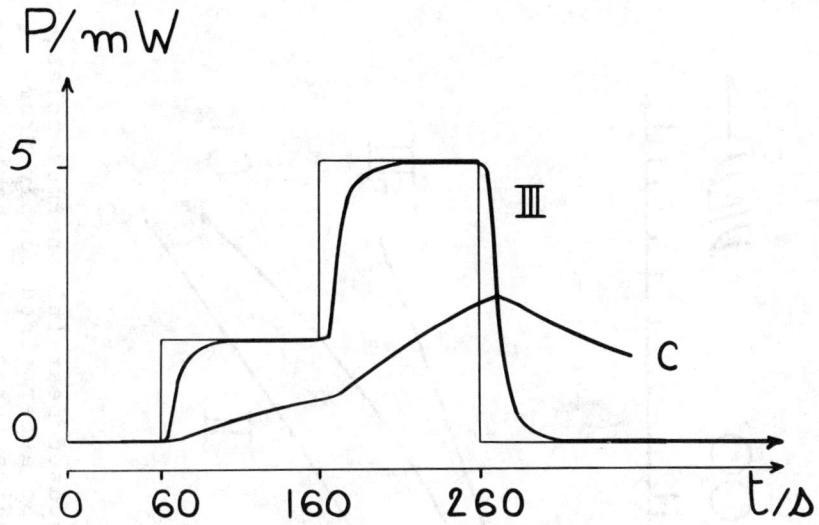

Figure 16 – By compensating the first zero of the transfer function, the overshoots observed on curve II (figure 14) disappear (curve III)

By integrating the response $s_2(t)$ obtained during the second step of inverse filtering, we get a signal which does not show a tendency to overshoot. In the specific case described, we have $\tau_1^* = 65$ seconds.

The positive effect of inverse filtering on the transfer function of the calorimetric device is shown in figure 17. The amplitude characteristic C clearly shows that the behaviour of the system cannot be ascertained simply from the two time constants τ_1 and τ_2, even at low frequencies (graph II). The introduction of a zero, however, results in a good description within a larger domain (graph III).

Figure 18 shows, in the two cases studied (heat source near and far from the thermocouples), the amplitude characteristic of the calorimetric device alone (graph C) and that of the same device associated with a two step inverse filtering (graph F) obtained after eliminating two poles and one zero or after eliminating two poles only.

We shall go on now to describe the results of a comparative study of harmonic analysis and digital inverse filtering (As we have already indicated, electronic

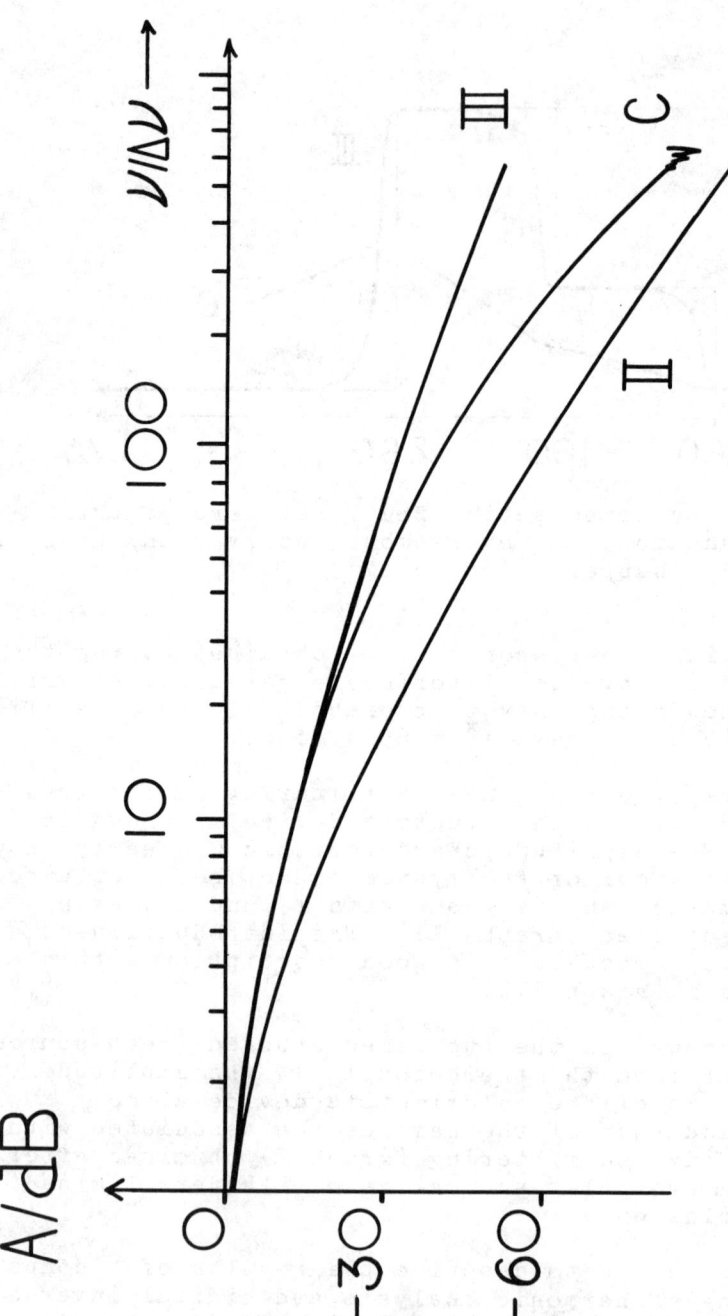

Figure 17 - As the source of the thermogenesis is located near the thermocouples, the amplitude characteristic C of the system cannot be ascertained simply from the two time constants τ_1 and τ_2 even at low frequencies (graph II). The introduction of a zero results in a good description within a large domain (graph III)

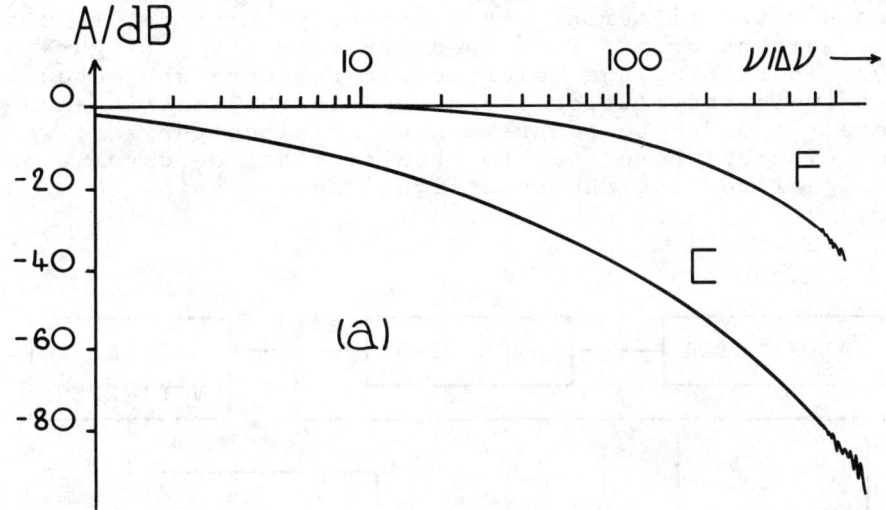

Figure 18a - Amplitude characteristic C of a typical heat conduction calorimeter when the source of the thermogenesis is located near the thermocouples. F represents the amplitude characteristic after compensating the first two time constant and the first zero of the transfer function

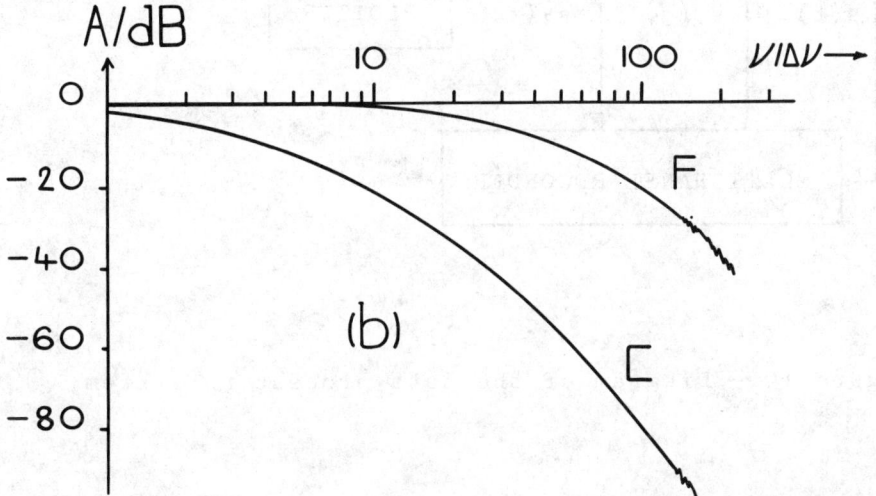

Figure 18b - Amplitude characteristic C of a typical heat conduction calorimeter when the source of the thermogenesis is located far from the thermocouples. F represents the amplitude characteristic after compensating the first two time constants

and digital filtering give virtually identical results).
The same calorimeter is used for this study (figure
19). Therefore, the calorimetric response subjected to
the two deconvolution techniques is the same in each
case. In order to secure a meaningful comparison, we
endeavoured, here too, to obtain, after deconvolution,
a signal bearing the same signal/noise ratio.

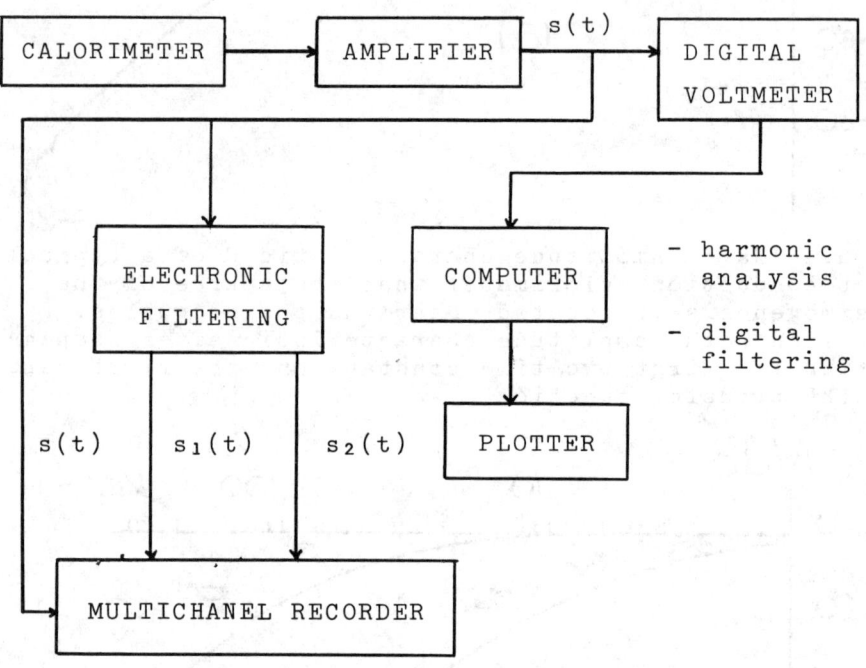

Figure 19 - Diagram of the data processing system

We should remember, first of all, that harmonic analysis gives a reproduced thermogenesis with an extra oscillation of frequency ν_c (cut-off frequency). To evacuate that fluctuation, we must perform a smoothing technique (integrating the signal over a period of time $1/\nu_c$ for example, figure 4b).

The results we shall subsequently be presenting will be those obtained after smoothing. It is clear from the quick decrease in both series of time constants $\{\tau_i\}$ and zeros $\{\tau_i^*\}$ that the two techniques give similar results (figures 20 and 21).

It should be also noted that inverse filtering invariably gives a delayed response because of a partial correction in phase. Harmonic analysis, on the other hand, will always give a response symmetrically distributed around the actual thermogenesis.

The results given by harmonic analysis and inverse filtering will be comparatively presented next in two special cases which clearly show how the interest provoked by the use of deconvolutive techniques in the field of heat conduction calorimetry is justified.

(B)

MISCELLANEOUS APPLICATIONS

We shall take as our first example the thermokinetic study of a solid—solid transformation by differential scanning microcalorimetry. The second example deals with measuring partial molar enthalpies of mixing (or partial molar excess enthalpies) by flow calorimetry. The data processing device has already been described.

I - STUDY OF A THERMOELASTIC PROCESS OF MARTENSITIC TYPE ($\beta \rightleftarrows \gamma'$)

The transformation takes place in a Cu-Zn-Al alloy under a programmed drop or rise in temperature. Figure 22 (a and b) shows two thermograms associated with a thermal cycle (cooling and heating the sample).

Thermogram a (exothermic) corresponds to the transformation $\beta \rightarrow \gamma'$ (when temperature drops from $-24.9°C$ to $-25.8\ °C$ at $0.225\ °C\ min^{-1}$). Thermogram b (endothermic) corresponds to the transformation $\gamma' \rightarrow \beta$ (when tempe-

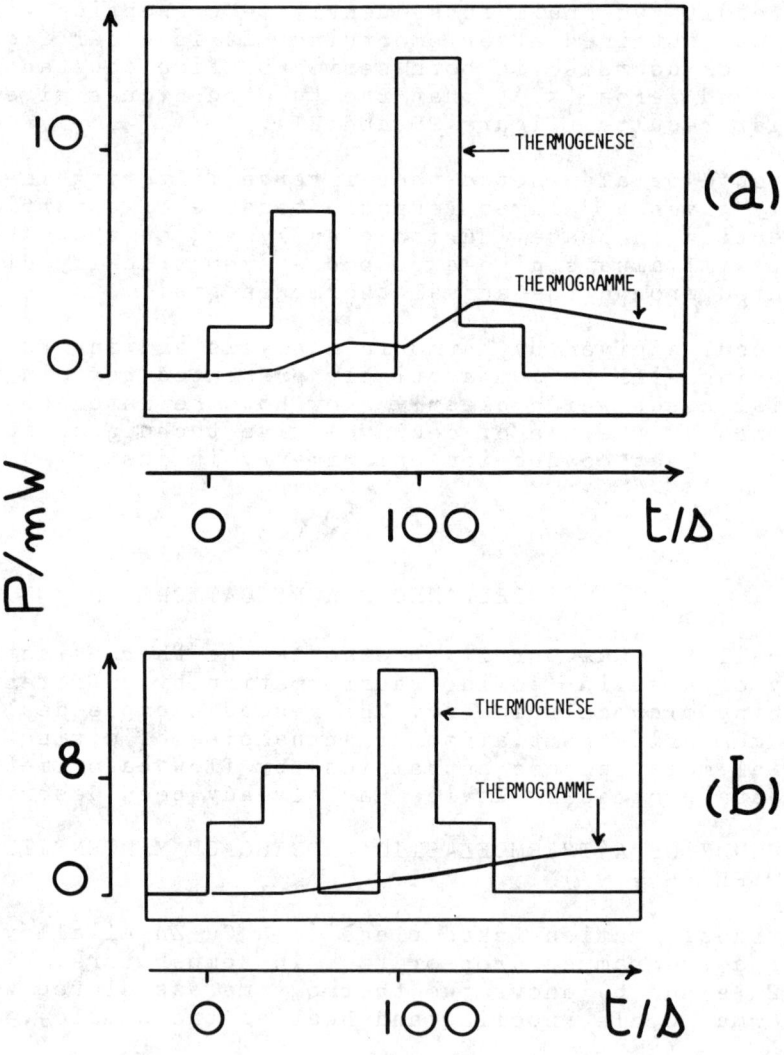

Figure 20 (a) and (b)

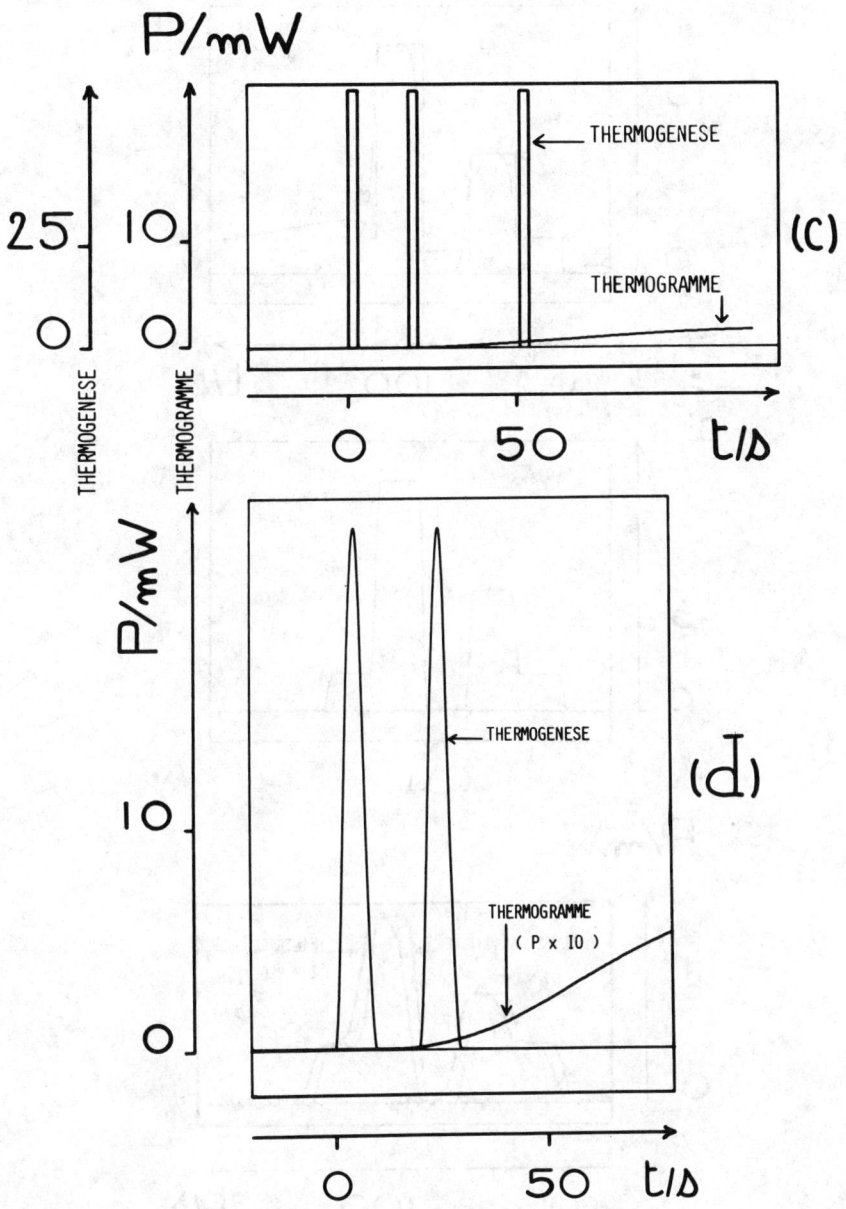

Figure 20 (c) and (d)

Figure 20 - Thermogenesis used for the comparative application of harmonic analysis and inverse filtering In the case (a), the source is located near the thermocouples

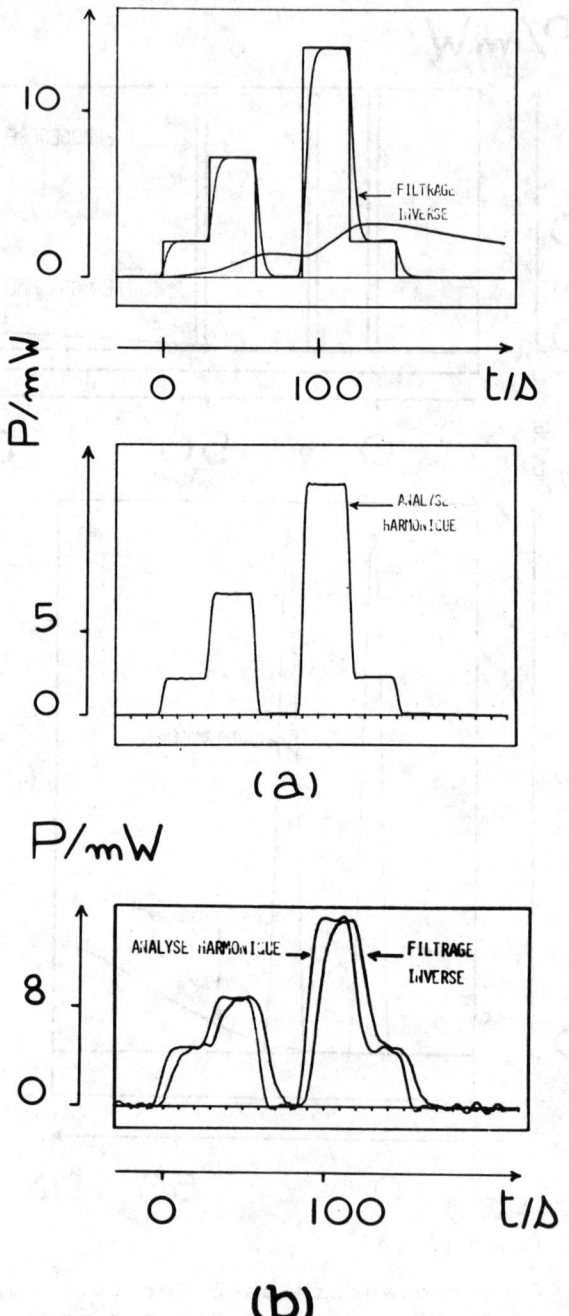

Figure 21 (a) and (b)

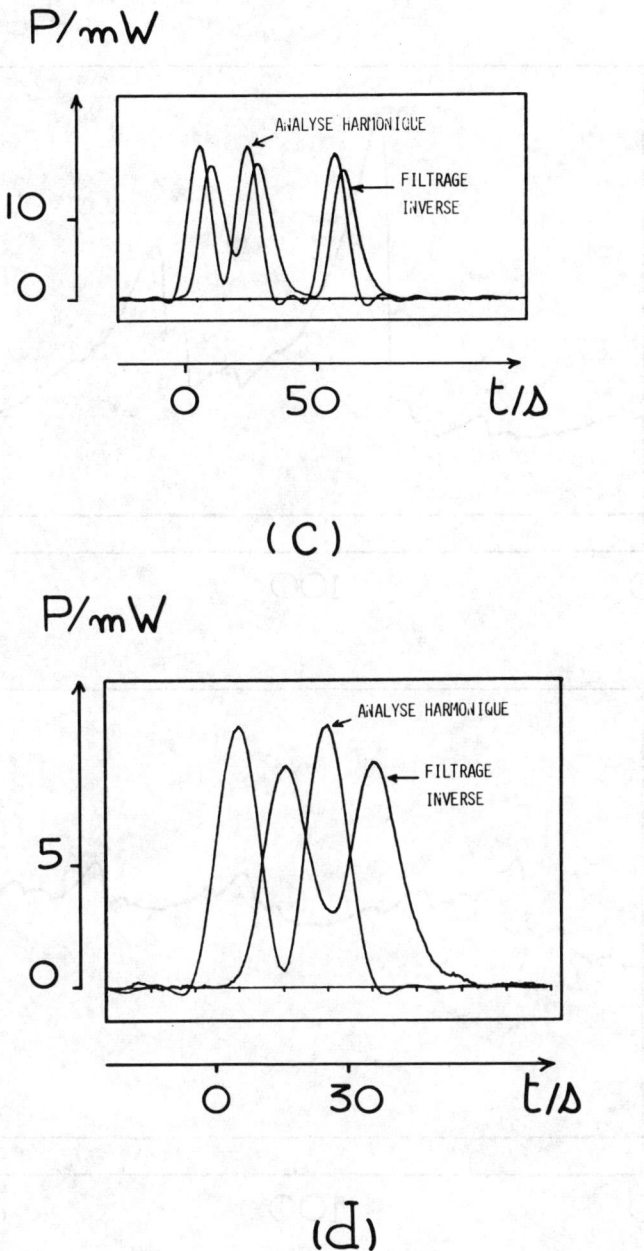

Figure 21 (c) and (d)
Figure 21 - Comparative application of harmonic analysis and inverse filtering

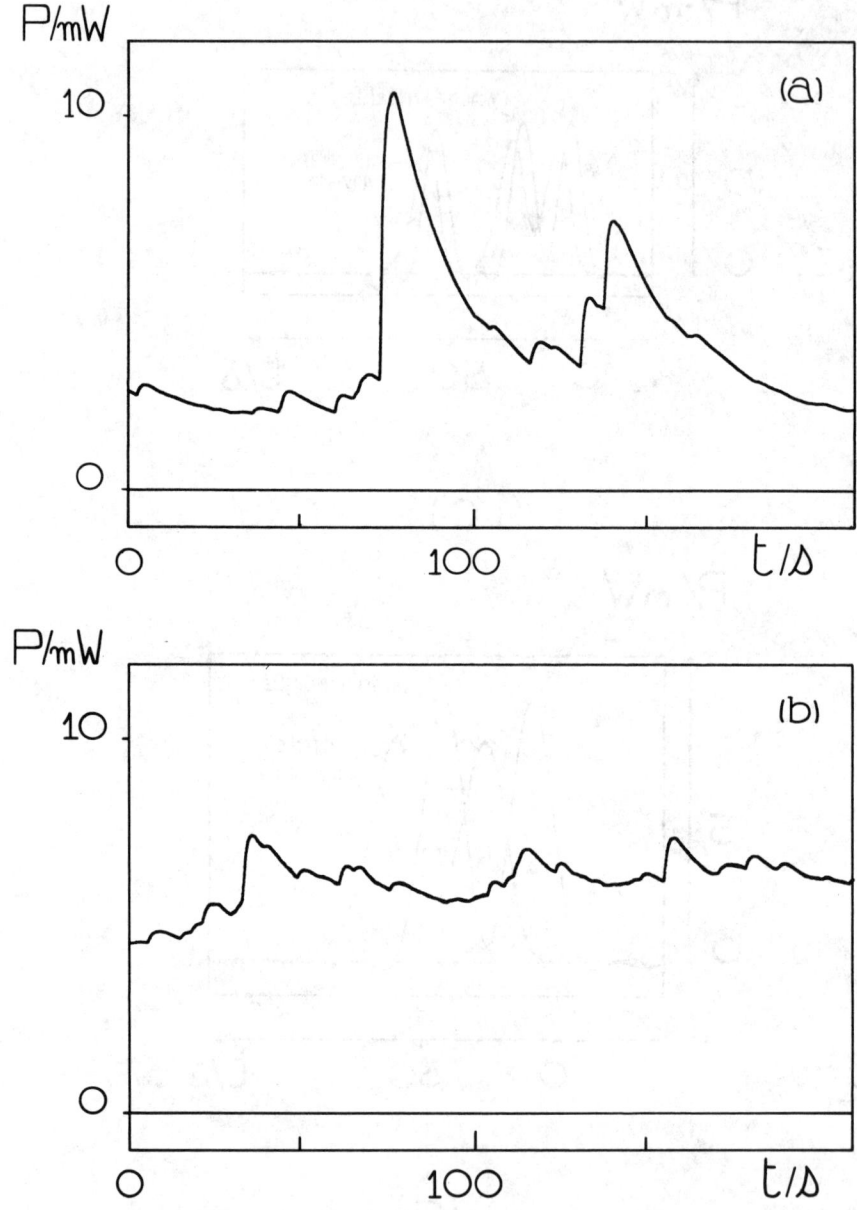

Figure 22 - Study of a solid-solid transition by differential scanning calorimetry. Two experimental thermograms ((a) and (b)), see text

rature rises at the same rate from -11.8°C to -11.1°C).

Figures 23a and 23b show the results given by inverse filtering compensating the first two poles of the instrumental transfer function (time constants 22.5 seconds and 0.75 second respectively). Figures 24a and 24b show the results given by harmonic analysis.

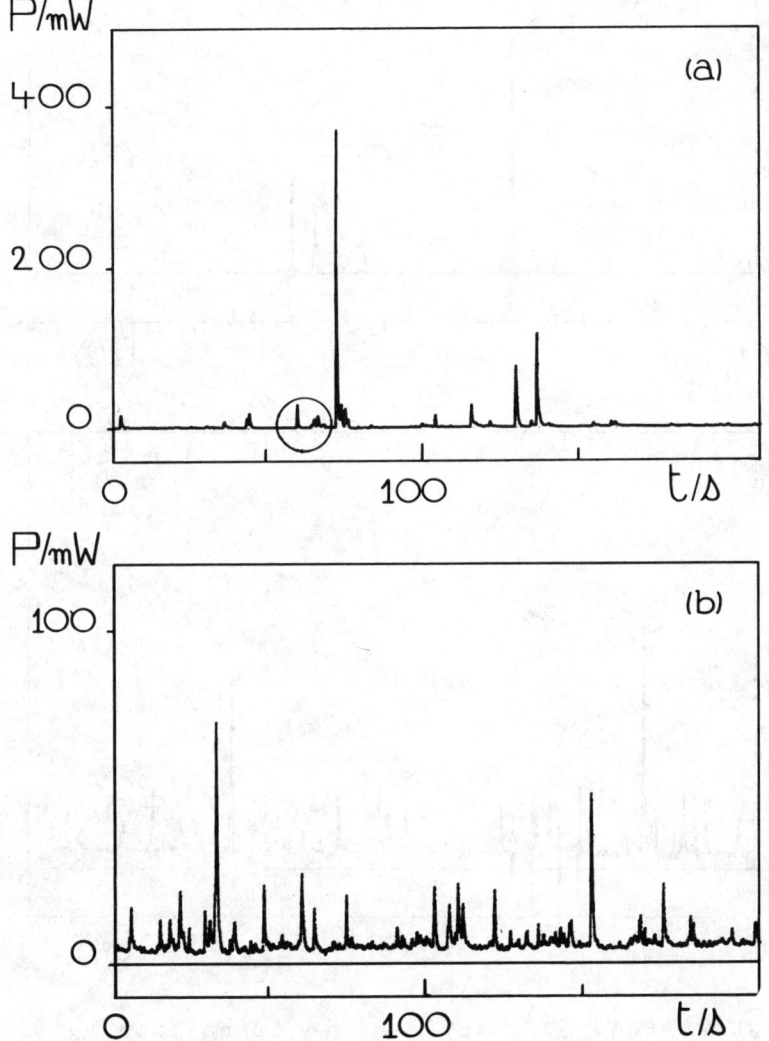

Figure 23 - Study of a solid-solid transition by differential scanning calorimetry. Deconvolution of the foregoing thermograms (figure 22(a) and (b)) by inverse filtering

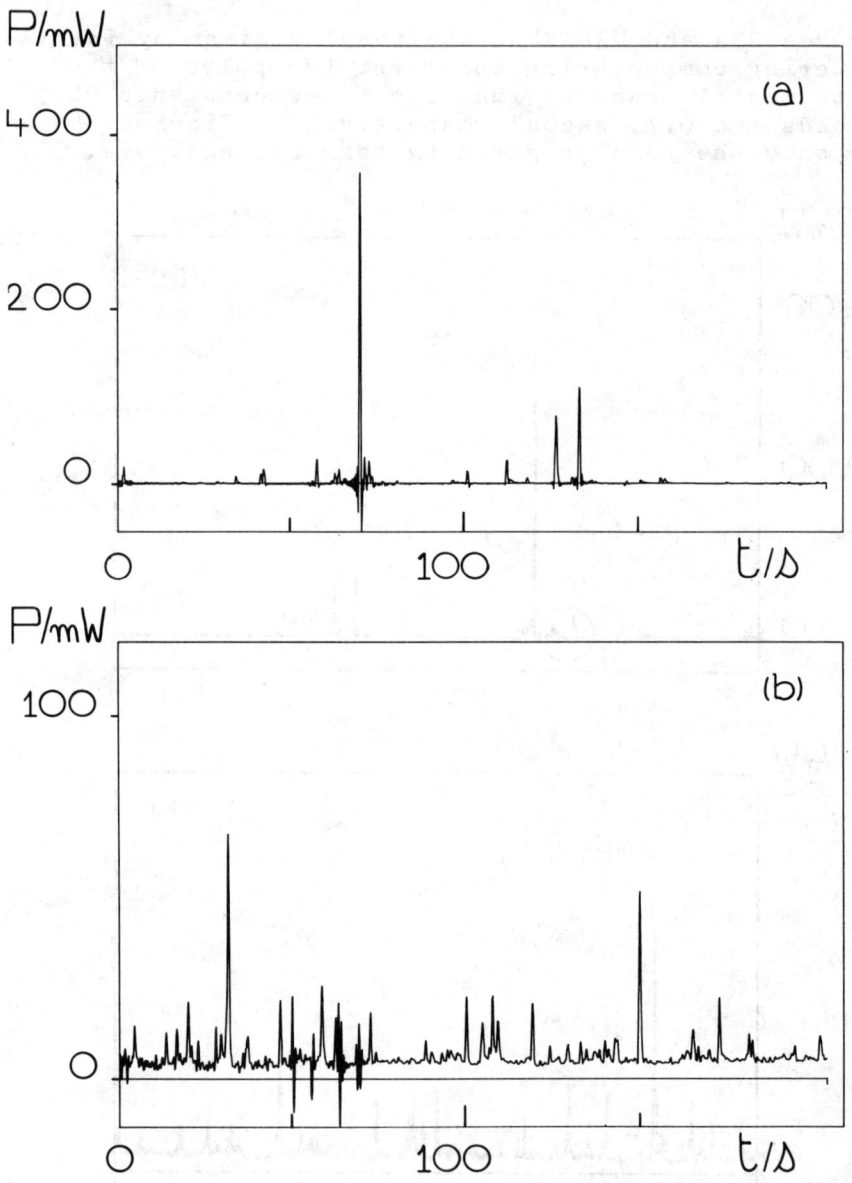

Figure 24 - Study of a solid-solid transition by differential scanning calorimetry. Deconvolution of the foregoing thermograms (figure 22(a) and (b)) by harmonic analysis

In figure 25, we have plotted the thermogenesis given both by inverse filtering (a) and by harmonic analysis (b) and corresponding to the area circled in the figure 23a.

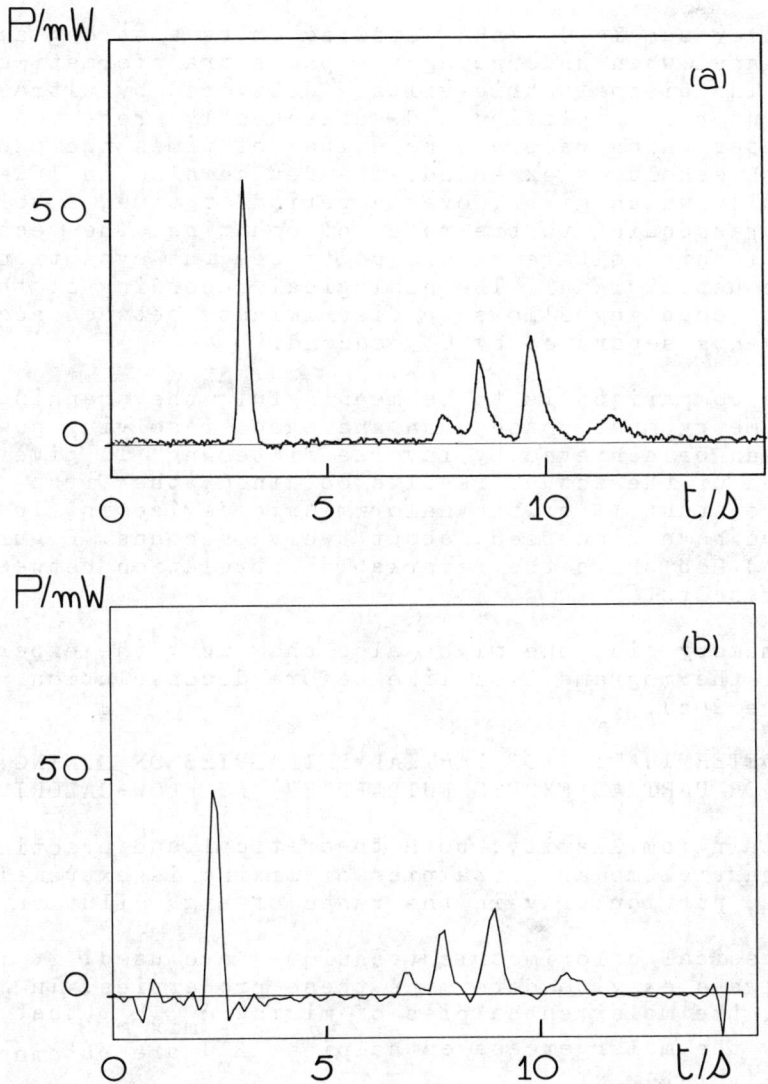

Figure 25 - Application of the inverse filtering (a) and harmonic analysis (b) to the deconvolution of a small detail circled on figure 23

When we look at the graphs representing the thermal power after deconvolution, we might be tempted to think that the very pronounced oscillations in the kinetics are simply due to background noise. But it can be shown that this is not the case.

The alloy sample Cu-Zn-Al studied emits a strong acoustic signal when undergoing the phase transformation (12). The piezoelectric voltage delivered by ultrasonic sensor is amplified and subsequently received by a counter which records the number of times the one volt threshold is exceeded. The device also includes a modulus which gives, over a period of time, a voltage corresponding to the rate of counting. The recording of this voltage allows us to see the evolution of the acoustic signal. The analogical recording of the rate of counting allows to discriminate between acoustic events separated by 0.5 second.

If the comparison is to be meaningful, the scanning calorimetry must possess an analogous resolving power. This can be achieved by inverse filtering. To give an idea of the actual results obtained, the first time constant τ_1 of the calorimetric device is, in the case we have discussed, about twelve seconds. Figure 26a and 26b shows the remarkable correlation between both results.

As a memory aid, one might also show what the experimental thermograms look like before deconvolution (figure 26c).

II - DETERMINATION OF PARTIAL ENTHALPIES OF MIXING (OR PARTIAL EXCESS ENTHALPIES) BY FLOW-CALORIMETRY

A certain familiarity, both theoretical and practical, with partial molar enthalpies of mixing is extremely useful, particularly in the range of high dilution (13).

If classical calorimetric techniques are used, it is not always easy to determine these properties. In most cases, the molar enthalpies of mixing $\Delta_{mix} H_m$ (that is to say the molar excess enthalpies H_m^E) are obtained directly (14).

Figure 27a shows an example of this function whilst figure 27b gives a representation of the corresponding partial molar enthalpies of mixing H_1^E and H_2^E for the benzene-cyclohexane system (15).

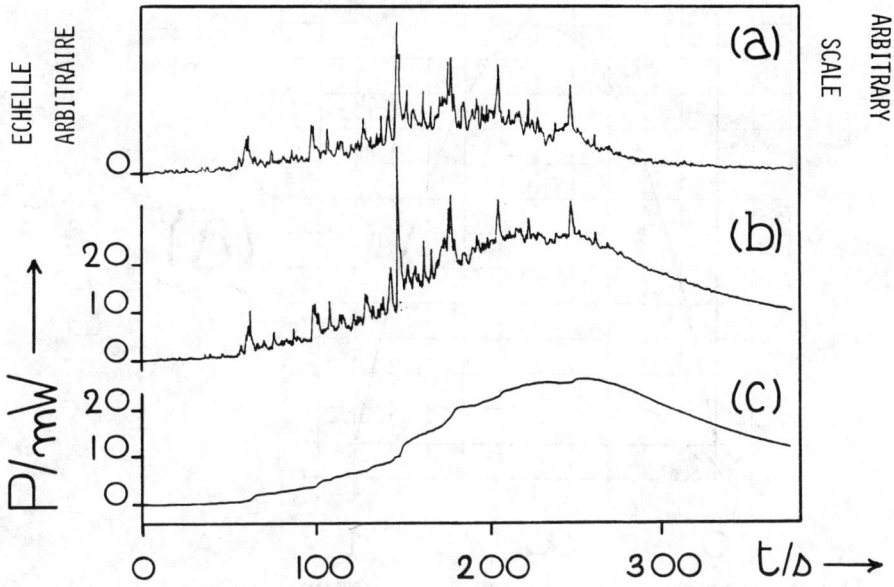

Figure 26 - Study of a solid-solid transition by acoustic emission (graph a) and differential scanning calorimetry (deconvolution by inverse filtering) graph b. (c) represents the experimental thermogram

The flow calorimeter allows us to obtain these results directly. The laboratory cell contains a certain quantity of (say) substance 2. Substance 1 is added at a constant rate. If the instrumental inertia were negligible, we would obtain the partial molar enthalpy of mixing H_1^E by calculating the ratio of the thermal power measured to the molar flow of substance 1.

It is clear from figure 27b that, in the range of high dilution, partial molar enthalpies of mixing change quickly v.s. composition of the system. This means that experimental thermograms do not give a good representation of the function H_1^E. This is confirmed by figure 28a. Benzene is the added compound 1 ; cyclohexane 2 is already inside the calorimetric cell. Deconvolution of the calorimetric response is therefore essential in this case too (16).

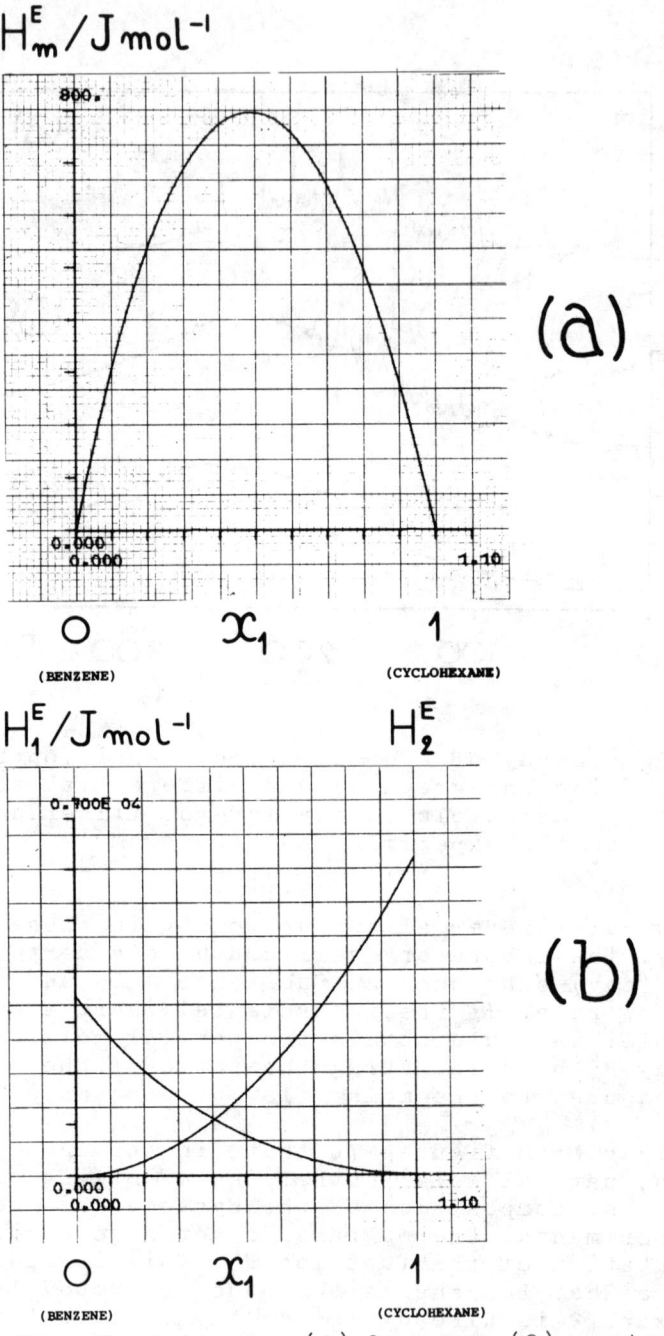

Figure 27 - Cyclohexane (1)-benzene (2) system
(a) molar excess enthalpy at 298,15 K (15)
(b) partial molar excess enthalpies at 298,15 K (15)

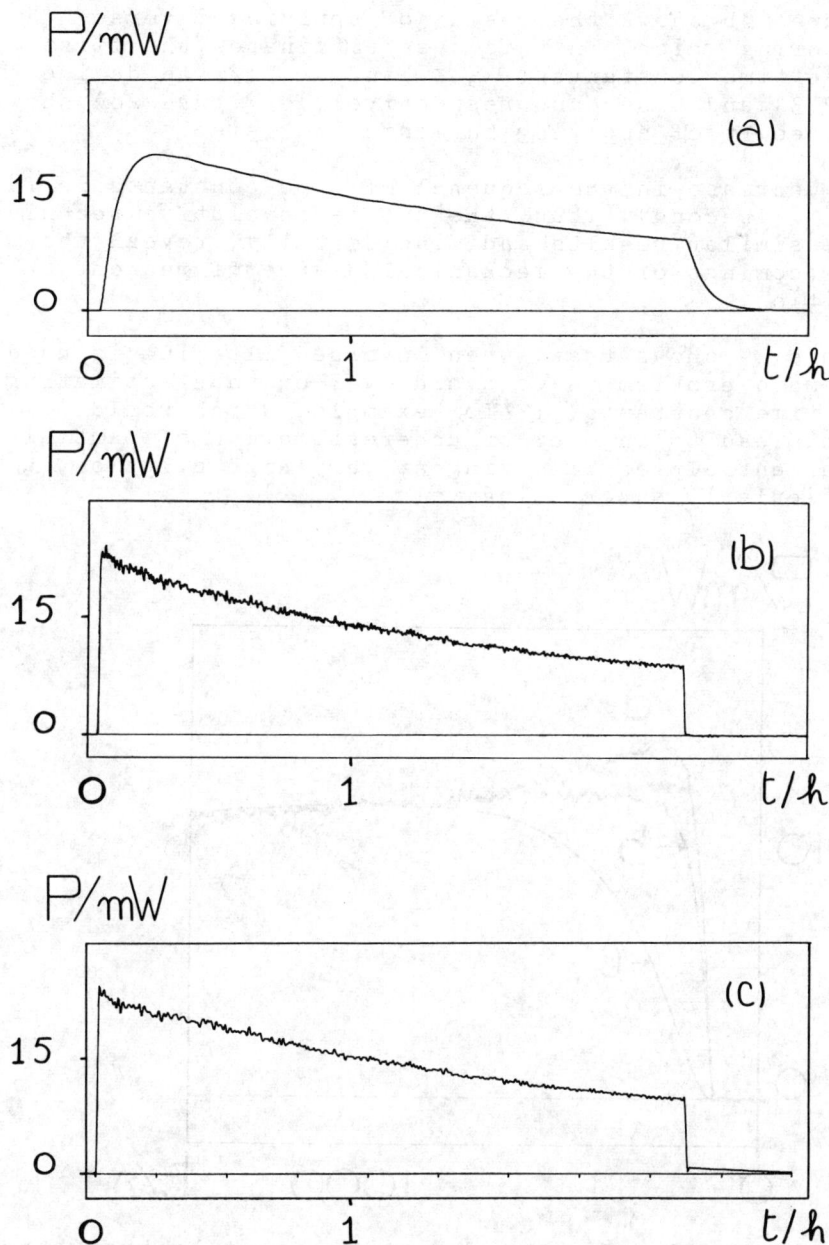

Figure 28 - Excess enthalpy measurement by heat conduction flow calorimetry
(a) a typical thermogram
(b) deconvolution by inverse filtering
(c) deconvolution by harmonic analysis

Figure 28b shows the result of applying inverse filtering which, in this case, eliminates the first three time constants τ_1, τ_2 and τ_3 of the device (222, 34 and 6 seconds respectively). Figure 28c shows the result of applying harmonic analysis.

The decrease in the sequence of time constants is fast ($\tau_4 \sim 1$ second). Thus, the two deconvolutive techniques give similar results and, incidentally, reveal the shortcomings of the mechanical device of reagent addition.

Figure 29 shows that, when inverse filtering is used, the main problem is to avoid over-or underestimating the time constants, τ_1 for example, which would, in turn, lead to an over-or underestimation of partial molar enthalpies of mixing at very high dilution, a particularly interesting value.

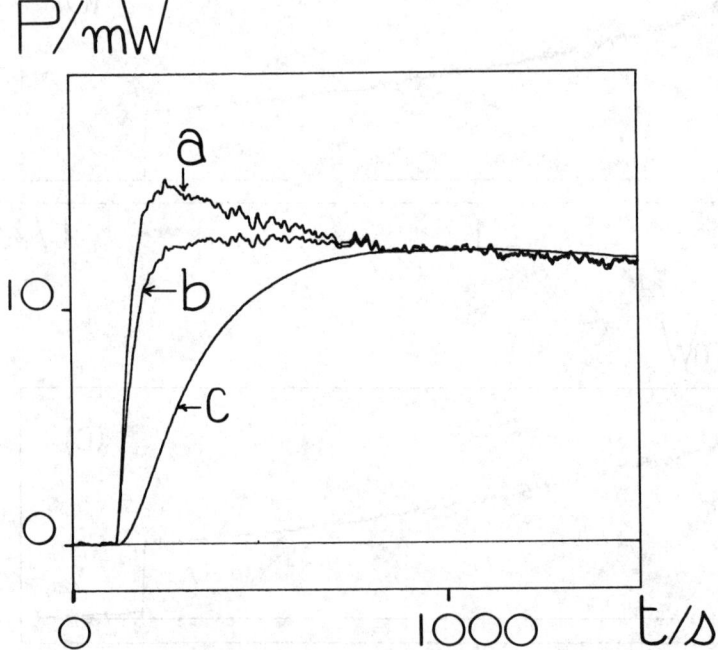

Figure 29 - Excess enthalpy measurement by heat conduction flow calorimetry. When inverse filtering is used, the main problem is to avoid over (graph a) or under (graph b) estimating the time constant τ_1 for example. C represents the experimental thermogram

We need only to look at figures 30a and 30b to appreciate how rewarding are the techniques just described.

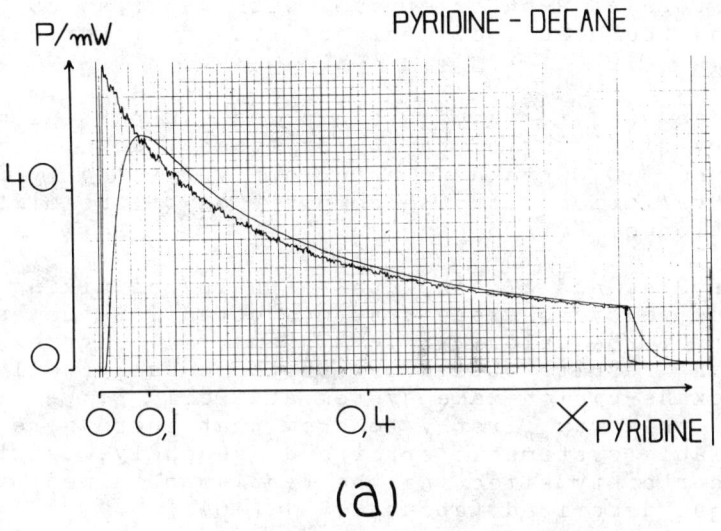

(a)

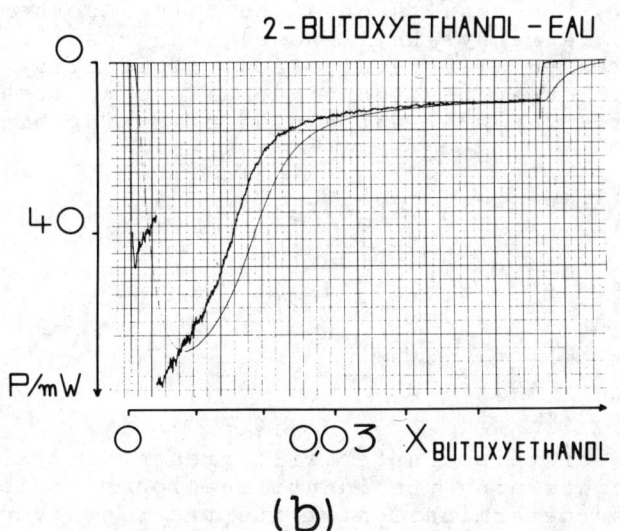

(b)

Figure 30 - Excess enthalpy measurement by heat conduction flow calorimetry. Typical results of the deconvolution by electronic inverse filtering
(a) pyridine-decane system (endothermic)
(b) butoxy 2 ethanol-water system (exothermic)

Figure 30a shows a rough thermogram obtained when pyridine is mixed with decane at 25 °C (17) and the result of deconvolution by electronic inverse filtering. Figure 30b shows the rough thermogram obtained from the mixture of 2-butoxyethanol with water at 25 °C (18) and the result of deconvolution by inverse filtering.

As we can see, it is possible to obtain partial molar enthalpy of mixing of the butoxyethanol down to molar fractions > 0.001 and to bring out very clearly its sudden variation v.s. the composition of the mixture (formation of "microphase").

Figures 27a and 27b show the enthalpy of mixing of the benzene-cyclohexane system adopted for the calibration of the calorimetric system (19). Figure 31 gives the results obtained with the equally well know hexane-cyclohexane system at 298,15 K. We were able in this way firstly to show that there were no appreciable systematic error and, secondly, to stress the uncertainty affecting the results obtained by the technique described (about 0.5%) (16).

Figure 32 shows the results obtained using another system (n heptane-methylethylketone), the results being already in the literature (20). It is clear from that figure that the technique allows to obtain both partial and integral enthalpies of mixing since we have

$$H_1^E = H_m^E - x_2 (dH_m^E/dx_2)$$

$$H_2^E = H_m^E + x_1 (dH_m^E/dx_2)$$

and

$$H_m^E = x_1 H_1^E + x_2 H_2^E$$

The final part of this chapter will present certain observations concerning the identification of calorimetric systems (determination of its transfer function).

III - CONCLUSION

We have attempted to show that deconvolution of the calorimetric response extends considerably the area of application of heat conduction calorimetry.

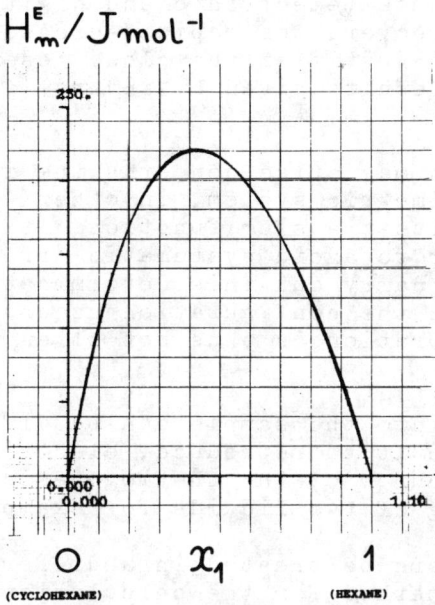

Figure 31 - Molar excess enthalpy for the hexane (1)-cyclohexane (2) system at 298,15 K. Upper graph our results (16), lower graph (ref. 19)

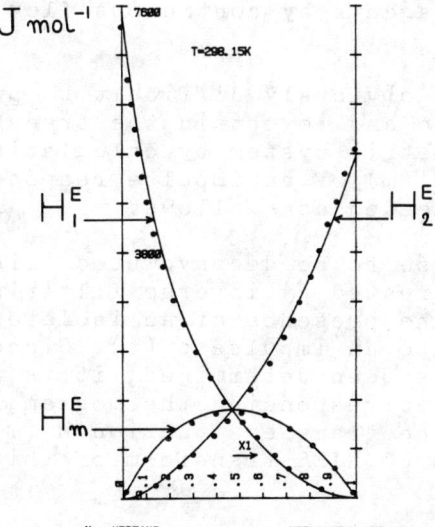

Figure 32 - Molar excess enthalpy of the system n heptane (2) - methylethylketone (1) at 298,15 K
..... results of KYHOHARA, HANDA and BENSON (20)
―――― our results (16)

Both filtering techniques (electronic and digital) give identical results. Moreover, the rapid decrease in the time constant sequence justifies the close resemblance observed between the results given by inverse filtering and harmonic analysis.

The only problem which may arise concerns the identification of the calorimetric system, in other words, the determination of its transfer function. It is obvious that, in order to avoid systematic error, the configuration used to carry out this determination must be identical with the configuration used subsequently. A JOULE calibration is thus not often appropriate.

Introducing a heater into the sample can be difficult. It is sometimes difficult to reproduce, electrically, the geometry of the source of the thermogenesis to be studied (the mixing of two liquids for example).

Each case must therefore be treated in a different way. The most appropriate method for the calculation of the transfer function must be found each time. To illustrate this, let us go back to the two applications already described
- a solid-solid transformation,
- a mixing of two reagents by continuous flow calorimetry.

In the first case, it is obviously difficult to put a heater into solid sample and to obtain the transfer function of the calorimetric system by calculating the FOURIER transform $H(\omega)$ of an impulse response $h(t)$. We therefore proceeded as follows.

The calorimetric response to be deconvoluted (figures 33a and 33b) is first treated by inverse filtering. We observe in graph b the presence of an isolated peak which corresponds to an impulse $\delta(t)$. Since the time constant τ_1 has been determined, it is possible to reconstruct the corresponding thermogram, figure 33c. To obtain the transfer function $H(\omega)$, we simply calculate the FOURIER transform of this impulse, figure 33d.

Since the location of the heat source has an effect on the instrumental transfer function, we must be careful to use a study-sample whose volume is small. The thermogenesis reproduced by harmonic analysis and represented in figure 24 shows oscillations which may be due to a failure to follow this requirement.

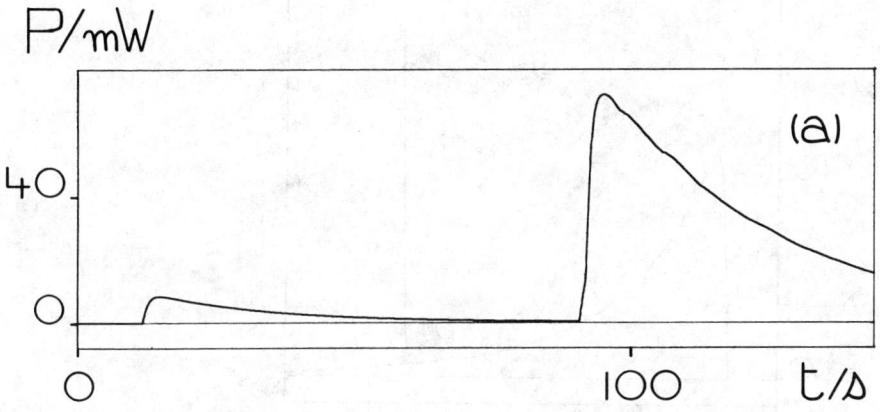

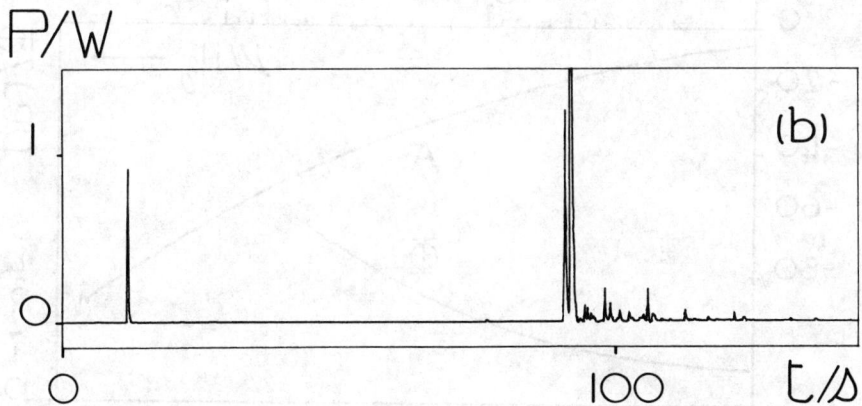

Figure 33 (a) and (b)

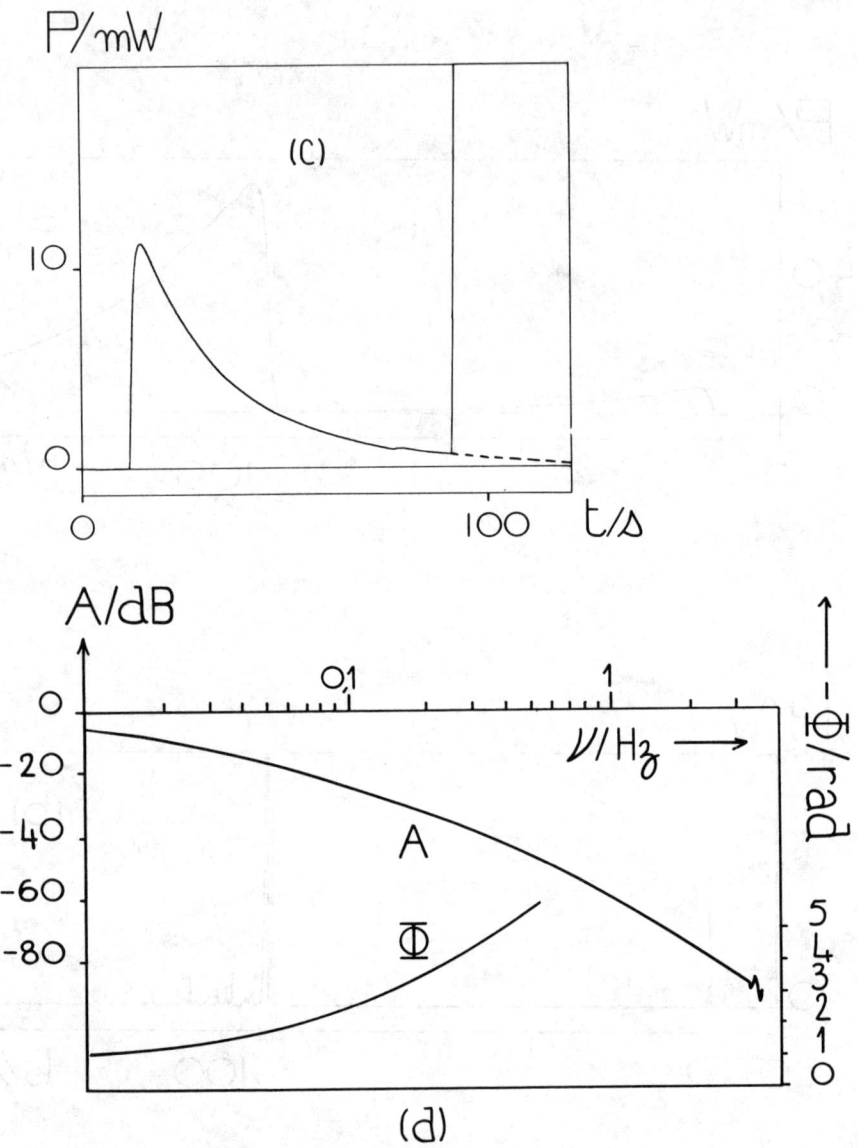

Figure 33 (c) and (d)

Figure 33 - Study of a solid-solid transition by differential scanning calorimetry (see text)
(a) part of an experimental thermogram
(b) deconvolution of the foregoing curve by inverse filtering
(c) reconstruction of the impulse response h(t)
(d) corresponding transfer function

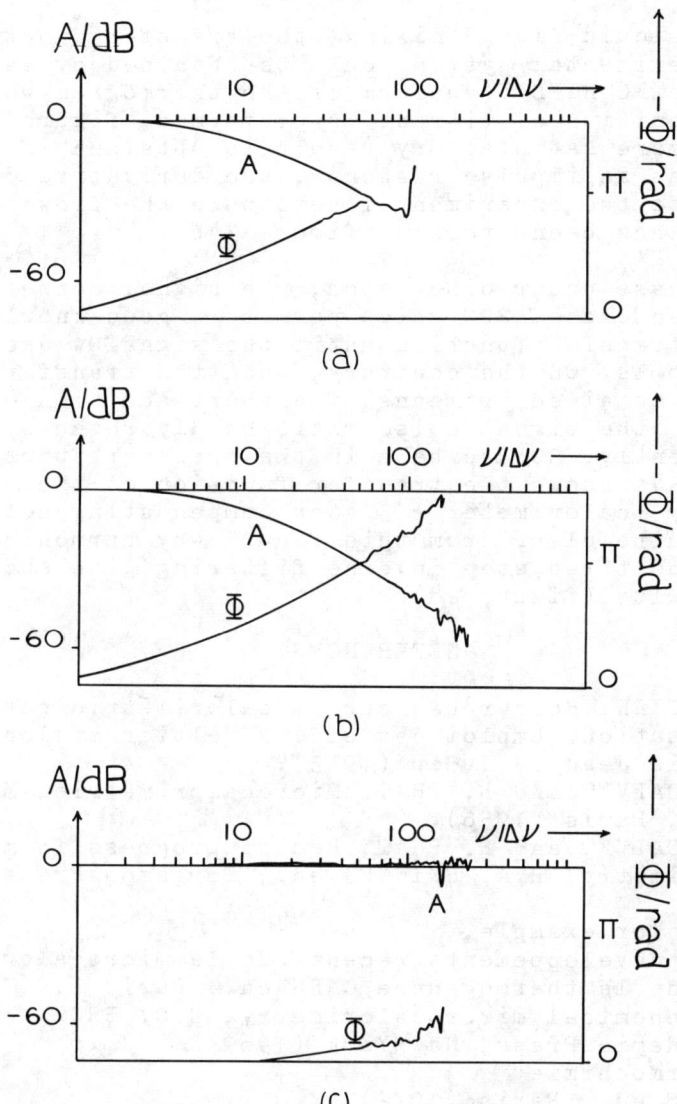

Figure 34 - Excess enthalpy measurement by heat conduction flow calorimetry
(a) transfer function by calculating the FOURIER transform of the calorimetric response to a short addition of reagent
(b) transfer function by calculating the FOURIER transform of the derivative of the return to the experimental zero once the flow of reagent has been stopped
(c) transfer function of the whole system {calorimeter + 3 step compensating network}

In the liquid-liquid mixing, the transfer function of
the experimental system could be obtained by calculating the FOURIER transform of the thermogram when a
short addition of titrant is performed, figure 34a.
A much more satisfactory result is obtained if we
adopt, as an impulse response, the derivative of the
return to the experimental zero once the flow of
reagent has been stopped, figure 34b.

In the case under discussion, the signal/noise ratio
will then exceed 200 which permits a good knowledge
of the transfer function until the signal weakens to
50 decibels. On the contrary, when the transfer function is obtained by means of a short addition of
titrant, the signal/noise ratio hardly reaches 20 or,
equivalently, 30 decibels in the transfer function.
Figure 34c shows the transfer function of the whole
assembly {calorimeter + 3 step compensating network}.
It is quite clear from this figure why harmonic analysis and three step inverse filtering give such similar results (figure 28).

REFERENCES

(1) A. TIAN, Recherches sur la calorimétrie par compensation. Emploi des effets Peltier et Joule,
Louis Jean ed., Gap (1933).
E. CALVET and H. PRAT, Microcalorimétrie, Masson
ed., Paris (1956).
E. CALVET and H. PRAT, Recent progress in microcalorimetry, H.A. SKINNER ed., Pergamon Press (1963).

(2) see for example,
Les développements récents de la microcalorimétrie
et de la thermogenèse, CNRS ed., Paris (1967)
Biochemical mircrocalorimetry, H.D. BROWN ed.,
Academic Press, New-York (1969)
Thermochimie
CNRS ed., Paris (1972)
Biological microcalorimetry, A.E. BEEZER ed.,
Academic Press, London (1980)

(3) E. CESARI, J. NAVARRO, V. TORRA, J.L. MACQUERON,
J.P. DUBES and H. TACHOIRE
Thermochim. Acta, $\underline{39}$, 73-80 (1980)

(4) F.M. CAMIA, J. Phys., $\underline{22}$, 271 (1961)
J. Phys., $\underline{23}$, 25 (1962)

(5) H. TACHOIRE in "Les développements récents de la
microcalorimétrie et de la thermogenèse", CNRS ed.,
Paris (1967)

(6) J. NAVARRO, E. ROJAS and V. TORRA, Rev. gén. Thermique, n° 143, 1137 (1973)
J. VAN BOKHOVEN and J. MEDEMA, J. Phys., E9, 123 (1976)
S. TANAKA, Thermochim. Acta, 25, 269 (1978)
E. CESARI, V. TORRA, J.L. MACQUERON and J. NAVARRO, An. Fis., 73, 300 (1977)

(7) O. ROSE, Thesis, Marseilles (1964)
Y. THOUVENIN, C. HINEN and A. ROUSSEAU in "Les développements récents de la microcalorimétrie et de la thermogenèse, CNRS ed., Paris (1967)
J.P. DUBES, M. BARRES and H. TACHOIRE, C.R. Acad. Sci., 283, 163 (1976)
J.P. DUBES, M. BARRES, E. BOITARD and H. TACHOIRE, Thermochim. Acta, 39, 63 (1980)
R. POINT, Thesis, Lyon (1978)
R. POINT, J.L. PETIT and P.C. GRAVELLE, Proceedings of "Journées de calorimétrie et d'analyse thermique", AFCAT, Turin (1978)
E. CESARI, V. TORRA, J.L. MACQUERON, R. PROST, J.P. DUBES and H. TACHOIRE, Proceedings of "Journées d'étude sur l'analyse et la déconvolution de la réponse instrumentale en spectroscopie et en calorimétrie, GTE (Société chimique de France), Cadarache (1980)
E. CESARI, V. TORRA, J.L. MACQUERON, R. PROST, J.P. DUBES and H. TACHOIRE, Thermochim. Acta, 53, 1 (1982)

(8) E. CESARI, V. TORRA, J.L. MACQUERON, R. PROST, J.P. DUBES and H. TACHOIRE, Thermochim. Acta, 53, 1 (1982)

(9) J.P. DUBES, M. BARRES, E. BOITARD and H. TACHOIRE, Thermochim. Acta, 39, 63 (1980)
J.P. DUBES and H. TACHOIRE, Thermochim. Acta, 51, 239 (1981)

(10) E. CESARI, V. TORRA, J.L. MACQUERON, R. PROST, J.P. DUBES and H. TACHOIRE, Thermochim. Acta, 53, 1 (1982)

(11) J.L. MACQUERON, P. FLEISCHMANN and H. MOULINIER in Proceedings of "Journées de calorimétrie et d'analyse thermique", AFCAT, Paris (1977)
A. PLANES, J.L. MACQUERON, M. MORIN and G. GUENIN, Phys. Stat. Sol., (a) 66, 717 (1981)

(12) G.R. SPEICH et R.M. FISCHER in Acoustic emission, ASTM, STP 500, Philadelphia (1972)

(13) see for example .
J.M. PRAUSNITZ, Molecular thermodynamics of fuid-phase equilibria, Prentice Hall, Englewood Cliffs (1969)

(14) see, for example,
M.L. Mc GLASHAN in Experimental thermochemistry, vol II, H.A. SHINNER ed., Interscience, New York (1962)
M.L. Mc GLASHAN, Chemical thermodynamics, Academic Press, New York (1979)

(15) R.H. STOKES, K.N. MARSH and R.P. TOMLINS, J. Chem. Thermodyn., $\underline{1}$, 211 (1969)

(16) R. KECHAVARZ, J.P. DUBES and H. TACHOIRE, Thermochim. Acta, $\underline{53}$, 39 (1982)

(17) A. AIT-KACI, J.Cl. MERLIN, R. KECHAVARZ, H.V. KEHIAIAN, J.P. DUBES and H. TACHOIRE, (to be published)

(18) Von ULFERT ONKEN, Bunsenges. physik. Chem., $\underline{63}$, n°2, 325 (1959)
GOPAL PATHAK, S.S. KATTI and S.B. KULKARNI, Ind. J. Chem., $\underline{3}$, 357 (1970)
R. KECHAVARZ, J.P. DUBES and H. TACHOIRE, (to be published)

(19) K.N. MARSH and R.H. STOKES, J. chem. Thermodyn., $\underline{1}$, 223 (1969)

(20) O. KIYOHARA, Y.P. HANDA and G.C. BENSON, J. chem. Thermodyn., $\underline{11}$, 453 (1979)

THE MEASUREMENT OF VAPOUR PRESSURE

A.S. Carson

Department of Physical Chemistry, University of Leeds,
Leeds LS2 9JT, U.K.

INTRODUCTION

Vapour pressure measurements may be undertaken for a variety of reasons, for example, the nature of a surface and its influence on evaporation and condensation may be studied as may also the diffusion coefficient of the evaporated species in air. Also information on differential vapour pressures is used to calculate molecular weights and the thermodynamic functions of solutions. From the point of view of the thermochemist, however, the temperature variation of a vapour pressure is measured so as to obtain indirectly an enthalpy of vaporisation (ΔH_v) or sublimation (ΔH_s). This may be combined with a standard enthalpy of formation to give a gas phase enthalpy of formation, which since it does not contain energy terms for the cohesive forces within the material, may be used to calculate bond enthalpies. Also vapour pressure measurements may be used to calculate K_p for a reaction and from K_p and its temperature variation, the standard free energy, enthalpy and entropy changes accompanying the reaction may be found. Further, if free energy functions or partition functions for the constituents are available from heat capacity or spectroscopic data, then an enthalpy change may be calculated from a single value of K_p. The two ways of handling K_p are often referred to as the "second law" and "third law" methods respectively.

The relationship between the enthalpy change and the temperature dependence of the saturated vapour pressure is given by the Clapeyron equation,

$$\frac{\Delta H}{T \Delta V} = \frac{dp}{dT}$$

where ΔV is the difference between the molar volumes of the coexisting phases at this pressure and at temperature T. This equation may be put into the form

$$\frac{\Delta H}{R} = \frac{p\Delta V}{RT}\left[-\frac{d \ln p}{d^{1}/T}\right]$$

where the expression in brackets is the experimentally measured quantity. This equation is most frequently used in a simplified form due to Clausius, in which the volume of the liquid or solid is ignored in comparison with the volume of the gas, which is assumed to be ideal. That is, $p\Delta V \sim pV_g \sim RT$ so that the equation becomes

$$-\frac{\Delta H}{R} = \frac{d \ln p}{d^{1}/T}$$

It is important, particularly in the case of a high saturated vapour pressure, to question the validity of these assumptions. A calculation in the case of benzene in which the molar volume of the liquid is included and the vapour is treated as a Van der Waals gas, shows that the percentage correction to ΔH_v is neglibible at 298 K (-0.05), but is appreciable at the boiling point, 350 K (-2.4). The main source of error in the simplified equation is the neglect of intermolecular forces in the vapour.

Vapour pressures may be measured in many different ways and the method chosen depends largely on the pressure range concerned, although there may be considerable overlap between instruments of each type. It is not possible in a single lecture to discuss all of these in detail; some will receive only a brief mention, others will be described more fully, but the main emphasis will be on the effusion method, which is the most important and widely used one for pressures below 10^{-2} Torr.[1] Two wide-ranging reviews are of particular interest (1,2).

Manometric

Manometers in various forms have been widely used in the pressure range 0.1-760 Torr (3). The apparatus may be very simple, as in the Smith-Menzies isoteniscope, in which the liquid and its vapour are contained in a bulb and isolated from the vacuum line by a U tube containing more of the same liquid. The pressure is measured directly by a Hg manometer. High temperatures require a different manometric fluid and liquid gold (4) has been used to measure the vapour pressures of alkali halides. Mercury must be protected from corrosive vapours either by using a fluorinated oil as a buffer or by sealing the material and its vapour inside a part of the apparatus which contains a spiral or a Bourdon spoon gauge. The gauge may be returned to a previously determined null point by adjusting the external pressure. The use of these gauges will increase the sensitivity of the equipment since an optical lever may be used, or the movement of a long pointer may be followed using a microscope.

An apparatus designed by Bradley (5) and used by him to measure the vapour pressures of monoclinic and cubic carbon tetrabromide, contained a Bourdon gauge and a pressure multiplier, in which the movement of mercury in a tube of 15 mm radius was transferred to that of a tube of 2.3 mm radius.

Metal bellows which are robust, have a small dead volume and which can be made corrosion resistant, can detect pressure changes of 10^{-3} Torr if the movement is followed optically or electrically.

Transpiration

In transpiration methods vapour is swept out by an inert carrier gas and collected by condensation or absorption. Since the saturated vapour pressure will be disturbed, experiments at different flow rates should be carried out and the results extrapolated ro zero rate of flow. High temperature transpiration techniques have been used to measure the vapour pressures of ionic salts in the temperature range 200-800°C (6). In a technique developed by Mackle (7), vapour, initially in equilibrium with a liquid, is isolated in a loop of constant volume and then is swept by the carrier gas into a gas-liquid chromatograph. The area under the peak on the recorder is proportional to the vapour pressure. The lower pressure limit of the transpiration method, possibly 10^{-4} Torr, depends on the precision with which the mass of the condensed or absorbed vapour may be estimated.

Ebulliometric

The vapour pressure of organic liquids in the range 20-760 Torr may be conveniently studied by the differential ebulliometric (9) method. Two ebulliometers are used, one containing the liquid being studied and the other a liquid - usually water - for which precise vapour pressure data are available. The reflux condensers of the ebulliometers are connected to a common stable pressure of inert gas and the vapour pressure at a measured temperature is obtained from the boiling temperature of the water. If precautions are taken to prevent the water "bumping" it can be used as a reference liquid down to 15 Torr.

Vibrating fibre

Particularly useful for corrosive vapours is the vibrating quartz fibre instrument. An apparatus described by Anderson (10) uses a bifilar suspension which has 1 mg iron attached to the bottom of the pendulum so that oscillations may be induced magnetically. The decay is followed by means of a photocell in front of which an illuminated grating is placed; each time the pendulum crosses a slit there is a fall in the output from the cell. The number of pulses produced in unit time is proportional to the amplitude of oscillation and the decay half-time t is related to the pressure and the molecular weight

$$pM^{\frac{1}{2}} = \frac{a}{t} - b \ .$$

a and b are constants, characteristic of the fibre; they may be found by direct calibration with a McLeod gauge. If a sail made from thin quartz sheet is included the method can be used for pressures down to 10^{-5} Torr.

Thermistor

The removal of heat from a hot wire has long been used as a measure of the gas pressure concerned (11). The use of thermistors for this purpose was first described by Becker *et al.* (12) in 1946 and in 1955 Engelsman (13) published an account of an improved experimental technique and a discussion of the underlying theory. The transfer of heat from a hot surface, temperature T_1 to a cold one at temperature T in the presence of a gas at a pressure p, under Knudsen conditions, depends on a radiation term [proportional to $(T_1^4 - T^4)$] a direct conduction term [proportional to $(T_1 - T)$] and a gas conduction term [proportional to $\frac{p}{T^{\frac{1}{2}}}(T_1 - T)$]. The thermistor forms one arm of a Wheatstone bridge and the voltage is adjusted until its temperature and resistance reach a selected fixed value. The power input is found by measuring the potential drop across the thermistor potentiometrically. The inconvenience of two thermostats, one for the sample and the other, at a higher temperature, for the thermistor is avoided by measuring the power required to keep the thermistor resistance constant at two temperatures T_1 and T_2, at a constant bath temperature T. Then

$$E = E_1 - E_2 = a(T_1^4 - T_2^4) + b(T_1 - T_2) + \frac{c(T_1 - T_2)p}{T^{\frac{1}{2}}}$$

which may be written as

$$E = E^o + \frac{cp}{T^{\frac{1}{2}}}$$

where E^o can be found from $E_1 - E_2$ at zero pressure (say 10^{-6} Torr). A plot of $\ln(E - E^o) + \frac{1}{2} \ln T$ against $\frac{1}{T}$ will yield ΔH. The fact that E^o appears to be independent of T indicates that the various theoretical assumptions, which have been made, are justified.

Engelsman studied trans-stilbene, cyclodecanol and cyclotetradecane in the pressure range 10^{-2}-10^{-3} Torr and Edwards and Kington (14) used essentially the same technique to obtain the enthalpy of sublimation and the vapour pressure of ferrocene, at pressures down to 5×10^{-3} Torr. This work was an important test of the validity of the method since a precise value of the enthalpy change was already available.

Knudsen effusion

This is the most important and extensively used method for pressures $< 10^{-2}$ Torr (15,16). Under ideal Knudsen effusion conditions, the rate at which a cell containing a sample of material will lose mass due to effusion through a small hole, is given by

$$W = pA \left(\frac{M}{2\pi RT}\right)^{\frac{1}{2}}$$

where A is the area of the hole,
p is the vapour pressure of the substance,
M is its molecular weight,
T is the temperature.

Several important criteria, some of which it is easy to overlook, must be satisfied if this equation is to be a valid one. It implies that an ideal vapour is effusing through an infinitely thin hole; the hole, of course, will not be of negligible thickness and non-specular collisions with the walls may return a molecule to the cell. A detailed theoretical study of the flow of gases through tubes of various shapes and lengths was carried out by Clausing (17) and the correction factor which he obtained for cylindrical tubes may be abbreviated for most purposes to

$$K_c = \frac{1}{1 + L/2r}$$

where L is the length and r the radius of the hole.

The extent to which the equilibrium vapour pressure within the cell is affected by the effusion loss must also be considered. If the surface area of a solid sample is S and the equilibrium vapour pressure is p^o at a temperature T, then the number of molecules condensing on the surface in unit time is

$$\frac{\alpha S p^o}{(2\pi m k T)^{\frac{1}{2}}}$$

where α is the condensation coefficient. Under steady state conditions this will also be the rate of loss from the surface. When the effusion hole is present the pressure falls to p, and the number of molecules escaping through the hole will be

$$\frac{K_c A p}{(2\pi mkT)^{\frac{1}{2}}}$$

This must equal the difference between the number leaving the surface and the number condensing upon it

$$\frac{s(p^o - p)}{(2\pi mkT)^{\frac{1}{2}}} \quad \text{so that} \quad p = p^o - \frac{K_c A}{\alpha s} p$$

More rigorous treatments obtain an equation (18) of the same form. If different hole sizes are used and p is plotted against $K_c A p$, $1/\alpha s$ may be obtained from the slope of the graph and the intercept on the pressure axis yields p^o. Normally the correction will be small since $\alpha \sim 1$ and $\frac{A}{S}$ will usually be very small. It is always worthwhile, however, to use holes of different sizes because other concealed sources of error may be revealed, for example, self-cooling of the sample, the hole diameter – mean free path ratio not satisfactory, or deviations from the assumed absolute vacuum on the low pressure side of the hole.

Corrections which must be applied if the absolute vapour pressure is to be obtained, will not affect the calculated enthalpy change unless they contain temperature dependent terms. In this case by considering $\frac{d \ln p^o}{d \, 1/T}$ it can be shown that

$$\Delta H^o = \Delta H_{observed} - \frac{R}{\frac{1}{AK_c} + \frac{1}{\alpha s}} \cdot \frac{d\left(\frac{1}{\alpha s}\right)}{d\left(\frac{1}{T}\right)}$$

The variation in αs with temperature may be found by using a series of holes of different diameters at two different temperatures. The correction to ΔH is very small and in general the method may be used with confidence even when the hole size makes the steady state pressure less than the saturated one.

The angular distribution of molecules leaving the hole is not quite cosine in form even for an ideal hole, but mirrors to some extent the geometry of the cell interior and the shape of the sample (16). The effect is of negligible importance unless the ratio of the radius of the hole to that of the cell is greater than 0.2.

Effusion under ideal Knudsen conditions assumes free molecular flow in which viscosity plays no part (19). This holds when the ratio $\frac{r}{\lambda}$ tends to zero, where r is the radius of the hole and λ the mean free path of the gas molecules. Knudsen believed from experimental observations that the assumption was valid for $\lambda/2r$ equal to 10. Other workers found it satisfactory for ratios less than 10, but opinion is very varied as to the value for which flow can be described by the free molecular model. In this respect the work by Edwards and Kington on ferrocene (14), in which λ is in the range 2r-10r, is particularly important. They obtained a value for the enthalpy of sublimation of 76.6 kJ mole^{-1} which appears to be unacceptably high when compared with the entropy of crystalline ferrocene, 216.23 J K^{-1} mole$^-$ (low temperature calorimetry) and gaseous ferrocene, 462.75 J K^{-1} mole^{-1} (spectroscopy) which yield an enthalpy of sublimation of 73.51 kJ mole^{-1}.

Hiby and Pahl (ref. 29, p.67) have shown that when $\frac{r}{\lambda}$ is not zero an additional term

$$\frac{1}{1 + \frac{K_2 r}{2}},$$

should be included in the expression, for p. $K_2 = 0.48$, is a function of L/r but it is not very sensitive to changes in L when L < r. Usually if one is only interested in the enthalpy of sublimation and not in the absolute value of the pressure, the expression chosen for p is not important since $\frac{d \ln p}{d\ 1/T}$, upon which ΔH depends, is unaffected by the choice. However in this case the term r/λ alters the slope and changes the enthalpy value to 73.34 kJ mole^{-1}. For the size of holes used in effusion work, in general, molecular streaming through the hole may be assumed for pressures less than 5×10^{-6} Torr but an upper limit of 10^{-2} Torr should be imposed on the effusion technique.

The Knudsen equation assumes that mass transport through the hole occurs only in the vapour phase. Winterbottom (20) discusses the situation where a molecule may condense close to the hole, migrate through the hole along the surface (surface diffusion) and re-evaporate from the external surface. The nature of the surface is important since the free energy of desorption is involved as well as the vibration frequency of the adsorbed molecule against the surface. The theory predicts that the effect will become more important as L, r and p decrease but usually the parameters for a particular molecule are not available and are almost impossible to assess. The effect will cause a systematic variation in vapour pressure with hole size although it may be impossible to disentangle this from, for example, unsaturation.

The effusion equation also assumes that there is a complete vacuum outside the cell; in practice so long as the external pressure is less than 10^{-6} Torr, there is no observable effect on the effusion rate.

The experimentally measured quantity in the effusion method

is the rate at which the cell is losing mass. Air can be admitted to the system and the cell removed to be weighed directly, but it is more usual to follow the loss of mass continuously. Bradley (21) designed extremely delicate silica, torsion vacuum microbalances in which the movement of the beam was followed by a travelling microscope and weight changes of 10^{-8} g could be detected. The effusion cell was made from soft glass, to the side of which had been fused a small piece of very thin platinum foil, which carried a 1 mm hole. Bradley has also studied polycrystalline beads, formed by dipping a very small glass bead into the molten material. The radius of the bead was measured by means of a travelling microscope and the area was calculated. The Langmuir equation was used which gives the amount of material evaporating in unit time from unit area of the surface as $\alpha p \left(\frac{M}{2\pi RT} \right)^{\frac{1}{2}}$. α, the evaporation coefficient may be found by comparing the rates of evaporation and effusion under the same conditions.

In recent years automatic, torsion vacuum microbalances (e.g. Sartorius Ltd., model 4104) have become available. The torque due to the decreasing weight of the effusion cell is continually balanced electromagnetically. The compensation current is a measure of the weight change and the current may be displayed on a galvanometer and chart recorder.

In studying very low vapour pressures (less than 10^{-6} Torr), various methods have been used to go beyond direct weighing. For example, the vapour can be made more detectable by labelling the compound with a radioactive isotope, and pressures as low as 10^{-12} Torr have been measured (22). The method has been used mainly for elements and simple compounds, oxides and halides, but some organometallic compounds have been studied, for example, mercury diphenyl (23) which was synthesised directly from active mercury (^{197}Hg, ^{203}Hg). A simple static apparatus was used initially, in which the equilibrium vapour pressure was established in a glass cylinder, adjacent to a scintillation counter. This worked very successfully for mercury (^{203}Hg) and similarly labelled mercuric chloride, a constant count rate which was reproducible at each temperature and easily removed by pumping was observed. However, mercury diphenyl itself was strongly adsorbed on the walls of the chamber producing a high and steadily increasing background count, which could not be removed by pumping and an effusion technique had to be used instead. The effusate was collected and counted on a cooled, silvered surface. Lead and tin tetraphenyls, labelled by direct exposure to tritium gas, were also examined (24). Everything seemed to be satisfactory but subsequent work by ourselves and other workers made it clear that, for reasons we have never been able to explain, the enthalpies of sublimation obtained were considerably in error.

Nesmeyanov and co-workers (25) have used an isotope exchange method to study the rates of evaporation of metals. Two specimens of different isotopic composition are placed in a vacuum chamber, heated by induction and a shutter is used to control the exchange between them. The underlying theory is complex since the exchange depends on the rate of evaporation from the active sample and the rate of diffusion into the inactive one.

A very sensitive method of detecting radioactive molecules in the vapour phase is by recording the tracks produced in photographic emulsion. Ausländer (26) has claimed that vapour pressures as low as 10^{-17} Torr may be determined by this method.

In general the active tracer method cannot easily yield accurate values of absolute vapour pressures since a definite mass of material must be identified with the activity of the effusate.

The use of the mass spectrometer as a sensitive vapour detector will be discussed in the next section.

The final quantity in the Knudsen equation which needs to be discussed is M, the atomic or molecular weight of the effusing species. If it is assumed that individual atoms or molecules are evaporating from the surface but, in fact, they are, for example, dimers, then the calculated vapour pressure will be incorrect, whereas the enthalpy of sublimation will be unaffected, although, of course, it will not refer to the monomeric species. The problem is more serious if several different species are present and the proportions vary with temperature, for example, it is of little value to know the total pressure in the system Ge, Ge_2 ---- Ge_8, or in the case of graphite, where species ranging from C_1 ---- C_5 exist, with enthalpies of sublimation varying from 715 to 971 kJ $mole^{-1}$. The same problem can arise with molecules, for example, the vapours of the alkali metalhalides contain dimers, trimers and some tetramers.

An effusion method, which does not depend on M, measures the rate of loss of momentum from the hole; this is recorded as a recoil force. The cell is designed with two holes, one on each side, equidistant from the suspension. Effusion causes a twist in the supporting fibre and the torsion angle so produced is measured. This torsional Knudsen effusion recoil method is known as the "torker" technique. Under ideal conditions the recoil force for an area A and a pressure p is $\frac{pA}{2}$. It has been shown (27,28) that the geometrical factor, f, which allows for the finite hole thickness is somewhat greater than K_c, because the force depends on the angular distribution of the effusing molecules as well as on their number, for example for $L/r = 1$, $K_c = 0.67$; $f = 0.73$. Therefore, $F = \frac{pAf}{2}$. If the two holes are a distance λ from the suspension, the torque will be $2F\lambda$ and if D is the torsional constant of the suspension, which is twisted through an angle θ

$$D\theta = 2F\lambda = pAf\lambda$$

so that $\quad p = \dfrac{D\theta}{Af\lambda} = k\theta$

k may be found from experiments with compounds of known vapour pressure and D may be found separately by measuring the period of oscillation of the suspension, with and without a ring weight hanging on the cell.

$$D = \frac{4\pi^2 I}{t_w^2 - t_s^2}$$

where I is the moment of inertia of the ring weight and t_w and t_s are the two periods of oscillation, with and without the weight respectively. This may be checked by calculating D from the relation

$$D = \frac{Ed^4}{32L}$$

where E is the torsional shear modulus for the suspension material and d and L are the diameter and length of the suspension.

Clearly an instrument which measures simultaneously both the mass loss from the cell and the torsional deflection would be of the greatest value since agreement between the vapour pressures measured by both methods would confirm that the assumed value of M was correct. An apparatus of this type was first developed by Wessel (29), Bradley and Cleasby (30) and more recently by McCreary (31) and Thorn and by Keiser and Kana'an (32). The apparatus used in Leeds is very similar in type. The metal vacuum jacket can be pumped down to 10^{-7} Torr and the effusion cell is supported by an aluminium rod, 35 cm in length, to which is attached 53 cm of 0.0025 cm diameter, tungsten wire. The wire is attached to the automatic vacuum microbalance (Sartorius Ltd., model 4104) and the mass loss is recorded continuously throughout the experiment. The aluminium rod carries a small mirror and light reflected from the mirror was detected by an automatic light spot follower (Photodyne Sefram Paris Ltd.) which recorded the torsional deflection. The calibration of the torsional wire followed the criteria suggested by Freeman and Gwinup (33). The oil diffusion pump was protected by a trap containing copper baffles, which were cooled by direct conduction along a liquid nitrogen cooled copper rod. A water-cooled spiral trap protected the balance. The pressure in the vacuum jacket was measured as close as possible to the cell, by means of a Penning gauge.

THE MEASUREMENT OF VAPOUR PRESSURE

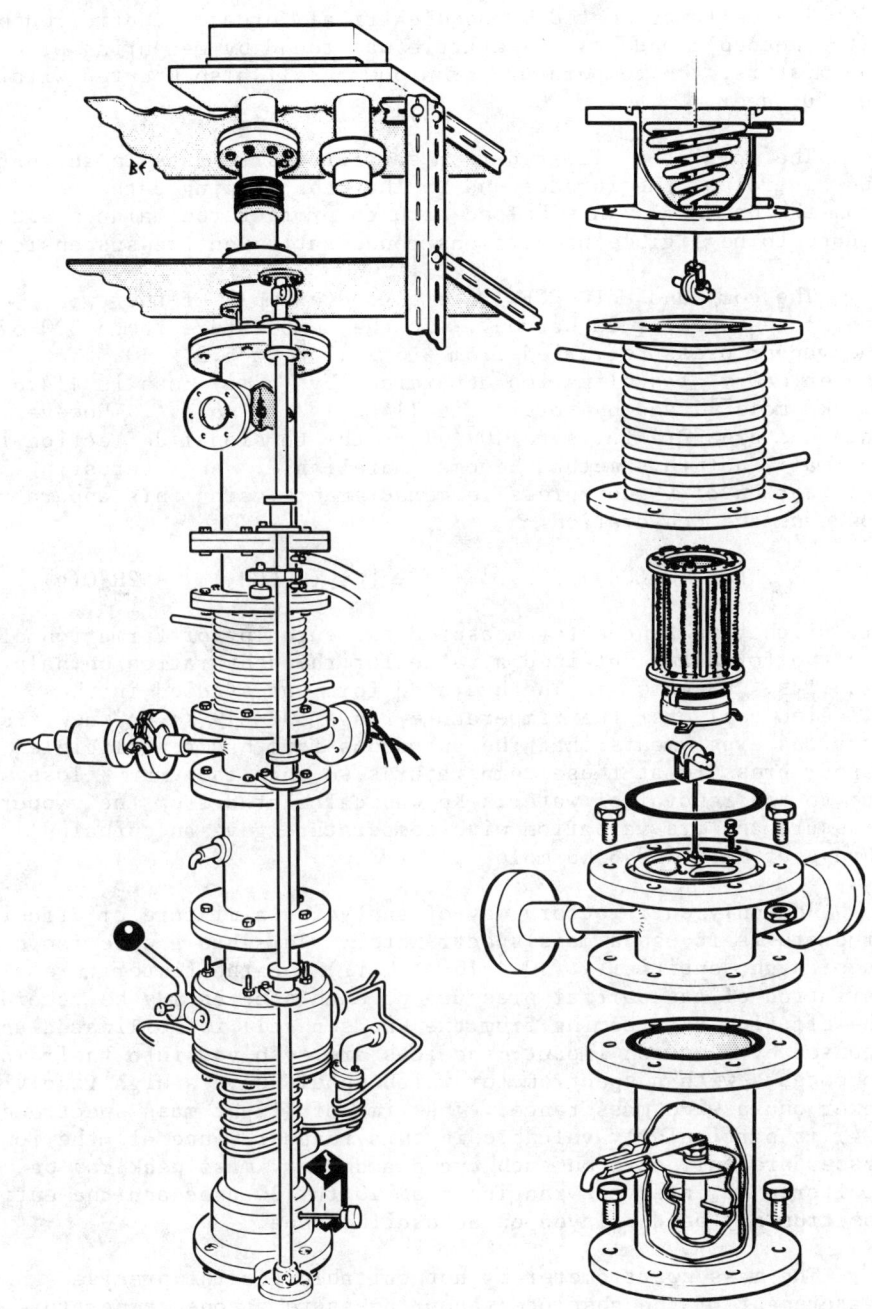

The Leeds combined mass-loss and torsion Knudsen effusion apparatus.
(Carson, Frankland, Morris and Laye)

The cell was heated by an electrical furnace, controlled by a thermocouple and its temperature was found by measuring with thermistors, the temperature of a dummy cell also mounted within the furnace.

The apparatus differed from earlier instruments in the nature and control of the furnace and in that the pumping path was away from the balance. This helped both to protect the balance and to reduce to negligible proportions condensation on the suspension.

The compound $Mo(O_2CCF_3)_4$ is a good example of the two approaches working in harmony. In the temperature range 320-370 K the vapour pressure varied from 4.6×10^{-6} to 5.3×10^{-4} Torr and the enthalpy of sublimation determined by mass loss is 114.6 ± 1.7 kJ mole^{-1}, and by torsion is 114.2 ± 2 kJ mole^{-1}. However, once the pressure falls to 10^{-7} Torr the torsional deflection is so small that this method becomes unreliable. An interesting application of vapour pressure measurements using this apparatus concerns the dehydration –

$$[Cr_2(O_2CCH_3)_4 \cdot 2H_2O] \longrightarrow [Cr_2(O_2CCH_3)_4] + 2H_2O(g) .$$

Dr. Pilcher in Manchester measured the enthalpy of formation of the two forms and obtained a value for the dehydration enthalpy of 94.3 ± 9.4 kJ mole^{-1}. The hydrated form was studied in the effusion cell over the temperature range 302-319 K; we knew from previous experiments that the anhydrous form had a negligible vapour pressure at these temperatures so that the mass loss was due to the removal of water. Kp was calculated from the vapour pressure and its variation with temperature gave an enthalpy change of 96.9 ± 8.3 kJ mole^{-1}.

The only satisfactory way of analysing a mixture of effusion products is to use a mass spectrometer. This has a wide range and is of high sensitivity (10^{-1}-10^{-10} Torr) and the temperature variation of the partial pressure of each species may be recorded. The effusing beam coming from the Knudsen cell is collimated and ionised by electron impact; the ions are resolved into their mass components with a spectrometer which should have a high resolving power and a wide mass range. The time-of-flight mass spectrometer (34) is particularly valuable in this respect since all the ion masses are collected on each cycle and every mass peak may be monitored at intervals ranging from 10 to 100 μsec and the entire spectrum may be displayed on an oscilloscope.

The mass spectrometer is not suitable for the precise measurement of the absolute vapour pressure at one temperature. This depends on the positive ion current for a particular species which in turn depends on the ion collection efficiency of the instrument and the ionisation cross-section of the species

concerned. Systematic uncertainties in these will make the
absolute vapour pressure uncertain but should leave the temperature
variation unaffected.

Malaspina and co-workers have used an interesting combination
of techniques. Liquid rubidium was studied (35) both in a conventional Knudsen cell and in a Bendix time-of-flight mass spectrometer containing a Knudsen source. The results were in good
agreement and confirmed that up to 551 K the vapour consists
almost entirely of monatomic species. Values of Kp for the
dissociation

$$Rb_2(g) = 2Rb(g)$$

were also obtained and the third-law method was used to calculate
the dissociation energy. The temperature range used was too small
for a reliable application of the second-law method.

They have also used an ingenious combination of Knudsen
effusion and direct calorimetry. The effusion cell and a matching
copy were placed in a Calvet high temperature, differential, micro
calorimeter. The enthalpy of sublimation was measured at a series
of temperatures and at the end of each experiment the cell was
removed and the mass loss was recorded. The cell could be
sealed by a silicon sphere, controlled externally and once a
constant base line was obtained the sphere was raised and the
thermogram of the evaporation was recorded. The vacuum was removed
by the addition of argon and the cell was weighed. Cadmium (36),
m and p-nitroaniline (37), 1,2- and 2,4-dihydroxy anthraquinones
(38), pyrene and 1,3,5-triphenylbenzene (39) were studied and in
each case there was excellent agreement between the two methods.

NOTE

[1] $Atm = 101.325 \text{ kPa}$; $Torr = \dfrac{101.325}{760} = 0.1333 \text{ kPa} \equiv 1 \text{ mm Hg}$

$Pa \equiv Nm^{-2} \equiv kg\ m^{-1}\ s^{-2}$

REFERENCES

1. Cooper, R. and Stranks, D.: 1966, 'Technique of Inorganic
   Chemistry' Vol. VI. Editors, Jonassen, H.B. and
   Weissberger, A. Interscience, New York.
2. Bradley, R.S.: 'Encyclopaedic Dictionary of Physics B81'
   Vol. 7. Editor, Thewlis, J. Pergamon Press, London.
3. Thomson, G.W.: 1949, 'Technique of Organic Chemistry' Vol. I
   Part 1 Chap. 5. Interscience, New York.
4. U.S. Atomic Energy Commission O.R.N.L.: 1960, 2933, 73.

5. Bradley, R.S. and Dury, T.: 1959, Trans. Faraday Soc. 55, 1844.
6. Sense, K.A., Snyder, M.T. and Clegg, J.W.: 1958, J. Phys. Chem. 58, 223.
7. Mackle, H., Mayrick, R.G. and Rooney, J.J.: 1960, Trans. Faraday Soc. 56, 115; 1964, 60, 817.
8. Mackle, H. and McClean, R.T.B.: 1964, Trans. Faraday Soc. 60, 817.
9. Osborn, A.G. and Douslin, D.R.: 1966, J. Chem. Eng. Data 11, 502.
10. Anderson, J.R.: 1958, J. Sci. Instru. 29, 1073.
11. Dushman, S.: 1949, 'Scientific Foundations of Vacuum Technique', Wiley, New York.
12. Becker, J.A., Green, C.B. and Pearson, G.L.: 1946, Trans. Am. Inst. Elect. Engrs. 65, 711.
13. Engelsman, J.J.: 1955, Doctoral Dissertation, The Free University of Amsterdam.
14. Edwards, J.W. and Kington, G.L.: 1962, Trans. Faraday Soc. 58, 1323.
15. Freeman, R.D.: 1967, 'The Characterisation of High Temperature Vapours' Chap. 7. Editor, Margrave, J.L. Wiley, New York.
16. Cater, E.D.: 1970, 'Techniques of Metals Research' Vol. IV, Part 1, Chap. 2A. Editor, Rapp, A. Interscience
17. Clausing, P.: 1932, Physik. 12, 961. (see also refs. 28, 29).
18. Ward, J.W. and Fraser, M.V.: 1968, J. Chem. Phys. 49, 3743.
19. Carman, P.C.: 1956, 'Flow of Gases through Porous Media', p.62. Butterworths, London.
20. Winterbottom, W.L.: 1968, J. Chem. Phys. 49, 106.
21. Bradley, R.S.: 1951, Proc. Roy. Soc. A 205, 553.
22. Ref. 1, p.54.
23. Carson, A.S., Stranks, D.R. and Wilmshurst, B.R.: 1958, Proc. Roy. Soc. 244A, 72.
24. Carson, A.S., Cooper, R and Stranks, D.R.: 1962, Trans. Faraday Soc. 58, 2125.
25. Nesmeyanov, A.N., Lozzachev, V.I. and Lebedev, N.F.: 1955, Dokl. Akad. Nauk. S.S.S.R. 102, 307.
26. Ausländer, J.S.: 1958, 'Inter. Conf. Peaceful Uses of Atomic Energy', 15, 1285.
27. Freeman, R.D. and Searcy, A.W.: 1954, J. Chem. Phys. 22, 762.
28. Schultz, D.A. and Searcy, A.W.: 1962, J. Chem. Phys. 36, 3099.
29. Wessel, G.: 1951, Z. Physik. 130, 539.
30. Bradley, R.S. and Cleasby, T.G.: 1953, J. Chem. Soc. 1681.
31. McCreary, T.R. and Thorn, R.J.: 1968, J. Chem. Phys. 48, 3290.
32. Keisen, D. and Kana'an, A.S.: 1969, J. Phys. Chem. 73, 4264.
33. Freeman, R.D. and Gwinup, P.D.: 1966, Rev. Sci. Inst. 37, 773.

34. Bowles, R.: 1969, 'High Temperature Studies of Inorganic Solids', 'Time-of-Flight Mass Spectrometry', p.211-226. Editors, Price, D. and Williams, J.E. Pergamon, London.
35. Piacente, V., Bardi, G. and Malaspina, L.: 1973, J. Chem. Thermodynamics 5, 219.
36. Malaspina, L., Gigli, R. and Bardi, G.: 1971, J. Chem. Thermodynamics 3, 827.
37. Malaspina, L., Gigli, R., Bardi, G. and de Maria, G.: 1973, J. Chem. Thermodynamics 5, 699.
38. Malaspina, L., Bardi, G. and Gigli, R.: 1973, J. Chem. Thermodynamics 5, 845.
39. Malaspina, L., Bardi, G. and Gigli, R.: 1974, J. Chem. Thermodynamics 6, 1053.

THE USE OF A SIMULTANEOUS TORSION AND MASS-LOSS EFFUSION APPARATUS
FOR DETERMINING VAPOUR PRESSURES AND ENTHALPIES OF SUBLIMATION

C.G. de Kruif

General Chemistry Laboratory, Chemical Thermodynamics
Group, State University of Utrecht, Transitorium III,
Padualaan 8, 3508 TB Utrecht, The Netherlands.

The use of two techniques based on two different properties of a
Knudsen gas provides information on both the vapour pressure and
the composition of the vapour. With pure substances the mutual
consistency of the two techniques can be tested. A description
of the fully automated set-up is given. The usefulness and
general applicability of the instrument is illustrated by the
results obtained for a number of different substances each of
which is representative for a group of closely related compounds.
It should be emphasised that part of these results could not have
been obtained using either technique on its own. We shall discuss the results for acetic acid, butanoic acid, L-alanine, TCNQ-TTF, 1,4-benzoquinone-hydroquinone, pp'-methane diphenyl isocyanate, theophylline and ammonium halides. Finally we shall discuss
the two lines of research in which we are engaged at present.

1. INTRODUCTION

During the sixties Kitaigorodsky[1] and Williams[2] and others
published their work on the calculation of physical properties
of molecular organic crystals by means of the atom-atom potential
method. Of fundamental importance for the development and extension of this method is the availability of both crystal structure data and enthalpies of sublimation, the latter being roughly
equal to the lattice energy U_{lat}: $\Delta H_{sub} \approx - U_{latt} - 2RT$. As for
the crystal structure data there is an abundance of numbers in
the Cambridge files and our group is closely connected with a
group of X-ray crystallographers at Utrecht. By contrast, however,
literature data on the enthalpies of sublimation are scarce and
in a number of cases have proved to be inconsistent.

We therefore decided in the early seventies to build an experimental set-up for measuring enthalpies of sublimation either by direct or indirect methods. Reviews of the available methods are given in the literature (see for instance Margrave[3] and Ambrose[4]). As direct calorimetric methods are difficult to operate at high or low temperatures and give only approximate information on the vapour pressure we opted for an indirect method i.e. we decided to calculate ΔH_{sub} or ΔH_{vap} from the temperature dependence of vapour pressure. Furthermore we wanted to obtain information about the composition of the evaporating species because we are also working on homogeneous mixtures of organic molecular crystals. Measuring the p-x diagram of such mixtures provides information about the excess behaviour of the solid phase, which in turn can be combined with results obtained with our adiabatic calorimeter developed by Van Miltenburg[5]. For these reasons we chose to develop a combined torsion and mass-loss effusion apparatus. A study of the literature showed three problems connected with the method: a) the measurement of the temperature of the sample, b) the mechanical properties (hysteresis, linearity) of the torsion wire, and c) the automation of the measurements. We solved these difficulties by using a bifilar suspension which enabled us to compensate the effusion torque electromagnetically and to measure the temperature of the cell directly from the resistance of a thermistor. An additional advantage of the compensation method is that the properties of the suspension wires do not affect the proportionality constant although they determine sensitivity. Moreover the method now becomes a null method and can easily be automated. By employing a specially modified electronic vacuum microbalance from the beam of which the torsion wires are suspended we could also automate the weighing and determine mass loss in situ. The only other experimental set-up we know of which is fully automated and employs simultaneous torsion and mass-loss effusion has been described by Edwards[6]. This set-up is used to measure high temperature vapour pressures; this in fact is the field in which most effusion apparatus is used. We however have used the apparatus almost exclusively for making vapour pressure measurements of low volatile organic substances.

Before discussing several typical examples of results obtained, we shall give a short description of the main features of the apparatus. For financial reasons automation was realized a few years after the main body of the set-up was put into operation. We shall give a detailed account of the automated measuring procedure, since this has not yet been published. Following the experimental part we shall discuss some results that have been published already and some unpublished results which have been obtained recently. Finally we shall give an account of the work in which we are currently engaged and describe the investigations we are planning and the developments which we hope to achieve in the near future.

2. EXPERIMENTAL

The principal features of the simultaneous torsion and mass-loss effusion set-up are as follows. The effusion cell is provided with two orifices 1 mm in diameter made in 6 μm platinum foil: the orifices are positioned 20 mm apart on opposite sides of the cell which resembles a garden sprinkler. The cell is screwed to the thermistor holder, see figure 1, with some beryllium oxide in order to provide a good heat contact. The thermistor is in series with the galvanometer coil 15 cm above the cell. The galvanometer coil, connected to the thermistor holder by a thin stainless steel tube, is at the end of two 75 cm long phosphor-bronze torsion bands which are suspended about 1 mm apart. This torsion system, which has a resolution of 10^{-3} to 10^{-4} Pa, is suspended from the beam of a vacuum microbalance (1 μgr resolution corresponds to 10 μV output). The balance was modified so that an electric current could be led over its beam to the two suspension bands without affecting the weighing circuits. These features of the measuring set-up have been described before.[7] Since the fully automated set-up has not yet been described, we give some details here. Figure 2 is a block scheme of the experimental set-up. A measurement is made as follows. After the effusion cell has been loaded with a sample weighing approximately 300 mgr it is screwed to the thermistor holder. The arrest mechanism is then released and the system evacuated to 133 Pa. From this moment on the whole measuring procedure is completely automatic and controlled by the HP-9825A desk-top calculator. At a system background pressure of 133 Pa the offset of the photodiodes amplifier, which detects the angular deflection of the torsion pendulum, is set to zero. Meanwhile the temperature of the oven surrounding the effusion cell is regulated so that the vapour pressure of the sample is approximately 0.08 Pa. Then the vacuum system is pumped to 10^{-3} Pa or lower and the current for the electromagnetic compensation of the effusion torque is switched on. After five minutes oscillations of the system are damped out (Focault damping) and the following voltmeter (6 digits Hp-3436A) readings are input into the calculator; V_1, voltage over a 10 KΩ standard resistor; V_2, voltage drop over the suspension wires + compensation coil + thermistor; V_3, output voltage of the Sartorius 4401 microbalance. In order to eliminate noise these readings are repeated 5 times, averaged and stored in the calculator together with elapsed time in 0.1 s units. The procedure is repeated 6 times at 30 s intervals. From the decrease that occurs in the output voltage of the microbalance in 30 s the mass loss dm from the effusion cell is calculated and converted to vapour pressure according to:

$$p_m = dm/dt \, A^{-1} \, (2 \pi R T/M_{form})^{\frac{1}{2}} \qquad (1)$$

where A is the orifice area and M_{form} is the formula mass. From the

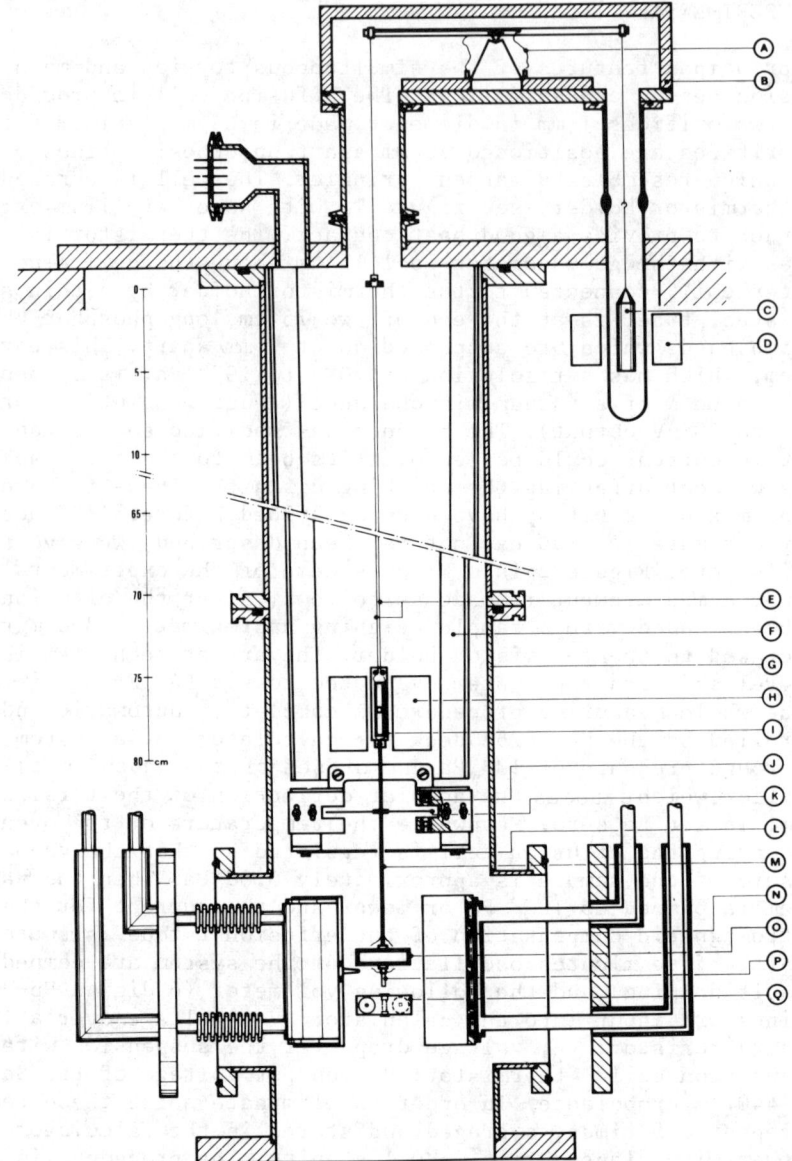

Figure 1. The torsion mass-loss effusion apparatus. The figure is drawn in elevation with some parts cut away. A gold ribbon; B, stainless-steel microbalance case; C, tare weight; D, H, permanent magnets; E, torsion bands; F, U-profile support; G, Stainless-steel vacuum tube; I, galvanometer coil; J, Ga-As diode; K, aluminium vane; L, radiation shield; M, stainless-steel tube; N, heater; O, heat resistor/shunt; P, aluminium thermistor holder; Q, effusion cell detached from the thermistor holder.

voltage over the suspension wires the resistance of the thermistor is determined, from which in turn the temperature of the cell is calculated. Multiplying the compensation current, measured from the voltage over the 10 kΩ resistor, by an apparatus constant yields the torsion pressure.

$$p_t = C \cdot I \tag{2}$$

Altogether we have in approximately 180 seconds 5 sets of values for temperature p_t and p_m. These data are printed and stored on a magnetic tape for later evaluation. The D/Al converter Hp-59501A is increased, which causes a temperature rise of 0.9 K for the oven and effusion cell. After the temperature of the cell has become practically constant (180 s) measurements are resumed as above. When the vapour pressure exceeds 1.2 Pa the sequence is stopped and a so-called run, consisting typically of 24 x 5 = = 120 singular data points in a temperature range of 22 K, is complete; a run takes 150 minutes. Then the temperature of the oven is lowered to its initial value, the vacuum pumps are shut off, the system pressurized to 133 Pa, the compensation current switched off, and the photodiode amplifier checked for offset and if necessary adjusted. Following this "zeroing" procedure, which takes 15 min, the next run is started. Meanwhile the experimental results are fitted to a "Clausius-Clapeyron equation".

With a load of about 300 mg in the cell we can make 8 or more runs, depending on the molecular mass of the evaporating substance. Usually these runs are made overnight. The next day the cell is refilled and turned 180° around the "long" horizontal axis. This results in the reversal of the effusion torque; then a new series of 8 runs is made. By measuring with "right" and "left" turn sense of the effusion torque it is possible to eliminate a small systematic error in the torsion effusion part. In addition an "independent" check is made on the first measurement (8 runs). Usually the samples are measured as delivered and then again after purification (zone refining, vacuum sublimation, recrystallisation). Generally there is no significant difference between these measurements. When there is doubt about the crystalline state of the sample, an X-ray powder diffraction pattern is recorded.

Usually the first run gives too high pressures because the sample contains impurities. During the next 5 or 6 runs vapour pressure is found to be a function of temperature only for pure non-decomposing substances. The average values of these runs are reported, unless stated otherwise. When the vapour pressure is not constant in time, this may be due either to impurities in the sample or to decomposition of the sample or to both. Non-volatile impurities manifest themselves in a steady decrease in vapour pressure, since they accumulate on and cover the evapora-

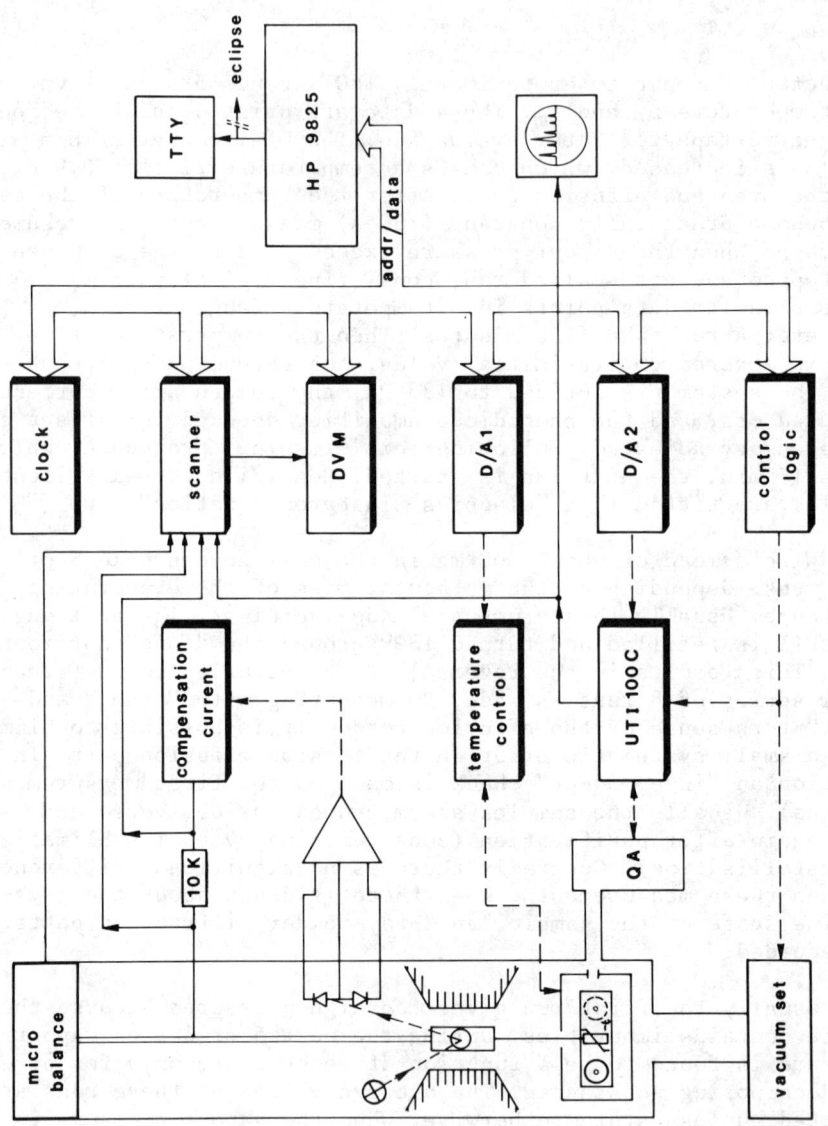

Figure 2. Block scheme of experimental set-up.

ting surface. This is confirmed by visual inspection of the
sample after several runs. In such cases additional purification
is carried out. Decomposition, dissociation or association of
the sample molecules is immediately noticed because then the
ratio p_m/p_t is significantly different from unity. It is unity
for pure substances. The apparent molecular mass of the vapour
species is calculated from (3,8,9)

$$M_{app} = (p_m/p_t)^2 M_{form} \text{ with } M_{app}^{\frac{1}{2}} \equiv \sum_i x_i M_i^{\frac{1}{2}} \qquad (3)$$

where x_i is the mole fraction of the i-th component in the gas
phase. When there is doubt about the composition of the vapour
(10,11) we employ a quadrupole gas analyzer which is able to
detect mass fragments with mass to charge ratios $\leq$ 300. The
ionizer of the QA is 15 cm away from the effusion cell and indirect line of sight through a slit in the oven surrounding the
effusion cell. The QA can also be operated by means of the
desk-top computer.

For temperature measurement we use four interchangeable
thermistors which are nominally 0.1, 1, 10 and 100 kΩ; these are
used for temperatures up to 200 K, 273 K, 350 K and 600 K
respectively. These thermistors are calibrated on IPTS-68.

3. CALIBRATION

The calibration constants of the apparatus can be obtained by
carefully measuring the geometry of the cell and by determining from torsion pendulum experiments the proportionality
constant of the electromagnetic compensation torque, using
either static[12] or dynamic[13] methods. When we built the
apparatus we determined the proportionality constants in that
very tedious way, since at that time there were no recommended
values for vapour pressures of organic substances in the sub
Pascal region. Since naphthalene can be used as a calibration
substance[14,15] we periodically check the performance of the
whole set-up with that substance. Recently we measured, using a
static method, the vapour pressure of benzoic acid[16] benzophenone[17] and trans stilbene[18]. These data are used to check
the set-up at higher temperatures. For a given cell we find that
calibration constants are independent of temperature and only a
function of geometry.

4. RESULTS

The experimental results are expressed in the coefficients of
the vapour pressure equation

$$R \ln (p/p^o) = -\Delta G^o(\theta)/\theta + \Delta H^o(\theta) (1/\theta - 1/T) \qquad (4)$$

in which ΔG^o and ΔH^o are the thermodynamic function changes occurring on evaporation and θ is a reference temperature mostly chosen midrange, $p_o = 1$ Pa. The other symbols have their usual meaning. In table 1 we give the coefficients of equation (4) for a number of substances each of which is representative for a group of compounds. It should be noted that the results for the substances listed in table 1 for which $p_m/p_t \ne 1$ should be interpreted as apparent values. We shall illustrate this by considering the following equilibria:

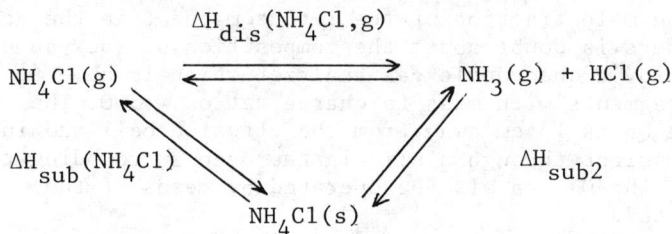

Figure 3. Phase reactions of NH_4Cl

The gas phase equilibrium is characterized by

$$K_p(T) = p(NH_3)\, p(HCl)\, p(NH_4Cl)^{-1} = b^2(1-b)^{-1}(1+b)^{-1} p_t \quad (5)$$

where b is the degree of dissociation of NH4Cl and p_t is the total pressure, which is just equal to the torsion effusion pressure. It can be shown[8,9] that b is found from:

$$b = (p_m/p_t - 1)\, \{(M_{NH_3}^{\frac{1}{2}} + M_{HCl}^{\frac{1}{2}})\, M_{NH_4Cl}^{-\frac{1}{2}} - p_m/p_t - 1\}^{-1} \quad (6)$$

Furthermore, differentiating equation (5) with respect to temperature yields relations between the "experimental" enthalpies of table 1 and $\Delta H_{sub}(NH_4Cl)$ and $\Delta H_{diss}(NH_4Cl,g)$:

$$-R\, d\ln(p_t)/d\, 1/T = \Delta H_t = b\,(1+b)^{-1}\{H_{dis}(NH_4Cl,g) - \Delta H_{sub}(NH_4Cl)\} + \Delta H_{sub}(NH_4Cl) \quad (7)$$

A similar relation holds for the slope of the mass-loss effusion "Clapeyron plot" and in principle the enthalpy changes defined above e.g. ΔH_{dis} can be evaluated.

A detailed discussion of the dimerization of the lower carboxylic acids is given in reference 8 and of the dissociation of the ammonium salts in reference 9.

We shall now discuss the results shown in table 1 giving special attention to peculiarities observed and the use that is made of the data obtained. Although it may be superfluous we would emphasize that the results presented below probably could not have been obtained with other methods.

Acetic Acid, Butanoic Acid.

These acids are members of the aliphatic carboxylic acids. We investigated[8,19,20] this homologous series from C1 up to C20 both by adiabatic calorimetry and by measuring vapour pressure in order to establish molar heat capacities and absolute entropies of solid, liquid and vapour according to a programme the purpose of which is to provide accurate thermodynamic data. From table 1 it can be seen that acetic acid is completely dimerized in the vapour phase $(p_m/p_t)^2 \approx 2$. As the number of C-atoms increases, this dimerization vanishes.

L-alanine

This substance is one of 14 amino acids and peptides which we investigated[21] in order to study the intermolecular interactions in hydrogen-bonded molecular crystals using the atom-atom potential method. Exceptionally in the case of l-alanine, literature data are available for ΔH_{sub}. A comparison of our results with those of Sabbah and Laffite[22] proved very satisfactory. In our investigation the use of the simultaneous techniques was especially valuable since a few substances showed decomposition, as could be seen from the ratio p_m/p_t but not from the separate techniques (because "vapour pressures" and "ΔH_{sub}" were practically constant). Hence the use of only one method would doubtless have led to an erroneous interpretation of the experimental results.

Tetracyanoquinodimethane – tetrathiofulvalene (TCNQ-TTF)

Since there is growing interest in the study of molecular interaction in organic donor-acceptor crystals with high electrical conductivity[23] and the literature[24] suggests that the complex evaporates as one entity we investigated the complex and its parent compounds. We established that the complex evaporates stoichiometrically but not as a whole. Dr. Govers[25] of our laboratory performed intermolecular energy and structure calculations which are in good agreement with the experimental results.

Benzoquinone – Hydroquinone (BQ-HQ)

De Wit et al.[26] from our laboratory calculated the molecular interactions of the (naphtho-)quinones, which also form stoichio-

	M_{form} gr.mol^{-1}	T_1 K	T_2 K	θ K	torsion effusion $\Delta G_t(\theta)$ J.mol^{-1}	$\Delta H_t(\theta)$ kJ.mol^{-1}	$p_t(\theta)$ Pa	mass-loss effusion $\Delta G_m(\theta)$ J.mol^{-1}	$\Delta H_m(\theta)$ kJ.mol^{-1}	$p_m(\theta)$ Pa	$p_m/p_t(\theta_t)$	M_{app}/M_{form}	b	evaporization
acetic acid	60.05	213	230	223.15	2015	69.9	0.34	1442	68.7	0.460	1.36	1.85	0.07	dimers
butanoic acid	88.10	238	255	248.15	1826	80.3	0.41	1483	80.2	0.487	1.18	1.40	0.39	partly dimers
L-alanine	89.09	407	426	414.08	3155	133.9	0.40	3089	131.7	0.41	1.02			stoichio-metric
TCNQ-TTF	204.3	401	428	410.57	3128	131.2	0.40	3162	128.8	0.40	0.99			
benzoquinone-hydroquinone (1:1)	180.10	308	325	312.91	2384	89.3	0.40	2410	87.8	0.40	0.99			
pp'-methane diphe-nyl isocyanate	250	338	362	349.53	2663	95	0.40	2751	93	0.39	0.97			non-stoichio-metric (BQ)
theophylline	180.2	404	434	421.27	3209	124.5	0.40	3245	127.6	0.40	0.99			
ammonium chlo-ride	53.49	337	364	352.02	2682	86.9	0.40	3645	86.3	0.283	0.72	0.52	0.85	NH_3+HCl
ammonium bicarbonate	79.06	262	280	270.56	2061	68.8	0.40	3263	67.4	0.234	0.59	0.34	0.85	NH_3+H_2O+CO_2

Table 1. Experimental results expressed in terms of the coefficients of equation (4). If $p_m/p_t \neq 1$ the values given are apparent values, see text.

metric complexes. From the vapour pressure measurements we evaluated enthalpies of sublimation of the pure substances and of the complexes as well as Gibbs energies and entropies of complexation[7]. It was also shown that in contrast to the TCNQ-TTF complex the quinone complexes do not evaporate congruently. The (naphtho-)hydroquinone is much less volatile than the corresponding quinone.

pp'-methane diphenyl isocyanate (pp'-MDI)

We were asked to determine the vapour pressure of this basic chemical for the production of polyurethanes. These polymers are used for parts of cars, for shoes, skiing boots, foams, paints and a number of elastomers. The maximum allowable concentration (MAC-value) of MDI in air was fixed at 0.02 ppm by the public authorities and some even recommended 0.005 ppm for this highly toxic (to the respiratory organs) chemical. This corresponds to a partial vapour pressure of approximately 2.10^{-3} Pa at $25°C$. Since we found, by extrapolating, that the vapour pressure at the melting point, i.e. at 41°C, is still 10^{-2} Pa which is about 0.1 ppm in air of 1 atm. it will be clear that protective measures are necessary when working with this chemical.

Theophylline

This is a pharmaceutical used to treat respiratory ailments. It is produced as a byproduct in the decaffeination of coffee beans. In cooperation with the pharmaceutical department we determined lattice energies and vapour pressures of both theophylline and its monohydrate in order to test whether experimentally found solubilities agreed with the theoretically predicted ones. Solubility and the associated kinetics are of course important parameters for drugs.

5. CURRENT RESEARCH

Within the framework of a programme that is financially supported by ZWO (organisation for scientific research in The Netherlands) we have been and are still working on some 30 substances which can be divided roughly into three groups according to whether they contain oxygen, sulphur or nitrogen. The molecules are all of the small "rigid body" type and the aim is to obtain a set of consistent parameters for intermolecular interaction potential functions. For all these substances we measured the enthalpies of sublimation, as only a few enthalpies are available in the literature. We also measured specific heat capacities either by adiabatic or DSC calorimetry. Heat capacity data for the vapour were calculated from spectroscopic data or estimated by Benson's method[27]. These data are needed for the temperature correction of ΔH_{sub} from the experimental temperature to the temperature where the crystal structure was determined, which is often liquid nitrogen temperature.

Very interesting too are the results obtained for the ammonium halides and several other ammonium salts. For instance the results for NH_4Cl show a dissociation in NH_3 and HCl in the vapour phase. Although our results are in a lower pressure region they are in the main consistent with the results obtained by Callanan et al.[28] Also very promising and interesting are the results obtained so far for the evaporation coefficient of, for instance, trans stilbene, benzoic acid, adamantane, benzophenone, urea and some other compounds. By employing both a "Langmuir" and the "Knudsen" effusion cell described above we obtain data on both "vacuum evaporation" and saturation vapour pressures in the same set-up. Preliminary results indicate an evaporation coefficient of unity. This is in sharp contrast to very scarce literature data, which, with a few exceptions all date back to before 1940. The discrepancies with literature can and must be ascribed to artefacts in earlier experiments, arising from phenomena such as surface cooling of the sample, incorrect saturation vapour pressure data and too low a pumping speed. We are working on a theoretical interpretation of the results obtained with the help of the so-called Terrace Ledge Kink model (TLK-model).

6. FUTURE DEVELOPMENTS

The research of our group at Utrecht will continue to concentrate on finding the relation between structure and thermodynamic properties of molecular organic (mixed) crystals. Since we have close links with a group of X-ray crystallographers and we are able to measure relevant thermodynamic quantities with for instance adiabatic calorimetry, isoperibol calorimetry and vapour pressure measurement apparatus (both static and dynamic) we hope to achieve our aims. We shall also study methods for calculating[29] phase diagrams and try to obtain accurate thermodynamic data. Special attention will be given to the properties of homogeneous mixed crystals prepared by "zone" levelling [30,31] These samples appear to show a sharp melting behaviour and contain crystals suitable for single crystal X-ray diffraction. We have already used the simultaneous torsion mass-loss effusion apparatus for determining saturation vapour pressures of substances which could be harmful to the environment. Future legislation is likely to insist on the provision of data on the volatility of industrial chemicals, insecticides and herbicides. Because only very few laboratories are equipped with suitable apparatus for determining these data we expect to become involved in this area too. In the not too distant future we may be concerned with developments in materials research. The apparatus described here can easily be adapted for high temperature vapour pressure measurements (inorganic substances) and used as a combined thermogravimetric DTA instrument.

ACKNOWLEDGEMENT

Some months before this conference an opportunity arose for me to change my main field of research from thermochemistry to the dynamics of colloidal solutions and I decided to accept the challenge. The work described in this paper is to some extent a review of the work done up till now in our thermodynamics group and will be continued. I should like to take this chance of thanking Prof. A.J.J. Sprenkels and Prof. A. Schuijff for their stimulating interest over the years. I have greatly enjoyed working with Dr. Harry Oonk and Dr. Kees van Miltenburg and look back with pleasure on our fruitful periode of cooperation.
I also wish to thank the many students who have participated in the research of our Chemical Thermodynamics Group. Mathy Tollenaar is thanked for making drawings and Hanneke Pellegrom for typing the manuscripts. Finally I am grateful to Koos Blok for his technical support and especially to Tjibbe Kuipers, who made the instruments, for his outstanding craftmanship and cooperation.

REFERENCES

1. Kitaigorodsky, A.I. *Molecular Crystals and Molecules*. Academic Press: New York, 1973.
2. Williams, D.E. *Acta Cryst.* 1972, A28, 629.
3. Margrave, J.L. *The Characteristics of High Temperature Vapours*. Wiley: New York, 1967.
4. Ambrose, D. *Chemical Thermodynamics; Specialists Periodical Reports*. McGlashan, M.L. Sen. Rep. The Chemical Society, Burlington House, London. 1971.
5. Schaake, R.C.F.; Offringa, J.C.A.; van den Berg, G.J.K.; van Miltenburg, J.C. *Rec. Trav. Chim. Pays Bas* 1979, 98, 408.
6. Edwards, J.G. *Rev. Sci. Instr.* 1979, 50, 374.
7. de Kruif, C.G.; van Ginkel, C.H.D. *J. Chem. Thermodynamics* 1977, 9, 725.
8. Calis-van Ginkel, C.H.D.; Calis, G.H.M.; Timmermans, C.W.M.; de Kruif, C.G.; Oonk, H.A.J. *J. Chem. Thermodynamics* 1978, 10, 183.
9. de Kruif, C.G. The vapour phase dissociation of ammonium salts. Submitted for publication.
10. de Kruif, C.G.; Govers, H.A.J. *J. Chem. Phys.* 1980, 73, 533.
11. de Kruif, C.G.; Smit, E.J.; Govers, H.A.J. *J. Chem. Phys.* 1981, 74, 5838.
12. de Kruif, C.G.; van Ginkel, C.H.D. *J. Phys. E.* 1973, 6, 764.
13. Pelino, M.; Viswanadham, P.; Edwards, J.G. *J. Phys. Chem.* 1979, 83, 2964.

14. de Kruif, C.G.; Kuipers, T.; van Miltenburg, J.C.; Schaake, R.C.F.; Stevens, G. *J. Chem. Thermodynamics* 1981, 13, 1081.
15. Ambrose, D.; Lawrenson, I.J.; Sprake, C.N.S. *J. Chem. Thermodynamics* 1975, 7, 1173.
16. de Kruif, C.G.; Blok, J.G. *J. Chem. Thermodynamics* 1982. 14, 201.
17. Molar heat capacities and vapour pressure of solid and liquid benzophenone. Submitted for publication.
18. To be published.
19. de Kruif, C.G.; Oonk, H.A.J. *J. Chem. Thermodynamics* 1979, 11, 287.
20. Schaake, R.C.F.; van Miltenburg, J.C.; de Kruif, C.G. *J. Chem. Thermodynamics* 1982, 00.
21. de Kruif, C.G.; Voogd, J.; Offringa, J.C.A. *J. Chem. Thermodynamics* 1979, 11, 651.
22. Ngauv, S.N.; Sabbah, R.; Laffite, M. *Thermochimica Acta* 1977, 20, 371.
23. Metzger, R.M. *Ann. N.Y. Acad. Sci.* 1978, 313, 145.
24. Chen, T.H.; Schechtman, B.N. *Thin Solid Films* 1975, 30, 173.
25. Govers, H.A.J. *Acta Cryst.* 1978, A34, 960.
26. de Wit, H.G.M.; van der Klauw, K.; Derissen, J.L.; Govers, H.A.J. *Acta Cryst.* 1980, A36, 490.
27. Benson, S.W. *Thermochemical Kinetics.* Wiley: New York, 1976.
28. Callanan, S.J.E.; Smith, N.O. *J. Chem. Thermodynamics* 1971, 3, 531.
29. Oonk, H.A.J. *Phase Theory.* Elsevier: Amsterdam. 1981.
30. Kolkert, W.J. *J. Chryst. Growth* 1975, 30, 213.
31. de Kruif, C.G.; van Genderen, A.C.G.; Bink, J.C.W.G.; Oonk, H.A.J. *J. Chem. Thermodynamics* 1981, 13, 457.

VAPORIZATION STUDIES BY KNUDSEN CELL - MASS SPECTROMETRY AND
THERMODYNAMIC STABILITY OF HIGH TEMPERATURE GASEOUS SPECIES

G.De Maria
Istituto di Chimica Fisica - Università di Roma,
Rome,Italy.

The Knudsen cell - mass spectrometric technique utilized for high temperature equilibrium studies is described together with the usual thermodynamic procedure followed in the determination of the dissociation energy of gaseous molecular species.The application of the technique to the determination of the activity in alloys is also described. Physico-chemical properties of different classes of homonuclear and heteronuclear species are discussed in the light of the chemical and physical stability of the molecules.Correlations between gaseous species formed in the vapor over compounds belonging to groups of refractory materials of particular interest in modern technology are also discussed.

1.INTRODUCTION

A fundamental problem in Chemistry is represented by the absolute calculation of chemical equilibrium.To the solution of the problem concur in a specific way a number of disciplines such as the Statistical Thermodynamics,the Molecular Spectroscopy and the Theoretical Chemistry.This problem,even though can be considered solved in principle on the basis of quantum mechanics treatment (Atomic Orbital and Molecular Orbital methods)it is in practice far from being successful for a large number of homonuclear and heteronuclear molecular species(1).Furthermore the complexity of the calculations involved requires,even in the case of simple chemical reactions,the use of highly expensive computer programs which in practice represents a severe limitation to progress in this field (2).

The basic relation $\Delta G° = -RT\ln K_a$ remains therefore an important reference point in treating the problem. The equation correlates the standard free energy of a chemical reaction with the equilibrium constant, namely with the activities of the reactants and products at the equilibrium and therefore with their fagacities. The latter, in ideal systems, are measured by the total or partial pressures of the components in the gaseous phase in thermodynamic equilibrium with the condensed system. All substances, by changing the temperature, can be vaporized and if this process is investigated under thermodynamic equilibrium, one can derives physico-chemical properties which are determinant to the solution of the problem. One can therefore easily understand why in the past an increasing number of physico-chemists have got interested in tensimetric measurements by developing a large variety of sophisticated experimental techniques (3).

2. KNUDSEN CELL - MASS SPECTROMETRIC TECHNIQUE

A number of experimental techniques have been utilized to carry out tensimetric determinations of various systems at high temperature. They are schematically reported in Fig. 1.

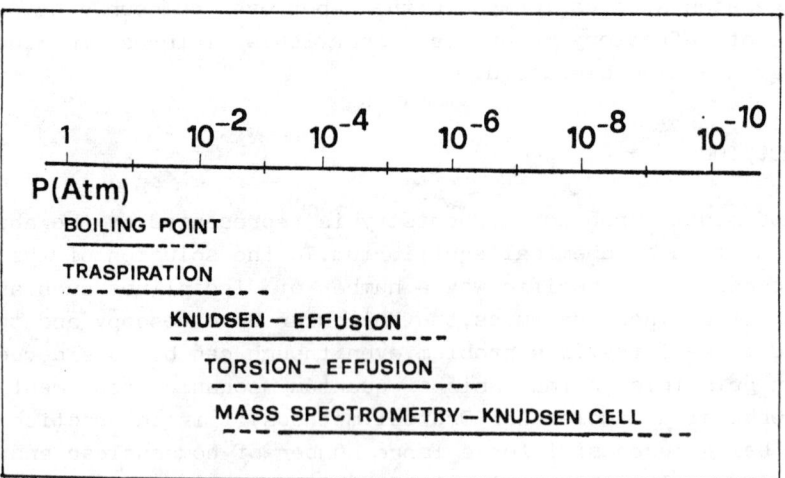

Fig. 1. Pressure range of operation of tensimetric methods.

For the determination of equilibrium pressures below 10^{-5} atm almost the totality of the methods devised utilizes, as a basic component, the Knudsen cell source coupled with a system of measurement either of the effects engendered by the effluent vapor or, by resorting to spectroscopic methods, of the intensity and composition of the molecular beam. Among these methods the coupling with a mass spectrometer has represented a uniquely powerful tool for the investigation of equilibrium systems especially at high temperature.

The first application of the mass spectrometric technique was made by Chupka and Inghram (4) at the University of Chicago. The apparatus, which can be considered a real standard Knudsen cell-mass spectrometric combination, has had minor changes in the course of the past years. A number of review papers have been published since them (5,6,7,8,9,10), an authentic testimony of the fertility and success of this technique which, even today, has no competitors in identifying all of the minor and major molecular species present in the equilibrium vapor and in measuring their partial pressures as a function of temperature, composition and time. The Knudsen effusion cell source assembly coupled with the ionization chamber is shown in Fig. 2.

A number of basic assumptions are to be satisfied in order to obtain accurate vapor pressure measurements of the constituent in the system (11). They relate specifically to Knudsen cell features and to the peculiarity of the coupled mass spectrometer. The vapor in the cell is supposed to be in thermal and mechanical equilibrium, that is to say collision density at all surface must be constant, nor pressure gradient from sample surface to orifice must exist. Chemical equilibrium between the gaseous and condensed phase must be present, even though this condition is not prejudicial for obtaining equilibrium data involving gaseous species only. No grain boundary or surface diffusion, nor chemical reactions with the container or diffusion through the wall must be operative. Cell orifices close to ideality (zero thickness) with constant geometry must be insured during the experiments and, in the case of non ideal orifice, accurate transmission probabilities (Clausing factor) must be available (12). Displacement of the cell should also be avoided during vaporization by appropriate mounting of the molecular source.

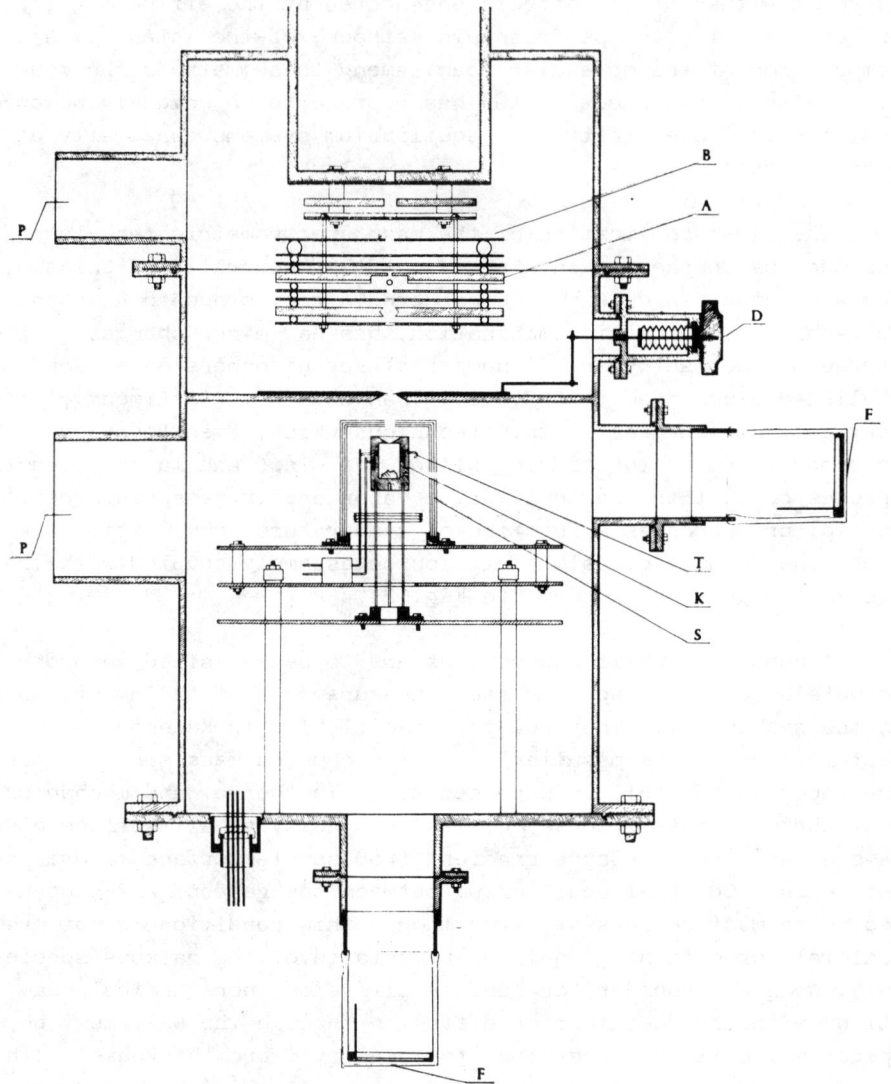

Fig. 2 - Knudsen effusion cell and ionization source assembly for thermodynamic equilibrium studies. K-Knudsen cell; T-bombarding filament; S-radiation shields; F-windows for temperature measurement; D-movable beam defining slit; A-ionization chamber; B-ion beam defining and accelerating system; P-to pumps.

2.1- Identification of Molecular Species

The identification of the molecular species is accomplished by this method in four steps:
- a) mass-to-charge ratio
- b) isotopic distribution
- c) appearance potential
- d) intensity distribution in the molecular beam

From the shape of the ionization curve one can, in some cases, reveal the occurrence of dissociative ionization processes and obtain an evaluation of the bond strength in the molecule. The intensity distribution in the molecular beam allows one to distinguish between species originating from inside the cell and species vaporizing from the surrounding shields. The effusion of the gas, in thermodynamic equilibrium with the condensed system, throughout a knife-edge orifice of area s, obeys to the Knudsen equation (13,14).

$$Z = \frac{1}{4} n \bar{c} s \qquad (2.1)$$

where Z is the number of molecules which effuse throughout the orifice s, n the equilibrium concentration of the gas in the cell in number of molecules per cm^3, $\bar{c}$ their mean velocity. The molecular beam issued spreads out according to the cosine law:

$$\frac{dN}{dt} = \frac{1}{4} n \bar{c} \sin\vartheta \cdot \cos\vartheta \cdot d\vartheta \cdot d\varphi \qquad (2.2)$$

where $\frac{dN}{dt}$ represents the number of particles travelling per unit time, in a solid angle having polar coordinates ϑ and φ, originated from unit surface orifice. In the Knudsen cell-ion source assembly represented in Fig. 2 the number of particles which will pass per unit time throughout the ionization chamber A is given by:

$$\frac{dN_i}{dt} = \frac{sn\bar{c}f}{4\pi L^2} \qquad (2.3)$$

where f is the area of the ion chamber slit and L its distance from the orifice of the cell K. If Δz is the depth of the ion chamber and Δt the time needed for a molecule to pass through the ion source, $\Delta z = \bar{c} \Delta t$. Using equation (2.3) and incorporating the gas low $P_i = n_i kT$ one obtains for the concentration N_s

in the ion chamber (in number of molecules per cm^3):

$$N_s = \frac{s \, P_i}{4 \pi L^2 k T} \qquad (2.4)$$

where P_i represents the equilibrium pressure of the species i in the Knudsen cell, T its temperature and k the Boltzman constant. The ion intensity I_i^+ of a species i measured by the mass spectrometer is related to the concentration N_s by means of the equation:

$$I_i^+ = \eta \, \sigma_i \, \gamma_i \, I^- \, l \, N_s \qquad (2.5)$$

where η is the efficiency of collection of the ions on the collector plate, σ_i the cross section, γ_i the electron multiplier gain, I^- the intensity current of the ionizing electron beam and l the length of its path in the ion-chamber. Combining this equation with the relation (2.4) one obtains:

$$P_i = \frac{4 \pi L^2 k}{\eta I^- l \, \sigma_i \, \gamma_i} \, I_i^+ T \qquad (2.6)$$

If I_i^+ is the measured isotope peak and a_i the corresponding isotopic abundance, the absolute partial pressure of a molecular species in the Knudsen cell is given by:

$$P = \frac{I_i^+ \, T}{S \, \sigma_i \, \gamma_i \, a_i} \qquad (2.7)$$

where S is the sensitivity factor of the mass spectrometer which takes care of the geometrical and electronic properties of the experimental set-up.

The equilibrium partial pressures are measured as a function of temperature. Each single species having a partial pressure between 10^{-12} and 10^{-3} atm is detected and measured in a wide temperature range, up to 2800°K. Limitation for higher pressure values and consequently higher temperatures is due to the condition imposed by the validity of the Knudsen effusion equation, namely, that the mean free path must be much larger than the diameter of the effusion hole. Interactions with the container may introduce severe limitations in the study of the condensed phase-gas phase equilibria. Several refractories are employed as Knudsen cell materials (Al_2O_3, Graphite, Ta, W, Mo and so on) and their choice depends on the nature of the system investigated. In applying relation (2.7) one is faced primarily with three fundamental problems: (i) the identification of the neutral progenitor of an ion; (ii) the evaluation of the sensitivity constant and (iii) the determination of the ionization cross section and multiplier gain. An excellent discussion of the topic is reported by F.E.Stafford (15) and by E.D. Cater (11). As a whole one can say that the evaluation of the above parameters introduce some degree of uncertainty in the final thermodynamic data; these are to be discussed case by case, but the practice show that in general these uncertainties are not so severes as one could be brought to think.

2.2. Activity Measurements by Knudsen Cell-Mass Spectrometric Technique

One of the peculiar application of the Knudsen cell-mass spectrometric technique has been the direct determination of the activity of components in alloys utilizing a twin-cell assembly coupled with a mass spectrometer (16,17). The activity a_i of a component in an alloy is, by definition, given by:

$$a_i = \frac{f_{i,x}}{f_o} \qquad (2.8)$$

where $f_{i,x}$ is the fugacity of the component i in the alloy with molar fraction x_i and f_o the fugacity of the component in the pure state. If thermodynamic equilibrium is insured in both cells at the same temperature, filled one with the pure constituent, the other with the alloy of well defined composition, then the activity is obtained from relation (2.7) as:

$$a_i = \frac{P_i}{P_o} = \frac{I_i^+}{I_o^+} \cdot \psi \qquad (2.9)$$

where I_i^+ is the ion intensity of the gaseous species in the equilibrium vapor over the alloy and I_o^+ the one corresponding to the pure component; $\psi = \frac{S_i}{S_o}$ is the ratio of the instrumental constant relative to the cells, which can be determined straightforwardly by filling both cells with the pure component. An extension of this method utilizing multiple rotating-cells has been also devised (18) and brought to a good development level (19). Fig.3 reports the activity of Ca and Ga in the Ca-Ga system as measured by this method. This technique has also been proved to be very useful for direct determination of the relative ionization cross-sections of elements (15,18).

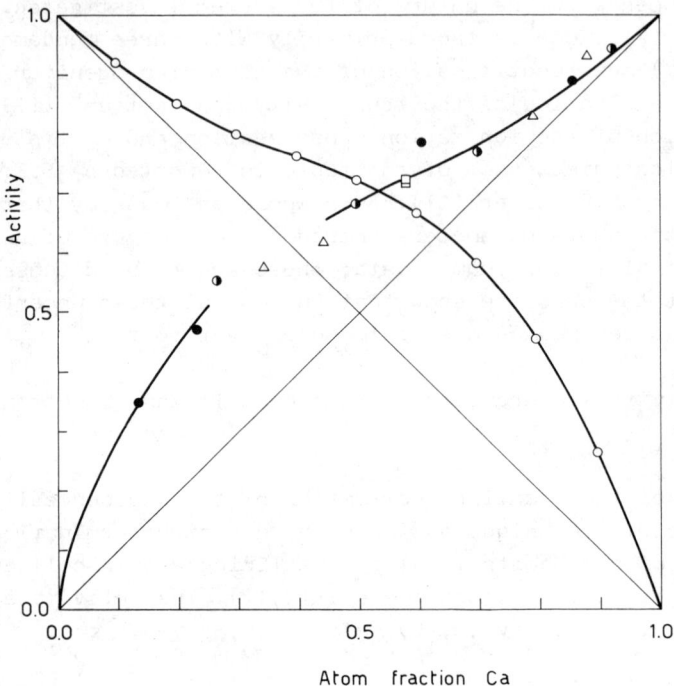

Fig.3 - Activities in the Ca-Ga system at an average temperature of 1270 K ●,◐ and △ are relative to data obtained by the multiple-rotating Knudsen -cell source. The square points are relative to data taken at 1200 K using a twin - Knudsen-cell source. The Ga activity curve was obtained using the Gibbs-Duhem equation.

2.3 - Thermodynamic treatment.

The most general and extensive application of the Knudsen cell mass spectrometric method has been the study of high temperature inorganic systems with the evaluation of the thermodynamic properties of the gaseous molecular species.

The thermodynamic stability of an homogeneous compound is simply defined by its standard free energy of formation. This, in many cases, gives also a measure of the "real" stability, that is to say of the stability referred to the normal environment conditions;

yet there are cases, particularly in the class of refractory compounds, where the real stability appears to be related to the different behaviour of evaporation of compounds. If one compares the thermodynamic stability of berillium oxide and magnesium oxide given by $\Delta G°_{298,f} = -581.7$ kJ mol^{-1} and $\Delta G°_{298,f} = -569.6$ kJ mol^{-1} respectively, one is brought to consider BeO slightly more stable than MgO. In practice MgO is much more stable than BeO and this can be explained by the different modes of evaporation of the two compounds. Berillium oxide vaporizes at high temperature giving rise as predominant species polymers of BeO principally the trimer, tetramer and pentamer (20). In contrast, magnesium oxide vaporizes primarily by decomposition to $Mg_{(g)}$ and $O_{2(g)}$. As a consequence of the differences in major species, in changing fron neutral environment to an oxygen atmosphere, the total equilibrium pressure of Berillium oxide remains nearly unchanged, while an high oxygen pressure greatly reduces the equilibrium partial pressure of $Mg_{(g)}$, conferring to the condensed phase a stability higher than that in an inert atmosphere.

A pecularity which appears immediately by examining the modes of evaporation of inorganic systems at high temperature is the fact that the composition of the saturated vapor phase over the condensed systems becomes more complex as the temperature increases. This very surprising generalization, sometime reported as Brewer's rule(21), can be explained on the basis of simple thermodynamic arguments.

Considering for simplicity the case of saturated equilibrium gaseous phase composed only of monomers and dimers, we can write the equations:

$$- RT \ln P_M = \Delta G°_M = \Delta H°_M - T \Delta S°_M \qquad (2.10)$$

$$- RT \ln P_D = \Delta G°_D = \Delta H°_D - T \Delta S°_D \qquad (2.11)$$

where the subscripts M and D stand for the monomeric and dimeric species. As a good approximation we can consider the entropy changes $\Delta S°_M$ and $\Delta S°_D$ to be the same; furthermore the term $T\Delta S°$ is somewat minor than $\Delta H°$, which in turn determines in a predominant way the value of $\Delta G°$ and therefore the equilibrium partial pressures. Since for a large majority of high temperature systems the conditions of interest are those for which the partial pressures are less than 1 atm., according to (2.10) and (2.11) the minor species will exhibit the higher partial enthalpies of sublimation. As a consequence the proprtion of the minor species in the equilibrium vapor phase will increase with temperature and at sufficiently high temperatures this species will become comparable with the one that predominates at low temperature (Fig.4)

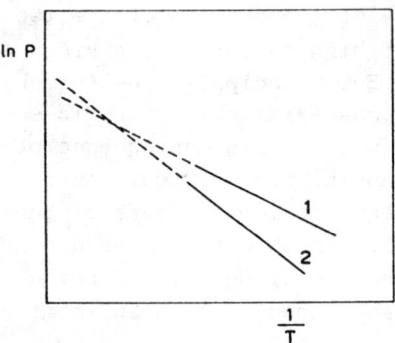

Fig.4 -Qualitative trend of monomeric and dimeric species versus 1/T in the saturated vapor.

These considerations can be generalized not only to systems that give rise to three or more species, such as monomers, dimers and polymers, but also to more complex vaporization reactions. One can therefore draw the important conclusion that a large variety of new and complex molecules are expected to be found in the gas phase in equilibrium with condensed systems at high temperatures.

The equilibrium partial pressures of the molecular species as measured by means of the experimental methods are utilized to obtain thermodynamic data of the specific molecule and of the condensed phase. The thermodynamic treatment of the data articulates essentially in two indipendent methods based respectively on second- and third-law of thermodynamics.

The so called second-law method is expressed by the Clausius-clapeyron or Van't Hoff equations for vapor pressure or equilibrium reactions respectively:

$$\frac{d \ln K_a}{d\left(\frac{1}{T}\right)} = -\frac{\Delta H_T}{R} \qquad (2.12)$$

where K_a is the equilibrium constant, R the gas constant and ΔH_T the heat of evaporation at the temperature T.

The third-law method isbased on absolute entropy evaluation of the reactants and, being

$$S_T^o = \frac{H_T^o - H_o^o}{T} - \frac{G_T^o - H_o^o}{T} \qquad (2.13)$$

it can be expressed in a more convenient formulation by the equation:

$$\Delta H_o^o = - RT \ln K_a - T\Delta\left\{\frac{G_T^o - H_o^o}{T}\right\} \qquad (2.14)$$

where ΔH_o^o, according to cases, represents a measure of:
a) - Cohesive energy as for $M_{(s)} \longrightarrow M_{(g)}$
b) - Dissociation energy ($=D_o^o$) as $M_{2(g)} \longrightarrow 2M_{(g)}$
c) - Bond energy for the process $MX_{2(g)} \longrightarrow M_{(g)} + X_{2(g)}$
d) - Atomization energy in the case of $MX_{2(g)} \longrightarrow M_{(g)} + 2X_{(g)}$

The free energy function is simply related to the partition function Q by:

$$\frac{G_T^o - H_o^o}{T} = - R\ln Q \text{ with } Q = Q_{tr} \cdot Q_{rot} \cdot Q_{vib} \cdot Q_{el},$$

corresponding respectively to the translational, rotational, vibrational and electronic contribution to the partition function. All these contributions can be calculated throughout the usual methods of statistical mechanics(22,23), taking into account an appropriate correction factor for the case of a non rigid rotor anharmonic oscillator. Even though the spectroscopic data(interatomic distances, vibration frequencies, force constants, anharmonicity constants, electronic states, etc.)implied in statistical mechanics calculations are obtained by various spectroscopic methods, one must point out that in the case of many high temperature species they are generally not known, nor, due to their very low abundance in the gaseous phase, easy to be measured by the conventional spectroscopic techniques. In these cases one can have resort to analogycal evaluations or to correlations rules as between vibration frequencies and interatomic distances in the same molecules, between vibration frequencies in gaseous species and in the condensed phase, etc. In all these cases some degree of uncertainty is introduced in the final thermodynamic data in addition to the uncertainties associated with the method utilized for the determination of the equilibrium partial pressures of the species. As a whole, due mainly to the random character of a part of such errors, the overall uncertainty is less severe of what one would be brought to think.

Exchange reactions in the gaseous phase are often utilized for measuring dissociation energies of high temperature molecules, by fixing in some cases the partial pressures in the vapor phase throughout an appropriate choice of a stable compound to which the element of interest is added(24,25).

3. CHEMICAL STABILITY OF GASEOUS MOLECULAR SPECIES

3.1 - Homonuclear Diatomic Molecules

A great effort has been made during the past years to the experimental detection and determinations of the dissociation energies of yet undiscovered homonuclear diatomic species, particularly the diatomic metals. A number of compilations have appeared on this subject(26,27,28,29) some dealing more specifically with experimental stability(30) and models for calculation of the dissociation energies(31). Among the homonuclear molecules are some which are little known but which are nontheless as stable as well known species; for example the dissociation energies of molecules such as Cu_2, Ag_2 and Au_2 (D_0^o= 190.4, 157.4 and 215.5 kJ mol^{-1} respectively) are comparable with the dissociation energies of the halogens Cl_2, Br_2 and I_2 (D_0^o= 239.0, 190.0 and 149.0 kJ mol^{-1} respectively). If one considers the ratio between the cohesive energies ΔH_0^o(vap.M) of the elements and the dissociation energies $D_0^o(M_2)$ of the related diatomic molecules $\alpha = \Delta H_0^o$(vap.M)/$D_0^o(M_2)$ one can recognize a noticeable difference. In general the α values for all elements plotted as a function of the atomic number can be grouped into three distinct, almost horizontal bands, normally denoted "categories"(32). Each category is composed of members with similar electronic structure. Elements of Groups VB(s^2p^3), VIB(s^2p^4) and VIIB(s^2p^5), with $0.5 < \alpha < 1.0$ belong to the first category; elements of Groups IA(s^1), IB($d^{10}s^1$), IIIB(s^2p^1), and IVB(s^2p^2) with $1.2 < \alpha < 2$, belong to the second category. The third category, with $\alpha > 5$, includes elements of groups IIA(s^2) and IIB($d^{10}s^2$). The different values evidentiate the diversity of the bonding nature.

A general thermodynamic treatment of the subject can be based on the following cicle:

$$M_{(s)} + M_{(g)} \xrightarrow{-\Delta H_o^o(\text{vap } M)} 2 M_{(s)}$$

$$\Delta H_o^o(\text{vap } M) \downarrow \qquad \qquad \downarrow \Delta H_o^o(\text{vap } M_2)$$

$$2 M_{(g)} \xrightarrow{-D_o^o(M_2)} M_2(g)$$

For the process:

$$M_{(s)} + M_{(g)} \longrightarrow M_2(g) \qquad (3.1)$$

one can write:

$$\Delta H_o^o(\text{vap } M_2) - \Delta H_o^o(\text{vap } M) = \Delta H_o^o(\text{vap } M) - D_o^o(M_2) \qquad (3.2)$$

which gives, taking into account the value:

$$\frac{R \, d\ln P(M_2)/P(M)}{d(1/T)} = \Delta H_o^o(\text{vap } M) \frac{1 - \alpha}{\alpha} \qquad (3.3)$$

For the elements with $\alpha < 1$ the relative proportion of M_2 is high at low temperatures but decreases with increasing temperature. Diatomic molecules of the elements of the first category are in fact well known at room temperature. When $\alpha > 1$ (bimetallic molecules) the proportion of diatomic molecules relative to monoatomic species in the saturated vapor increases with increasing temperature. One can denote these molecules as high-temperature molecules.

A basis for a discussion of the possibilities of detecting the remaining unknown diatomic molecules is represented by the free energy function of equation (3.1):

$$\Delta G_o^o = -RT \ln P(M_2)/P(M) = \Delta H_o^o(\text{vap } M_2) - \Delta H_o^o(\text{vap } M) - T\Delta\{(G_T^o - H_o^o)/T\}$$
$$= \Delta H_o^o(\text{vap } M) \{(\alpha - 1)/\alpha\} - T\Delta\{(G_T^o - H_o^o)/T\} \qquad (3.4)$$

3.2 - Heteronuclear Diatomic Molecules.

This important class of molecules which in theory should count approximately 4000 members, has gained in recent years a growing interest. In addition to the mass-spectrometric method(24,33,34,35), matrix isolation technique(36), optical spectroscopy(37), magnetic circular dichroism spectroscopy(38) have been applied to investigate a number of new heteronuclear species, with particular enphasis to the bimetallic molecules. The basic thermodynamic approach to the problem follows a treatment analogous to that applied to the heteronuclear diatomic molecules. For these species a parameter $a = 0.5 \Delta H_0^o$ (at AB)/D_0^o (AB) can be defined, where ΔH_0^o (at AB) represents the cohesive energy of the solid AB and D_0^o (AB) the dissociation energy of the heteronuclear gaseous species. The process of vaporization may be different for different substances; nonetheless it can be shown (39) that in all cases the free energy of the reaction is related to the ratio between the partial pressures of the heteronuclear species and that of the monoatomic or diatomic species of one of the two components, giving a straighforward measure of the "chemical stability".

Table I - Values of a^a for Alkali Halides

Element	F	Cl	Br	I
Li	0.83	0.85	0.92	1.08
Na	0.88	0.89	0.96	1.11
K	0.80	0.84	0.89	1.01
Rb	0.78	0.82	0.87	0.98
Cs	0.74	0.78	0.82	0.91

$^a a = 0.5 \Delta H_{298}^o$ at. MX /D_{298}^o (MX).

Also for these compounds one can allocate the systems into two categories with $0.5 < a < 1$ and $a > 1$. To the first category belong heteronuclear diatomic molecules of Group IVB-VIB and IIIB-VIIB, isosteric with homonuclear molecules of group VB. This category is enriched by the typically ionic alkali halides whose a values ranges from 0.7 to 1.1, with a slight monotonic trend with increasing halogen atomic number (Table I).
These molecules, by analogy with what has been discussed for the homonuclear species, must be considered as "low-temperature molecules". A large part of others systems probability exibit a values > 1. Therefore most of these molecules are difficult to detect and needs special experimental arrangements for their observation (33,34). Table II reports the dissociation energies of III_B-V_B molecules at 0°K in kJ mol^{-1}.

Table II- Dissociation Energy of III_B-V_B Molecules at 0°K in kJ mol^{-1}

AlP=212.6±12.6	GaP=230.2±10.5	InP=195.0±8.3	TlP=205±12.6
AlAs=199.9±7.1	GaAs=209.7±1.2	InAs=197.9 ±10.0	TlAs 194.1±14.6
AlSb=212.6±21.0	GaSb=189.2±12.6	InSb=148.1±10.5	TlSb=123.0±10.5
AlBi= ---	GaBi=154.8±16.7	InBi= ---	TlBi=117.2±12.6

AlP- ref.40; AlAs- ref.41; AlSb and TlP- ref.42; GaP- ref.43; GaAs- ref.33; GaSb- ref.44; GaBi ref.45; InP- ref.46; InAs- Piacente,V. Gigli, R. (1982), J. Chem.Phys. 77 p.91; InSb- ref.48; TlAs- ref.49; TlSb- ref.50; TlBi- ref.39.

A critical assessment of the thermodynamic properties of the chalcogenide family, with particular emphasis to the dissociation energies of the gaseous species,are reported in the paper by Drowart (24).

3.3. Vaporization of Oxides

This group of compounds, due to the important role that plays in ceramic science and in general in manifold applications of the modern technology, has been the object of many investigations at high temperature (24,51,52). The high temperature investigation of the equilibrium vapor over these systems constituted, since the beginning, sort of surprise by revealing the existence of unusual oxidation states such as molecules LiO, Ba_2O, Ba_2O_3, Al_2O, AlO and polymers as observed in the vapors over CrO_3, MoO_3, WO_3 and BeO. A numbers of spectroscopic techniques have been utilized for the investigation of gaseous metal oxides (24,52) but, especially for gaseous dioxide species, a more intense effort is auspicable.

From analysis of the vaporization data one can readily distinguish three different modes of vaporization:
1. Vaporization by dissociation to the elements
2. Vaporization to molecular oxides or by dissociation to monoxides and oxygen
3. Vaporization to polymers

Most of the refractory oxides belong to the second category, in which can be included also members of Group IIIB elements and the rare-earth series not included in Table III.

Table III- Vapor Pressure, Melting Point and Predominant Vapor Composition for a Series of Refractory Oxides[a]

Oxide	Temperature (°K) for Vapor Pressure (Torr)			Melting Point (°C)	Predominant Vapor Composition
	10^{-6}	10^{-3}	1		
Li_2O	1175	1466	1825	(1700)	Li_2O, elements, LiO
BeO	1862	2300	2950	2530	BeO, $(BeO)_n$, elements
MgO	1600	1968	2535	2800	Elements, MgO
CaO	1728	2148	2795	2580	Elements
SrO	1600	1897	2247	2430	SrO, elements
BaO	1358	1694	2198	1923	BaO
B_2O_3	1090	1337	1734	450	B_2O_3
Al_2O_3	1910	2339	3000	2015	Elements, Al_2O, AlO
Y_2O_3	2100	2523	3040	—	YO, O, O_2
La_2O_3	1820	2239	2754	2315	LaO, O, O_2
TiO	1618	1968	2489	1750	TiO
TiO_2	1800	2203	2825	1640	TiO_2, TiO, O_2
ZrO_2	2060	2512	(3048)	2700	ZrO_2
HfO_2	2270	2748	3443	2810	HfO_2
ThO_2	2061	2512	3192	3050	ThO_2
VO	1663	2075	(2594)	—	VO
MoO_2	1368	1654	2004	—	MoO_3, MoO_2, $(MoO_3)_2$
MoO_3	762	878	1038	795	$(MoO_3)_3$, $(MoO_3)_4$, $(MoO_3)_5$
WO_2	1641	1954	(2317)	—	WO_2, WO_3
WO_3	1138	1409	(1531)	1473	$(WO_3)_3$, $(WO_3)_4$, $(WO_3)_5$
UO_2	1754	2165	2786	2176	UO_2
PuO_2	1722	2133	(2622)	—	PuO_2
MnO	1384	1758	2223	1650	Elements
FeO	1413	1774	2239	1420	Elements
NiO	1300	1629	2046	2090	Elements
CoO	1368	1683	2138	1800	Elements
ZnO	1002	1230	1592	(1800)	Elements

[a] From De Maria, G. (ref.51)

Trends in the stability of gaseous oxides have been discussed by Searcy (43) and it was also emphasized the fact that oxides of the right hand side of the periodic table display a general decrease in stabilities of the oxides of the left-hand side of the periodic table rise with atomic number.

Ternary oxide molecules have been more recently object of investigation. A number of complex species have been detected over systems such as Eu-W-O (54); Eu-Mo-O (55); Eu-Cr-O; Eu-P-O (56). Atomization energies are reported in Table IV.

Table IV- Ternary Oxide Molecules

System	Condensed phase	Species (M)[a]	ΔH°_o, atom(M) (kJ mol^{-1})	$-\Delta H^\circ_{o,f}$ (M) (kJ mol^{-1})
Eu-W-O[b]	Eu_2O_3 in W or Eu_2O_3+$EuPO_4$ in W	$EuWO_4$	2990 ± 27	986 ± 28
		$EuWO_3$	2383 ± 29	625 ± 30
		Eu_2WO_5	4017 ± 46	1594 ± 46
		EuW_2O_7	5340 ± 38	1748 ± 40
Eu-Mo-O[c]	Eu_2O_3 in Mo or Eu_2O_3 + $EuPO_4$ in Mo	$EuMoO_4$	2774 ± 25	962 ± 25
		$EuMoO_3$	2159 ± 33	594 ± 33
Eu-Cr-O[d]	Cr + Eu_2O_3 or Cr_2O_3 in W or Ta	EuCrO	877 ± 35	63 ± 35
		$EuCrO_2$	1401 ± 36	344 ± 37
Eu-P-O[e]	Eu_2O_3 + $EuPO_4$ in W or Mo	$EuPO_2$	1502 ± 25	506 ± 25
		$EuPO_3$	2019 ± 23	778 ± 23

(a) Typical temperature interval of the study: 1900-2300°K; (b) ref.44; (c) ref.45; (d) Balducci, G., De Maria, G., Gigli, G. and Guido, M. (1977), Fifth Int. Conf. Chemical Thermodynamics, Ronneby, Sweden, poster session; (e) ref. 46.

These species may play a significant role in transport phenomena connected with nuclear oxides and fission products. Europia is assuming an increasing interest in Nuclear power plant; it is one of the few materials suitable for use as neutron absorber in fast reactor control and shut-off rods.

3.4. Vaporization of Carbides

This class of compounds, whose technological interest is certainly not minor than that of the oxides, have received so far an attention from high temperature researchers not proportionate to their importance. This is due probably to the high refractory properties of these materials which require severe limiting conditions in the experimental investigations. Nevertheless a number of systems have been already investigated in the past by mass-spectrometric technique, revealing in some cases the complexity of the gaseous phase (57) and in others (58) the formation of MeC_4 species of high physical stability, to which, more recently (59) have followed others to the MeC_5 and MeC_6, summarized in table IV.

Table IV- Examples of the complexity of the vapour in equilibrium over the metal carbide +C condensed phase (usually MC_2+C for Sc, Y, La, Ce, U, Th; or MC+C for Ti, Zr, U, Hf). Temperature interval: 2300 - 2800 K

M	
	YC_2, YC_4, YC_3, YC, YC_5, YC_6
La	LaC_2, LaC_4, LaC_3
Ce	CeC_2, CeC_4, CeC_3, CeC, CeC_5, CeC_6
U	UC_2, UC_4, UC, UC_5, UC_3, UC_6
Th	ThC_2, ThC_4, ThC, ThC_3, ThC_5, ThC_6

Even though there has been in the past an interesting investigation of the partial thermodynamic properties throughout the homogeneity region of non stoichiometric carbides such as UC up to UC_2 (50) the large majority of the mass-spectrometric data obtained are relative to the carbon richest carbide-phase (51,52). In other cases observation of the species were made in a gas phase over a not well defined condensed phase.

Gaseous metal dicarbides are the most frequently observed molecular species and their bond energies are correlated to the corresponding dissociation energies of the gaseous monoxide by the

Chupka–De Maria's rule (63,64) Table V.

Table V– Comparison of the bonding energies of gaseous monoxides and gaseous dicarbides (in kJ mol^{-1}) (values in parenthesis are estimated)

M	D_0^o (M–O)	D_0^o(M–C$_2$)	D_0^o(M–O) – D_0^o(M–C$_2$)
B	799	636	163
Al	506	515	–9
Sc	670	565	105
Y	703	632	71
La	795	669	126
Ce	791	678	113
Pr	749	632	117
Nd	707	603	104
Sm	569	(485)	(84)
Eu	468	(410) or 539	(58) or –71
Gd	711	632	79
Tb	703	628	75
Dy	602	556	46
Ho	615	556	59
Er	607	565	42
Yb	356	(376)	—
Lu	695	607	88
Si	795	695	100
Ge	653	615	38
Ti	661 or 602	565	96 or 37
V	619	569	50
Zr	757	569	188
Hf	791	670	121
Cr	436	443	–7
U	757	607	87
Th	866 or 845	724	142 or 121
Ru	510 or 477	523	–13 or –46
Rh	406	439	–33
Os	590	573	—
Ir	(351)	548	(–197)
Pt	368	552	–184
Tm	556	498	58

From this rule the relative concentration of the dicarbide species with respect to the metal atom can be estimated at a given temperature from knowledge of the dissociation energy of the corresponding monoxide.

The pressure-independent reation $M(g) + 2C_{(s)} \rightarrow MC_{2(g)}$ is combined with the enthalpy of sublimation of Carbon to C_2 molecules to obtain the expression:

$$\frac{P_{MC_2}}{P_M} = -\frac{A - D_0^o(MO)}{RT} - \frac{1}{R}\Delta\left\{\frac{G_T^o - H_0^o}{T}\right\} \qquad (3.5)$$

The constant A, equal to 900.0 kJ mol^{-1}, incorporates the enthalpy of sublimation of Carbon to $C_{2(g)}$ and the average difference between $D_0^o(MO)$ and $D_0^o(MC_2)$ 75.3 kJ mol^{-1} (51). An extension of this rule to the gaseous dioxide and tetracarbide molecules is also apparent from inspection of the data reported in table VI.

Table VI- Comparison of bond energies for dioxides and tetracarbides (in kJ mol^{-1})
(Values in parenthesis are estimated)

M	D_0^o (O-M-O)	D_0^o (C_2-M-C_2)	
Sc	(1280±92)	1218±21	(62±92)
Y	(1351±92)	1272 or 1330	(79±92)
La	(1519±92)	1330±17	(189±92)
Ce	1439±21	1351±21	88±29
Pr	(1415±92)	1310±21	(105±92)
Nd	1330±84	1297±42	33±92
Dy	(1192±113)	1297±21	(−105±113)
Ho	1285±105	1213±13	72±105
Lu	(1289±105)	1255±42	(34±113)
Ti	1310±21	1218±21	92±29
Zr	1443±42	1289±29	154±50
Hf	1506±84	1347±25	159±88
U	1423±29	1364±25	59±38
Th	1632±25	1402±25	230±33

The "inversion temperature points" plotted for the carbon-rare earth group in Fig. 5 are of interest for chosing appropriate operative conditions for spectroscopic investigations of gaseous dicarbide species (65).

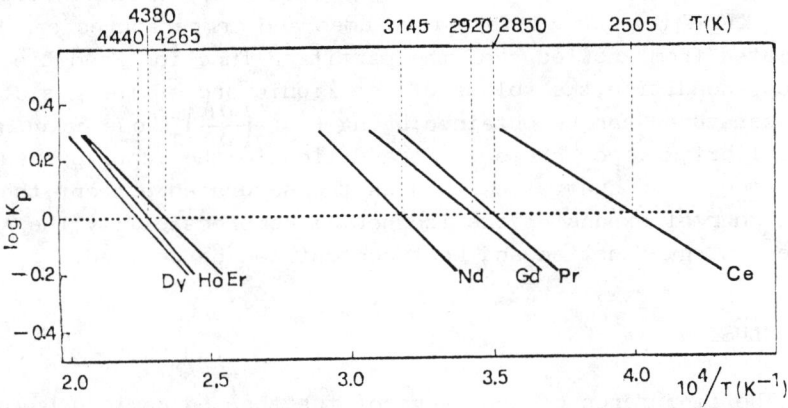

Fig. 5- Temperature dependence of evaluated K for reaction: $M_{(g)} + 2C_{(s)} \longrightarrow MC_{2(g)}$, in proximity of "inversion temperatures"

3.5 - Critical point data prediction of nuclear fuels

One of the present problem in nuclear material science is the assessment of equation of state of nuclear fuels. Important contributions have been made in this field by Ohse et al(66,67) and by Potter et al(68).

By applying the Significant Structure Theory(SST) model(69) the relative importance of the complex species over the UC system at very high temperature-pressure region can be evaluated(70) from knowledge of the nature and thermodynamic properties of the species involved.

The partition function for the different gaseous species is being written as:

$$f_g^{N\frac{V-V_s}{V}} = \left\{ \frac{(2\pi Mm_i KT)^{3/2}}{h^3} (V-V_s) f_{int} \right\}^{N_i/N_i!} \quad (3.6)$$

where $f_{i,int} = f_{el,i} \cdot f_{rot,i} \cdot f_{vib,i}$ is the partition function related to the motion with respect to the center of mass of the species i, N_i is the number of particles of the species i. A system of 17 equations is being considered;in this connection 14 indipendent equilibria have been selected and the relative equilibrium constants have been evaluated at each temperature of interest by means of third-law procedure.The value of the work function $A = -KT\ln f_{(v,T)}$ at different volumes and temperatures can be calculated from knowledge of the partition function.From the common slope condition,the volume of the liquid and of the gas at various temperatures can be obtained;being $P = -\left(\frac{\partial A}{\partial v}\right)_T$ one calculates the equilibrium vapor pressure.In addition to the vapor pressure various thermodynamic quantities such as the saturated entropy,the isobaric thermal expansion,the isothermal compressibility,the standard free energy function and heat content can be derived.

CONCLUSION

The importance of knowledge of the thermodynamic behavior of inorganic chemical compounds at high temperature does not merely derive from the variety of informations that can be obtained , extending from ascertaining the nature of high-temperature reactions to establishing the nature and energetics of chemical binding of the gaseous molecular species,but is also connected with the impact on a number of disciplines such as Geochemistry(72), Astrophysics(73) and Cosmology. One can enphasize the fact that nowday among astrophysicists is generally recognized that condensation from a primival nebula has been the primary process which led to the formation of the solid bodies in the solar system. Investigations on the vaporization process of lunar samples at high temperature(74) were aimed to give a contribution to such a problem,while the enlargement of the panorama of known high-temperature gaseous species has permitted to investigate in a more comprehensive way the composition of circumstellar atmospheres(73) A number of high temperature molecules have been already observed in the sun and stars.The molecule SiC_2,one of the predominant species in the equilibrium vapor over SiC(75),was recognized to be the emitter of the astronomically observed Merril-Sanford bands and the SiO molecule identified in the M-supergiant star Orionis is in full agreement with the large abundance of this molecule predicted by the thermodynamic calculations.Certainly this close interconnection between researches in high themperature chemistry and Astrophysics constitute one of the most attractive aspect of high temperature chemistry.

BIBLIOGRAPHY

1. Mulliken,R.S.-1975,Selected papers of R. Mulliken-Ramsay,D.A. and Hinze,J. ed.- Chicago:University of Chicago Press.
2. Mulliken,R.S. - 1967,Science,157,p.13
3. Margrave,J.L. - 1967,The Characterization of High-Temperature Vapors-J.Wiley & Sons,Inc.NewYork
4. Chupka,W.A. and Inghram,M.G.- 1955,J.Phys.Chem.59,p.100
5. Inghram,M.G. and Drowart,J.,1959,in "Proceedings of the International Conf.on High-Temperature Technology",Asilomar, California,Mc Graw-Hill,N.Y.
6. Margrave,J.L.- Ann.Rev.Phys.Chem.10,p.457
7. Gilles,P.W. -1961,Ann.Rev.Phys.Chem.12,p.355
8. Drowart,J and Goldfinger,P.- 1962,Ann.Rev.Phys.Chem.13,p.459
9. Grimley,R.T.- 1967,in "The Characterization of High Temperature Vapors" - Margrave,J.L.,ed.J.Wiley,N.Y. p.195
10. De Maria,G.and Balducci,G.- 1972,in MTP International Review of Science-Physical Chemistry Series One-Skinner,H.A. ed., 10,p.209
11. Cater,E.D.- 1979,in"Characterization of High Temperature Vapors and gases"- Hastie,J.W.ed.-NBS Special Public.561,p.3
12. Grimley,R.T. and Forsman,J.A.- ibidem p.211
13. Knudsen,M.- 1909,Ann.Physik,28,p.75
14. Knudsen,M.- 1950,Kinetic Theory of Gases,3rd ed. Metuen and Co., Ltd,London.
15. Stafford,F.E.- 1971,High Temperature-High Pressure,3 p.213
16. Pattoret,A,Smoes,S.and Drowart,J.- 1966,Thermodynamics,Vienna IAEA,I,p.377
17. Cameresi,G.G.,De Maria,G.,Gigli,R and Piacente,V.:1967,Ric.Sci. 37,p.1092
18. De Maria,G.and Piacente,V.- 1972,Bull.Soc.Chim.Belges,81,p.155
19. Chatillon,C,Allibert,M.and Pattoret,A.-1979 in "Characterization of High Temperature Vapors and Gases",Hastie,J.W.ed.,NBS Spec.Publ.561,p.185
20. Chupka,W.A.,Berkowitz,J.and Giese,C.F.- 1959,J.Chem.Phys.30,827
21. Brewer,L.- 1950 in "Chemistry and Metallurgy of Miscellaneous Materials:Thermodynamics,Nat.Nucl.En.- Mc Graw Hill,N.Y.
22. Fowler,R.H.and Guggenheim,E.A.- "Statistical Thermodynamics", Cambridge University Press
23. Mayer,J.E.and Goepert-Mayer,M.- 1950,"Statistical Mechanics", J.Wiley & Sons,Inc.N.Y.

24. Drowart,J.- 1969 in Proc.of the Intern.School on Mass Spectrometry,Ljubijama,Yugoslavia
25. Hildebrand,D.L.- in "Characterization of High Temperature Vapors and Gases" Hastie,J.W.,ed. -NBS Spec.Publ.561,p.171
26. Gaydon,A.G.- 1968,"Dissociation Energies and Spectra of Diatomic Molecules",Chapman and Hall,London,3rd Ed.
27. Drowart,J.- 1967 in "Phase Stability in Metals and Alloys,ed. Rudman,P.S.,Stringer,J. and Jaffer,R.I.,Mc Graw-Hill,N.Y.
28. Drowart,J.and Goldfinger,P.- 1967,Agew Chem.,$\underline{79}$,589
29. Gurvich,L.V.,Karachevstev,G.V.,Kondratyev,V.N.,Lebedev,Y.A., Mendredev,V.A.,Potapov,V.K. and Khodeev,Y.S.- 1974, "Bonf Energies,Ioniz.Potentials and Electron Affinities" Nauka,Moscow
30. Gingerich,K.A.- 1980,Faraday Symp.Chem.Soc.,Skinner,H.A.ed.$\underline{14}$,109
31. Brewer,L.and Winn,J.S.,ibidem,p.126
32. Verhagen,G.,Stafford,F.E.,Goldfinger,P.and Ackerman,M.- 1962, Trans.Faraday Soc.,$\underline{58}$,p.1926
33. De Maria,G.,Malaspina,L.and Piacente,V.- 1970,J.Chem.Phys.$\underline{52}$,1019
34. De Maria,G.,Malaspina,L.and Piacente,V.- 1972,J.Chem.Phys.$\underline{56}$,1978
35. Zmbov,K.F.,Wu,W.H.anf Ihle,H.R.,1977,J.Chem.Phys.$\underline{67}$,p.4603
36. Montano,P.A.- 1980 in Faraday Symp.Chem.Soc.,$\underline{14}$,p.79
37. Shulze,W.and Abe,H.- ibidem p.87
38. Grinter,R.,Armstrong,S.,Jayasooriya,U.A.,McComly,J.,Norris,D. and Springall,J.P.,ibid.p.94
39. Colin,R.and Goldfinger,P.- 1964,in "Condensation and Evaporation of Solids",Gordon and Breach,N.Y.
40. De Maria,G.,Gingerich,K.A.,Malaspina,L.and Piacente,V.- 1966, J.Chem.Phys.$\underline{44}$,2531
41. Piacente,V.- 1979,J.Chem.Phys.$\underline{76}$,2912
42. Piacente,V. and Balducci,G.- 1977 in Advances in Mass Spectr. $\underline{7}$,p.626, Heyden & Sons,London
43. Piacente,V. and Gingerich,K.A.- 1971,High Temper.Science $\underline{6}$,214
44. Piacente,V.and Balducci,G.- 1974,High Temp.Science $\underline{6}$,254
45. Piacente,V.and Desideri,A.- 1972,J.Chem.Phys.$\underline{57}$,2213
46. Piacente,V.and Balducci,G.- 1975,"Dynamic Mass Spectr.$\underline{4}$,p.295 Heyden & Sons,London
47. Piacente,V. and Gigli,R.- J.Chem.Phys. $\underline{77}$,p.198
48. De Maria,G.,Drowart,J. and Inghram,M.G.- 1959,J.Chem.Phys.$\underline{31}$,1076
49. Piacente,V.and Malaspina,L.- 1972,J.Chem.Phys. $\underline{56}$,1780
50. Balducci,G.,Ferro,D. and Piacente,V.,submitted to High Temp.Sci.
51. De Maria,G.- 1968 in "Chemical and Mechanical Behavior of Inorganic Materials",p.81,Wiley Interscience,N.Y.

52. De Maria,G.- 1975, "The role of high temperature molecules in transition phenomena" in'The flames as reactions in flow¦ p.39, Padua- CNR
53. Searcy,A.W.- 1962,in "Progress in Inorganic Chemistry,$\underline{3}$,p.49, Interscience,N.Y.
54. Balducci,G.,Gigli,G.and Guido,M.- 1977,J.Chem.Phys.$\underline{67}$,p.147
55. Balducci,G.,Gigli,G. and Guido,M.- 1979,J.Chem.Phys.$\underline{70}$,p.3146
56. Balducci,G.,Gigli,G. and Guido,M - 1977,High Temper.Sci.$\underline{9}$,149
57. Drowart,J,De Maria,G.and Inghram,M - 1958,J.Chem.Phys.$\underline{29}$,1015
58. Balducci,G.,Capalbi,A.,De Maria,G.and Guido,M.-1965,J.Chem.Phys. $\underline{43}$,p.2136
59. Gingerich,K.A. in "Characterization of High Temperature Vapors and Gases", Hastie,J.W.Ed.-NBS Spec.Publ.561 p.289
60. Storms,E.K.- 1966,Thermodynamics,Vienna,IAEA,$\underline{1}$,p.309
61. Balducci,G.,Capalbi,A,De Maria,G.and Guido,M.-1969,J.Chem.Phys. $\underline{50}$,p.1969
62. De Maria,G.- in "Symposium on Dynamics of Chemical Reactions", 1966,Padua-CNR,p.121
63. Chupka,W.A.,Berkowitz,J.,Giese,C.F.and Inghram,M.G.-1958,J.Phys. Chem.$\underline{62}$,p.611
64. De Maria,G.,Balducci,G.Capalbi,A.and Guido,M.- 1967,in Proceed. of the British Ceramic Soc.,Stoke-on-Trent,England $\underline{8}$,127
65. De Maria,G.- 1970,in Recent Developm.in Mass Spectroscopy,Univ. of Tokyo Press,p.1132
66. Fisher,E.A.,Kinsman,P.R. and Ohse,R.W.- 1976,J.Nucl.Mater.$\underline{59}$,125
67. Ohse,R.W.,Babelot,J.F.,Cercignani,C.,Kinsman,P.R.,Long,K.A., Magill,J.and Scotti,A.- 1979,J.Nucl.Materials $\underline{80}$,p.232
68. Browning,P.,Gillan,M.J.andPotter,P.E.- 1978,Rev.Int.Hautes Temp. Rèfract.$\underline{15}$,p.333
69. Ree,T.S.,Eyring,H.and Perkins,R.-1965,J.Phys.Chem.$\underline{69}$,p.3322
70. Gigli,G,Guido,M.and De Maria,G.-1981,J.Nucl.Materials,$\underline{98}$,p.35
71. Ohse,R.W.,private communication
72. De Maria,G. and Piacente,V.-1969,Atti Accad.Naz.Lincei,$\underline{47}$,525
73. Tsuji,T.- 1973,Astron.and Astrophys.$\underline{23}$,p.411
74. De Maria,G.,Balducci,G.,Guido,M.and Piacente,V.- 1971,in Proc. of the Second Lunar Science Conference,$\underline{2}$,p.1367
75. Drowart,J.,De Maria,G. and Inghram,M.G.-1958,J.Chem.Phys.$\underline{29}$,1015
76. Kleman,B- 1956,Astrophys.J.,$\underline{12}$,162

TRANSPIRATION MASS SPECTROMETRY--A NEW THERMOCHEMICAL TOOL

J. W. Hastie and D. W. Bonnell

High Temperature Processes Group
National Bureau of Standards
Washington, DC 20234

ABSTRACT

   Classical vaporization methods such as transpiration and Knudsen or Langmuir effusion have been limited because they do not establish the molecular identity of transport species or because low pressures are necessary to make effusion measurements. We have developed a new technique--Transpiration Mass Spectrometry (TMS)--that overcomes both of these limitations by combining the basic features of transpiration and molecular beam mass spectrometry. With this technique, it is possible to sample reactive gases directly from high-temperature (to 1500 °C), high-pressure (to 10 atm) atmospheres for quantitative characterization with a mass spectrometer. The accuracy of thermochemical data obtained by the TMS method is competitive with that of established lower dynamic range techniques. Examples of application to vaporization of complex silicate slags, glasses, and minerals are considered. Implications and precautions resulting from cooling effects during the sampling process are also discussed.

PART I: DEVELOPMENT OF METHOD[1]

1. Introduction

   1.1 <u>Technological background</u>. Many high temperature processes of technological importance rely upon, or are adversely affected by, chemically active heterogeneous subsystems. One such case involves chemical interaction of a multicomponent high temperature (e.g., ∼ 500-2000 K) high pressure (e.g., ∼ 0.1-100 atm) gas mixture with a solid or liquid substrate. This interaction frequently leads to the formation of intermediate species containing elements of both the gaseous and substrate materials. These reaction intermediates can be transported in the presence of temperature, concentration, or momentum (e.g., forced convection) gradients. Transport to another regime of temperature, pressure, or concentration can lead to a reaction-reversal or the introduction of secondary processes. In many instances, this changing chemistry along gradients result in a deposition of intermediate species with the overall result that the initial substrate has been physically transported to another part of the system.

Such materials transport can have either an undesirable or beneficial effect on the system of interest. Examples in modern technology include, respectively:

(1) Hot corrosion of gas turbines, jet engines, rockets, coal gasifiers, magnetohydrodynamic channels, coal fired boilers, and numerous pyrometallurgical systems.

(2) Controlled material transport for production of crystals and films, extractive metallurgy, flame inhibition and fire extinguishment, combustion modification such as smoke and antiknock control with additives, and regenerative lamp cycles.

A detailed recent discussion of these and other examples of high temperature materials transport has been given elsewhere (1).

In order to understand these heterogeneous processes for development of new or improved control strategies, it is necessary to define at a molecular level the transport mechanisms and particularly the reaction intermediates. To date, extensive reliance has been placed on speculation and empirical observations based solely on macroscopic variables defining the initial and final state of the system. For instance, current understanding of hot corrosion derives from metallurgical examination of the clean and corroded materials under ambient conditions. Extrapolation of these observations to the actual conditions of usage requires many assumptions about the corrosion mechanisms. However, detailed mechanistic information is practically nonexistent. A further inaccuracy occurs because standard test procedures (e.g., use of laboratory burner rigs) simulate only part of the actual system. To properly relate test results to practical usage would require a knowledge of scale-up factors derived from a detailed mechanistic description of both the test and the end-use systems.

Information on reaction mechanisms, including the intermediate transport species, has been practically nonexistent heretofore because of the measurement problems associated with these extreme conditions. Very limited molecular-specific information can be obtained using optical spectroscopic methods at high temperature and pressure. In principle, the broad spectrum of gaseous and vaporized species present could be identified and their spatial and temporal concentrations determined by the high pressure sampling mass spectrometric (HPMS) technique. This molecular beam technique, which uses a sonic nozzle for sample extraction, has been amply demonstrated by us (2) and others for <u>homogeneous</u> systems, including flames. A summary of recent work in this area has been given by Stearns, et al. (3). For <u>heterogeneous</u> systems, the gas is saturated by condensible species. In this case, the nozzle probes used to interface the hot gas and the mass spectrometer become part of the reaction system and severe limitations arise. Conventional use of the HPMS technique (homogeneous sytems) has required probe temperatures to be considerably less than for the sample system. With saturated gases, this leads to condensation of inorganic species at the probe tip and results in physical blockage of the small entrance orifice or in corrosive loss of the probe.

We have developed a procedure for avoiding this difficulty, in addition to providing other advantages. Basically, the sample probe and skimmer (to a lesser degree) are maintained isothermal by an external heat source. Maintenance of a steady state (and constant pressure) between gas and substrate requires a continuous replenishment of the gas extracted by the sampling process. This leads naturally to a transpiration procedure, where the input flow rate matches the gas extraction rate through the small-orifice probe. The technique therefore combines the basic features of both the transpiration (4) and the HPMS methods.

1.2 <u>Scope of application</u>. In the following sections we describe this technique of transpiration mass spectrometry (TMS) and its application to the characterization of high temperature vapors. We have demonstrated that it is a nonperturbing method, at least for the equilibrium systems studied, and this is most likely the case for many nonequilibrium systems. Sample perturbation is always an important concern with probe methods. The method extends the dynamic range of classical vaporization and vapor transport techniques by many orders of magnitude; at least five orders with respect to Knudsen effusion mass spectrometry. We believe that TMS should be applicable as a <u>quantitative</u> measurement tool for laboratory simulations of most of the technological transport systems mentioned above and eventually may be applicable to plant-scale systems. Examples of application to alkali vapor transport thermochemistry are considered in Part II.

With regard to systems of initial academic interest, this technique should be useful for producing novel high temperature species in a spectroscopically cool form, resulting from free jet expansion cooling. This application would greatly extend the utility of thermally sensitive molecular beam spectroscopic methods. In particular, electron diffraction, microwave, laser fluorescence, Raman, photoelectron, and photoionization spectroscopy, as well as matrix isolation methods, would benefit from a coupling with the TMS technique. The extended high pressure range should also allow for the study of basic thermodynamic properties of novel complexes and adducts of the type suggested indirectly from gas-solid solubility studies [e.g., see Hastie, pp. 126-148; pp. 73-87 (1)]. Future extension to even greater pressures should permit definitive molecular characterization of systems in their critical state and greatly supplement equation of state studies and the fundamental understanding of fluids. Controlled formation of clusters, or large molecular aggregates ($\sim 10^5$ amu), is also possible using the TMS approach.

2. Apparatus

2.1 <u>The mass spectrometer system</u>. Figure 1 shows a schematic of the initial mass spectrometer (stage II) and vacuum system layout with the transpiration apparatus attached (stage I). Technical details may be found elsewhere (see note 1).

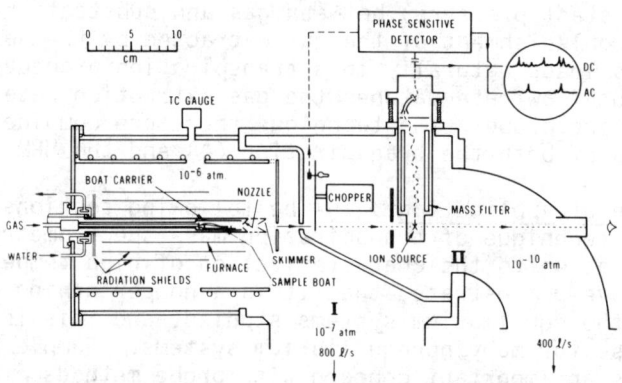

Figure 1. Detailed schematic of assembled transpiration reactor-mass spectrometer system.

2.2 <u>The transpiration reactor</u>. The transpiration reactor is located in stage I of the overall assembly (see Figure 1). Essential features of this reactor include: a sample container or boat, a boat carrier, a thermocouple for temperature measurements, a carrier gas inlet system, and a gas extraction system or probe. The boat carrier allows for boat removal from the reactor without need for a complete disassembling of the transpiration system. Molecular beam sonic probes are typically conical nozzles with design details determined by reasonably well established gas dynamic criteria, as outlined in Section 3. However, for highly reactive systems, we also found it desirable to develop a relatively more robust capillary probe, at the possible expense of sampling fidelity. Below we describe the construction details of these probes and the transpiration assembly in general. Figure 2 shows a schematic view of a typical transpiration reactor (Pt construction material), including details of the assembled boat carrier and boat.

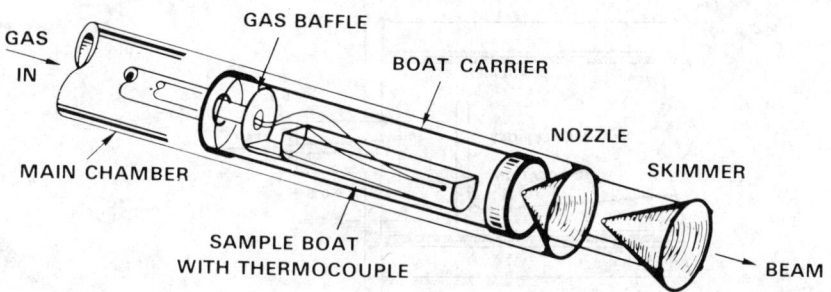

Figure 2. Transpiration reactor showing internal details.

Details of the nozzle-skimmer arrangement are shown in figure 3.

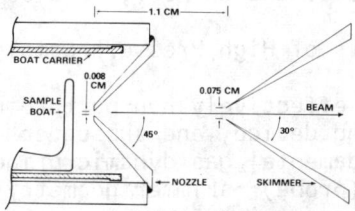

Figure 3. Schematic of conical sampler construction details and relative location of internal components. Forward edge of skimmer is attached to outer edge of chamber exterior with 3 thin (0.2 cm thick) platinum struts (not shown in Figure). For clarity, not all dimensions are drawn to scale.

A capillary sampler, shown in figure 4, was fabricated to be interchangeable with the conical nozzle-skimmer assembly, Note the absence of a skimmer with the capillary method. This omission was possible since beam velocity distribution measurements (5) with short channels (length to orifice diameter ratio of 2) and no skimmer have indicated formation of a well developed isentropic expansion leading to a high Mach number supersonic beam.

Our longer channel (corresponding ratio of 30) capillary also produces a good quality supersonic beam (see Section 3).

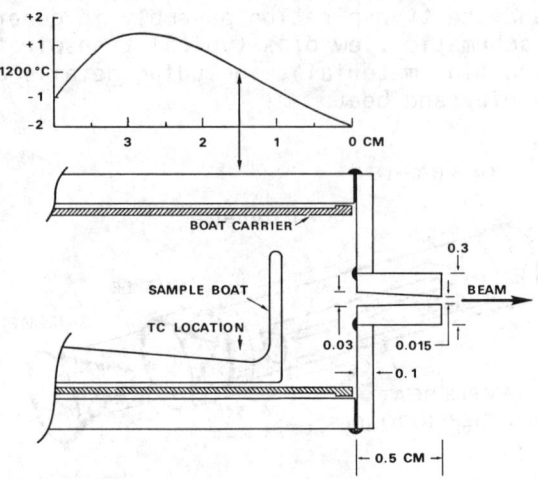

Figure 4. Schematic of capillary sampler showing construction details and relative location of internal components. Channel not drawn to scale for clarity. A typical temperature profile, obtained along the central axis of the chamber, is also shown.

3. Gas Dynamic Aspects of High Pressure Sampling

The design of an effectively non-perturbing high pressure molecular beam sampling device, and subsequent data analysis, relies heavily on fundamental gas dynamic principles. In deciding on an optimum nozzle (probe)--skimmer geometry, consideration must be given to system pumping capacity, source pressure and temperature, desired beam intensity, and the characteristic thermodynamic and dynamic properties of the sample gas. We have discussed this subject in greater detail elsewhere (see note 1).

For normal gas-gas and condensed-gas reactions involving uncharged species, reaction times are typically greater than 10^{-5} s. As continuum flow times of 10^{-7} s are typical for modern sampling systems, including ours, the criterion for minimal sampling perturbation should apply for most molecular systems of interest. In the absence of chemical reaction kinetic data, however, it is desirable to provide experimental verification of sampling fidelity. This has been done for the TMS system, as described in Section 4.

3.1 **Beam flux and sensitivity**. Greene, et al. (6) showed that the flux through the skimmer of an isentropically expanding gas from a supersonic nozzle (i.e., conical probe) has the functional form of effusive flow multiplied by a function dependent only on the gas specific heat and the Mach number at the skimmer. Experimentally it has also been shown (7) that, to a good approximation, the dilute components of a gas mixture undergoing supersonic expansion are accelerated with the characteristics of the solvent carrier gas. In the absence of mass separation effects these results suggest that the usual mass spectrometer relationship, based on effusive flow, can be applied to the supersonic jet sampling process. That is,

$$P_o = kIT_o$$

where I is the observed ion intensity,

T_o the source temperature,
k the effective mass spectrometer sensitivity constant, and
P_o the source pressure.

As will be shown by our data, the functional dependence of beam flux on specific heat and Mach number can be empirically absorbed into the k factor when this "constant" is obtained from the observed ion intensity for the carrier gas at known conditions of P_o and T_o. In practice, k is measured at every temperature and pressure of an experimental run. The k's obtained show a minor variation with temperature and pressure, reflecting small changes in the expansion process including mass separation effects.

3.2 **System mass discrimination**. In order to make quantitative measurements of species partial pressures, it was necessary to determine the magnitude of possible mass discrimination effects in our system. One such effect is inherent in the free-jet expansion process and results from differences in the perpendicular velocity component for various beam species leading to radial diffusion and depletion of low mass species in the beam. In practice, this effect is superimposed on the normal mass discrimination effects of the quadrupole mass analyzer and electron multiplier.

In combination, the various mass discrimination effects are taken into account by calibration with a commercially prepared standard gas mixture analyzed against gravimetric standards. This gas mixture contains ∼ 1 percent of each of the rare gases, helium, neon, argon, krypton, and xenon in nitrogen and has been analyzed to 0.01 percent of the gas composition. The overall mass discrimination function, as obtained in the temperature and pressure region of operation and normalized to the carrier gas (N_2), is used to correct ion intensity and, hence, pressure data (see Section 4).

A typical plot of normalized ion intensities versus mass number, at several temperatures and pressures, is given in figure 5. The normalization of ion intensities, I_m, for species m, to the nitrogen carrier gas (using the $^{29}N_2$ isotopic species) is made using the expression

$$I_m(\text{norm}) = \frac{I_m}{f_m \cdot A_m \cdot \sigma_m} \frac{f_{29} \cdot A_{29} \cdot \sigma_{29}}{I_{29}},$$

where f_m is the mole fraction,

A_m the isotopic abundance, and

σ_m the ionization cross section at 30 eV.

The reference discrimination curve (see effusive curve of Figure 5) was obtained under effusive conditions at a pressure of less than 10^{-5} atm using a large orifice cell. This curve may be used as a measure of the mass discrimination effects present in the quadrupole filter and detector.

Figure 5 shows a small enhancement (∼ 10 percent) of higher mass species in supersonic beams which agrees qualitatively with the arguments outlined earlier, considering an early onset of collisional "freezing" and the relatively large skimmer-orifice separation used. The effective coincidence of the mass separation curves for the conical and capillary probes (see Figure 5) suggests that both probe types are effectively isentropic in their expansion characteristics.

To account for the combined mass discrimination effects of mass separation in the expanding jet and quadrupole mass filter discrimination, a sensitivity factor, $S_{Na^+(NaCl)}$, for example, was obtained from

$$S_{Na^+(NaCl)} = \left(\frac{S_h - S_e}{S_e}\right)_{58\ amu} + (S_e)_{23\ amu}$$

where h and e refer to the high pressure and effusion curves of figure 5, respectively. It should be noted that the experimental mass discrimination function is essentially fixed for a given set of quadrupole mass filter operating parameters. However, when the mass spectrometer is re-tuned by varying the ion source and resolution settings a new discrimination function is obtained.

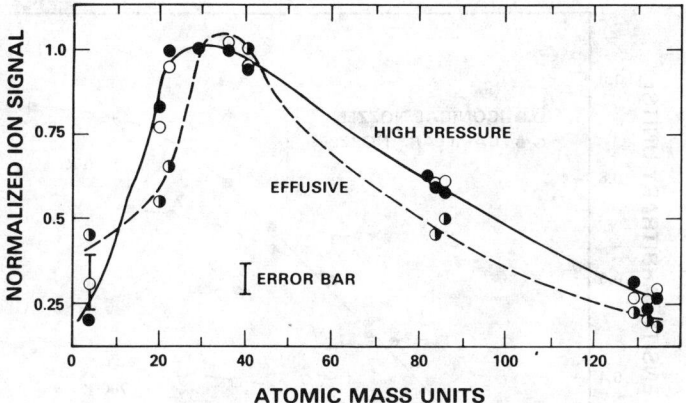

Figure 5. Mass discrimination behavior obtained using a standard gas mixture. Open circles--conical sonic orifice sampler at 1065 K and 0.6 atm, closed circles--capillary sampler at 800 K and $< 10^{-5}$ atm. Error bar represents worst case for variations in ion signal; most abundant isotope of each species normally exhibits variations less than half this amount. An exception is ^{20}Ne, where contributions from $^{40}Ar^{2+}$ and the narrow peak width limit the signal to noise ratio. Ion intensities measured at 30 eV ionizing electron energy and normalized with respect to $^{29}N_2^+$.

3.3 <u>Pressure operating characteristics</u>. Figure 6 shows the behavior of the detected ion signal as a function of source pressure. For this instrument, the expansion ratio is essentially constant, set by the available pumping speed, and the turnover of the unit slope portion of the curves, marked by arrows on figure 6, occurs consistently at an observed pressure of $P_I \sim 2 \times 10^{-6}$ atm. The fact that this turnover occurs for both the capillary (no skimmer) and conical sampling system at the same stage I pressure, independent of temperature, suggests the effect is a post-expansion scattering phenomenon dependent primarily on the mean free path. This argument is also consistent with the sensitivity of the turnover point onset to the skimmer alinement relative to the nozzle.

It appears possible, as Green, et al. (6) showed, to operate at pressures where scattering is significant and still maintain sampling fidelity; our data also indicate this (see Sections 4 and 5). However, the majority of our experimental runs were made in a pressure range where scattering was relatively low, i.e., within the linear region of the intensity-pressure curves such as those shown in figure 6.

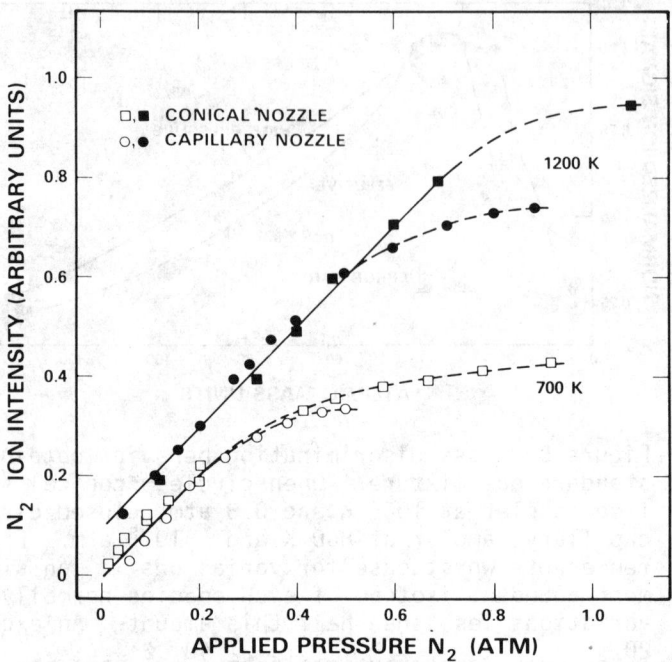

Figure 6. Dependence of carrier gas ion intensity on applied pressure and temperature. The 1200 K data have been offset vertically +0.1 div for clarity. Solid curves are lines of unit slope and zero intercept in accord with the theoretical relationship. Arrows indicate where stage I pressure exceeds 2×10^{-6} atm. The capillary curve deviates from linearity at a lower pressure because the total flow rate is higher through the relatively large carillary orifice.

4. The NaCl System: Results and Discussion

This system was chosen to test the validity and accuracy of the transpiration mass spectrometric technique, since the literature data for NaCl is extensive, reasonably self-consistent, and has been critically evaluated and formatted as thermochemical tables by JANAF (8).

Between about 800 to 1600 K the significant reactions in the equilibrium NaCl vaporization process are

$$NaCl(s) = NaCl(g) \qquad R1$$
$$NaCl(l) = NaCl(g) \qquad R2$$
$$2NaCl(s) = (NaCl)_2(g) \qquad R3$$
$$2NaCl(l) = (NaCl)_2(g) \qquad R4$$

$$3\text{NaCl(s)} = (\text{NaCl})_3(\text{g}) \qquad \text{R5}$$
$$3\text{NaCl(l)} = (\text{NaCl})_3(\text{g}). \qquad \text{R6}$$

Reactions R1 - R4 can be expressed in homogeneous form as

$$2\text{NaCl(g)} = (\text{NaCl})_2(\text{g}), \qquad \text{R7}$$

and similarly

$$3\text{NaCl(g)} = (\text{NaCl})_3(\text{g}). \qquad \text{R8}$$

This variety of reactions, involving both homogeneous and heterogeneous equilibria should provide a stringent test of the TMS technique, and particularly where the second law method of data analysis is used.

4.1 <u>NaCl mass spectral fragmentation data</u>. Typical mass spectral fragmentation data obtained by us and others are summarized in Table 1.

Table 1. Mass spectral ion intensities[a] for the NaCl system.

Na^+	NaCl^+	Na_2Cl^+	R^b	Cell	T(K)	eV	Source
1.00	0.49	0.88	2.0	Cu-Knudsen	968	70	(9)
1.00	0.73	0.91	1.4	Cu-Knudsen	~960	20	(10)
1.00[c]	0.66	0.66	1.5	Pt-Knudsen	957	30	This work
1.00[d]	0.04	0.08	25.0	Pt-transp.[e]	1312	30	This work
1.00[g]	0.03	0.14	33.3	Pt-transp.[f]	1360	30	This work

[a] Normalized to I_{Na^+}, and corrected for Cl-isotope fractions.
[b] $R = I_{\text{Na}^+}/I_{\text{NaCl}^+}$.
[c] Equivalent to a signal intensity of 920 µV ($10^7 \Omega$).
[d] Equivalent to a signal intensity of 1.8×10^4 µV.
[e] Carrier gas N_2 at 0.54 atm; $I_{29_{\text{N}_2}} = 1.00$ ($=1.8 \times 10^4$ µV) and $k_{\text{N}_2} = 1.66 \times 10^{-10}$ µV/atm K.
[f] Transpiration weight loss run. Note S factors differ to e. $k_{\text{N}_2} = 3.7 \times 10^{-11}$ µV/atm K (0.65 atm).
[g] Equivalent to a signal intensity of 1.9×10^4 µV.

The transpiration mass spectrometric data for the vapors over solid NaCl (i.e., at temperatures more compatible with Knudsen effusion data) showed no measurable $NaCl^+$ ion. However, while the 23 amu Na^+ ion is at a low noise position, $NaCl^+$ at 58 and 60 amu is relatively noisy (an order of magnitude greater than for 23 amu) and R values of about two cannot be ruled out on this basis. But, well into the liquid range, where the ion signals at both 23 and 58 amu are strong and relatively noise-free, we observe R ~ 26 ± 4 where error is standard deviation for data above 1170 K. This ratio varies slightly between experimental runs. The average ratio above 1200 K is quite constant within a variable temperature run and shows no significant variation from ~ 1200-1500 K, even with carrier gas pressures varying over the range 0.35-0.8 atm.

It appears that the high $Na^+/NaCl^+$ ion intensity ratio, with respect to those characteristic of Knudsen effusion measurements, is associated with the electron impact ionization process and the relatively cool nature of the TMS molecular beam. We believe that the Franck-Condon electron impact transition to the $NaCl^+$ potential curve is sensitive to vibrational exitation and hence temperature. That is, at low temperatures (TMS) transitions above the dissociation limit ($NaCl^+ \rightarrow Na^+ + Cl$) are more favorable.

4.2 <u>NaCl vaporization and pressure calibration</u>. The effective temperature insensitivity of the $Na^+/NaCl^+$ ratio, R, together with auxiliary appearance potential data (not presented here as it basically agrees with earlier work), leads to the conclusion that both Na^+ and $NaCl^+$ may be used as a measure of the NaCl partial pressure without significant interference from electron impact fragmentation of the $(NaCl)_2$ dimer species. The similar T-dependence of Na^+ and $NaCl^+$, but not Na_2Cl^+, also supports this claim, as do similar data reported in the literature. Typical partial pressure data for NaCl derived on this basis are shown in figure 7. This plot represents a number of sets of data obtained using either Ar or N_2 as carrier gases, with total pressures varying from 0.35-0.75 atm and 0.4-0.8 atm, respectively. Two different conical orifice samplers of similar, but not identical, construction were used. Also, the data plotted in the liquid range include pressures calculated from both Na^+ and $NaCl^+$ ion intensities. The essential concordance between these various data sets support the ion-precursor assignments and all other aspects of the procedure adopted for converting ion intensities to partial pressures. Note also in figure 7 the good agreement between the experimental and literature melting points (T_m), indicative of accurate temperature measurement.

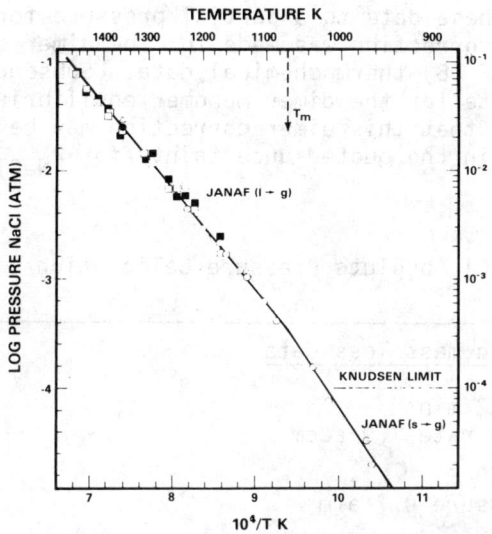

Figure 7. Vapor pressure curves for NaCl (s and l) obtained by TMS. Open squares--$P_{NaCl}(Na^+)$ in N_2 carrier gas, total pressure 0.4-0.8 atm, at 30 eV. Closed squares--$P_{NaCl}(NaCl^+)$ in N_2 carrier gas total pressure 0.4-0.8 atm, at 30 eV. Open circles--$P_{NaCl}(Na^+)$ in Ar carrier gas, total pressure 0.35-0.75 atm, at 70 eV. The open triangle and closed circle data points were obtained using the integrated ion intensity (b) and classical transpiration (a) methods, respectively (see Section 4.2). All other points were obtained using the ion intensity comparison method (c). Solid curve represents data from the JANAF compilation (8). The broken line labeled Knudsen limit was calculated assuming a typical orifice diameter of 0.1 cm.

There are three methods by which the mass spectral ion intensity data were converted to absolute pressures in the present study:

(a) classical transpiration,
(b) integrated ion intensity-time method, and
(c) reference gas ion intensity comparison.

With method (a) the salt mass loss is measured (gravimetrically) for a known volume of transpiration carrier gas at a fixed temperature and total pressure. Typical data are given in table 2. To convert these data to a partial pressure for the NaCl species, a small correction was made for the dimer species $(NaCl)_2$ using the JANAF (8) thermochemical data. Subsequent analysis of our own data for the dimer-monomer equilibrium reaction will indicate that this dimer correction may be too high but is still within the quoted uncertainty for P_{NaCl} given in table 2.

Table 2. NaCl Absolute Pressure Calibration

transpiration mass loss data:

> time 172 min
> N_2 flow rate 7.9 sccm
> T = 1360 K
> N_2 pressure 0.7 atm
> NaCl mass loss 0.193 gm

auxiliary input data:

> average mol wt 72[a]
> $P_{Na_2Cl_2} / (P_{Na_2Cl_2} + P_{NaCl}) = 0.23$[a]

derived pressures:[b]

$$P_{NaCl_{total}} = [n_{NaCl}/(n_{NaCl} + n_{N_2})]P_{total} = 2.96 (\pm 0.6) \times 10^{-2} \text{ atm}$$

$$P_{NaCl} = 2.28 (\pm 0.4) \times 10^{-2} \text{ atm}$$

[a] Derived from JANAF (8) data and also self consistent with independent Knudsen and TMS data from the present study.

[b] n refers to moles of gas transported.

Method (b) is an adaptation of the integrated ion intensity method often used with Knudsen effusion cells (11). In the present case use is made of equation 4.1 and the relationship

$$k_{NaCl} = R(\tfrac{n}{V})\Delta t / \Sigma I \Delta t$$

where R is the Universal gas constant, n the number of moles of salt transported as NaCl species, and V the volume of carrier gas transported for time t and constant temperature T.

Method (c) utilizes the known pressure and ion intensity of the carrier gas (Ar or N_2), but relative mass discrimination factors and ionization cross sections are needed. For the relatively straightforward case of Na^+ as a measure of NaCl, ion intensities were converted to partial pressures as follows. The partial pressure of NaCl is given by,

$$P_{NaCl}(Na^+) = k_{NaCl} \cdot I_{Na^+} \cdot T, \qquad (4.1)$$

where k_{NaCl} is a constant which takes into account mass spectrometer sensitivity (k'), mass discrimination (S) and ionization cross section (σ) effects; I_{Na^+} is the measured Na^+ ion intensity at the sample temperature, T. The constant k_{NaCl} can be derived from a knowledge of the carrier gas pressure and ion intensity, i.e.,

$$P_{Ar} = k_{Ar} \cdot I_{Ar^+} \cdot T, \qquad (4.2)$$

and

$$k_{Ar} = P_{Ar} / (I_{Ar^+} \cdot T) \qquad (4.3)$$

then

$$k_{NaCl} = k_{Ar} \left(\frac{\sigma_{Ar}}{\sigma_{NaCl}}\right) \left(\frac{S_{Ar}}{S_{NaCl}}\right) \left(\bar{S}\right)^{-1}. \qquad (4.4)$$

The term $\bar{S}$ (≤ 1) is an empirical correction factor discussed below.

For the case of $NaCl^+$ as a measure of NaCl, an average R value, denoted as $\bar{R}$, was used as a scaling factor (necessary for the total ionization cross section argument to be valid) to avoid overwhelming the $NaCl^+$ variations by the large Na^+ contribution. The $P_{NaCl}(NaCl^+)$ data were obtained from the expression

$$P_{NaCl}(NaCl^+) = [(1 + \bar{R}) \cdot I_{NaCl^+}] k_{NaCl} \cdot T. \qquad (4.5)$$

Figure 7 shows the pressures derived from this expression to be equivalent to those using Na^+ ion intensity data [and Eq. (4.1)].

Values of σ for the reference gases N_2 and Ar are known from the literature (12). However, no such data exist for the high temperature species NaCl and $(NaCl)_2$. Expressions of the type 4.4 can be used to obtain ratios of ionization cross sections. Values of σ, relative to Ar, were measured for N_2, O_2, SO_2 and NaCl using the TMS technique and the results are summarized in table 3. The data for N_2 and O_2 agree with the literature results within the combined experimental errors. Also the measured σ's for SO_2 and NaCl seem reasonable in comparison with data for electronically similar species. To derive σ_{NaCl}, k_{NaCl} was obtained using expression 4.1 and the value of P_{NaCl} obtained by transpiration [method (a)].

Table 3. Ionization Cross Sections [πa_0^2 units (12)][a]

	30 eV		70 eV	
Ar	--	$(2.6)^b$	--	(4.0)
N_2	1.48	(1.2)	3.0	(2.8)
O_2	1.26	(1.0)	2.8	(2.8)
SO_2	1.3		--	
Na	--	(4.6 ± 0.3)	--	(4.0 ± 0.2)
NaCl	1.00	(± 0.3)^c	--	
$NaCl_2$	1.5^d			

[a] Experimental values determined from expressions of the type 4.4 using Ar as a reference σ_j.

[b] Literature values of Kieffer and Dunn (12) given in parentheses. Typical uncertainty ± 0.2.

[c] Based on $\bar{S} = 0.6$.

[d] Determined from literature empiricism for dimer to monomer cross section ratio of ~ 1.5 [Meyer and Lynch (13)]; probable uncertainty ± 0.2.

The empirical factor $\bar{S}$ contained in expression 4.4 corrects for a differential gas scattering effect. This effect which is characteristic of our TMS system, arises as follows. In the working pressure range of our system, the beam number density at the skimmer is only four times higher than for the background. Since the differential pump aperture does not completely limit the solid angle viewed by the ion source to beam species, some of the carrier gas (N_2 or Ar) admitted arises from the scattered background (in region I) and not the beam itself. This would not be the case for the condensible salt species. As the ion intensities of salt species are normalized against the carrier gas signal [by expressions (4.3) and (4.4)], as part of the pressure calibration procedure, they would be underestimated by the degree that scattered carrier gas competes with beam gas. Thus, in treating ion signals of condensible species relative to the carrier gas signal, we need to include an empirically determined multiplying factor of about 1.67 ($\bar{S}$ = 0.6) as a calibration constant, e.g., in expression 4.4. This factor is essentially invariant with total pressure since the ratio of background to beam pressure remains constant for our unthrottled pumping system. We have applied this correction factor to all <u>condensible</u> species partial pressures included in this study.

An independent test and verification of this correction procedure will be demonstrated for the Na_2SO_4 system in Section 5. We should emphasize, however, that even without this correction the agreement between our data and the JANAF (8) values is well within that usually expected for vapor pressures of complex high temperature systems.

Below the melting point, the sublimation data (Figure 7) fit the least squares expression:

$$\log P_{NaCl}(atm) = 7.8(\pm 1.6) - 11970(\pm 1500)T.$$

It follows that the enthalpy and entropy of sublimation (reaction R1) are, respectively,

ΔH_s = 54.8 ± 7 kcal/mol and

ΔS_s = 35.6 ± 8 cal/deg mol. at 1000 K.

For the liquid region, the corresponding data-fit is given by:

$$\log P_{NaCl}(atm) = 4.85(\pm 0.3) - 8820(\pm 200)/T,$$

and for reaction R2,

ΔH_v = 40.4 ± 0.9 kcal/mol and

ΔS_v = 22.2 ± 1.4 cal/deg mol at 1290 K.

The enthalpy of sublimation compares favorably with the JANAF (8) selection of 51.6 ± 2.4 kcal/mol where the uncertainty represents the standard deviation in the second law data analyzed by JANAF(8). For the liquid region, the literature data are based on total pressure measurements (i.e., not molecular specific) which have been corrected for dimer contribution by extrapolating mass spectrometric data obtained over solid NaCl. JANAF (8) selects $\Delta H_v = 42.7 \pm 3.5$ kcal/mol and $\Delta S_v = 24.2 \pm 0.5$ cal/deg mol at 1290 K for reaction R2. Both sets of data are in very good agreement.

The key uncertainty in the present data arises from the cross section and $\bar{S}$ terms in expression 4.4. However, the agreement between NaCl pressures obtained by the methods (a) - (c) at 1360 K and a subsequent check on $\bar{S}$ using Na_2SO_4 strongly support the pressure calibration procedure and hence the entropy data.

In view of the sparsity of molecular specific measurements in the liquid range by other investigators, the general agreement between JANAF (8) and our data is excellent. Further, this appears to be the first report of measurements in the temperature range from the melting point to 1250 K. This result demonstrates an additional advantage of TMS over the classical effusion (e.g., Knudsen effusion mass spectrometry) and total pressure (e.g., conventional transpiration) methods in its ability to bridge the $10^{-4} - 10^{-3}$ atm gap in existing partial pressure measurement capability at high temperature. The temperature interval of > 450 K covered by our measurements is exceptionally large, and more recent measurements have been made over an even wider range.

4.3 <u>NaCl dimerization and test for homogeneous equilibrium</u>. The NaCl sublimation and vaporization reactions represented by the data in figure 7 provide a convincing test of the attainment of heterogeneous equilibrium, for condensible vapor species, in the transpiration reactor and their faithful sampling into the mass spectrometer. However, in the absence of additional experimental insight, it is conceivable that the sampling process could perturb homogeneous vapor phase chemical equilibria. A sensitive test of the system fidelity in this regard can be made using the known equilibrium represented by reaction R7.

It is well established from the literature, and our results support this, that the Na_2Cl^+ ion may be used as a measure of the dimer species $(NaCl)_2$. Our appearance potential and temperature dependence data indicate that Na_2Cl^+ is a fragment ion of $(NaCl)_2$ and that any Na^+ or $NaCl^+$ arising from dimer electron impact fragmentation is negligible (for present purposes). The equilibrium constant for the dimerization process (R7) is then given by:

$$K_d = \frac{I_{Na_2Cl^+}}{(I_{Na^+} \text{ or } I^*_{NaCl^+})^2 k_{Na_2Cl_2} \cdot T} \quad (4.6)$$

where $k_{Na_2Cl_2}$ includes the relative ionization cross section, mass spectrometer sensitivity, differential gas scattering and beam segregation factors for the monomer and dimer species, as discussed previously. The term $I^*_{NaCl^+}$ refers to the scaled $NaCl^+$ ion intensity, i.e., as in expression 4.5. In practice, where I_{Na^+} and $I_{Na_2Cl^+}$ are used as measures of monomer and dimer, respectively, $k_{Na_2Cl_2} = 4.0 \, k_{N_2}$.

Typical data sets for K_d as a function of temperature are given in figure 8. These data were obtained using both the conical and the capillary-type samplers. Note the satisfactory agreement between the various K_d data sets and the JANAF (8) literature curve, indicating good control over the various effects discussed earlier as contributing errors in converting ion intensity to partial pressure, or equivalent equilibrium constant, data. Both the I_{Na^+} [as per Eq. (4.1)] and the scaled $I^*_{NaCl^+}$ [as per Eq. (4.5)] signals were used as measures of P_{NaCl}. The least squares line shown in figure 8 is an average of individual least squares fits to each data set, giving

$$\log K_d = 10640 \, (\pm 400)/T - 6.97 \, (\pm 0.4).$$

The calculated enthalpy of dimerization (reaction R7) at 1265 K is then $\Delta H_d = -48.7 \, (\pm 1.8)$ kcal/mol, which is in good agreement with the selected JANAF (8) value of -47.1 kcal/mol, particularly since the corresponding literature data span a difference of more than 7 kcal/mol [see JANAF (8)]. Our data for the conical probe were obtained at carrier gas pressures varying over the range 0.4-0.6 atm and gas flow rates in the range 3-10 sccm. The capillary probe data, were obtained at pressures of 0.3-0.5 atm and gas flow rates in the range 15-40 sccm. That there is agreement between these data sets over a range of flow rates suggests saturated flow conditions in the transpiration reactor.

The isentropic expansion model discussed earlier (see Section 3.3) for the capillary sampler suggested that a temperature decrease of a few percent might occur in the early gas expansion phase along the capillary, while the velocity in the channel was

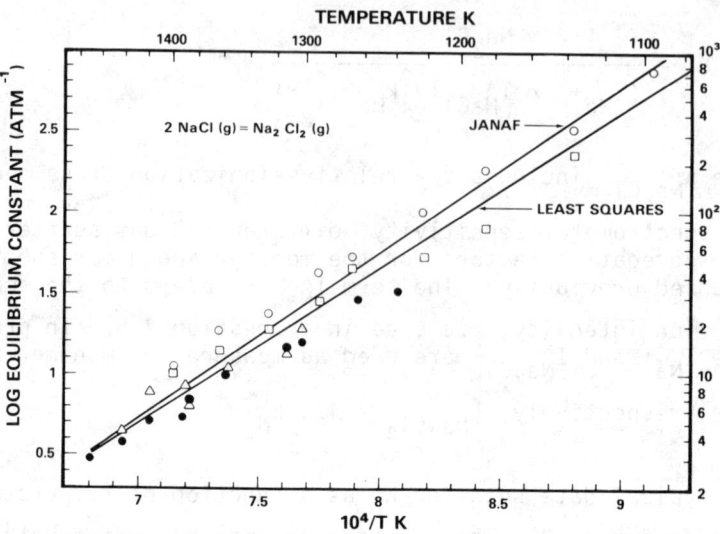

Figure 8. TMS data, at 30 eV with N_2 carrier gas for NaCl monomer-dimer equilibrium. Open circles and squares using capillary sampler at 0.3-0.5 atm with Na^+ and $NaCl^+$ (scaled--see text) as measures of P_{NaCl}, respectively. Closed circle and open triangle obtained with conical sampler, 0.4-0.6 atm, using Na^+ and $NaCl^+$ (scaled--see text), respectively [see Eq. (4.6) in main text].

still low enough that a re-adjustment of simple gas equilibria might be kinetically feasible. However, the generally good agreement between the capillary and conical samplers, and between the JANAF (8) literature data, suggests that during the early expansion period the isothermal model is a better description of capillary flow. That is, isentropic expansion dominates only when the remaining residence time in the capillary is a few microseconds and the flow velocity approaches unit Mach number. Direct computer modeling is required to predict more definitively the early flow conditions for this type of sampling orifice. However, the empirical approach adopted seems to satisfy present requirements.

4.4 <u>Comparison of Knudsen effusion and TMS-NaCl dimerization data</u>. The various Knudsen effusion and TMS data sets have been replotted in the form of K_d (R7), as shown in figure 9. Note (in Figure 9) that our Knudsen effusion data extrapolate almost perfectly to the liquidus data curve obtained by TMS. Thus both the Knudsen effusion and TMS data sets appear to be more consistent with each other than with the JANAF (8) data, though the difference is not disturbingly great and is almost within the combined data

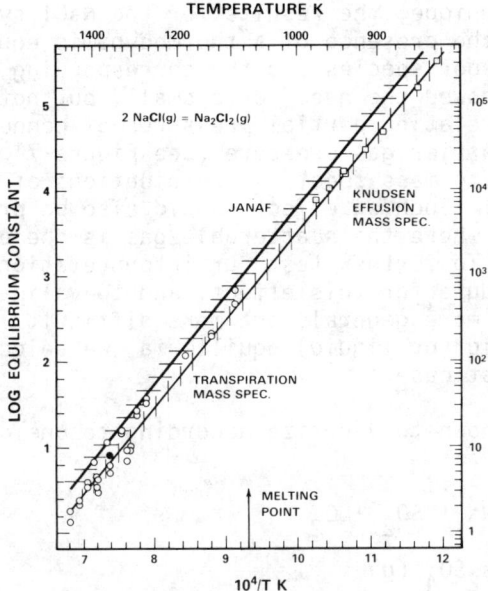

Figure 9. Comparison of various data sets for NaCl dimerization (reaction R7). Horizontal and vertical bars indicate (to a good approximation) the slope uncertainties for the JANAF (8) and experimental curves, respectively (see text). The closed circle data point was obtained using calibration method (a).

uncertainties (see Figure 9). Such agreement between widely differing measurement techniques clearly establishes the TMS method as reliable and accurate, and capable of extending the pressure range of conventional effusion methods by four orders of magnitude or greater. The temperature range is also naturally extended by such a high pressure sampling capability.

5. The Na_2SO_4 System

5.1 Background. Most practical systems adaptable to characterization by the TMS technique are heterogeneous, involving equilibria between reactive gases and volatile substrates. Experience with Knudsen effusion mass spectrometric studies of gas-solid (or liquid) systems has indicated a need to exercise great care in achieving, and then demonstrating, thermodynamic equilibrium. On the other hand, considerable experience has shown that equilibria involving only condensible vapor species and their substrates are generally attainable in Knudsen cells.

With the TMS technique, the results for the NaCl system clearly demonstrated the presence of a thermodynamic equilibrium between condensible vapor species and the corresponding substrate. However, these data showed the need for a small, but noticeable, correction factor in relating partial pressures of condensible NaCl species to the carrier gas pressure (see Figure 7). This effect was attributed to mass spectral contributions of non-beam scattered carrier gas. Such an effect should also be present in heterogeneous systems where the scatterable gas is one of the reaction components. To further test our interpretation and data correction procedure for this effect, and to validate the TMS technique for the more general, but more difficult, case of heterogeneous gas-solid (or liquid) equilibria, we selected the Na_2SO_4 system as a test case.

This system is known to vaporize according to one or more reactions:

$$Na_2SO_4(s, l) = 2Na + SO_2 + O_2 \qquad \text{R9}$$

$$Na_2SO_4(s, l) = Na_2SO_4 (g) \qquad \text{R10}$$

$$Na_2SO_4(s, l) = Na_2O \text{ (soln or gas)} + SO_2 + 1/2\ O_2 \qquad \text{R11}$$

$$SO_3 = SO_2 + 1/2\ O_2 \qquad \text{R12}$$

Historically, there has been considerable disagreement among researchers concerning the relative contribution of each reaction to the overall vaporization process. It now appears that container reactions (catalytic or otherwise), temperature, sample treatment, and gas composition (e.g., external O_2 or SO_2 present), all contribute to the overall process defined by reactions R9-R12. For the high temperature (> 1360 K) conditions of interest to the present study, we can readily eliminate reaction R12 as a contributing process.

For most conditions, reaction R9 is usually predominant. This is a very convenient reaction to test TMS sampling of gas-condensed phase equilibria. In addition to carrying out temperature dependence studies for the isolated system, it is possible to externally control the concentration of product species SO_2 and O_2 and provide an isothermal mass action (or Le Chatelier effect) test for equilibrium. With our particular design of external gas supply multiple mass flow meter/controller system we were able to introduce either SO_2, O_2 or their mixtures at concentrations, with respect to the N_2 or Ar carrier gas, in the range of ∼ 0.02-4 percent. Usually we maintained a constant O_2 pressure and varied the SO_2 concentration. The capability for varying the SO_2 partial pressure over a range of two orders of magnitude was more than adequate for a mass action test of equilibrium sampling.

5.2 **Closed system temperature dependence.** Under closed system conditions, i.e., no external reactive gas load applied, we observed reaction R9 as the predominant process in the temperature range ca. 1400-1700 K. The measured temperature dependence of the equilibrium constant,

$$K_p(R9) = P_{Na}^2 \cdot P_{O_2} \cdot P_{SO_2} \tag{5.1}$$

is shown in figure 10, where an excellent agreement with the JANAF (8) curve is indicated. The right-hand side axis represents an equivalent partial pressure of sodium calculated from the observed $K_p(R9)$ using $P_{Na} = [4K_p(R9)]^{1/4}$. This is a useful representation of $K_p(R9)$ for making comparisons with the literature data.

The partial pressures were obtained using the established relation,

$$P_i = k_i I_i^+ T$$

for conversion of the ion intensities for Na^+ (23 amu), SO_2^+ (64 amu), and O_2^+ (32 amu). k_i was derived from k_{Ar} or k_{N_2} at each temperature T where

$$k_{Ar} = \frac{P_o \cdot f_i}{I_{36_{Ar}} \cdot T} \text{ and } k_{N_2} = \frac{P_o \cdot f_i}{I_{29_{N_2}} \cdot T},$$

P_o is the source pressure of carrier gas (Ar; N_2),
$I_{36_{Ar}}$ the ion intensity of the argon isotope at 36 amu,
$I_{29_{N_2}}$ the ion intensity of the nitrogen isotope at 29 amu,
$f_i = 0.0034$ is the ^{36}Ar isotope fraction of argon gas, and
$f_i = 0.0072$ is the $^{29}N_2$ isotope fraction of N_2 gas.

A typical data point for figure 10 is given by the experimental parameters: T = 1563 K, P_{Ar} = 0.81 atm, $I_{36_{Ar}}$ = 26 x 10³, I_{Na} = 2100, I_{SO_2} = 500 and I_{O_2} = 2300 µV. These data agree within experimental error with the JANAF $K_{Na_2SO_4}$ data if a

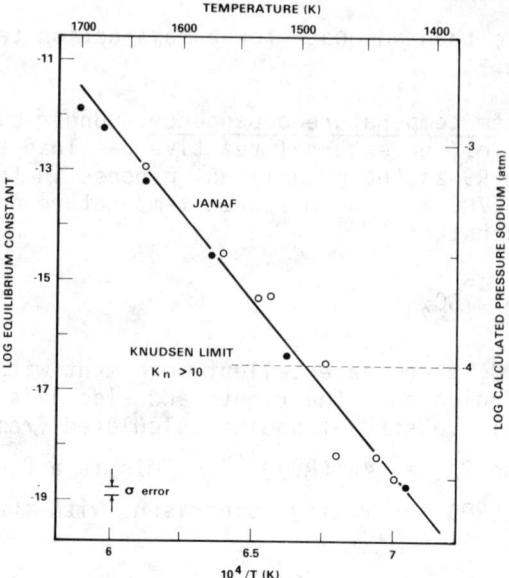

Figure 10. Temperature dependence of the equilibrium constant for reaction $Na_2SO_4(l) = 2Na + SO_2 + O_2$ from TMS observations of Na^+, SO_2^+ and O_2^+. Data taken at 30 eV ionizing electron energy using Ar carrier gas at pressures of 0.4 atm (closed circles) and 0.5-0.8 atm (open circles). The solid line is taken from JANAF (8). The Knudsen limit was estimated for a cell with ~ 0.1 mm orifice diameter. The right-hand side axis has been calculated, assuming congruent vaporization, as

$$P_{Na} = [4K_p(R9)]^{1/4}.$$

scattering factor of $\bar{S} = 0.6$ ($\pm\ 0.06$) is assumed. That this factor has the same magnitude as for the NaCl study tends to establish its use as a universal correction factor for our particular TMS system.

The constants k_i are then derived as follows. For the case of k_{SO_2}, use can be made of the system response to a known external pressure of SO_2. A linear response of the mass spectrometer system to known concentrations of SO_2 was found, thereby yielding a direct relationship between observed intensity and true pressure, i.e.:

$$\frac{P_{SO_2}(obs)}{P_{SO_2}(appl)} = \frac{k_{N_2} \cdot I_{SO_2}^+ \cdot T}{k_{SO_2} \cdot I_{SO_2}^+ \cdot T} = \frac{1}{1.55} ;$$

hence, $k_{SO_2} = 1.55\, k_{N_2}$ (at 30 eV). With the mass discrimination function of figure 5, the cross section data of table 3, and the scattering correction (multiplication factor of 1.67) obtained from the NaCl TMS data, the equilibrium constant is then

$$K_p(R9) = 4.8(k_{Ar}T)^4 \, (I_{Na^+})^2 \cdot I_{SO_2^+} \cdot I_{O_2^+} \quad . \tag{5.2}$$

The data represented in figure 10 were obtained at various argon carrier gas pressures P_o, in the range of 0.4-0.8 atm. Separate least squares fits to the low (0.4 atm) and high (0.5-0.8 atm) P_o data points yield, respectively (in terms of Na):

$\Delta H_v = 69 \pm 1$ and 71 ± 4 kcal/mol

$\Delta S_v = 27.5 \pm 1.5$ and 28.1 ± 4 cal/deg mol (for reaction R9).

The good fits within and between various total pressure-ranges is indicative of saturated sampling conditions since the carrier gas flow rate is porportional to P_o at constant T. Also, the excellent agreement with the JANAF (8) data for both ΔH_v and ΔS_v (evident in Figure 10) is consistent with a non-perturbing sampling process.

5.3 <u>Open system, isothermal variable SO_2 conditions</u>. The effect of variable SO_2 pressure on the equilibrium constant for reaction R9 at near isothermal conditions (1465 ± 7 K) is shown in figure 11. Note that the least squares curve has zero slope [required by definition of $K_p(R9)$] and is located very close to the JANAF (8) curve. There are two contributing factors to the data scatter observed in figure 11. First, during the runs, the VARIAC driven furnace system allowed a temperature variation of ± 7 K, due partly to fluctuations in line power and to interaction of variable background SO_2 with the thin-foil tantalum furnace. The second factor resulted from a slow response of $I_{SO_2^+}$ to changes in applied P_{SO_2} from high to low SO_2 concentrations. This slow response is believed to result from SO_2 dissovling, and being retained, in the liquid Na_2SO_4. In spite of these effects, the agreement between the TMS data and the JANAF (8) results is very good and the invariance of $K_p(R9)$ with applied SO_2 pressure is clear evidence of thermodynamic equilibrium and a faithful sampling process.

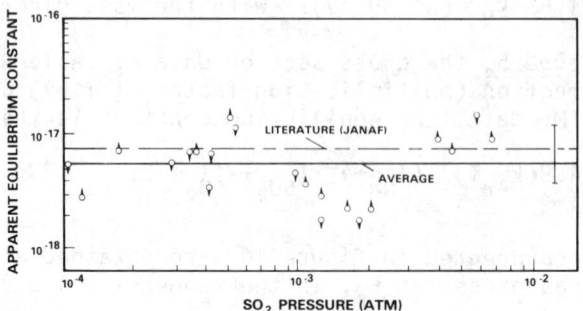

Figure 11. Isothermal equilibrium behavior of $K_p(R9)$ with variable SO_2 partial pressure at 1475 (± 7) K, using N_2 carrier gas at 0.5 atm and 30 eV ionizing electron energy. Error bar on JANAF (8) value (broken line) represents the change in $K_p(R9)$ for the experimental temperature variation of ± 7 K. The solid line is the average of all measurements; a two parameter fit yields a similar result with a statistically insignificant pressure dependent term. Arrows on data symbols indicate the direction of change of applied SO_2 pressure. Three symbols with double arrows denote the turning points of the two cycles of variation in SO_2.

PART II: APPLICATION TO ALKALI VAPOR TRANSPORT THERMOCHEMISTRY[2]

1. Introduction

Vapors containing alkali metal species have diverse implications to high temperature processes (14). Potential new applications of alkalies in combustion systems include--their vapor phase catalytic action in smoke reduction [see Haynes et al. (15) and the review of Hastie (1)], their liquid phase catalysis of coal gasification (16), and their role as electron sources for magnetohydrodynamic (MHD) combustion systems. In most combustion systems, however, their presence is undesirable. This is particularly true in fossil energy systems.

More efficient coal utilization can be realized with combined power plant cycles. For instance, the post combustion gases of a conventional combustor or an advanced MHD system can be further utilized to drive a gas or steam turbine. However, the sustained durability of downstream turbine or heat exchanger components requires minimal transport of corrosive fuel impurities. Control of mineral-derived impurities is also required for environmental protection. For the special case of open cycle-coal fired MHD systems, the thermodynamic activity of potassium is much higher in the seeded combustion gas (plasma) than in common coal minerals and slags. This results in the loss of plasma seed by slag absorption and is of critical concern to the economic feasibility of MHD.

Empirical experience with conventional coal-fired power plants has indicated minerals containing alkali metal (Na, K), sulfur- and chlorine-bearing species to be the most aggressive fuel components leading to fire-side or hot corrosion (17). Species containing these elements appear to act synergistically in degrading alloy or ceramic materials. The mechanisms by which such species are released from their mineral source, transported, and deposited are not known, though the literature contains numerous speculative schemes [see p. 216 (1)]. Rational development of new control strategies, such as gas clean-up or the use of fuel additives, requires a clear understanding of the role played by the active fuel impurities. For instance, new control systems based on scavenging (e.g., absorption of alkali by glass or other oxide media) or chemical modification of the active inorganic impurities will need as design criteria information, such as species identity, concentration profiles, dew points, thermodynamic reactivity, nucleation and absorption rates and diffusivities. Such data will also be pertinent to minimization of seed-slag interaction in MHD systems.

Previous attempts to define the mode of release and transport of fuel impurities have largely been unsuccessful, owing mainly to a lack of knowledge concerning species identities. This has resulted from the inability of molecular specific measurement techniques to function under the combined aggressive conditions of high temperature, high pressure, and high chemical reactivity. The Transpiration Mass Spectrometric (TMS) technique is, however, ideally suited to these conditions. Examples of alkali-containing systems, analyzed by TMS, are presented as follows. These include hydroxide, halide, glass, carbonate, mineral, synthetic, and actual coal slag materials. In some of these examples comparison is made with data obtained by the classical Knudsen effusion mass spectrometric (KMS) technique. A detailed account of this method and the apparatus used has been given elsewhere by Plante (11). Comparisons between KMS and TMS data are particularly useful as the cell residence times, $\sim$ 0.04 and 20 s respectively, differ considerably. Agreement between these two techniques almost certainly ensures establishment of thermodynamic equilibrium.

2. KOH(ℓ) Vaporization

The thermodynamic stability of KOH in the vapor phase can be obtained from the Second and Third Law analyses of KOH(ℓ) vaporization as the thermodynamic functions for the liquid phase are reasonably well established. JANAF (8) has evaluated the various disparate sets of KOH vaporization data but with considerable uncertainty. Much of the difficulty associated with obtaining reliable thermodynamic data for this system arises from its reactivity with container materials, the presence of carbonate impurity, and the coexistence of dimers and monomers. Previous studies have also been hampered by decomposition to K and H_2O. In the present work, using the TMS technique, this decomposition was suppressed by addition of H_2O to the carrier gas.

We have obtained extensive data for the KOH(ℓ) system, which will be presented in a formal publication elsewhere (18). Representative data for KOH(ℓ) are presented here in comparison with other recent results not considered by JANAF (8). Species partial pressure data are summarized in figure 12. Note that the KOH species data are in good agreement with JANAF (8), as might be expected. However, there is no agreement between workers regarding the $(KOH)_2$ species, except that the <u>relative</u> amounts of dimer to monomer found in the present study agree quite well with the KMS results of Gusarov and Gorokhov (19). When the monomer and dimer partial pressures are summed, the total pressures are about a factor of two greater than the JANAF (8) data.

3. Dolomite Vaporization

Transpiration mass spectrometry (TMS) and Knudsen effusion (KMS) experiments were carried out on an NBS Standard dolomitic limestone, SRM 88a. Typical data are shown in figure 13. Note that these alkali pressures exceed the turbine tolerance limit of $\sim 1 \times 10^{-7}$ atm. From the slope of figure 13, a value of 58 kcal/mol is derived for the apparent vaporization enthalpy of K. As the TMS and KMS data are essentially identical and as the cell residence times for the two techniques vary by several orders of magnitude, we consider the system to be at thermodynamic equilibrium.

4. Vaporization in The Soda-Lime-Silica Glass System

4.1 <u>Background</u>. A commercially common soda-lime-silica glass has been considered as an absorbing medium for removing fly ash particulates in combustion gas streams (20,21). However, a possible limitation with this application is the release of alkali from the glass into the gas stream. Glass also has some common features with coal slag, and the basic thermodynamic data derived from glass-combustion gas studies will benefit our basic understanding of slag-gas interactions.

Description of alkali release, or retention, by glass requires accurate Na_2O activity data at various temperatures and

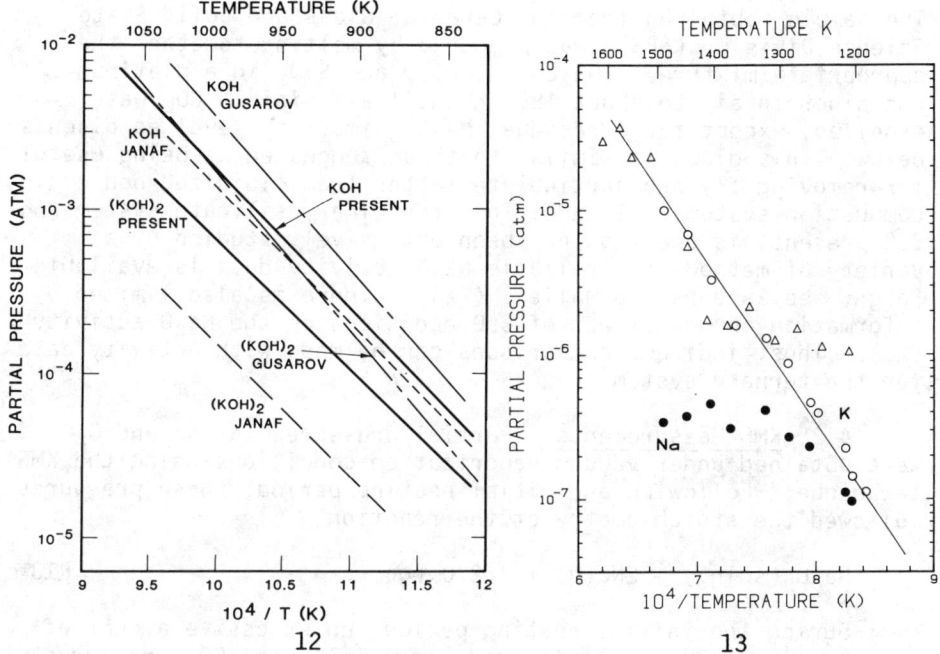

Figure 12. Partial pressure data for KOH and $(KOH)_2$ over liquid KOH, obtained by TMS. Curves labelled Gusarov refer to data of Gusarov and Gorokhov (19). JANAF curves are from (8).

Figure 13. Partial pressures of K and Na as a function of reciprocal temperature for dolomitic limestone. The solid curve represents the best fit to the Knudsen K pressure data (open circles); triangles represent TMS K-pressures; closed circles represent KMS Na-pressures; and open squares and dashed line are solution model predictions. Dolomite composition (wt.%) was as follows: SiO_2 (1.2), Al_2O_3 (0.19), Fe_2O_3 (0.28), CaO (30.1), MgO (21.3), Na_2O (0.01), K_2O (0.12), CO_2 (46.6).

glass compositions. However, thermodynamic activity data for glass systems are surprisingly sparse. No critical analysis, for instance of the type represented by JANAF Thermochemical Tables (8), has been made on the available data. Glass activity data for a common system can vary by several orders of magnitude, or more, depending on the measurement method used.

In the present study, we have utilized the KMS and TMS methods to obtain vaporization and activity data for a glass of initial composition (wt.%): Na_2O(17), CaO(12), and SiO_2(71).

The sample, obtained from the Ceramic, Glass and Solid State Science Division (NBS), was prepared by melting together the appropriate mixture of Na_2CO_3, $CaCO_3$, and SiO_2 in a platinum container in air to about 1800 K until all visible CO_2 was expelled, except for a residual Na_2CO_3 impurity level as discussed below. This glass is similar to those suggested as being useful for removing fly ash particulate matter from fluidized bed coal combustion systems. In addition, the binary silicate system (no CaO present) is one that has been extensively studied by a variety of methods and reliable Na_2O activity data is available [e.g., see Sanders and Haller, (22)]. There is also limited information on the effect of CaO additions on the Na_2O activity (23). Thus, indirect comparisons can be made with activity data for the ternary system.

4.2 <u>KMS Measurements</u>. Partial pressures for Na and O_2 were obtained under vacuum vaporization conditions using the KMS technique. Following an initial heating period, these pressures followed the stoichiometry of the reaction,

$$Na_2O \text{ (soln.)} = 2Na(g) + 1/2\ O_2(g) \ . \qquad \text{R13}$$

During the initial heating period, an excessive amount of Na, as well as CO_2, was observed. Initially the CO_2 pressures approximated those for O_2 but decreased to a negligible level ($CO_2/O_2 < 0.1$) in subsequent experimental runs. Two processes appeared to be controlling the release of CO_2,

$$Na_2CO_3 \text{ (soln.)} = 2Na + 1/2\ O_2 + CO_2 \quad \text{and} \qquad \text{R14}$$

$$Na_2CO_3 \text{ (soln.)} = Na_2O \text{ (soln.)} + CO_2 \ , \qquad \text{R15}$$

with the former predominating in the early and lower temperature phase of the experimental runs, and the latter at higher temperatures and later observation times.

Figure 14 summarizes Na partial pressure curves for two experimental runs. Experimental details have been presented elsewhere (20) and the data points basically followed the curve (labeled PLANTE) given in figure 14. Several data sets from the literature are indicated for comparison. Neudorf and Elliott (23) measured Na_2O activities in the binary silicate solution, as well as the effects of CaO on the Na_2O activity, using an emf method. We have extrapolated their data to our experimental conditions based on the effects of Na_2O and CaO content on the Na_2O activity. The data point of Cable and Chaudhry (24) was obtained by a classical transpiration method under conditions where surface segregation effects were negligible. Similarly, the data point of Sanders et al. (25) represents a stirred-melt transpiration experiment where surface depletion is also unlikely.

The curve of Argent et al. (26) represents Knudsen effusion mass spectrometric data (without beam modulation).

4.3 **TMS Measurements**. Glass vaporization in a N_2 atmosphere was monitored using the TMS technique. Representative Na-partial pressure curves are given in figure 14. Note that these partial pressures are more than an order of magnitude greater than those obtained by the KMS technique. Ideally, both sets of data should coincide. We believe that the explanation for this apparent discrepancy is as follows.

Under the conditions of the KMS experiments, the rate of alkali removal was about an order of magnitude greater than for TMS. During the initial phase of each type of experiment, excessive amounts of CO_2 and Na were released. Only when about one percent of the glass Na_2O was depleted did the excess CO_2 and Na become negligible in the KMS experiments. As this level of alkali depletion was never reached during the TMS experiments, we believe that these latter data correspond to the anomalously high Na pressures found in the early phase of the KMS experiments. These high alkali pressures can be attributed to the presence of unreacted Na_2CO_3 impurity in the original glass samples, even though care was taken to avoid this in the glass preparation. Residual carbonate impurity is a common problem with glass experimentation [e.g., see (24)].

We calculate from the time-integrated CO_2 and excess Na signals that the initial concentration of impurity Na_2CO_3 was 0.45 wt.%. From the relative amounts of Na, O_2, and CO_2 released prior to vaporization from the silicate itself, two types of impurity-related vaporization processes appear to be present, as represented by reactions R14 and R15. Comparison of the TMS and KMS data indicates that the activity of Na_2O (solution), produced by reaction R15, is substantially greater than that for the silicate-bound Na_2O characteristic of the pristine glass. Apparently, the Na_2O produced *in situ* by carbonate decomposition is not readily incorporated into the silicate matrix, at least on the time scale of the vaporization measurements. Formally, we can consider Na_2CO_3 as a solute, in a metastable glass solution, and with an activity defined by reaction R14, as discussed in the following Section. Alternatively, one could argue that under the higher vaporization rate conditions typical of KMS, surface depletion of alkali led to the relatively low Na-pressures observed, e.g., see (24). However, no significant isothermal time dependent vaporization was noted on the several minute time scale of individual KMS measurements. Also, the KMS pressures are greater than the stirred-melt data of Sanders et al., (25) which would not be the case in a surface-alkali depleted system.

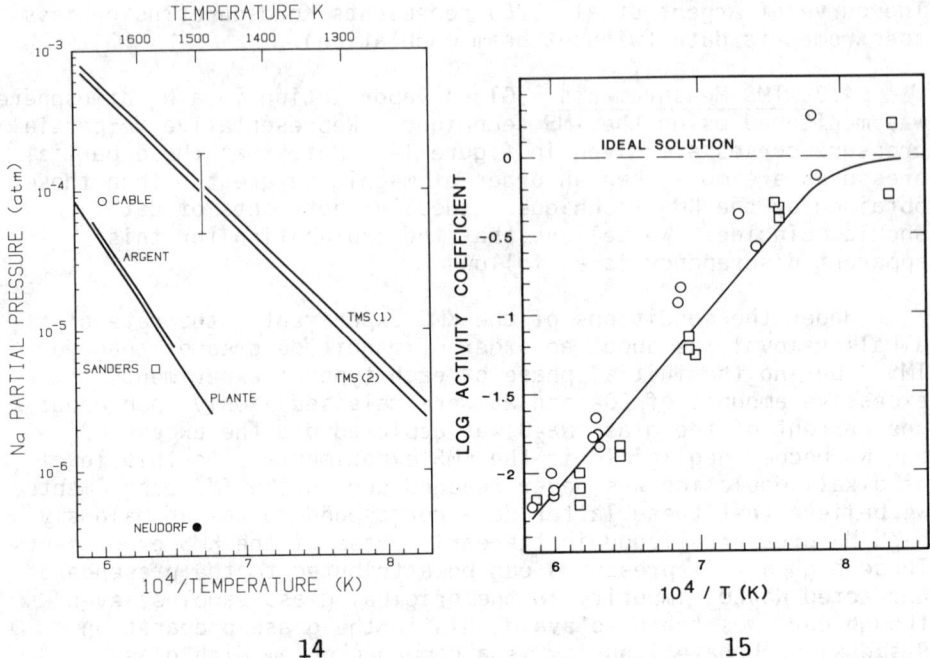

Figure 14. Comparison of glass melt Na partial pressure data obtained by various workers (see text) for compositions (wt.%) similar to $Na_2O(16)$, $CaO(12)$, and $SiO_2(72)$. KMS(1), 17.0 to 16.7 and KMS(3), 15.6 to 13.3 wt.%: TMS(1), 17.0 and TMS(2), 16.9 wt.%.

Figure 15. Activity coefficient data (TMS) for Na_2CO_3 (0.45 wt.%) in glass (see caption 14 for glass composition). Open circles and squares refer to run-chronology of increasing and decreasing temperature, respectively.

4.4 Na_2CO_3 in Glass. We have interpreted the anomalously high alkali vapor pressures of the TMS experiments described above in terms of impurity Na_2CO_3 decomposition in a glass solution. By monitoring the release of CO_2, and integrating over time, we have determined the mole fractions of Na_2CO_3 present at the various measurement temperatures and times. Hence, from the observed partial pressures for reaction R14, and the corresponding reference state values (8), we can calculate Na_2CO_3 activity coefficient data, as shown in figure 15. These data appear to be thermodynamically reasonable and tend to support the alkali carbonate impurity interpretation of alkali vaporization differences between the KMS and TMS experiments. We can likewise argue that the data of Cable and Chaudhry (24), shown in figure 14, also appear to suffer from this impurity problem, even though they also took precautions to eliminate residual carbonate during the glass synthesis process.

Future studies should be pursued under controlled doping conditions and in atmospheres containing CO_2 and O_2. The known synergistic effect of CO_2 on O_2-solubility in silicate melts at very high gas pressures has, in fact, been interpreted in terms of Na_2CO_3 formation in solution. Effects of this type could significantly enhance alkali vapor transport in practical combustion systems.

5. MHD Channel Slag (K_1) Vaporization

Detailed TMS and KMS studies were made of vapor transport over a high liquidus temperature ($\sim$ 1700 K) potassium-enriched coal slag with initial composition (wt.%): K_2O (19.5), Al_2O_3 (12.1), Fe_2O_3 (14.3), CaO (3.8), MgO (1.0), SiO_2 (46.8), Na_2O (0.5). This slag sample was obtained by combustion of Illinois No. 6 coal with additional potassium added to the combustor [see Hastie et al. (27)]. Note that this slag composition lies between those of the "Eastern" and "Western" coal-types. For identification purposes, this slag is given the designation K_1. X-ray diffraction data indicated that the bulk of the slag potassium was present as the compound $KAlSiO_4$. TMS analysis indicated that about two percent of the slag potassium was present in relatively volatile form, mainly K_2SO_4 and K_2CO_3.

5.1 <u>Identity of Volatile Species</u>. The as-received potassium-enriched coal slag was subjected to a series of heating cycles (runs) in nitrogen carrier gas. During the initial heating cycle, mass spectral scans, obtained using the TMS technique, revealed many volatile species in addition to the expected K and Na species. The following species were positively identified: H_2O, CO_2, SO_2, O_2, K, and Na. Some of the other low-intensity ion signals can be very tentatively assigned to the species (some hypothetical): KO or KOH, KS or KSH, SiS, SiSH, H_2S, H_2SO_4, and KSiO. From JANAF (8), we can expect to see KOH under these conditions. Some of these more minor species may result from slag occlusions and metastable phases and most likely do not represent an equilibrium release from the slag. Following this initial heating cycle, the only significant slag vapor species were K and O_2, and these were present in the approximate stoichiometric ratio expected for K_2O decomposition.

5.2 <u>Initial Species Partial Pressure--Temperature Dependence</u>. The initial volatiles showed a non-monotonic variation of partial pressure with temperature, as shown in figure 16. These volatiles constitute only a few percent of the total slag components and are not representative of the bulk slag composition. However, they do provide a sufficiently high flux of alkali (Na, K) and

SO_2 to be a potential source of corrosion in downstream MHD components. The high initial partial pressures of SO_2, CO_2, K, and Na are indicative of the presence of alkali sulfate and carbonate in the slag. An additional contribution to low temperature (T < 1300 K) alkali release could result from the high H_2O content leading to the formation of volatile hydroxide species (KOH). However, no definitive hydroxide signals were observed. Note that at T > 1400 K, the potassium pressures fall below those expected from $KAlO_2$, but that the SO_2, CO_2, and H_2O pressures are still relatively high. Apparently, at this stage, the K produced by sulfate and carbonate decomposition is retained in the bulk slag. After further heating, the sample was virtually depleted of Na, SO_2, and CO_2; H_2O also continued to fall-off in pressure to a negligible level. Following this initial clean-up period, the sample showed a more normal vaporization behavior and representative data are given in figure 17.

5.3 Potassium Partial Pressure--Temperature Dependence.

The variation of K-partial pressure was followed over a wide range ($\sim$ 1150 to 1820 K) using both the TMS and KMS techniques. Usually, the O_2 partial pressure tracked with the K data (but not always in stoichiometric proportion) indicating the main vaporization process to be:

$$K_2O(slag) = 2K + 1/2\ O_2 , \qquad\qquad R16$$

with possible secondary contributions, as discussed later. At temperatures corresponding to K partial pressures of 10^{-5} atm, or less, the rate of loss of K_2O from the bulk was sufficiently low that the data represent constant composition conditions. At higher temperatures, and vaporization rates, the data are greatly modified by the effects of changing slag composition.

Representative vaporization data are given in figure 17. By combining data from both the TMS and KMS methods we were able to cover a wide range of temperature and K_2O mole fraction. The rate of vapor transport differs appreciably for the two methods, with KMS losing about three times as much material as for TMS as equal K pressure. Thus the bulk slag composition changes more rapidly during a KMS experiment. The N_2 carrier gas used for the TMS experiments also contained a small, but significant, amount of O_2 (typically $\sim$ 5 x 10^{-5} atm). This caused a suppression of the K-vapor pressure by a reversal of reaction R16. When this effect is taken into account, using the observed thermochemical data for reaction R16, good agreement is found between the TMS and KMS data, e.g., see the comparison temperature (1561 K) and composition point in figure 17. That such an agreement is possible is a strong indication of thermodynamic equilibrium, at least with respect to K transport, as species residence times in Knudsen cells and transpiration tubes differ appreciably.

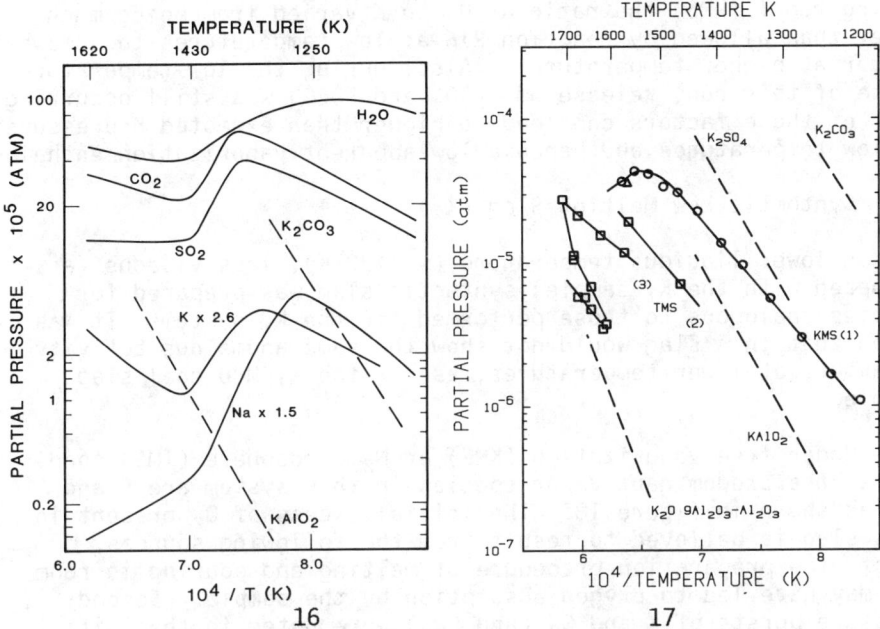

Figure 16. Partial pressure variation of initial volatiles (K_2O, 19.5 to 19.1 wt.%) as a function of temperature (and time) for the K_1 slag (liquidus temperature ∼ 1700 ± 30 K) using the TMS approach (run 1). Conditions: 0.5 atm N_2, capillary probe. Dashed comparison curves represent K-pressures over the K_2CO_3 and $KAlO_2$ phases.

Figure 17. Variation of K-partial pressure with temperature for K_2O contents of 19.1 to 17.8 wt.% (run 1--open circles, KMS), 18.17 to 18.07 wt.% (run 2--open squares, TMS) and 18.07 to 18.0 (run 3, open squares, TMS). Corresponding pressure curves (literature data) for the phases $K_2CO_3(l)$, $K_2SO_4(l)$, $KAlO_2(s)$, and $K_2O \cdot 9Al_2O_3$--$Al_2O_3(s)$ are given for comparison. The triangular point at 1561 K is a TMS data point corrected for the presence of O_2 (see main text). Each curve represents a separate experimental run. Usually, data were obtained for successive increases in temperature, except for run 3--TMS where the arrows indicate the run chronology, i.e., T increasing or decreasing. TMS conditions, 0.5 atm N_2, capillary probe.

Several processes appear to be contributing to the relatively shallow slopes of the curves presented in figure 17, which under equilibrium conditions represent enthalpies of vaporization.

During run 1 (KMS) the ratio of O_2 to K varied from being much lower than allowed by reaction R16 at low temperatures to somewhat larger at higher temperatures. Also, during the low temperature phase of this run, release of K_2SO_4 and K_2CO_3 was still occurring. Each of these factors can lead to higher than expected K-pressures at low temperatures and hence a low apparent vaporization enthalpy.

6. Synthetic Low Melting Slag (K_2)

A lower liquidus temperature (~ 1480 K), less viscous (as compared with the K_1 sample) synthetic slag was prepared for studies analogous to those performed for the K_1 system. It was hoped that this slag would not show the same anomalous activity behavior, at lower temperatures, as for the K_1-MHD coal slag sample.

Under free vaporization (KMS) or N_2-atmosphere (TMS) conditions, the predominant vapor species in this system are K and O_2, as shown in figure 18. The initial excess of O_2 present in this slag is believed to result from the following sources. First, the preparation procedure of melting and pouring in room air may have led to oxygen absorption by the sample. Second, pressure bursts of K and O_2 (and CO_2) were noted in the initial phase of the TMS experiments and particularly near the liquidus temperature, e.g., at 1500 K in figure 18. This effect is attributed to K_2CO_3 impurity. Third, reduction of Fe_2O_3 to Fe_3O_4, with release of excess O_2, is favorable at these temperatures and, in fact, has a similar temperature dependence to that of the initial O_2 data shown in figure 18. Using the JANAF (8) thermochemical data for Fe_3O_4 and Fe_2O_3, it is possible to predict O_2 partial pressures for given condensed phase activities. On this basis, the initial experimental data of figure 18 at 1600 K, for instance, are consistent with ~ 50 percent and 25 percent Fe_2O_3 reduction for the KMS and TMS experiments, respectively.

The KMS data were obtained using the integrated ion intensity-weight loss method of pressure calibration, taking into account the additional weight loss due to Fe_2O_3 reduction. Calibration of the TMS data, on the other hand, was made using the relative ionization cross section approach. The apparent difference between the KMS and TMS data, indicated in figure 18, is related to the problem of additional sources of O_2 already mentioned. That is, the TMS data were obtained at an earlier stage of the sample history, where the high O_2 pressure depresses the K-pressure by the mass-action effect. In fact, if the data are converted to K_2O activities, the KMS and TMS data are in satisfactory agreement. Such agreement is good evidence of system thermodynamic equilibrium. At a later phase of the KMS experiments, the O_2/K pressures were of the correct stoichiometry for $K_2O(\ell)$ decomposition.

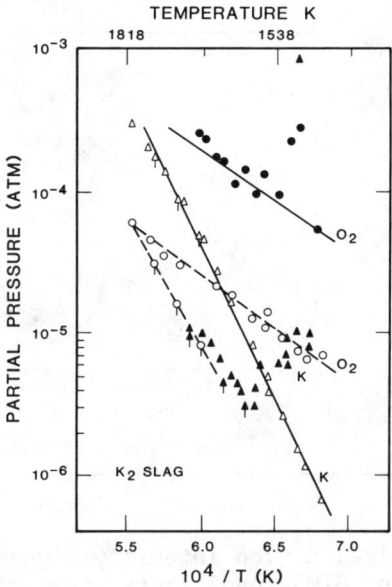

Figure 18. Vaporization of K and O_2 from the K_2 slag. Open circles--O_2 (KMS), closed circles--O_2 (TMS), open triangles--K (KMS), closed trianges--K (TMS). Chronological order of data taken with increasing temperature except for ticked data points where the temperature was decreasing.

A second type of time dependent phenomenon was observed for this slag using the TMS method, as shown in figure 19. Once an isothermal condition was achieved, the K-pressure decreased with time. This result could be taken as evidence of surface depletion of K (and O_2) from the sample, due to the bulk diffusion rate being too small relative to the surface vaporization rate. Also, this effect was found to be much less pronounced at higher temperatures where the diffusion rates are higher. However, this initial interpretation (27) no longer seems reasonable. The order-of-magnitude greater vapor transport rates for KMS vs TMS experiments would indicate lower apparent activities in the former case due to surface depletion effects. However, in practice the KMS activities are somewhat higher. Hence, this time dependent phenomenon is attributed to the combined effects of K_2CO_3 impurity decomposition, as noted previously for the analogous glass system, and, to FeO_x reduction, as noted with the "Eastern" slag data. That this latter effect was more apparent in the TMS, versus the KMS, experiments can be attributed to the much higher transport rates and shorter FeO_x reduction times for the latter case.

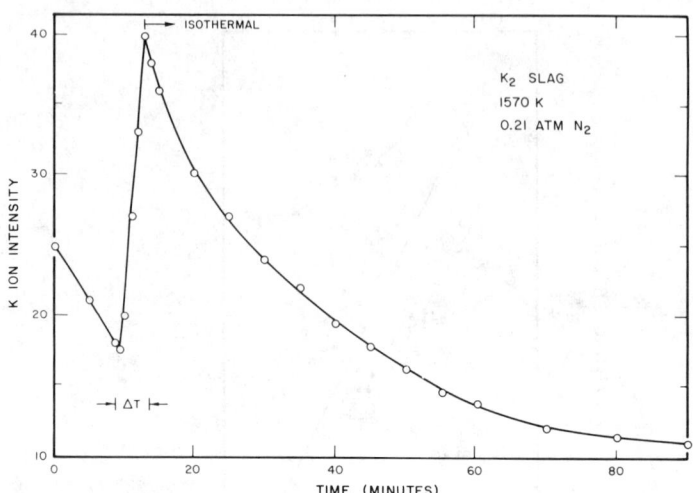

Figure 19. Typical K$^+$ ion intensity signal-decay with time for K$_2$ slag. TMS conditions: temperature, 1570 K; N$_2$ carrier gas pressure, 0.21 atm; capillary nozzle.

7. Salt-Slag Alkali Exchange

The common disposition of alkali in coal minerals is Na as NaCl and K as K$_2$O--bound in a low-activity silicate phase. Thus, during coal conversion, Na is expected to be released to the vapor phase more readily than K. However, the possibility of NaCl-K$_2$O (slag) interaction to produce KCl-Na$_2$O (slag) could greatly enhance K-release to the vapor phase. Also, in MHD slags, where about 20 wt.% K$_2$O content is possible, the problem of recovering this lost seed could likewise be resolved through replacement by NaCl. The feasibility of such an exchange process was tested by a TMS monitoring of the vapor phase over the system, NaCl + K$_1$ slag (19.4 wt.% K$_2$O). Details of this study will be given elsewhere (28), but the main observations are as follows.

When a thin layer of powdered NaCl was present on the surface of the K$_1$ slag, a rapid exchange reaction occurred near the melting point of NaCl, i.e.,

$$NaCl(\ell) + K_2O \text{ (slag)} = KCl(\ell) + Na_2O \text{ (slag)} \ .$$

This result is demonstrated in figure 20, where the observed partial pressures of NaCl and KCl are expressed in thermodynamic activity form. Note the marked decrease in NaCl activity and concomitant increase in KCl activity just above the melting point of NaCl. After a heating period of about 50 min., the NaCl sample was virtually depleted, as was the KCl product. Insufficient salt was present in the initial mixture to convert all the available K_2O to KCl. However, 90 percent of the initial NaCl was converted to Na_2O (slag) with stoichiometric release of KCl. About six percent of the available K_2O was converted to KCl vapor, and we expect that nearly complete removal of K_2O from the slag would have been possible if sufficient NaCl was present. The remaining ten percent NaCl was lost by vaporization before, and during, the exchange process. During the isothermal, constant activity, phase of the exchange process (20 to 40 min. region of fig. 20), a potassium vapor transport enhancement factor of,

$$\frac{P_{KCl}}{P_K \text{ (no NaCl)}} \sim 10^4$$

was observed. Also, during this period, the high KCl activity suggests formation of an essentially ideal solution of KCl-NaCl, as well as the establishment of thermodynamic equilibrium. Note the near unit NaCl activity in the initial phase of the experiment (fig. 20), which confirms the calibration factors used to convert mass spectral ion intensities to partial pressures and reflects establishment of thermodynamic equilibrium. Additional study of this exchange process is in progress.

8. Heterogeneous Reactive Gas Systems

In the previous sections, we have considered alkali vapor transport from condensed phase systems in the absence of external influences, such as reactive gases. However, some of the component gases of combustion systems, such as H_2O, HCl, SO_2, O_2, CO, and H_2, can be expected to significantly modify alkali vapor transport through mass action effects or formation of new molecular species. Some representative cases are considered as follows.

8.1 <u>Synthetic Slag (K_2) - H_2O - H_2 System</u>. In order to extend the vapor transport conditions in slag systems to a reducing hydrous environment similar to that present in coal gasification, a series of TMS and KMS measurements were made using H_2 or H_2O as the initial reactant gas. With the TMS system, compositions of H_2-N_2-H_2O up to 10 vol % H_2 were attained prior to hydrogen-induced corrosive loss of the transpiration reactor.

As H_2 was introduced to the slag system, the O_2 concentration decreased and K and H_2O increased, as expected for the process,

$K_2O(slag) + H_2 = 2K + H_2O.$ R17

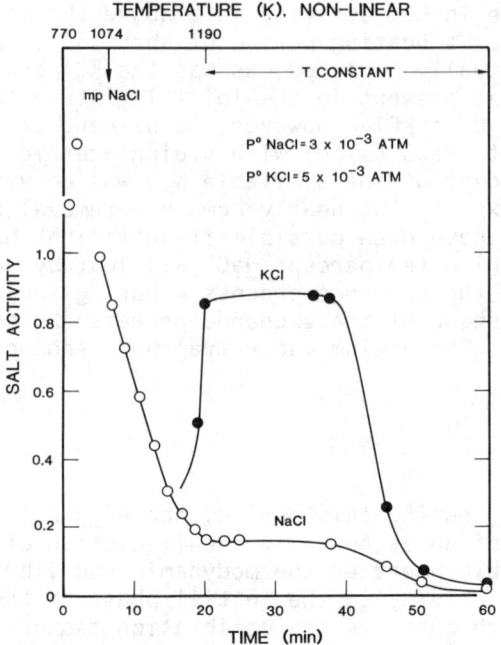

Figure 20. Thermodynamic activities (TMS data) for NaCl and KCl in the K_1 slag-alkali exchange process. The indicated reference state partial pressures were obtained from JANAF (26).

Typical TMS data are given in figure 21 where the H_2 partial pressure was varied over the range 10^{-4} to 10^{-2} atm. Note the pronounced hysterisis effect for increased versus decreased H_2 and H_2O-content. Though not shown here, this effect is also present in the K_2O activity data, as calculated from the observed K and O_2-pressures. Hence the system is not at thermodynamic equilibrium. From the established equilibrium constants for reaction R16 with K_2O (slag) and K_2O (pure liquid), together with the measured K_2O activity data, we calculate, K_p R17 = 209 at 1650 K. The corresponding experimental value, obtained from the measured partial pressures of K, H_2, and H_2O, is K_p R17, obs. = 4.2 at 1650 K. Thus the system is far from equilibrium.

Similar TMS experiments were performed but with H_2O as the added reactant and a non-reducing atmosphere. An unexpected K-pressure dependence on H_2O was found, as shown in figure 22. No hysterisis effects were observed in this case. A similar, though less pronounced (factor of four less effect on K-pressure), H_2O-induced K vaporization effect was noted in the more acidic and more viscous MHD (K_1) slag sample (27).

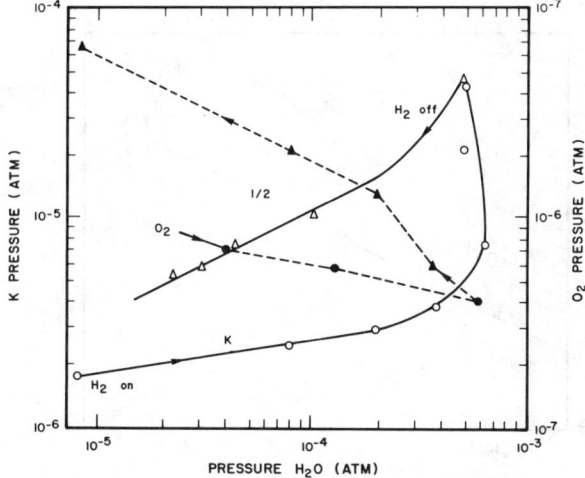

Figure 21. Variation of K-pressure (open symbols) and O_2-pressure (closed symbols) with H_2O-pressure for K_2 slag in the presence of added H_2. TMS conditions: temperature, 1655 K; N_2 carrier gas pressure, 0.18 atm; capillary nozzle. Arrows indicate run chronology.

For the H_2O-pressure and temperature conditions used, KOH should have formed according to the process,

$$K_2O \text{ (slag)} + H_2O = 2KOH. \hspace{2cm} \text{R18}$$

However, no KOH was observed in the TMS mass spectra. We also established the K^+ precursor as atomic K, from the pure KOH data and appearance potential measurements. A higher temperature study, using the KMS method, did show the expected formation of KOH in the presence of added H_2O. However, the KOH-pressures were about an order of magnitude below predicted equilibrium values, even though the correct H_2O pressure dependence was found.

The apparently anomalous H_2O-induced increase in K-pressure can be explained as follows. Literature water solubility data for aluminate and silicate melts [e.g., see (29)] suggest solubilities of at least several hundred ppm for our experimental conditions. Various acid-base reaction mechanisms have been suggested to explain water solubility in silicate melts, as summarized by Turkdogan (30). For basic melts, H_2O acts as an acid and enhances the silicate network structure, and vice-versa for acid (high silica) melts. Though apparently not previously recognized, these structural changes should be reflected in the

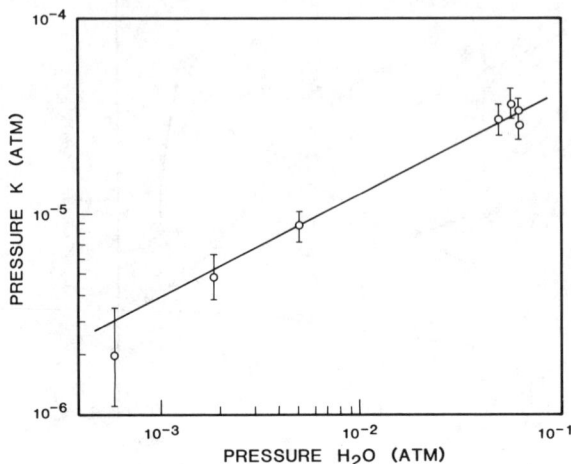

Figure 22. Isothermal (1610 K) dependence of K-pressure on H_2O-pressure for the K_2 slag with added H_2O. TMS conditions: N_2 carrier gas pressure, 0.21 atm; capillary nozzle.

alkali activity data. Thus, we can reasonably expect an activity increase when water is incorporated into the silicate matrix of the relatively basic K_2 slag. Reaction sets of the type,

$$H_2O + O^{2-} \text{ (slag)} = 2(OH^-)$$

$$K_2O \text{ (slag)} = 2K + 1/2\ O_2, \text{ and}$$

$$O_2 + 2e^- = 2O^{2-},$$

would be consistent with the observed one-half power dependence of $\log P_K$ on $\log P_{H_2O}$ (see fig. 22). These structural changes

should be reflected in viscosity data. The water-solubility viscosity enhancement effects noted by Brower et al. (31) for a similar slag are consistent with the present activity trends. For more basic systems a decreased alkali activity is possible. The recent observations of Gray (32), where water vapor decreased alkali vaporization rates in low silica glasses, could be interpreted in this manner.

We believe that a similar water vapor solubility enhancement of alkali vapor transport is possible in soda-lime-silica glass systems, and work is in progress to verify this. Some of the disparities between various glass vaporization studies may well result from variations in water content and, hence, alkali activities. The common explanation for water vapor enhanced alkali vapor transport over silicates has revolved around formation of volatile NaOH (33) and KOH (34) species. However, no direct test for the presence of these species has been made, and the possibility of water vapor enhancement of atomic Na and K transport exists in these systems.

Mention should also be made of the possible effect of H_2O dissociation to yield vapor phase H_2, which was suggested by Horn et al. (35) as a factor in ceramic degradation. However, in the present system there is an additional source of O_2 and, at thermodynamic equilibrium, K should be H_2O-independent except for the noted H_2O-solubility effect.

9. Notes

1. Based on material presented by the authors at the 10th Materials Research Symposium on Characterization of High Temperature Vapors and Gases held at NBS, Gaithersburg, Maryland, September 18-22, 1978; see NBS SP-561 (1979), p. 357.

2. Additional material may be found in the reports NBSIR 80-2178 (1980) and NBSIR 81-2279 (1981) available from NTIS.

10. References

(1) Hastie, J. W.: 1975, High Temperature Vapors: Science and Technology (Academic Press, New York).

(2) Hastie, J. W.: 1975, Intl. J. Mass Spectrom. Ion Phys. 16, p. 89 and Hastie, J. W.: 1973, Combust. Flame, 21, p. 49.

(3) Stearns, C. A., Kohl, F. J., and Fryburg, G. C.: 1979, 10th Materials Research Symposium on Characterization of High Temperature Vapors and Gases, J. W. Hastie, ed., NBS SP-561.

(4) Mertin, U. and Bell, W. E.: 1967, The Characterization of High Temperature Vapors, Margrave, J. L., ed., (Wiley, New York) p. 91.

(5) Kantrowitz, A. and Grey, J.: 1951, Rev. Sci. Instru. 22, p. 328.

(6) Greene, F. T., Brewer, J., and Milne, T. A.: 1964, J. Chem. Phys. 40, p. 1488.

(7) Abuaf, N., Anderson, J. B., Andres, R. P., Fenn, J. B., and Marsden, D. G. H.: 1967, Science, 156, p. 997.

(8) JANAF Thermochemical Tables, 2nd ed.: 1971, NSRDS-NBS 37, (U. S. Govt. Printing Office, Washington, D.C.)

(9) Feather, D. H. and Searcy, A. W.: 1971, High Temp. Sci. 3, p. 155.

(10) Milne, T. A. and Klein, H. M.: 1960, J. Chem. Phys. 33, p. 1628.

(11) Plante, E. R.: 1979, "Vapor Pressure Measurements of Potassium Over K_2O-SiO_2 Solutions by a Knudsen Effusion Mass Spectrometric Method," in Characterization of High Temperature Vapors and Gases, Hastie, J. W. ed. NBS-SP-561, (U.S. Govt. Printing Office, Washington, DC) p. 265.

(12) Kieffer, L. J. and Dunn, G. H.: 1966, Rev. Mod. Phys. 38, p. 1.

(13) Meyer, R. T. and Lynch, A. W.: 1973, High Temp. Science, 5, p. 192.

(14) Stwalley, W. C., and Koch, M. E.: 1980, Opt. Eng. 19, p. 71.

(15) Haynes, B. S., Jander, H., and Wagner, H. G.: 1978, Symp. (Int.) Combust., 17th, (The Comb. Inst., Pittsburgh, PA) p. 1365.

(16) Gangwal, S. K., and Truesdale, R. S.: 1980, Energy Res. 4, p. 113.

(17) Rapp, R., ed.: 1981, International Conference on High Temperature Corrosion, San Diego, CA, March 2-6, NACE, Proceedings in press.

(18) Hastie, J. W., Bonnell, D. W., and Zmbov, K.: 1981, "Transpiration Mass Spectrometric Analysis of $KOH(\ell)$ and $KCl(\ell)$ Vaporization," to be published.

(19) Gusarov, A. V. and Gorokhov, L. N.: 1968, Russ. J. Phys. Chem. 42, p. 449.

(20) Hastie, J. W., Plante, E. R., Bonnell, D. W., and Horton, W. S.: 1980, "Molecular Basis for Release of Alkali and Other Inorganic Impurities From Coal Minerals and Fly Ash," Report to DOE Morgantown, WV.

(21) Gatti, A., Goldstein, H. W., McCreight, L. R., and Semon, H. W.: 1980, "Feasibility Study of Coal Slag Based Glasses for Hot Gas Clean-up," Report FE-2068-32 to DOE (February).

(22) Sanders, D. M. and Haller, W. K.: 1979, "A High Temperature Transpiration Apparatus for the Study of the Atmosphere Above Viscous Incongruently Vaporizing Melts," p. 111, ibid, (3).

(23) Neudorf, D. A., and Elliott, J. F.: 1980, Met. Trans., 11B, p. 607.

(24) Cable, M., and Chaudhry, M. A.: 1975, Glass Tech., 16, p. 125.

(25) Sanders, D. M., Blackburn, D. H., and Haller, W. K.: 1976, J. Am. Ceram. Soc., 59, p. 366.

(26) Argent, B. B., Jones, K., and Kirkbride, B. J.: 1980, "Vapors in Equilibrium with Glass Melts," in The Industrial Use of Thermochemical Data, Barry, T. I., ed., (The Chemical Society, London, UK) p. 379.

(27) Hastie, J. W., Bonnell, D. W., Plante, E. R., and Horton, W. S.: 1980, "Vaporization and Chemical Transport Under Coal Gasification Conditions," NBSIR 80-2178.

(28) Hastie, J. W., Plante, E. R., and Bonnell, D. W.: 1981, "Slag-Alkali Halide Exchange Reactions," to be published.

(29) Schwerdtfeger, K., and Schubert, H. G.: 1978, Met. Trans. 9B, p. 143.

(30) Turkdogan, E. T.: 1980, Physical Chemistry of High Temperature Technology," (Academic Press, New York).

(31) Brower, W. S., Waring, J. L., and Blackburn, D. H.: 1980, "Slag Characterization: Viscosity of Synthetic Coal Slag in Steam," NBSIR 80-2124.

(32) Gray, W. J.: 1980, Radioactive Waste Management, 1, p. 147.

(33) Sanders, D. M. and Haller, W. K.: 1977, J. Amer. Ceram. Soc., 60, p. 3.

(34) Charles, R. J.: 1967, J. Amer. Ceram. Soc., 50, p. 631.

(35) Horn, F. L., Fillo, J. A., and Powell, J. R.: 1979, J. Nucl. Mater., 85, p. 439.

APPENDIX A:

TEMPERATURE DEPENDENT ELECTRON IMPACT FRAGMENTATION IN HIGH TEMPERATURE MOLECULAR BEAMS

High temperature mass spectrometry normally assumes the electron impact process to be temperature independent, since kT << eV. For the alkali halides (MX), where fragmentation of molecular ions is extensive, the M^+/MX^+ ion intensity ratio is normally temperature insensitive by Knudsen Effusion Mass Spectrometry (KMS). However, in one known exception, Akishin et al. (1) observed a 20 percent variation in the ratio $Cs^+/CsCl^+$ over the temperature range 800 K to 900 K. This effect can be explained in terms of Franck-Condon overlap for the molecule → molecule-ion transition. Bloom et al. (2), in explaining extended curvature in the toe region of ionization efficiency curves of alkali metal fragment ions, invoked the Franck-Condon principle to postulate approximate functions for various states of CsCl leading to ionization. They suggested that the Franck-Condon region of vertical transition overlap for ground state CsCl intersects the repulsive edge of the lowest bound state for the molecule-ion. As the potential curve for the molecule ion $CsCl^+$ has a very shallow minimum, the ratio $Cs^+/CsCl^+$ resulting from this transition will be particularly sensitive to the location of the CsCl potential curve and hence temperature.

More recently, Dronin and Gorokhov (3,4) have modeled the CsCl ionization process using a similar concept and reported good agreement with the experimental results of Akishin et al. (1). However, details of their calculation are unavailable. A key aspect of their theory is that alkali halide molecules in effusive beams have a significant population of excited vibrational and rotational states. Typically, for CsCl at 800 K, only 32 percent of the molecules are in the v = 0 state, and the rotational population peaks in a broad maximum near J = 60. This temperature-sensitive thermal population of upper states affects the extent of Franck-Condon overlap and hence the degree of fragmentation as a function of temperature. However, there are no other reports in the literature, to our knowledge, of an observed temperature dependent fragmentation in alkali halide systems. Feather and Searcy (5) reported no detectable temperature dependence in the ratio $Na^+/NaCl^+$ over the temperature range 850 to 1050 K. Table A.1 summarizes fragment to parent ion intensity ratio data for certain alkali chlorides and hydroxides.

In the application of Transpiration Mass Spectrometry (TMS) to alkali halide systems, anomalously high ratios for M^+/MX^+ are observed (6). This can be interpreted in terms of the above model

Table A.1
Fragmentation Ratios of Selected Alkali Halides/Hydroxides

Molecule	R = M^+/MX^{+a}	Temperature K	eV	Reference
NaCl	1.7	~ 950	75	(12)
	1.4	~ 940	20	(15)
	1.0 and 0.8	850 to 1050	70	(5)
	1.5	952	30	(6)
	25 to 33	1090 to 1440^b	30	This work
KCl	6.0	~ 900	75	(12)
	60 to 85	990	30	This work
CsCl	86	~ 830	75	(12)
	154	861	50	(13)
	110	850	20	(14)
	86; 70; 67	800; 828; 900	90	(1)
	310 to 135	980 to 1440^b	30	This work
KOH	0.43	623 to 670	50	(16)
	1.1	650 to 950	100	(17)
	2.0 ± 0.3	990^b	30	This work

^{a}Assumes no contribution to M^+ or MX^+ from M_2X^+ or higher polymers; for X = halide and M > Li, it has been concluded (18) that M^+ and MX^+ have the same molecular origin. Literature data are from KMS experiments where $T_b = T_o$. In the present study, with supersonic expansion conditions $T_b \ll T_o$.

^{b}Sample temperature. Calculated beam temperature < 100 K due to expansion.

by noting that the beam temperature by TMS is much lower than the source gas temperature. Translational, rotational, and, in many cases, vibrational cooling occurs during the isentropic expansion process. Thus the true temperature of molecules in the beam is very low.

The isentropic relation for the initial to final temperature ratio in the expansion process is,

$$\frac{T_o}{T_b} = 1 + \left[\frac{\gamma - 1}{2}\right]M^2 \qquad (A.1)$$

where T_o is source temperature, T_b the beam temperature, γ the gas heat capacity ratio, C_p/C_v, and M the local Mach number, defined as the ratio of the local flow velocity, v, to the local speed of sound, c. Ashkenas and Sherman (7) relate the center line flow Mach number to the downstream distance to orifice diameter ratio z/d_o for values of $z/d_o > 4$ as

$$M = A(\gamma) \left[\frac{z}{d_o}\right]^{(\gamma - 1)} \quad (A.2)$$

Knuth (8) has suggested that $A(\gamma)$ in this regime be given by

$$A(\gamma) = \left[\frac{2.2}{[\gamma(\gamma - 1)]^{1/2}}\right]^{(\gamma - 1)/2} \left[\frac{\gamma + 1}{\gamma - 1}\right]^{(\gamma + 1)/4} \quad (A.3)$$

The sudden freeze approximation (1,9,10) allows the definition of a terminal Mach number. The terminal Mach number, M_T, is that value which the flow Mach number approaches asymtotically at infinitely large values of z/d_o. This value can be expressed in general (Stearns et al., 1979) as

$$M_T = \left[\frac{\gamma}{\gamma - 1}\right] \left[\frac{2}{A(\gamma)}\right]^{1/\gamma} \left[\frac{8}{\gamma\pi}\right]^{(\gamma - 1)/2\gamma} \left[\frac{Kn_o}{\varepsilon}\right]^{(1 - \gamma)/\gamma}$$

$$\quad (A.4)$$

$$= \left[C_1(\gamma)\right] \frac{Kn_o}{\varepsilon}^{(1 - \gamma)/\gamma}.$$

where ε is a collision effectiveness parameter ($0 \leq \varepsilon \leq 1$) and Kn_o the Knudsen number, defined as the ratio of gas mean free path length λ_o to orifice diameter d_o. The mean free path is given as

$$\lambda_o = (\sqrt{2} \pi \sigma^2 n_o)^{-1}, \quad (A.5)$$

where σ is collision diameter and n_o is the source gas number density. To reasonable accuracy, $\varepsilon \cong 0.3$ for all gases. Collision dominated flow is considered to cease when the expansion temperature no longer follows the isentropic relation. At this point the Mach number, M_1 is given by

$$M_1(d_o, P_o, T_o, \gamma) = \frac{M_T}{\gamma} \quad . \tag{A.6}$$

Thus $T_o/T_b = f(d_o, P_o, T_o, \gamma)$ can be calculated from the macroscopic observables.

By adjusting the source gas pressure, various beam temperatures can be obtained using Eq. (A.1) with $M = M_1$ and assuming complete translational, rotational, and vibrational relaxation (6). Figure A.1 shows the relationship of T_b to P_o for typical source gas conditions.

With the above approach, beam temperatures have been calculated for the TMS experiments. Typical results for NaCl are given in Figure A.2 in comparison with the KMS molecular effusion data. Note the smooth transition from low (TMS) to high (KMS) temperatures. The curve is qualitatively very similar to that obtained for the population ratio of ground to vibrationally excited states from a thermal Boltzmann distribution. Similar results are found for KCl. However, in contrast, the KOH system yields similar K^+/KOH^+ data by the TMS and KMS methods, as shown in Table A.1.

To summarize, it is clear that more attention should be given to possible temperature dependent electron impact fragmentation in KMS and TMS thermochemical studies of high temperature species. As a first approximation, we can expect those species with low stability molecular ions to be more susceptible to this effect. Molecular-ion stabilities have been summarized elsewhere (11).

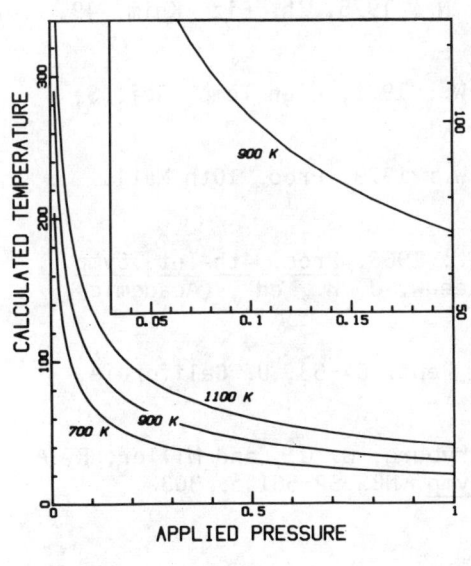

Figure A.1. T_b vs P_o (atm) and T_o for $d_o = 0.00762$ cm, $\gamma = 7/5$ (e.g., N_2), $\varepsilon = 0.3$ and $\sigma = 5.5 \times 10^{-8}$ cm.

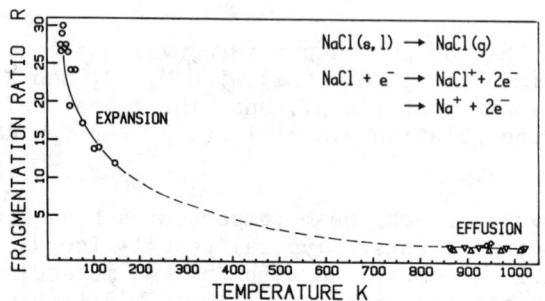

Figure A.2. Variation of R with T_b.

References to Appendix A

1. Akishin, P. A., Gorokhov, L. N., and Sidorov, L. N.: 1960, Proc. Acad. Sci. USSR, [Dok. Phys. Chem.] 135, 1, p. 1001.

2. Bloom, H., Hastie, J. W., and Morrison, J. D.: 1968, J. Phys. Chem. 72, p. 3041.

3. Dronin, A. A. and Gorokhov, L. N.: 1972, Teplofiz. Vys. Temp. 10, p. 49.

4. Dronin, A. A. and Gorokhov, L. N.: 1975, Zh. Fiz. Khim. 49, p. 798.

5. Feather, D. H. and Searcy, A. W.: 1971, High Temp. Sci. 3, p. 155.

6. Bonnell, D. W. and Hastie, J. W.: 1979, Proc. 10th Matls. Res. Symp. NBS SP-561 p. 357.

7. Ashkenas, H. and Sherman, F. S.: 1966, Proc. 4th Int. Symp. on Rarefield Gas Dynamics, deLeeuw, J. H., ed., (Academic Press, New York) p. 84.

8. Knuth, E. L.: 1964, Dept. Eng. Rept. 64-53, U. California, Los Angeles.

9. Stearns, C. A., Kohl, F. J., Fryburg, G. C., and Miller, R. A.: 1979, Proc. 10th Matls. Res. Symp. NBS SP-561 p. 303.

10. Anderson, J. B. and Fenn, J. B.: 1965, Phys. Fluids, 8, p. 780.

11. Hastie, J. W. and Margrave, J. L.: 1968, Fluorine Chem. Rev. 2, p. 77.

12. Berkowitz, J. and Chupka, W. A.: 1958, J. Chem. Phys. 29, p. 653.

13. Milne, T. A. and Klein, H. M.: 1960, J. Chem. Phys. 33, p. 1628.

14. Bloom, H. and Hastie, J. W.: 1966, Aust. J. Chem. 19, p. 1003.

15. Bloom, H. and Hastie, J. W.: 1968, J. Chem. Phys. 49, p. 2230.

16. Gusarov, A. V. and Gorokhov, L. N.: 1968, Zh. Fiz. Khim. 42, p. 860.

17. Schoonmaker, R. C. and Porter, R. F.: 1959, J. Chem. Phys. 31, p. 830.

18. Hastie, J. W. and Margrave, J. L.: 1969, High Temp. Sci. 1, p. 481.

THERMODYNAMIC ACTIVITY AND VAPOR PRESSURE MODELS FOR SILICATE SYSTEMS INCLUDING COAL SLAGS

J. W. Hastie, D. W. Bonnell, E. R. Plante and W. S. Horton

High Temperature Processes Group
National Bureau of Standards
Washington, DC 20234

ABSTACT

A new modeling approach is described for thermodynamic predictions of multicomponent, multiphase high temperature silicate systems including coal slags. The model, which attributes negative deviations from ideal solution behaviour to the formation of complex liquids and solids, is demonstrated for quarternary systems containing K_2O, Al_2O_3, CaO, and SiO_2. Good agreement between the model predictions and experimental vapor pressure data is found.

INTRODUCTION

Silicate systems, including slags, are an important component in the development of new coal-fired energy systems. These include magnetohydrodynamics (MHD), pressurized fluidized bed combustion, coal gasification, and the extension of current boiler technology to higher temperatures and lower grade coals. Problems of hot corrosion, loss of alkali seed (in MHD) and slag control require basic thermochemical data and predictive models of vapor transport and deposition for alkali and other slag/inorganic components. Such models would also be applicable to other phase transformation predictions, such as phase separations and solubilities, by virtue of the thermodynamic interrelationships between vapor pressure, activity and the partial and integral thermodynamic functions.

In a previous paper we described, and demonstrated, a solution model for binary and ternary systems containing K_2O, Al_2O_3 and SiO_2 (1). This model is unique among the many models developed for activity predictions in silicate and other condensed phase systems in that no prior knowledge is required for the mixture of interest. Here, we apply the model to quarternary systems containing K_2O, Al_2O_3, CaO and SiO_2. As we have shown earlier (2), the thermodynamic properties of many coal slags can, to a good approximation, be represented by these four components or the Na-analogue.

The basis of the present solution model is outlined here as follows. We attribute large negative deviations from ideal thermodynamic activity behaviour to the formation of stable complex liquids (and solids) such as K_2SiO_3, $KAlSiO_4$,... The free energies of formation (ΔG_f) are either known or can be estimated for these component liquids (and solids). By minimizing the total system free energy one can calculate the equilibrium composition with respect to these components. Thus, for instance, the mole fraction of K_2O present ($X^*_{[K_2O]}$) in equilibrium with K_2SiO_3, and other complex liquids (and solids) containing K_2O, is known. As we have shown previously for the ternary systems, the component activites can, to a good approximation, be equated to these mole fraction quantities. From this assumption of Ideal Mixing of Complex Phases (IMCP) it also follows that potassium partial pressures can be obtained from the relationship

$$P_K = 2 \cdot X^*_{[K_2O]} K_p^{0.4},$$

where K_p is the stoichiometric dissociation constant for pure K_2O (liquid or solid) to K and O_2. In the following discussion we test the model by comparing predicted P_K data determined in this manner with experimental values. Thermodynamic activities and phase compositions are also calculated using this model. The experimental K-pressure data were obtained by Knudsen effusion mass spectrometry as discussed in detail elsewhere (2).

THE DATA BASE

The SOLGASMIX computer program (3) used for calculation of the equilibrium composition and hence activities utilizes a data base of the type given in table 1. The coefficients to the ΔG_f equation were obtained by fitting ΔG_f vs T data available in JANAF (4), Robie et al (5), Barin and Knacke (6), Rein and Chipman (7) and Kelley (8). The elemental reference state data are those given by JANAF (4). In some cases no literature data were available and we estimated functions in the manner described earlier (1). These cases can be identified in table 1 as those where only two coefficients are listed. Note also in table 1 the asterisk liquid compound $CaAl2Si4O12^*$ (i.e., $CaO \cdot Al_2O_3 \cdot 4SiO_2$) is a component not previously recorded and for which we have

THERMODYNAMIC ACTIVITY AND VAPOR PRESSURE MODELS

Table 1. Thermodynamic Data Base for K_2O, CaO, Al_2O_3, SiO_2 System

$$\Delta G_f(T) \text{Joules} = \underline{1} \cdot T^{-1} + \underline{2} \cdot T^0 + \underline{3} \cdot T^1 + \underline{4} \cdot T^2 + \underline{5} \cdot T^3 + \underline{6} \cdot T \ln T$$

LIQUIDS	coeff 1	coeff 2	coeff 3	coeff 4	coeff 5	coeff 6
Al2O3	-5.1021425+008	1.1897489+006	-2.0394746+004	-1.2145752+000	9.2059900-005	2.8258033+003
K2O	.0000000	-4.5469820+005	2.2537340+002	.0000000	.0000000	.0000000
SiO2	.0000000	-8.9144570+005	7.5239100+001	.0000000	.0000000	1.2545960+001
K2Si03	-8.0014313+007	-7.7409560+005	-6.3254273+003	-3.3633172-001	1.8964800-005	9.1118181+002
K2Si2O5	-8.5294893+007	-7.1201679+006	-6.6582289+003	-4.1297721-001	2.6558900-005	9.9499587+002
K2Si4O9	.0000000	-4.4413160+006	9.7068800+002	.0000000	.0000000	.0000000
KAlO2	-6.6031322+007	-4.0771820+005	-6.7718013+003	-5.2394735-001	4.6888800-005	9.9161504+002
KAlSiO4	.0000000	-2.0919864+006	4.1232080+002	.0000000	.0000000	.0000000
KAlSi2O6	.0000000	-3.0213582+006	5.7392890+002	.0000000	.0000000	.0000000
CaAl2Si4O12*	.0000000	-6.4348874+006	1.2504930+003	.0000000	.0000000	.0000000
Al6Si2O13	.0000000	-5.9751009+006	-9.0003420+003	-1.2179684+000	1.4569980-004	1.5426376+003
K2Al18O28	.0000000	-1.1569249+007	2.9974060+003	.0000000	.0000000	.0000000
CaO	2.4693906+010	-1.2859793+008	9.0366898+005	5.1437253+001	-3.9681461-003	-1.2271341+005
CaSiO3	-1.7364559+009	6.1170297+006	-4.5878183+004	-2.0348609+000	1.2279700-004	6.0915231+003
CaAl2O4	-9.2730135+007	-1.3377933+006	-1.0090940+004	-9.2414711-001	9.3646380-005	1.5131449+003
CaAl2Si2O8	-3.5566419+009	1.2196323+007	-1.0203235+005	-4.9938669+000	3.3924038-004	1.3696191+004
CaAl2Si2O7	-1.8291860+009	4.7189539+006	-5.5962010+004	-2.9590080+000	2.1644338-004	7.6046680+003
Ca12Al14O33	-6.4911094+008	-1.2462817+007	-7.0221678+004	-6.4690298+000	6.5552466-004	1.0592014+004

SOLIDS	coeff 1	coeff 2	coeff 3	coeff 4	coeff 5	coeff 6
Al2O3	-8.5337210+006	-1.6525534+006	.0000000	-2.6404848-002	1.9218494-006	4.6927662+001
K2O	.0000000	-5.7326594+005	5.6414644+003	1.0812991+000	-1.7184390-004	-8.9711854+002
CaO	.0000000	-6.6007023+005	1.2222659+002	.0000000	.0000000	.0000000
K2Al18O28	.0000000	-1.6137018+007	3.2444966+003	.0000000	.0000000	.0000000
SiO2	-1.7364559+009	6.7988567+006	-4.5963590+004	-2.0348609+000	1.2279700-004	6.0915231+003
KAlSiO4	.0000000	-2.2003285+006	4.6587600+002	.0000000	.0000000	.0000000
KAlSi2O6	.0000000	-3.1188960+006	6.2371850+002	.0000000	.0000000	.0000000
Ca2Al2Si2O7	.0000000	-4.0318970+006	8.1153500+002	.0000000	.0000000	.0000000
CaSiO3	-1.7364559+009	6.0555265+006	-4.5844263+004	-2.0348609+000	1.2279700-004	6.0915231+003
Ca2SiO4	-1.7364559+009	5.3766162+006	-4.5743437+004	-2.0348609+000	1.2279700-004	6.0915231+003
Ca3Al2O6	-9.2730135+007	-2.7391466+006	-9.8163742+003	-9.2414711-001	9.3646380-005	1.5131449+003
Ca12Al14	-6.4911094+008	-1.3151094+007	-6.9823559+004	-6.4690298+000	6.5552466-004	1.0592014+004
CaAl2O4	-9.2730135+007	-1.4164061+006	-1.0049127+004	-9.2414711-001	9.3646380-005	1.5131449+003
CaAl4O7	-1.8546027+008	-2.1510420+006	-2.0210781+004	-1.8482942+000	1.8729276-004	3.0262898+003
Al6Si2O13	-3.7511022+009	1.1381955+007	-1.2239894+005	-6.8421631+000	5.2653314-004	1.6722481+004
CaAl2Si2O8	-3.5566419+009	1.2123507+007	-1.0199181+005	-4.9938689+000	3.3924038-004	1.3696191+004
Ca2Al2Si2O7	-1.8291860+009	4.6252803+006	-5.5911791+004	-2.9590080+000	2.1644338-004	7.6046680+003

Computer notation used, e.g., 5.1+008 = 5.1×10^8

estimated thermodynamic data. Many of the compounds listed in table 1 are common mineral phases such as mullite (Al6Si2Ol3), kaliophilite (KAlSiO4), leucite (KAlSi2O6), feldspar (KAlSi3O8), gehlenite (Ca2Al2SiO7) and β-alumina (K2Al18O28).

RESULTS

Table 2 lists the compositions of synthetic coal slag systems selected as test or demonstration cases of the model. These compositions were derived from those of actual coal slags as follows. We assume MgO and CaO to be equivalent on a molar basis and similarly for Fe_2O_3 and Al_2O_3. Alternatively, Fe_2O_3 can be assumed to be inert relative to the other components. We expect the true effect of Fe_2O_3 to fall between these two extreme cases. These assumptions are supported by our earlier (2) experimental observations on actual (containing MgO and Fe_2O_3) and synthetic ternary slag systems.

Table 2. Slag Compositions (wt%).

	K_2O	Al_2O_3	CaO	SiO_2
"Simplified Western"	2-14	30	30	38-27
	14	10-40	30	46-16
K_1 (Fe_2O_3 = Al_2O_3)	11.1	17.0	8.0	63.9
K_1 (Fe_2O_3 inert)	12.0	10.4	8.6	69.0
K_2 (Fe_2O_3 = Al_2O_3)	4.2	12.2	30.1	53.5
K_2 (Fe_2O_3 inert)	4.5	7.6	31.6	56.3
"Simplified Rosebud"	1	26	22	51
	1	36	2	61
	1	37	0	62

The "Simplified Western" slag system contains compositions typical of Western type coals but enriched in K_2O to simulate MHD channel slags [see Hastie et al, (2)]. The K_1 slag was extracted from an MHD test channel, K_2 is a synthetic slag, and the "Simplified Rosebud" system represents an average ash composition for Rosebud coal with various levels of added CaO. Tables 3-7 list representative activity and phase composition data for these systems, as well as the explicit ΔG_f values (derived from table 1) used in the calculations. Note in tables 3 and 4 that activities of solids are also listed. However, no solid phases are present under these conditions. Only those solids near unit activity are close to their point of precipitation which occurs at unit activity.

Figures 1-6 show the predicted K-pressures as a function of composition or temperature. Note in figures 1 to 3, the excellent agreement with our experimental data. For these cases the experimental data are based on identical compositions, i.e., no assumptions regarding MgO and Fe_2O_3 were required. For the K_1 and K_2 systems where the experimental samples did contain MgO and Fe_2O_3, agreement with the model is still good as shown in figures 4 and 5. No corresponding experimental data exist for the Rosebud system but we estimate the model K-pressures to be as accurate as can be determined experimentally. Note in figure 5 the dramatic effect of CaO content on K-pressure. Higher CaO concentration levels lead to increased K_2O activities and K-pressures. This result is quite consistent with our earlier experimental observations on "Eastern" and "Western" type slags (2). As can be seen in table 7, CaO removes SiO_2 and Al_2O_3 through formation of stable $CaSiO_3(\ell)$, $CaO \cdot Al_2O_3 \cdot 4SiO_2(\ell)$, $Ca_2SiO_4(s)$ and $2CaO \cdot Al_2O_3 \cdot SiO_2(s)$. This, in turn, results in less $K_2SiO_3(\ell)$ and $KAlSiO_4(\ell)$ formation and a higher $K_2O(\ell)$ activity. The model also predicts a substantial decrease in the liquidus temperature with added CaO, as shown in figure 6. For the one case where an experimental liquidus temperature determination was made (see figure 5), the model prediction is in reasonable agreement with experiment.

More detailed comparisons between model and experimental results for the K_1 and K_2 systems are unwarranted owing to the neglect of Fe_2O_3 and MgO in the model. However, our earlier experimental studies have shown Fe_2O_3 to be the least significant component in determining the K_2O activities, provided the K_2O oxygen dissociation pressure exceeds that of Fe_2O_3. Also, the fact that the model, which is certainly validated for the non Fe-containing "Simplified Western" system (see figures 1 and 2), shows satisfactory agreement with experiment for the K_1 and K_2 systems, supports this viewpoint. We are currently developing an Fe-species data base (as well as Mg-species) for more precise model calculations on actual coal slag compositions.

In summary, it appears that the present model, which is based on an ideal mixing of complex phases approximation, provides activity and hence vapor pressure data to an accuracy comparable with experiment. The model has been validated for binary, ternary and quarternary systems containing K_2O, CaO, Al_2O_3 and SiO_2, including simplified coal slag compositions. Similar validations have also been made recently for Na_2O-containing glasses and molten salts, as discussed elsewhere (9). Thus we believe that this approach can be applied to virtually any high temperature liquid-solid-gaseous system exhibiting negative deviations from ideal-solution behaviour, referenced to normal components such as K_2O, CaO,.... In principle, though not indicated here, all the solution properties are determined by this approach including integral and partial enthalpies and entropies of mixing. Phase diagrams can also be constructed using this approach though we should caution that the liquidus location is particularly sensitive to the accuracy of the data base thermodynamic functions, some of which are estimates only.

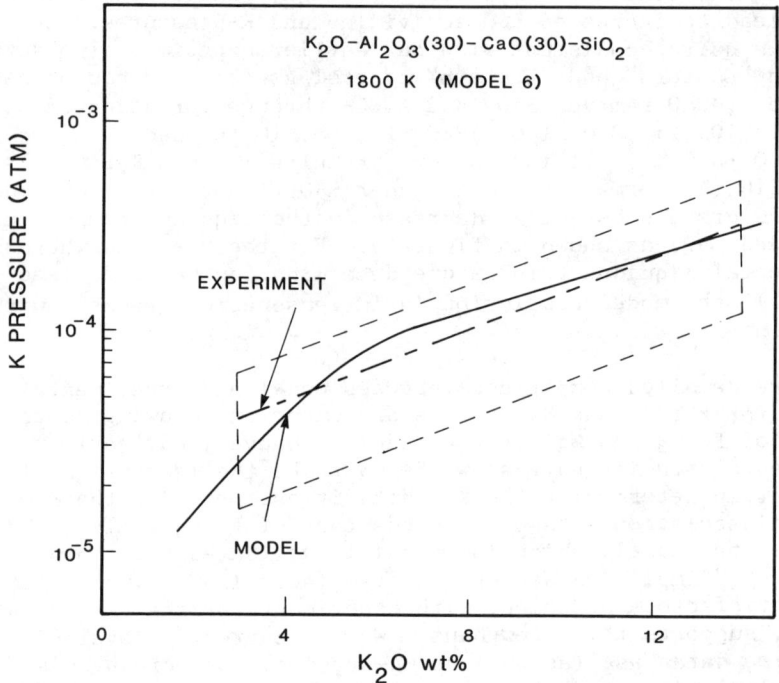

Figure 1. Comparison of model (solid curve) and experimental (broken curve) K-pressure data as a function of total K_2O content (wt%) for the "Simplified Western" synthetic coal slag system. The dashed curves indicate the upper and lower limits in experimental uncertainty.

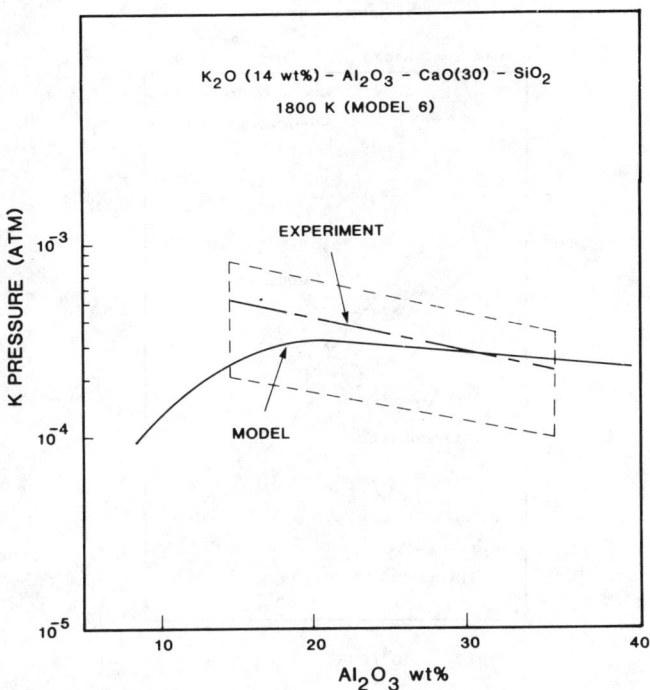

Figure 2. Comparison of model (solid curve) and experimental (broken curve) K-pressure data as a function of total Al_2O_3 content (wt%) for the "Simplified Western" synthetic coal slag system.

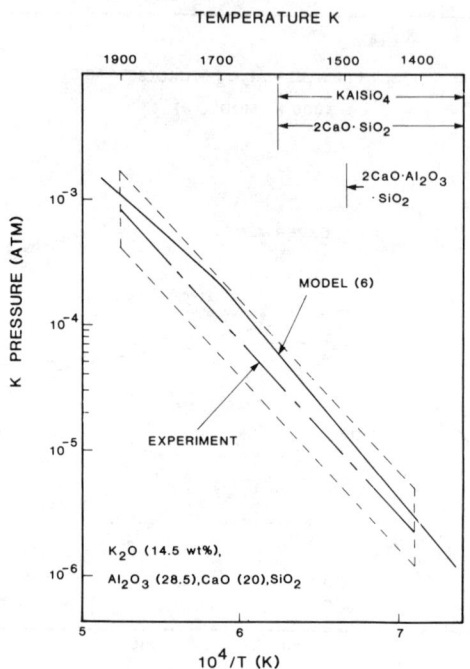

Figure 3. Comparison of model (solid curve) and experimental (broken curve) K-pressure data as a function of reciprocal temperature, for the "Simplified Western" slag system. Compounds listed are solid precipitates formed over the temperature interval indicated.

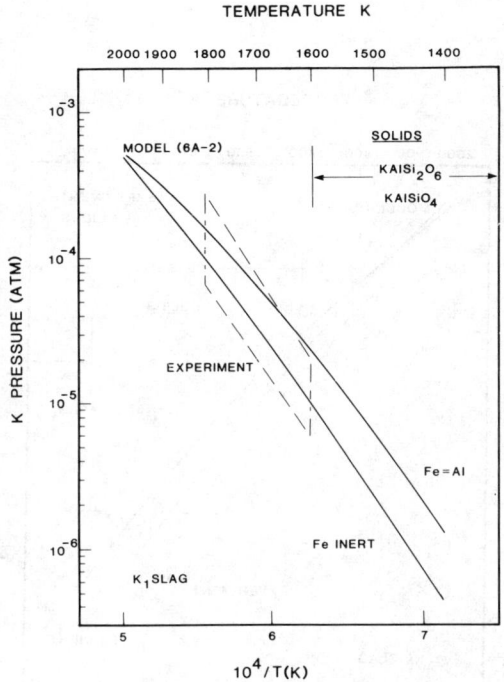

Figure 4. Comparison of model (solid curves) and experimental (broken curve) K-pressure data as a function of reciprocal temperature for the K_1 slag. See table 2 for gross compositions. Indicated solid phases correspond to the Fe = Al case.

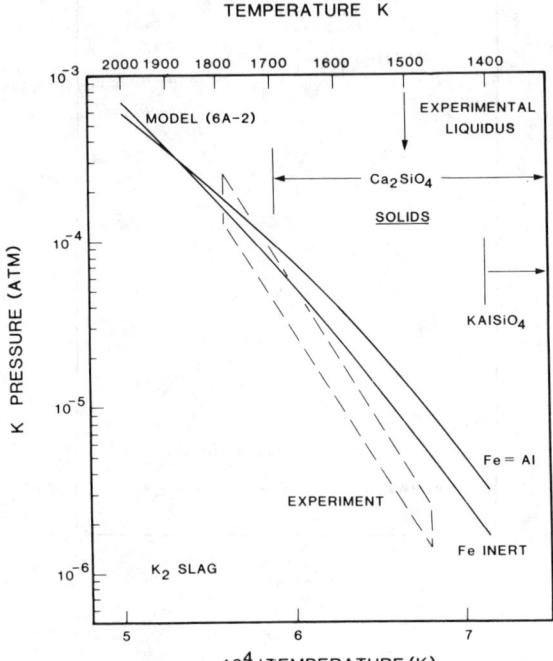

Figure 5. Comparison of model (solid curves) and experimental (broken curve) K-pressure data as a function of reciprocal temperature for the K_2 slag. See table 2 for gross compositions. Indicated solid phases correspond to the Fe = Al case.

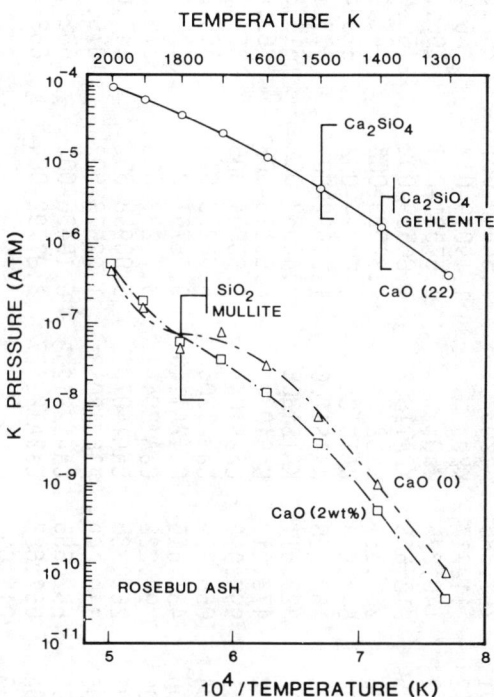

Figure 6. Model K-pressures as a function of reciprocal temperature for synthetic Rosebud coal ash compositions. The CaO content (in wt%) is indicated in parentheses for each curve. See table 2 for gross compositions.

Table 3. Model Activity Data for "Simplified Western" Type Coal Slag at 1800 K

LIQUIDS	ACTIVITY	$-\Delta G_f(T)$ kcal/mol	SOLIDS	ACTIVITY	$-\Delta G_f(T)$ kcal/mol
Al2O3	1.78474-001	257.386	Al2O3	7.54940-001	262.546
K2O	7.58123-012	11.717	K2O	1.19019-012	5.095
SiO2	5.72664-002	140.235	CaO	1.67521-004	105.177
K2SiO3	1.02079-005	212.663	K2Al18O28	1.86539-002	2461.015
K2Si2O5	5.87255-006	361.151	SiO2	7.32023-002	141.113
K2Si4O9	3.63963-008	643.898	KAlSiO4	9.47894-002	325.465
KAlO2	2.69706-003	162.268	KAlSi2O6	1.31532-001	477.102
KAlSiO4	4.26768-002	322.611	KAlSi3O8	3.42052-003	614.513
KAlSi2O6	7.74998-002	475.210	CaSiO3	4.53115-003	267.437
CaAl2Si4O12*	6.23400-001	999.996	Ca2SiO4	3.50604-005	386.324
Al6Si2O13	7.24322-004	1065.719	Ca3Al2O6	5.06273-010	595.818
K2Al18O28	4.88640-006	2431.515	Ca12Al14O33	9.78757-034	3208.295
CaO	1.81221-005	97.222	CaAl2O4	3.71238-003	379.810
CaSiO3	4.39733-003	267.330	CaAl4O7	6.00958-003	645.083
CaAl2O4	2.96787-003	379.009	Al6Si2O13	1.02573-002	1075.199
CaAl2Si2O8	6.16715-003	682.555	CaAl2Si2O8	6.10436-003	682.519
Ca2Al2SiO7	3.11490-004	657.687	Ca2Al2SiO7	3.87792-004	658.471
Ca12Al14O33	6.50148-033	3215.068			

Gross composition in wt%: Al2O3(30), CaO(30), SiO2(38), K2O(2). See figure 1.
1 cal = 4.184 Joule.

Table 4. Model Activity Data for "Simplified Western" Type Coal Slag at 1800 K

LIQUIDS	ACTIVITY	$-\Delta G_f(T)$ kcal/mol	SOLIDS	ACTIVITY	$-\Delta G_f(T)$ kcal/mol
Al2O3	7.43698-007	257.386	Al2O3	3.14581-006	262.545
K2O	2.08643-009	11.717	K2O	3.27550-010	5.095
SiO2	3.99271-001	140.235	CaO	1.37592-003	105.177
K2SiO3	1.95869-002	212.663	K2Al18O28	1.94455-048	2461.015
K2Si2O5	7.85644-002	361.151	SiO2	5.10379-001	141.113
K2Si4O9	2.36697-002	643.898	KAlSiO4	2.23805-002	325.465
KAlO2	9.13340-005	162.268	KAlSi2O6	2.16526-001	477.102
KAlSiO4	1.00763-002	322.611	KAlSi3O8	3.92589-002	614.513
KAlSi2O6	1.27579-001	475.210	CaSiO3	2.59478-001	267.437
CaAl2Si4O12*	5.04178-002	999.996	Ca2SiO4	1.64905-002	386.324
Al6Si2O13	2.54756-018	1065.719	Ca3Al2O6	1.16890-012	595.818
K2Al18O28	5.09376-052	2431.515	Ca12Al14O33	2.01238-060	3208.295
CaO	1.48845-004	97.222	CaAl2O4	1.27056-007	379.810
CaSiO3	2.51815-001	267.330	CaAl4O7	8.57057-013	645.083
CaAl2O4	1.01576-007	379.009	Al6Si2O13	3.60771-017	1075.199
CaAl2Si2O8	1.02604-005	682.555	CaAl2Si2O8	1.01560-005	682.519
Ca2Al2SiO7	6.10493-007	657.687	Ca2Al2SiO7	7.60039-007	658.471
Ca12Al14O33	1.33674-059	3215.068			

Gross composition wt%: Al_2O_3(10), CaO(30), SiO_2(46), K_2O(14). See figure 2.

Table 5. Model Activity Data for K_1 Slag at 1800 K

LIQUID COMPONENT	ACTIVITY	$-\Delta G_f(T)$ kcal/mol
Al2O3	7.931-003	271.513
K2O	4.046-009	22.491
SiO2	5.923-002	148.892
K2SiO3	2.696-002	232.350
K2Si2O5	1.604-002	389.517
K2Si4O9	2.222-005	690.301
KAlO2	1.313-002	173.921
KAlSiO4	2.149-001	342.322
KAlSi2O6	4.037-001	502.647
CaAl2Si4O12*	2.223-001	1059.776
Al6Si2O13	6.800-008	1120.204
K2Al18O28	1.763-015	2574.818
CaO	1.271-004	102.284
CaSiO3	3.190-002	280.295
CaAl2O4	9.251-004	397.158
CaAl2Si2O8	2.056-003	717.787
Ca2Al2SiO7	7.044-004	689.141
Ca12Al14O33	3.157-032	3361.945

See Table 2 for gross composition, where $Fe_2O_3 = Al_2O_3$ and MgO = CaO has been assumed.

Table 6. Model Activity Data for K_2 Slag at 1600 K

LIQUIDS	ACTIVITY	$-\Delta G_f(T)$ kcal/mol
Al2O3	2.173-005	271.513
K2O	1.697-009	22.491
SiO2	4.415-002	148.892
K2SiO3	8.176-002	232.350
K2Si2O5	4.872-002	389.517
K2Si4O9	4.752-005	690.301
KAlO2	9.127-004	173.921
KAlSiO4	1.862-002	342.322
KAlSi2O6	2.996-002	502.647
CaAlSi4O12*	2.364-001	1059.776
Al6Si2O13	2.384-016	1120.204
K2Al18O28	1.294-036	2574.818
CaO	1.281-003	102.284
CaSiO3	5.370-001	280.295
CaAl2O4	4.327-005	397.158
CaAl2Si2O8	1.111-004	717.787
Ca2Al2SiO7	9.165-004	689.141
Ca12Al14O33	4.046-036	3361.945

See Table 2 for gross composition, where Fe_2O_3 = inert and MgO = CaO is assumed. A small amount of Ca_2SiO_4 solid precipitated.

Table 7. Model Activity Data for CaO Enriched "Simplified Rosebud" Coal Ash at 1600 K

LIQUIDS	MOLES	ACTIVITY	$-\Delta G_f(T)$ kcal/mol	SOLIDS	MOLES	ACTIVITY	$-\Delta G_f(T)$ kcal/mol
Al2O3	6.49308-003	3.68800-003	285.903	Al2O3	.00000	6.26492-002	293.783
K2O	4.31168-010	2.44899-010	33.264	K2Al18O28	.00000	1.64941-006	2771.195
SiO2	5.74513-003	3.26317-003	157.473	KAlSiO4	.00000	5.71445-001	370.003
K2SiO3	2.54535-002	1.44573-002	256.445	KAlSi2O6	.00000	5.18995-002	536.731
K2Si2O5	1.86973-003	1.06199-003	422.581	Ca2SiO4	1.75584-001	1.00000+000	424.511
KAlO2	1.96860-002	1.11814-002	185.659	Ca3Al2O6	.00000	9.55358-004	658.811
KAlSiO4	5.72508-002	3.25179-002	362.030	Ca12Al14O33	.00000	2.94528-012	3543.227
KAlSi2O6	8.36351-003	4.75039-003	530.079	Al6Si2O13	.00000	1.27193-008	1200.693
CaAl2Si4O12*	1.18487+000	6.72994-001	1119.546	Ca2Al2SiO7	1.90320-001	1.00000+000	726.001
Al6Si2O13	2.47930-012	1.40822-012	1175.352	K2O	.00000	7.83675-011	30.094
K2Al18O28	1.49187-014	8.47365-015	2718.095	CaO	.00000	4.39461-002	116.862
CaO	1.60424-003	9.11190-004	106.079	SiO2	.00000	4.31994-003	158.254
CaSiO3	1.81569-001	1.03129-001	292.636	KAlSi3O8	.00000	7.94309-005	692.098
CaAl2O4	3.03973-002	1.72653-002	415.753	CaSiO3	.00000	3.43824-001	295.986
CaAl2Si2O8	7.40754-004	4.20741-004	752.220	CaAl2O4	.00000	9.68669-002	420.551
Ca2Al2SiO7	2.36505-001	1.34332-001	720.416	CaAl4O7	.00000	1.03068-002	715.807
Ca12Al14O33	6.76853-017	3.84446-017	3511.939	CaAl2Si2O8	.00000	1.67225-003	756.059

See Table 2 for gross composition (case CaO 22 wt%)

REFERENCES

(1) Hastie, J. W., Horton, W. S., Plante, E. R., and Bonnell, D. W., Thermodynamic Models of Alkali Vapor Transport in Silicate Systems, IUPAC Conf., Chemistry of Materials at High Temperatures, Harwell, U. K., August 1981; High Temp. High Press, in press.

(2) Hastie, J. W., Plante, E. R. and Bonnell, D. W.: 1982, Alkali Vapor Transport in Coal Conversion and Combustion Systems. ACS Symposium Series, 179, p. 543-600, Gole, J. L., and Stwalley, W. C., eds. Metal Bonding and Interactions in High Temperature Systems with Emphasis on Alkali Metals (see also NBSIR 81-2279).

(3) Eriksson, G.: 1975, Chemica Scripta, 8, p. 100.

(4) JANAF: 1971, Joint Army, Navy, Air Force Thermochemical Tables, 2nd Ed. NSRDS-NBS 37. See also later supplements for 1971-1981.

(5) Robie, R. A., Hemingway, B. S. and Fisher, J. R.: 1979, Thermodynamic Properties of Minerals and Related Substances at 298.15 K and 1 Bar (10^5 Pascals) Pressure and at Higher Temperatures", Geol. Survey Bull 1452 (Washington, DC: U. S. Govt. Printing Office).

(6) Barin, I., Knacke, O.: 1973, "Thermochemical Properties of Inorganic Substances" (New York: Springer Verlag).

(7) Rein, R. H. and Chipman, J.: 1965, Trans. Metall. Soc. AIME, 233, p. 415.

(8) Kelley, K. K.: 1962, U. S. Bur. Mines Rep. Invest., No. 5901.

(9) Hastie, J. W., Bonnell, D. W. and Plante, E. R.: 1982, Thermodynamic Models of Glass Vaporization, American Ceramic Society 84th Annual Meeting, Cincinnati, Ohio.

METAL-LIGAND HEATS AND RELATED THERMODYNAMIC QUANTITIES BY TITRATION CALORIMETRY

James J. Christensen

Department of Chemical Engineering
and the Thermochemical Institute
Brigham Young University
Provo, Utah 84602, USA

Titration calorimetry provides a means to evaluate the thermodynamic quantites log K, ΔH, and ΔS for a wide range of metal-ligand reactions in solution. These quantities give information concerning extent of reaction, bond energies, and solvent interactions, respectively, which provide the basis for understanding metal-ligand reactions. Heat of reaction data for metal-ligand interactions are not numerous and many of the reported values were obtained from K versus 1/T plots making them of questionable validity (1).

In this lecture it will be shown how titration calorimetric data in the form of temperature versus volume (moles) of titrant added can be analyzed to calculate the log K, ΔH, and ΔS values of the reactions taking place. The evaluation of log K and ΔH values from calorimetric data for reactions in solution involves four steps: (1) the experimental determination of the gross heat, Q, liberated in the reaction vessel as a function of titrant added; (2) the calculation of all correction terms for heat effects occurring in the reaction vessel other than those due to the chemical reaction(s) of interest; (3) evaluation of heat effects contributed from reactions other than the ones for which log K and ΔH values are to be evaluated; and (4) the calculation of the energy changes due to the reactions in question and the evaluation of log K and ΔH values. For simplicity and completeness the procedures given here apply to measurements obtained from a continuous titration, isoperibol calorimeter, but with slight modifications they can be applied to either incremental titration or isothermal titration calorimetry data. (2-5).

The principles presented here apply equally to studies in either aqueous or nonaqueous solvents for a wide range of interactions such as proton ionization, metal-ligand interactions, oxidation-reduction reactions, and adduct formation as the heat produced by interactions in solution is dependent only on the types and quantities of species involved. The use of titration calorimetry for the determination of log K, ΔH and ΔS for metal-ligand binding is illustrated for several chemical systems (6-10).

1. TITRATION CALORIMETRY

1.1. Thermograms

The thermogram for a typical calorimetric run is given in Figure 1. The initial or lead (region a) and final or trail

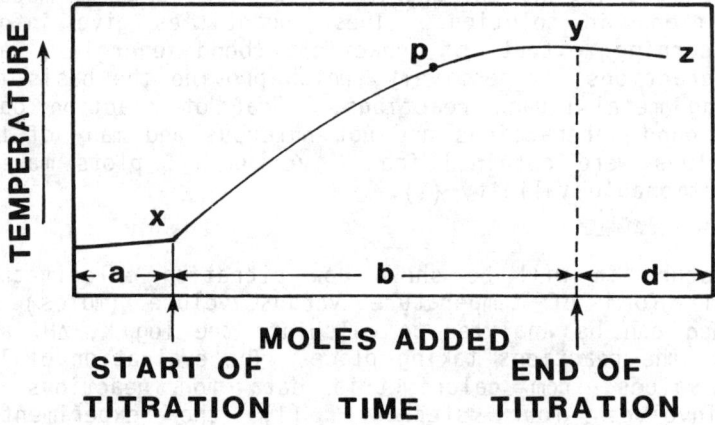

Figure 1. Typical thermogram showing temperature versus time trace for lead (region a), trail (region d) and reaction (region b) periods.

(region d) periods are measures of the nonchemical heat effects due to stirring, resistance heating of the thermistor, and heat losses to the surroundings in these periods. Titrant is added in region b with the resulting temperature rise being due to nonchemical heat effects, dilution of the titrant and titrate, and chemical interactions as the titrant and titrate are mixed. In actual practice, the time at point x is not the same as that at which the titrant buret is turned on because a small air space is usually left at the tip of the titrant delivery tube to prevent premixing of the solutions. Point x may be found directly if a stripchart recorder is used to record the data. If the data are in the form of digital output at set time intervals, then point x must be found by interpolation of the data points or from independent calibration.

The following data can be evaluated from the thermogram: the slopes of the initial (S_i, °C/sec.) and final (S_f, °C/sec) periods, the temperatures of the reaction vessel and contents (T_1, T_2, T_3 ..., T_p, ... T_m) at the reaction times (t_1, t_2, t_3,...,t_p, ..., t_m). A set of Q_p values corresponding to the T_p values can be calculated from the thermogram. Q_p is the quantity of heat which would be evolved by the system if reaction were started at temperature T_x and the temperature after reaction were reduced to the initial temperature, T_x. These Q_p values represent the total heat produced in the reaction vessel from points x to p and must be corrected for all heat effects other than those due to the reaction of interest before they can be used to calculate equilibrium constants. These corrections are detailed in the next section.

1.2. Calculation of corrected $Q_{C,p}$ values

When the data corrections described in the preceeding sections have been made the data from any of the instruments will have been reduced to an array of Q_p vs n_p values where Q_p is the heat produced and n_p the moles titrated or mixed at data point p. The Q_p values will include the contributions from nonchemical energy terms, temperature difference between titrant and titrate, dilution of titrant and from reactions other than the ones for whch thermodynamic calculations are to be made. Nonchemical contributions ($Q_{HL,p}$) to the total energy change measured in the reaction vessel include those energy quantities associated with stirring of the solution, heat losses between the reaction vessel and its surroundings, and resistance heating of the thermistor. The contribution due to the temperature difference between the titrant and titrate ($Q_{TC,p}$) results from the titrant's being introduced into the reaction vessel at a temperature different from the temperature of the contents of the reaction vessel at the beginning of the titration (T_x). The contribution due to dilution of the titrant ($Q_{D,p}$) arises because of chemical changes such as solvation, hydrolysis, and ion paring occuring as the titrant is added to the titrate. Other reactions than the ones of interest occuring in the calorimeter will give rise an energy contribution ($\Delta H_R \Delta n_R$) to the total energy change. Proceedures for calculating the above energy effects are described in the literature in great detail (2-5).

The heat change due to the reactions of interest associated with the addition of reactants corresponding to point p, is determined by subtracting from the overall heat term, Q_p, the heat effects given by the previously described correction terms

$$Q_{C,p} = Q_p - Q_{HL,p} - Q_{TC,p} - Q_{D,p} - \Delta H_R \Delta n_R \qquad (1)$$

where the last term above refers to those reactions occurring other than the ones for which the ΔH values are to be determined. The $Q_{C,p}$ value is a function only of the K and ΔH values for the reactions of interest. For isothermal calorimeters $Q_{HL,p} = 0$.

1.3. Calculation of ΔH values

The addition of titrant to the titrate solution produces one or more reactions where the extent of the reaction(s) and the energy produced are related to the corresponding equilibrium constant(s) and enthalpy change(s) for the reaction(s). The equations relating the heat produced, the equilibrium constant(s), and enthalpy change(s) for the reaction(s) are generally complex. It is convenient to express the relationship among these quantities for the general case of n reactions occurring in the reaction vessel in the form given by equation 2

$$Q_{C,p} = \sum_{i=1}^{n} \Delta H_i \Delta n_{i,p} \qquad (2)$$

where $\Delta n_{i,p}$ is the change in the moles of product i formed from point x to point p and is a function of the equilibrium constant for reaction i.

In general, the best values of ΔH are calculated by a least squares analysis of equation 2. The error square sum over the m data points is given by equation 3.

$$U(\Delta H_i) = \sum_{p=1}^{m} (Q_{C,p} - \sum_{i=1}^{n} \Delta n_{i,p} \Delta H_i)^2 \qquad (3)$$

where the subscript p is over all the data points and the subscript i is over all the reactions being studied. The best values for ΔH for a given run are those which minimize $U(\Delta H_i)$, that is, those values which satisfy equation 4

$$\partial U(\Delta H_i)/\partial \Delta H_k = 0 = \sum_{p=1}^{m} Q_{C,p} \Delta n_{k,p} - \sum_{p=1}^{m} \Delta n_{k,p} \sum_{i=1}^{n} (\Delta n_{i,p} \Delta H_i) \qquad (4)$$

where (k = 1,2,...,n). The n expressions given by equation 4 are all homogeneous first-order linear equations in the ΔH_i

values and are easily solved.

1.4. Calculation of K values

The equilibrium constant for a given reaction can be determined by titration calorimetry if the magnitudes of K and ΔH for the overall reaction taking place in the reaction vessel are within certain limits. The family curves presented in Figure 2a shows that increased overall curvature of the thermogram is obtained with decreasing values of K (ΔH is assumed to be constant); in other words, the shape of a given curve is a function of the K value.

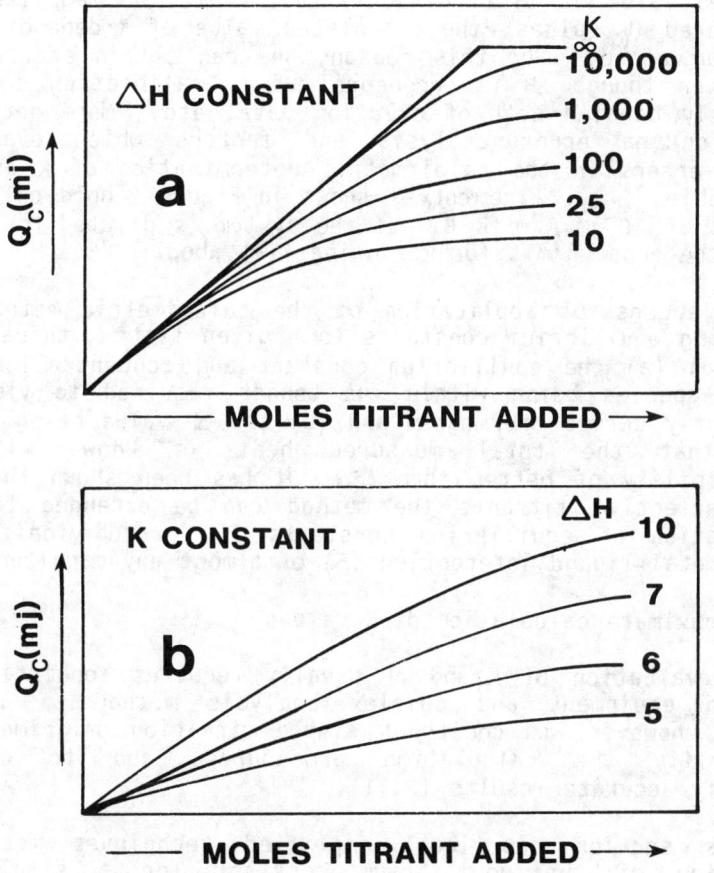

Figure 2. Dependence of calorimetric titration data on K and ΔH for the reaction A + B = C.

The curves for systems with K values greater than approximately 10^4 differ only slightly from one another; hence it is difficult to make accurate calculations of K in these cases. For reactions with K values less than 10, very little reaction takes place and hence very little heat is evolved. In addition, the shape of the curve is little affected by incremental decreases in K for K values either greater than 10^4 or less than approximately 10. The magnitude of the Q_c value is directly related to ΔH in that is ΔH changes there is a corresponding change in Q_c as illustrated in Figure 2b. It follows, that in order to obtain data which can be used to calculate reliable K and ΔH values, the lower the K value is, the higher the ΔH must be. It is emphasized that, while the calculated value of ΔH is directly related to the magnitude of the measured Q_c values, the calculated value of K depends only on the curvature. For this reason, one can obtain accurate K values even though ΔH is erroneous due to calibration errors, incorrectly handled heat of dilution data, etc. More detailed information on error analysis and factors which evaluate possible errors in the caloirmetric determination of K and ΔH is available (2-5). The curves shown in Figure 2 hold only for titration of 10^{-2}M A with B. If the system is diluted to 10^{-4}M A, then the upper limit for measuring K is about 10^6.

The successful application of the calorimetric method of determining equilibrium constants to a given system, therefore, depends on (a) the equilibrium constant and concentrations of reacting species being within the bounds required to yield a sufficiently curved thermogram and (b) the ΔH value being large enough that the total measured heat is known with a reproducibility of better than 5%. It has been shown that by use of selective titrants the method can be extended to the determination of equilibrium constants for proton ionization (4) and metal-ligand interaction (5) of almost any magnitude.

1.5. Approximate calculation of K values

Precise evaluation of K and ΔH usually requires sophisticated titration equipment and complex analysis methods. It is possible, however, to construct simple titration calorimeters, to simplify the calculation procedures, and to obtain moderately accurate results (3,11).

This section describes simplified techniques for the calculation of the equilibrium constant for a simple pH independent reaction using titration calorimetric data. Assume that the reaction producing the thermogram in Figure 2 is for the association of A with B to yield AB. The heat, corrected for all extraneous heat effects, due to the reaction from the start of the titration to any point, p, on the thermogram will be $Q_{C,p}$. This quantity is related by equation 5 to the number of moles of AB formed

$$Q_{C,p} = \Delta H (\Delta n_p) \tag{5}$$

where ΔH is the change in enthalpy for the reaction and Δn_p is derived from the concentration of each species present in the reaction vessel at point p, which necessitates knowing the value of the equilibrium constant for the reaction. However, in this case, the K value for the formation of AB is not known, and therefore Δn_p is not known. If one assumes a value for K, a corresponding value of Δn_p can be calculated A value for ΔH is then obtained at a given point p using equation 5. This operation is carried out for each of the chosen points on the thermogram, $Q_{C,1}$, $Q_{C,2}$, $Q_{C,3}$, ..., $Q_{C,n}$ using the same assumed value of K. Because ΔH and K are constant for a given reaction at a constant ionic strength, μ, and temperature, T (T and μ vary only slightly during the course of reaction), it is to be expected that the calculated ΔH value will be the same for each point. If such is not the case, a new value of K is chosen and ΔH values are calculated again. The process is repeated until a K value is found such that the ΔH values are the same at each point p throughout the run. When this occurs, the proper values for K and ΔH will have been determined. The value for ΔG is determined by the relationship $\Delta S = (\Delta H - \Delta G)/T$.

To illustrate the above procedure, the following experimental conditions and concentrations have been assumed for the reaction $A + B = AB$. A is the titrant, B is the titrate, and four data points taken at four minute intervals are used in the calculations. The titrant, 0.200 M A, is titrated into 100.00 cm^3 of 0.100 M B at a constant rate of 0.600 cm^3 min^{-1} giving a total of 2.40, 4.80, 7.20 and 9.60 cm^3 of titrant delivered at the end of each interval. After the appropriate heat corrections are made, it is determined that the total heat of reaction at the end of each time interval is: $Q_{C,1} = 1.04$ J, $Q_{C,2} = 1.81$ J, $Q_{C,3} = 2.37$ J and $Q_{C,4} = 2.77$ J. the concentrations of the various species in the calorimeter at the end of each time interval can be calculated using equations 6-10,

$$K = [AB]_p / [A]_p [B]_p \tag{6}$$

$$C_{A,p} = [A]_p + [AB]_p \tag{7}$$

$$C_{B,p} = [B]_p + [AB]_p \tag{8}$$

$$K = [AB]_p / (C_{A,p} - [AB]_p)(C_{B,p} - [AB]_p) \tag{9}$$

$$K[AB]_p^2 - (KC_{B,p} + KC_{A,p} + 1)[AB]_p + KC_{A,p}C_{B,p} = 0 \tag{10}$$

where the bracketed quantities represent the equilibrium concentrations of the species indicated. $C_{A,p}$ and $C_{B,p}$ represent the analytical or total concentrations of A and B irrespective of the species they are combined with, these values are known at each point p. For example, at the end of the first time interval 2.40 cm^3 of titrant has been added to the reaction vessel, and $C_{A,1} = (0.200M)(2.40/102.40) = 4.69 \times 10^{-3}$M, similarly, $C_{B,1}$ is calculated to be 9.76×10^{-3}M. In calculating the equilibrium concentrations of the species present after each time interval, a value of K must be assumed. In particular, $[AB]_p$ is calculated from equation 9, using the assumed K value and the analytical concentrations of A and B. Equations 7 and 8 may then be used to calculate the equilibrium concentrations of A and B, respectively. In our example, only the value of $[AB]_p$ is necessary to complete the calculations. The moles of AB formed, Δn_p, can be calculated from the relation $\Delta n_p = [AB]_p V_p$ where V_p is the volume of solution in the reaction vessel at the end of the time interval in question. A value for ΔH is then calculated using equation 5. In our example two values of K were chosen, 200 and 100, and the resulting species concentrations and ΔH values at the end of each time interval are given in Table 1. Values of ΔH were constant when K was chosen to be 100, but not when K was taken to be 200. Therefore, for this hypothetical case, K = 100, and ΔH = 5.0 kJ mol^{-1}. An actual determination would require that activity coefficients be incorporated into the calculations, that many more data points be taken from the curve, and that smaller increments be used in estimating K. (Incidentally, if K is accurately known, the ΔH value for the reaction is determined by solving equation 10 for $[AB]_p$, converting it to Δn_p, and calculating ΔH from equation 5).

It should be stressed that the above method is useful only for the determination of approximte values of K and ΔH.

1.6. Evaluation of K values by a least squares technique

The method of calculating the heat, $Q_{C,p}$ (equation 1), released by the reaction for which the K values are to be determined has been described. The mathematical relationships among the heat produced, the equilibrium constant(s), and the enthalpy chang(s) for the reaction(s) are given in equation 2. A simple method of solving equation 2 for K and ΔH was outlined previously for the reaction A + B = AB (n = 1).

In general, the best values for K_i and ΔH_i are those which minimize $U(K_i, \Delta H_i)$ (see equation 3), that is, those values which satisfy equations 11 and 12.

Table 1. Calculation of K and ΔH values from titration calorimetric data

QUANTITY, UNITS		4 min (p = 1)	8 min (p = 2)	12 min (p = 3)	16 min (p = 4)
$C_{A,p}$, mol dm^{-3}		4.69×10^{-3}	9.16×10^{-3}	1.34×10^{-2}	1.75×10^{-2}
$C_{B,p}$, mol dm^{-3}		9.76×10^{-3}	9.54×10^{-3}	9.33×10^{-3}	9.12×10^{-3}
[AB], mol dm^{-3}	(K^a=200)	2.74×10^{-3}	4.57×10^{-3}	5.67×10^{-3}	6.31×10^{-3}
[AB], mol dm^{-3}	(K^a=100)	2.04×10^{-3}	3.46×10^{-3}	4.42×10^{-3}	5.06×10^{-3}
n_{AB}, mol	(K^a=200)	2.80×10^{-4}	4.79×10^{-4}	6.08×10^{-4}	6.92×10^{-4}
n_{AB}, mol	(K^a=100)	2.09×10^{-4}	3.63×10^{-4}	4.74×10^{-4}	5.55×10^{-4}
Q_C, J		1.04	1.81	2.37	2.77
ΔH, J mol^{-1}	(K^a=200)	3729	3790	3897	4011
ΔH, J mol^{-1}	(K^a=100)	5000	5000	5000	5000

^{a}Correct values of K and ΔH are 100 dm^{3}mol^{-1} and 5000 J mol^{-1}, respectively.

$$\frac{\partial U(K_i, \Delta H_i)}{\partial K_k} = 0 = \sum_{p=1}^{m} (Q_{C,p} \Delta n_{k,p} - \sum_{p=1}^{m} \Delta n_{k,p} \sum_{i=1}^{n} (\Delta n_{i,p} \Delta H_i)) \quad (11)$$

$$\frac{\partial U(K_i, \Delta H_i)}{\partial K_k} = 0 = \sum_{p=1}^{m} (Q_{C,p} - \sum_{i=1}^{n} (\Delta n_{i,p} \Delta H_i)) \frac{\partial}{\partial K_k} (\Delta n_{i,p} \Delta H_i) \quad (12)$$

where $(k = 1, 2, \ldots, n)$. The n equations given by equation 11 are all homogeneous first-order linear equations in the ΔH_i values and may be solved easily if K_i values, and therefore $\Delta n_{i,p}$ values are known. The n equations given by equation 12 are nonlinear in K_i values and must be solved either by trial and error or by some iterative technique. A complete and accurate solution of equations 11 and 12 involves five steps: (a) assumption of initial K_i values; (b) calculation of the concentration of each species in the reaction vessel at each data point using the assumed K_i values; (c) calculation of the best ΔH_i values corresponding to the K_i values chosen; (d) evaluation of the K_i and ΔH_i values to establish how well they fit the experimental data; (e) recalculation of the quantities in steps b, c, and d, using new K_i values until the best set of K_i and ΔH_i values is found. These five steps have been discussed in detail (3-5).

1.7. Typical values of Log K and ΔH values from titration calorimetry

Examples of log K and ΔH values for proton- and metal ion-ligand association obtained by use of the calorimetric titration procedure are given in Table 2 together with comparative values determined by other methods (6). The excellent agreement of these results with literature values indicates that a wide range of K values can be accurately determined by this method.

Our experience shows that calorimetry can be used effectively and confidently for the simultaneous determination of equilibrium constants and enthalpy changes provided that appropriate log K and ΔH values, and titrant and reactant concentrations exist (12). For example, if log K for the reaction is 2 and the concentration of titrant is 0.5 F, then ΔH must be at least 0. kcal/mole to give an error of <0.05 log K units.

2. APPLICATIONS OF CALORIMETRY TO SPECIFIC SYSTEMS

The use of titration calorimetry to investigate the thermodynamic of metal and organic cations binding with ligands

Table 2. Comparison of K and ΔH values determined by calorimetric titration with those determined by conventional procedures

Reaction	-log K	Method[a]	ΔH (kcal./mole)	Method
$HSO_4^- = H^+ + SO_4^{2-}$	1.97±0.03 1.99	I II	-4.9±0.2 -5.2±0.01	I V
$HPO_4^{2-} + OH^- = PO_4^{3-} + H_2O$	-1.6±0.03 -1.625±0.01	I IV	-9.1±0.4 -9.9±0.5	I VI
$ad = H^+ + ad^-$	12.35±0.03 12.5	I II	9.7±0.2	I
$cyt = H^+ + cyt^-$	12.15±0.05 12.16	I II	11.5±0.1	I
$Hmet = H^+ + met^-$	3.76±0.04 3.74	I II	4.92±0.15 4.96	I V
$Hpy^+ = H^+ + py$	5.17±0.02 5.18	I III	4.98±0.04 4.80±0.01	I I
$Him^+ = H^+ + im$	6.99±0.02 6.99	I II	8.78±0.03 8.79	I V
$Hthma^+ - H^+ + thma$	8.03±0.37 8.075	I II	11.39±0.04 11.38	I V

Table 2. (cont.) Comparison of K and ΔH values determined by calorimetric titration with those determined by conventional procedures

Reaction	-log K	Method[a]	ΔH (kcal./mole)	Method
$Ag^+ + py = Agpy^+$	-2.04±0.06 -2.00	I II	-4.6	I
$Ag^+ + 2py = Agpy_2^+$	-4.09±0.05 -4.11	I II	-11.25±0.09	I
$Cu^{2+} + py = Cupy^{2+}$	-2.50±0.02 -2.52	I III	-4.02±0.08	I
$Cu^{2+} + 2py = Cupy_2^{2+}$	-4.30±0.05 -4.38	I III	-8.86±0.1	I
$Cu^{2+} + 3py = Cupy_3^{2+}$	-5.16±0.06 -5.69	I III	-16.1±0.6	I
$Cu^{2+} + 4py = Cupy_4^{2+}$	-6.04±0.1 -6.54	I III	-21.5±1.5	I

[a] I, calorimetric determination of equilibrium constant and enthalpy change; II, hydrogen electrode (potentiometric measurements); III, glass electrode (pH measurements); IV, spectrophotometry; V, pK as f(T); VI, calorimetry.

met⁻, aniline-m-sulphonate ion; py, pyridine; im, imidazole; thma, tris(hydroxymethyl)methylamine; ad, adenosine; cyt, cytosine.

in solution is illustrated in the following four examples.

2.1. Calorimetric determination of Log K_i, ΔH_i, and ΔS_i values for the interaction of thiourea with $Hg(CN)_2$ in water-formamide solvents at 25°C.

Log K_i, ΔH_i, and ΔS_i (i = 1,2) values have been determined calorimetrically at 25°C for the consecutive reaction of thiourea (Tu) with $Hg(CN)_2$ to form $Hg(CN)_2Tu_2$ in formamide water solvents (7). The values of log K_i, ΔH_i, and ΔS_i (i = 1,2) for reactions (1) and (2) in formamide-water solvents are given in Table 3.

$$Hg(CN)_2 + Tu = Hg(CN)_2Tu \tag{1}$$

$$Hg(CN)_2Tu + Tu = Hg(CN)_2Tu_2 \tag{2}$$

Examination of the thermodynamic values in Table 3 reveals the striking fact that the values do not vary significantly as the ratio of formamide to water in the solvent is increased and in general the results show little deviation from those values determined in pure water (13). This leveling nature of formamide towards the thermodynamic quantities for the interaction of Tu with $Hg(CN)_2$ confirms previous views that formamide is a water-like solvent.

2.2. Thermodynamics of formation of 18-crown-6 complexes with arenediazonium and anilinium salts in methanol at 25°C

As part of a systematic investigation of the bonding of organic cations to crown ethers, we have determined log K, ΔH, and TΔS values for formation of 18-crown-6 complexes with simple and substituted arenediazonium and anilinium salts in methanol at 25°C by titration calorimetry (8). We find that complexation of arenediazonium salts is much more sensitive to steric factors due to aromatic substitution than is that of anilinium salts, and is very sensitive to electronic factors as well.

Table 4 lists log K, ΔH, and TΔS values for the interaction of 18-crown-6 with the diazonium and anilinium cations studied. The values reported represent the average of results from 3-6 independent determinations. As the log K data illustrate, binding of 18-crown-6 to diazonium ions is invariably weaker than is that to corresponding anilinium ions. This difference is due to must stronger entropic opposition to reaction in the diazonium case.

X-Ray crystallographic data show that NH_4^+ binds to dicyclohexano-18-crown-6 via hydrogen bonds to the ether oxygen

Table 3. Log K_i, ΔH_i and ΔS_i values for the stepwise interaction of Tu with $Hg(CN)_2$ in formamide-water solvent mixtures at 25°C and $\mu=0$.

Formamide (%, w/w)	Log K_1	Log K_2	ΔH_1 (kcal/gmole)	ΔH_2 (kcal/gmole)	ΔS_1 (cal/gmol degree)	ΔS_2 (cal/gmol degree)
0	1.97±0.06	0.58±0.04	-1.5±0.1	-7.9±0.2	4.0	-23.8
20	2.08±0.06	0.56±0.10	-1.4±0.1	-9.1±0.4	4.8	-28
40	1.95±0.05	0.57±0.11	-1.7±0.1	-9.2±0.4	3.2	-28
60	2.01±0.08	0.60±0.09	-1.7±0.1	-8.7±0.4	3.5	-26
80	2.02±0.08	0.64±0.11	-1.6±0.1	-7.8±0.4	3.8	-23
100	2.02±0.08	0.64±0.11	-1.5±0.1	-7.4±0.4	4.0	-22

Table 4. Log K, ΔH, and TΔS for formation of 18-crown-6 complexes with aremadiazonium and anilinium salts in methanol at 25°C.

Cation	Anion	Log K	ΔH[a]	TΔS[a]
Ph-N≡N+	BF$_4^-$	2.37±0.04	-14.0±0.5	-10.8
p-MeC$_6$H$_4$-N≡N+	BF$_4^-$	2.26±0.02	-10.03±0.28	-6.95
o-MeC$_6$H$_4$-N≡N+	BF$_4^-$	b	b	---
oo'-Me$_2$C$_6$H$_3$-N≡N+	BF$_4^-$	c	c	---
p-Et$_2$NC$_6$H$_4$-N≡N+	BF$_4^-$	d	d	---
PhNH$_3^+$	Br$^-$	3.80±0.03	-9.54±0.14	-4.36
p-MeC$_6$H$_4$NH$_3^+$	Br$^-$	3.82±0.04	-9.92±0.22	-4.71
mm'-Me$_2$C$_6$H$_3$NH$_3^+$	I$^-$	3.74±0.02	-9.07±0.06	-3.97
o-MeC$_6$H$_4$NH$_3^+$	Br$^-$	2.86±0.03	-7.59±0.15	-3.69
oo'-Me$_2$C$_6$H$_3$NH$_3^+$	Br$^-$	2.00±0.05	-5.65±0.27	-2.92

[a] kcal mol^{-1}. [b] No measurable heat other than heat of dilution, indicating that log K is very small and/or ΔH is close to zero. [c] Cation decomposes too rapidly to allow measurement. [d] Heat is produced, but quantity is too small to allow calculation of thermodynamic parameters.

atoms. It is likely that the anilinium ion behaves similarly sitting on top of the macrocycle (Figure 3b) rather than inserting itself into the ring cavity as does the diazonium ion[3] (Figure 3a). Thus, the diazonium group brings the entire aromatic system closer to the macrocyclic ligand in the complexed form than does an ammonium group. For this reason, one would expect a larger decrease in complex stability as a result of ortho-substitution on the benzene-diazonium ion than on the anilinium ion since such substitution would more effectively block the approach of the substituted diazonium cation to the ligand. The data in Table 4 support this idea. Specifically, the addition of just one ortho methyl group to the benzenediazonium ion effectively eliminates reaction altogether, whereas the stability of the anilinium complex drops approximately one log K unit upon each addition of a methyl group adjacent to the ammonium group.

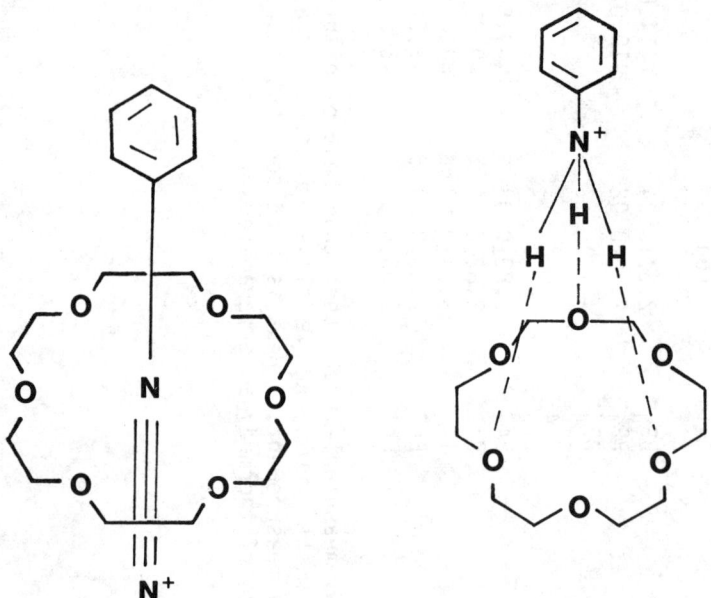

Figure 3. The structures of 18-crown-6 complexes of the benzenediazonium ion (left) and the anilinium ion (right).

2.3. Thermodynamic origin of the macrocyclic effect in crown ether complexes of Na^+, K^+, and Ba^{2+}

As compared to the noncyclic analogue pentaglyme, 18-crown-6 complexes of Na^+, K^+, and Ba^{2+} have superior stabilities in methanol and methanol/water mixtures. This macrocyclic effect amounts to 3-4 orders of magnitude in the equilibrium formation constant. One of the objectives of this study was to determine whether the macrocyclic effect found in polyether complexes of Na^+, K^+, and Ba^{2+} was the result of either enthalpic or entropic factors or a combination of both factors. This was accomplished by comparing log K, ΔH, and TΔS values for the reaction of these metal cations with cyclic 18C6 with corresponding values for PG complexes and pentaethylene glycol (PEG) complexes (9). It was hoped that such experiments would

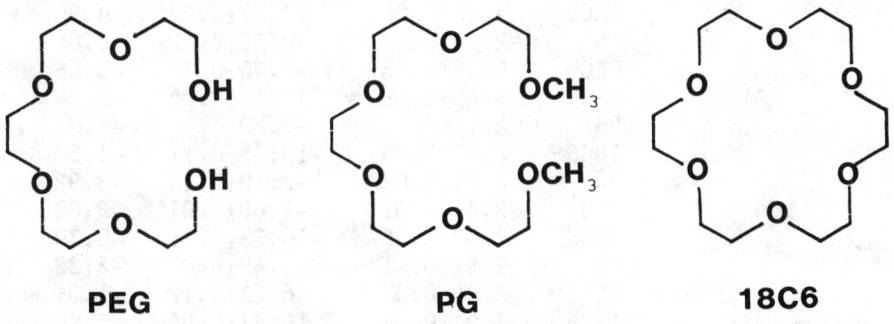

PEG   **PG**   **18C6**

help in understanding the nature of the polyether-metal cation interaction and give some information about the macrocyclic effect in general. Log K, ΔH, and TΔS values for the reactions of Na^+, K^+ and/or Ba^{2+} with 18C6, PG, and PEG in 90 wt% methanol and 99 wt% methanol are given in Table 5.

The data in Table 5 indicate that the macrocyclic effect in the five systems studied is primarily the result of favorable enthalpic factors, not entropic factors. The Na/18C6/PG (100% MeOH), K/18C6/PG (99% MeOH), and Ba/18C6/PEG (99% MeOH) comparisons all show negligible entropy contributions (Δ(TΔS) ≤ 0.5 kcal/mol) to the macrocyclic effect. In the K/18C6/PEG (99% MeOH) system, the data show that entropy changes actually work against (Δ(TΔS) = -1.4 kcal/mol) complexation by the cyclic polyether; yet, the highly favorable enthalpy of complexation of K^+ by 18C6 overcomes these entropy changes and is large enough to cause a sizable macrocyclic effect. Even in the one system, Ba/18C6/PG (99% MeOH), where entropy changes (Δ(TΔS)) do contribute to the

Table 5. Log K, ΔH, and TΔS values for reactions of Na^+, K^+ and Ba^{2+} with various ligands in water/methanol mixtures at 25°C

cation	solvent[a]	ligand	log K	ΔH[b]	TΔS[b]
Na^+	90	18C6	3.66±0.02	-6.64±0.07	-1.65
	99	18C6	4.33±0.02	-8.11±0.05	-2.20
	100	18C6	4.36±0.02	-8.36±0.37	-2.41
		PG	1.44±0.05	-4.02±0.28	-2.96
K^+	90	18C6	5.35±0.25	-11.77±0.05	-4.47
		PG	1.95±0.01	-7.00±0.08	-4.34
		PEG	1.91±0.01	-4.57±0.02	-1.96
	99	18C6	6.05±0.05	-13.21±0.07	-4.96
		PG	2.27±0.02	-8.15±0.12	-5.05
		PEG	2.05±0.03	-6.36±0.10	-3.56
	100	18C6	6.06±0.03	-13.41±0.06	-5.14
		PG	2.1	-8.70	-5.8
Ba^{2+}	90	18C6	6.56±0.09	-10.33±0.11	-1.38
		PG	2.33±0.03	-7.10±0.16	-3.92
		PEG	3.45-0.01	-7.60±0.01	-2.89
	99	18C6	7.03±0.06	-10.38±0.15	-0.79
		PG	2.51±0.01	-5.65±0.05	-2.23
		PEG	3.96±0.08	-6.71±0.12	-1.31
	100	18C6	7.04±0.08	-10.41±0.06	-0.80
		PG	2.3	-5.55	-2.4

[a] Given as wt % methanol in methanol/water mixture. [b] kcal/mol.

macrocyclic effect, 77% of this effect is the result of enthalpic factors. thus, the expected entropy contribution to the macrocyclic effect resulting from less unfavorable configurational entropy changes in the cyclic ligand was not in fact observed, and if configurational entropy changes do play a role at all, they are masked by other thermodynamic factors.

2.4. Anomalous stability sequence of lanthanide(III) chloride complexes with 18-crown-6 in methanol. Abrupt decrease to zero from Gd^{3+} to Tb^{3+}

Although the preparation of solid complexes of lanthanide(III) metal cations with cyclic polyethers has been reported, thermodynamic data describing these reactions in solution are not available. Furthermore, contradictory results have been presented as to whether or not complexes are formed between the latter members of the lanthanide series and the crown-6 members

Table 6. Log K, ΔH,[a] and $T\Delta S$[a] values for the interaction of 18-crown-6 with several trivalent rare earth chlorides in methanol at 25°C and $\mu = 0.005$

	Log K	ΔH	$T\Delta S$
La^{3+}	3.29±0.03	2.81±0.04	7.30
Ce^{3+}	3.57±0.20	2.54±0.13	7.41
Pr^{3+}	2.63±0.28	4.46±0.40	8.05
Nd^{3+}	2.44±0.16	4.77±0.26	8.10
Sm^{3+}	2.03±0.07	3.67±0.04	6.44
Eu^{3+}	1.84±0.14	3.06±0.14	5.57
Gd^{3+}	1.32±0.12	3.73±0.15	5.53
(Tb^{3+}, Dy^{3+}, Ho^{3+}, Er^{3+}, Tm^{3+}, Yb^{3+}, Lu^{3+}, UO_2)[b]			

[a] Kilocalories/mole.

[b] No measurable heat other than heat of dilution was found for the mixing of these cations with the ligand.

of the cyclic polyether family. It has been suggested that trivalent rare earth-cyclic polyether interactions could be useful in the separation and purification of lanthanides. Similar application might also be made to the series of actinides, some elements of which form stable tripositive ions of comparable chemical behavior. In addition, isotopic separation of elements from either series by cyclic polyethers may be feasible. Isotopes of Ca^{2+} have already been separated by a solvent extraction technique employing dicyclohexo-18-crown-6. We therefore embarked on a study of the thermodynamics of interaction of the hydrated lanthanide(III) chlorides with cyclic polyethers using a titration calorimetric technique (10).

Log K, ΔH, and $T\Delta S$ values for the reaction in methanol of La^{3+}, Ce^{3+}, Pr^{3+}, Nd^{3+}, Sm^{3+}, Eu^{3+}, and Gd^{3+} with 18-crown-6 are given in Table 6. A plot of the Log K values measured vs. the reciprocal of metal ion radius is given in Figure 4. Identical calorimetric titrations with Tb^{3+}, Dy^{3+}, Ho^{3+}, Er^{3+}, Tm^{3+}, Yb^{3+}, Lu^{3+}, UO_2^{2+}, and Th^{4+} produced no measurable heat other than heat of dilution. Three features of these results are particularly significant. (i) No heat of reaction is observed with the post-Gd^{3+} lanthanide cations. (ii) All reaction enthalpies in Table 6 are positive and, thus, observed stabilities are entropic in origin. (iii) With increasing atomic number, the stabilities of the complex formed decrease, rather than increase as is the case with the trivalent lanthanide complexes of most other ligands.

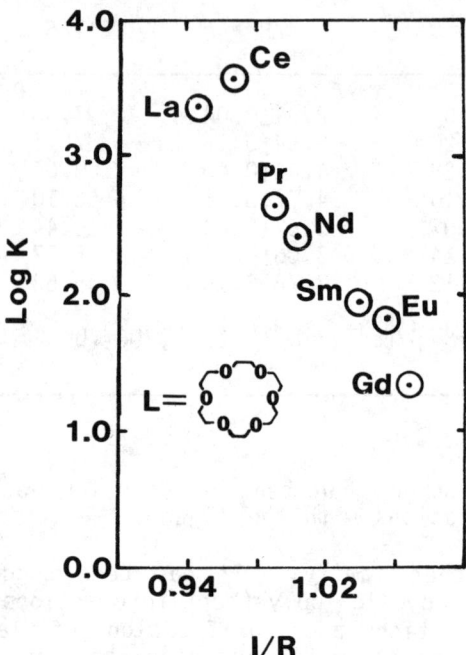

Figure 4. Log K for the reaction $M^{3+} + L = ML^{3+}$ (L = 18-crown-6) in methanol at 25°C in presence of Cl^- vs. reciprocal of cation radius (Å^{-1}). No heat of reaction was observed with Tb^{3+}, Dy^{3+}, Ho^{3+}, Er^{3+}, Tm^{3+}, Yb^{3+}, or Lu^{3+}.

REFERENCES

(1) Christensen, J.J., Eatough, D.J., Izatt, R.M.: 1975, "Handbook of Metal Ligand Heats," 2nd Ed., Marcel Dekker, Inc., New York.
(2) Eatough D.J., Christensen, J.J., Izatt, R.M.: 1974, "Experiments in Thermometric Titrimetry and Titration Calorimetry," Brigham Young University Press, Provo.

(3) Christensen, J.J., Ruckman, J., Eatough, D.J., and Izatt, R.M.: 1972, Thermochimica Acta 3, p. 203.
(4) Eatough, D.J., Christensen, J.J., Izatt, R.M.: 1972, Thermochimica Acta 3, p. 219.
(5) Eatough, D.J., Izatt, R.M., Christensen, J.J.: 1972, Thermochimica Acta, 3, p. 233.
(6) Christensen, J.J., Rytting, J.H., Izatt, R.M.: 1979, J. Chem. Soc. (A), p. 862.
(7) Izatt, R.M., Bartholomew, C.H., Morgan, C.E., Eatough, D.J., Christensen, J.J.: 1971 Thermochimica Acta 2, p. 313.
(8) Izatt, R.M., Lamb, J.D., Rossiter, B.E., Izatt, N.E., Christensen, J.J., Haymore, B.L.: 1978, J.C.S. Chem. Comm., p. 386.
(9) Haymore, B.L., Lamb, J.D., Izatt, R.M., Christensen, J.J.: 1982, Inorg. Chem., 21, p. 1598.
(10) Izatt, R.M., Lamb, J.D., Christensen, J.J., Haymore, B.L.: 1977, J. Am. Chem. Soc. 99, p. 8344.
(11) Martin, C.J., Marini, M.A.: 1979, in "Critical Reviews in Analytical Chemistry," Vol. 8, CRC Press, pp. 221-286 and 407-408.
(12) Christensen, J.J., Wrathall, D.P., Oscarson, J.L., Izatt, R.M.: 1968, Analyst. Chem., 40, p. 1713.
(13) Izatt, R.M., Eatough D.J., Christensen, J.J.: 1968, J. Phys. Chem., 72, p. 2720.

MEASUREMENT OF COMPLEXING CONSTANTS AND ENTHALPIES OF COMPLEXING BY BATCH CALORIMETRY

Michael H. Abraham

Department of Chemistry, University of Surrey, Guildford, Surrey, U.K.

The technique of batch calorimetry is a useful alternative to that of titration calorimetry for the simultaneous determination of $K°$ and $\Delta H°$ values in homogeneous solution. Examples are given of the application of the batch calorimetric technique to acid-base reactions in nonaqueous solvents, and to the complexation of cations by crown ethers. It is shown how measurements in the latter system can yield, through suitable thermodynamic cycles, enthalpies and entropies of transfer from one solvent to another of cations complexed by crown ethers.

1. INTRODUCTION

The determination, not only of standard enthalpies of reaction, but also of equilibrium constants by calorimetry is a particularly convenient and general technique. It is convenient because the same set of experiments will yield both $\Delta H°$ and $K°$, and it is general because the technique does not rely on any particular molecular structure of the species involved - the only requirement is that there be a measureable enthalpy change on reaction. By far the most widespread calorimetric method is that of titration calorimetry, developed by Christensen, Izatt, and their coworkers (1,2). However, the very success of this method has tended to obscure the alternative method of batch calorimetry, and is the purpose of this article to illustrate the application of the latter method method to a number of systems.

Consider a simple equilibrium (1), taking place in homogeneous solution, and characterised by an equilibrium constant, $K°$, and a standard enthalpy of reaction, $\Delta H°$. If a and b are the initial

$$A + B \rightarrow P \qquad (1)$$

concentrations of A and B, and x the concentration of product formed, then $K°$ is given by eqn (2), where the activity coefficient terms correct the observed

$$K° = \frac{x}{(a-x)(b-x)} \cdot \frac{f^x}{f^{a-x} f^{b-x}} \qquad (2)$$

equilibrium constant to that referred to a standard set of conditions (e.g. all species at zero concentration). If $K°$ is very large, then on mixing solutions of A and B, essentially 100% P is formed and the observed enthalpy change, Q, may then be used to

Table 1. Percentage reaction $A + B \rightarrow P$

mol ℓ^{-1}		\multicolumn{4}{c}{$K°$}			
a	b	1	10^2	10^4	10^6
1	10^{-1}	49	99		
10^{-1}	10^{-2}	9	90		
10^{-2}	10^{-3}		49	99	
10^{-3}	10^{-4}		9	90	
10^{-4}	10^{-5}			49	99
10^{-5}	10^{-6}			9	90

calculate $\Delta H°$. If $K°$ is not too large, then values of a and b may be chosen to give only partial reaction, see Table 1, and it is this situation that is necessary for the calorimetric determination of $\Delta H°$ and $K°$. The limits of $K°$ that can be obtained in this way depend on the sensitivity of the calorimetric system. In Table 2 are given values of the observed enthalpy of reaction for the case of eqn (1), in which $\Delta H° = 5000$ cal mol^{-1}. Although values of a and b can be chosen to yield only partial reaction with $K°$ as high as 10^6, the observed enthalpy change then becomes too small to measure. However, the method of batch calorimetry has been used to determine K° values in systems where K° is as low as unity and as high as 10^5 in value.

2. ACID-BASE REACTIONS WITH LOW K° VALUES

Drago and his coworkers (3-5) have used the batch method to obtain quite low equilibrium constants in a

Table 2. Enthalpy of reaction in cal if $\Delta H° = 5000$ cal mol^{-1}

mol ℓ^{-1}		K°			
a	b	1	10^2	10^4	10^6
1	10^{-1}	25	50		
10^{-1}	10^{-2}	0.45	4.5		
10^{-2}	10^{-3}		0.25	0.50	
10^{-3}	10^{-4}		5×10^{-3}	0.045	
10^{-4}	10^{-5}			2×10^{-3}	5×10^{-3}
10^{-5}	10^{-6}			5×10^{-5}	5×10^{-4}

number of systems involving interaction of acids and bases, and in Table 3 are given results (5) of two batch experiments on the m-fluorophenol/pyridine system. The values of ΔH are calculated from the observed enthalpy change as though 100% reaction had taken place, and are related to the standard enthalpy change since $x_1/b_1 = -3976/\Delta H°$, and $x_4/b_4 = -6057/\Delta H°$. Now if the activity coefficients in eqn (2) are set equal to unity,

$$2x_i = \left(a_i + b_i + \frac{1}{K°}\right) - \left[\left(a_i + b_i + \frac{1}{K°}\right)^2 - 4a_i b_i\right]^{\frac{1}{2}} \quad (3)$$

for $i = 1$ and 4, and the expressions in x_i/b_i and eqn (3) may be solved to yield the unique solution that $K° = 32$ mol^{-1} ℓ and $\Delta H° = -6410$ cal mol^{-1}. Of course, in the expressions for x_i/b_i, the basic assumption that the enthalpy of reaction is directly proportional to the fraction of product formed has been used.+

Table 3. Results of two batch experiments on the m-fluorophenol/pyridine system in solvent 1,2-dichloroethane (5)

Expt	mol ℓ^{-1}		cal mol^{-1}
1	a_1 = 0.0614	b_1 = 0.0156	ΔH_1 = -3976
4	a_4 = 0.5389	b_4 = 0.0159	ΔH_4 = -6057

+It should be noted that in some systems, for example those involving ion-exchange, this basic assumption does not hold, see for example, (6). However these ion-exchange processes are not homogeneous solution systems.

When a number of batch experiments have been carried out a graphical solution may be obtained by calculating sets of corresponding ΔH and K values for each experiment, as listed in Table 4. Drago and coworkers have shown that if ΔH is plotted against $1/K$ for each batch experiment, there results a series of curves, the intersection of which gives $\Delta H°$ and $1/K°$ (i.e. $K°$). Although this method is useful as a preliminary scan of the results, the point (or area) of intersection has to be estimated by eye, and as shown in Figure 1 this is not so easy to do.

Table 4. Analysis of all results for the m-fluorophenol/pyridine system (5) $-\Delta H$ in kcal mol^{-1}

Expt	mol ℓ^{-1}		1/K			
	a	b	0.0400	0.0309	0.0250	0.0200
1	0.0614	0.0156	6.99	6.35	5.93	5.55
2	0.1306	0.0165	6.70	6.33	6.09	5.87
3	0.4016	0.0151	6.43	6.30	6.21	6.14
4	0.5389	0.0159	6.51	6.41	6.34	6.28
		$\sigma(\Delta H)$	0.24	0.05	0.17	0.32
	Average $-\Delta H$		6.66	6.35	6.14	5.96

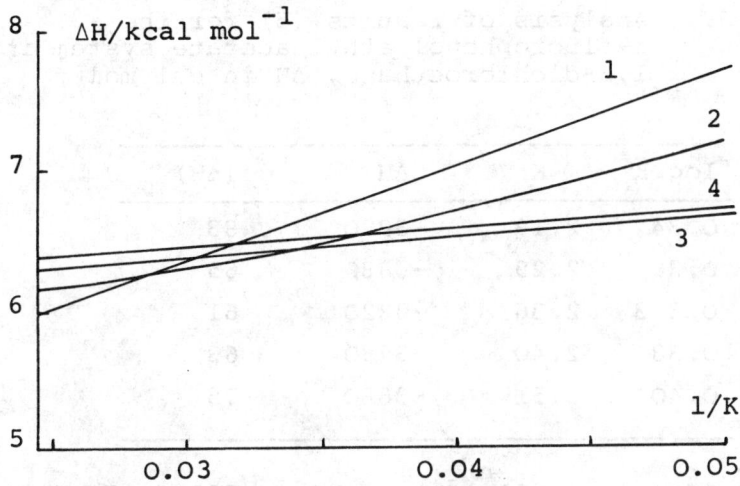

Figure 1. Plot of ΔH against 1/K for the four batch experiments in the m-fluorophenol/pyridine system, see Table 4.

A better method is again to calculate corresponding ΔH and K values but then for each value of K to determine the standard deviation of ΔH, as shown in Table 4. The taken values of ΔH° and K° are those for which σ(ΔH) is a minimum, viz -6.35 kcal mol^{-1} for ΔH° and 32 mol^{-1} ℓ (1/K = 0.0309). The 95% confidence level for ΔH° is ±0.07 kcal mol^{-1}, and the corresponding level for K° is ±3 mol^{-1} ℓ, the latter estimated through the plot of σ(ΔH) against K. One of the smallest equilibrium constants obtained by Nozari and Drago (5) is that for the m-fluorophenol/ ethyl acetate system, and the results of our statistical analysis are in Table 5. The final calculated values of ΔH° = (-3.82±0.07) kcal mol^{-1} and K° = (2.36±0.07) mol^{-1} ℓ agree quite well with those originally reported (5), viz -3.7±0.2 kcal mol^{-1} and 2.5±0.1 mol^{-1} ℓ respectively. As noted by Drago and his co-workers (4,5), for equilibria involving nonelectrolytes there seems no alternative than to set the activity coefficients in eqn (2) equal to unity, and also to assume zero enthalpies of dilution.

Table 5. Analysis of results (5) for the m-fluorophenol/ethyl acetate system in 1,2-dichloroethane, ΔH in cal mol^{-1}

log K	K	ΔH	$\sigma(\Delta H)$
0.34	2.19	-3990	83
0.36	2.29	-3880	65
0.373	2.36	-3820	61
0.38	2.40	-3780	63
0.40	2.51	-3680	75

$K° = 2.36 \pm 0.07$ (95% CL) $\Delta H° = -3820 \pm 70$ (95% CL)

3. M^+/18-CROWN-6 REACTIONS WITH HIGH $K°$ VALUES

We have used the statistical analysis of batch experiments to obtain much larger equilibrium constants than those found by Drago and co-workers, the systems we have studied being of the type shown in eqn (4) where the ligand, L, is 18-crown-6. The most convenient procedure is to measure the enthalpy change, Q, on breaking an ampoule containing a solution of the crown ether in a given solvent (say methanol) into a

$$M^+ + L \rightarrow LM^+ \qquad (4)$$

solution of the metal ion in the given solvent. In separate experiments the enthalpy change on breaking the ampoule of crown ether into pure solvent, Q_L, is measured so as to obtain the corrected enthalpy Q_C. Note that Q_L includes not only the enthalpy of dilution of the crown ether but also the actual enthalpy of ampoule breaking.

L(soln in ampoule) + $M^+(b)$
$\quad\rightarrow L(a-x) + M^+(b-x) + LM^+(x)$ $\quad Q$ (5)
L(soln in ampoule) $\rightarrow L(a)$ $\quad Q_L$ (6)
$L(a) + M^+(b) \rightarrow L(a-x) + M^+(b-x) + LM^+(x)$

$$Q_C = Q - Q_L \qquad (7)$$

Now in eqn (2), the activity coefficient of the ligand at concentration $(a-x)$ may be taken as unity, but the ionic activity coefficients f^x and f^{b-x} can be evaluated either from the limiting or extended Debye-Huckel expressions (8) and (9). In these expressions A and B are the Debye-Huckel constants, I is the concentration M^+ or LM^+, and å is the ion-size parameter. One of the systems we have studied, Li^+/18C6 in methanol, required quite high electrolyte concentrations, between 8×10^{-3} and 2×10^{-2} mol ℓ^{-1}, and so we analysed this system in some detail. In Table 6 is a preliminary survey using the statistical method to fix the approximate values of $K°$ and $\Delta H°$, and in

$$\log f = -A\sqrt{I} \qquad (8)$$

$$\log f = -A\sqrt{I}/(1+B\text{å}\sqrt{I}) \qquad (9)$$

Table 7 are the final results obtained with different activity coefficient corrections. As might be expected for an equilibrium (4) in which there is no net change in electrical charge, these corrections are quite small and within the experimental error. Of course, for less polar solvents than methanol, the activity coefficient corrections would be larger, and also in less polar solvents there might arise complications because of ion association between M^+ and the counter-anion and between LM^+ and the counter-anion.

Table 6. Analysis of results on the Li^+/18C6 system in methanol

log K	ΔH/cal mol^{-1}	$\sigma(\Delta H)$
1.6	-740	120
1.8	-560	65
1.9	-500	51
2.0	-450	46
2.1	-410	47
2.2	-380	51
2.4	-340	60

In addition to evaluating the activity coefficients in eqn (2), it is also possible to correct the enthalpy change Q_C to zero concentration as in eqns (10) to (16).

Table 7. Activity coefficient corrections in the $Li^+/18C6$ system in methanol

å(Li^+)/Å	å(Li^+18C6)Å	log K°	ΔH°/cal mol^{-1}
3	6	2.03	-430
2	4	2.03	-440
2	3	2.03	-440
limiting law		2.01	-470
no corrections		2.00	-390

$$L(a) + M^+(b) \rightarrow L(a-x) + M^+(b-x) + LM^+(x) \quad Q_C \quad (10)$$
$$L(a) \rightarrow L(o) \quad Q_1 \quad (11)$$
$$L(a-x) \rightarrow L(o) \quad Q_2 \quad (12)$$
$$M^+(b) \rightarrow M^+(o) \quad Q_3 \quad (13)$$
$$M^+(b-x) \rightarrow M^+(o) \quad Q_4 \quad (14)$$
$$LM^+(x) \rightarrow LM^+(o) \quad Q_5 \quad (15)$$

$$Q° = Q_C - Q_1 + Q_2 - Q_3 + Q_4 + Q_5 \quad (16)$$

Corrections Q_1 and Q_2 tend to cancel out, and in any case are very small indeed. The enthalpy of dilution term Q_3 also tends to cancel out the term (Q_4+Q_5), but it is possible to calculate the effects due to Q_3, Q_4, and Q_5 using the Debye-Huckel limiting expression for enthalpies of dilution, or a much more complicated expression relating to the extended expression. For the $Li^+/18C6$ system in methanol, we find that over the range of concentrations studied, see Table 8, the correction to ΔH° due to the three enthalpy of dilution terms average 478 (Q_3), -474(Q_4), and -72(Q_5) in cal mol^{-1}, leading to a total correction of only -68 cal mol^{-1} when calculated through the extended expression. The corresponding total correction is -70 cal mol^{-1} when calculated through the limiting law. Since these corrections are theoretical ones, and since they are of the same order as the error in ΔH° (±43 cal mol^{-1}, 95% confidence level), it does not seem worthwhile to include them. As mentioned in connection with activity coefficients, the corrections in less polar solvents may be much larger in magnitude.

Table 8. Calculated total corrections to $\Delta H°$ due to enthalpies of dilution for $M^+ + L \rightarrow LM^+$ in methanol

System	Electrolyte concentration mol ℓ^{-1}	Extended law cal mol^{-1}	Limiting law
$Li^+/18C6$	8×10^{-3} to 6×10^{-2}	-68	-70
$Rb^+/18C6$	2×10^{-4} to 1×10^{-3}	-24	-25

One of the highest equilibrium constants we have been able to determine by batch calorimetry is that between Rb^+ and 18C6 in methanol. For this reaction we reduced the initial concentrations of reactants to $(2 \times 10^{-4} - 2 \times 10^{-3})$ mol ℓ^{-1} for the electrolyte and to 1×10^{-4} mol ℓ^{-1} for the ligand and were able to measure $K°$ as 1.9×10^5, molar scale. However, the plot of $\sigma(\Delta H)$ against log K, see Figure 2, is much flatter than the corresponding plot for complexing of Na^+ ($K° = 3.1 \times 10^4$) indicating that at the reactant concentrations used, the value of $K°$ is approaching the upper limit that is measureable. Indeed the plot for complexing of K^+ (where $K° = 1.1 \times 10^6$) is so flat that we were not able to determine $K°$ in this system. An advantage of the low electrolyte concentrations used, is that corrections for activity coefficients and enthalpies of dilution become insignificant in these methanolic solutions. In Table 8 are results of calculations on enthalpies of dilution, and we confirmed that corrections for activity coefficients had no effect whatsoever on the computed value of $K°$.

As mentioned above, with very large $K°$ values, the a and b values chosen to produce measureable enthalpy changes will lead to 100% reaction, effectively. In such cases, $\Delta H°$ can be obtained but not $K°$; a plot of $\sigma(\Delta H)$ against K or log K gives rise to a flat or almost horizontal curve instead of one with a pronounced minimum, see Figure 2. Thus in our hands, the upper limit of values of $K°$ that can be measured by the batch method is about 10^5.

Christensen and Izatt and their co-workers (7,8) have studied by titration calorimetry a number of complexing reactions that we have investigated by the

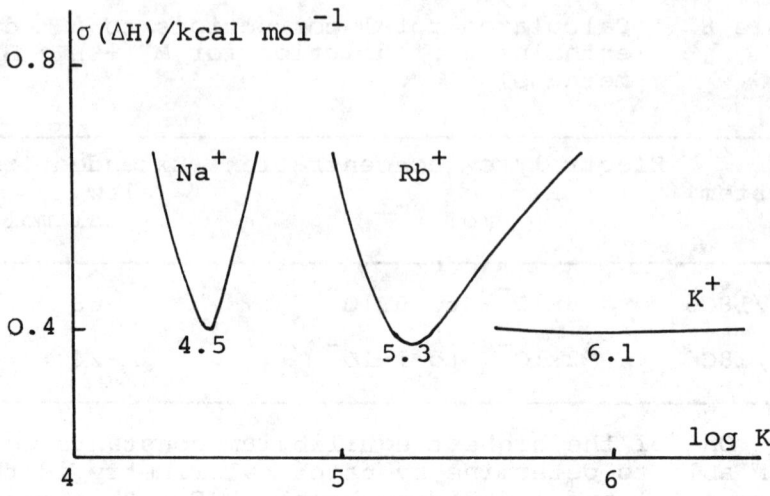

Figure 2. Plots of ΔH against log K for the complexing of M⁺ions with 18C6 in methanol.

batch calorimetric method, and in Table 9 is a comparison of the two sets of results. In general, there is good agreement between the log K° and ΔH° values obtained by the two methods.

Table 9. Comparison of log K° and ΔH° by titration and batch calorimetry, ΔH° in kcal mol^{-1}

With 18C6 in methanol	Titration (7,8) log K°	ΔH°	Batch log K°	ΔH°
Li^+	-	-	2.03	-0.44
Na^+	4.36	-8.4	4.49	-7.66
K^+	6.06	-13.41	-	-13.00
Rb^+	5.32	-12.09	5.27	-12.20
Cs^+	4.79	-11.29	4.69	-12.27
NH_4^+	4.27	-9.27	4.23	-10.38
Ag^+	4.58	-9.15	4.72	-9.66
Average error	±0.04	±0.13	±0.16	±0.31

4. THERMODYNAMICS OF TRANSFER OF M^+18C6 COMPLEXES

Having to hand the log K° and ΔH° values for complex formation between M^+ ions and 18C6 in methanol, we took the opportunity to combine these ΔH° values in methanol, ΔH_M°, with the corresponding values in water, ΔH_W°, (9) to obtain enthalpies of transfer of the complex ions M^+18C6 through (10) a thermochemical cycle.

$$
\begin{array}{ccc}
M^+(aq) + L(aq) & \xrightarrow{\Delta H_W^\circ} & LM^+(aq) \\
\Delta H_t^\circ(M^+) \downarrow \quad \downarrow \Delta H_t^\circ(L) & & \downarrow \Delta H_t^\circ(LM^+) \\
M^+(MeOH) + L(MeOH) & \xrightarrow{\Delta H_M^\circ} & LM^+(MeOH)
\end{array}
$$

The only unknown in the cycle is $\Delta H_t^\circ(LM^+)$ where L=18C6 and this can be evaluated through eqn (17). The determined values of $\Delta H_t^\circ(LM^+)$ for the M^+18C6

$$\Delta H_t^\circ(LM^+) = \Delta H_M^\circ - \Delta H_W^\circ + \Delta H_t^\circ(M^+) + \Delta H_t^\circ(L) \qquad (17)$$

complexes are given in Table 10, together with values for the cryptate complex ions M^+222 where 222 is the 222 cryptand ligand, and values for simple M^+ and R_4M^+ ions (11). The values for the M^+18C6 complexes are quite close to those for the M^+222 complexes, suggesting that in solution these two types of complex ion interact with the solvent in quite similar ways.

Table 10. Values of ΔH_t° for complex ions for transfer from water to methanol, kcal mol^{-1} (10), at 298 K

Na^+	-4.9	Na^+18C6	3.3	Na^+222	5.5	Me_4N^+	0.3
K^+	-4.5	K^+18C6	2.3	K^+222	4.1	Et_4N^+	2.2
Rb^+	-3.7	Rb^+18C6	1.5	Rb^+222	4.1	Pr_4N^+	3.8
Cs^+	-3.3	Cs^+18C6	1.8	Cs^+222	3.9	Bu_4N^+	4.9
Ag^+	-5.0	Ag^+18C6	1.1	Ag^+222	1.2	Ph_4As^+	-0.4

An exactly equivalent cycle can be constructed in terms of Gibbs energy and of entropy (12) and in Table 11 are the standard entropies of transfer of complex and simple ions from water to methanol (11, 12). Again, values for the M^+18C6 complexes resemble those for the M^+222 complexes both in sign and in magnitude.

Table 11. Values of ΔS_t^o for complex ions for transfer from water to methanol, cal K^{-1} mol^{-1} (12), at 298 K

Na^+	-23	Na^+18C6	16	Na^+222	26	Me_4N^+	-5
K^+	-23	K^+18C6	14	K^+222	26	Et_4N^+	7
Rb^+	-21	Rb^+18C6	9	Rb^+222	23	Pr_4N^+	20
Cs^+	-19	Cs^+18C6	10	Cs^+222	15	Bu_4N^+	34
Ag^+	-23	Ag^+18C6	8	Ag^+222	6	Ph_4As^+	17

The effect of complexing the M^+ ions by either 18C6 or cryptand 222 is to convert the hydrophilic M^+ ion (ΔH_t^o negative, ΔS_t^o negative) into a typical hydrophobic ion (ΔH_t^o positive, ΔS_t^o positive), compare the thermodynamics of transfer of M^+18C6 or M^+222 with those for the known hydrophobic ions Et_4N^+ or Pr_4N^+ or Bu_4N^+. For both types of complex, there must be substantial interaction between the CH_2 groups at the exterior of the complex with surrounding water and methanol molecules. There must also be interaction (direct or indirect) between the solvent molecules and the central metal ions, however, because within each series of complexes the ΔH_t^o and ΔS_t^o values depend on the actual central metal ion present.

REFERENCES

(1) Christensen, J. J., Eatough, D. J., Ruckman, J., and Izatt, R. M.: 1972, *Thermochimica Acta* 3, 203. Eatough, D. J., Christensen, J. J., and Izatt, R. M.: 1972, *Thermochimica Acta* 3, 219. Eatough, D. J., Izatt, R. M., and Christensen, J. J.: 1972, *Thermochimica Acta* 3, 233.

(2)   Christensen, J. J.: 1980, *Bioenergetics and Thermodynamics: Model Systems*, Ed. A. Braibanti, Reidel Publishing Co., Holland.

(3)   Bolles, T. F., and Drago, R. S.: 1965, *J. Am. Chem. Soc.* 87, 5025.

(4)   Epley, T. D., and Drago, R. S.: 1967, *J. Am. Chem. Soc.* 89, 5770.

(5)   Nozari, M. S., and Drago, R. S.: 1972, *J. Am. Chem. Soc.* 94, 6877.

(6)   Irving, R. J., Abraham, M. H., Salmon, J. E., Marton, A., and Inczédy, J.: 1977, *J. Inorg. Nuclear Chem.* 39, 1433.

(7)   Lamb, J. D., Izatt, R. M., Swain, C. S., and Christensen, J. J.: 1980, *J. Am. Chem. Soc.* 102, 475.

(8)   Izatt, R. M., Lamb, J. D., Izatt, N. E., Rossiter, B. E. Jr., Christensen, J. J., and Haymore, B. L.: 1979, *J. Am. Chem. Soc.* 101, 6273.

(9)   Izatt, R. M., Terry, R. E., Haymore, B. L., Hansen, L. D., Dailey, N. K., Avondet, A. G., and Christensen, J. J.: 1976, *J. Am. Chem. Soc.* 98, 7620.

(10) Abraham, M. H., Danil de Namor, A. F., Ling, H. C., and Schulz, R. A.: 1980, *Tetrahedron Letters*, 21, 961.

(11) Abraham, M. H., Danil de Namor, A. F., and Schulz, R. A.: 1980, *J.C.S. Faraday I.* 76, 869.

(12) Abraham, M. H., and Ling, H. C.: 1982, *Tetrahedron Letters*, 23, 469.

METAL LIGAND BOND STRENGTHS AND BONDING THEORY

C.T. Mortimer

Chemistry Department, University, Keele,
Staffordshire, U.K., ST5 5BG.

ABSTRACT

Metal-ligand mean bond dissociation energies have been calculated in those transition-metal complexes for which we have values of ΔH_f^o(complex, cryst.) and a measured or reliably estimated value for ΔH(sublimation). The complexes are of two general types, ML_nX_2, where L is a ligand molecule and X is a halogen, and ML_n, where L is a ligand radical, such as the pentanedionato group. For the octahedral complexes, ML_4Cl_2, the values of the gas-phase mean bond dissociation energies, $\overline{D}$(M-L), follow the trend Mn<Fe<Co<Ni>Cu>Zn, with a maximum value at Ni. For the tetrahedral complexes of Group II metal, ML_2X_2, the trend of $\overline{D}$ values is Zn>Cd>Hg, whilst for the square planar pentanedionato complexes, ML_2, the order of $\overline{D}$(M-L) values is Mn>Co>Ni>Cu>Zn. A simple molecular orbital description (the Angular Overlap Model) of the complexes is consistent with, but not predictive of these trends in $\overline{D}$(M-L) values. A comment is made on the variation of $\overline{D}$(Co-pyridine) values, with substitution on the pyridine ring, in terms of the bonding theory.

1. INTRODUCTION

Transition metal complexes, in which the metal is bonded to ligand atoms of Groups V and VI, are of great importance in certain biological processes. One thinks readily of haemoglobin, where Fe-N bonds hold the iron atom in and above the nitrogen atoms of the porphyrin ring; of nitrogenase, the enzyme involved in nitrogen fixation on plants, where Mo-S-Fe bonds are found; and of DNA polymerase, the function of which in the replication of DNA molecules depends on the formation of Zn-O bonds. In therapy, where transition metals are involved, cure depends on the formation of metal-ligand bonds. Thus, the use of certain platinum compounds in anti-tumour therapy results from the formation of Pt-N and Pt-O bonds between platinum and the bases cytosine and adenine. Removal of excess copper in the treatment of Wilson's disease is achieved efficiently by use of glycyl-glycylhistidine N-methylamide, to which copper coordinates through both Cu-N and Cu-O bonding. For a better understanding of processes of this type, we need more information about the strengths of bonds between metals and these Group V and VI atoms.

Few thermochemical data have been reported for these complexes, and even for simpler model systems, such as the amino-acid complexes of transition metals, the available information refers mainly to reactions in solution. Reviews of these data are available (1,2). However, without values for the enthalpies of formation of the gaseous complexes, it is not possible to calculate metal-ligand bond dissociation energies. The purpose of this paper is to show how these dissociation energies are derived and to examine the extent to which trends in these values are consistent with simple bonding theory of these compounds.

The structure of the two complexes $Co(2Mepy)_2Cl_2$, where 2Mepy is the 2-methylpyridine molecule, and $Ni(pd)_2$, where pd is the pentanedionato radical, are shown.

The enthalpies of the gas-phase dissociation reactions (1) and (2) are measures of the strengths of the metal-ligand mean bond dissociation energies in these particular complexes and we write $2\bar{D}(Co-N) = \Delta H(1)$, and $4\bar{D}(Ni-O) = \Delta H(2)$.

$$Co(2Mepy)_2Cl_2(g) \rightarrow CoCl_2(g) + 2(2Mepy)(g) \quad (1)$$

$$Ni(pd)_2(g) \rightarrow Ni(g) + 2pd(g) \quad (2)$$

In general, the mean bond dissociation energies, $\bar{D}(M-L)$, can be derived from the enthalpy, ΔH, of a gas-phase dissociation reaction

$$ML_nX_m(g) \rightarrow MX_m(g) + nL(g)$$

and can be calculated from the relationship

$$n\bar{D}(M-L) = \Delta H = \Delta H_f^o(MX_m, g) + n\Delta H_f^o(L, g) - \Delta H_f^o(ML_nX_m, g).$$

Values for $\Delta H_f^o(MX_m, g)$ and $\Delta H_f^o(L, g)$ are often available, so that it is necessary only to obtain $\Delta H_f^o(ML_nX_m, g)$. Since the majority of these complexes are solid, two separate experiments are carried out, determination of (i) the enthalpy of formation of the crystalline complex, $\Delta H_f^o(ML_nX_m, c)$, and (ii) the enthalpy of sublimation, $\Delta H_{sub}(ML_nX_m)$. In some cases it is possible to avoid determination of $\Delta H_f^o(ML_nX_m, c)$, for example in a direct measurement of the enthalpy of the following thermal decomposition reaction.

$$Co(2Mepy)_2Cl_2(c) \rightarrow CoCl_2(c) + 2(2Mepy)(g) \quad (3)$$

In this case the only additional datum needed to obtain $\bar{D}(Co-N)$

is $\Delta H_{sub}\{Co(2Mepy)_2Cl_2\}$ since the value for $\Delta H_{sub}(CoCl_2)$ is known. In the sections which follow, we summarise methods by which calorimetric measurements have been made on crystalline transition metal complexes, list the enthalpies of sublimation, measured or estimated, and provide values for the mean metal-ligand bond dissociation energies.

2. ML_nX_2 COMPLEXES, WHERE L IS A LIGAND MOLECULE

2.1 Differential Scanning Calorimetry. This technique is especially suited to measurement of the enthalpies, ΔH_d, of thermal decomposition reactions of the type shown in the lower line of the enthalpy cycle. On raising the temperature, the crystalline complex dissociates to crystalline MX_2, generally

$$\begin{array}{ccc} ML_2X_2(g) & \xrightarrow{\Delta H_g} & MX_2(g) + 2L(g) \\ \uparrow \Delta H_{sub}(ML_2X_2) & & \uparrow \Delta H_{sub}(MX_2) \\ ML_2X_2(c) & \xrightarrow{\Delta H_d} & MX_2(c) + 2L(g) \end{array}$$

a metal halide, and gaseous ligand L. A number of transition-metal complexes have been studied by use of this technique. Those shown in Table 1 have been selected because values of $\Delta H_{sub}(ML_2X_2)$ are also known. Since values for $\Delta H_{sub}(MX_2)$ are also available, the enthalpy, ΔH_g, of the gas-phase dissociation reaction, the upper line of the cycle, can be derived and values for $\overline{D}(M-L)$ calculated. Results for some cobalt benzothiazole and benzoxazole complexes are also to be found in the paper on "Differential Scanning Calorimetry" by C.T. Mortimer. It is important to note that the Δh_d values refer to temperatures well above 298 K. Where heat capacity data for ML_2X_2, MX_2 and L are known, it is possible to refer the ΔH_d value to 298 K. There is some difficulty in obtaining heat capacity data for the complexes ML_2X_2 up to the appropriate temperatures because of the onset of decomposition. The values quoted for $\Delta H_{sub}(ML_2X_2)$ refer to temperatures between 298 K and the decomposition

Table 1

Enthalpies of decomposition of methyl- and halogen-substituted pyridine complexes, values in kJ mol^{-1}

Complex	$\Delta H_d{}^{\underline{b}}$	$\Delta H_{sub}{}^{\underline{c}}$	$\overline{D}$(M-N)
Co(py)$_2$Cl$_2{}^{\underline{a}}$	119.7 ± 2.1 (3)	(100 ± 10)$^{\underline{d}}$	127 ± 6
Co(py)$_2$Br$_2$	114.2 ± 1.7 (3)	(87 ± 4)$^{\underline{d}}$	124.6 ± 3
Co(2Mepy)$_2$Cl$_2$	109.6 ± 2.1 (3)	86.6 ± 3.8 (4)	128.5 ± 3
Co(2Mepy)$_2$Br$_2$	88.7 ± 3.8 (3)	69.5 ± 2.9 (4)	120.6 ± 3
Co(2Clpy)$_2$Cl$_2$	115.1 ± 1.3 (4)	101.2 ± 6.7 (4)	124.0 ± 4
Co(2Brpy)$_2$Cl$_2$	126.4 ± 1.3 (4)	120.5 ± 4.6 (4)	120.0 ± 3
Co(2Brpy)$_2$Br$_2$	129.3 ± 1.7 (4)	100.8 ± 2.1 (4)	125.3 ± 2
Co(3Clpy)$_2$Cl$_2$	134.7 ± 2.5 (4)	95.8 ± 3.3 (4)	136.5 ± 3
Co(3Brpy)$_2$Cl$_2$	134.3 ± 1.3 (4)	77.0 ± 4.2 (4)	145.7 ± 3
Ni(3Clpy)$_2$Cl$_2$	142.2 ± 2.5 (5)	72.0 ± 7.1 (5)	158.1 ± 5
Ni(3Brpy)$_2$Cl$_2$	147.3 ± 2.1 (5)	76.1 ± 5.0 (5)	158.6 ± 4
Cu(3Brpy)$_2$Cl$_2$	146.4 ± 1.7 (5)	56.9 ± 5.4 (5)	141.8 ± 3

(py) pyridine

$\underline{a}$ blue tetrahedral form of Co(py)$_2$Cl$_2$
$\underline{b}$ ML$_2$X$_2$(c) → MX$_2$(c) + 2L(g), at 480 - 575 K
$\underline{c}$ ML$_2$X$_2$(c) → ML$_2$X$_2$(g), at 240 - 317 K
$\underline{d}$ Values in parenthesis estimated

temperature.

Using this approach it is clearly not necessary, when deriving values for $\bar{D}(M-L)$, to obtain a value for the enthalpy of formation of the crystalline complex, although this could be obtained from the relationship

$$\Delta H_f^o(ML_2X_2, c) = \Delta H_f^o(MX_2, c) + \Delta H_f^o(L, g) - \Delta H_d.$$

This has not been done because (a) there are insufficient data to refer values of ΔH_d and ΔH_{sub} to 298 K and (b) of the ligands listed, only the enthalpy of formation of pyridine is available from the literature.

2.2 Solution Calorimetry. An alternative to measuring the enthalpy of the dissociation of a complex ML_nX_2 is to determine the enthalpy of the reaction in which it forms from the metal halide and ligand. This is conveniently done by using a precision solution calorimeter to measure the enthalpies of solution of metal halide, ligand and complex in some suitable solvent, $\Delta H(a)$, $\Delta H(b)$ and $\Delta H(c)$. The enthalpy, ΔH_R is then

$$\begin{array}{ccccc}
MX_2(c) & + & nL(c \text{ or } \ell) & \xrightarrow{\Delta H_R} & ML_nX_2(c) \\
\downarrow \Delta H(a) & & \downarrow \Delta H(b) & & \downarrow \Delta H(c) \\
\text{solvent} & & \text{solution A} & & \text{solvent} \\
\downarrow & & \downarrow & & \downarrow \\
\text{solution A} & & \text{solution B} & & \text{solution B}
\end{array}$$

given by $\Delta H_R = \Delta H(a) + \Delta H(b) - \Delta H(c)$.

S.J. Ashcroft (6) has determined the enthalpies of the formation reactions, ΔH_R, of the thiourea complexes $M(tu)_4Cl_2$, by use of 0.1 M HCl as solvent for the reactants and products. He has determined the enthalpies of sublimation of the complexes and of thiourea by use of a thermogravimetric analytical technique. Data are shown in Table 2; values for ΔH_f^o (complex, c), and ΔH_{sub} (complex), together with the derived $\bar{D}(M-S)$ values, which are plotted in Figure 1. This figure also shows values of

Table 2

Enthalpies of formation of thiourea complexes (6), values in kJ mol^{-1}

$M(tu)_4Cl_2$ [a]	ΔH_R [b]	ΔH_f^o(complex, c)	ΔH_{sub}(complex)	$\overline{D}$(M-S)
$Mn(tu)_4Cl_2$	-(31.4 ± 0.4)	-(865.9 ± 1.0)	133.1 (390 K)	122.7
$Fe(tu)_4Cl_2$	-(46.0 ± 0.8)	-(741.0 ± 1.3)	109.6 (385 K)	126.1
$Co(tu)_4Cl_2$	-(44.4 ± 0.8)	-(710.1 ± 1.3)	128.9 (370 K)	131.1
$Ni(tu)_4Cl_2$	-(44.3 ± 0.2)	-(702.5 ± 0.8)	73.6 (430 K)	147.9
$Zn(tu)_4Cl_2$	-(50.2 ± 0.4)	-(818.5 ± 1.2)	90.4 (365 K)	120.9
$Cd(tu)_4Cl_2$	-(29.3 ± 2.8)	-(774.0 ± 3.0)	73.3 (290 K)	128.0
$Hg(tu)_4Cl_2$	-(4.1 ± 5.0)	-(579.9 ± 5.0)	101.7 (375 K)	92.0
tu	-		93.7 (360 K)	-

[a] (tu) thiourea $(NH_2)_2C=S$
[b] $MCl_2(c) + 4tu(c) \rightarrow M(tu)_4Cl_2(c)$

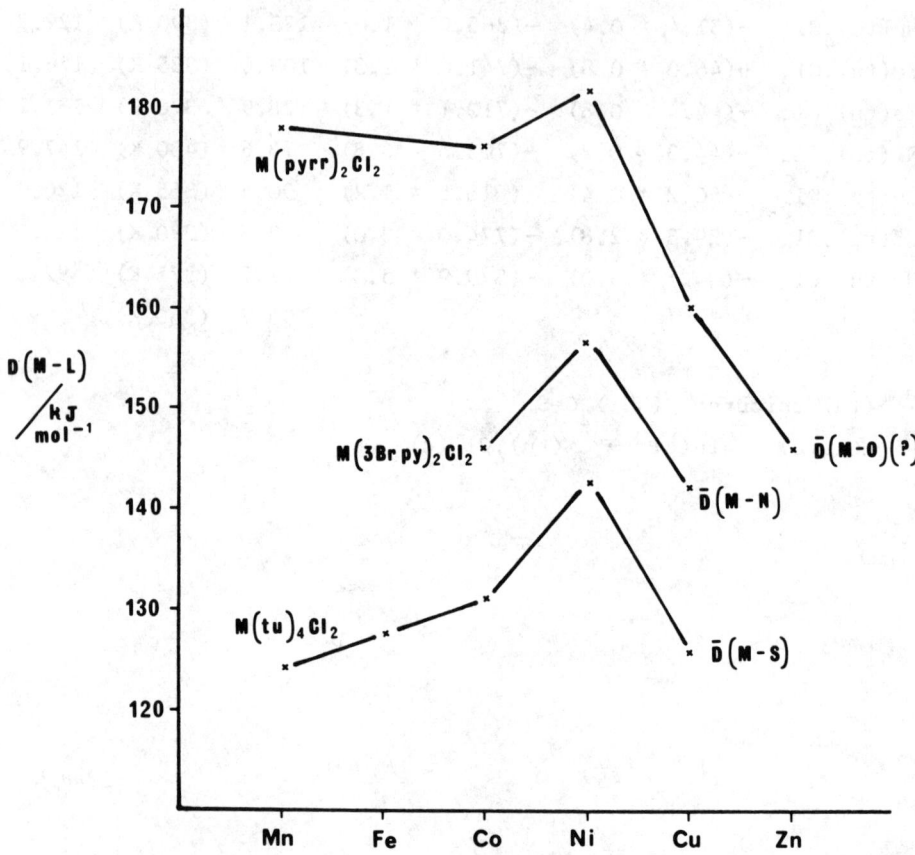

Figure 1. $\bar{D}(M-S)$ in $M(tu)_4Cl_2$, $\bar{D}(M-N)$ in $M(3Brpy)_2Cl_2$ and $\bar{D}(M-O)$ in $M(pyrr)_2Cl_2$

Table 3

Enthalpies of formation of 2-pyrrolidone complexes (7), values in kJ mol^{-1}

Complex[a]	ΔH_R [b]	ΔH_f^o(complex, c) [c]	$\bar{D}$(M-O)(?)
Mn(pyrr)$_2$Cl$_2$	-(70.3 ± 0.4)	-(1155.6 ± 2.0)	178 ± 3
Co(pyrr)$_2$Cl$_2$	-(50.2 ± 0.3)	-(966.7 ± 2.0)	176 ± 3
Ni(pyrr)$_2$Cl$_2$	-(49.5 ± 0.3)	-(958.8 ± 2.0)	182 ± 3
Cu(pyrr)$_2$Cl$_2$	-(58.3 ± 0.2)	-(802.4 ± 2.0)	160 ± 3
Zn(pyrr)$_2$Cl$_2$	-(74.4 ± 0.2)	-(1093.5 ± 2.0)	146 ± 3

[a] (pyrr) 2-pyrrolidone

[b] MCl$_2$(c) + 2(pyrr)(c) → M(pyrr)$_2$Cl$_2$(c)

[c] based on ΔH_f^o(pyrr, c) = -(302 ± 1) kJ mol^{-1}

$\bar{D}$(M-N) for the complexes M(3Brpy)$_2$Cl$_2$, where M = Co, Ni and Cu, together with $\bar{D}$(M-O) for the pyrrolidone complexes M(pyrr)$_2$Cl$_2$.

Data for these latter compounds are shown in Table 3. They were obtained by M.S. Barvinok and L.G. Lukina (7), who measured the enthalpies of solution of the complexes, metal halides and ligands in 2 M HCl solution. It is uncertain whether pyrrolidone bonds to the metal via an oxygen or a nitrogen atom. The mean metal-ligand bond dissociation energy for the cobalt complex, 176 ± 3 kJ mol^{-1} is considerably greater than either the $\bar{D}$(Co-N) values for the pyridine and benzothiazole complexes (120 to 146 kJ mol^{-1}) or the $\bar{D}$(Co-O) values in the benzoxazole complexes of cobalt (128 to 148 kJ mol^{-1}). Possibly, the pyrrolidone molecule is bidentate, with both the nitrogen and oxygen atoms bonding to give a six-coordinate metal complex.

The shape of the plot for the octahedral M(tu)$_4$Cl$_2$ complexes can be explained in terms of the Angular Overlap Model, a simplified molecular orbital description of the bonding in transition metal complexes, which considers only the interaction between metal d-orbitals and ligand group-orbitals (8) as shown on the adjacent diagram. Along the series Mn to Zn, the energy level of the d orbitals will be lowered, which results in a better match with the ligand group orbitals and stronger M-L bonding. However, along this series of metals, this effect is offset by the increasing number of d electrons, which reside in antibonding orbitals. The relative energies of the antibonding orbitals are shown in Figure 2. The antibonding electrons in the Fe^{II}(d^6), Co^{II}(d^7), Ni^{II}(d^8) and Cu^{II}(d^9) complexes are not spherically distributed between the σ^* and π^* antibonding orbitals, but occupy the lower energy π^* orbitals preferentially. As a result the antibonding energy is less, per electron, than for the Mn^{II}(d^5) or Zn^{II}(d^{10}) cases. This "stabilization" energy, E, is given by the expression $E = \Delta_o/5[3n_{\sigma^*} - 2n_{\pi^*}]$, where Δ_o is the energy separation between the σ^* and π^* orbitals, which

ML_6 complexes: interaction of metal 'd' orbitals with ligand group orbitals.

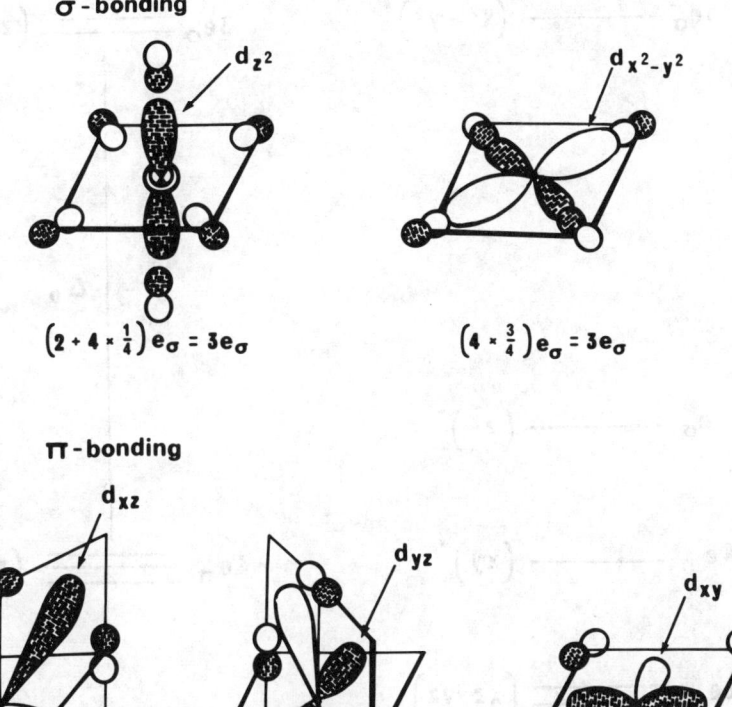

σ - bonding

d_{z^2} $d_{x^2-y^2}$

$\left(2 + 4 \times \frac{1}{4}\right)e_\sigma = 3e_\sigma$ $\left(4 \times \frac{3}{4}\right)e_\sigma = 3e_\sigma$

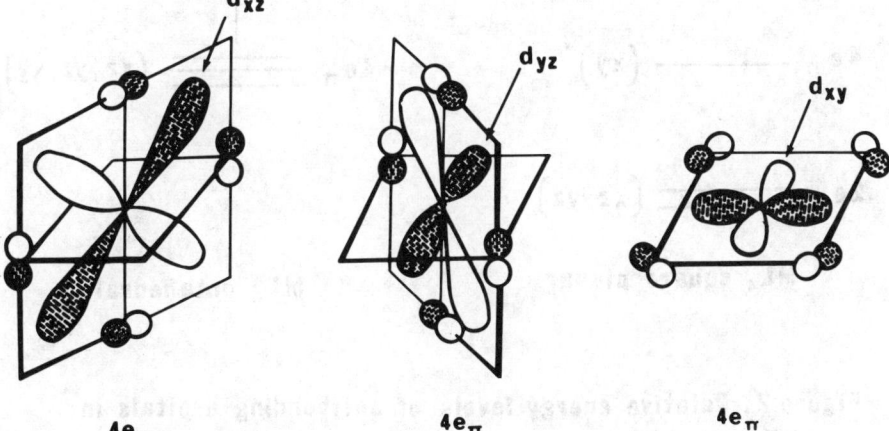

π - bonding

d_{xz} d_{yz} d_{xy}

$4e_\pi$ $4e_\pi$ $4e_\pi$

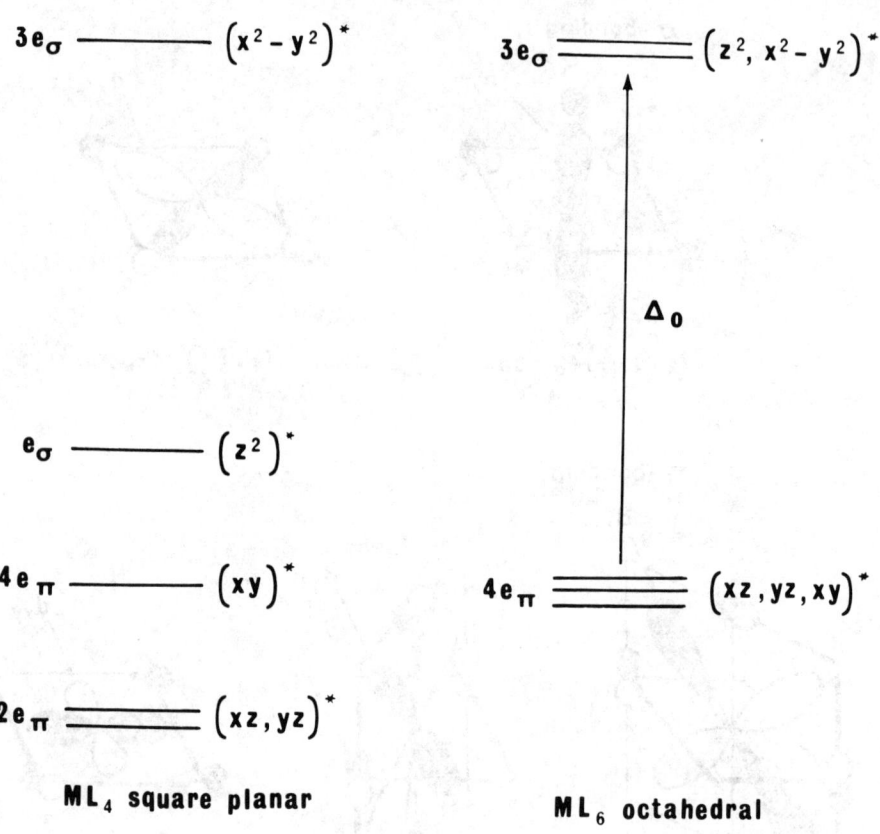

Figure 2. Relative energy levels of antibonding orbitals in Angular Overlap Model description of ML_4 square planar and ML_6 octahedral complexes.

are occupied by a number $n_{\sigma*}$ and $n_{\pi*}$ of electrons, respectively. For high spin complexes, this stabilization energy rises to a maximum at Ni and then falls to Cu, which is consistent with the experimental values of $\overline{D}(M-S)$ and $\overline{D}(M-N)$.

C. Airoldi, A.P. Chagas and co-workers (9-13), at the Universidade Estadual de Campinas, Sao Paulo, Brazil, have measured the enthalpies of formation, from metal halide and ligand, of a number of zinc, cadmium and mercury complexes. The ligands were N-(2-pyridyl)acetamide, (pa), which probably bonds to the metal through the pyridine N atom, some phosphine oxides, (tepo), (tppo), diacetamide, da, which is a bidentate ligand, and tetramethylurea, (tmu), all of which form metal-oxygen bonds. Data for the complexes $M(pa)_n X_2$ are shown in Table 4.

It has not been possible to determine the enthalpies of sublimation of any of these metal complexes, because they dissociate when heated to temperatures at which it is possible to measure their vapour pressures. Airoldi et al. make the assumption that $\Delta H_{sub}^{298}(complex) = \Delta H_{sub}^{298}(ligand)$ in order to calculate a value for ΔH_g, the gas-phase formation of complex from metal halide and ligand. There is some justification for this assumption. Thus, the values of ΔH_{sub} for the complexes $M(tu)_4 Cl_2$ lie in the range 73 to 133 kJ mol, compared with $\Delta H_{sub}(tu) = 94$ kJ mol^{-1}. The values of ΔH_{sub} for the metal complexes of the substituted pyridines, benzothiazoles and benzoxazoles also lie in the band 70 to 125 kJ mol^{-1} (with the exception of $Cu(3Brpy)_2 Cl_2$ for which we have $\Delta H_{sub} = 57$ kJ mol^{-1}) although the values of $\Delta H_{sub}(ligand)$ are not available for comparison. Moreover, for complexes $ML_n X_2$, with n ligands L, the error in the calculated value of $\overline{D}(M-L)$ is only $1/nth$ of the error in the estimated value of $\Delta H_{sub}(complex)$.

The derived values for $\overline{D}(M-N)$ and $\overline{D}(M-O)$ in the metal chloride complexes are plotted in Figure 3. For comparison, values of $D(M-S)$ from the thiourea complexes $M(tu)_4 Cl_2$ are also shown.

Table 4

Enthalpies of formation of N-(2-pyridyl)acetamide complexes (9), values in kJ mol^{-1}.

Complex[a]	ΔH_R [b]	ΔH_f^o(complex, c)	$\bar{D}$(M-N)
Zn(pa)$_2$Cl$_2$	-(72.8 ± 0.6)	-(785 ± 22)	163
Zn(pa)$_2$Br$_2$	-(76.5 ± 0.8)	-(702 ± 22)	155
Zn(pa)$_2$I$_2$	-(74.4 ± 0.8)	-(580 ± 22)	156
Cd(pa)$_2$Cl$_2$	-(35.0 ± 0.7)	-(724 ± 22)	160
Cd(pa)$_2$Br$_2$	-(36.7 ± 0.8)	-(650 ± 22)	152
Cd(pa)$_2$I$_2$	-(40.2 ± 0.4)	-(541 ± 21)	145
Hg(pa)$_2$Cl$_2$	-(22.9 ± 1.1)	-(544 ± 23)	106
Hg(pa)Br$_2$	-(10.7 ± 0.7)	-(330 ± 22)	119

[a] (pa) N-(2-pyridyl)acetamide ⟨pyridyl⟩-NHCOCH$_3$

[b] MX$_2$(c) + n(pa)(c) → M(pa)$_n$X$_2$(c)

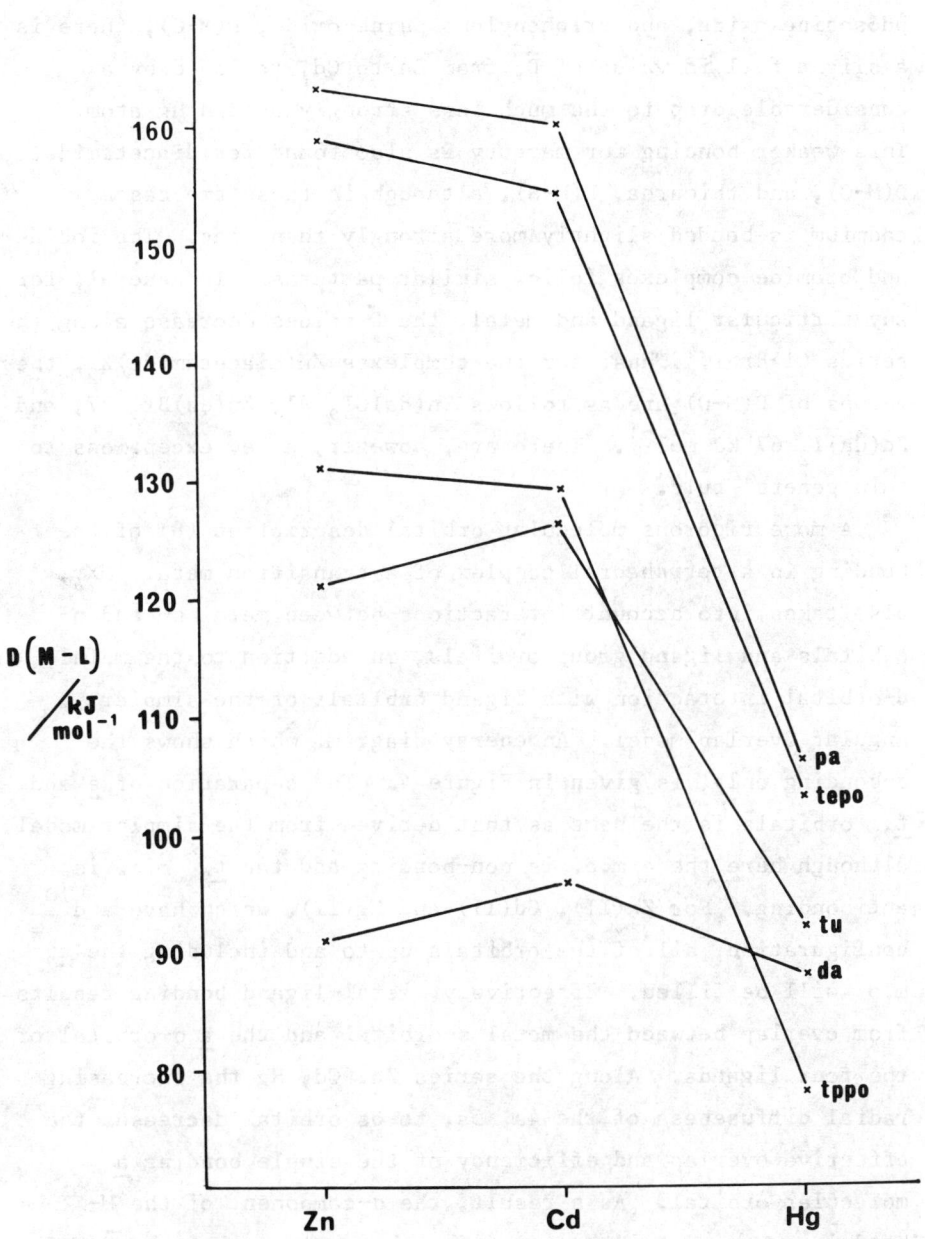

Figure 3. $\bar{D}(M-S)$ in $M(tu)_4Cl_2$; $\bar{D}(M-N)$ in $M(pa)_2Cl_2$ and $\bar{D}(M-O)$ in $M(tepo)_2Cl_2$, $M(tppo)_2Cl_2$ and $M(da)Cl_2$.

For the complexes of N-(2-pyridyl)acetamide, D(M-N), triethylphosphine oxide, and triphenylphosphine oxide, D(M-O), there is a slight fall in value of $\bar{D}$, from Zn to Cd, followed by a considerable drop to the much less strongly bonded Hg atom. This weaker bonding for mercury is also found for diacetamide, D(M-O), and thiourea, D(M-S), although in these two cases cadmium is bonded slightly more strongly than zinc. The iodide and bromide complexes follow similar patterns. In general, for any particular ligand and metal, the $\bar{D}$ values decrease along the series Cl>Br>I. Thus, for the complexes Zn(diacetamide)X_2, the values of D(M-O) are as follows Zn(da)Cl_2 91, Zn(da)Br_2 77, and Zn(da)I_2 67 kJ mol^{-1}. There are, however, a few exceptions to this general rule.

A more rigorous molecular orbital description (8) of the bonding in a tetrahedral complex of a transition metal, MX_4, also takes into account interactions between metal s- and p-orbitals and ligand group orbitals, in addition to the metal d-orbital interaction with ligand orbitals of the simpler Angular Overlap Model. An energy diagram, which shows the σ-bonding only, is given in Figure 4. The separation of $\underline{e}$ and $\underline{t_2}^*$ orbitals is the same as that derived from the simpler model, although here the $\underline{e}$ m.o. is non-bonding and the $\underline{t_2}^*$ m.o. is antibonding. For Zn(II), Cd(II) and Hg(II), which have a d^{10} configuration, all of the orbitals up to and including the $\underline{t_2}^*$ m.o. will be filled. Effectively, metal-ligand bonding results from overlap between the metal s-orbital and the $\underline{a}$ σ-orbital of the four ligands. Along the series Zn, Cd, Hg the increasing radial diffuseness of the 4s, 5s, to 6s orbital decreases the effective overlap and efficiency of the single bonding $\underline{a}$ molecular orbital. As a result, the σ-component of the M-X bond is expected to decrease. The observation that the $\bar{D}$(M-X) values decrease along this series implies that it is the σ-bonding which is dominant in these molecules.

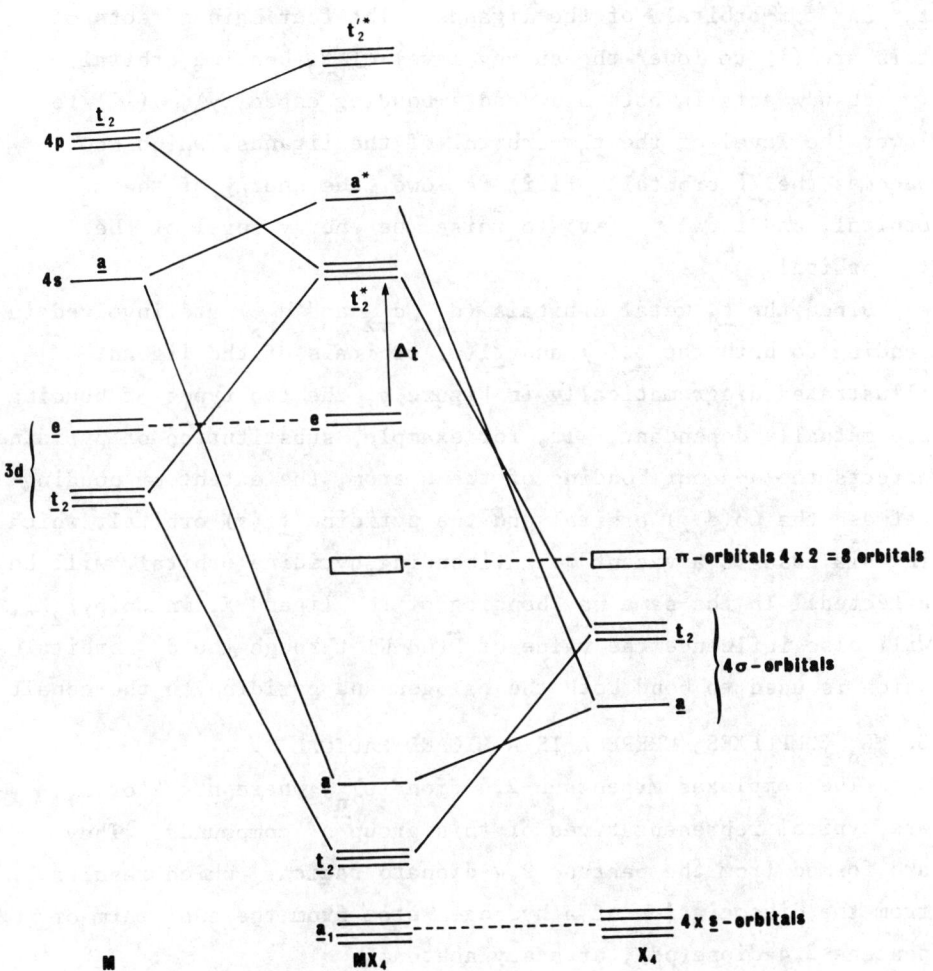

Figure 4. Molecular orbital energy diagram of tetrahedral complex MX_4, σ-bonding only.

Incorporation of π-bonding between ligands and metal (Figure 5) results in a lowering of the $\underline{e}$ non-bonding m.o., which now becomes weakly bonding. Also, there is interaction between the $\underline{t_2}^*$ and $\underline{t_2}$π-orbitals of the ligands. The four main effects of this are (i) to lower the energy level of $\underline{t_2}$ bonding orbital (which now acts in both a σ- and π-bonding capacity), (ii) to lower the level of the $\underline{t_2}$π-orbital of the ligands, which now becomes the $\underline{t_2'}$ orbital, (iii) to lower the energy of the $\underline{t_2}^*$ orbital, and finally (iv) to raise the energy level of the $\underline{t_2}'^*$ orbital.

Since the $\underline{t_2}$ metal orbitals (d_{xz}, d_{yz} and d_{xy}) are involved in bonding to both the $\underline{t_2}(\sigma)$ and $\underline{t_2}(\pi)$ orbitals of the ligand, illustrated diagramatically in Figure 6, the two types of bonding are mutually dependant. If, for example, substitution of pyridine affects the σ-donor bonding of the N atom, the extent of bonding between the Co(d_{yz}) orbital and the pyridine $\underline{t_2}(\pi)$ orbital, which in this case is a vacant π* antibonding pyridine orbital, will be affected. In the same way bonding of the ligand X, in $Co(py)_2X_2$, will also influence the value of $\overline{D}(Co-N)$ through the d_{yz} orbital which is used to bond both the halogen and pyridine to the cobalt.

3. ML_n COMPLEXES, WHERE L IS A LIGAND RADICAL

The complexes M(pentane-2,4-dionato)$_{\underline{n}}$, where $\underline{n}$ = 2 or 3, are typical representatives of this group of compounds. They are formed from the pentane-2,4-dionato radical, which results from the dissociation of a hydrogen atom from the enol form of pentane-2,4-dione(pd), or acetylacetone.

keto form   enol form   radical   + H

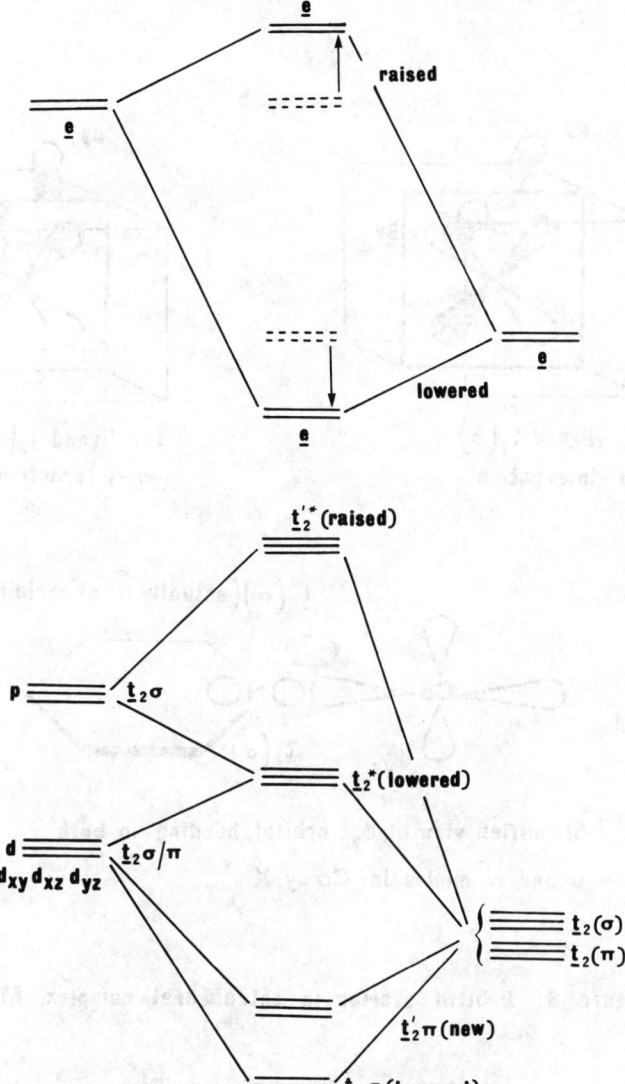

Figure 5. Molecular orbital energy diagram of tetrahedral complex MX_4, π bonding only.

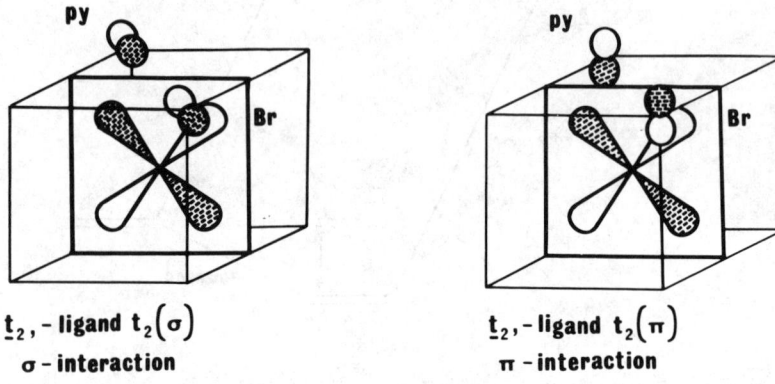

$\underline{t}_2$, - ligand $t_2(\sigma)$
σ - interaction

$\underline{t}_2$, - ligand $t_2(\pi)$
π - interaction

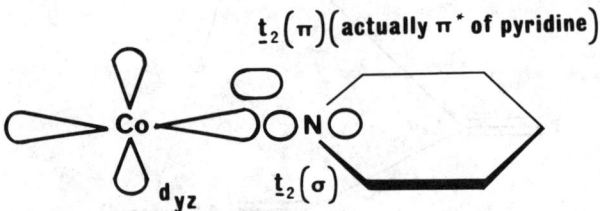

Simplified view of d_{yz} orbital bonding in both σ and π modes in $Co\,py_2\,X_2$

Figure 6. Orbital overlap in tetrahedral complex MX_4.

3.1 Dionato complexes.

Metal dionato complexes, ML_2 and ML_3, decompose into metal and free ligand in sulphuric, hydrochloric, or perchloric acid, although complexes of low water solubility react slowly. For these particular complexes the rate of reaction can be increased by using a mixture of hydrochloric acid and 1,4-dioxan.

This approach to the determination of enthalpies of formation of these complexes has been used by a number of separate groups of workers; under the leadership of R.J. Irving, University of Surrey; M.A.V. Ribeiro da Silva, University of Oporto, Portugal; E. Geira, University of Wroclaw, Poland; J.O. Hill, La Trobe University, Bundoora, Victoria, Australia; G. Pilcher and H.A. Skinner, University of Manchester.

The enthalpies, ΔH_R, of the following reactions were obtained by measuring the enthalpies of solution of ML_2 in the solvent. Then, in a fresh sample of solvent, the successive

$$ML_2(c) + [H_2SO_4, 53.54H_2O](\ell) \rightarrow (53.54-n)H_2O(\ell) + 2HL(\text{stand.state}) + MSO_4 \cdot nH_2O(c) \qquad (4)$$

$$ML_2(c) + 2[HCl, 11.6OH_2O](\ell) \rightarrow (23.20-n)H_2O(\ell) + 2HL(\text{stand.state}) + MCl_2 \cdot nH_2O(c) \qquad (5)$$

enthalpies of solution of the reaction products were measured. The solvent used was either $[H_2SO_4, 53.54H_2O]$, $[HCl, 11.6OH_2O]$, $[HClO_4, 3.57H_2O]$, or a mixture of 25% by volume $[HCl, 11.6OH_2O]$ and 75% of 1,4-dioxan.

The procedure is illustrated in more detail for the particular case of bis(pentane-2,4-dionato)copper (II), $Cu(pd)_2$ (14).

$$[H_2SO_4, \quad \quad \quad \Delta H_R \quad \quad \quad CuSO_4 \cdot$$
$$53.54H_2O](\ell) + Cu(pd)_2(c) \xrightarrow{\Delta H_R} 48.54H_2O(\ell) + 2Hpd(\ell) + 5H_2O(c)$$

$$\quad | \quad \quad \quad | \quad \quad \quad \quad \quad \quad | \quad \quad \quad | \quad \quad \quad |$$
$$\Delta H(a) \quad \Delta H(b) \quad \quad \quad \quad \quad \Delta H(c) \quad \Delta H(d) \quad \Delta H(e)$$
$$\downarrow \quad \quad \quad \downarrow \quad \quad \quad \quad \quad \quad \downarrow \quad \quad \quad \downarrow \quad \quad \quad \downarrow$$

Solvent $[H_2SO_4, 53.54H_2O]$ $\xrightarrow{\Delta H(f)}$ Solvent $[H_2SO_4, 53.54H_2O]$
Final solution A_2 $\quad\quad\quad\quad\quad\quad\quad\quad$ Final solution B_3

The following enthalpies of solution were determined. Provided that

Solvent + $[H_2SO_4, 53.54H_2O](\ell)$ $\xrightarrow{\Delta H(a)}$ Solution A_1
Solution A_1 + $Cu(pd)_2(c)$ $\xrightarrow{\Delta H(b)}$ Solution A_2
Solvent + $H_2O(\ell)$ $\xrightarrow{\Delta H(c)}$ Solution B_1
Solution B_1 + $Hpd(\ell)$ $\xrightarrow{\Delta H(d)}$ Solution B_2
Solution B_2 + $CuSO_4 \cdot 5H_2O(c)$ $\xrightarrow{\Delta H(e)}$ Solution B_3

the precise stoichiometry is realised and solution A_2 is the same as solution B_3, then the value of $\Delta H(f)$ is zero and we can write

$$\Delta H_R = \Delta H(a) + \Delta H(b) - 48.54 \Delta H(c) - 2\Delta H(d) - \Delta H(e)$$

Since care has been taken in setting up the stoichiometry of the reaction so that the sulphuric acid used in the reaction is of the same concentration as the solvent, $[H_2SO_4, 53.54H_2O]$, then $\Delta H(a) = 0$. The value $\Delta H(c) = -0.121$ kJ mol^{-1} is calculated from published data (15). By measuring the values $\Delta H(b) = -21.75$ kJ mol^{-1}, $\Delta H(d) = +6.19$ kJ mol^{-1}, $\Delta H(e) = 22.61$ kJ mol^{-1} we obtain $\Delta H_R = 50.6 \pm 0.4$ kJ mol^{-1}. Using the relationship $\Delta H_f^o\{Cu(pd)_2, c\} = 2\Delta H_f^o(Hpd, \ell) + \Delta H_f^o(CuSO_4 \cdot 5H_2O) - \Delta H_f^o[H_2SO_4,$ in $53.54] - 5\Delta H_f^o(H_2O, \ell) - \Delta H_R$ and incorporating enthalpies of formation we obtain the value $\Delta H_f^o\{Cu(pd)_2, c\} = -(760.7 \pm 4)$ kJ mol^{-1}. Data for these compounds are given in more detail in the paper "Thermochemistry of Metal β-diketonates" by M.A. Ribeiro da Silva.

The mean metal-oxygen bond dissociation energies in the metal dionato complexes ML_2 and ML_3, $\bar{D}(M-O)$, corresponding to the gas-phase dissociation reactions

$$ML_2(g) \rightarrow M(g) + 2L(g) \text{ and}$$

$$ML_3(g) \rightarrow M(g) + 3L(g)$$

can be calculated from the relationships

$$\overline{D}(M-O) = {}^1/4\{\Delta H_f^o(M,g) + 2\Delta H_f^o(L,g) - \Delta H_f^o(ML_2,g)\} \text{ and}$$

$$\overline{D}(M-O) = {}^1/6\{\Delta H_f^o(M,g) + 3\Delta H_f^o(L,g) - \Delta H_f^o(ML_3,g)\}.$$

To calculate these $\overline{D}(M-O)$ values, we require $\Delta H_f^o(M,g)$ and $\Delta H_f^o(L,g)$. We use the $\Delta H_f^o(L,g)$ values; pentanedionato (pd) -235.7, 1-phenyl-1,3-butanedionato (pbd) -117.3 and 1,3-diphenyl-1,3-propanedionato (dppd) -5.3 kJ mol^{-1}. The values which we have used for $\Delta H_f^o(ML_2,g)$ and $\Delta H_f^o(ML_3,g)$ have been derived elsewhere (16).

A plot of $\overline{D}(M-O)$ values is shown in Figure 7. Values for $M(pd)_3$ complexes indicate a fall in the value of $\overline{D}(M-O)$ from Cr and Mn. The higher value for $\overline{D}(Fe-O)$ in $Fe(pd)_3$ may be due to this complex being low spin. For the complexes $M(pd)_2$, the values fall along the series Mn>Co>Ni>Cu>Zn, which is contrary to the normal plot of enthalpies of complexation in solution, where the bond strength increases from Mn to a maximum at Ni or Cu, before falling to Zn. Substitution of a CH_3 group in pd by a C_6H_5 group to give pbd increases the value of $\overline{D}(M-O)$, considerably for Ni, less so for Cu and to a yet smaller extent for Zn complexes. The value of $\overline{D}(Ni-O)$ is marginally lowered by substitution of a CH_3 group in pd by $(CH_3)_2CH$ in the Ni complex (not shown in Figure 8). The trend in these values is consistent with the progressive filling of antibonding orbitals shown in Figure 2.

3.2 Diethyldithiocarbamate complexes. In the bis(diethyldithiocarbamato)nickel (II) and copper (II) complexes, all four of the sulphur atoms are assumed to bond to the metal.

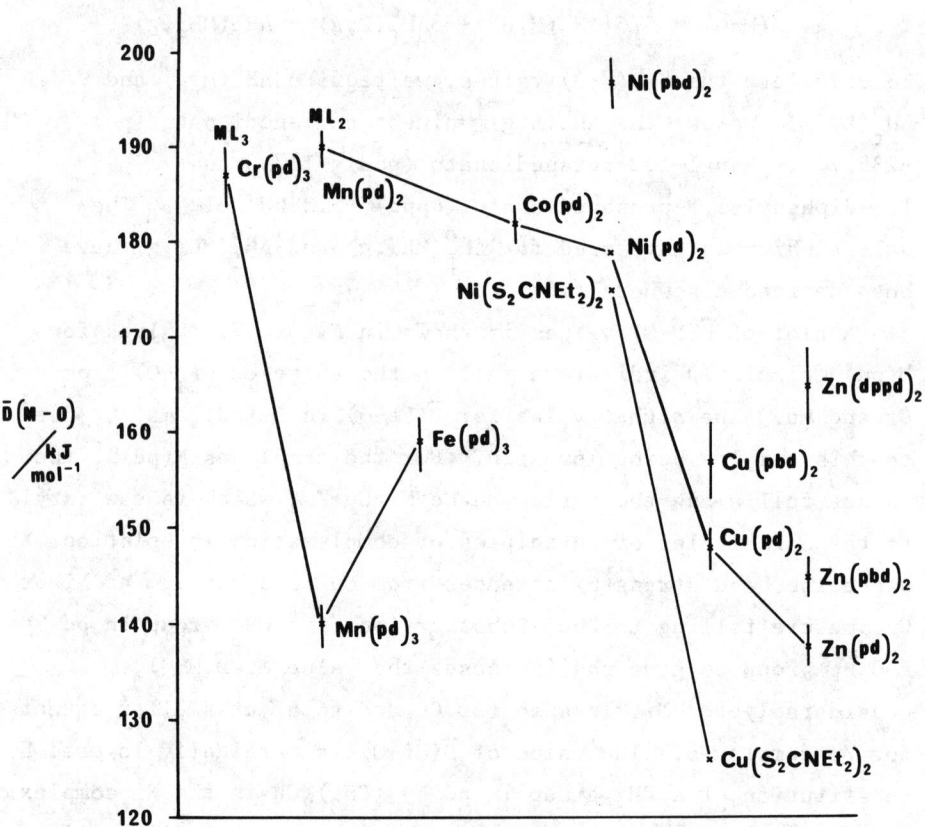

Figure 7. $\bar{D}(M-O)$ in dionato and thiocarbamato complexes.

$$\text{Et}_2\text{NC} \begin{array}{c} \diagup\!\!\!\!\!\!\!\diagdown \text{S} \quad \text{S} \diagup\!\!\!\!\!\!\!\diagdown \\ \diagdown\!\!\!\!\!\!\!\diagup \text{S} \quad \text{S} \diagdown\!\!\!\!\!\!\!\diagup \end{array} \text{M} \begin{array}{c} \text{CNEt}_2 \end{array}$$

K.J. Cavell, J.O. Hill and R.J. Magee (17) have determined the enthalpies of formation of these complexes, by measuring the enthalpies, ΔH_R, of the following ligand replacement reactions.

$$1/3\{Ni(pd)_2\}_3(c) + 2[NEt_2H_2][S_2CNEt_2](c) \rightarrow Ni(S_2CNEt_2)_2(c) + 2NEt_2H(\ell) + 2Hpd(\ell) \quad (6)$$

$$Cu(pd)_2(c) + 2[NEt_2H_2][S_2CNEt_2](c) \quad Cu(S_2CNEt_2)_2(c) + 2NEt_2H(\ell) + 2Hpd(\ell) \quad (7)$$

In these reactions, the pentane-2,4-dionato group in the complex $M(pd)_2$ is replaced by the diethyldithiocarbamato group to give $M(S_2CNEt_2)_2$, where M is Ni or Cu. For the nickel complex dioxan was used as solvent, whilst for the copper complex dimethylformamide was used. Using a direct calorimetric technique involving differential scanning calorimetry K.J. Cavell et al. (18) obtained ΔH_{sub} values $Ni(SCNEt_2)_2$ 102.6±1.5 kJ mol^{-1} and $Cu(S_2CNEt_2)_2$ 116.2±1.3, which are much higher than those obtained by G. D'Ascenzo and W.W. Wendlandt (19) based on isoteniscope vapour pressure/temperature data, viz, $Ni(SCNEt_2)_2$ 61.1±1.7 and $Cu(SCNEt_2)_2$ 87±1.7 kJ mol^{-1}. Incomplete outgassing of the sample appears to give low results by the isoteniscope method.

<u>Table 5</u>

Enthalpies of formation of diethyldithiocarbamato complexes (17), values in kJ mol^{-1}

Complex	ΔH_R(298 K)	ΔH_f^o(complex, c)	ΔH_{sub}	$\overline{D}$(M-S)
$Ni(S_2CNEt_2)_2$	94.5 ± 2.0	-105.5 ± 20	102.6 ± 1.5	175 ± 18
$Cu(S_2CNEt_2)_2$	97.6 ± 1.1	-5.4 ± 18	116.2 ± 1.3	123 ± 17

In order to calculate the enthalpy of the following gas-phase dissociation reaction, a value for $\Delta H_f^o(S_2CNEt_2, g)$ is required. A value for the enthalpy of formation of gaseous

$$M(S_2CNEt_2)(g) \rightarrow M(g) + 2S_2CNEt_2(g)$$

diethyldithiocarbamic acid has been calculated by Cavell et al. (17), using the Allen (20) bond-energy scheme, as $\Delta H_f^o(acid, g) = 35.13$ kJ mol^{-1}. The S-H bond dissociation energy D(S-H) for diethyldithiocarbamic acid is assumed to be 316 ± 25 kJ mol^{-1}, some 60 kJ mol^{-1} less than D(HS-H) = 376.6 kJ mol^{-1} in H_2S, because D(O-H) = 433 ± 10 kJ mol^{-1} in CH_3COOH (21) is 60 kJ mol^{-1} less than D(HO-H) = 498.3 kJ mol^{-1} in H_2O. This leads to $\Delta H_f^o(S_2CNEt_2, g) = 133 \pm 25$ kJ mol^{-1}. Using this value leads to those for $\overline{D}$(M-S) in Table 5, and shown in Figure 7. The lower value of $\overline{D}$(Cu-S) than $\overline{D}$(Ni-S) is consistant with the placing of an additional electron on the $(x^2-y^2)^*$ antibonding orbital.

REFERENCES

1. Ashcroft, S.J. and Mortimer, C.T., 1970 "Thermochemistry of Transition Metal Complexes", Academic Press, London.
2. Christensen, J.J., Eatough, D.J. and Izatt, R.M., 1975, "Handbook of Metal Ligand Heats", 2nd. edn., Dekker.
3. Beech, G., Mortimer, C.T. and Tyler, E.G., 1967, J. Chem. Soc., p.925.
4. Mortimer, C.T. and McNaughton, J.L., 1974, Thermochim Acta, 10, p.125.
5. Mortimer, C.T. and McNaughton, J.L., 1974, Thermochim Acta, 10, p.207.
6. Ashcroft, S.J., 1970, J. Chem. Soc., p.1020.
7. Barvinok, M.S. and Lukina, L.G., 1981, Zhur. Neorg. Khim., 26, p.253.
8. Purcell, K.F. and Kotz, J.C. 1977, "Inorganic Chemistry", Saunders, London, p.533.
9. Airoldi, C. and Chagas, A.P., 1979, Thermochim. Acta, 33, p.371.

10. Queiroz, J.C., Airoldi, C. and Chagas, A.P., 1978, J. Chem. Soc. Dalton, p.1102.

11. Queiroz, J.C., Airoldi, C. and Chagas, A.P., 1981, J. Inorg. Nucl. Chem., 43, p.1207.

12. Airoldi, C., Chagas, A.P. and Filho, M.N., 1981, J. Inorg. Nucl. Chem., 43, p.89.

13. Airoldi, C., Chagas, A.P. and Assunção, F.A., 1980, J. Chem. Soc. Dalton, p.1823.

14. Ribeiro da Silva, M.A.V. and Reis, A.M.M.V., 1979, Bull. Chem. Soc. Japan, 52, p.3080.

15. Technical Notes 270-3 and 270-4, 1968, 1969, National Bureau of Standards, Washington, D.C.

16. Burkinshaw, P.M. and Mortimer, C.T., 1982, Coord. Chem. Rev., in press.

17. Cavell, K.J., Hill, J.O. and Magee, R.J., 1980, J. Chem. Soc. Dalton, pp. 763 and 1638.

18. Cavell, K.J., Hill, J.O. and Magee, R.J., 1979, Thermochim Acta, 34, p.155.

19. D'Ascenzo, G. and Wendlandt, W.W., 1969, J. Thermal Anal., 1, p.423.

20. Allen, T.L., 1959, J. Chem. Phys., 31, p.1039.

21. Cook, K.D. and Taylor, J.W., 1979, Int. J. Mass Spectrom. Ion Phys., 30, p.93.

THERMOCHEMISTRY OF β-DIKETONES AND METAL-β-DIKETONATES.
METAL-OXYGEN BOND ENTHALPIES

Manuel A. V. Ribeiro da Silva

Department of Chemistry, Faculty of Sciences
University of Oporto, 4000 PORTO, PORTUGAL

The thermochemical data available for standard enthalpies of formation and for standard enthalpies of sublimation or vaporization, both for β-diketones and metal-β-diketonates, are reviewed, and the mean homolytic metal-oxygen bond enthalpies in the metal complexes are derived. The results are discussed and, whenever appropriate, trends in data are suggested.

1. INTRODUCTION

The β-diketones are bidentate ligands which appear to form a complex with virtually every metal and metalloid in the periodic table. Metal β-diketonates were known as early as 1887, but the nature of the coordinate bond was not completely understood until 1945 when aromaticity within the diketonate ring was proposed. Many metal-β-diketonates were rapidly recognised as inner complexes and several such complexes were demonstrated to be volatile and to be soluble in organic solvents.

The thermochemical study of metal-β-diketonates originated in 1962 when the combustion enthalpy of the majority of the first row transition metal acetylacetonate complexes was reported. This was pioneering work in that these data were used to derive thermochemical homolytic and heterolytic bond dissociation enthalpies, the later ones used to show a classical double periodic variation with metal atomic number as predicted by the simple crystal field theory. The experimental work involved has been shown subsequently to be of very low accuracy, but calculations remain the principal method in use for the derivation of the mean metal-ligand bond dissociation enthalpies. Since then the thermochemistry of both β-diketones and metal-β-diketonate complexes has been investigated progressively and extensively over the last two decades. More thermochemical data are available for metal-β-diketonate complexes than for any other group of complexes containing metal-oxygen coordinate bonds. Thus it is possible to classify such thermochemical data into well defined categories and to identify several trends within each category.

Since the preparation of acetylacetone and similar β-dicarbonyl compounds in the later half of last century, organic chemists have had considerable interest in their properties and reactions. The best known phenomenon of β-dicarbonyls is their participation in tautomeric equilibrium of keto and enol structures, which is a most important feature of β-diketones since it is the enolate form which complexes with metals (fig. 1).

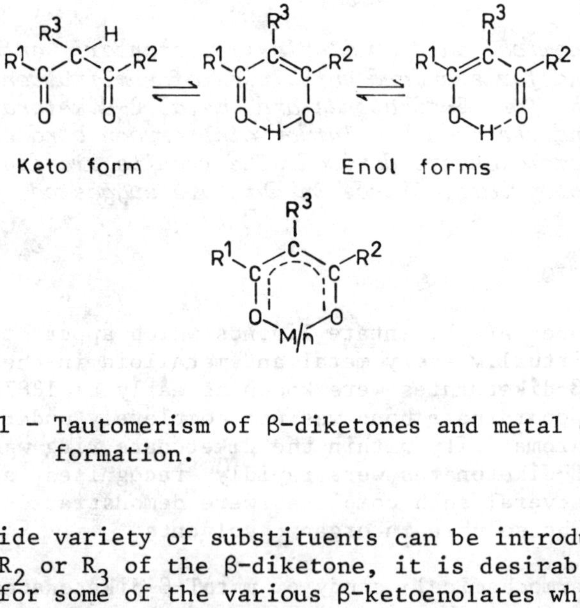

Fig. 1 - Tautomerism of β-diketones and metal complex formation.

Since a wide variety of substituents can be introduced at positions R_1, R_2 or R_3 of the β-diketone, it is desirable to use abbreviations for some of the various β-ketoenolates which have been extensively studied. Unfortunatly, a unique system of abbreviations has not yet been agreed on, and consequently the same compound appears in the literature under different abbreviations. Throughout this paper, the β-diketones will be generally abbreviated as "Hlig", "H" being the enolic hydrogen atom, and "lig" the abbreviation for the rest of the structure. The abbreviations used are resumed in table 1, together with the systematic names and the trivial names.

β-diketones are weak acids, stronger than phenol (except the

Table 1 – Nomenclature and abbreviations of β-diketones

Formula *	Systematic name	Trivial name	Abbreviation
$CH_3COCH_2COCH_3$	2,4-Pentanedione	Acetylacetone	Hacac
$CF_3COCH_2COCH_3$	1,1,1-Trifluoro-2,4-pentanedione	Trifluoroacetylacetone	Htfac
$CF_3COCH_2COCF_3$	1,1,1,5,5,5-Hexafluoro-2,4-pentanedione	Hexafluoroacetylacetone	Hhfac
$(CH_3)_2CHCOCH_2COCH(CH_3)_2$	2,6-Dimethyl-3,5-heptanedione	Diisobutyrylmethane	Hdibm
$(CH_3)_3CCOCH_2COCH_2CH_3$	2,2-Dimethyl-3,5-heptanedione	Pivaloylpropionylmethane	Hpiprm
$(CH_3)_3CCOCH_2COCH(CH_3)_2$	2,2,6-Trimethyl-3,5-heptanedione	Isobutyrylpivaloylmethane	Hibpm
$(CH_3)_3CCOCH_2COC(CH_3)_3$	2,2,6,6-Tetramethyl-3,5-heptanedione	Dipivaloylmethane	Hdpm
$C_6H_5COCH_2COCH_3$	1-Phenyl-1,3-Butanedione	Benzoylacetone	Hbzac
$C_6H_5COCH_2COC_6H_5$	1,3-Diphenyl-1,3-Propanedione	Dibenzoylmethane	Hdbzm

* Expressed as the keto form

Table 2 - Some pK_a values at 298.15 K

Compound	pK_a		
Acetylacetone	9.00 (1,2)	8.77 (3)	8.99 (4)
Trifluoroacetylacetone			6.79 (4)
Hexafluoroacetylacetone			5.35 (4)
Dipivaloylmethane	11.77 (5)		
Tropolone	6.95 (6,7)	6.92 (7)	
Methyltropolone	7.92 (7)		
Benzoylacetone			8.74 (4)
Phenol	10.0 (7)		
Acetic Acid	4.8 (7)		
Benzoic Acid	4.2 (7)		

dipivaloylmethane) but weaker than acetic acid and benzoic acid. Table 2 resumes the relative acidities of these ligands, by means of their pK_a values. The 3-proton can be lost easily and give the enolate anion which has a five-atoms π network extending over the two oxygens and three non-terminal carbon atoms. Hence six electrons occupy the resulting π-type molecular orbitals, the enolate anion having a delocalized symmetric structure.

Besides the usual bidentate way of bonding to the metals (through the two oxygen atoms, forming a chelate ring) the β-diketones can be monodentate and can also bridge atoms in a number of ways (8). Chrystallographic studies (9) for numerous octahedral acetylacetonate complexes show that all coordinate enolate anions have the same symmetry, and that in each complex the six M-O bond distances are equal, within the experimental error.

Various aspects of the structure, bonding and other properties of these compounds have been treated in several review articles (10-16)

Besides the academic interest that β-diketones provide for the chemists, these compounds have become quite useful, not only as tools in chemistry, but also where practical applications are concerned. Possible applications exist in the areas of:
- solvent extraction (rare earths, lanthanides and actinides);
- fractional separation of metals;
- mass spectrometric determination of a metal in a mixture, using an integrated current method;
- spectrometric determination of uranium;
- fluorometric determination of metals;
- fluorescence production;
- inhibitors of aluminium in alkaline media;

- solvent for dissolution of various metals;
- liquid organometallic lasers;
- convenient methods of research in Physical Organic Chemistry as a typical polar reaction to elucidate structural effects or acid-base catalysis;
- elucidation of structural arrangements of different atoms in inorganic molecules;
- control of soil-borne fungi;
- determination of δ-aminolevulinic acid in urine;
- insecticides;
- catalysts for polymerization;
- etc.

2. STANDARD ENTHALPIES OF FORMATION OF β-DIKETONES

The enthalpies of formation of β-diketones are of key importance in determining the enthalpies of formation of metal β-diketonate co-ordination complexes from reaction-calorimetric studies.

The enthalpy of formation of acetylacetone was firstly determined in 1957 by Nicholson (17) from the enthalpy of combustion in oxygen: the amount of reaction was calculated from the mass of the sample contained in a glass ampoule and the observed energy of combustion was corrected for the presence of about 2 per cent of xylene impurity. Hacking and Pilcher (18) redetermined this enthalpy of combustion, in which the sample was contained in a plastic bag and the amount of reaction was calculated from the mass of carbon dioxide produced. The standard enthalpies of formation of four others methyl-substituted heptane-3,5-diones, were measured by static bomb calorimetry, by Ferrão et al (19) using the same technique.

The enthalpy of combustion of crystalline benzoylacetone was firstly measured by Farrar and Jones (20) on a commercial product and redetermined on a purified sample by Ferrão et al (19).

The available values for the standard enthalpies of formation of β-diketones are resumed in Table 3.

There are no experimental values for the standard enthalpies of formation of trifluoroacetylacetone and hexafluoroacetylacetone. However, the values can be estimaded on the basis of the existing experimental values for other compounds, by using a Group Scheme; the estimated values are also listed in Table 3, between brackets.

At 298.15 K, the liquid acetylacetone is the equilibrium mixture: (0.186 keto + 0.814 enol) (22). The enthalpies of enolization have been reported; in the liquid phase, -11.3 ± 0.4 kJ mol^{-1}

Table 3 – Standard enthalpies of formation of β-diketones

β-diketone	ΔH_f^o(ℓ or c)/kJ mol^{-1}	Ref.
Hacac	−423.9±1.5	17
	−425.5±1.0	18
Hbzac	−339.7±4.6	20
	−335.1±2.8	19
Hdbzm	−224.9±1.8	21
Hdpm	−587.7±3.8	19
Hpiprm	−527.1±2.2	19
Hdibm	−526.8±2.0	19
Hibpm	−568.1±2.0	19
Htfac	(−1051.0±5.0)	Estimated
Hhfac	(−1673.9±5.0)	Estimated

(22), and in the gaseous phase, −10.0±0.8 kJ mol^{-1} (23). The enthalpy of vaporization of the enol form was derived by Irving and Wadsö (24) from the enthalpy of vaporization of the equilibrium mixture, ΔH_v^o(Hacac, enol)=43.2±0.1 kJ mol^{-1}. These data allow the calculations of ΔH_f^o(Hacac), both in the keto and enol forms listed below:

	ΔH_f^o(ℓ)/kJ mol^{-1}	ΔH_f^o(g)/kJ mol^{-1}
Hacac, keto	−416.3±1.1	−374.4±1.3
Hacac, enol	−427.6±1.1	−384.4±1.3

The keto-enol equilibrium compositions of several liquid β-diketones at 298.15 K have been determined by n.m.r. spectroscopy, but the enthalpy of enolization has been deduced only for acetylacetone. The enol tautomer is the major constituent (> 93 per cent) in the methyl-substituted heptane-3,5-diones (25,26) so there will be insignificant error introduced in deriving ΔH_f^o(enol,ℓ) from the observed ΔH_f^o(ℓ) by assuming that ΔH^o(enolization) is unchanged from the known value for acetylacetone. The enthalpies of vaporization of the enol forms have been reported (27) and table 4 lists the enthalpies of formation of the enol forms in the liquid and gaseous states. It is not possible to predict accurately ΔH_f^o(enol,g) by application of a simple bond-energy scheme because of the difficulty of estimating the delocalization energy and the intramolecular hydrogen-bonding energy in the enol form. An indirect estimate of ΔH_f^o(enol,g) can be made by deriving a bond-energy scheme to predict ΔH_f^o(keto,g), calibrated against ΔH_f^o(acetylacetone,keto,g) previously reported, and using the group bond-energy terms given by Cox and Pilcher (28). By assuming a constant gaseous enthalpy of enolization, −10.0 kJ mol^{-1}, values of ΔH_f^o(enol,g) may then be obtained. These estimates are given in table 4 for comparison with the experimental values.

Table 4 - Enthalpies of formation of enol forms at 298.15 K (all values in kJ mol^{-1})

	Hpiprm	Hdibm	Hibpm	Hdpm
% of enol(ℓ)	93.4 (25)	95.8 (25)	96.0 (26)	98.0 (25)
ΔH_f^o(enol,ℓ)	-527.8±2.3	-527.3±2.1	-568.4±2.1	-587.9±3.9
ΔH_f^o(enol,g,obs.)	-470.9±2.3	-470.4±2.1	-510.7±2.1	-528.4±3.9
ΔH_f^o(enol,g,calc.)	-478.5	-475.3	-502.6	-529.8

In view of the assumptions made, the agreement with the observed values seems reasonable.

Hubbard et al (29) measured the standard enthalpy of formation of tropolone (2-hydroxycyclohepta-2,4,6-trien-1-one), hereafter abbreviated as Htrop, by static bomb combustion calorimetry, and found the value ΔH_f^o(Htrop,c)=-239.24±0.88 kJ mol^{-1}. From this value, it is possible to estimate an identical parameter for the 4-methyltropolone, as ΔH_f^o(H-4Metrop,c)=-275.7±2.9 kJ mol^{-1}.

3. STANDARD ENTHALPIES OF VAPORIZATION OR SUBLIMATION OF β-DIKETONES

The available values for the standard enthalpies of vaporization of liquid β-diketones have been measured by a direct calorimetric method, using the Wadsö vaporization calorimeter (30). The values of the molar enthalpies of vaporization, listed in table 5, refer

Table 5 - Standard Enthalpies of Vaporization (298.15 K) of some β-diketones

β-Diketone	Normal $t_{b.p.}$/°C	Enol %	ΔH_v^o/kJ mol^{-1}
Hacac	140	81.4 (22)	41.8±0.2 (24)
Hhfac	70	100 (32)	30.6±0.1 (27)
Htfac	107	97 (32)	37.2±0.2 (27)
Hdibm	199-201 *	95.8 (25)	56.1±0.2 (27)
Hpiprm	203-204 *	93.4 (25)	56.9±0.1 (27)
Hibpm	206-208 *	96.0 (26)	57.7±0.2 (27)
Hdpm	214-216 *	98.0 (25)	59.5±0.1 (27)

* - Reduced to normal pressure using Dreisbach Tables (31)

to the isothermal vaporization (298.15 K) of the pure liquid (keto-enol) to the real gas (keto-enol) formed under its saturated vapour pressure. With the exception of acetylacetone, neither the keto-enol composition of the vapour nor the enthalpy of enolization of these compounds is known. As previously stated, in view of the high percentage of enol form in the liquid equilibrium, even assuming complete conversion to enol in the vapour state, it is calculated that the values for the vaporization of pure liquid to real gas for all compounds in table 5, with the exception of acetylacetone, are the same as for liquid (100% enol) to gas (100% enol) within the experimental error quoted. For acetylacetone, this value is calculated (24) as +43.2 kJ mol^{-1}.

A plot of the standard enthalpies of vaporization versus the normal boiling point of the β-diketones (Fig. 2) shows that all seven fall on a smooth weakly convex curve corresponding to the equation

$$\Delta H_v^o / kJ\ mol^{-1} = 20.29 + 13.18 \times 10^{-2} (t_{b.p.} /^o C) + 23.64 \times 10^{-5} (t_{b.p.} /^o C)^2 \quad (1)$$

Irving and Ribeiro da Silva (27) plotted some of Wadsö's (33-37) results (Fig. 2) together with the results obtained for the β-diketones, showing that the β-diketone curve lies very clo-

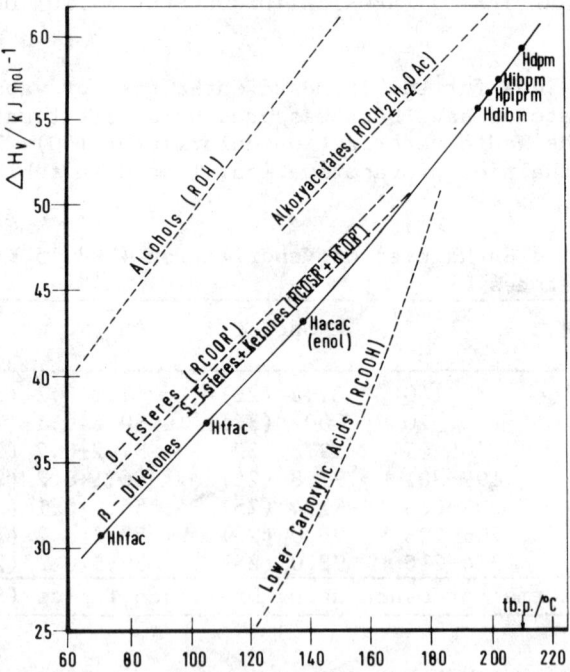

Fig. 2 - Correlation between standard enthalpies of vaporization and normal boiling points.

sely to the families of slightly associated compounds, suggesting that β-diketones have a very low or even non-existent degree of association. This is to be expected, as the enolic hydrogen of the β-diketones forms an intramolecular hydrogen bond with the ketonic oxygen. If Wadsö's (33) relationship

$$\Delta H_v^o / kJ\ mol^{-1} = 20.9 + 0.172(t_{b.p.}/^oC) \qquad (2)$$

is used to calculate the heats of vaporization of β-diketones, reasonable results are obtained, the maximum variations occurring at the two extremes, Hhfac being 2.5 kJ mol^{-1} more exothermic, and Hdpm 1.7 kJ mol^{-1} less exothermic than the experimental determined values.

Irving and Ribeiro da Silva also showed (27) that for β-diketones the increment of a methylene group in a secondary carbon increases the enthalpy of vaporization by +0.9 kJ mol^{-1} and the increment of a methylene group in a tertiary carbon, increases the enthalpy of vaporization by +1.7 kJ mol^{-1}. These increments are remarkably close (38) to those of the much more thoroughly studied aliphatic ketone series.

The enthalpy of sublimation of benzoylacetone has been reported by Aihara (39) from vapour pressures measurements using a viscosity gauge, as ΔH_{subl}^o(Hbzac, c)=83.8±0.8 kJ mol^{-1}. This result, combined with ΔH_f^o(Hbzac, c)=-335.1±2.8 (19), leads to ΔH_f^o(Hbzac, g)=-251.3±2.9 kJ mol^{-1}. Estimation of ΔH_f^o(Hbzac, enol, g), as described in part 2 of this paper, gives -253.9 kJ mol^{-1}, suggesting that in the gaseous state this compound will exist predominantly in the enol form.

The enthalpy of sublimation of dibenzoylmethane has been reported by Wood and Jones (40) as 75.3 kJ mol^{-1}, from measurements of vapour pressure using the isoteniscope technique of Truemper. This value seems too low if compared with an identical parameter for benzoylacetone, which probably is due to a systematic error in the experimental method used.

Jackson et al (41) reported the standard enthalpy of sublimation of tropolone as ΔH_{subl}^o(Htrop, c)=84.1±0.4 kJ mol^{-1}, from vapour pressure measurements by the Knudsen-cell method.

4. STANDARD ENTHALPIES OF FORMATION OF CRYSTALLINE METAL-β-DIKETONATES

The first values reported for standard enthalpies of formation of metal-β-diketonates were obtained by static bomb calorimetry in the early sixties. Jones and co-workers (40,42-45) and Kawasaki (46) reported the first results. The experimental work involved

Table 6 - Early work on thermochemistry of $[M(\beta\text{-dik})_n]$

$[M(\beta\text{-dik})_n]$	$\Delta H_f^o(c)$ kJ mol^{-1}	ΔH_{sub}^o kJ mol^{-1}	$\bar{D}(M\text{-}O)$ kJ mol^{-1}	Ref
$[Sc(acac)_3]$	-1776±15	49.8±2.5	255±4	44
$[V(acac)_3]$	-1649±30	44.4±2.5	263±4	44
$[Cr(acac)_3]$	-1480±14	27.6±2.9	218±4	44
$[Mn(acac)_3]$	-1360±12	77.8±0.8	172±4	44
$[Fe(acac)_3]$	-1486±15	65.3±1.7	218±4	44
$[Co(acac)_3]$	-1364±18	75±5	197±4	44
$[Ni_3(acac)_6]$	-4284	69	431	43
$[Cu(acac)_2]$	-774	63	176	42
$[Al(acac)_3]$	-1749	-	-	46
$[Fe(bzac)_3]$	-1054±8	45.6±0.4	292±3	45
$[Fe(dbzm)_3]$	-752±8	31.8±1.3	215±6	45

has later been shown to be of very low accuracy (47), probably because insufficient attention was paid to controlling the combustion with the auxiliary aid and mainly in defining the completeness of the combustion and the nature of the combustion products. Table 6 contains some results of the early work on combustion of metal-β-diketonates, together with the reported entalpies of sublimation and mean metal-oxygen bond dissociation enthalpies.

In 1966, Irving (48) published the first article on the determination of the standard enthalpy of formation of a crystalline metal-β-diketonate, $[Al(acac)_3]$, by solution and reaction calorimetry.

The standard enthalpies of formation of metal-β-diketonates, $[M(\beta\text{-dik})_n]$, are obtained from the enthalpy change which occurs as a result of the reaction

$$n\, H\beta\text{-dik}(\ell \text{ or } c) + MX_n \cdot xH_2O(c) + yH_2O(\ell) \underset{\leftarrow}{\overset{\Delta H_r}{\rightarrow}}$$

$$\underset{\leftarrow}{\overset{}{\rightarrow}} [M(\beta\text{-dik})_n](c) + nHX \cdot (\frac{x+y}{n}) H_2O(\ell) \qquad (3)$$

The enthalpy change of the reaction, ΔH_r, is measured indirectly in an isothermal-jacketed solution-calorimeter by reaction in a non-volatile solvent, in which all the components are soluble. If

Table 7 - Standard Enthalpies of Formation of Metal-Acetylacetonates

Complex	Method	$\Delta H_f^o(c)^\dagger$/kJ mol^{-1}	Ref
[Ca(acac)$_2$]	Sol.	-1310.4±4.2	50
[Cr(acac)$_3$]	Sol.	-1564.8±8.9	51
[Mn(acac)$_2$]	Sol.	-1036.8±3.3	50
[Mn(acac)$_3$]	Sol.	-1378.9±3.7	52
[Fe(acac)$_3$]	Sol.	-1314.7±3.0	53
[Co$_4$(acac)$_8$]	Sol.	-3494.5±8.4	50
	Sol.	-3471.9±9.2	54
[Ni$_3$(acac)$_6$]	Sol.	-2699.9±6.3	55
	Sol.	-2582.9±6.2	56
[Cu(acac)$_2$]	Sol.	-782.4±2.4	50,57
	Sol.	-809.9±1.3	58
[Zn(acac)$_2$]	Sol.	-932.6±2.9	50
[Be(acac)$_2$]	Sol.	-1250.0±2.2	59
[Al(acac)$_3$]	Sol.	-1797.8±3.7	48
	Comb.	-1793.3±2.0	47
[Ga(acac)$_3$]	Sol.	-1487.5±5.2	49
	Comb.	-1476.0±4.5	47
[In(acac)$_3$]	Comb.	-1405.7±3.6	47
[Mo(acac)$_3$]	Sol.	-1327±6	60,61

† Some of the values have been recalculated using more modern auxiliary data.

equilibrium is rapidly reached from either side, the difference between enthalpies of solution of all products and reactants, in the required stoicheiometric ratio, gives the enthalpy of reaction. The method introduced by Irving has since been used by different workers for measuring standard enthalpies of formation of a great number of crystalline metal-β-diketonates, using appropriate solvents. Table 7 reports the available literature values for metal acetylacetonates.

Cavell and Pilcher (47) in 1977 measured the enthalpies of combustion the tris(acetylacetonates) of Al^{III}, Ga^{III} and In^{III}, by static combustion calorimetry, and derived the standard enthalpies of formation (values in table 7),which are in good agreement with the values from reaction calorimetry reported by Irving et al

Table 8 - Standard Enthalpies of Formation of Metal-β-Diketonates and Metal-Tropolonates †

Ligand	$\Delta H_f^o(c)/kJ\ mol^{-1}$			
	Ni^{II}	Cu^{II}	Be^{II}	Al^{III}
Htfac			-2498±10a	-3696±15b
Hdibm		-998.0±4.4c		
Hpiprm		-995.5±4.7c		
Hibpm		-1078.5±4.4c		
Hdpm	-1776.7±7.8d	-1111.5±7.8c	-1552.0±7.8a	-2255±12b
Hbzac	-1976±13§e			
	-1895±17§f	-591.4±5.8g	-1013.1±5.7h	-1483.6±8.9h
Htrop		-414.8±1.8i	-855.2±1.8a	-1262.7±2.8j
HMetrop				-1316.7±9.2j

† All values have been determined by reaction calorimetry and most of them have been recalculated using more modern auxiliary data

§ These values refer to the trimer $[Ni_3(bzac)_6]$

| a=Ref. 62 | b=Ref. 63 | c=Ref. 58 | d=Ref. 56 | e=Ref. 64 |
| f=Ref. 65 | g=Ref. 66 | h=Ref. 67 | i=Ref. 68 | j=Ref. 69 |

(48,49), giving confidence in the reaction calorimetric method.

The modern literature values for standard enthalpies of formation of crystalline metal-β-diketonates, are resumed in table 7 (acetylacetonates) and table 8 (other β-diketonates and tropolonates). These is also a published value for $\Delta H_f^o\{[Zn(bzac)_2],c\}$ = = -747.7±4.2 kJ mol^{-1} (50).

5. STANDARD ENTHALPIES OF SUBLIMATION OF METAL-β-DIKETONATES

The standard enthalpies of sublimation of coordination compounds are of fundamental importance as it is necessary to refer the standard enthalpies of formation to the ideal gas state, in order to remove the intermolecular forces and so calculate the strengths of the coordination bonds.

Several methods have been used to measure enthalpies of sublimation of metal β-diketonates, but unfortunately the values derived for the same complex by different techniques are generally inconsistent, and so large discrepancies exist between reported enthalpies of sublimation for metal—β—diketonate complexes. For example, the published ΔH_{sub} values for tris(acetylacetonato)aluminium(III) range from 19.2 kJ mol^{-1} (70) obtained by the isoteniscopic method by Berg and Truemper to 129 kJ mol^{-1} (71) obtained by differential scanning calorimetry by Beech and Lintonbon, with several values scattered between these extremes: 24.3 kJ mol^{-1} (72), 47 kJ mol^{-1} (73), 105 kJ mol^{-1} (74), 111 kJ mol^{-1} (74), 120 kJ mol^{-1} (75) and 121.7 kJ mol^{-1} (76). For tris(acetylacetonato)-iron(III), values range from 19 kJ mol^{-1} (70), again obtained by the isoteniscopic method, to 121 kJ mol^{-1} (74) obtained by sublimation bulb technique, with seven other values between these extremes (44, 45, 71, 72, 74, 77 and 78). A similar situation can be found for most of the other β-diketonates, suggesting that this area is widely open for further research.

In view of the present situation, it is necessary, at the moment, to choose sensible values among the published ones. It seems likely that the sublimation enthalpies reported by Jones et al as well as the isoteniscopic ones from Berg and Truemper are too low, even if compared with the accepted values for the ligands, probably due to sistematic errors in the determinations. For instance, the isoteniscopic method, as modified by Berg and Truemper (70), failed to give good results using benzoic acid as a calibrating substance! In these circunstances, in the present article we chose mainly values of standard enthalpies of sublimation derived from the Knudsen Effusion Technique, when available, and values from the "vacuum sublimation drop microcalorimetric method" of Skinner et al (79). Although this later method is a simple and easy one, it yields results that appear to be reliable to ±4 kJ mol^{-1} (80).

Table 9 - Standard Enthalpies of Sublimation of Metal-β-Diketonates

Ligand	ΔH^o_{sub}/kJ mol^{-1}					
	Cr^{III}	Fe^{III}	Co^{III}	Cu^{II}	Be^{II}	Al^{III}
Hacac	†123±3[a]			*110±4[b]	†94±1[a]	†120±3[a]
Htfac	*117±4[d]	*138±4[d]	*114±4[d]	*115±4[d]	†88±4[a]	†108±2[a]
Hhfac	*112±4[d]			*111±6[b]		
Hdibm				*118±4[b]		
Hpiprm				*123±4[b]		
Hibpm				*126±4[b]		
Hdpm	*133±4[d]	*146±4[d]	*126±4[d]	*123±7[b]	†102±3[a]	†119±3[a]
Hbzac	*186±4[d]	*200±4[d]		*160±4[b]	*142±4[c]	*194±4[c]

† Measured by the Knudsen Effusion Technique
* Measured by the Skinner Vacuum Sublimation Drop Technique (79)
a=Ref. 75   b=Ref. 58   c=Ref. 67   d=Ref. 81

The values choosen from these two techniques (Table 9) seem to be consistent and show a trend (Fig. 3) which can be useful to estimate standard enthalpies of sublimation not yet measured.

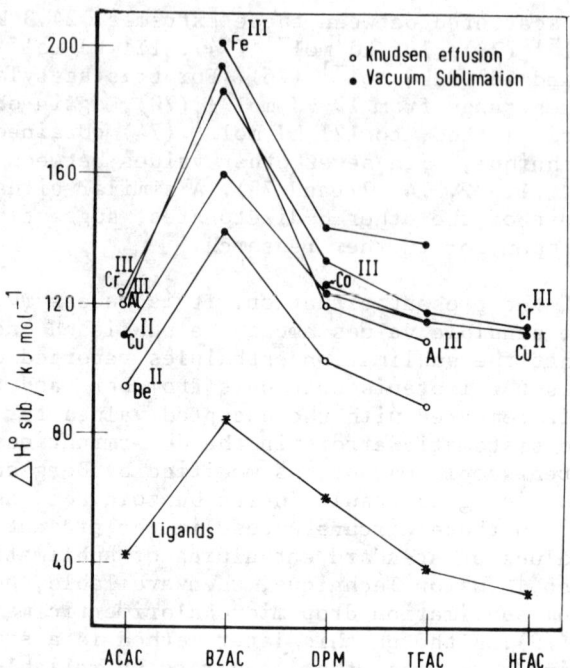

Fig. 3 - Enthalpies of sublimation of metal-β-diketonates and enthalpies of vaporization(*) of β-diketones

6. THE MEAN METAL-OXYGEN BOND DISSOCIATION ENTHALPIES

From the standard enthalpies of formation of the crystalline metal-β-diketonates, $\Delta H_f^o\{[M(\beta dik)_n],c\}$ and the corresponding standard enthalpies of sublimation, $\Delta H_{sub}^o\{[M(\beta dik)_n],c\}$, the standard enthalpies of formation of the gaseous metal-β-diketonates, $\Delta H_f^o\{[M(\beta dik)_n],g\}$ are derived.

From the molar enthalpies of dissociation, $\Delta H_{f.r.}$, of the gaseous molecules into metal atoms and ligand radicals (equation 4)

$$[M(\beta dik)_n](g) \xrightarrow{\Delta H_{f.r.}} M(g) + n\, \beta dik \cdot (g) \qquad (4)$$

the mean metal-oxygen homolytic bond dissociation enthalpies, $\langle D \rangle(M-O)$, can be derived. Since the oxygen atoms in metal β-diketonates are equivalent (9), $\langle D \rangle(M-O)$ are calculated as

$$\langle D \rangle(M-O) = \Delta H_{f.r.}/2n \qquad (5)$$

with

$$\Delta H_{f.r.} = \Delta H_f^o(M,g) + n\Delta H_f^o(H-\beta dik,g) - n\Delta H_f^o(H,g) - \Delta H_f^o\{[M(\beta dik)_n],g\} + n<D>(O-H,enol) \quad (6)$$

Where $<D>(O-H,enol)$ corresponds to the molar enthalpy of the dissociation of the enolic hydrogen

$$H-\beta dik(g) \longrightarrow H(g) + \beta dik^\bullet(g) \quad (7)$$

There are no measured values for the molar enthalpy of dissociation of the enolic hydrogen from β-diketones. For the eight aliphatic alkanols C_1-C_4, $D(O-H)$ is within the range 427-431 kJ mol^{-1} (82), and so Irving and Ribeiro da Silva (118) estimated the value $<D>(O-H,enol) = 418\pm20$ kJ mol^{-1}, since probably this value does not vary considerably from one β-diketone to another (83). Later, Cavell and Pilcher (47) estimated 365 kJ mol^{-1}, on the basis that a more realistic comparison would be with $D(C_6H_5O-H) = 352$ kJ mol^{-1} and as the more extensive delocalisation in phenol, $D(C_6H_5O-H)$ should be slightly less than $D(O-H,enol)$. They also noted that for acetone $D(H-CH_2COCH_3) = 385$ kJ mol^{-1} (82) and this would be an upper value since the stabilization would be larger for the radical acac$^\bullet$ than for the radical $CH_3COCH_2^\bullet$. The estimate of Cavell and Pilcher seems low as in phenol the formation of the radical is stabilized by delocalisation and in the case of β-diketones the enolic forms are extensively stabilized both by delocalisation and by the intramolecular hydrogen bonding. It should be pointed that Cavell et al (61) suggested later a value of 400 kJ mol^{-1}. In this work, for comparison reasons, the value $<D>(H-\beta dik) = 418\pm20$ kJ mol^{-1} will be considered.

The mean metal-oxygen homolytic bond dissociation enthalpies for metal β-diketonates have been calculated, most of them being registered in table 10. For Ga(III), Mo(III) and In(III), there are only values for the acetylacetonates, respectively, 183±10 kJ mol^{-1} (47,49), 209±10 kJ mol^{-1} (60,61) and 162±10 kJ mol^{-1} (47).

The available results allow several conclusions to be drawn:

(a) - The mean metal-oxygen bond dissociation enthalpies, for the same metal with different ligands, are almost the same within the associated uncertainties. (The average values of $<D>(M-O)$ for the same metal are in the last row of table 10);

(b) - For the cobalt and manganese β-diketones $<D>(Co^{II}-O)$ is larger than $<D>(Co^{III}-O)$ and $<D>(Mn^{II}-O)$ is larger than $<D>(Mn^{III}-O)$, which can be assigned to the increase in electronic density in the metal;

(c) - If one thinks in terms of interatomic distances, despite

Table 10 - Mean Metal-Oxygen Bond Dissociation Enthalpies in Metal β-Diketonates

$<D>(M-O)/kJ\ mol^{-1}$

	Cr^{III}	Mn^{II}	Mn^{III}	Fe^{III}	Co^{II}	Co^{III}	Ni^{II}	Cu^{II}	Be^{II}	Al^{III}
Hacac	214 ± 10^a		164 ± 10^b	174 ± 10^c	202 ± 11^d		202 ± 10^e	167 ± 10^h	278 ± 10^j	$242\pm10^{i,m}$
Hbzac	207 ± 11^p		163 ± 10^p	175 ± 10^p	201 ± 10^p	164 ± 10^p	203 ± 11^f	167 ± 10^g	273 ± 10^k	250 ± 10^k
Htfac	221 ± 11^p	212 ± 11^p		176 ± 11^p	204 ± 11^p	177 ± 11^p		164 ± 11^p	$277\pm11^\ell$	246 ± 10^n
Hhfac	238 ± 11^p	217 ± 11^p		180 ± 11^p	203 ± 11^p		203 ± 11^p	163 ± 11^p		
Hdpm	209 ± 10^p		170 ± 10^p	183 ± 10^p		171 ± 10^p	202 ± 11^e	168 ± 10^h	$280\pm10^\ell$	247 ± 10^n
Hdibm								169 ± 10^h		
Hpiprm								168 ± 10^h		
Hibpm								168 ± 10^h		
Average	218	214	166	178	202	171	202	167	277	246

a=Ref. 51; b=Ref. 52; c=Ref. 53; d=Ref. 54; e=Ref. 56; f=Ref. 65; g=Ref. 66; h=Ref. 58;
i=Ref. 47; j=Ref. 59; k=Ref. 67; ℓ=Ref. 62; m=Ref. 48; n=Ref. 63; p=Ref. 84.

the fact that not all of the molecular structures are known, the results seem consistent with the fact that for the different metal-β-diketonates of the same metal the M-O bond length is almost constant (9);

(d) – The average value for $<D>$(Cr-O) is 218 kJ mol^{-1}. For the chromium(III)-β-diketonates, whose crystal structures are known (85,86), the average Cr-O bond length is 195.4 pm. Entering this value in the suggested correlation (87) $<D>$(Cr-O)=f$[r$(Cr-O)$]$, one estimates $<D>$(Cr-O)=220 kJ mol^{-1}, in excellent agreement with the value for tris(β-diketonates) chromium(III);

(e) – The almost constant values for $<D>$(M-O) suggest that, for a certain metal, the mean metal-oxygen bond dissociation enthalpy is not affected by the structure of the ligand, and so, as the metal is completely surrounded by the oxygen atoms of the ligands, it must be totally oxidized, i.e, in a degree of oxidation similar to the one of the corresponding metal-oxide. So a linear correlation $<D>$(M-O, complex, g)=f{$<D>$(M-O,oxide,g)} is to be expected, where $<D>$(M-O, oxide,g) is ΔH^{o}_{diss} for reaction (8)

$$MO(g) \xrightarrow{\Delta H^{o}_{diss}} M(g) + O(g) \qquad (8)$$

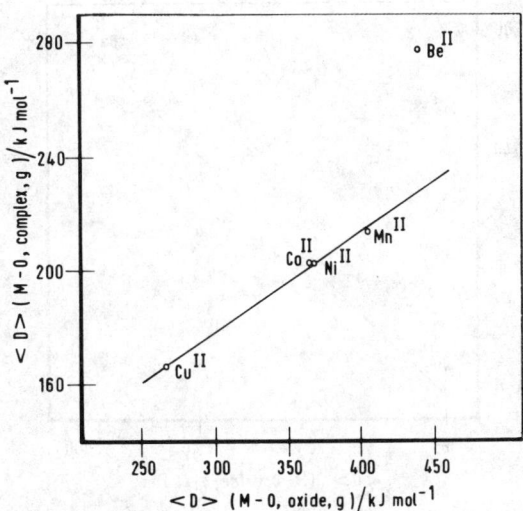

Fig. 4 – Variation of $<D>$(M-O, complex g) with $<D>$(M-O, oxide, g)

The above correlation is represented in fig. 4, and is excellent except for Be(II), for which there is no explanation.

(f) - It might be expected (88) that, the metal-oxygen bonds in the metal-β-diketonates are more similar to the corresponding bonds of the crystalline metal oxides, as in the oxides with polymeric structures the metal presents a coordination number and a configuration equal to the one in the complexes. So, if we consider the process

$$M_xO_y(c) \xrightarrow{\Delta H^o_{diss}} xM(g) + yO(g) \qquad (9)$$

and define the parameter

$$<D>(M-O, oxide, c) = \frac{\Delta H^o_{diss}}{\text{coord. number of the metal}} \qquad (10)$$

a plot of $<D>(M-O, complex, g)$ versus $D(M-O, oxide, c)$ should be a straight line. These plots, for bis(β-diketonate)metal (II) (see, fig. 5) and for tris(β-diketonate)metal(III), (see fig. 6) show excellent correlations.

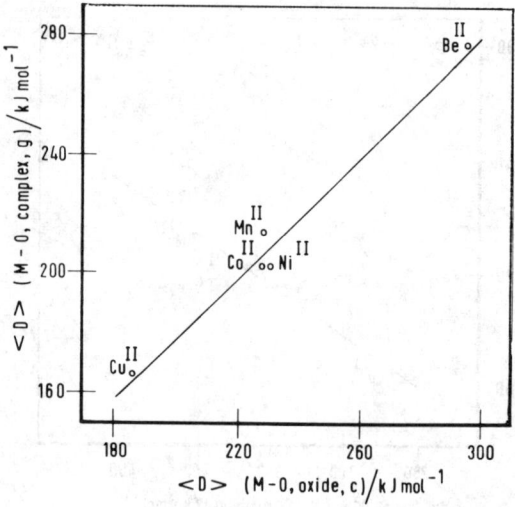

Fig. 5 - Variation of $<D>(M^{II}-O, complex, g)$ with $<D>(M^{II}-O, oxide, c)$

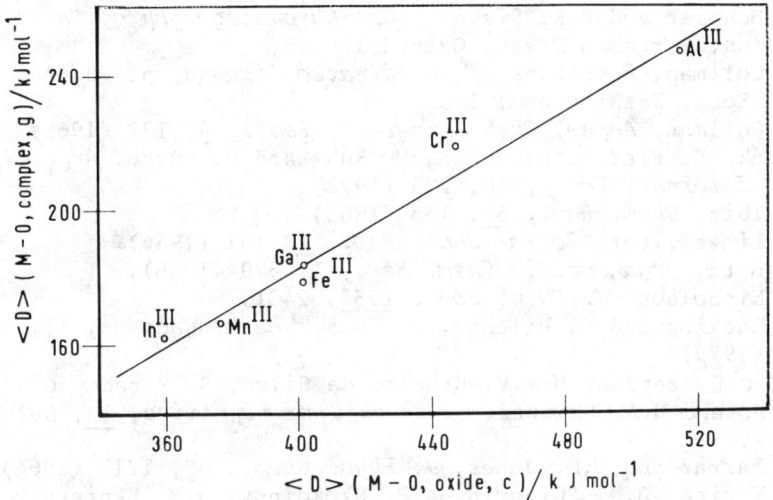

Fig. 6 - Variation of <D>(M^{III}-O, complex, g) with <D>(M^{III}-O, oxide, c).

ACKNOWLEDGMENTS

The author thanks NATO, Research Grant 1709, for partial financial support and the Instituto Nacional de Investigação Cientifica, Lisboa, for support given to a research project of the Chemistry Research Center, Oporto University. Thanks are also due to Junta Nacional de Investigação Cientifica e Tecnológica, for partial financial support under research contract nº 220.80.32.

REFERENCES

1. M.L. Eidinoff, *J. Am. Chem. Soc.*, 67, 2072 (1945)
2. R.P. Bell, E. Gelles and E. Moller, *Proc. Roy. Soc. (London)*, A198, 308 (1949).
3. F. Hashimoto, J. Tanaka and S. Nagakura, *J. Mol. Spectry.*, 10, 401 (1963).
4. S.P. Patel, *Ph.D. Thesis*, University of Surrey, 1973.
5. G.A. Guter and G.S. Hammond, *J. Am. Chem. Soc.*, 78, 5166 (1956).
6. J.W. Cook, A.R. Gibb, R.A. Raphael and A.R. Somerville, *Chemistry & Industry*, 1950, 427.
7. P.L. Pauson, *Chem. Rev.*, 55, 9 (1955).
8. D.W. Thompson, *Structure and Bonding*, 9, 27 (1972).
9. E.C. Lingafelter and R.L. Braun, *J. Am. Chem. Soc.*, 88, 2951 (1966).

10. R.W. Moshier and R.E. Sievers, *Gas Chromatography of Metal Chelates*, Pergamon Press, Oxford, 1965.
11. J.P. Collman, *Reactions of Coordinated Ligands*, p. 78, Am. Chem. Soc., Washington, 1963.
12. J.P. Collman, *Angew. Chem. Internat. Edit.*, 4, 132 (1965).
13. B. Bock, K. Flatau, H. Junge, M. Kuhr and H. Musso, *Angew. Chem. Internat. Edit.*, 10, 225 (1971).
14. J. Selbin, *Chem. Rev.*, 65, 153 (1965).
15. E.C. Lingafelter, *Coord. Chem. Rev.*, 1, 151 (1966).
16. F. Bonati, *Organometal. Chem. Rev.*, 1, 379 (1966).
17. G.R. Nicholson, *J. Chem. Soc.*, 1957, 2431.
18. J.M. Hacking and G. Pilcher, *J. Chem. Thermodynamics*, 11, 1015 (1979).
19. M.L.C.C.H. Ferrão, M.A.V. Ribeiro da Silva, S. Suradi, G. Pilcher and H.A. Skinner, *J. Chem. Thermodynaxics*, 13, 567 (1981).
20. D.T. Farrar and M.M. Jones, *J. Phys. Chem.*, 68, 1717 (1964).
21. M.P. Kozina, D.N. Shigorin, A.P. Skoldinov, S.M. Skuratov, *Dokl. Akad. Nauk SSSR*, 160, 135 (1965).
22. L.W. Reeves, *Can. J. Chem.*, 35, 1351 (1957).
23. H.J. Bernstein and J. Powling, *J. Am. Chem. Soc.*, 73, 4353 (1951).
24. R.J. Irving and I. Wadsö, *Acta Chem. Scand.*, 24, 589 (1970).
25. H. Koshimura, J. Saito and T. Okubo, *Bull. Chem. Soc. Japan*, 46, 632 (1973).
26. G.K. Schweitzer and E.W. Benson, *J. Chem. Eng. Data*, 13, 452 (1968).
27. R.J. Irving and M.A.V. Ribeiro da Silva, *J. Chem. Soc. Dalton*, 1975, 798.
28. J.D. Cox and G. Pilcher, *Thermochemistry of Organic and Organometallic Compounds*, Academic Press, London, 1970.
29. W.N. Hubbard, C. Katz, G.B. Guthrie Jr and G. Waddington, *J. Am. Chem. Soc.*, 74, 4456 (1952).
30. I. Wadsö, *Acta Chem. Scand.*, 20, 536 (1966).
31. R.R. Dreibach, *Pressure-Volume-Temperature Relationships of Organic Compounds*, Handbook Publishers Inc. Ohio, 3rd edn, 1952.
32. J.L. Burdett and M.T. Rogers, *J. Am. Chem. Soc.*, 86, 2105 (1964).
33. I. Wadsö, *Acta Chem. Scand.*, 20, 544 (1966).
34. I. Wadsö, *Acta Chem. Scand.*, 23, 2061 (1969).
35. P.B. Howard and I. Wadsö, *Acta Chem. Scand.*, 24, 145 (1970).
36. J. Konicek and I. Wadsö, *Acta Chem. Scand.*, 24, 2612 (1970).
37. K. Kusano and I. Wadsö, *Acta Chem. Scand.*, 25, 219 (1971).
38. M.A.V. Ribeiro da Silva and R.J. Irving, *Rev. Port. Quim.*, 20, 36 (1978).
39. A. Aihara, *Bull. Chem. Soc. Japan*, 32, 1242 (1959).
40. J.L. Wood and M.M. Jones, *J. Inorg. Nucl. Chem.*, 29, 113 (1967).
41. W. Jackson, T.S. Hung and H.P. Hopkins Jr, *J. Chem. Thermody-*

namics, <u>3</u>, 347 (1971).
42. M.M. Jones, B.J. Yow and W.R. May, *Inorg. Chem.*, <u>1</u>, 166 (1962).
43. J.L. Wood and M.M. Jones, *J. Phys. Chem.*, <u>67</u>, 1049 (1963).
44. J.L. Wood and M.M. Jones, *Inorg. Chem.*, <u>3</u>, 1553 (1964).
45. D.T. Farrar and M.M. Jones, *J. Phys. Chem.*, <u>68</u>, 1717 (1964).
46. Y. Kawasaki, T. Tanaka and R. Okawara, *Technol. Reports Osaka Univ.*, <u>13</u>, 217 (1963).
47. K.J. Cavell and G. Pilcher, *J. Chem. Soc. Faraday I*, <u>73</u>, 1590 (1977).
48. J.O. Hill and R.J. Irving, *J. Chem. Soc. (A)*, <u>1966</u>, 971.
49. R.J. Irving and G.W. Walter, *J. Chem. Soc. (A)*, <u>1969</u>, 2690.
50. W. Kakolowicz and E. Giera, *4eme Conf. Int. Therm. Chim.*, Montpellier, <u>1</u>, 73 (1975).
51. J.O. Hill and R.J. Irving, *J. Chem. Soc. (A)*, <u>1967</u>, 1413.
52. J.O. Hill and R.J. Irving, *J. Chem. Soc. (A)*, <u>1968</u>, 3116.
53. J.O. Hill and R.J. Irving, *J. Chem. Soc. (A)*, <u>1968</u>, 1052.
54. R.J. Irving and M.A.V. Ribeiro da Silva, *J. Chem. Soc. Dalton*, <u>1981</u>, 99.
55. W. Kakolowicz and E. Giera, *Thermochimica Acta*, <u>32</u>, 19 (1979).
56. R.J. Irving and M.A.V. Ribeiro da Silva, *J. Chem. Soc. Dalton*, <u>1978</u>, 399.
57. W. Kakolowicz and E. Giera, *Roczniki Chemii, Ann. Soc. Chim. Polonorum*, <u>47</u>, 1817 (1973).
58. M.A.V. Ribeiro da Silva, M.D.M.C. Ribeiro da Silva, A.P.S.M.C. Carvalho, M.J. Akello and G. Pilcher, *J. Chem. Thermodynamics*, in press (1983).
59. R.J. Irving and M.A.V. Ribeiro da Silva, unpublished results.
60. G. Pilcher, K.J. Cavell, D.C. Garner and S. Parkes, *J. Chem. Soc. Dalton*, <u>1978</u>, 1311.
61. K.J. Cavell, J.A. Connor, G. Pilcher, M.A.V. Ribeiro da Silva, M.D.M.C. Ribeiro da Silva, H.A. Skinner, Y. Virmani and M.T. Zafarani-Moattar, *J. Chem. Soc. Faraday I*, <u>77</u>, 1585 (1981).
62. R.J. Irving and M.A.V. Ribeiro da Silva, *J. Chem. Soc. Dalton*, <u>1977</u>, 413.
63. R.J. Irving and M.A.V. Ribeiro da Silva, *J. Chem. Soc. Dalton*, <u>1976</u>, 1940.
64. W. Kakolowicz and E. Giera, *Thermochimica Acta*, <u>32</u>, 19 (1979).
65. M.A.V. Ribeiro da Silva and A.M.M.V. Reis, *Thermochimica Acta*, <u>55</u>, 89 (1982).
66. M.A.V. Ribeiro da Silva and A.M.M.V. Reis, *Bull. Chem. Soc. Japan*, 52, 3080 (1979).
67. M.A.V. Ribeiro da Silva and A.M.M.V. Reis, *J. Chem. Thermodynamics*, in press (1983).
68. M.A.V. Ribeiro da Silva and R.J. Irving, *Bull. Chem. Soc. Japan*, <u>50</u>, 734 (1977).
69. R.J. Irving and M.A.V. Ribeiro da Silva, *J. Chem. Soc. Dalton*, <u>1975</u>, 1257.
70. E.W. Berg and J.T. Truemper, *J. Phys. Chem.*, <u>64</u>, 487 (1960).
71. G. Beech and R.M. Lintonbom, *Thermochimica Acta*, <u>3</u>, 97 (1971).

72. S.V. Volkov, E.A. Mazurenko, Zh.N. Bublik, *Svoistva Primen β-Diketonatov Met.*, [*Mater. Vses. Sem.*], 3rd, 1977 (Publ. 1978), 119-22, V.I. Spitsyn (editor), Izd Nauka, Moscow. [C. A., 91, 79727c (1979)].
73. R. Teghil, D. Ferro, L. Bencivenni and M. Pelino, *Thermochimica Acta*, 44, 213 (1981).
74. J. Sachinidis and J.O. Hill, *Thermochimica Acta*, 35, 59 (1980).
75. H. Naghibi-Bidokhti, *Ph.D. Thesis*, University of Surrey, 1977.
76. R.J. Irving and M.A.V. Ribeiro da Silva, *Proceedings of 3rd International Conference on Chemical Thermodynamics*, Baden bei Wien, 1, 2/11 (1973).
77. T.P. Melia and R. Merrifield, *J. Inorg. Nucl. Chem.*, 32, 2573 (1970).
78. S.J. Ashcroft, *Thermochimica Acta*, 2, 512 (1971).
79. F.A. Adedeji, D.L.S. Brown, J.A. Connor, M. Leug, M.I. Paz--Andrade and H.A. Skinner, *J. Organometallic Chem.*, 97, 221 (1975).
80. G. Pilcher and H.A. Skinner, in *The Chemistry of Metal-Carbon Bond*, F.R. Hartley and S. Patai (editors), Chapter 2, page 43, John Wiley & Sons, Ltd, 1982.
81. M.A.V. Ribeiro da Silva *et al*, unpublished results.
82. J.A. Kerr, *Chem. Rev.*, 66, 465 (1966).
83. C.A. Coulson, Private Communication (1973).
84. M.A.V. Ribeiro da Silva and M.L.C.C.H. Ferrão, unpublished results.
85. B. Morosin, *Acta Cryst.*, 19, 131 (1965).
86. B.G. Thomas, M.L. Morris and R.L. Hilderbrandt, *Inorg. Chem.*, 17, 2901 (1978).
87. K.J. Cavell, C.D. Garner, J.A. Martinho Simões, G. Pilcher, H. Al-Samman, H.A. Skinner, G. Al-Tekhin, I.B. Walton e M.T. Zafarani-Moattar, *J. Chem. Soc. Faraday I*, 77, 2927 (1981).
88. S. Murata, M. Sakiyama and S. Seki, *6th. International Conference on Chemical Thermodynamics*, Merseburg, G.D.R., Poster Paper nº 21 (1980).

THERMOCHEMISTRY OF METAL-POLYAMINE COMPLEXES

P. Paoletti

Istituto di Chimica Generale ed Inorganica,
Università di Firenze, Firenze (Italy)

It is now widely known that polyamine ligands form very stable complexes with transition metal ions. This stability is measured by the formation constants of these complexes and it is made up of two main contributions: enthalpic (ΔH^o) and entropic (ΔS^o). The entropic contribution to the complex formation is always positive and favourable, the enthalpic contribution is very often negative and favourable as well. Much effort has been made in order to obtain reliable values of the thermodynamic functions ΔH^o and ΔS^o. In recent years, commercially available accurate solution calorimeters have made it possible to measure and publish many ΔH^o and ΔS^o values. In particular we have studied the systems formed by tetraaza-cycloalkanes (macrocycles) and some transition metals.

1. INTRODUCTION

This lecture is concerned with both known and unknown aspects of the thermodynamics and thermochemistry of complexes in solution. Column 1 of Table 1 contains values which are known and which can be predicted whereas Column 2 contains values which can only be predicted as good approximations.

Table 1. Known and Unknown on Thermodynamics and Thermochemistry of Complex-Formation

KNOWN	UNKNOWN
A metal reacts with a ligand in a given solvent. (criteria: Hard and Soft theory-donicity numbers)	Speciation problems: number and stoichiometry of species.
Prediction of log K, ΔH^o, ΔS^o. (criteria: Irving and Williams series. E and C equation. Cobble equation........)	Formation of protonated and/or hydroxo compounds.

Can we say when a metal ion and a ligand are mixed in a particular solvent whether they give rise to the formation of metal complexes or not? To do this, we can apply various criteria which range from the hard/soft concept to the donicity number of the solvent. Thus we are able to say that an amine will form complexes with transition metals in aqueous solution but not with lanthanides. At the same time we can say that lanthanides will form complexes with amines in a solvent such as acetonitrile. Again we know that the alkali metals will form complexes with ligands which contain oxygen donors such as, for example, polyalcohols or crown ethers in a solvent such as methanol. Furthermore, we are now sometimes able to predict the value of the stability constant of some complexes to a good level of approximation. Important rules govern such a procedure and one of the most famous is the Irving-Williams series which allows us to predict the value of the stability constant of an iron(II) complex provided we know the values for manganese(II) and cobalt(II). In other cases very interesting correlations have been found with non-thermodynamic features. Thus, we have found a linear relationship (1) between the heat of formation and the maximum in the absorbance of the electronic spectrum. Obviously, I will only quote those criteria which can be used ultimately to predict thermodynamic aspects relevant to the formation of complexes.

Thus, for example, it is known that some crown ethers form both
1:1 and also 4:3 complexes with lanthanide nitrate in acetonitrile. The reason why is not known. In other words, the problem is
one of considerable speculation, involving both the number and
the stoichiometry of the various species which can be formed by a
metal ion and a ligand. For example, triethylenetetramine (trien)
forms 1:1 complexes in aqueous solution with all divalent 3d metal
ions from chromium(II) to zinc(II) but with nickel(II) it also
forms 2:3 and 1:2 complexes. Once more, we do not know the reason
for this behaviour. As well as normal complexes one can also form
complexes in solution which contain molecules of either the
solvent or of ions produced by the dissociation of the solvent.
The most common examples are protonated and hydroxo complexes.
Protonated complexes are easily formed with polydentate ligands
which produce highly stable species by virtue of their sufficiently basic donor atoms. Thus for the two triamines 1,4,9-triazanonane and 1,4,7-triazaeptane, (2) the former forms a monoprotonated
complex more easily for at least three reasons: (Fig. 1)

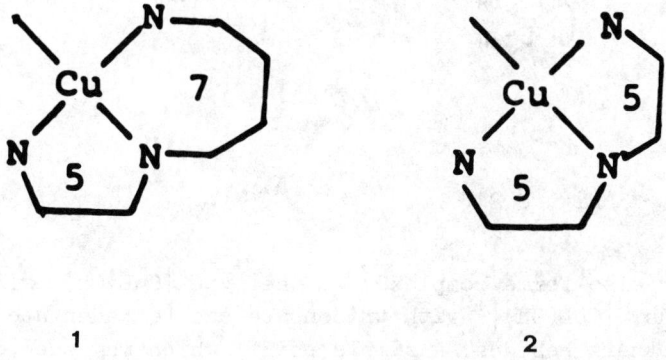

Figure 1. Protonated complexes are formed more easily
by a ligand such as 1,4,9-triazanonane(1),
rather than by 1,4,7-triazaeptane (2). It is
much easier to break a 7 membered chelate
ring than a more stable 5 membered one.

1) it has a more basic nitrogen atom;
2) seven-membered rings are less stable than five-membered rings and are broken more easily;
3) the positive charge on the amino group will be at a greater distance from the charge on the metal.

The tendency to form hydroxo complexes is linked with the tendency for the same metal ion to form soluble hydroxides during alkaline hydrolysis. Thus copper forms, amongst others, the stable species $|Cu_2(OH)_2|^{2+}$ and this then combines with 1 or 2 molecules of a diamine (3) which can be substituted in various places (Fig.2)

Thermodynamic functions for the olation and hydrolysis of CuL^{2+} complex in $0.5M-KNO_3$ at $25.0\ °C$

			dmen	admn	tmen
Olation $2[CuL]^{2+} + 2\ OH^- \rightleftharpoons [Cu_2(OH)_2L_2]^{2+}$	$-\Delta G°$ (kcal mol^{-1}) $-\Delta H°$ (kcal mol^{-1}) $\Delta S°$ (e.u.)		20·49(1) 11·2(1) 31(1)	21·03(2) 11·2(3) 33(1)	20·81(3) 9·4(2) 38·3(8)
Hydrolysis $[CuL]^{2+} + 2\ OH^- \rightleftharpoons [Cu(OH)_2L]$	$-\Delta G°$ (kcal mol^{-1}) $-\Delta H°$ (kcal mol^{-1}) $\Delta S°$ (e.u.)		11·33(2) 7·4(4) 13(2)	11·99(5) 5·9(5) 20(2)	12·44(3) 5·6(1) 23·0(4)

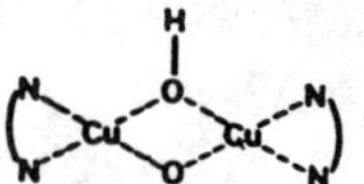

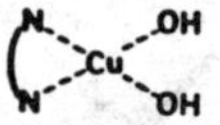

Figure 2

Copper also forms complexes of the type $|CuL(OH)_2|$ with bidentate ligands, $|CuL(OH)|^+$ with tridentate and tetradentate ligands and many others besides. A simple point which may seem obvious is that if the ligand is completely able to wrap itself around the metal, then the metal will become inaccessible both to water molecules and to hydroxide ions. An example can be found with the tripod-like ligand triaminoethylamine (tren) which forms a complex with zinc which is presumably five-coordinate with a trigonal bipyramidal geometry

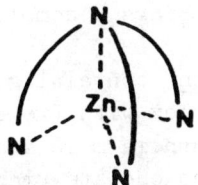

Figure 3. The probable stereochemistry of the Zinc(II) complex with the tripod-like ligand tpt in aqueous solution. The metal ion is completely encapsulated by the ligand in a tetrahedral stereochemistry. In consequence of this, the complex $|Zn\ tpt|^{2+}$ does not show the tendency to hydrolysis observed in analogous Zinc(II) complexes.

The arms of this ligand are too short to completely enclose the metal and on hydrolysis this complex forms the species $|ZnL(OH)|^{+}$. However if we elongate the arms by inserting a CH_2 group forming the ligand tpt (Fig. 3), this ligand forms a complex in which the zinc is in a tetrahedral cavity in the center of the ligand. This complex cannot be hydrolysed (4). Another interesting example with reference to hydrolysis can be found with the ligand 1,4,8,11-tetramethyl-1,4,8,11-tetraazacyclotetradecane (TMC) which forms complexes with various metals. Strangely, the copper complex cannot be hydrolised while the nickel complex can.
We think that the copper, which is the smaller of the two metals, might sit completely in the center of the cavity of the macrocycle, while nickel remains slightly above the plane of the macrociclic ring, as seen from its X-ray structure, and thus is far more accessible to the small hydroxo ligand, OH^{-}. (5)

2. CHELATE EFFECT.

I am now going to speak about chelate complexes. Much has been said and written on the chelate effect, which is now receiving renewed interest for two mains reasons:
1) Macrocyclic complexes with their very high stability are seen as some sort of "superchelates".
2) The discovery of new bidentate ligands which can complex

across 'trans' position rather than across 'cis' positions. Most people know of the chelate effect. For a given co-ordination number the more chelate rings a complex contains the more stable it will be.

Thus complexes with trien containing three chelate rings are more stable than the complex with two ethylenediamines which has only two and which in turn is more stable than that with four ammonias which contains no chelate rings at all. (Fig. 4)

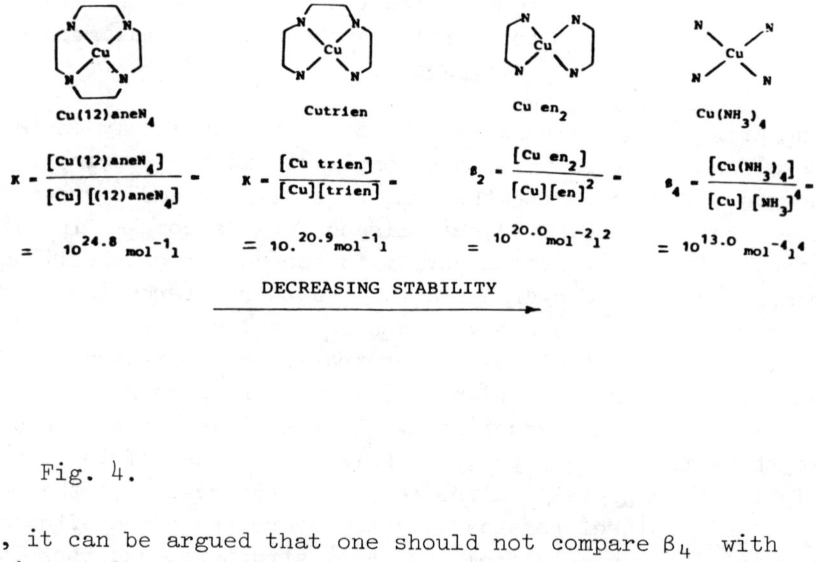
Cu(12)aneN4
Cutrien
Cu en2
Cu(NH3)4
K = [Cu(12)aneN4] / [Cu] [(12)aneN4] = 10^24.8 mol^-1 l
K = [Cu trien] / [Cu][trien] = 10^20.9 mol^-1 l
β2 = [Cu en2] / [Cu][en]^2 = 10^20.0 mol^-2 l^2
β4 = [Cu(NH3)4] / [Cu] [NH3]^4 = 10^13.0 mol^-4 l^4
DECREASING STABILITY

Fig. 4.

However, it can be argued that one should not compare β_4 with β_2 and with K since the dimensions of these stability constants are different, namely $mol^{-4}l^4$, $mol^{-2}l^2$ and mol^{-1} l, respectively. It is agreed that the magnitude of the chelate effect depends on the choice of units for the concentration used to express the unit standard state. Normally, the chelate effect is expressed as the difference in the logarithms of the two stability constants. If, instead of using mol l^{-1} we use mmol.l^{-1}, the chelate effect will vary by a factor of 10^3, or multiples of this. If, instead we use mole fractions the chelate effect can be seen to diminish or even vanish. If we take the complex with trien as a standard and add another chelate ring, we obtain a macrocyclic complex

(Fig. 4) which is more stable than trien complex. In this case, the problem of the standard states does not exist because the two stability constants have the same dimensions.Hence we have reason to believe that, following the direction shown by the arrow, we will have complexes which are always less stable. It is from this that the conclusion has been drawn that the macrocyclic effect enhances the chelate effect.

3. MACROCYCLIC EFFECT WITH TETRAAZA-CYCLOALKANES

It is now time to examine the thermodynamic origins of the macrocyclic effect. At the present time we have sufficient experimental data to be able to conclude that the entropy of formation of a macrocyclic complex, at least with tetraazacycloalkanes, is always larger (that is more positive) than that with a corresponding open-chain ligand.The entropy is thus more favourable. To this end we have studied a series of ligands of the type (Fig. 5) with n = 2,3,4,5.

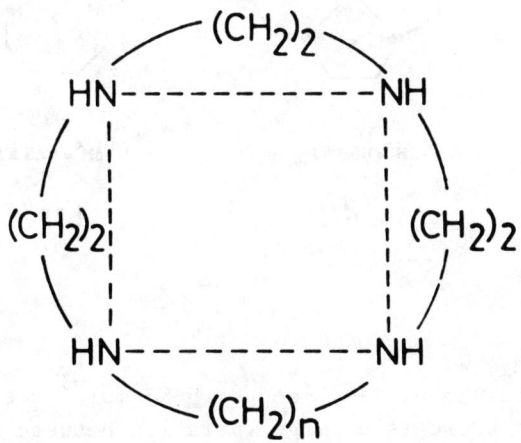

n = 2, 3, 4, 5
tetra-azocycloalkanes

Figure 5.

Here we find that at least as far as n = 4 the entropy of
formation of the complexes with copper is again more positive
than the reference $|Cu(trien)|^{2+}$. This is due to the fact that
the open-chain ligand must lose more entropy in wrapping itself
around the metal than is the case with the cyclic ligand.
On the other hand, with regard to the enthalpy, we can find
examples both of instances where the enthalpy term is favourable
and where it is not. In fact, the enthalpy of formation of a
tetraaza macrocyclic complex depends on numerous factors.
Let us take as an example the complexes of copper with the
three ligands illustrated in the figure (Fig. 6).

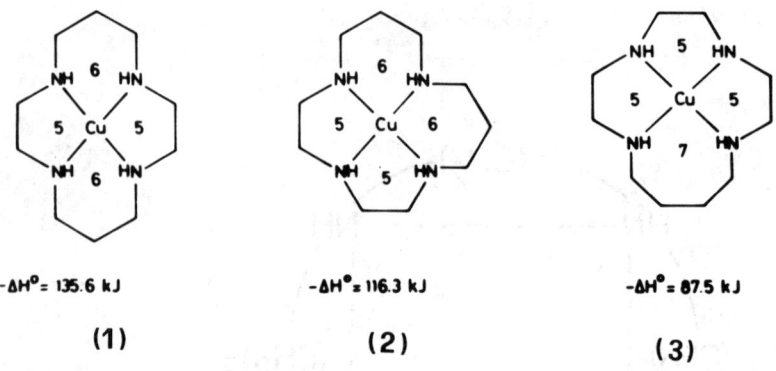

Figure 6.

The values of $-\Delta H$ are 135.6, 116.3 and 87.5 KJ mol^{-1} respectively (6). We see that there is a large variation between the
first and last. This is most likely to be due to the energy
which must be expended by the less symmetrical ligands in
arranging their donor atoms in the most favourable positions
to coordinate to the metal. This conformational energy is
primarily associated with the presence of hydrogen-hydrogen
interactions as can be seen by making accurate molecular models.
Another factor is the strength of the metal-nitrogen bond which
is likevise connected with the above considerations.

It is clear, for example, that in the complexes (2) and (3) the co-ordinate bonds will be weaker than in complex (1). Another interesting aspect of the formation of complexes with polyazacycloalkane ligands concerns the existence of endo and exo forms of the uncoordinated ligand. (Fig. 7).

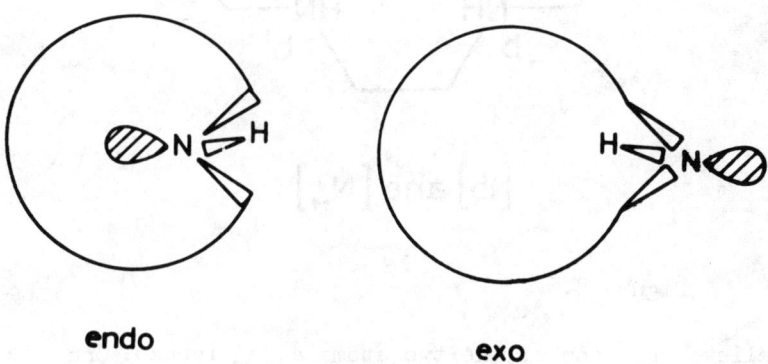

endo   exo

Figure 7.

In the <u>endo</u> form the lone pair points towards the centre of the ring while in the <u>exo</u> form outwards from the ring. It may be that the <u>endo</u> form is also stabilised by the formation of intramolecular hydrogen-bonds. It should be evident that the <u>exo</u> form must be transformed into the <u>endo</u> form in order to form metal complexes. One would expect that this transformation would be endothermic. However, reliable methods do not exist which confirm the presence of <u>exo</u> and <u>endo</u> forms in solution.
We have determined successive protonation parameters (pk, ΔH and ΔS) for some tetraazacycloalkanes and have advanced some hypotheses about the existence of <u>exo</u> and <u>endo</u> forms. For cyclam, we do believe that all the nitrogen atoms are endo, while for the ligand |15| $aneN_4$ shown in the Figure 8,

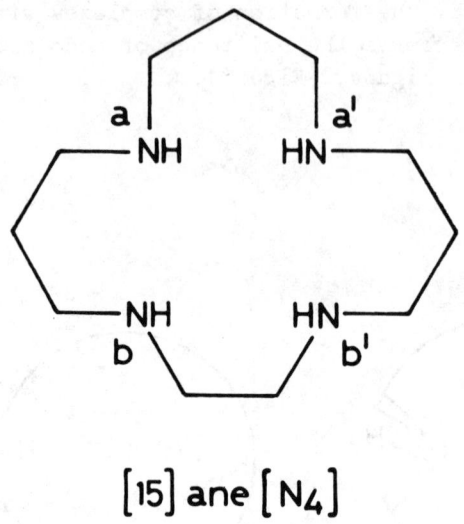

[15] ane [N4]

Figure 8.

we believe that one of the two atoms a, a' is exo. One can then visualise that the transformation exo → endo will produce a situation involving ring-strain which will have the following consequences:
1) the X-Ray structure of the complex |Cu(15)aneN4|(ClO4)2| shows that the 5-membered ring has a stable gauche conformation while the two outside 6-membered rings have stable chair conformations unlike the middle 6-membered ring which is flattened into a sofa conformation which contains the atoms C-N-Cu-N-C in a plane and the sixth atom, (which is a carbon atom) outwith this plane. In our view this signifies the existence of strain.
2) The value of $-\Delta H$ is much lower than the value for cyclam, 110.9 compared with 135.6 kJ mol^{-1}
3) The strained ring has the tendency to open and this favours the formation of a monoprotonated complex ($K_H = 5 \cdot 1 \times 10^2$)

Recently we have been studying macrocyclic complexes of nickel, again with tetraazacycloalkanes. This work is complicated by various factors:
1) the complexes are both slow to form and to decompose;
2) in solution blue, paramagnetic trans-diaquo species exist in equilibrium with the yellow, diamagnetic square-planar form.

More rarely, cis-diaquo species are also formed.
This all combines to make the determination of the stability
constants somewhat difficult. Until recently, there was only
the spectrophotometric determination carried out by Margerum in
1974 on the nickel-cyclam complex and two of its C-methyl
derivatives. (9) We have developed a Batch-wise potentiometric
technique which has allowed us to determine the formation
constants of the two complexes with the ring-sequence 5,5,5,7
and 5,6,6,6. The first exists as a blue-yellow equilibrium;
the second only in the pure blue form.
Both these complexes form monoprotonated species $|NiHL|^{3+}$ which
are presumably octahedral. This creates a problem: a ring
exists in these protonated complexes which, because of the
NH_2^+ group will be ten-membered. There are two possibilities:
either the complex is meridional and hence the two donor atoms
at the extremities of the ten-membered ring will be <u>trans</u> or
the complex must rearrange and adopt a facial configuration.
I believe that the latter is more likely. In fact, from models
one sees that it should be possible to maintain the octahedral
structure with three water molecules co-ordinated in a <u>fac-</u>
arrangement. Furthermore, the positive charge on the nitrogen
will be kept at a greater distance from the Ni^{2+} cation.
With regard to studies on the macrocyclic effect I would like
to mention two problems: we are comparing complexes with ligands
which are closed and open. The closed ligands have nitrogens
which are all secondary while the open ligands have two
terminal nitrogen atoms which are primary. To avoid this we
are now studying a series of linear tetraamines which have been
mono-methylated on the two terminal nitrogens. As yet, we do
not have sufficient data to draw definitive conclusions but
it seems that there will not be such great differences as will
cause us to alter our ideas profoundly on the thermodynamic
origins of the macrocyclic effect.

Another problem is raised in macrocyclic complexes with
different sized rings. Thus, for example, a complex with the
ring sequence 5,5,5,6 can be compared with complexes with the
sequence 5,5,5 or 5,6,5 or even 6,5,5. The most stable is
definitely the second and it thus seems most appropriate to
use it as a reference rather than the others.

All these determinations have been made possible by the
introduction of microcalorimetric techniques using both flow
and batch modes. Thus has allowed us to study these systems
in dilute solution and also to follow slow reactions. As an

example we have followed the destruction of the nickel-cyclam complex with cyanide in an LKB 10700 Batch microcalorimeter and in this case the reaction was complete only after five hours.(10)

Thus I hope that I have been able to demonstrate that the introduction of some of the new techniques I have discussed during this lecture has made it possible to obtain new and interesting data relating to the thermodynamic origins of the macrocyclic effect and that this should open the gate to more such work being carried out in the future.

GENERAL REFERENCES

Aschroft, S. J., and Mortimer, C. T.; 1975 *"Thermochemistry of Transition Metal Complexes"*, Academic Press.

Smith, R. M., and Martell, A. E.: 1975 *"Critical Stability Constants"*, Vol. 2, Amines. Plenum Press.

Christensen, J. J., and Izatt, R. M.: 1970 *"Handbook of Metal Ligand Heats and Related Thermodynamic Quantities"*, Marcel Dekker Inc.

Christensen, J. J., Hansen, L. D., and Izatt, R. M.: 1976 *"Handbook of Proton Ionization Heats and Related Thermodynamic Quantities"*, Wiley Interscience Pub.

Paoletti, P.: 1980 *"The formation of Metal Complexes: the known and the unknown"*, Adv. Mol. Relaxation Processes, 18, p. 73

Paoletti, P.: *"Encircling of 3d-Metal Ions by Polyazamacrocycles: Thermodynamic Aspects"*, Pure & Appl. Chem., 52, p. 2433.

REFERENCES

(1) Lever, A.B.P., Paoletti, P., and Fabbrizzi, L.: 1979, "Thermodynamic and Spectroscopic Parameters in Metal Complexes: Extension of a Linear Relationship to Nickel(II) Derivatives", Inorg. Chem., 18, p. 1324

(2) Barbucci, R., Paoletti, P., and Vacca, A.: 1975, "Stability of Some Transition Metal Ion Complexes with a Linear Aliphatic Triamine Potentially Forming a Five-Membered Chelate Ring fused with a Seven-Membered Chelate Ring: 1,4,9-Triazanonane (2,4-tri). II; Inorg. Chem., 14, p. 302.

(3) Barbucci, R., Fabbrizzi, L., Paoletti, P., and Vacca, A.: 1972, "Thermodynamic of Complex Formation in Aqueous Solution. Reactions of Copper(II) with Ethylenediamine, NN'-dimethylethylenediamine and NN-dimethylethylenediamine: log K, ΔH and ΔS Values", J. Chem. Soc., p. 740.

(4) Vacca, A., and Paoletti, P.: 1968, "Reactions of Tris-(3-aminopropyl)-amine with Protons and Some Bivalent Transition Metal Ions: ΔH°, ΔS and Visible Spectra, J. Chem. Soc., A, p. 2378

(5) Micheloni, M., Paoletti, P., Bürki, S., and Kaden, A.: 1982, "59. Metal Complexes with Macrocyclic Ligands. XVI^1. Spectrophotometric and Thermodynamic studies of Solvents and Unidentate Ligands Interaction with the Pentacoordinate Co^{2+}, Ni^{2+}- and Cu^{2+}- complexes of 1,4,8,11-Tetramethyl-1,4,8,11-Tetraazacyclotetradecane", Helv. Chim. Acta, 65, 2, p.587-594

(6) Micheloni, M., Paoletti, P., Poggi, A., and Fabbrizzi, L.: 1982, "Co-ordinating Tendencies of a 14-membered Tetra-aza Macrocycle which forms a seven-membered chelate ring", J. Chem. Soc., Dalton, p. 61.

(7) Micheloni, M., Paoletti, P., and Vacca, A.: 1978 "Solution Chemistry of Macrocycles. Part 2. Enthalpic and Entropic Contribution to the Proton Basicity of Cyclic Tetra-amine Ligands: 1,4,8,11-Tetra-azacyclotetradecane, 1,4,8,12-Tetra-azacyclopentadecane, and 1,4,8,11-Tetra-methyl-1,4,8,11 - Tetra-azacyclotetradecane", J. Chem. Soc., Perkin II, p. 945; Bartolini, M., Bianchi, A., Micheloni, M., and Paoletti, P.: 1982, "Solution Chemistry of Macrocycles. Part 3. Synthesis and Thermodynamics of Protonation of Some Tetra-aza-macrocycles", J. Chem. Soc., Perkin II, in press.

(8) Anichini, A., Fabbrizzi, L., Paoletti, P., and Clay, R.M.: 1978, "A Microcalorimetric study of the Macrocyclic Effect. Enthalpies of formation of Copper(II) and Zinc(II) complexes with some Tetra-aza Macrocyclic Ligands in Aqueous Solution", J. Chem. Soc., Dalton, p. 577

(9) Hinz, F.P., and Margerum, D.W.: 1974, "Ligand Solvation and the Macrocyclic effect. A study of Nickel(II)-tetramine complexes", Inorg. Chem., 13, p. 2941.

(10) Fabbrizzi, L., Paoletti, P., and Clay, R.M.: 1978, "Microcalorimetric Determination of the Enthalpy of a Slow Reaction: Destruction with Cyanide of the Macrocyclic (1,4,8,11-Tetraazacyclotetradecane)nickel(II) Ion" Inorg. Chem., 17, p. 1042.

THERMOCHEMISTRY OF MAIN-GROUP ORGANOMETALLIC COMPOUNDS

Geoffrey Pilcher

Department of Chemistry,
University of Manchester,
Manchester M13 9PL, U.K.

Recent reviews of the thermochemistry of organometallic compounds were made by Skinner[1](1964), Cox and Pilcher[2](1970), Pilcher[3](1975), Pedley and Rylance[4](1977), and Pilcher and Skinner[5](1982) : and in these, the individual experimental values with references can be traced.

Of approximately 300 measurements of the enthalpies of formation of main-group organometallic compounds; 50% have been determined by combustion calorimetry, 90% of these by the static-bomb method, the remainder by using the rotating-bomb.

ENTHALPIES OF COMBUSTION

Static-bomb combustion calorimetry requires the solution of three experimental problems: (i) To achieve complete combustion of the organic ligand, which can be tested by measuring the CO_2 produced; (ii) To determine the extent of oxidation of the metal requiring chemical analysis; and (iii) To determine the crystalline form of the metal oxide produced : X-ray powder photographs can mislead because a small degree of crystallinity is often sufficient to give a clear diffraction pattern.

The validity of the static-bomb method for compounds of any particular element should be confirmed by an alternative procedure.

The static-bomb method has been shown to be satisfactory for compounds of Zn, Hg, B, and Sn, as seen in the following comparisons where the reaction calorimetry values can be considered to be reliable.

$$\Delta H_f^o(298K)/kJ\ mol^{-1}$$

	Static-bomb	Reaction calorimetry
$ZnMe_2(\ell)$	+26.8 ± 6.3	+22.6 ± 8.4
$ZnEt_2(\ell)$	+20.5 ± 2.9	+22.2 ± 8.4
$HgMe_2(\ell)$	+59.8 ± 0.4	+56.1 ± 3.3
$HgEt_2(\ell)$	+27.2 ± 0.8	+30.5 ± 3.8
$BEt_3(\ell)$	−194.5 ± 15.5	−189.1 ± 5.0
$B(\underline{n}-C_6H_{13})_3(\ell)$	−488.7 ± 2.9	−486.6 ± 9.2

Although there is no independent experimental verification of the static-bomb method for tin compounds; this method has given a consistent set of enthalpy of formation data when tested against bond-energy schemes.

The static-bomb method is probably satisfactory for compounds of P, As, Sb, and Bi, but there are no independently determined values nor are there sufficient data to test for consistency using bond-energy schemes.

The static-bomb method has been shown to be unsatisfactory for compounds of Cd, Al, Si, Ge and Pb as seen in the following comparisons where the reaction calorimetry and rotating-bomb values are considered to be reliable.

$$\Delta H_f^o(298K)/kJ\ mol^{-1}$$

	Static-bomb	Reaction calorimetry
$CdMe_2(\ell)$	+88.3 ± 12.5	+69.9 ± 1.3
$CdEt_2(\ell)$	+90.4 ± 3.8	+60.7 ± 1.7
$Al(n-C_3H_7)_3(\ell)$	−251.0 ± 15.9	−297.5 ± 18.8
		Rotating-bomb
$SiO_2(c)$	−859.4 ± 2.2	−910.9 ± 4.5
$GeEt_4(\ell)$	−189.7 ± 3.3	−210.5 ± 6.7
$PbMe_4(\ell)$	−24.7 ± 5.4	+98.3 ± 3.8
$PbEt_4(\ell)$	+219.7 ± 5.0	+53.1 ± 3.8

Static-bomb measurements for Al and Si compounds should be rejected : probably the oxide formed envelops the burning compound resulting in incomplete combustion. Two independent sets of static-bomb data for Al compounds differ by ca. 100 kJ mol^{-1} and although the larger enthalpies of combustion must be less subject to error due to incomplete combustion, none of these values can be considered to be reliable.

The problem with static-bomb combustions of Ge compounds lies in defining the state of the germanium oxide formed, and the enthalpies of transition between the solid forms are large,

$$GeO_2(am.) \rightarrow GeO_2(c,hex.) \quad \Delta H = -15.7 \text{ kJ mol}^{-1}$$

$$GeO_2(c,hex.) \rightarrow GeO_2(c,tet.) \quad \Delta H = -25.4 \text{ kJ mol}^{-1}$$

The static-bomb combustions of Pb compounds were made by experts in this field with careful analyses of the combustion products, nevertheless the results showed very large deviations from those obtained by rotating-bomb calorimetry.

Rotating-bomb calorimetry has been applied successfully to compounds of Si, Ge, Pb, P, As, Bi, S and Se; this method has the advantage that independent confirmation of the results is not required. Prior to combustion, a solution of a suitable solvent is placed in the bomb; after combustion the bomb is rotated to wash thoroughly the crucible, its contents, and the bomb walls to produce a final system consisting of a homogeneous solution in equilibrium with the gaseous phase. The final system can be thermodynamically defined. Generally, a measurement and a comparison experiment are required to derive ΔH_f^o. To illustrate for Si compounds, e.g. hexamethyldisiloxane was mixed with trifluoromethylbenzene and water added to the bomb. After combustion and rotation, a final solution of hexafluorosilicic acid was produced according to,

$$5CF_3Ph(\ell) + (Me_3Si)_2O(\ell) + 49.5O_2(g) + 406H_2O(\ell) \rightarrow$$
$$41CO_2(g) + 2\{H_2SiF_6 + 1.5HF)(212H_2O)\}(\ell)$$

For the comparison experiment, a mixture of silicon and poly(vinylidene fluoride) was burned to produce a final solution of the same composition. By subtraction, the unknown enthalpy of formation of the final bomb solution is eliminated, and $\Delta H_f^o(Me_3Si)_2O,(\ell)$ determined relative to elemental silicon.

For Ge compounds two techniques have been used. Aqueous HF in the bomb will dissolve the GeO_2 formed by combustion of $GeEt_4$; or aqueous NaOH in the bomb will dissolve both the GeO_2 and the CO_2. Comparison experiments were made by burning benzoic acid and at the same time dissolving $GeO_2(c,hex.)$ to form a final solution of the same composition; hence by subtraction the enthalpy of reaction

$$GeEt_4(\ell) + 14 O_2(g) \rightarrow GeO_2(c,hex.) + 8CO_2(g) + 10H_2O(\ell)$$

was determined.

For Pb compounds, the bomb solution was aqueous HNO_3 containing arsenious oxide to ensure that the dissolved lead was in the Pb^{2+} oxidation state. For comparison, hydrocarbon oil was burned and $Pb(NO_3)_2(c)$ dissolved to produce a final solution of the same composition, hence $\Delta H_f^o(PbEt_4,\ell)$ was determined from the enthalpy of reaction

$$PbEt_4(\ell) + 13.5 O_2(g) + 2(HNO_3 \text{ in } 30H_2O)(\ell) \rightarrow$$
$$8CO_2(g) + 11H_2O(\ell) + Pb(NO_3)(c)$$

For Sb and Bi compounds, aqueous NaOH was placed in the bomb to dissolve the As_2O_3 and the CO_2 produced (a small correction was required for formation of arsenate). The enthalpy of solution of the CO_2 was determined by burning benzoic acid with the same initial solution in the bomb. Hence $\Delta H_f^o(AsPh_3,c)$ was determined from the enthalpy of reaction

$$AsPh_3(c) + 22.5 O_2(g) + 43(NaOH.12.93 H_2O)(\ell) \rightarrow$$
$$18CO_2(g) + (40NaOH.NaAsO_3.565 H_2O)(\ell)$$

Some Se compounds have been measured with water in the bomb as this easily dissolves the SeO_2 formed, $\Delta H_f^o(SePh_2,c)$ was determined from the enthalpy of reaction

$$SePh_2(c) + 15.5 O_2(g) + 396H_2O(\ell) \rightarrow$$
$$12CO_2(g) + (SeO_2.401H_2O)(\ell)$$

The enthalpy of solution of $SeO_2(c)$ was measured separately and comparison experiments were made by burning benzoic acid with a solution of SeO_2 in the bomb.

ENTHALPIES OF REACTION

(1) Enthalpies of hydrolysis are often simple to measure by reaction calorimetry but it is best if a homogeneous solution is produced rather than a precipitate which may introduce errors due to uncertain crystal structure and to surface adsorption, e.g. the results of hydrolysis of $CdMe_2$ in acid were preferred

over those in neutral solution

$$CdMe_2(\ell) + 2H_2O(\ell) \rightarrow Cd(OH)_2(c,ppt.) + 2CH_4(g)$$

$$CdMe_2(\ell) + H_2SO_4 \cdot 100H_2O(\ell) \rightarrow CdSO_4 \cdot 100H_2O(\ell) + 2CH_4(g)$$

If the reaction of the liquid compound with liquid water is violent, or explosive, the reaction can be moderated by bubbling nitrogen saturated with water vapour though the liquid : this was done for $AlEt_3(\ell)$.

(2) Enthalpies of halogenation are obviously important for determining the enthalpies of formation of halogeno-organometallic compounds, e.g. $\Delta H_f^\circ(Me_3SnBr,\ell)$ was determined from the enthalpy of reaction

$$SnMe_4(\ell) + Br_2(g) \rightarrow Me_3SnBr(\ell) + MeBr(g)$$

Nitrogen saturated with bromine vapour was passed into a reaction vessel containing $SnMe_4(\ell)$ within a reaction calorimeter : the amount of reaction was determined from the weight loss of the bromine evaporator. The enthalpy of formation of hexamethyldistannane was then determined from the enthalpy of reaction

$$Me_3SnSnMe_3(\ell) + Br_2(\ell) \rightarrow 2Me_3SnBr(\ell)$$

(3) Enthalpies of hydroborination were used to determine enthalpies of formation of boron alkyls from the enthalpy of reaction

$$6RCH = CH_2(\ell) + B_2H_6(g) \rightarrow 2(RCH_2CH_2)_3B(\ell)$$

A solution of the alkene in diglyme was placed in a reaction vessel within a calorimeter and a measured volume of $B_2H_6(g)$ admitted through a sintered glass disc covered with mercury. The results were in agreement with the most recent combustion studies.

(4) Enthalpies of redistribution reactions In redistribution reactions the chemical bonds change in relative position about a single centre, but do not change in number or in formal character. Such reactions have been useful for organomercury and organotin compounds, e.g. the enthalpy of the reaction

$$HgMe_2(\ell) + HgCl_2(c) \rightarrow 2MeHgCl(c)$$

was determined from the enthalpy of reaction in solution and the enthalpies of solution of reactants and products and used to obtain $\Delta H_f^\circ(MeHgCl,c)$. All the enthalpies of formation of alkylmercury halides were obtained in this way.

The enthalpies of redistribution of tin tetramethyl and tin tetrachloride were measured by mixing $SnMe_4(\ell)$ and $SnCl_4(\ell)$ in a reaction vessel within a calorimeter according to,

$$SnMe_4(\ell) + SnCl_4(\ell) \rightarrow$$

$$\text{mixture}(Me_3SnCl, Me_2SnCl_2, MeSnCl_3)$$

The relative amounts of products in the final mixture were determined by glc, and these could be varied by changing the initial ratio of the amounts of reactants. Hence ΔH_f° for all the tin chloroalkyls could be determined from a series of experiments.

THERMOCHEMICAL BOND STRENGTHS

In recent years, the experimental uncertainties, particularly in the enthalpies of formation of organic radicals have been greatly reduced so it is becoming more necessary from an experimental point of view to be precise in defining the commonly used terms for bond strengths. It has of course always been desirable from a theoretical point of view to be precise about definitions.

For a diatomic molecule, the spectroscopic <u>dissociation energy</u>, D_0° is measured from the lowest energy level of the molecule and corresponds to ΔU° for the reaction,

$$AB(g,0K) \rightarrow A(g,0K) + B(g,0K)$$

and at the absolute zero, $\Delta U^\circ = \Delta H^\circ$.

The thermochemist usually uses ΔH° for the reaction,

$$AB(g,298K) \rightarrow A(g,298K) + B(g,298K)$$

and this should be called the <u>bond dissociation enthalpy</u>. The spectroscopic value at 298K corresponds to $\Delta U^\circ(298K)$ for the dissociation, hence

$$D_{298}^\circ = \Delta U^\circ(298) = \Delta H^\circ(298) - RT = \Delta H^\circ(298) - 2.48 \text{ kJ mol}^{-1}$$

$D_{298}^\circ > D_0^\circ$ because ΔC_v° for dissociation will be positive : the maximum difference will be $3RT = 7.5$ kJ mol^{-1}. It is necessary to be careful as indiscriminate comparisons of spectroscopic and thermochemical dissociation energies and enthalpies can lead to unaccounted discrepancies of up to 10 kJ mol^{-1}.

For a polyatomic molecule MR_n (R=atom or radical), for the disruption reaction at 298K,

$$MR_n(g, 298K) \rightarrow M(g, 298K) + nR(g, 298K)$$

$$\Delta H^o(\text{disrupt.}) = \Delta H^o_f(M,g) + n\Delta H^o_f(R,g) - \Delta H^o_f(MR_n,g)$$

$$\Delta U^o(\text{disrupt.}) = \Delta H^o_f(\text{disrupt.}) - nRT.$$

If the (M-R) bonds are equivalent, then the <u>mean bond dissociation energy</u> = $\Delta U^o(\text{disrupt.})/n$.

$$\Delta H^o(\text{disrupt.})/n = \text{mean bond dissociation enthalpy} =$$

$$\bar{D}(M-R)$$

The difference between the mean bond dissociation enthalpy and the mean bond dissociation energy at 298K will be RT = 2.48 kJ mol^{-1}.

For the atomization of any molecule at 298K according to,

$$\text{Molecule}(g) \rightarrow \text{Atoms}(\text{ground state}, g)$$

$$\Delta H^o_a = \Sigma \Delta H^o_f(\text{atoms}, g) - \Delta H^o_f(\text{compound}, g)$$

$$= \Sigma \bar{E} + \text{Stabilization energy} - \text{Strain energy}$$

so that conventionally, the enthalpy of atomization is regarded as a measure of the total chemical binding energy in the molecule. $\bar{E}$ is called the bond energy; it ought to be called the bond enthalpy or bond enthalpy contribution. The values $\bar{E}(M-X)$ depend on the rules of the scheme used to divide up ΔH^o_a, hence $\bar{E}(M-X)$ values must never be confused or compared with $\bar{D}(M-X)$ values.

Mean Bond Dissociation Enthalpies in Metal Alkyls

Mean bond dissociation enthalpies can be derived for many alkyl derivatives of main-group metals, especially the methyl, ethyl and phenyl derivatives. As

$$\bar{D}(M-R) = (1/n)(\Delta H^o_f(M,g) + n\Delta H^o_f(R,g) - \Delta H^o_f(MR_n,g))$$

the enthalpies of formation of the gaseous metals, the radicals as well as those for the gaseous alkyls are required. The enthalpies of formation of the gaseous main-group metals are given in Table 1.

H	218.00 ± 0.01	S	276.98 ± 0.25	Cd	110.0 ± 0.4
Li	160.7 ± 1.7	K	89.1 ± 0.8	In	243 ± 8
Be	324 ± 5	Ca	177.8 ± 0.8	Sn	301.2 ± 1.7
B	560 ± 12	Cu	337.6 ± 1.2	Sb	264 ± 8
C	716.67 ± 0.44	Zn	130.42 ± 0.20	Te	193 ± 8
N	472.68 ± 0.40	Ga	288.7	Cs	78.2 ± 1.3
O	249.17 ± 0.10	Ge	377 ± 13	Ba	177.8
Na	107.9 ± 0.4	As	289 ± 13	Au	369.4 ± 3.8
Mg	147.1 ± 0.8	Se	206.7 ± 4.2	Hg	61.38 ± 0.04
Al	329.7 ± 4.0	Rb	81.6 ± 4.2	Tl	179.9 ± 4.2
Si	450 ± 8	Sr	143.6 ± 4.2	Pb	195.2 ± 0.8
P	316.5 ± 1.0	Ag	284.9 ± 0.8	Bi	207.1 ± 4.2

Table 1. Enthalpies of formation of gaseous metals/kJ mol^{-1} (298K).

and the enthalpies of formation of the radicals, ΔH_f^o(R,g,298K)/kJ mol^{-1} : R=Me, + 146.3 ± 0.6; R=Et, + 108.2 ± 4.3; R=Ph, + 325.1 ± 4.3 were used. Table 2 lists the mean bond dissociation enthalpies in metal methyl, ethyl and phenyl derivatives.

$ZnMe_2$	186.4	$ZnEt_2$	145.0		
$CdMe_2$	148.5	$CdEt_2$	110.6		
$HgMe_2$	130.0	$HgEt_2$	102.7	$HgPh_2$	160.1
BMe_3	373.9	BEt_3	344.5	BPh_3	468.4
$AlMe_3$	283.3	$AlEt_3$	272.6		
$GaMe_3$	256.4	$GaEt_3$	224.9		
$InMe_3$	169.4				
CMe_4	367.3	CEt_4	345.3	CPh_4	404.8
$SiMe_4$	320.2	$SiEt_4$	287.1	$SiPh_4$	352.2
$GeMe_4$	258.2	$GeEt_4$	242.9	$GePh_4$	308.8
$SnMe_4$	226.4	$SnEt_4$	194.7	$SnPh_4$	257.2
$PbMe_4$	161.1	$PbEt_4$	129.6	$PbPh_4$	196.4
NMe_3	311.8	NEt_3	296.7	NPh_3	373.7
PMe_3	285.5	PEt_3	230.2	PPh_3	325.9
$AsMe_3$	238.4	$AsEt_3$	185.8	$AsPh_3$	285.4
$SbMe_3$	223.5	$SbEt_3$	179.8	$SbPh_3$	267.8
$BiMe_3$	150.5	$BiEt_3$	105.3	$BiPh_3$	193.9
OMe_2	362.9	OEt_2	358.7	OPh_2	423.8
SMe_2	303.6	SEt_2	293.5	SPh_2	348.0
		$SeEt_2$	240.3	$SePh_2$	285.3

Table 2 Mean bond dissociation enthalpies at 298K in kJ mol^{-1}

Skinner's comments made in 1964 are still valid : (i) $\bar{D}$-Me > $\bar{D}$(M-Et) by ca. 20 kJ mol^{-1}, (ii) $\bar{D}$(M-Ph) > $\bar{D}$(M-Me) by ca. 40 kJ mol^{-1}, (iii) $\bar{D}$(M-R) falls as M descends a particular B group (an opposite trend is apparent for A group elements). When $\bar{D}$(M-R) is plotted against ΔH_f^o(M,g), both A and B group elements follow the same pattern of $\bar{D}$(M-R) increasing with ΔH_f^o(M,g) as shown in Figures 1 and 2.

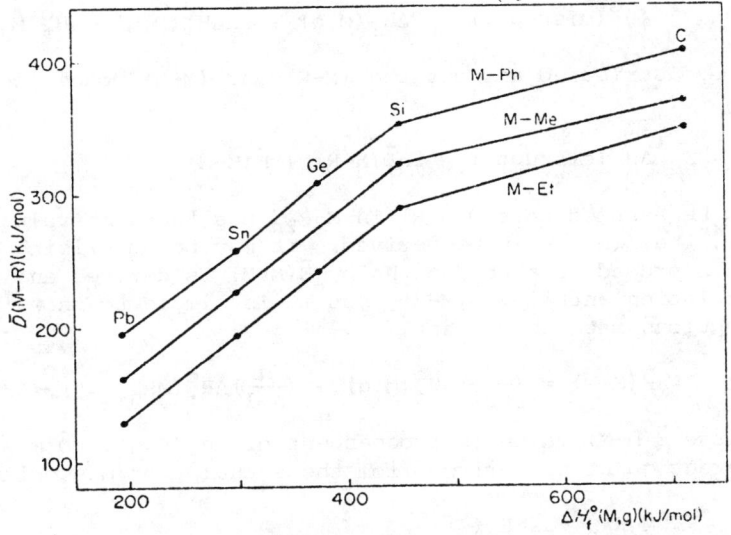

Figure 1. Plots of $\bar{D}$(M-Me), $\bar{D}$(M-Et), and $\bar{D}$(M-Ph) versus ΔH_f^o(M,g) for C, Si, Ge, Sn and Pb.

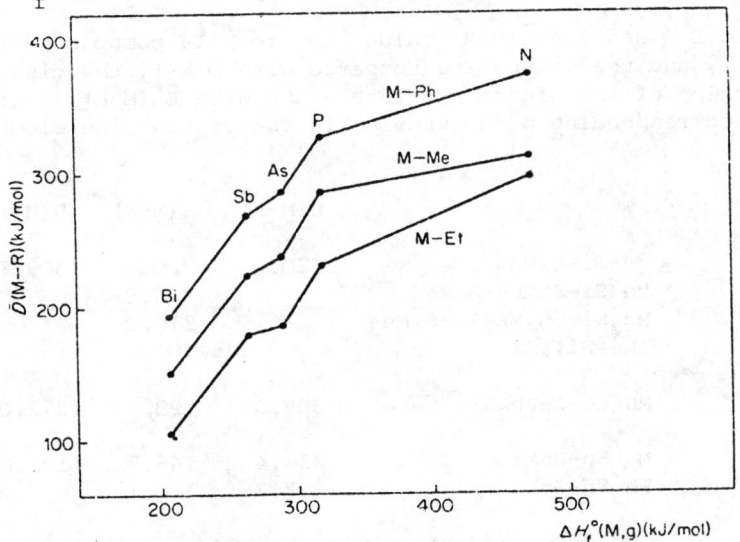

Figure 3. Plots of $\bar{D}$(M-Me), $\bar{D}$(M-Et), and $\bar{D}$(M-Ph) versus ΔH_f^o(M,g) for N, P, As, Sb, and Bi.

Metal-metal bond enthalpies in M_2R_{2n} molecules.

For the disruption of an M_2R_{2n} molecule which contains a metal-metal bond,

$$M_2R_{2n}(g) \to 2M(g) + 2nR(g)$$

the enthalpy of disruption,

$$\Delta H^o(\text{disrupt.}) = 2\Delta H_f^o(M,g) + 2n\Delta H_f^o(R,g) - \Delta H_f^o(M_2R_{2n},g)$$

can be distributed amongst the (M-R) and (M-M) bonds according to,

$$\Delta H^o(\text{disrupt.}) = 2n\bar{D}(M-R) + \bar{E}(M-M)$$

If it be assumed that $\bar{D}(M-R)$ in M_2R_{2n} has the same value as in MR_{n+1}, the $\bar{E}(M-M)$ can be derived. It may be surprising that by this procedure a bond enthalpy $\bar{E}(M-M)$ is derived and not the dissociation enthalpy $\bar{D}(M-M)$, but it is simple to show from the assumptions made above that,

$$\bar{E}(M-M) = (\frac{2}{n+1})\Delta H_f^o(M,g) + (\frac{2n}{n+1})\Delta H_f^o(MR_{n+1},g) - \Delta H_f^o(M_2R_{2n},g)$$

i.e. the $\bar{E}(M-M)$ value is independent of $\Delta H_f^o(R,g)$. $\bar{D}(M-M)$ is a different quantity derived from the enthalpy of dissociation,

$$M_2R_{2n} \to 2MR_n(g)$$

$$\bar{D}(M-M) = 2\Delta H_f^o(MR_n,g) - \Delta H_f^o(M_2R_{2n},g)$$

Table 3 lists the $\bar{E}(M-M)$ values in Group IV compounds of the type M_2R_{2n}, and the values are compared with $D(M_2)$, the dissociation enthalpy of the diatomic molecule and with $E^*(M-M)$, a value for the corresponding bond enthalpy in the crystalline element.

	$\bar{D}(M-R)$	$\bar{E}(M-M)$	$D(M_2)$	$E^*(M-M)$
$Me_3Si-SiMe_3$	320.2	220.0	309.6	225.0
$Me_3Si-SiMe_2-SiMe_3$		214.8		
$Me_3Si-(SiMe_2)_2-SiMe_3$		210.0		
$(Me_3Si)_4Si$		180.7		
$Ph_3Ge-GePh_3$	308.0	193.7	272.0	188.5
$Me_3Sn-SnMe_3$	226.4	148.7	191.6	150.6
$Ph_3Sn-SnPh_3$	257.2	160.1		

Table 3. $\bar{E}(M-M)$ values at 298K in kJ mol^{-1}.

The crystalline elements have diamond structures, hence $E^*(M-M)$ = $\frac{1}{2}\Delta H^o(\text{sub.})$ of the crystalline element. It is seen that in all cases $E(M-M)$ is quite close to $E^*(M-M)$ but is considerably less than $D(M_2)$, the dissociation enthalpy of the diatomic molecule. The trend in $E(M-M)$ for the silicon methyls suggests that when sufficient experimental data become available for compounds of this type, application of a modern bond-energy scheme would be satisfactory.

Bond energy schemes for organometallic compounds.

By equating the enthalpy of atomization to the total chemical binding energy according to,

$$\Delta H_a^o = \Sigma \bar{E} + \text{Stabilisation energy} - \text{Strain energy}$$

it is clear that bond energies should only be derived from ΔH_a^o values for compounds for which it is reasonable to assume exceptional stabilisation or strain are absent.

The three modern bond-energy schemes, the Laidler scheme, the Group method and the Allen scheme are equivalent[2]; the reason for this equivalence is that these schemes rest on the assumption, made either implicitly or explicitly, that the energy of a particular bond is constant provided the nearest neighbours of the bond are the same. Each scheme produces identical results if the parameters are chosen in accordance with the equivalence relations : hence it is necessary only to apply one scheme. As

$$\Delta H_a^o = \Sigma \Delta H_f^o(\text{atoms,g}) - \Delta H_f^o(\text{compound,g})$$

it is easy to see that any bond-energy scheme which is additive in contributions to ΔH_a^o must have a counterpart additive in contributions to $\Delta H_f^o(g)$.

Here we select the Laidler scheme giving parameters in Table 4 to estimate $\Delta H_f^o(g)$. The parameters previously derived for hydrocarbons are given first. The Laidler scheme for alkanes recognises one (C-C) bond energy and three (C-H) bond energies, $E(C-H)_p$, $E(C-H)_s$ and $E(C-H)_t$ in $-CH_3$, $-CH_2-$ and $>CH-$ groups respectively. C_d refers to a doubly bound carbon atom, so $E(C_d-H)_2$ is for the (C-H) bond in $=CH_2$ and $E(C_d-H)_1$ for the (C-H) bond in $\overset{H}{=}C\overset{}{\underset{C}{}}$. For benzene derivatives, the parameters include the delocalisation energy, e.g. for benzene itself,

$$\Delta H_a = 6E(C_b-C_b) + 6E(C_b-H)$$

For the metal alkyls, MR_n, $E(M-C)$ is taken as constant and $E(C-H)$ involving the C atom of (M-C) depends on the degree of substitution giving rise to $E(C-H)_p^M$, $E(C-H)_s^M$ and $E(C-H)_t^M$.

$E(C-C)$	-0.13	$E(C=C)$	$+158.57$	$E(C_b-C_b)$	$+37.70$
$E(C-H)_p$	-14.10	$E(C_d-H)_2$	-26.61	$E(C_b-H)$	-23.85
$E(C-H)_s$	-10.21	$E(C_d-H)_1$	-23.85	$E(C_b-C)$	-14.48
$E(C-H)_t$	-7.11	$E(C_d-C)$	-19.33		
$E(C-H)_p^M$	-14.10 for all metals.				
$E(C-Zn)$	$+67.45$	$E(C-Si)$	-19.05	$E(C-P)$	$+8.60$
$E(C-H)_s^{Zn}$	$+1.72$	$E(C-H)_s^{Si}$	-2.44	$E(C-H)_s^P$	-8.68
$E(C-Cd)$	$+95.15$	$E(C-Ge)$	$+24.63$	$E(C-As)$	$+46.50$
$E(C-H)_s^{Cd}$	-0.04	$E(C-H)_s^{Ge}$	-11.30	$E(C-H)_s^{As}$	$+7.33$
$E(C-Hg)$	$+89.35$	$E(C-Sn)$	$+37.50$	$E(C-Sb)$	$+53.00$
$E(C-H)_s^{Hg}$	-4.81	$E(C-H)_s^{Sn}$	-4.29	$E(C-H)_s^{Sb}$	$+2.83$
$E(C-H)_t^{Hg}$	$+15.66$	$E(C-H)_t^{Sn}$	$+14.96$		
$E(C_b-Hg)$	$+88.75$	$E(C_d-Sn)$	$+24.60$	$E(C-Bi)$	$+107.10$
		$E(C_b-Sn)$	$+20.55$	$E(C-H)_s^{Bi}$	$+3.62$
$E(C-B)$	$+1.37$	$E(Sn-Sn)$	$+1.90$		
$E(C-H)_s^B$	-4.29				
$E(C-H)_t^B$	$+22.41$	$E(C-Pb)$	$+76.35$		
		$E(C-H)_s^{Pb}$	-3.27		
$E(C-Ga)$	$+28.43$				
$E(C-H)_s^{Ga}$	-3.23				

Table 4. Laidler parameters for estimation of $\Delta H_f^o(g)/kJ\ mol^{-1}$ at 298K.

In Table 5, the Laidler parameters are applied to tin compounds: these are chosen for illustration because the experimental data are reliable.

	$\Delta H_f^\circ(g)$ (obs.)	$\Delta H_f^\circ(g)$ (calc.)	Δ
Me_4Sn	-19.2 ± 2.1	-19.2	0.0
Me_3SnEt	-29.5 ± 3.0	-27.9	-1.6
$Me_3Sn(i-Pr)$	-46.8 ± 4.8	-46.8	0.0
$Me_3Sn(t-Bu)$	-67.1 ± 6.2	-104.2	37.1
Et_4Sn	-44.9 ± 3.3	-54.0	9.1
$(n-Pr)_4Sn$	-144.4 ± 5.7	-136.2	-8.2
$(n-Bu)_4Sn$	-219.2 ± 4.2	-218.4	-0.8
$Me_3Sn(CH=CH_2)$	91.7 ± 13.4	91.7	0.0
Me_3SnPh	113.1 ± 5.2	113.1	0.0
Me_3SnCH_2Ph	82.8 ± 5.7	107.0	-24.2
Ph_4Sn	572.7 ± 5.6	510.0	62.7
$(Me_3Sn)_2$	-26.9 ± 8.4	-26.9	0.0
$(Ph_3Sn)_2$	849.7 ± 16.2	766.9	82.8

Table 5. Calculated and observed $\Delta H_f^\circ(g)$ values in kJ mol^{-1} at 298K for organotin compounds.

For those compounds in which steric strain is expected, $Me_3Sn(t-Bu)$, Ph_4Sn, $(Ph_3Sn)_2$, the observed $\Delta H_f^\circ(g)$ indicates the compound is less stable energetically than indicated by the calculated value. The deviation for Me_3SnCH_2Ph may arise because the secondary (C-H) bonds are subject to the influence of the Sn atom and the Ph group and allowance cannot be made for this.

Steric strain is present in $(Ph_3Sn)_2$ relative to $(Me_3Sn)_2$, but in Table 3 it is seen that E(Sn-Sn) is larger in $(Ph_3Sn)_2$ than in $(Me_3Sn)_2$. This contradiction arises because in deriving E(Sn-Sn) from ΔH°(disrupt.) of $(Ph_3Sn)_2$, $\bar{D}$(Sn-Ph) was taken from ΔH°(disrupt.) of $SnPh_4$, which is affected by steric strain whereas the bond-energy term for E(Sn-Sn) in Table 4 was derived from $\Delta H_f^\circ((Me_3Sn)_2,g)$ and the value for $E(C_b-Sn)$ derived from $\Delta H_f^\circ(Me_3SnPh,g)$ which is less likely to be affected by steric strain.

This shows the danger of considering mean bond dissociation enthalpies or bond energies as measures of bond strength : one must always consider the assumptions made in the derivations and the structures of the molecules and radicals involved.

Reviews of Organometallic Thermochemistry.

1. H.A. Skinner, Adv.Organometallic Chem. 2, 49 (1964).

2. J.D. Cox and G. Pilcher, "Thermochemistry of Organic and Organometallic compounds", Academic Press, London (1970).

3. G. Pilcher, Chapter 2, "Thermochemistry and Thermodynamics", Phys.Chem.Ser.2, Vol.10, Int.Rev.Sci. Butterworths, London (1975).

4. J.B. Pedley and J. Rylance, "Sussex-NPL Computer Analysed Thermochemical Data, Organic and Organometallic Compounds". University of Sussex, Brighton, U.K. (1977).

5. G. Pilcher and H.A. Skinner, Chapter 2, "The Chemistry of the Metal-Carbon Bond", J.Wiley and Sons Limited, (1982).

EXPERIMENTAL THERMOCHEMISTRY OF TRANSITION-METAL
ORGANOMETALLIC COMPOUNDS

Geoffrey Pilcher

Department of Chemistry,
University of Manchester,
Manchester M13 9PL, U.K.

The number of reported enthalpies of formation of
organometallic compounds of the transition metals has risen
dramatically during the last decade, from ca. 12 values in 1970
to over 150 in 1982. The reason for the increase is that some
special techniques for determining these enthalpies of formation
have been developed by three groups of workers, in Gorky, U.S.S.R.,
in Manchester U.K. and in Lisboa, Portugal.

Static-bomb combustion calorimetry has been applied by
I.B. Rabinovich and V.I. Tel'noi in Gorky, U.S.S.R. Attempts to
burn organometallic compounds in samples of normal size, ca. 1 g,
in oxygen at 30 atm. usually results in explosion depositing
unburned material on the interior walls of the bomb. The Russian
workers overcame this problem by burning samples consisting of
ca. 10% of the organometallic compound with ca. 90% of an auxiliary
material, e.g. benzoic acid, hydrocarbon oil, paraffin wax. The
auxiliary seems to promote completeness of combustion by moderating
the violence of reaction rather than acting as a simple combustion
aid. The price paid is in a reduced precision of the measured
value because the compound contributes less than 10% to the measured
enthalpy change.

Complete combustion of the organic component of the sample
was established by measuring the CO_2 produced. The oxide product
was formed at a high temperature hence the surface was fused and
hydration of the oxide did not occur, but the extent of
oxidation of the metal required careful analysis, e.g. for the
following dicyclopentadienyl derivatives, the enthalpies of
combustion were reported for,

$$(C_5H_5)_2Cr(c) + \frac{53}{4}O_2(g) \rightarrow 10CO_2(g) + 5H_2O(\ell) + \tfrac{1}{2}Cr_2O_3(c)$$

$$(C_5H_5)_2Mn(c) + \frac{79}{6}O_2(g) \rightarrow 10CO_2(g) + 5H_2O(\ell) + \tfrac{1}{3}Mn_3O_4(c)$$

$$(C_5H_5)_2Fe(c) + \frac{79}{6}O_2(g) \rightarrow 10CO_2(g) + 5H_2O(\ell) + \tfrac{1}{3}Fe_3O_4(c)$$

$$(C_5H_5)_2Co(c) + 13O_2(g) \rightarrow 10CO_2(g) + 5H_2O(\ell) + CoO(c)$$

$$(C_5H_5)_2Ni(c) + 13O_2(g) \rightarrow 10CO_2(g) + 5H_2O(\ell) + NiO(c)$$

but in the experiment the compositions of the solid products were:-

Cr_2O_3(100%) : MnO(10-24%) + Mn_3O_4 : Fe_2O_3(7-14%) + Fe_3O_4 : Co(7-25%), CoO(54-71%), Co_3O_4(14-25%) : Ni(16-50%) + NiO

Hence thermal corrections to the measurements were needed to derive the enthalpies of combustion as given.

There are few instances of compounds studied by this combustion method also being studied by an alternative method. One example is bisbenzene chromium; the combustion work gave $\Delta H_f^\circ(Cr(C_6H_6)_2, c) = +146 \pm 8$ kJ mol^{-1}, in agreement with the value derived from the enthalpy of decomposition, $+142.3 \pm 8.4$ kJ mol^{-1}.

A selection of the compounds studied by combustion calorimetry is given below where Cp = cyclopentadienyl, B = benzene:-

Cp_2Cr, Cp_2Mn, Cp_2Fe, Cp_2Co, Cp_2Ni,

Cp_3Se, Cp_3Y, Cp_3La, Cp_3Tm, Cp_3Yb,

Cp_2Ti, Cp_2TiMe_2, Cp_2TiPh_2, $Cp_2Ti(CH_2Ph)_2$,

Cp_2TiCl_2, $CpTiCl_3$, Cp_2MoCl_2, Cp_2WCl_2,

B_2Cr, B_2CrCl, B_2CrBr.

The Russian combustion work has been reviewed by the main authors, V.I. Tel'noi and I.B. Rabinovich, Usp.Khim. 46, 1337 (1977).

Mean Bond Dissociation Enthalpies in Metal Cyclopentadienyls.

For $MCp_n(g) \rightarrow M(g) + nCp(g)$

$$\bar{D}(M-Cp) = \tfrac{1}{n}[\Delta H_f^\circ(M,g) + n\Delta H_f^\circ(Cp, g) - \Delta H_f^\circ(MCp_n, g)]$$

From recent ion-cyclotron resonance studies $\Delta H_f^o(Cp, g) = + 264.4 \pm 9.0$ kJ mol^{-1}, and the following $\Delta H_f^o(M, g)$ values are required.

Mg	147.1 ± 0.8	Y	424.7 ± 0.8	Rh	556.5 ± 4.2
Sc	381.6 ± 1.3	La	431.0 ± 0.4	Pd	380.7 ± 4.2
Ti	470.7	Pr	372.8 ± 1.3	Hf	619.2
V	514.6	Tm	247.3 ± 0.8	Ta	786.6 ± 4.0
Cr	396.6 ± 4.2	Yb	151.9 ± 0.4	W	859.9 ± 4.6
Mn	279.1	Zr	608.4 ± 1.7	Re	783 ± 8
Fe	416.3 ± 4.2	Nb	724 ± 8	Os	783 ± 8
Co	425.1	Mo	658.1 ± 2.1	Ir	665 ± 8
Ni	430.1	Ru	640 ± 8	Pr	565.7 ± 4.2

TABLE 1. $\Delta H_f^o(M, g)$/kJ mol^{-1} at 298K.

MgCp$_2$	272.6	TiCp$_2$	505.9
ScCp$_3$	363.8	VCp$_2$	420.0
YCp$_3$	385.9	CrCp$_2$	340.3
LaCp$_3$	337.0	MnCp$_2$	265.6
PrCp$_3$	343.0	FeCp$_2$	351.7
TmCp$_3$	326.1	CoCp$_2$	323.5
YbCp$_3$	339.7	NiCp$_2$	300.9

TABLE 2. $\bar{D}$(M-Cp)/kJ mol^{-1} at 298K.

In figure 1, these values are plotted against $\Delta H_f^o(M, g)$ and the straight line drawn through points considered the most reliable.

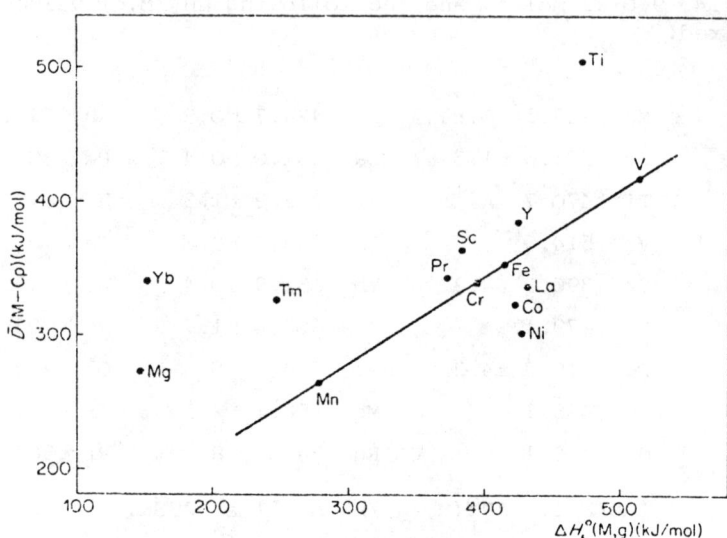

Figure 1. Plot of $\bar{D}$(M-Cp) versus ΔH_f°(M,g) for metal cyclopentadienyl derivatives.

Although the points for the rare-earth compounds show considerable scatter, it is not yet established that the combustion method is satisfactory for these compounds. $\bar{D}$(Co-Cp) and $\bar{D}$(Ni-Cp) lie below the line and this is expected because in $CoCp_2$ one electron occuppies an antibonding orbital and in $NiCp_2$ there are two such antibonding electrons. The point for $TiCp_2$ is doubtful as solid $TiCp_2$ has an unusual structure and the structure in the gaseous state may not be comparable with those of the other molecules.

Figure 1 shows that in spite of the considerable experimental difficulties, the Russian combustion work appears to be quite successful with the possible exception of the rare-earth compounds.

The Calvet high temperature microcalorimeter has been applied in Manchester. A small quantity, 3-5 mg, of the compound at room temperature contained in a glass capillary was dropped into a glass reaction vessel situated within the microcalorimeter held at a higher temperature, typically 150-350°C. Several types of reaction could be studied, e.g.

(a) Thermal decomposition, as

$$Cr(CO)_6(c, 298) \rightarrow Cr(c, 533K) + 6CO(g, 533K)$$

(b) **Iodination**, carried out by first dropping I_2(c) into the reaction vessel,

$$Cr(CO)_6(c, 298K) + I_2(g, 533K) \rightarrow CrI_2(c, 533K) + 6CO(g, 533K)$$

The observed enthalpies can be corrected to the enthalpies of reaction at 298K from the heat capacities of the reactants and products (literature or estimated values) and the temperature difference.

The thermal decomposition value is subject to error because of the exothermic adsorption of CO on the Cr metal film, whereas the iodination value does not suffer in this way. There is however, the problem of analysis of the non-stoichiometric transition-metal iodide and the estimation of its enthalpy of formation but fortunately this usually leads to a relatively small uncertainty. The superiority of the iodination method is shown by the following comparison,

	$\Delta H_f^o(Cr(CO)_6, c, 298K)/kJ\ mol^{-1}$
Microcalorimeter decomposition	-932.6 ± 7.3
Microcalorimeter iodination	-980.3 ± 6.3
Macro-scale determination	-978.2 ± 2.1

(c) **Bromination** is faster than iodination so can be carried out at lower temperatures; it has been used for $RMn(CO)_5$ compounds, e.g.

$$PhMn(CO)_5(c, 298) + \frac{3}{2}Br_2(g, 403) \rightarrow MnBr_2(c, 403) + 5CO(g, 403) + PhBr(g, 403)$$

(d) **Sublimation**. Enthalpies of sublimation can be measured by dropping ca. 3 mg of the compound into a reaction vessel in microcalorimeter held at a temperature below that for decomposition; then subliming the solid by applying a vacuum. The calorimeter and procedure were calibrated by using substances of known enthalpies of sublimation, e.g. I_2, naphthalene, benzoic acid.

A small selection of compounds studied by high-temperature microcalorimetry showing the different types of compound is given below:-

(i) Carbonyls, both mononuclear and polynuclear, e.g.

$$Cr(CO)_6, Mn_2(CO)_{10}, Fe_3(CO)_{12}, Co_4(CO)_{12}, Ru_6(CO)_{16}$$

(ii) Arenes, e.g. $Cr(C_6H_6)_2$, $Mo(C_6H_6)_2$, $Cr(naphthalene)_2$

(iii) Arene carbonyls, e.g. $(C_6H_6)Cr(CO)_3$, $(C_6Me_6)Cr(CO)_6$

(iv) Pyridine and methylcyanide complexes, e.g. $Py_3W(CO)_3$, $Py_2Cr(CO)_4$, $(CH_3CN)_3Mo(CO)_3$.

(v) Alkyl manganese carbonyls, $RMn(CO)_5$ with $R=CH_3$, CF_3, C_6H_6 etc.

(vi) PF_3 complexes, e.g. $Ni(PF_3)_4$, $Cr(PF_3)_6$.

The microcalorimetric method has a wide range of application with the limitation that the ligands produced from the decomposition should be stable and unreactive at the calorimeter temperature.

To illustrate the range of application of the microcalorimeter we can consider some recent studies. $RMn(CO)_5$ compounds have been studied mainly by the bromination reaction but also iodination and some partial thermal decompositions have played their part e.g. sublimation of acetylmanganese pentacarbonyl at about 60°C is straightforward, but on raising the temperature to 100°C, although a normal sublimation appears to occur, the enthalpy rises sharply and the process occurring in the calorimeter is,

$$CH_3COMn(CO)_5 (c, 25°C) \rightarrow CH_3Mn(CO)_5 (g, 100°C) + CO(g, 100°C)$$

A list of ΔH_f^o's for $RMn(CO)_5$ compounds in the solid and gaseous states is given in Table 3.

	$\Delta H_f^o(c)$	$\Delta H_f^o(g)$	$D(R-Mn(CO)_5)$
$CH_3Mn(CO)_5$	-813 ± 4	-753 ± 5	153 ± 5
$CF_3Mn(CO)_5$	-1464 ± 4	-1386 ± 4	172 ± 7
$C_6H_5Mn(CO)_5$	-675 ± 5	-590 ± 7	170 ± 11
$C_6H_5CH_2Mn(CO)_5$	-726 ± 8	-642 ± 8	87 ± 12
$CH_3COMn(CO)_5$	-997 ± 7	-897 ± 10	119 ± 12
$CF_3COMn(CO)_5$	-1587 ± 5	-1508 ± 5	
$C_6H_5COMn(CO)_5$	-848 ± 5	-725 ± 6	89 ± 10
$ClMn(CO)_5$	-1009 ± 8	-918 ± 10	294 ± 10
$BrMn(CO)_5$	-964 ± 4	-876 ± 5	242 ± 6
$IMn(CO)_5$	-912 ± 5	-834 ± 5	195 ± 6

HMn(CO)$_5$	−778 ± 10	−740 ± 10	213 ± 10
(CO)$_5$MnMn(CO)$_5$	−1667 ± 4	−1585 ± 5	94

TABLE 3. ΔH_f^o(RMn(CO)$_5$)/kJ mol^{-1} at 298K

The right hand column of table 3 lists the dissociation enthalpy, D(R-Mn(CO)$_5$) based on a value of 94 kJ mol^{-1} for the dissociation enthalpy of the (Mn-Mn) bond in Mn$_2$(CO)$_{10}$. Any change in this D(Mn-Mn) will affect the absolute values of D(R-Mn(CO)$_5$) but not the differences between them. The values of D(Mn-R) are compared with D(H-R) in figure 2; the difference between the curves is roughly constant until R=halogen, when the increased ionicity of the (Mn-R) bond compared with the (H-R) bond would be expected to increase D(Mn-R) relative to D(H-R).

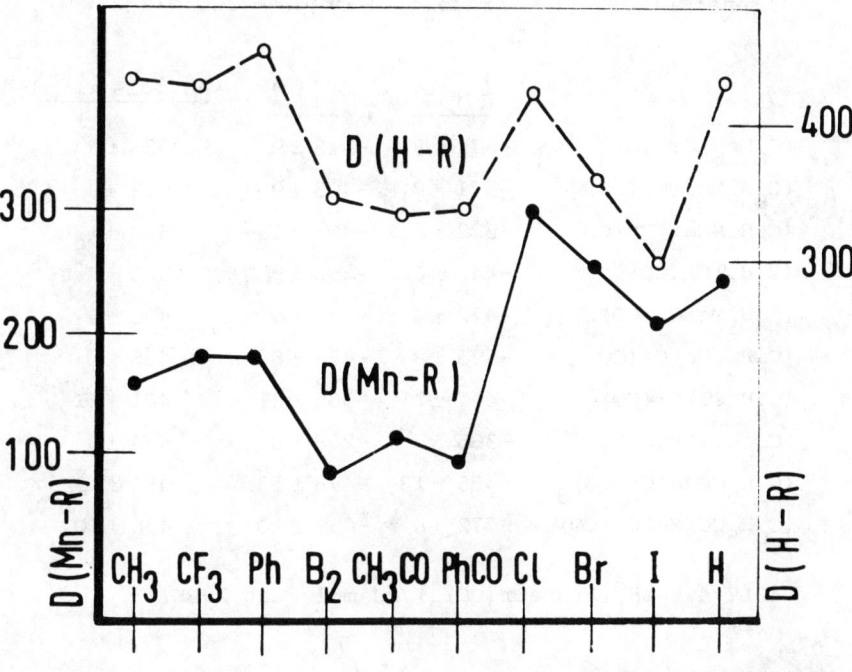

Figure 2. D(Mn-R) compared with D(H-R)/kJ mol^{-1}.

The results can also be used to throw light on the thermodynamic feasibility of carbonylation reactions. For

$$CH_3Mn(CO)_5(c) + CO(g) \rightarrow CH_3COMn(CO)_5(c)$$

$\Delta H_r^o = -54$ kJ mol^{-1}. If we assume the major contribution to ΔS_r^o is the loss of translational and rotational entropy of CO(g), then $\Delta S_r^o = -150$ JK^{-1} mol^{-1}, so that at 300K, $\Delta G_r^o = -9$ kJ mol^{-1}. In accord with this indication of thermodynamic feasibility, this carbonylation reaction occurs. If however, we consider

$$CF_3Mn(CO)_5(c) + CO(g) \rightarrow CF_3COMn(CO)_5(c)$$

$\Delta H_r^o = -12$ kJ mol^{-1} and by making the same assumption concerning ΔS_r^o, then at 300K, $\Delta G_r^o = +33$ kJ mol^{-1}, and in fact direct carbonylation of $CF_3Mn(CO)_5$ has never been carried out.

A series of <u>arenechromium tricarbonyls</u> have been studied by the iodination method, and table 4 lists the enthalpies of formation in the solid and gaseous phases and also the enthalpy of disruption according to,

$$(Arene)Cr(CO)_3(g) \rightarrow Arene(g) + Cr(g) + 3CO(g)$$

	$\Delta H_f^o(c)$	$\Delta H_f^o(g)$	ΔH^o(disrupt.)
$(C_6Me_6)Cr(CO)_3$	-671 ± 8	-548 ± 9	526 ± 9
$(C_6H_3Me_3)Cr(CO)_3$	-571 ± 8	-463 ± 9	512 ± 9
$(C_6H_5NMe_2)Cr(CO)_3$	-522 ± 8	-404 ± 13	503 ± 13
$(C_6H_6)Cr(CO)_3$	-443 ± 8	-352 ± 9	500 ± 4
$(C_6H_5Me)Cr(CO)_3$	-473 ± 4	-380 ± 5	496 ± 5
$(C_6H_5OMe)Cr(CO)_3$	-593 ± 8	-489 ± 8	486 ± 8
$(C_6H_5Cl)Cr(CO)_3$	-467 ± 21	-365 ± 21	481 ± 21
$(C_{10}H_8)Cr(CO)_3$	-365 ± 7	-258 ± 8	474 ± 8
$(C_6H_5COMe)Cr(CO)_3$	-585 ± 13	-478 ± 13	456 ± 13
$(C_6H_5CO_2Me)Cr(CO)_3$	-772 ± 8	-659 ± 10	436 ± 10

TABLE 4. $\Delta H_f^o(\text{AreneCr}(CO)_3)/$kJ mol^{-1} at 298K.

It is simple to show that

$$\Delta H^o(\text{disrupt.})\,[(Arene)Cr(CO)_3 - (Benzene)Cr(CO)_3]$$
$$= D[(Arene)-Cr(CO)_3] - D[(Benzene-Cr(CO)_3]$$

The differences in these dissociation energies is shown in Figure 3. $D[(Arene) - Cr(CO)_3]$ increases relative to $D[(Benzene) - Cr(CO)_3]$ when there are substituents which release electrons to

the benzene ring whereas the position is reversed when the
substituents are electron-withdrawing. It would be an
oversimplification to ascribe the effect of the substituent
groups solely upon the strength of the (Cr-Arene) bond. The mean
bond dissociation enthalpies for $Cr(arene)_2$ are $Cr(benzene)_2$
165 ±5, $Cr(hexamethylbenzene)_2$ 155 ±7 kJ mol^{-1}. The question of
whether the strengthening is to be regarded as being in the
(Cr-Arene) or in the (Cr-CO) bonds cannot be answered by thermochemical arguments alone.

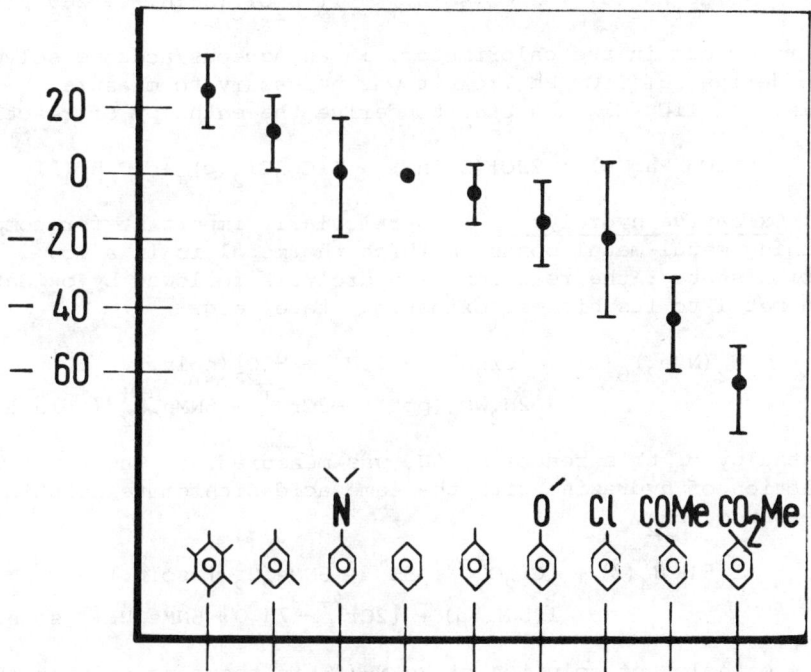

Figure 3. $D(Arene-Cr(CO)_3) - D(Benzene-Cr(CO)_3)/kJ\ mol^{-1}$.

The solution-reaction calorimeter has been used in Manchester
mainly for compounds containing metal-metal multiple bonds and in
Lisboa for studies on compounds containing metal-carbon bonds.

An all glass unsilvered Dewar calorimeter of capacity ca.
150 cm^3 has been used : it is fitted with a stirrer, electrical
calibration heater and an ampoule breaking device. Temperatures
are measured with either a quartz-thermometer or a thermistor.
Provision is made for carrying out reactions in an inert
atmosphere, necessary for compounds spontaneously affected by air.

The main types of reaction studied have been:

(i) <u>Hydrolysis</u> : many hydrolyses are simple to carry out,

e.g. $TaMe_5$(isopentane, soln.) + $2.5H_2O(\ell)$ + excess(aq.ether)

$\rightarrow$ 0.5 Ta_2O_5(ppt.) + $5CH_4$(isopentane, ether)

It is often necessary to measure the enthalpies of solution of reactants and products in order to derive the required enthalpy of formation, e.g. the reaction

$TiCp_2Ph_2$(c) + 2HCl(soln.) $\rightarrow$ $TiCp_2Cl_2$(soln.) + $2C_6H_6$(soln.)

was carried out in the calorimeter, in an aqueous/acetone solvent and to derive $\Delta H_f^o(TiCp_2Ph_2$, c) it was necessary to measure ΔH(soln.) of $TiCp_2Cl_2$ and C_6H_6 to derive the enthalpy of reaction,

$TiCp_2Ph_2$(c) + 2HCl(soln.) $\rightarrow$ $TiCp_2Cl_2$(c) + $2C_6H_6(\ell)$

(ii) <u>Oxidative-hydrolysis</u> is particularly important for compounds containing metal-metal bonds in which the metal is in a low oxidation state : the reaction is hydrolysis followed by oxidation of the metal to its highest oxidation state, e.g.

$W_2(NMe_2)_6$(c) + $[Cr_2O_7^{2-}$ + $14H^+$ + $H_2O]$(soln.)
$\rightarrow$ $2H_2WO_4$(ppt.) + $2Cr^{3+}$ + $6NMe_2H_2^+$ (soln.)

the enthalpy of this reaction, ΔH_1 was measured. The enthalpy of reaction of hydrazine with the same acid dichromate solution was measured, ΔH_2,

$1.5 N_2H_4(\ell)$ + $[Cr_2O_7^{2-}$ + $8H^+$ + $6NMe_2H_2^+]$(soln.) $\rightarrow$
$1.5 N_2$(g) + $[2Cr^{2+}$ + $7H_2O$ + $6NMe_2H_2+]$(soln.)

and the enthalpy of solution of $Me_2NH(\ell)$ in the same solvent ΔH_3,

$6NMe_2H(\ell)$ + $6H^+$(soln.) $\rightarrow$ $6[NMe_2H_2^+]$(soln.)

Then, $\Delta H_1 - \Delta H_2 - \Delta H_3$ gives the enthalpy of reaction

$W_2(NMe_2)_6$(c) + $1.5N_2$(g) + $8H_2O(\ell)$ $\rightarrow$ $2H_2WO_4$(ppt.)
+ $1.5N_2H_4(\ell)$ + $6NMe_2H(\ell)$

from which $\Delta H_f^o(W_2(NMe_2)_6$, c) was derived without requiring the enthalpies of formation of dichromate or chromic ions in solution.

For the tetraacetate derivatives of dimolybdenum (II), chromium (II) etc., acidic ferric chloride was used as the oxidising agent, and the enthalpy of formation was determined from the enthalpy of reaction,

$$Mo_2(OAc)_4(c) + 8FeCl_3(c) + 8H_2O(\ell) + NaCl(c) \rightarrow$$
$$2Na_2MoO_4(c) + 8FeCl_2(c) + 12HCl(in\ 7.97H_2O)(\ell)$$
$$+ 4AcOH(\ell)$$

The enthalpy of reaction was determined by measuring the enthalpy of solution of each reactant and product successively in the calorimetric solvent, so that the final solution from dissolution of the reactants was of the same composition as that from dissolution of the products.

Halogen-abstraction reactions were introduced in Lisboa for determining the enthalpies of formation of compounds containing metal-hydrogen bonds, e.g.

$$MoCp_2H_2(c) + 2CCl_4(\ell) \rightarrow MoCp_2Cl(c) + 2CHCl_3(\ell)$$

Carbon tetrabromide has also been used for this type of reaction, but at present there is the disadvantage that the enthalpies of formation of the halogen compounds used are not well established.

Metal-Metal Multiple Bonds.

Multiple bonds between transition metal atoms have become well known and it is an interesting question whether from thermochemical measurements, realistic estimates of the strengths of these bonds can be made. Some of the arguments and difficulties can be illustrated by considering the metal-metal quadruple bond. All the recently determined enthalpies of formation of compounds containing metal-metal quadruple bonds are listed in table 5.

	$\Delta H_f^o(c)$	$\Delta H^o(sub)$	$\Delta H_f^o(g)$
$Mo_2(OAc)_4$	-1970.7 ± 8.4	165 ± 5	-1805.7 ± 10.0
$MoCr(OAc)_4$	-2113.9 ± 6.4	$[165 \pm 5]$	-1948.9 ± 8.0
$Mo_2(OAc)_2(pd)_2$	-1808.4 ± 8.9	$[163 \pm 4]$	-1645.4 ± 10.0
$Cr_2(OAc)_4$	-2297.5 ± 6.6	314 ± 27	-1984 ± 28
$Cr_2(OAc)_4 \cdot 2H_2O$	-2875.4 ± 6.7		
$Mo_2(mhp)_4$	-754.0 ± 9.0	157 ± 3	-597 ± 10
$Mo_2(OAc)_2(mhp)_2$	-1367 ± 12	161 ± 4	-1206 ± 13
$Cr_2(mhp)_4$	-948 ± 9	150 ± 4	-798 ± 10
$Cr_2(dmp)_4$	-961 ± 22	$[192 \pm 8]$	-769 ± 23

TABLE 5. Enthalpies of Formation of Compounds containing metal-metal quadruple bonds (kJ mol^{-1} at 298K).

In Table 5, (OAc) = acetate : (pd) = 2,4 pentanedionate, (mhp) = 2 methyl-6-hydroxypyridyl, (dmp) = 2,6-dimethoxyphenyl.

To deduce values for $E(M\equiv M)$, some assumptions must be made concerning the transferability of bond energies between molecules. Consider the redistribution reaction,

$$0.5Mo_2(mhp)_4(g) + 0.5Mo_2(OAc)_4(g) \rightarrow Mo_2(mhp)_2(g)$$

$$\Delta H_r^o = -4.6 \pm 15.0 \text{ kJ mol}^{-1}$$

In the event that $E(Mo\equiv Mo)$ is constant, or that in the mixed compound it is the mean of the values for the other compounds, this redistribution being effectively thermoneutral indicates that transfer of ligand binding energies should be permissible in such cases.

It is also interesting to consider the redistribution,

$$0.5Mo_2(OAc)_4(g) + 0.5Cr_2(OAc)_4(g) \rightarrow MoCr(OAc)_4(g)$$

$$\Delta H_r^o = -54 \pm 17 \text{ kJ mol}^{-1}$$

If it be accepted that ligand binding energies are transferable, then this redistribution indicates that $E(Mo\equiv Cr)$ is not the mean of $E(Mo\equiv Mo)$ and $E(Cr\equiv Cr)$.

Consider the disruption of the following molecules,

$$Mo_2(OAc)_4(g) \rightarrow 2Mo(g) + 4(OAc)(g)$$
$$Mo_2(OAc)_2(pd)_2(g) \rightarrow 2Mo(g) + 2(OAc)(g) + 2(pd)(g)$$
$$Mo(pd)_3(g) \rightarrow Mo(g) + 3(pd)(g)$$

It was assumed that,

(i) $E(Mo\equiv Mo)$ was the same in $Mo_2(OAc)_4$ and in $Mo_2(OAc)_2(pd)$: the bond lengths $R(Mo Mo)$ in the crystals are close, 2.093 and 2.192Å respectively;

(ii) $\bar{D}(Mo-O)$ for the (pd) groups are the same in $Mo(pd)_3$ and in $Mo_2(OAc)_2(pd)_2$.

By applying these assumptions, it is easy to show that,

$$E(Mo\equiv Mo) = \frac{2}{3}\Delta H_f^o(Mo, g) + \Delta H_f^o(Mo_2(OAc)_4, g)$$
$$- 2\Delta H_f^o(Mo_2(OAc)_2(pd)_2, g) + \frac{4}{3}\Delta H_f^o(Mo(pd)_3, g)$$

and as $\Delta H_f^o(Mo(pd)_3, g) = -1202.0 \pm 7.8$, then

$$E(Mo\equiv Mo) = 321 \text{ kJ mol}^{-1}$$

The value so derived is unambiguous, but it does depend on the assumptions made and of these, the third is the weakest, as it is in fact assuming that D(Mo-O) for a (pd) group in independent of the oxidation state of Mo.

By making an additional reasonable assumption, that

$$D(Mo-O)_{pd} - D(Cr-O)_{pd} = D(Mo-O)_{OAc} - D(Cr-O)_{OAc}$$

then unambiguous expressions can also be derived for the metal-metal bonds in $MoCr(OAc)_4$ and in $Cr_2(OAc)_4$, which will involve additionally $\Delta H_f^o(Cr(pd)_3, g) = -1431.0 \pm 6.9$ kJ mol^{-1}, to give

$$E(Mo\equiv Cr) = 234 \text{ kJ mol}^{-1}$$
$$E(Cr\equiv Cr) = 43 \text{ kJ mol}^{-1}$$

The bond length r(Cr-Cr) in $Cr_2(OAc)_4(c)$ is one of the longest of the proposed (Cr≡Cr) bonds and the very small value for E(Cr≡Cr) may reflect this. It is important to realize that the values derived are also functions of the assumptions made. The difference between E(Mo≡Mo), E(Mo≡Cr) and E(Cr≡Cr) may be realistic but the absolute values should be regarded as tentative. If however, the values derived above are not in accord with your expectations concerning the chemical nature of these compounds, there is the opportunity to produce a different set of assumptions which could lead to different values.

Metal Cyclopentadienyl Derivatives (Cp_2MR_2)

Several derivatives of this general type have been studied by the static-bomb combustion method in Gorky, and by reaction calorimetry in Lisboa. The results are of interest in that they pose one of the fundamental problems of thermochemistry in a direct manner : in what way is it reasonable to divide the enthalpy of disruption of Cp_2MR_2 between the (M-Cp) and the (M-R) bonds.

Consider the following disruptions;

$Cp_2TiCl_2(g) \rightarrow 2Cp(g) + TiCl_2(g)$ $\Delta H = 556 \pm 13$ kJ mol^{-1}
$CpTiCl_3(g) \rightarrow Cp(g) + TiCl_3(g)$ $\Delta H = 229 \pm 16$ kJ mol^{-1}

the enthalpies would lead to $\bar{D}(Cp-TiCl_2) = 278 \pm 12$ and $D(Cp-TiCl_3) = 229 \pm 16$ kJ mol^{-1}. These values differ considerably from each other and the linear plot of figure 1 would place $D(Cp-Ti)$ at about 383 kJ mol^{-1}, a much higher value still.

The (Ti-Cl) bond length in $TiCl_4$ is 2.21 ± 0.03 Å and is close to that in $[Cp_2TiCl_2]$, 2.24 ± 0.01 Å, so it is reasonable to assign $\bar{E}(Ti-Cl)$ the same value in these two compounds. For $TiCl_4$, $\Delta H_a^o = 1722 \pm 5$ kJ mol^{-1}, hence $\bar{E}(Ti-Cl) = 430.5 \pm 1.3$ kJ mol^{-1}. For $[Cp_2TiCl_2]$, $\Delta H_a^o = 10329 \pm 10$ kJ mol^{-1}, and the assignment of $\bar{E}(Ti-Cl) = 430.5$ leaves 9469 ± 10 kJ mol^{-1} for the Cp_2Ti fragment of the molecule. Applying the same argument to $[CpTiCl_3]$ for which $\Delta H_a^o = 6017 \pm 13$ kJ mol^{-1}, leaves 4727 ± 13 kJ mol^{-1} for the CpTi fragment of the molecule. These "in molecule" ΔH_a^o values correspond to "in molecule" $\Delta H_f^o(g)$ values of 351 ± 10 kJ mol^{-1} for Cp_2Ti and 420 ± 13 for CpTi leading to $\bar{D}(Cp-Ti) = 326 \pm 8$ kJ mol^{-1} in $[Cp_2TiCl_2]$ and $D(Cp-Ti) = 318 \pm 16$ kJ mol^{-1} in $[CpTiCl_3]$. Some "in molecule" $\bar{D}(M-Cp)$ values calculated in this manner are listed in Table 6.

	$\Delta H_f^o(g)$	ΔH_a^o	$\bar{E}(M-Cl)$	$\Delta H_f^o(MCp_2)$	$\bar{D}(M-Cp)$
Cp_2TiCl_2	-266 ± 9	10326	430	351 ± 10	328 ± 8
Cp_2ZrCl_2	-433 ± 4	10631	489	302 ± 8	418 ± 8
Cp_2HfCl_2	-429 ± 3	10638	496	306 ± 8	421 ± 8
Cp_2MoCl_2	5 ± 5	10243	304	370 ± 15	409 ± 12
Cp_2WCl_2	34 ± 5	10416	347	485 ± 8	450 ± 10

TABLE 6. $\bar{D}(Cp-M)$ in $[Cp_2MCl_2]$ kJ mol^{-1} at 298K.

Similar arguments are applied to the data in Table 7 to determine $\bar{D}(M-X)$ in (Cp_2MX_2) molecules : in cases of discrepancies the experimental data from reaction calorimetry are preferred over those from the combustion method.

	$\Delta H_f^o(g)$	$\Delta H_f^o(Cp_2M)$	$\Delta H_f^o(X)$	$\bar{D}(M-X)$
Cp_2TiMe_2	134 ± 12	351	146	255 ± 8
Cp_2TiPh_2	382 ± 12	351	325	315 ± 12
$Cp_2Ti(CH_2Ph)_2$	280 ± 10	351	179	215 ± 15
Cp_2ZrMe_2	37 ± 3	302	146	278 ± 8
Cp_2ZrPh_2	368 ± 12	302	325	292 ± 8
Cp_2MoH_2	303 ± 6	370	218	252 ± 8
Cp_2MoBr_2	109 ± 19	370	112	242 ± 12
Cp_2MoI_2	170 ± 9	370	107	207 ± 9
Cp_2MoMe_2	354 ± 6	370	146	154 ± 8
Cp_2WH_2	311 ± 6	485	218	305 ± 5
Cp_2WBr_2	112 ± 18	485	112	298 ± 14
Cp_2WI_2	162 ± 9	485	107	268 ± 6
Cp_2WMe_2	359 ± 6	485	146	209 ± 5

TABLE 7. $\bar{D}(M-X)$ in $[Cp_2MX_2]$ kJ mol^{-1} at 298K.

The $\bar{D}(M-X)$ values fall into a reasonable pattern but we cannot yet regard all these experimental data as being free from systematic error, and more work is required to improve the quality of data for these compounds.

The mean dissociation enthalpies do include the reorganisation energies of the fragments and radicals produced on dissociation hence these values are of significance in discussing reactivity. To correlate bond strengths with structural data, mean bond enthalpy terms are more useful, and one approach is to make allowance for reorganisation energies.

The problem is to derive or make allowance for the reorganisation energy of the radical as it changes its configuration from that existing in the molecule to that of the free radical. The first requirement is knowledge of the structures or at least reasonable estimates of the structures involved. This will depend on the specific molecule under consideration, and the arguments are too detailed to expand here but the recent publications of the Lisboa School, A.R. Dias, J.A. Martinho Simoes et al merit careful study as they have initiated a promising approach to this problem.

Conclusions.

The rapid growth in our knowledge of the enthalpies of formation of transition-metal organometallic compounds has given rise to many interesting problems, both experimental and in the interpretation of the data. To make progress in this area it is important that;

(i) There is good collaboration between the thermochemist and the inorganic chemist.

(ii) There is also a requirement to develop techniques for making measurements on smaller samples, as many interesting compounds will only ever be available in small amounts. The Calvet high-temperature microcalorimeter has been of value in this regard but there remains a need for further development of combustion calorimetry and reaction calorimetry to reduce the size of samples needed for measurement.

THERMOCHEMISTRY OF MOLECULAR TRANSITION METAL COMPOUNDS

J.A. Connor

University Chemical Laboratory,
Canterbury, Kent, England.

Results of calorimetric measurements made on binary alkyl, arene and carbonyl metal complexes as well as σ-alkyl and π-alkene and arene metal carbonyl compounds are reviewed. Some of the limitations on transferability of bond enthalpy contributions are described. The information obtained is applied to the examination of simple models of reactions (exchange, oxidative addition, reductive elimination, alkyl migration) thought to play a role in catalytic processes. The enthalpy contributions of simple and multiple metal-metal bonds are surveyed to indicate problems of interpretation.

INTRODUCTION

Calorimetric methods have been reviewed in this volume (1). It is important to emphasize the constraints which determine the type of thermochemical measurements which can be made on transition metal organometallic compounds. The most important is the limitation of quantity; many compounds are routinely available in small (less than ten grams) quantities - unless cost is not a problem, so that micro-scale methods of measurement are required. The second constraint relates to the identification of the products of any process occurring in the calorimeter; this may be difficult where small (milligram) samples are used. Where possible, attempts should be made to compare results obtained by two independent methods. Much of our own work with a high temperature microcalorimeter has involved both the measurement of the heat of thermal decomposition of a substance and also the measurement of the heat of reaction of the same substance with a halogen (bromine or iodine).

Binary Metal Compounds

(a) <u>Homoleptic alkyls</u>. The term bond enthalpy is often rather ill-defined. Recently the definitions have been precisely restated (2), and will not be reviewed here. The confusion can be indicated with reference to the example of zirconium(IV) alkyls. The mean dissociation enthalpy ($\bar{D}$-value) refers to the process

$$ZrR_4(g) \rightarrow Zr(g) + 4 R (g) \qquad (i)$$

which includes the value of $\Delta H_f^o(R,g)$ which, in turn, will take account of the fact that $D(R-H)$ is sensitive to the nature of R. The mean enthalpy (E-value) refers to a reaction such as

$$ZrR_4(g) + R'OH(g) \rightarrow Zr(OR')_4(g) + RH(g) \qquad (ii)$$

and assumes a constant value for $E(R-H)$. The result of this is that values of $D(M-R)$ and $E(M-R)$ will be similar if $D(R-H)$ is close to $E(R-H)$ (chosen as the value in methane, 413 kJ mol^{-1}) e.g., for [$Zr(CH_2CMe_3)_4$], ($\bar{D}(Zr-C)$ = 227 kJ mol^{-1}. $\bar{E}(Zr-C)$ = 226 kJ mol^{-1}; $D(Me_3CCH_2-H)$ = 417 kJ mol^{-1}, but the values are very different if $D(R-H) \not\simeq 413$ kJ mol^{-1}, e.g. for $Zr(CH_2Ph)_4$ $\bar{D}(Zr-C)$ = 252 kJ mol^{-1}; $\bar{E}(Zr-C)$ = 310 kJ mol^{-1} $D(PhCH_2-H)$ = 365 kJ mol^{-1}). The increase in the value of $\bar{D}(M-X)$ (M = Ti, Zr, Hf; X = CH_2CMe_3, NEt_2, $OCHMe_2$, F) as the electronegativity of the ligating atom in X changes (X = C < N < O < F) indicates the increasing importance of the ionic contribution to the M-X bond. The effect of the metal within one subgroup is less significant, but in general $\bar{D}(Ti-X) < \bar{D}(Zr-X) \leq \bar{D}(Hf-X)$ (3). The effect of the metal is more significant when values of $\bar{D}(M-X)$ (M = Hf, Ta, W) are compared which show $\bar{D}(W-X) < \bar{D}(Ta-X) < \bar{D}(Hf-X)$ (e.g. $D(M-CH_3)$ M=W, 159; M=Ta, 261; M=Hf, 330 (estimate) kJ mol^{-1}) (4).

(b) <u>Carbonyls</u>. A problem of interpretation in the derivation of values of bond enthalpy contributions is epitomized by the values of $\bar{D}(W-C)$ in [WMe_6] (ΔH_f^o, g 772 kJ mol^{-1}; $\bar{D}$ = 159 kJ mol^{-1}) (5) and in [$W(CO)_6$] (ΔH_f^o, g -884 kJ mol^{-1}; $\bar{D}$ = 178 kJ mol^{-1}) (6). The close similarity of these values of $\bar{D}$ for two compounds which are chemically so very dissimilar can be traced to the values of the enthalpy of formation of the ligands (CH_3, + 145.6; CO, -110 kJ mol^{-1}), without consideration of valence state reorganisation energies.

The standard enthalpies of formation of gaseous polynuclear metal carbonyls have been interpreted in two contrasting ways (7). The first assumes a simple two centre electron pair bond description of the structure in terms of M-CO terminal (T) bonds, M-CO bridging (B) bonds and metal-metal (M) bonds. The enthalpy of disruption, ΔH_D, of the compound [$M_m(CO)_n$]

$$\Delta H_D = \underline{m}\Delta H_f^o[M,g] + \underline{n}\Delta H_f^o[CO,g] - \Delta H_f^o[M_{\underline{m}}(CO)_{\underline{n}},g] \qquad (iii)$$

can be related to the bonds and their enthalpy contributions in the structure of $[Fe_2(CO)_9]$ for example as $\Delta H_D = 6T + 6B + M$. Similar expressions can be derived for $[Fe(CO)_5]$ and for $[Fe_3(CO)_{12}]$. It is possible to construct a set of empirical approximations which relate the bond enthalpy contributions (B,M,T) to the enthalpy of atomisation of the metal $\Delta H_f^o(M,g)$ and the coordination number of the bulk metal, $\underline{z}$, from the solution to the simultaneous equations in ΔH_D, $\overline{B}$, M and T, which are

$$2T \sim 3M \sim 4B \sim 6\Delta H_f^o(M,g)/\underline{z} \qquad (iv)$$

The second method for the interpretation of ΔH_D proposes an empirical logarithmic relationship between the length, d(M-M), and the enthalpy contribution, M, of a metal-metal bond in the bulk which can be shown to have the form $M = A[d(M-M)]^{-4.6}$ (8). The values of T and B are then derived from ΔH_D using an electron pair bond description. This model necessarily presumes that a long bond also makes a small contribution to ΔH_D (e.g. D(Mn-Mn) in $[Mn_2(CO)_{10}] = 35$ kJ mol^{-1} because d(Mn-Mn) = 2.904Å), but has the advantage that it distinguishes between the enthalpy contributions of the Fe-Fe bonds in $[Fe_3(CO)_{12}]$ (d(Fe-Fe) 2.56, 2.68Å, D(Fe-Fe) 65, 52 kJ mol^{-1} respectively) which the electron pair bond description does not allow. This model is not able to deal with alternative descriptions of polynuclear metal carbonyls which suggest, for example, that metal-metal interaction occurs in the manner of a superexchange process (no direct M-M interaction/overlap) through a bridging CO ligand when this is present (9), whereas the electron pair bond description can be adjusted to allow for this. On the other hand, the electron pair bond description is unable to account for face bridging (μ_3) CO ligands and other more unusual ligand/metal cluster interactions in these systems.

Finally, is it known from a study of the heterometallic complexes $[Fe_2Ru(CO)_{12}]$ and $[FeRu_2(CO)_{12}]$ that the mean enthalpy contribution D(Fe-Ru), is greater than the arithmetic mean of the two components, $\frac{1}{2}[D(Fe-Fe) + D(Ru-Ru)]$ whichever method of evaluation is used (10).

(c) <u>Arenes</u>. The mean bond dissociation enthalpies D(Cr-arene) in a series of $[Cr(\eta-arene)_2]$ complexes (arene = C_6H_6 ($\overline{D} = 165$ kJ mol^{-1}), mesitylene ($\overline{D} = 151$ kJ mol^{-1}), C_6Me_6 ($\overline{D} = 155$ kJ mol^{-1}) indicate weaker binding to chromium by the substituted benzenes than by benzene itself, the weakest bond being in $[Cr(\eta-naphthalene)_2]$ (D = 145 kJ mol^{-1}) (11). This conflicts with the view that increasing methyl substitution leads to stabilization of the Cr-(η-arene) bond. Empirically, there is a strong correlation (0.997) : $\Delta H_f^o[Cr(\eta\text{-arene})_2,g] = 2.03 \Delta H_f^o(\text{arene},g) + 88$ kJ mol^{-1}. Calculations of the enthalpy change in arene displacement reactions:

$$[Cr(\eta\text{-arene}_A)_2,g] + 2 \text{ arene}_B(g) \rightarrow [Cr(\eta\text{-arene}_B)_2,g] + 2 \text{ arene}_A(g)$$

suggest that naphthalene is readily displaced from [Cr(η-C$_{10}$H$_8$)$_2$] by monocyclic arenes and that benzene displaces any other arene from [Cr(η-arene)$_2$]. The magnitude of the contribution of the metal, M to D(M-(η-arene)) is significant, as shown by the increase (12) from Cr (165) to Mo (247) and W (304 kJ mol^{-1}) when the ligand is benzene (Cr, Mo) or toluene (W).

Values of $\Delta H_f^o[M(\eta\text{-C}_5H_5)_2,c]$ have been determined by two groups (4,13). Although the agreement between the two sets of results is not particularly good in detail, they are consistent in showing the influence of the metal atom on the mean bond dissociation enthalpy, $\overline{D}(M\text{-C}_5H_5)$. These have been calculated

Table 1
Enthalpy of formation of gaseous [M(η-C$_5$H$_5$)$_2$] (13) and mean bond dissociation enthalpy $\overline{D}(M\text{-C}_5H_5)$. Values in kJ mol^{-1}.

M	$\Delta H_f^o[M(C_5H_5)_2,g]$	$\overline{D}(M\text{-C}_5H_5)$
V	182	431
Cr	249	339
Mn	270	272
Fe	231	358
Co	275	341
Ni	334	312

using $\Delta H_f^o[C_5H_5,g] = (264.4 \pm 9.2)$ kJ mol^{-1} recently determined (14) by ion cyclotron double resonance spectroscopy. The weakening of the metal ligand bond in cobaltocene and nickelocene relative to ferrocene can be explained in terms of the presence of electrons in the antibonding e_{1g} molecular orbital.

Organic Derivatives of Metal Carbonyls

(a) Decacarbonyldimanganese. The interpretation of the enthalpy of formation of [Mn(CO)$_5$R] compounds depends on the value of $\Delta H_f^o[Mn(CO)_5,g]$, which is linked to the dissociation enthalpy of the Mn-Mn bond in [Mn$_2$(CO)$_{10}$]. Experimental determinations of this quantity vary between 79 and 154 kJ mol^{-1}. A value of D(Mn-Mn) = 94 kJ mol^{-1} has been chosen for reasons given in detail elsewhere (15); this choice corresponds to

$\Delta H_f^o[Mn(CO)_5,g] = -(745.5\pm2.5)$ kJ mol^{-1}.

Table 2
Enthalpy of formation of gaseous [Mn(CO)$_5$R] and
bond enthalpy contribution D(Mn-R).
All values in kJ mol^{-1}.

R	ΔH_f^o,g	D(Mn-R)
Me	-(753±4)	153± 5
CH$_2$Ph	-(642±8)	87±12
Ph	-(590±7)	170±11
CF$_3$	-(1386±4)	172± 7
COMe	-(897±10)	129±12
COPh	-(724± 5)	89±10
COCF$_3$	-(1508±6)	147±11
H	-(740±10)	213±10

The values of D(Mn-R) decrease in passing along the series
CF$_3$ ~ Ph > Me > CH$_2$Ph and follow the trend observed in D(R-CH$_3$).
It is worth recalling that CH$_3$ and Mn(CO)$_5$ are isolobal
fragments. Whereas D(R-CH$_3$) is greater for R = Ph than for
R = CF$_3$, they are almost equal for R = Mn(CO)$_5$ which might be
interpreted as a relative strengthening of the Mn-CF$_3$ bond.
The enthalpy of the carbonylation reaction (v) at 298 K shows
that there is a dependence on R. These reactions involve the

$$[Mn(CO)_5R](c) + CO(g) \rightarrow [Mn(CO)_5COR](c) \quad (v)$$

loss of the translational contribution to the entropy of CO
absorbed by reaction and this corresponds to a change (TΔS) of
45 kJ mol^{-1}, so that the free energy change ΔG (v) remains
negative for R = Ph, Me but is positive for R = CF$_3$. In practice,
carbonylation of [Mn(CO)$_5$CH$_3$] is readily accomplished but
[Mn(CO)$_5$CF$_3$] resists carbonylation even under high pressures of
CO. Decarbonylation of [Mn(CO)$_5$COCF$_3$] by warming the solid
under reduced pressure is the preferred route to [Mn(CO)$_5$CF$_3$].
Metal-formyl complexes have been considered as possible intermediates in the initial stages of metal-catalyzed reactions
between CO and H$_2$ to form oxohydrocarbons (Fischer-Tropsch
reaction). Using information for D(X-R) and ΔH_f^o(R·,g) (X = H,
Me; R = H, Me, Ph, CHO, COMe, COPh) and values of D(Mn-R), it is
possible to estimate ΔH_f^o[Mn(CO)$_5$CHO,g] ~ -830 kJ mol^{-1} and then
to calculate the enthalpy change ΔH(vi) for the reaction, as ca.
+20 kJ mol^{-1} which is made even less favourable when the entropy

$$[Mn(CO)_5H](g) + CO(g) \rightarrow [Mn(CO)_5CHO](g) \quad (vi)$$

contribution of CO is included.

The enthalpy values in Table 2 can be used (15)

to calculate the enthalpy change in intermolecular elimination reactions which produce alkanes, aldehydes and alcohols as follows:

$$Mn(CO)_5H + Mn(CO)_5R \rightarrow RH + Mn_2(CO)_{10} \quad (vii)$$
$$Mn(CO)_5H + Mn(CO)_5COR \rightarrow RCHO + Mn_2(CO)_{10} \quad (viii)$$
$$3Mn(CO)_5H + Mn(CO)_5COR \rightarrow RCH_2OH + 2Mn_2(CO)_{10} \quad (ix)$$

In each case the enthalpy change is favourable to reaction; for example, in the case $R=CH_3$, ΔH (vii) = -160 kJ mol^{-1}, ΔH (viii) = -114 kJ mol^{-1}, ΔH (ix) = -288 kJ mol^{-1}. The enthalpy change for the intramolecular elimination of ethene from [Mn(CO)$_5$Et] can be calculated as ca. 100 kJ mol^{-1}, assuming that $D(Mn-C_2H_5)$ is the same as in [Mn(CO)$_5$Me]. Oxidative addition of alkyl and acyl halides to $Mn_2(CO)_{10}$ giving [Mn(CO)$_5$X] and either [Mn(CO)$_5$R] or [Mn(CO)$_5$COR] is calculated to be nearly thermoneutral.

(b) <u>Pentacarbonyliron</u>. Substitution of carbon monoxide in Fe(CO)$_5$ by olefinic ligands causes r(Fe-CO) to decrease from 1.82Å to a value in the range 1.76–1.79Å. Oxidation to iron(II) results in a further decrease in r(Fe-CO) to 1.72–1.76Å, which can be accounted for by the smaller radius of iron(II). The iron-carbon distances in all complexes containing organic π-bonded ligands (which contain sp^2-hybridised carbon atoms in the free ligand) are effectively constant and the carbon radius is that of an sp^3-hydridised atom irrespective of whether the iron atom is formally iron(0) or iron(II). This result is reflected in an interpretation of the bond enthalpy contributions of organic ligands in these complexes (16). If the ligands are considered as <u>n</u> electron donors where n = 2 (monoolefin), 4 (enyl, diene) and 6 (dienyl), the enthalpy contributions of these ligands per electron <u>pair</u>, 2 $\overline{D}$(Fe-L)/n is ca. 100 kJ mol^{-1} and does not change very much from one organic ligand to another.

Table 3
Iron-ligand bond enthalpy contributions for <u>n</u>-electron donor ligands.

ligand	n	D(Fe-L)	D/n
C_2H_4	2	97	49
C_3H_5	4	192	48
C_6H_8	4	192	48
C_5H_5	6	358	59

There is evidence for a constant bond enthalpy contribution per electron pair donated in organic derivatives of molybdenum carbonyl, which is similar (ca. 100 kJ mol^{-1} per electron pair) to that for the iron complexes, but as usual more detailed information is required.

(c) **Hexacarbonylchromium.** An important general question concerns the transferability of bond enthalpy contributions from one system to another. For example, the transference of $\bar{D}$(Cr-arene) and $\bar{D}$(Cr-CO) from [Cr(η-arene)$_2$] and from Cr(CO)$_6$ to [Cr(CO)$_3$(η-arene)]. If transferability is perfect, then the enthalpy change for the redistribution reaction $\Delta H(x) = 0$

$$[Cr(\eta\text{-arene})_2](g) + Cr(CO)_6(g) \rightarrow 2[Cr(CO)_3(\eta\text{-arene})](g) \quad (x)$$

and deviations from this, given as $-[\Delta H(x)]/2 = \Delta$(Cr-arene) + 3Δ(Cr-CO) where Δ represents a change in the bond enthalpy concerned, indicate the extent to which transferability is not justified. The experimental values of ΔH_f^o[Cr(CO)$_3$(η-arene)](g) taken with other evidence which indicates that both (Cr-CO) and (Cr-arene) bonds in these complexes have greater multiple bond character than in the binary complexes, suggests that both Δ(Cr-CO) and Δ(Cr-arene) are positive and less than 15 kJ mol^{-1}. This variation is within the error of the experiment. Transferability is seen to be valid in the systems considered to date (11).

The enthalpy change in the arene displacement reactions:

$$[Cr(CO)_3(\eta\text{-arene}_A)](g) + \text{arene}_B(g) \rightarrow [Cr(CO)_3(\eta\text{-arene}_B)](g)$$
$$+ \text{arene}_A(g) \quad (xi)$$

can be calculated and a selection of the results is shown in Table 4.

Table 4
Enthalpy change for arene displacement reaction, $\Delta H(xi)$/kJ mol^{-1}.

Cr(CO)$_3$(η-arene)$_A$	arene$_B$			
	$C_{10}H_8$	C_6H_6	$C_6H_3Me_3$	C_6Me_6
$C_{10}H_8$	-	-(26±12)	-(38±12)	-(52±15)
C_6H_6	+(26±12)	-	-(12±13)	-(26±16)
$C_6H_3Me_3$	+(39±12)	+(12±13)	-	-(14±16)
C_6Me_6	+(53±15)	+(26±16)	+(14±16)	-

Studies of arene displacement reactions from [Cr(CO)$_3$(η-arene)] complexes in a donor solvent (THF) at high temperatures have shown that the position of equilibrium in reaction (xi) is not very temperature dependent but is sensitive to the nature of the arene. The thermochemical measurements support this.

Multiple Metal-to-Metal Bonds

The central problem is one of interpretation: what is understood by the formal oxidation state of a metal in a binuclear or polynuclear metal complex in which, additionally, the bonds between some (or all) of the metal atoms have an order greater than one? A system suitable for the determination of the bond enthalpy contribution of a multiple bond between two metal atoms, $D(M \stackrel{x}{=} M)$, should be free of any ligand bridging the metal-metal bond and the metal should form a mononuclear binary complex with the same ligand. The heats of formation of the compounds $Mo(NMe_2)_4$, $W(NMe_2)_6$, and $[M_2(NMe_2)_6]$ (M=Mo,W) were determined (18) by various calorimetric methods. From these data it is possible to determine $\overline{D}(Mo^{IV}-NMe_2)$ and $\overline{D}(W^{VI}-NMe_2)$ directly and, from comparison with the meagre information available for other binary compounds of the type $M'X_n$ (M'=Zr, Hf, Ta, Mo, W; X=F, Cl, Br, OMe, NMe_2, Me; n = 4,5,6 as appropriate), it is possible to estimate values of $D(M-NMe_2)$ for other formal oxidation states of M. In the case of $[M_2(NMe_2)_6]$, the enthalpy of disruption, $\Delta H_D = 6\, D(M-NMe_2) + D(M \stackrel{3}{=} M)$. The arguments for recognizing the metal atom as four coordinate and thus comparable to $M(NMe_2)_4$ have been detailed elsewhere (18), they rely principally on structural information. From this premise, with $D(Mo-NMe_2) = (255.4 \pm 5)$ kJ mol^{-1} and $D(W-NMe_2) \sim (295 \pm 5)$ kJ mol^{-1}, values of $D(Mo \stackrel{3}{=} Mo) = (396 \pm 18)$ and $D(W \stackrel{3}{=} W) = (558 \pm 20)$ kJ mol^{-1} have been derived.

Other complexes containing $(M \stackrel{x}{=} M)$ bonds in which this bond is bridged by a ligand such as acetate (in $[Mo_2(OAc)_4]$) or alkoxide (in $Mo_2(OR)_8$), present worse problems of interpretation. It has been suggested that there is a logarithmic relation between the bond length and bond enthalpy contribution of the metal-metal bond: $E(Mo\stackrel{4}{=}Mo) = A[r(Mo-Mo)]^{-4.29}$ with $\log A = 12.64$ (19). This leads to $D[Mo\stackrel{4}{=}Mo] = 489$ kJ mol^{-1} in $Mo_2(OAc)_4$, $D(Mo\stackrel{3}{=}Mo) = 378$ kJ mol^{-1} in $[Mo_2(OPr^i)_6]$ and $Mo\stackrel{2}{=}Mo = 220$ kJ mol^{-1} in $[Mo_2(OPr^i)_8]$ and Mo-Mo = 75 kJ mol^{-1} in $[Mo_2(CO)_6(\eta-C_5H_5)_2]$. Further work is necessary to explore the limitations of this bond length/bond enthalpy approach. In these complexes it has the particular advantage that it allows an interpretation to be made. This interpretation is not unique as shown by the fact that it proposes that a long bond must have a low bond enthalpy contribution whereas, for example, it is proposed on the basis of kinetic measurements (20) that the long Mn-Mn bond in $Mn_2(CO)_{10}$ (2.90Å vs 2.74Å in manganese metal) may have an enthalpy contribution ca. 150 kJ mol^{-1} which is more than half ΔH_{vap} for manganese metal (284.5 kJ mol^{-1}).

The author wishes to record his indebtedness to Professor H.A. Skinner and Dr. G. Pilcher and the many other people at Manchester whose names are mentioned in the references. He thanks the Science and Engineering Research Council for support of this work over many years.

REFERENCES

1. Pilcher, G.: in this volume.
2. Pilcher, G., Skinner, H.A., in: Chemistry of the Metal-Carbon Bond, Hartley, F.R., Patai, S. (editors). Wiley. 1982, pp. 43-75.
3. Lappert, M.F., Patil, D.S., Pedley, J.B.: J.C.S. Chem. Comm., 1975, pp. 830-831.
4. Connor, J.A.: Topics Curr. Chem., 1977, 71, pp. 71-110.
5. Adedeji, F.A., Connor, J.A., Skinner, H.A., Galyer, L., Wilkinson, G.: J.C.S. Chem. Comm., 1976, pp. 159-160.
6. Barnes, D.S., Pilcher, G., Pittam, D.A., Skinner, H.A., Todd, D.: J. Less Common Metals, 1974, 38, pp. 53-58.
7. Connor, J.A. in: Transition Metal Clusters, Johnson, B.F.G., editor, Wiley, 1980, pp. 345-389.
8. Wade, K.: Inorg. Nucl. Chem. Letters, 1978, 14, p. 71; Housecroft, C.E., Wade, K., Smith, B.C.: J.C.S. Chem. Comm., 1978, p. 765.
9. Heijser, W., Baerends, E.J., Ros, P.: Faraday Symposium, 1980, p. 14.
10. Baev, A.K., Connor, J.A., El-Saied, N.I., Skinner, H.A.: J. Organometallic Chem., 1981, 213, pp. 151-156.
11. Connor, J.A., Martinho-Simoes, J.A., Skinner, H.A., Zafarani-Moattar, M.T.: J. Organometallic Chem., 1979, 179, pp. 331-356.
12. Connor, J.A., El-Saied, N.I., Martinho-Simoes, J.A., Skinner, H.A.: J. Organometallic Chem., 1981, 212, pp. 405-410.
13. Chipperfield, J.R., Sneyd, J.C.R., Webster, D.E.: J. Organometallic Chem., 1979, 178, pp. 177-189.
14. DeFrees, D.J., McIver, R.T., Hehre, W.J.: J. Amer. Chem. Soc., 1980, 102, p. 3334.
15. Connor, J.A., Zafarani-Moattar, M.T., Bickerton, J., El-Saied, N.I., Suradi, S., Carson, R., Al Takhin, G., Skinner, H.A.: Organometallics, 1982, 1, pp. 1166-1174.
16. Connor, J.A., Demain, C.P., Skinner, H.A., Zafarani-Moattar, M.T.: J. Organometallic Chem., 1979, 170, pp. 117-130.
17. Mahaffy, C.A.L., Pauson, P.L.: J. Chem. Research (S) 1979, p. 126.
18. Adedeji, F.A., Cavell, K.J., Cavell, S., Connor, J.A., Pilcher, G., Skinner, H.A., Zafarani-Moattar, M.T.: J.C.S. Faraday 1, 1979, 75, pp. 603-613.
19. Connor, J.A., Skinner, H.A.: A.C.S. Symposium Ser., 1981, 155, pp. 197-205.
20. Halpern, J.: Acc. Chem. Res., 1982, 15, pp. 238-244; Poë, A.J.: A.C.S. Symposium Ser., 1981, 155, pp. 135-166.

MEASUREMENT OF ENTHALPIES OF SOLUTION OF ELECTROLYTES

Michael H. Abraham

Department of Chemistry, University of Surrey,
Guildford, Surrey, U.K.

The measurement of enthalpies of solution of electrolytes by batch calorimetry is discussed, with particular reference to extrapolations to zero electrolyte concentration in order to obtain values of ΔH_s°. Corrections that may have to applied for the enthalpy of ampoule-breaking and the enthalpy of dilution are listed, and problems connected with electrolytes that undergo ion-association are mentioned. Examples are given of various extrapolations to zero electrolyte concentration that may be used, and tests for self-consistency of measured ΔH_s° values for a series of electrolytes in a given solvent, through the calculation of enthalpies of transfer, are discussed. It is shown that in terms of enthalpy, electrolytes of the alkali halide series are more stable in nonaqueous solvents than in water; possible reasons for this behaviour are discussed.

1. GENERAL CONSIDERATIONS

Enthalpies of solution are now known for a large number of uni-univalent electrolytes in water and in a number of the more polar organic solvents (1). There is need, however, for measurements on new solvent systems as well as for further accurate measurements on previously studied systems. For example, since the evaluation by Parker (2) in 1965 on ΔH_s° for sodium chloride in water at 298 K, there have been several more studies, but there seems to be not the degree of consistency in the new values that might be expected for such a system, see Table 1.

Table 1. Enthalpies of solution of NaCl(cryst) in water, cal mol^{-1} at 298 K

ΔH_s°	Date	Author	Ref.
928	1965	Parker	(2)
930	1971	Somsen and Weeda	(3)
990±30	1972	Arnett et al.	(4)
928±1	1976	Bury et al.	(5)
907	1977	Dadgar et al.	(6)
908	1978	Taniewska-Osinska et al.	(7)
933±16	1981	Abraham and Ling	(8)

In solvents where ion-association is small (e.g. water, formamide, and DMSO), provided due care is taken over the purity of the salts and solvent, the main problems in batch calorimetric determination of ΔH_s° values for the electrolytes are the corrections to be made for the enthalpy of ampoule-breaking and for the enthalpy of dilution of the electrolyte from the final concentration in the calorimetric vessel to zero concentration. The ampoule-breaking correction will include the actual mechanical enthalpy of breaking and the enthalpy of evaporation of the solvent into the vapour space of the partially empty ampoule. The latter can be calculated from the enthalpy of vaporisation of the solvent at 298 K, together with the corresponding vapour pressure. In Table 2 are given the calculated enthalpies of evaporation of solvents into a 1 cm^3 vapour space at 298 K, together with the ampoule-breaking corrections that have been determined experimentally by breaking 1 cm^3 empty ampoules.

The calculated and found values are close enough to suggest that the mechanical enthalpy of breaking is quite small, and that the main contribution arises from evaporation of the solvent, as calculated. Although the enthalpies of ampoule-breaking are only

Table 2. Enthalpies of breaking 1 cm^3 empty ampoules at 298 K

solvent	calc/cal	found/cal
water	0.0134	0.0091, 0.0106
methanol	0.0587	0.0724
ethanol	0.0321	0.0316
1-propanol	0.0122	0.0115
DMF	0.0025	
acetonitrile	0.0357	
acetone	0.0938	
1,2-DCE	0.0356	

around 0.02 cal, they may become as large as the actual enthalpy of solution of an electrolyte when the latter is dissolved to yield a very dilute solution. Thus for solution of a solid electrolyte in water with ΔH_s° = 5000 cal mol^{-1} to yield 100 cm^3 of 10^{-3} mol ℓ^{-1} solution, solvent evaporation into a 1 cm^3 vapour space amounts to some 3% of the actual enthalpy of solution. In Table 3 are listed a number of comparisons of this type, and it is clear that for many solvents it may not be practical to carry out batch measurements of enthalpies of solution of electrolytes to below, say, a concentration of 5x10^{-4} mol ℓ^{-1} unless the ampoule-breaking correction can be circumvented.

Table 3. Enthalpy of breaking 1 cm^3 empty ampoules in 100 cm^3 solvent as a % of the enthalpy of solution ΔH_s° = 5000 cal mol^{-1}

concentration mol ℓ^{-1}:	10^{-2}	10^{-3}	10^{-4}	10^{-5}
water	0.3	2.7	27	270
methanol	1.2	11.7	117	1170
DMF	0.1	0.5	5	50
acetone	1.9	18.8	188	1880

In order to obtain ΔH_s° values for solution at zero concentration, rather than to attempt measurements at very low concentration with difficulties due to the presence of small quantities of impurities as well as to the ampoule-breaking effect, it is more practical to restrict measurements to a range of final concentration between, say, 1×10^{-2} to 5×10^{-4} mol ℓ^{-1} and to extrapolate the measurements to zero concentration. Either the obtained ΔH_s^{OBS} values (after correction for ampoule-breaking effects) may be plotted against $\sqrt{I}$, where I is the concentration of unassociated uni-univalent electrolyte, and extrapolated to I=0, or the ΔH_s^{OBS} values (again after application of the ampoule-breaking correction) may be corrected through the Debye-Huckel expression for the enthalpy of dilution from concentration I to zero concentration:

$$\Delta H_s^\circ = \Delta H_s^{OBS} + \Delta H_D \qquad (1)$$

The enthalpy of dilution, ΔH_D, is given generally by eqn (2) in which the term $\partial \ln f/\partial T$ may be evaluated either by the extended Debye-Huckel expression (3) or

$$\Delta H_D = \frac{4}{3} RT^2 \frac{\partial \ln f}{\partial T} \qquad (2)$$

by the limiting law (4).

$$\log f = -A\sqrt{I}/(1+B\mathring{a}\sqrt{I}) \qquad (3)$$

$$\log f = -A\sqrt{I} \qquad (4)$$

In the latter case, the equation for ΔH_D may be formulated as eqn (5), or more simply as eqn (6), where α_e is the coefficient of thermal expansion of the solvent. Values of D for a number of the more common solvents are in Table 4, together with the Debye-Huckel coefficients, A and B in eqns (3) and (4); values of ε and $\partial\varepsilon/\partial T$ were taken from references (9) and (10).

$$\Delta H_D^{LIM}/\text{cal mol}^{-1} = \frac{2.8831\times10^8}{\varepsilon^{3/2}} \left(\frac{1}{298.15} + \frac{1}{\varepsilon}\cdot\frac{\partial\varepsilon}{\partial T} + \frac{\alpha_e}{3}\right)\sqrt{I} \qquad (5)$$

$$\Delta H^{LIM}/\text{cal mol}^{-1} = D\sqrt{I} \qquad (6)$$

Table 4. Values of the Debye-Huckel coefficients A, B and D

solvent	A/ $\ell^{1/2}$ mol$^{-1/2}$	B/Å^{-1} $\ell^{1/2}$ mol$^{-1/2}$	D/cal $\ell^{1/2}$ mol$^{-3/2}$
water	0.5109	0.3290	-480
methanol	1.9006	0.5098	-3530
ethanol	2.9533	0.5904	-5580
1-propanol	3.8324	0.6441	-10160
$(CH_2OH)_2$	1.5311	0.4743	
formamide	0.3093	0.2783	-750
NMF	0.1439	0.2157	
DMF	1.5934	0.4807	-1590
DMSO	1.1113	0.4263	+1280
MeCN	1.6394	0.4853	-840
PhNO$_2$	1.7249	0.4936	
Acetone	3.8212	0.6434	-2830
1,2-DCE	10.8318	0.9106	-15100

It should be emphasised that the calculated enthalpies of dilution, ΔH_D^{LIM} or ΔH_D^{EXT}, obtained through the limiting-law or extended law, are theoretical only. In the case of water, the range of validity of these corrections can be assessed, for example through the careful work of Bury et al. (5) on the enthalpy of dilution of sodium chloride. In Table 5 are given the observed values for dilution from the given concentration to zero concentration, together with those calculated from the limiting law, eqn (6) with D = -480, and from the extended law, eqns (2) and (3) with å = 2Å.

As the solvents become less polar, the coefficients A and B increase considerably in magnitude. Because the numerical value of D depends on a balance between the positive term $\alpha_e/3$ and the negative term $(1/\varepsilon)\partial\varepsilon/\partial T$, the magnitude of D does not depend so evenly on the solvent dielectric constant but does again tend to be large in the less polar solvents. Furthermore, as the solvent dielectric constant decreases, so does the range of validity of the various Debye-Huckel expressions.

Table 5. Observed (5) and calculated enthalpies of dilution in cal mol^{-1} of sodium chloride in water, from a given concentration to zero concentration at 298 K

Concentration/mol ℓ^{-1}	ΔH_D^{OBS}	ΔH_D^{LIM}	ΔH_D^{EXT} (å=2Å)
0.1000	-83.3	-151.8	-118.8
0.0500	-70.4	-107.3	-89.6
0.0100	-38.3	-48.0	-44.0
0.0050	-29.6	-33.9	-31.8
0.0010	-14.2	-15.2	-14.7
0.00050	-10.2	-10.7	-10.5
0.00025	-7.3	-7.6	-7.5
0	0	0	0

However, for the more polar solvents in which ion-association is very small, the enthalpy of dilution correction is small, at least provided that the final concentration, I, in eqns (5) and (6), is not larger than about 10^{-3} mol ℓ^{-1}. Thus for these solvents, as mentioned above, it is usual to obtain ΔH_s^o either just by extrapolation of ΔH_s^{OBS} against $\sqrt{I}$ or by applying the theoretical correction for ΔH_D. It must be mentioned that in some cases, these theoretical corrections are not well known, since they depend on small differences between the (positive) coefficient of thermal expansion of the solvent and the (negative) coefficient of the dielectric constant.

2. ION-ASSOCIATION OF ELECTROLYTES

For many of the solvents listed in Table 3, ion-association must be taken into account, the electrolyte dissolving not only as the dissociated pair of ions, eq (7), but also as an associated ion-pair, eq (8). As suggested by Wu and Friedman (11), the observed

$$M^+X^-(cryst) \xrightarrow{\Delta H_s} M^+(soln) + X^-(soln) \qquad (7)$$

$$M^+X^-(cryst) \rightarrow M^+X^-(soln) \qquad (8)$$

$$M^+(soln) + X^-(soln) \xrightarrow{\Delta H_A} M^+X^-(soln) \qquad (9)$$

enthalpy of solution may be related to the required ΔH_s term through eq (10), in which α is the degree of

$$\Delta H_s^{OBS} = \Delta H_s^\circ + (1-\alpha)\Delta H_A \qquad (10)$$

dissociation of the electrolyte. Thus a plot of ΔH_s^{OBS} against α, on extrapolation to $\alpha=1$ will yield ΔH_s°. Calculation of the degree of dissociation at each final concentration in a series of batch experiments is itself not trivial. For the relevant equilibrium (9) we have that

$$K_A = x/(a-x)^2 f_\pm^2 \qquad (11)$$

where K_A is the ion-pair association constant, x is the concentration of ion-pair and a is the total formal concentration of electrolyte. The activity coefficient of the ion-pair is taken as unity, but that of the dissociated electrolyte must be evaluated either from the limiting or extended equation, (12) or (13).

$$\log f_\pm = -A\sqrt{(a-x)} \qquad (12)$$

$$\log f_\pm = -A\sqrt{a-x}/(1+B\mathring{a}\sqrt{a-x}) \qquad (13)$$

Knowing K_A and a, it is possible to evaluate $f_\pm$ and x by a series of iterative calculations, and then to obtain $\alpha = (a-x)/a$.

The determination of ΔH_s° for Pr_4NI in 1-propanol may be taken as an example of a system in which the electrolyte is associated, but with not too large a value of K_A (440 ℓ mol^{-1}). Details (12) of a series of batch measurements are in Table 6; ΔH_s^{OBS} refers to the observed enthalpy of solution after correction for ampoule breaking. Since the activity coefficient and ΔH_D expressions refer only to the dissociated electrolyte, values of α were first calculated through eqns (11) and (13) with $\overset{\circ}{a}$ = 4.2 Å, and the corrections applied only to the dilution from concentration $(a-x)$ to zero. Two series of corrections were used, one through the limiting equation (6) and one through the extended equation (2) and (3), and plots of the various ΔH_s values made against α. For this system, it makes little difference to the final result whether the ΔH_s^{OBS} values themselves are plotted against α (ΔH_s° = 6.91 kcal mol^{-1}) or whether the corrected values are used (ΔH_s° = 6.85 or 6.83 kcal mol^{-1}). Indeed, a simple plot of ΔH_s^{OBS} against the square root of the total final concentration yields ΔH_s° = 6.99 kcal mol^{-1}, very close to the other obtained values. For this particular system, a value of ΔH_s° = 6.72 kcal mol^{-1} may be calculated from various single-ion values of ΔH_t° obtained from other measurements (12).

Table 6. Enthalpies of solution of Pr_4NI in 1-propanol, kcal mol^{-1} at 298 K; K_A = 440 (12)

Concentration/mol ℓ^{-1}	α	ΔH_s^{OBS}	ΔH_s^{LIMD}	ΔH_s^{EXTD}
1.80x10^{-3}	0.770	6.26	5.88	5.93
1.33x10^{-3}	0.802	6.26	5.93	5.96
1.09x10^{-3}	0.822	6.42	6.12	6.15
4.82x10^{-4}	0.894	6.60	6.39	6.40
-	1	6.91	6.85	6.83
0	-	6.99	-	-

Our general experience is that for electrolytes that are not too associated, ΔH_s° obtained from plots of ΔH_s^{OBS} (i.e. after correction for ampoule-breaking) against the square root of the final concentration is usually very close to the value obtained from plots of ΔH_s^{LIMD} or ΔH_s^{EXTD} against α. This is sometimes the case also for very associated electrolytes, as shown by results (13) for Bu_4NBPh_4 in 1,2-dichloroethane

($K_A = 1.71 \times 10^3$), Table 7.

Table 7. Enthalpies of solution of Bu_4NBPh_4 in 1,2-dichloroethane, kcal mol^{-1} at 298 K; $K_A = 1.71 \times 10^3$ (13)

Concentration/mol ℓ^{-1}	α	ΔH_s^{OBS}	ΔH_s^{LIMD}	ΔH_s^{EXTD}
3.93×10^{-3}	0.729	2.80	1.99	2.20
3.28×10^{-3}	0.737	2.68	1.94	2.13
2.21×10^{-3}	0.757	2.59	1.97	2.11
1.33×10^{-3}	0.788	2.72	2.23	2.32
6.39×10^{-4}	0.838	2.59	2.24	2.29
-	1	2.41	2.75	2.55
0	-	2.51	-	-

However with Bu_4NClO_4 in the same solvent, Table 8, there is a noticeable difference between ΔH_s^o obtained through plots of ΔH_s^{OBS} against the square root of the concentration (-0.54 kcal mol^{-1}), and that obtained from the ΔH_s^{EXTD} vs α plot (-1.20 kcal mol^{-1}). Since measurements on various other electrolytes lead to single ion ΔH_t^o values from which ΔH_s^o (Bu_4NClO_4) = -1.14 kcal mol^{-1} may be deduced (13), it seems that in this particular case the simple plot of ΔH_s^{OBS} against the square root of the concentration leads to a ΔH_s^o value that is somewhat in error.

Table 8. Enthalpies of solution of Bu_4NClO_4 in 1,2-dichloroethane, kcal mol^{-1} at 298 K; $K_A = 6.41 \times 10^3$ (13)

Concentration/mol ℓ^{-1}	α	ΔH_s^{OBS}	ΔH_s^{LIMD}	ΔH_s^{EXTD}
9.14×10^{-3}	0.305	0.25	-0.55	-0.35
5.10×10^{-3}	0.333	0.09	-0.53	-0.41
2.54×10^{-3}	0.381	-0.06	-0.53	-0.45
1.04×10^{-3}	0.464	-0.18	-0.51	-0.47
4.72×10^{-3}	0.556	-0.45	-1.22	-0.70
-	1	-1.59	(-2.06)	-1.20
0	-	-0.54	-	-

With measurements on nonaqueous solvents, the effect of adventitious water must always be considered. We have found it useful to determine the water content of the given solvent both before the enthalpy of solution measurement and afterwards as well (the latter by sampling the solvent in the calorimetric vessel after reaction), and in addition to carry out measurements on batches of solvent to which known quantities of solvent have been added. In Table 9 are results of some ΔH_s determinations in 1,2-dichloroethane to which water has deliberately been added (8). Rather surprisingly, the effect of added water is quite small, although, to be sure, ΔH_s in dry 1,2-dichloroethane and in water-saturated 1,2-dichloroethane will differ by over 1 kcal mol^{-1}.

Table 9. The effect of water on ΔH_s for Bu$_4$NI (final concentration 7.7x10^{-4} mol ℓ^{-1}) in 1,2-dichloroethane (8)

| Percent water (w/w) | | ΔH_s^{OBS} |
Added	Total	kcal mol^{-1}
-	(0)	(2.79)
0	0.015	2.73
0.041	0.056	2.47
0.081	0.096	2.16
0.122	0.137	1.74
0.163	0.178	1.52
	0.188 (saturated)	

3. TREATMENT OF THE DATA

Having obtained a set of ΔH_s° values for a series of electrolytes in a given solvent, it is useful to test the self-consistency of the measurements. First of all, enthalpies of transfer from water to the given solvent may be calculated from $\Delta H_t^\circ = \Delta H_s^\circ$ (in solvent) - ΔH_s° (in water). Secondly, since all these measurements refer to completely dissociated ions, the value for a given cation (say) must be independent of the counter-anion, and vice-versa. In Table 10 is an illustration of a test for self-consistency that can be carried out using ΔH_t° values; the final column of $\Delta H_t^\circ(I^- - C\ell O_4^-)$ values is constant to about ±0.5 kcal mol^{-1}. Of course, any error in the values of ΔH_t°

will include not only that in ΔH_s°(ethanol) but also any error in ΔH_s°(water).

Table 10. Enthalpies of transfer from water to ethanol in kcal mol^{-1} at 298 K. Test for self-consistency (14).

Salt	ΔH_s°(ethanol)	ΔH_s°(water)	ΔH_t°	$\Delta H_t^\circ(I^- - ClO_4^-)$
KI	0.70	4.86	-4.16	> 0.63
KClO$_4$	7.41	12.20	-4.79	
NaI	-5.77	-1.80	-3.97	> 0.24
NaClO$_4$	-0.89	3.32	-4.21	
Bu$_4$NI	8.64	3.53	5.11	> 0.52
Bu$_4$NClO$_4$	7.10	2.51	4.59	

The test for self-consistency shown in Table 10 is cumbersome to use if ΔH_t° values are available for a large number of electrolytes. It is then more convenient to designate a value of ΔH_t° to some particular ion and then to construct a set of single-ion ΔH_t° values. For the purpose of a test for self-consistency, any ΔH_t° value may be assigned to any particular ion, but it is useful from the point of view of discussions on ion-solvent interactions to attempt to assign single-ion ΔH_t° values that are felt intuitively to be near to the actual "absolute" values. One such assignment sets $\Delta H_t^\circ(Ph_4As^+$ or $Ph_4P^+) = \Delta H_t^\circ(Ph_4B^-)$ and although it is impossible to prove the correctness of this assumption, it is generally felt to yield chemically reasonable single-ion values. The use of single-ion values in order to test for self-consistency of ΔH_t° values is illustrated by the set of results in Table 11 for transfer from water to ethanol, obtained from values of ΔH_s°(ethanol), and hence ΔH_t°, on 21 uni-univalent electrolytes (14).

The single-ion values are then used to regenerate the $\Delta H_t^\circ(M^+ + X^-)$ values and are adjusted until the total deviations in the observed and calculated $\Delta H_t^\circ(M^+ + X^-)$ values are minimised. Application to the iodide and perchlorate salts is illustrated in Table 12. For the six electrolytes studied, the observed ΔH_t° values are self-consistent within ±0.1 kcal mol^{-1}, and for

Table 11. A set of single-ion values of enthalpies of transfer from water to ethanol, in kcal mol^{-1} at 298 K (14).

Ion	ΔH_t^o	Ion	ΔH_t^o
Li^+	-4.17	Cl^-	2.44
Na^+	-3.91	Br^-	1.30
K^+	-4.11	I^-	-0.10
Me_4N^+	0.04	ClO_4^-	-0.62
Bu_4N^+	5.21	Ph_4B^-	0.01
Ph_4As^+	0.01		
Ph_4P^+	0.01		

fifteen electrolytes to which the test can be applied, values are self-consistent to within 0.08 kcal mol^{-1} (95% confidence level) or 0.10 kcal mol^{-1} (average deviation). Use of the single-ion method thus enables limits of self-consistency to be calculated for the entire set of electrolytes studied.

Table 12. Test for self-consistency of enthalpies of transfer from water to ethanol (14) using a single-ion assumption, kcal mol^{-1} at 298 K

Salt	ΔH_t^o (obs)	ΔH_t^o (calc)
KI	-4.16	-4.21
$KClO_4$	-4.79	-4.73
NaI	-3.97	-4.01
$NaClO_4$	-4.23	-4.53
Bu_4NI	5.11	5.11
Bu_4NClO_4	4.59	4.59

4. DISCUSSION

There are now available reasonably self-consistent sets

of enthalpies of transfer of uni-univalent electrolytes from water to many of the common nonaqueous solvents (1)(12-15), as well as less complete data for transfers from water to various other nonaqueous solvents and to aqueous organic mixtures. As an example of values for a typical alkali halide electrolyte, in Table 13 are listed enthalpies of transfer of (K^+ + Br^-) from water to a number of nonaqueous solvents. Rather surprisingly, all the given values are negative, so that (K^+ + Br^-) is more stable enthalpically in these nonaqueous solvents than in water. The very large solvation enthalpy of the gaseous electrolyte into water, -156.7 kcal mol^{-1}, cannot only be attributed to interactions of the type K^+...OH_2 and Br^-...HOH because the solvation enthalpy is even larger (-163.2 kcal mol^{-1}) in 1,2-dichloroethane where such interactions cannot take place. On the other hand, the ΔG_t^o values are quite well behaved and in general become more positive in value as the solvent becomes less polar. As seen from the listed ΔS_t^o values, it is the very negative entropies of transfer that lead to the positive ΔG_t^o values.

Table 13. Values of ΔH_t^o, ΔG_t^o, and ΔS_t^o for transfer of (K^+ + Br^-), mol fraction scale at 298 K.

Solvent	ΔG_t^o/kcal mol^{-1}	ΔH_t^o/kcal mol^{-1}	ΔS_t^o/cal K^{-1} mol^{-1}
water	0	0	0
methanol	4.0	-3.4	-25
ethanol	6.7	-2.8	-32
PC	6.6	-2.0	-29
DMSO	1.8	-7.5	-31
DMF	4.7	-8.5	-44
Acetone	10.2	-4.9	-51
1,2-DCE	13.7	-6.5	-68
gas phase (1 atm)	142.3	156.7	48

It is possible to account qualitatively and even semi-quantitatively for these observations in terms of electrostatic theories of ionic solvation. The dielectric constant of the solvent surrounding an ion is generally considered to gradually decrease from the bulk dielectric constant, ε_o, at a distance of about 6Å from the centre of a univalent ion to a rather low value of around 2 at the surface of the ion. Abraham and Liszi (9,10,16) have attempted to mimic such a dielectric profile in a mathematically amenable form by constructing a stepwise function in which an ion of given crystallographic radius is surrounded by a solvent layer of dielectric constant ε_ℓ and of thickness, r, the radius of a solvent molecule. Outside this layer is the bulk solvent of dielectric constant ε_o. Not only does such a simple model fit closely to calculated dielectric profiles, but the derived electrostatic equations can be solved to yield mathematically simple expressions for the electrostatic Gibbs energy ΔG_E, entropy ΔS_E, and thence enthalpy ΔH_E. Combination of the electrostatic term with a corresponding neutral term, obtained from values for solvation of the gaseous inert gas of the same size as the ion in question leads to calculated values for the various solvation parameters of a gaseous ion. Using values of ε_ℓ, $\partial \varepsilon_\ell/\partial T$, ε_o, $\partial \varepsilon_o/\partial T$, and r given by Abraham and Liszi (9,10), the ΔG_E, ΔS_E, and ΔH_E values for gaseous ($K^+ + Br^-$) have been calculated (Table 14), with reference to a number of solvents. The neutral term, also given in Table 14, may then be added and the total compared with the observed values. The difference, $\Delta = [\Delta H^{OBS} - (\Delta H_E + \Delta H_N)]$ should be zero if the calculations did indeed include all the interactions that take place; similar Δ-values may be obtained in terms of Gibbs energy or entropy. As found by Abraham and Liszi (9,10) the Δ-values for ($K^+ + Br^-$) in the aprotic solvents are essentially zero, so that no specific interactions need to be invoked. The various observed ΔG_s^o, ΔS_s^o and ΔH_s^o values for solvation of gaseous ($K^+ + Br^-$) in aprotic solvents can be accounted for by the nonspecific neutral term and the general electrostatic continuum term. This is also the case for the ΔG_s^o term in the hydroxylic solvents, so that the calculated solvation free energy, $\Delta G_E + \Delta G_N$, becomes more negative as the solvents become more polar, exactly as observed. In other words, $\Delta G_t^o(K^+ + Br^-)$ for transfer from water to the other solvents is both calculated and observed to be positive.

For water, however, $(\Delta H_E + \Delta H_N)$ is far more negative than the observed solvation enthalpy by some

18 kcal mol^{-1}, with a corresponding Δ-value of no less than 58 cal K^{-1} mol^{-1} in terms of entropy. According to Abraham and Liszi (17,18) these specific terms arise because of a second disordered layer of solvent outside the first solvation layer. In this disordered layer, hydrogen bonds are broken, thus giving rise to an extra positive entropic contribution and a corresponding extra positive enthalpic contribution. Just as in the melting of ice at 273 K, these contributions almost exactly balance, so that there is very little net effect on the Gibbs energy of solvation. Thus for water, although the ΔH_E or ($\Delta H_E + \Delta H_N$) term is more negative than in the other solvents, the incursion of the anomalous contribution from the disordered layer results in ΔH_s° for solvation in water being actually less negative than in much less polar solvents. For methanol, the same effect is present but to a much less extent.

Table 14. Calculations on the solvation of ($K^+ + Br^-$) gaseous ions, in kcal mol^{-1} or cal K^{-1} mol^{-1} at 298 K

	1,2-DCE	acetone	methanol	water
ΔG_E	-137.0	-141.5	-147.6	-151.3
ΔG_N	9.1	8.9	9.7	12.8
Δ	-0.7	0.5	-0.4	1.2
ΔG_s^{OBS}	-128.6	-132.1	-138.3	-142.3
ΔS_E	-95	-70	-64	-48
ΔS_N	-25	-25	-34	-58
Δ	4	-4	25	58
ΔS_s^{OBS}	-116	-99	-73	-48
ΔH_E	-165.2	-162.4	-166.8	-170.6
ΔH_N	1.6	1.3	-0.3	-4.6
Δ	0.4	-0.5	7.0	18.5
ΔH_s^{OBS}	-163.2	-161.6	-160.1	-156.7

It should be noted that although the presentation in Table 14, in terms of the combination ($K^+ + Br^-$), avoids any difficulties due to single-ion assumptions, it does mask any specific cation-solvent interactions, for example. Hence the results in Table 14 should be regarded, for the moment at least, as a qualitative and perhaps semi-quantitative account only. These calculations, however, do provide some explanation of the otherwise rather peculiar ΔH_t^0 values shown in Table 13.

(1) Krishnan, C. V., and Friedman, H. L.: 1976, in *Solute-Solvent Interactions, Vol. 2*, Ed. J. F. Coetzee and C. D. Ritchie, Marcel Dekker, New York.

(2) Parker, V. B.: 1965, *Thermal Properties of Aqyeous Uni-univalent Electrolytes*, U.S. National Bureau of Standards, NSRDS-NBS2, Washington, D.C.

(3) Somsen, G., and Weeda, G.: 1971, *Rec. Trav. Chim.* 90, 81.

(4) Arnett, E. M., Ko, H. C., and Minasz, R. J.: 1972, *J. Phys. Chem.* 76, 2474.

(5) Bury, R., Mayaffre, A., and Chemla, M.: 1976, *J. Chim. Phys.* 73, 935.

(6) Dadgar, A., and Taherian, M. R.: 1977, *J. Chem. Thermodynamics* 9, 711.

(7) Taniewska-Osinska, S., Pietrzynska, B., and Logwinienko, R.: 1980, *Canad. J. Chem.* 58, 1584.

(8) Abraham. M. H., Ling, H. C., and Schulz, R. A.: unpublished work.

(9) Abraham, M. H., and Liszi, J.: 1978, *J.C.S. Faraday I* 74, 1604.

(10) Abraham, M. H., and Liszi, J.: 1978, *J.C.S. Faraday I* 74, 2858.

(11) Wu, Y.-C., and Friedman, H. L.: 1966, *J. Phys. Chem.* 70, 501.

(12) Abraham, M. H., Danil de Namor, A. F., Schulz, R. A.: 1977, *J. Solution Chem.* 6, 491.

(13) Abraham, M. H., Danil de Namor, A. F., Schulz, R. A.: 1976, *J. Solution Chem.*, 8, 529.

(14) Abraham, M. H., Ah-Sing, E., Danil de Namor, A. F., Hill, T., Nasehzadeh, A., and Schulz, R. A.: 1978, *J.C.S. Faraday I* 74, 359.

(15) Cox, B. G., Hedwig, G. R., Parker, A. J., and Watts, D. W.: 1974, *Australian J. Chem.* 27, 477.

(16) Abraham, M. H., Liszi, J., and Kristóf, E.: 1982, *Australian J. Chem.* in the press.

(17) Abraham, M. H., and Liszi, J.: 1980, *J.C.S. Faraday I* 76, 1219.

(18) Abraham, M. H., Liszi, J., and Papp, E.: 1982, *J.C.S. Faraday I* 78, 197.

HYDROPHOBIC HYDRATION OF ALKYLAMMONIUM BROMIDES IN AQUEOUS
MIXED SOLVENTS

G. Somsen

Department of Chemistry, Free University,
De Boelelaan 1083, 1081 HV Amsterdam,
The Netherlands

The enthalpies of solution and of transfer of a number of different
alkylammonium bromides in several aqueous mixed solvents will be
discussed. For comparison some enthalpies of solution in nonaqueous mixed solvents are included. The results give information
on the hydrophobic hydration of the alkylammonium ions in water
and the way in which this hydration is influenced by the presence
of cosolvents. Attention will be given to the influence of the
size and the nature of the (substituted) alkyl groups, to the
influence of the temperature and to the role of the cosolvent. In
addition the applicability of a clathrate-like hydration model
will be discussed.

The energetic aspects of solute-solvent interactions can be
studied by calorimetric measurement of the enthalpies of solution
of a certain solute in various solvents. Differences between the
enthalpies of solution in the solvents, i.e. the enthalpies of
transfer, reflect directly the changes in solvation of a solute.
These changes may be due to alterations in the solute-solvent
interactions but also to a change in mutual interactions between
solvent molecules in the neighbourhood of the solute molecule. The
solvents can be pure liquids or mixtures of liquids. In the latter
systems it is possible to change the solvent properties gradually
by the use of two component mixtures with varying composition.
    In this contribution I will focus the attention on the
solvation of tetraalkylammonium bromides in mixtures of water (W)
and N,N-dimethylformamide (DMF). Tetraalkylammonium ions are
well-known "hydrophobic solutes". This means that they are
solvated in a particular way when they are dissolved in water,
which is called "hydrophobic hydration" (1). It involves a shift

in the spatial and orientational correlations between the water molecules in the vicinity of the solute molecules towards a more hydrogen bonded arrangement. In nonaqueous solvents and especially non-hydrogen bonded solvents it is absent. For that reason we have mainly used DMF as cosolvent. DMF is aprotic, its properties are close to those of an ideal dipolar liquid and it has proved to be a suitable reference solvent.

Hydrophobic hydration is also present in mixtures of water and other liquids. However, its influence reduces considerably when the amount of nonaqueous cosolvent increases. Hence, mixtures of water and nonaqueous cosolvents can serve as systems to study hydrophobic hydration. We have done this in a systematic way with regard to a number of tetraalkylammonium bromides in mixtures of water and DMF.

ENTHALPIES OF SOLUTION AND ENTHALPIES OF TRANSFER

All enthalpies of solution discussed in this paper have been measured calorimetrically. Details about the calorimetric procedure, the way in which the measurements were performed and the purification of the solutes and solvents can be found in the original publications (2-6). Each enthalpy of solution is reduced to the standard state and hence equal to the value at infinite dilution. Unless indicated otherwise, all data refer to 298.15 K.

The peculiar solvation behaviour of tetraalkylammonium bromides can be demonstrated by a comparison of their enthalpies of transfer, ΔH_{tr}, with those of alkali bromides. Table I shows that the enthalpies of transfer of both alkali and tetraalkylammonium bromides from DMF to two nonaqueous solvents of different nature, i.e. dimethylsulphoxide, DMSO, aprotic, and N-methyl-

Table I

Enthalpy of transfer (in kJ mol^{-1}) of alkali and tetraalkylammonium bromides from DMF to DMSO, NMF and water at 25 °C

solute	DMF → DMSO	DMF → NMF	DMF → water
LiBr	+6.0	+14.3	+28.7
NaBr	+6.0	+13.5	+30.3
kBr	+5.0	+12.8	+36.2
RbBr	+5.6	+13.6	+35.6
CsBr	+5.0	+13.5	+33.8
Me$_4$NBr	-3.6	+11.0	+ 8.2
Et$_4$NBr	+4.2	+11.7	- 3.2
Pr$_4$NBr	+6.1	+10.9	-14.2
Bu$_4$NBr	+8.3	+10.0	-20.9

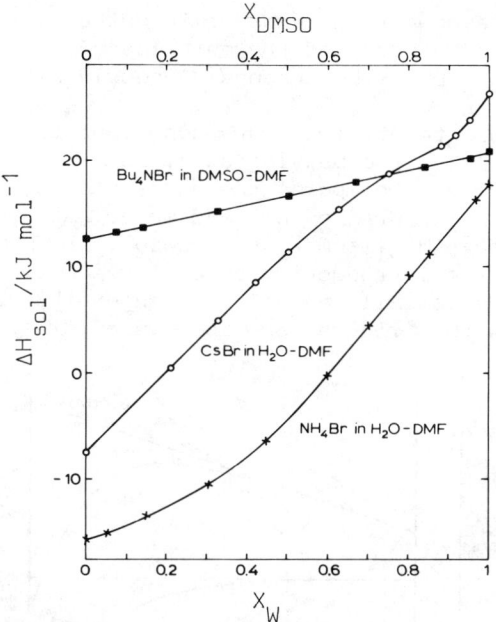

Figure 1. Enthalpies of solution of some bromides in mixed solvents

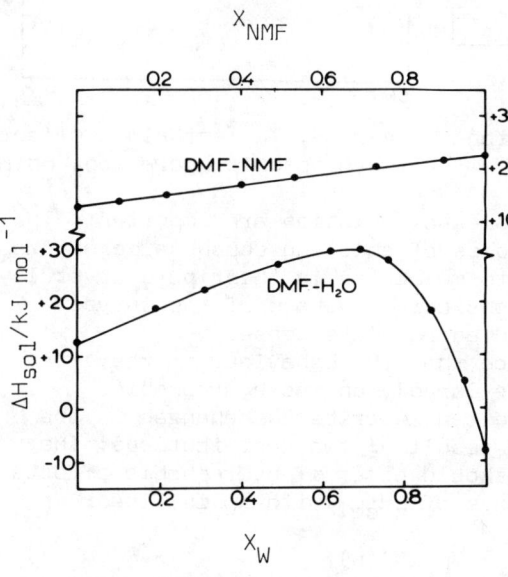

Figure 2. Enthalpies of solution of Bu_4NBr in DMF + NMF and DMF + water

formamide, NMF, strongly hydrogen bonded, are almost equal. On the other hand the enthalpies of transfer from water to DMF differ considerably. For the alkali bromides ΔH_{tr} is virtually constant, but for the tetraalkylammonium bromides ΔH_{tr} deviates from this constant value and more so if the alkyl group is larger. These deviations can be attributed to the hydrophobic hydration of the R_4N^+ ions. Hydrophobic hydration leads to an exothermic contribution to the enthalpy of transfer to water and it can be estimated from Table I that in the case of Bu_4NBr this contribution amounts to ca. -55 kJ mol^{-1}.

The difference between alkali bromides and tetraalkylammonium bromides is also present in aqueous mixed solvents. Figure 1 shows that the enthalpies of solution, ΔH_{sol}, of CsBr and NH_4Br in mixtures of DMF and water change gradually from one pure solvent to the other and almost proportionally to the solvent composition as expressed by the mole fraction of water X_W. The same is true for the enthalpies of solution of Bu_4NBr in the nonaqueous mixture DMSO + DMF. However, the enthalpies of solution of Bu_4NBr in mixtures of water and DMF change completely different. Figure 2 shows that the endothermic enthalpy of solution of Bu_4NBr in pure DMF changes linearly with X_W in DMF + water mixtures up to a more endothermic value at $X_W = 0.7$. At higher mole fraction of water a strong exothermic shift in ΔH_{sol} occurs down to an exothermic enthalpy of solution in pure water. On the other hand the enthalpy of solution of Bu_4NBr in the mixture DMF + NMF changes linearly with solvent composition.

That it is highly reasonable to attribute this behaviour to the Bu_4N^+ ion, or more explicitly to the butyl groups, is demonstrated by figure 3. This figure compares the enthalpies of solution of NH_4Br and $BuNH_3Br$ in mixtures of DMF and water. At low water content the curves relating ΔH_{sol} and X_W are closely parallel. In the more waterrich mixtures the introduction of a single butyl group affects the solvation substantially and also in an exothermic sense. It will be clear that in this figure (and in several others)

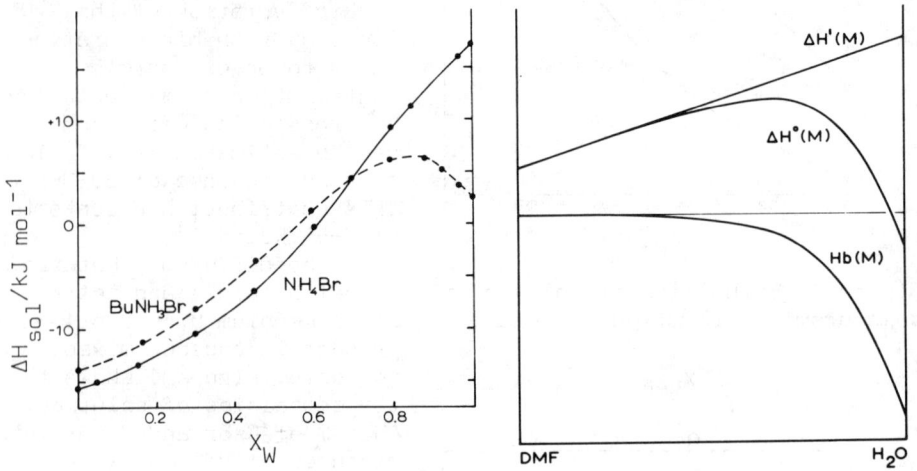

Figure 3. Enthalpies of solution of NH_4Br and $BuNH_3Br$

Figure 4. The enthalpic effect of hydrophobic hydration, Hb(M)

only the trends in the enthalpies of solution are important. The absolute values of the enthalpies of solution depend also on the interactions between the solute molecules in their pure crystalline state. Consequently we consider the shape of the curves only and not their location in an absolute sense.

Starting from the evidence that the behaviour of the enthalpies of solution depends largely on the hydrophobic character of the R_4N^+ ions, we can describe the changes in the enthalpies of solution as the result of two contributions. The first is the solvation which should occur if hydrophobic effects would be absent. Then the change of ΔH_{sol} with X_W is linear as expressed by $\Delta H'(M)$ in figure 4:

$$\Delta H'(M) = (1-X_W) \Delta H^O(DMF) + X_W \Delta H'(W) \qquad (1)$$

The second contribution is the result of hydrophobic hydration. It is exothermic, maximal in pure water and decreases strongly with decreasing water content. It is represented by Hb(M) in figure 4. The sum of both contributions leads to a curve which is shaped as found experimentally and which can be represented by the equation:

$$\Delta H^o(M) = (1-X_W) \Delta H^o(DMF) + X_W \Delta H'(W) + Hb(W) \quad (2)$$

Since in pure water:

$$\Delta H'(W) = \Delta H^o(W) - Hb(W) \quad (3)$$

we can write:

$$\Delta H^o(M) = (1-X_W) \Delta H^o(DMF) + X_W \Delta H^o(W) + Hb(M) - X_W Hb(W) \quad (4)$$

$Hb(W)$ is the enthalpic effect of hydrophobic hydration in pure water and we will use this measure frequently in the following discussions.

SIZE AND NATURE OF THE ALKYL GROUPS

Figure 5 gives the changes of the enthalpies of solution in DMF + water mixtures for five tetraalkylammonium bromides. Only the curves for Pr_4NBr, Bu_4NBr and Pen_4NBr are comparable to the simplified model presented in figure 4. For these compounds the exothermic shifts, and consequently the values of $Hb(W)$, increase from Pr_4NBr to Pen_4NBr, indicating that Pen_4N^+ ions are more hydrophobic than Pr_4N^+ ions. In addition the estimated value of $Hb(W)$ for Bu_4NBr is approximately four times that for $BuNH_3Br$ which can be evaluated from figure 3. This suggests that the alkyl groups are solvated more or less independently and contribute additively to the total hydrophobic effect. We will return on this point later.

For Et_4NBr the shape of the curve is somewhat different from that of Pr_4NBr. Presumably this is due to the influence of the Br^- ion and/or the influence of the charge of the R_4N^+ ion. Certainly the relative effect of the latter will be larger for smaller R_4N^+ ions. Indeed, the curve of Me_4NBr is close to those for alkali bromides, showing that hydrophobic effects do not longer dominate the influence of the charge.

The question of the independency of the solvation of the different alkyl groups can be investigated by studying asymmetric alkylammonium bromides. One example is already presented in figure 3. Other cases are given in figure 6, where we present the results of measurements on Me_3BuNBr and $MeBu_3NBr$ and compare them with the parent compounds

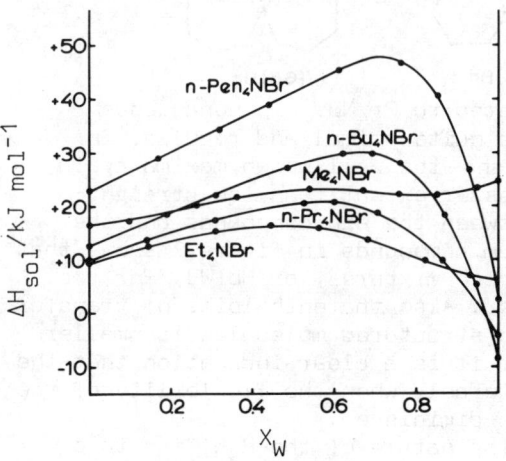

Figure 5. Enthalpies of solution of different R_4NBr

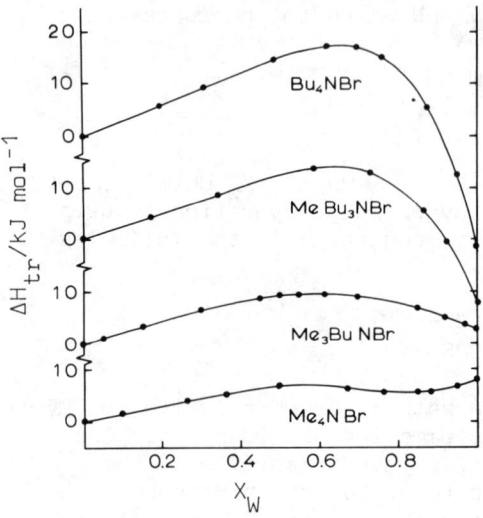

Figure 6. Enthalpies of transfer of some asymmetric R_4NBr

Me_4NBr and Bu_4NBr. The shape of the curves changes gradually from Me_4NBr to Bu_4NBr. A numerical analysis of the experimental results reveals that the curves of the asymmetric compounds can be calculated from those of the parent compounds by application of simple additivity rules. Consequently, the results clearly substantiate the independent solvation and hydrophobic hydration of the different alkyl groups in the solute molecules.

It has been argued (7,8) that the hydrophobic character of an alkyl group is also due to the ability of the alkyl chain to adopt different conformations. In this context we have studied the influence of the flexibility of the alkyl group by means of the so-called azonia-spiro-alkane bromides. These compounds can be regarded as tetraalkylammonium salts in which the side chains are interconnected pairwise. When the resulting cyclic structure contains 4 C atoms we use the abbreviation [4.4]Br. This compound can be compared with Et_4NBr.

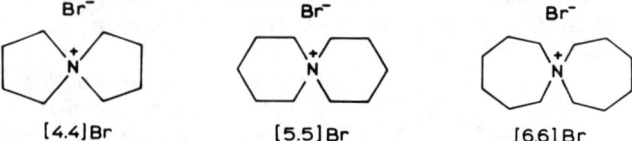

In the same way [6.6]Br is related to Pr_4NBr. In nonaqueous mixtures these compounds behave quite normal and regular. In mixtures of DMF and water they show the well-known maxima again (figure 7), though to a much lesser extent than the straight-chain R_4NBr. The comparison between the azo compounds and the corresponding tetraalkylammonium compounds in figure 7 shows that the exothermic shift in water-rich mixtures, or Hb(W), for the cyclic compounds is much smaller. Also the enthalpies of transfer from DMF to water of the cyclic structured molecules is smaller or even endothermic. Altogether it is a clear indication that the hydrophobicity decreases considerably when the flexibility of the side chains in the R_4N^+ ions is diminished.

It is possible to affect the nature of the R_4N^+ ion to a larger extent by means of substitutions in the alkyl groups. We have done this by introducing terminal hydroxy groups in Et_4NBr. It appears that subsequent introduction of several terminal OH

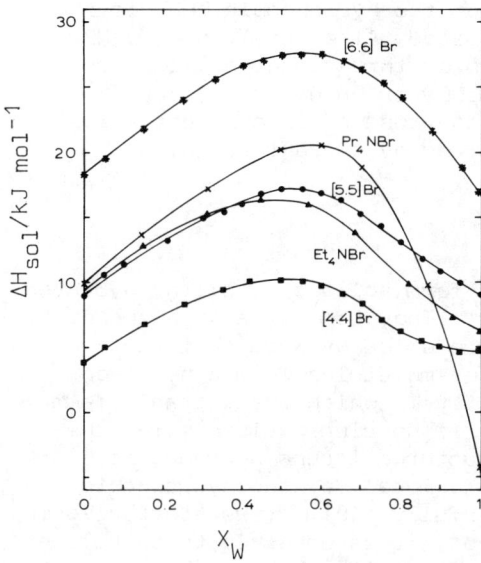

Figure 7. Comparison of ΔH_{sol} of azo compounds with normal tetraalkylammonium bromides

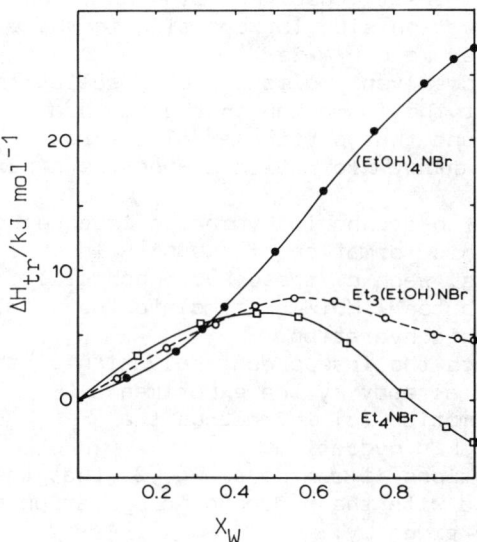

Figure 8. Enthalpies of transfer from DMF for hydroxy-substituted compounds

groups changes the solvation behaviour gradually to that of a non-hydrophobic compound. Figure 8 visualized by means of a comparison of enthalpies of transfer that the $Et_3(EtOH)N^+$ ion is already less hydrophobic than Et_4NBr, whereas the compound with four terminal OH groups, $(EtOH)_4NBr$ is not hydrophobic at all. The curve for the latter compound is close to that for CsBr. Also in this case the idea of independent solvation of the different groups in the ion is supported by the results given in figure 8. The exothermic shift shown by $Et_3(EtOH)NBr$ is approximately 75% of the exothermic shift of Et_4NBr. Results on $Et_2(EtOH)_2NBr$ and $Et(EtOH)_3NBr$, not represented in figure 8, lead to similar conclusions.

From these and previous results it is possible to estimate the effects of individual alkyl groups by considering differences between selected enthalpies of solvation. For example the hydrophobic effect of a single butyl group can be evaluated from a curve representing the difference $\Delta H_{sol}(BuMe_3NBr) - \frac{3}{4}\Delta H_{sol}(Me_4NBr)$ as a function of X_W. In the same way the effect of three ethyl groups can be determined from the differences $Et_3(EtOH)NBr - \frac{1}{4}(EtOH)_4NBr$. Along these lines it is possible to obtain values for the hydrophobic effect of one alkyl group on the basis of the enthalpies of solution of different compounds. They appear to be similar. Hence,

it hardly makes any difference for the hydrophobic behaviour of
an alkyl group whether it is isolated (like butyl in $BuMe_3NBr$
and $BuNH_3Br$) or combined with two or three other similar or
dissimilar groups (butyl in Bu_3MeNBr or Bu_4NBr; ethyl in
$Et_3(EtOH)NBr$ or Et_4NBr). Again, the idea of independent solvation
of the different groups is supported by these results.

A CLATHRATE-LIKE HYDRATION MODEL

Several tetraalkylammonium salts form solid crystalline hydrates
with a well-defined stoechiometry. They contain a high amount of
water. Their structure has been revealed by X-ray diffraction (9).
In such hydrates a number of water molecules form a hydrogen-
bonded network surrounding a cavity in which the tetraalkylammonium ion resides. Although there is no clear evidence for the
existence of rigid clathrate structures around hydrophobic
solutes in water, a clathrate-like model for the hydrophobic
hydration has been applied succesfully (10). If we start also in
our study from a clathrate concept, it is possible to calculate
the enthalpic contribution associated with the hydrophobic
hydration of a R_4N^+ ion in aqueous solvent mixtures, Hb(M), in
the following way.
(a) The clathrate-like cage around a symmetrical R_4N^+ ion is
    divided into four subunits, one for each alkyl group, such that
    there are n water molecules in each subunit.
(b) In a mixture of water and an "inert" cosolvent like DMF, the
    probability that a given solvation site (or position in the
    subunit) is occupied by a water molecule is X_W.
(c) The presence of one or more cosolvent molecules on a solvation
    site in a subunit will prevent the formation of the clathrate-
    like hydration structure around that particular alkyl group
    only. It does not affect the subunit formation around the other
    alkyl groups.
(d) The enthalpic contribution of hydrophobic hydration in aqueous
    solvent mixtures is due to this formation of subunits. In the
    case of symmetrical R_4N^+ ions, each of these hydrophobically
    hydrated alkyl groups contributes $\frac{1}{4}$ Hb(W) per mol to the total
    enthalpic effect of hydrophobic hydration.
Assumptions (a) and (c) represents the independent solvation of the
different alkyl groups suggested already by the experimental
results in figures 6 and 8. Assumption (c) introduces the
cooperative character of hydrophobic hydration.
   As a result of these assumptions it can be shown (2) that the
enthalpic contribution associated with the hydrophobic hydration in
mixtures of water and DMF can be given by:
$$Hb(M) = X_W^n Hb(W) \tag{5}$$
Combination with eq. (4) gives:
$$\Delta H^o(M) = (1-X_W)\Delta H^o(DMF) + X_W\Delta H^o(W) + (X_W^n - X_W)Hb(W) \tag{6}$$
In this way the enthalpy of solution of a hydrophobic compound

in mixtures of water and DMF can be expressed in the enthalpies of solution in the pure liquid components by means of two adjustable parameters, Hb(W) and n. However, it must be kept in mind that this approach is an oversimplification of the real situation and does not pretend to give an unique representation of the experimental data. We have shown (6) that it does lead to different results when one of the basic assumptions (i.e. assumption b) was modified. However, it appeared that only the values of n are affected. In good approximation the values of Hb(W) are independent of the model. Consequently Hb(W) is the most significant parameter. The physical meaning of the values of n are doubtful, although trends in n seem to bear some significance indeed.

Application of eq. (6) to the relevant experimental results has proved to be very succesful. Least squares analysis of the experimental data leads to a good fit with eq. (6), especially for the compounds with larger alkyl groups. An impression is given in Table II which lists the ensuing values of the parameters Hb(W) and n and the mean deviations between calculated and experimental values of the enthalpies of solution. This Table shows that for the different R_4NBr Hb(W) increases with the size of the alkyl

Table II

The enthalpic effect of hydrophobic hydration, Hb(W), the number, n, of hydrating water molecules per alkyl group and the mean deviation, δ, between calculated and experimental enthalpies of solution in DMF-water mixtures of several hydrophobic compounds

solute	Hb(W)/kJ mol^{-1}	n	δ/kJ mol^{-1}
Et_4NBr	-29.3	(2.3)[a]	-
Pr_4NBr	-39.6	4.2	0.17
Bu_4NBr	-52.8	6.4	0.23
Pen_4NBr	-58.0	8.6	0.37
[4.4]Br	-23.8	(1.9)[a]	-
[5.5]Br	-25.2	(2.3)[a]	-
[6.6]Br	-27.3	2.8	0.18

[a] based on data with X_W <0.75 only, due to poor fit for data with large X_W

groups. It can be shown that the height of the maxima in figure 5 is closely related to the value of Hb(W). Also the value of n increases with the size of the alkyl group. However, the magnitudes of n are far too small to be physically realistic. For the cyclic compounds eq. (6) applies only satisfactory to the results of [6.6]Br. Its value for Hb(W) clearly shows its decreased hydrophobicity. Poor fits are obtained with the results for Et_4NBr, [4.4]Br and [5.5]Br. In these cases, a reasonable estimation is possible for the values of Hb(W) only, not for those of n. This is meanly due to the shape of the curves at higher mole fraction of water. Apparently the clathrate-like hydration model is unable to describe the behaviour of these compounds satisfactorily. This is not surprising since the model neglects the influence of charge and Br^- ion completely. In addition the geometrie shape of the spiro ions is hardly compatible with a clathrate-like environment in the sense of the model. Indeed, solid clathrate hydrates of these compounds have never been identified.

INFLUENCE OF THE TEMPERATURE

Measurements of the enthalpies of solution of Bu_4NBr in mixtures of water and DMF show that between 5 and 55 °C the curves relating ΔH_{sol} and X_W have the same general shape as that given in figure 2. At six temperatures in this interval all enthalpies of solution increase from $X_W = 0$ to a value of X_W around 0.7. Then they decrease substantially to a lower value in pure water. However, the magnitude of the exothermic shift becomes smaller with increasing temperature. Consequently, the hydrophobic effect decreases with increasing temperature. At each temperature the experimental results can be described well by eq. (6). The results of a least squares analysis in terms of this equation is given in Table III. It appears that the parameter Hb(W) decreases from -65 kJ mol^{-1}

Table III

The enthalpic effect of hydrophobic hydration, Hb(W), the number, n, of hydrating water molecules per butyl group and the mean deviation, δ, between calculated and experimental values of ΔH_{sol} of Bu_4NBr in DMF + water mixtures of different temperatures

T/K	Hb(W)/kJ mol^{-1}	n	δ/kJ mol^{-1}
278.15	-65.6	6.0	0.38
288.15	-59.7	6.1	0.31
298.15	-52.8	6.4	0.23
308.15	-46.3	6.4	0.19
318.15	-40.5	6.5	0.12
328.15	-34.4	6.7	0.15

at 5 °C to -34 kJ mol^{-1} at 55 °C. An approximate extrapolation suggests that it will be zero around 130 °C. The temperature change does not affect the value of n. In terms of the model this means that the number of hydrating water molecules remains constant. Apparently, the change in Hb(W) is not due to a reduction of the number of hydrogen bonds in the clathrate-like

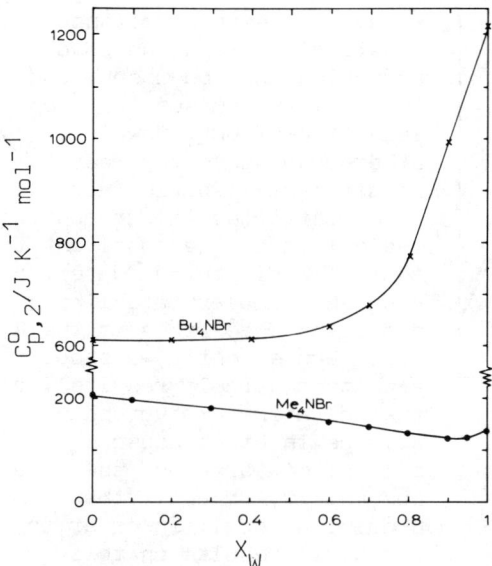

Figure 9. Partial molar heat capacities of Me_4NBr and Bu_4NBr in DMF + water mixtures.

subunit, but rather to a loosening of the hydrogen bonds leading to a smaller hydrogen-bonding energy at higher temperatures. This may mean that the subunit becomes more flexible at higher temperatures.

By means of the so-called "integral enthalpy of solution" method (11) it is possible to calculate partial molar heat capacities from enthalpies of solution at different temperatures. For Bu_4NBr the results are shown in figure 9. In pure DMF $C_{p,2}^o$ is approximately + 600 $JK^{-1} mol^{-1}$. It remains rather constant up to $X_W = 0.6$ and starts to rise with higher X_W to a much higher value in pure water: around + 1200 $JK^{-1} mol^{-1}$. In contrast to this large change, $C_{p,2}^o$ of Me_4NBr does not depend very much on the mole fraction of water. Generally the magnitude of the partial molar heat capacity at infinite dilution can be regarded as being the sum of two contributions (12,13). The first is the intrinsic heat capacity of the solute particle, $C_{p,2}(intr)$, which is the heat capacity when solute-solvent interactions are absent. The second contribution is the result of the influence of the solute on the solvent, $C_{p,2}(solv)$. In this approach the values of $C_{p,2}^o$ in figure 9 reflect the change in $C_{p,2}(solv)$. The constant value of $C_{p,2}^o$ of Bu_4NBr up to $X_W = 0.6$ is very close to the intrinsic heat capacity of Bu_4NBr. Hence, in these mixtures $C_{p,2}(solv) = 0$ and hydrophobic hydration is absent. Beyond $X_W = 0.6$ $C_{p,2}(solv)$ becomes significant and in pure water its value is ca. 600 $JK^{-1} mol^{-1}$. This result can be easily reconciled with the hydration model of the previous section. Using the observation that n is virtually independent of temperature, the model equation (eq. 6) can be applied to derive an equation which relates the partial molar heat capacity of a hydrophobic solute in a solvent mixture with the heat capacities in the pure solvents and the parameter n (14).

THE INFLUENCE OF THE COSOLVENT

In the discussions of the preceding sections we did not give much attention to the role of the cosolvent in the aqueous mixtures.

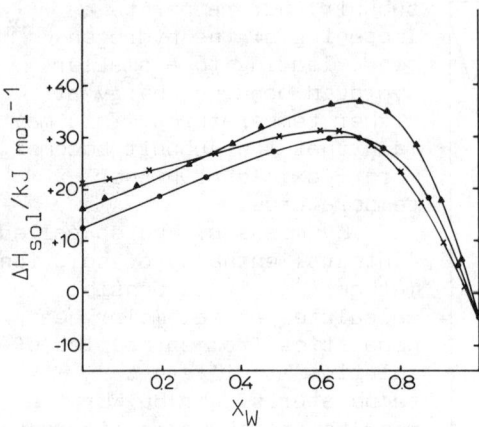

Figure 10. Enthalpies of solution of Bu$_4$NBr in binary solvents
●, DMF-H$_2$O; ×, DMSO-H$_2$O; ▲, DMA-H$_2$O

We only stressed that specific interactions with the cosolvent molecules must be absent and that was the reason for the selection of DMF. However, other cosolvents may meet this requirement also. In fact, we have restricted in our approach the effect of the cosolvent to that of a diluent of water. Also in the clathrate-like hydration model this is the case. Consequently we should expect comparable results for the solvation of hydrophobic solutes in other aqueous solvent mixtures and the same possibility to apply the hydration model. In order to check our results on this particular point we have measured enthalpies of solution of Bu$_4$NBr in mixtures of water with DMSO and N,N-dimethylacetamide (DMA). Both nonaqueous solvents are aprotic. Specific interactions between the solvent molecules in the pure liquid state are small as demonstrated by the Kirkwood correlation factors which are close to unity. The experimental enthalpies of solution of Bu$_4$NBr in three aqueous solvent systems are shown in figure 10. The usual pattern is followed in all three systems. However, there are differences. They emerge clearly when we apply the hydration model approach and analyze the enthalpies of solution in terms of eq. (6). Values of the parameters Hb(W) and n, together with the mean deviations of the curves are given in Table IV. The mean deviations show that the description of the experimental data with the hydration model is less accurate in water-DMSO and

Table IV

The enthalpic effect of hydrophobic hydration, Hb(W), the number, n, of hydrating water molecules per alkyl group and the mean deviation, δ, between calculated and experimental enthalpies of solution of Bu$_4$NBr in mixtures of water and three aprotic solvents

solvent system	Hb(W)/kJ mol^{-1}	n	δ/kJ mol^{-1}
DMF-water	-52.8	6.4	0.23
DMSO-water	-51.5	5.1	1.12
DMA-water	-59.9	7.0	0.64

water-DMA mixtures than in water-DMF mixtures. In addition, the values of the parameters Hb(W) and n are not, as the model requires, completely independent of the nature of the cosolvent. However, in view of the approximative character of the model, the results are not unreasonable and confirm a general picture.

When protic cosolvents like formamide or N-methylformamide are used, the deviations from the model description become much more pronounced. The same is true when preferential solvation occurs, for example in mixtures of water with acetonitrile and ethylene carbonate. Then strong deviations are present in mixtures with a low water content, probably due to a preferential solvation of the solute particles by water. However, in all aqueous mixtures the same exothermic shift of the enthalpies of solution of hydrophobic solutes is found in the water-rich mixtures, indicating the strong influence of the hydrophobic hydration in pure water and highly aqueous solvent mixtures [14].

ACKNOWLEDGEMENT

Figures 5 and 10 are reprinted with permission of the American Chemical Society from ref. 4.

REFERENCES

1. "Water, a Comprehensive Treatise", Franks, F.: editor, Plenum Press, New York, 1973, vol. 1, 2 and 3.
2. De Visser, C., and Somsen, G.: 1974, J. Phys. Chem. 78, pp. 1719-1722.
3. Heuvelsland, W.J.M., and Somsen, G.: 1976, J. Chem. Thermodynamics 8, pp. 873-880.
4. Heuvelsland, W.J.M., De Visser, C., and Somsen, G.: 1978, J. Phys. Chem. 82, pp. 29-32.
5. Heuvelsland, W.J.M., De Visser, C., Somsen, G., LoSurdo, A., and Wen-Yang Wen: 1979, J. Solution Chem. 8, pp. 25-34.
6. Heuvelsland, W.J.M., Bloemendal, M., De Visser, C., and Somsen, G.: 1980, J. Phys. Chem. 84, pp. 2391-2395.
7. Wen-Yang Wen and Kaatze, U.: 1977, J. Phys. Chem. 81, pp. 177-181.
8. LoSurdo, A., Wen-Yang Wen, Jolicoeur, C., and Fortier, J.-L.: 1977, J. Phys. Chem. 81, pp. 1813-1817.
9. Wen-Yang Wen, in: "Water and aqueous solutions", Horne, R.A.: editor, Wiley-Interscience, New York, 1972, pp. 613-661.
10. Mastroianni, M.J., Pikal, M.J., and Lindenbaum, S.: 1972, J. Phys. Chem. 76, 3050-3057.
11. Criss, C.M., and Cobble, J.W.: 1961, J. Am. Chem. Soc. 83, 3223-3228.
12. Philip, P.R., and Desnoyers, J.E.: 1972, J. Solution Chem. 1, 353-367.

13. De Visser, C., and Somsen, G.: 1973, J. Chem. Soc. Faraday Trans. I, 69, 1440-1447.
14. Heuvelsland, W.J.M.: .Thesis, Free University, Amsterdam, 1980.

SOLVATION OF ALCOHOLS, AMINES, UREAS AND AMIDES IN MIXED SOLVENTS

G. Somsen

Department of Chemistry, Free University,
De Boelelaan 1083, 1081 HV Amsterdam,
The Netherlands

The solvation and (hydrophobic) hydration of a number of organic model compounds of different type in mixed solvent systems is studied by means of enthalpies of solution and enthalpies of transfer. The solvents are mostly mixtures of water and N,N-dimethylformamide. In this way information can be obtained about the peculiar solvation of several organic molecules in water. The compounds studied comprise aliphatic alcohols; primary, secondary, tertiary and cyclic amines; alkyl substituted ureas and several (substituted) amides. It appears that in good approximation the different parts of the molecules are independently solvated. The experimental results will be discussed also in terms of a clathrate-like hydration model.

Quite some time ago it has been recognized that the solvation of organic solutes in water is different from that in other solvents (1). The cause of this difference is the presence of weakly interacting alkyl groups in most organic compounds. The different behaviour manifests itself also in the thermodynamic properties of solutions of organic compounds. In this contribution we will focus the attention on the enthalpies of solution. Organic solutes bearing alkyl groups are hydrated in pure water in a peculiar way. Around the alkyl groups a (dynamic) structure is induced which is different from that in bulk water. Generally spoken, the amount of hydrogen bonding between the water molecules is increased. This type of solvation is called "hydrophobic hydration". It is present also in mixtures of water and other solvents, but only when the water content of the mixture is high. The way in which it changes with the water content of the mixtures provides information on the nature of hydrophobic hydration. Since specific short range structural effects between solvent molecules are absent in non-hydrogen-bonded or aprotic solvents we have studied solvation effects mainly in mixtures of water and N,N-dimethylformamide (DMF) and used the latter as a reference solvent. As solutes we have selected a number of organic "model" compounds of different type: aliphatic alcohols, amines (primary, secondary as well as tertiary amines), substituted ureas and substituted amides. Data on the solvation of these model compounds may be used to evaluate similar data on the solvation of other and larger organic molecules. The solution properties of large organic molecules are assumed to be important. For instance, the solvation of the alkyl groups in proteins may play a role in the stabilization of their three-dimensional (tertiary) structure in water.

The enthalpies of solution discussed in this paper have been determined with an LKB 8700 calorimetric system. The experimental procedure and test of the calorimeter have been reported (2). Details about the purification of the solutes and solvents and the primary data can be found in the original papers (3-5). The enthalpies of solution given in this contribution have been obtained by application of necessary corrections to the primary data and can be considered as standard enthalpies of solution, i.e. enthalpies of solution of infinite dilution. They all refer to a temperature of 298.15 K.

ALCOHOLS

The variation in the solvation of a certain compound in different solvents is reflected directly by the changes in the enthalpies of solution or the enthalpies of transfer from a chosen reference solvent, in our case DMF. The results in Table I show that the

Table I

Enthalpies of solution, ΔH_{sol}, (in kJ mol^{-1}) of n-BuOH in different mixed solvents

DMF + DMSO		DMF + NMF	
X_{DMSO}	ΔH_{sol}	X_{NMF}	ΔH_{sol}
0	+3.03	0	+3.03
0.121	+2.89	0.101	+3.10
0.300	+2.97	0.300	+3.20
0.502	+3.09	0.500	+3.27
0.700	+3.44	0.701	+3.22
0.900	+3.82	0.900	+3.21
1	+4.12	1	+3.12

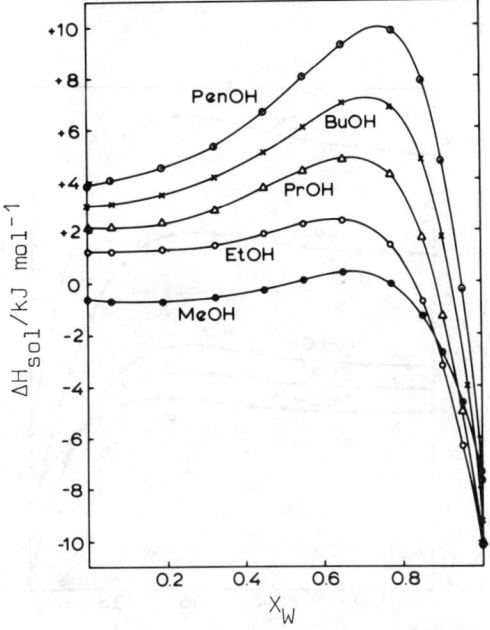

Figure 1. Enthalpies of solution of normal alcohols in DMF + water

enthalpies of solution of butanol (n-BuOH) in mixtures of DMF and an organic cosolvent change to a good approximation proportionally to the mole fraction of the cosolvent. The nature of the cosolvent seems to be of minor importance. The enthalpies of solution in mixtures of DMF and aprotic dimethylsulphoxide (DMSO) as well as those in mixtures of DMF and strongly hydrogen bonded N-methylformamide (NMF) show only small changes, mostly smaller than 1 kJ mol^{-1}. Probably the solvation of n-BuOH changes very gradually from one pure solvent to the other.

In contrast to the behaviour in nonaqueous solvents, the enthalpies of solution of n-BuOH in mixtures of DMF and water do vary much more, and do not change proportionally to the mole fraction of water, X_W. Figure 1 shows that the dependence on solvent composition is very pronounced and even characteristic for all primary alcohols which we have investigated. The enthalpies of solution increase gradually in mixtures with X_W smaller than 0.6, but as soon as X_W becomes larger, substantial changes occur and the enthalpies of solution shift in an exothermic sense towards values which are much more negative in pure water. This effect becomes more pronounced for alcohols with a larger number of carbon atoms. From our results on isomeric alcohols (see figures 2 and 3) it follows that the exothermic shift for n-BuOH and 2-methylpropanol (i-BuOH) is

Table II

Enthalpies of transfer, ΔH_{tr}, from DMF to water (in kJ mol^{-1}) of several alcohols

solute	ΔH_{tr}(DMF → W)	solute	ΔH_{tr}(DMF → W)
MeOH	− 6.75	i-BuOH	−12.40
EtOH	−11.46	s-BuOH	−16.63
n-PrOH	−12.31	t-BuOH	−20.81
i-PrOH	−16.11	n-PenOH	−11.53
n-BuOH	−12.28	t-PenOH	−21.35

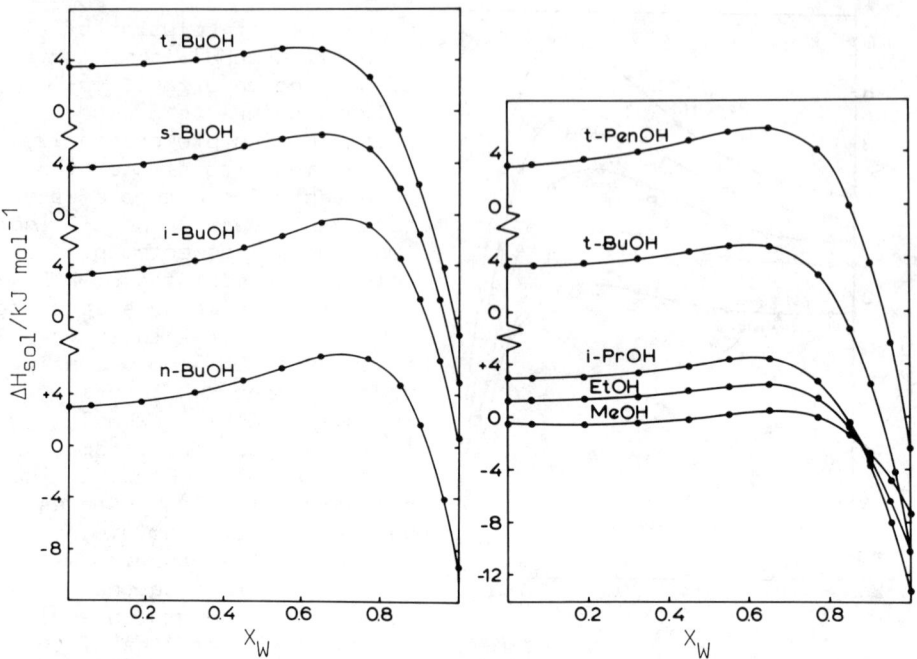

Figures 2 and 3. Enthalpies of solution of isomeric butanols (left) and of alcohols of different type (right) in DMF + water

approximately equal, but that it is larger for secondary and tertiary butanol (s-BuOH and t-BuOH, respectively). This is reflected also in the enthalpies of transfer, ΔH_{tr}, from DMF to water. Table II shows that the secondary and tertiary alcohols exhibit substantially more negative enthalpies of transfer than the corresponding primary ones, but that the difference between n-BuOH and i-BuOH is negligible. When we consider the enthalpies of solution as a function of the number of carbon atoms, we see that ΔH_{tr} of the series MeOH, EtOH, i-PrOH and t-BuOH depends linearly on this number. The same is true for the series n-PrOH, s-BuOH and t-PenOH. Apparently the subsequent substitutions of α-H atoms by methyl groups cause equal changes (ca. 4.7 kJ mol^{-1}) in the enthalpies of transfer. This effect is much larger than that of any β-substitution. Due to these regularities it is possible to relate the enthalpies of the non-primary higher alcohols to those of the lower primary ones. For example, in the case of t-PenOH the following equation applies:

$$\Delta H_{tr}(\text{t-PenOH}) = \Delta H_{tr}(\text{n-PrOH}) + 2[\Delta H_{tr}(\text{EtOH}) - \Delta H_{tr}(\text{MeOH})] \qquad (1)$$

It appears that a similar equation can be used to represent the enthalpies of transfer from DMF to mixtures of DMF and water. We conclude that for the alcohols the differences in solvation are mainly caused by the independent solvation of each alkyl group attached to the α-C-atom.

Table III

The enthalpic effect Hb(W) of hydrophobic hydration (in kJ mol^{-1}) the number n of water molecules involved in the hydration and the mean deviation δ (in kJ mol^{-1}) between calculated and experimental values of ΔH_{sol} for several alcohols

solute	Hb(W)	n	δ	solute	Hb(W)	n	δ
MeOH	-8.1	8.9	0.18	i-BuOH	-18.6	9.3	0.48
EtOH	-13.4	7.7	0.24	s-BuOH	-21.2	8.1	0.28
n-PrOH	-16.4	8.5	0.30	t-BuOH	-24.1	7.0	0.29
i-PrOH	-18.8	7.2	0.27	n-PenOH	-19.3	11.2	0.47
n-BuOH	-18.0	10.1	0.37	t-PenOH	-26.4	7.8	0.30

The independent solvation of the different parts of a molecule and especially that of alkyl groups is one of the basic assumptions of a solvation model which can be applied to the solvation, or more specifically the hydrophobic hydration, of organic compounds (2,6,7). This clathrate-like hydration model is described in more detail in our first contribution to this course. It leads to an equation relating the enthalpies of solution of hydrophobic compounds in aqueous solvent mixtures to those in the pure solvent components, the mole fraction of water, the enthalpic contribution of hydrophobic hydration in pure water, Hb(W), and the number of water molecules hydrating one alkyl group, n,:

$$\Delta H_{sol}(M) = (1-X_W)\Delta H_{sol}(DMF) + X_W \Delta H_{sol}(W) + (X_W^n - X_W)Hb(W) \qquad (2)$$

We have applied this equation to the experimental results of the alcohols, using Hb(W) and n as adjustable parameters. The fit results in values presented in Table III, together with the mean deviations, δ, between calculated and experimental values of $\Delta H_{sol}(M)$. Table III shows that the experimental values of ΔH_{sol} can be represented by eq. (2) reasonably. Like the enthalpies of transfer, Hb(W) for secondary and tertiary alcohols is more negative than that for primary alcohols. In fact Hb(W) follows the same pattern as the enthalpies of transfer with respect to the number of carbon atoms. This is visualized in figure 4. Hb(W) increases with the number of C-atoms. For n-alcohols the increase levels off for longer chains.

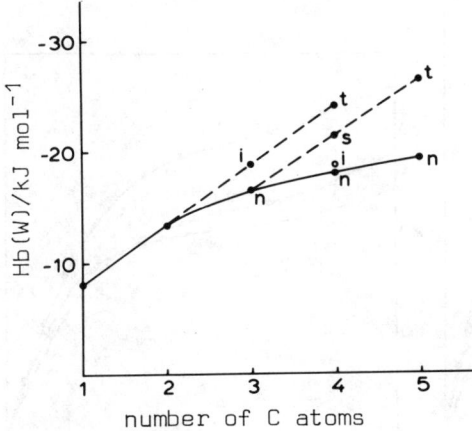

Figure 4. The enthalpic effect of hydrophobic hydration, Hb(W), of alcohols as a function of the number of carbon atoms

However, for alcohols with an increasing number of α-C atoms the relation is linear in good approximation. Again, the independent hydration of the alkyl groups is stressed. The values of n in Table III refer to alkyl groups attached to the α-C atoms of the alcohols. In some cases (s-BuOH and t-PenOH) n represents an average over different alkyl groups. For this reason values of n for secondary and tertiary alcohols are lower than those for the corresponding primary ones. For i-PrOH and t-BuOH n is related to the hydration cage around a methyl group. In consistency with the model they are very close.

AMINES

In alcohols the polar group is always an OH group, which remains virtually outside the part of the molecule consisting of alkyl groups. In contrast to this, the polar part of amines can be changed by substitution of the N hydrogen atoms leading to a more enclosed polar group in secondary and tertiary amines. For this reason, we have studied several primary amines: n-butylamine (n-BuNH$_2$), t-butylamine (t-BuNH$_2$), n-hexylamine (n-HexNH$_2$); a number of secondary amines: diethylamine (Et$_2$NH), di-n-propylamine (n-Pr$_2$NH), di-i-propylamine (i-Pr$_2$NH), di-n-butylamine (n-Bu$_2$NH), n-butylmethylamine (n-BuMeNH); and two tertiary amines: triethylamine (Et$_3$N) and tri-n-propylamine (n-Pr$_3$N). In addition we have determined enthalpies of solution of two cyclic amines: pyrrolidine, (CH$_2$)$_4$NH, and perhydroazepine, (CH$_2$)$_6$NH. All data have been

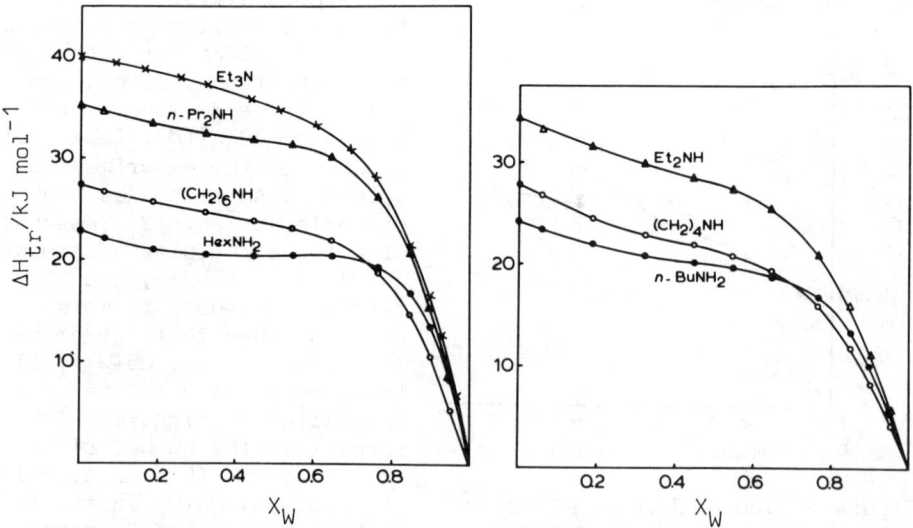

Figures 5 and 6. Enthalpies of transfer from water to mixtures of DMF + water for amines with 6 C-atoms (left) and 4 C-atoms (right)

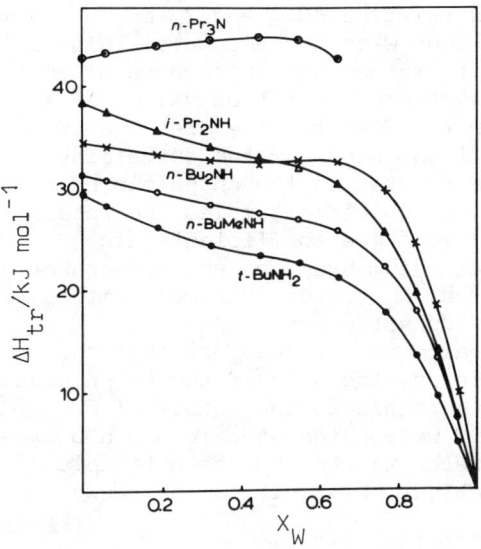

Figure 7. Enthalpies of transfer from water to mixtures of DMF + water for different amines

corrected for partial hydrolisis in water and aqueous mixtures.

In the nonaqueous solvent mixture NMF + DMF the enthalpy of solution of Et_3N changes approximately linear with solvent composition as expressed by the mole fraction of the components. As with the alcohols we consider this gradual change as an indication for the absence of specific solvation effects. In mixtures of DMF and water the enthalpies of solution decrease considerably from such linear dependence. In figures 5, 6 and 7 we have plotted the enthalpies of transfer, ΔH_{tr}, from water to mixtures of DMF + water for different amines with respect to the mole fraction of water, X_W.

They show the same trends as the enthalpies of solution. In figure 5 we present curves of mono-, di-, tri- and cyclic amines with 6 C atoms. All curves show the typical feature of hydrophobic hydration: after a more or less linear change of ΔH_{tr} at values of X_W below 0.6, an exothermic shift occurs for larger values of X_W. Hence, for X_W smaller than 0.6 the solvent mixture behaves essentially nonaqueous. A direct comparison of the enthalpies of transfer of the different amines, as has been done for the alcohols, is hampered by the different nature of the polar amino group in primary, secondary and tertiary amines. Contrary to the alcohols the number of protons varies and this will affect the value of ΔH_{tr}. However, in spite of the decreasing number of protons in the series $HexNH_2$, $n-Pr_2NH$, Et_3N, values of ΔH_{tr} increase from primary to secondary and tertiary amines. A similar conclusion can be drawn from figure 6 with regard to amines with 4 C atoms. Hence the effect of two or three ethyl groups is much larger than that of one butyl or hexyl group, respectively. Comparing $t-BuNH_2$ and $i-Pr_2NH$ (figure 7) with $n-BuNH_2$ (figure 6) and $n-Pr_2NH$ (figure 5) leads to the conclusion that ΔH_{tr} is also enhanced by branching of the alkyl groups. We have found similar influences with the alcohols. Finally the cyclic amines $(CH_2)_4NH$ and $(CH_2)_6NH$ show smaller exothermic shifts than the corresponding dialkylamines. This effect has been observed for cyclic ammonium bromides (8) and can be connected with the reduced hydrophobicity of cyclic compounds due to their limited ability to adopt favourable conformations in liquid water.

Except for n-Pr$_3$N the curves relating ΔH_{tr} and X_W do not show the endothermic maxima which we found with the alcohols. This difference is not fundamental. Rather than the appearance of an endothermic maximum, the strong exothermic shift at values of X_W larger than 0.6 is the characteristic feature. The presence of a maximum is largely due to the limiting slope of the enthalpies of transfer if $X_W \to 0$. This slope is related to the enthalpic pair interaction coefficients between the solute molecules and water molecules in the solvent DMF (9,10). These coefficients are strongly positive for tetraalkylammonium bromides and consequently give rise to maxima in the ΔH_{sol} vs. X_W curves. For most amines they are negative and thus maxima do not occur.

Another difference with the alcohols is the fact that the amines do not show the same simple relationship of the enthalpies of transfer from water to DMF with regard to the number of C atoms. The reason will be clear if we consider the relation between the enthalpies of transfer from water to DMF and the enthalpic contribution of hydrophobic hydration, Hb(W). It can be shown that

$$\Delta H_{tr}(W \to DMF) = -Hb(W) - \left(\frac{\partial \Delta H_{sol}}{\partial X_W}\right)_{X_W \to 0} \quad (3)$$

in which the last term represents the limiting slope of ΔH_{sol} with respect to X_W in DMF-rich mixtures. For the alcohol series MeOH, EtOH, i-PrOH, t-BuOH this limiting slope is very small. Consequently:

$$\Delta H_{tr}(DMF \to W) = -\Delta H_{tr}(W \to DMF) \cong Hb(W) \quad (4)$$

sothat the enthalpies of transfer show the same trend as (and are nearly equal to) the values of Hb(W). For the amines the limiting slopes are not negligible. Hence, the enthalpies of transfer differ from Hb(W) and do not show a regular trend.

We have analyzed our data in terms of the clathrate-like hydration model with the aid of eq. (2). The resulting values of the parameters Hb(W) and n are presented in Table IV. As for the alcohols the values of Hb(W) of the di- and trialkylamines are the sum of contributions of separate alkyl groups (compare n-BuNH$_2$ with n-Bu$_2$NH; Et$_2$NH with Et$_3$N). This agrees with the model assumption of independent solvation of the alkyl groups. Also values of Hb(W) for the cyclic amines (CH$_2$)$_4$NH and (CH$_2$)$_6$NH are smaller than those of the corresponding dialkylamines Et$_2$NH and n-Pr$_2$NH. We have found earlier that cyclic molecules seem to be less compatible with a clathrate-like structure in paper (8). For the alkyl groups the value of Hb(W) increases with the number of C atoms. The increase is linear for alkyl groups formed by subsequent substitution of α-H atoms by methyl groups, i.e. in the series Me, Et, i-Pr and t-Bu. However, the CH$_2$ increment is smaller than that for alcohols. Apparently the nature of the polar group influences the magnitude of Hb(W). Since the values of Hb(W) of n-alkyl groups with more than two C-atoms show the same CH$_2$ increment as the alcohols, the influence of the polar group does

Table IV

The enthalpic effect Hb(W) of hydrophobic hydration (in kJ mol^{-1}), the number n of hydrating water molecules per alkyl group and the mean deviation δ (in kJ mol^{-1}) between calculated and experimental values of ΔH_{tr} for several amines

solute	Hb(W)	n	δ
Et$_2$NH	-22.3	6.0	0.40
Et$_3$N	-31.6	6.2	0.13
n-Pr$_2$NH	-28.7	7.1	0.33
n-Pr$_3$N	(-45.0)[a]	-	-
i-Pr$_2$NH	-27.8	6.3	0.44
n-BuNH$_2$	-15.7	8.4	0.36
n-Bu$_2$NH	-31.8	8.3	0.54
t-BuNH$_2$	-16.8	6.8	0.53
HexNH$_2$	-18.3	11.8	0.45
(CH$_2$)$_4$NH	-15.1	6.4	0.44
(CH$_2$)$_6$NH	-20.4	6.5	0.30
Me[b]	- 8.1	5.0	0.06

[a] Estimated; [b] Me = n-BuMeNH - ½ n-Bu$_2$NH

not extend beyond the second C-atom. Probably the difference between the alcohols and the amines in this respect is due to a difference in hydrophobic hydration of the alkyl groups adjacent to the polar groups.

UREAS AND AMIDES

In this contribution we will use the following abbreviations for the different ureas and amides which we have investigated: urea: U; methylurea; MU; 1,1 dimethylurea: 1,1-DMU; 1,3-dimethylurea: 1,3-DMU; tetramethylurea: TMU; ethylurea: EU; formamide: FA; N-methylformamide: NMF; N,N-dimethylformamide: DMF; acetamide: AA; N,N-dimethylacetamide: DMA; N-n-butylacetamide: NBA; butyramide: BA.

The results of the measurements of the enthalpies of solution of the different compounds are represented by means of the enthalpies of transfer from water to mixtures of DMF + water in figures 8 and 9. The curves in these figures show the characteristic behaviour of hydrophobic compounds, except urea (U) and formamide (FA). Before analyzing them in more detail, we will focus the attention to the enthalpies of transfer from water to pure DMF. When we compare ΔH_{tr}(W → DMF) of U and MU, we see that the introduction of one methyl group causes a shift of +9.67 kJ mol^{-1}. This is higher than the effect of introducing a second methyl group on the same N atom to form 1,1-DMU, which is +7.31 kJ mol^{-1}.

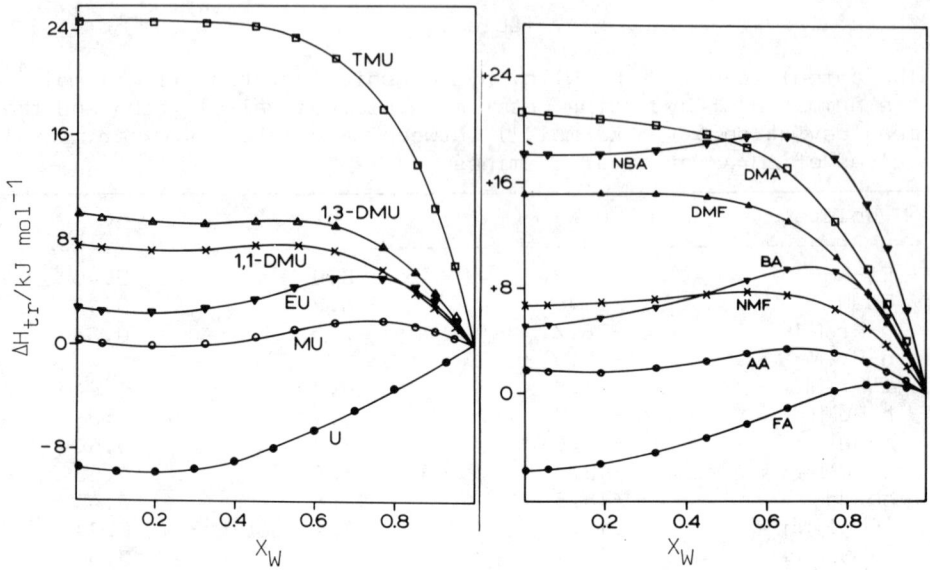

Figures 8 and 9. Enthalpies of transfer from water to mixtures of DMF + water for (substituted) ureas (left) and amides (right)

However, introduction of a second methyl group on the other N atom (to form 1,3-DMU) results in a shift (+9.88 kJ mol^{-1}) which is close to that of the first methyl group. Apparently the NH$_2$ and NHCH$_3$ groups on either side of the molecule are solvated independently. This should imply that the enthalpy of transfer of TMU will be larger than that of U by 9.67 + 9.88 + 7.31 + 7.31 = 34.17 kJ mol^{-1}. Experimentally we find 34.13 kJ mol^{-1}. This additivity applies in good approximation also to the enthalpies of transfer from water to the mixtures. For the amides, the introduction of a first methyl group into FA and AA (to form NMF and NMA) is accompanied by a change in ΔH_{tr}(W → DMF) of +12.72 and +12.60 kJ mol^{-1}, respectively. Again this is higher than the effect of a second N-methyl substitution (8.40 and 6.95 kJ mol^{-1}, respectively). Since the amides are not symmetrical, introduction of a methyl group at the other side of the molecule will give different results. Indeed the difference in ΔH_{tr}(W → DMF) between AA and FA is +7.68 kJ mol^{-1}. Similar lower values are found for the differences between NMA and NMF (+7.56 kJ mol^{-1}) and between DMA and DMF (+6.11 kJ mol^{-1}). Hence, the effect of C-substitution is smaller than that of N-substitution. This is in accordance with the fact that on C-substitution an aprotic CH group is replaced by an aprotic CCH$_3$ group, while on N-substitution a protic NH group becomes an aprotic NCH$_3$. The combined results for the amides substantiate also the idea of independent solvation of the different parts of the molecule.

The application of the clathrate-like hydration model to the

Table V

The enthalpic effect, Hb(W), of hydrophobic hydration (in kJ mol^{-1}), the number n of hydrating water molecules per alkyl group and the mean deviation δ (in kJ mol^{-1}) between calculated and "corrected" experimental values of ΔH_{tr} for ureas and amides

solute	"correction"	Hb(W)	n	δ
MU	$-\frac{3}{4}$U	-7.1	4.0	0.07
1,1-DMU	$-\frac{1}{2}$U	-13.0	3.5	0.16
1,3-DMU	$-\frac{1}{2}$U	-13.2	4.3	0.16
TMU	–	-24.6	5.1	0.08
EU	$-\frac{3}{4}$U	-10.7	5.5	0.09
FA	$-\frac{1}{2}$U	-3.7	(8.6)	0.05
NMF	$-\frac{1}{4}$U	-10.9	4.9	0.04
DMF	–	-16.1	4.2	0.07
AA	$-\frac{1}{2}$U	-8.0	4.4	0.09
DMA	–	-19.3	4.1	0.09
NBA	$-\frac{1}{4}$U	-22.6	7.1	0.16
BA	$-\frac{1}{2}$U	-14.9	6.6	0.14

data presented in figures 8 and 9 meets a problem. The model assigns deviations from a linear dependence on solvent composition exclusively to hydrophobic effects. But the parent compound U deviates considerably from linear behaviour. Hence, we must apply corrections for the deviation of this parent compound in order to analyze the results for the substituted ureas. We have reason to believe that the deviation of U is mainly due to its NH protons. Consequently we have assumed that the deviation consists of four additive and equal NH contributions. We have applied these contributions to the results for the substituted ureas (with an exception for TMU) and have analyzed the resulting values in terms of eq. (2). Values of the resulting model parameters are given in Table V. Since it can be shown (5) that the same corrections for NH contributions can be applied for the amides, Table V contains also values of the model parameters for the amides resulting from a least-squares analysis with regard to eq. (2). The values of Hb(W) for the methylsubstituted ureas indicate that in good approximation the hydrophobic hydration of the methyl groups does not depend on the solvation of neighbouring groups in the molecule. Per methyl group there are small differences indeed. But they have limited significance in view of the approximations due to the correction procedure. In FA the hydrophobic effect of the CH group appears to be -3.7 kJ mol^{-1}. A comparable value for the CH group, -3.8 kJ mol^{-1}, is obtained by taking the difference between $\frac{1}{2}$Hb(W) for TMU and Hb(W) for DMF. The Hb(W) values for AA and BA refer to a methyl and a n-propyl group, respectively, as C-substituents. Values for other alkyl groups can be obtained by conbining the results of Table V together

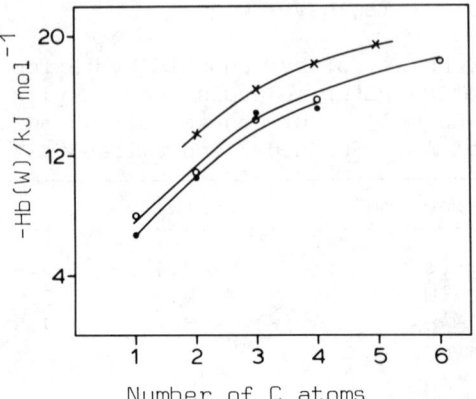

Figure 10. Variations of Hb(W) with the number of carbon atoms for alcohols (×), amines (o) and ureas + amides (•)

with a correction for the effect of NH protons (5). In this way values of Hb(W) can be obtained for methyl, ethyl, n-propyl and n-butyl groups. Their magnitude increases with the number of C atoms, as with the alcohols and amines. Figure 10 shows that the behaviour of n-alkyl groups in the three classes of compounds is similar. The CH_2 increments for compounds with more than 2 C atoms are approximately equal. However, the increments from C1 to C2 compounds are different and larger for alcohols than for amines and ureas + amides. Probably this is due to the polar group, which has some influence on the first neighbouring methyl group. This influence does not reach beyond this group. Hence in all cases the difference in values of Hb(W) is due to a difference in hydrophobic hydration of the CH_2 groups adjacent to the polar groups.

ACKNOWLEDGEMENT

Figures 1 and 2 are reprinted from ref. 3 with permission of Academic Press Inc. (London) Ltd.; figures 5, 6 and 7 are reprinted from ref. 4 with persmission of Plenum Publishing Corporation, New York; and figures 8, 9 and 10 are reprinted from ref. 5 with permission of the Royal Society of Chemistry.

REFERENCES

1. Frank, H.S., and Evans, M.W.: 1945, J. Chem. Phys. 13, pp. 507-532.
2. Heuvelsland, W.J.M., De Visser, C., and Somsen, G.: 1978, J. Phys. Chem. 82, pp. 29-32.

3. Rouw, A.C., and Somsen, G.: 1981, J. Chem. Thermodynamics 13, pp. 67-76.
4. Rouw, Aart C., and Somsen, Gus: 1981, J. Solution Chem. 10, pp. 533-547.
5. Rouw, A.C., and Somsen, G.: 1982, J. Chem. Soc. Faraday Trans. I (in press).
6. De Visser, C., and Somsen G.: 1974, J. Phys. Chem. 78, pp. 1719-1722.
7. De Visser, C., Heuvelsland, W.J.M., and Somsen, G.: 1975, J. Solution Chem. 4, pp. 311-318.
8. Heuvelsland, W.J.M., De Visser, C., Somsen, G., LoSurdo, A., and Wen Yang Wen: 1979, J. Solution Chem. 8, pp. 25-34.
9. Friedman, H.L., and Krishnan, C.V.: 1973, J. Solution Chem. 2, pp. 119-138.
10. Desnoyers, J.E., Perron, G., Avédikian, L., and Morel, J.-P.: 1976, J. Solution Chem. 5, pp. 631-644.

MICROCALORIMETRY AS AN ANALYTICAL TOOL IN BIOLOGY

Ingemar Wadsö

Thermochemistry Laboratory, University of Lund, Chemical Center, Box 740, S 220 07, Lund, Sweden

HISTORICAL BACKGROUND

As early as 200 years ago calorimetry was applied on biological systems (1, 2). At that time Crawford in Scotland and Lavoisier and Laplace in France measured heat production in small animals and were able to draw conclusions about the nature of animal respiration. During the following 100 years, there was not much progress in the field, but toward the end of the 19th century combustion calorimetry was developed and was employed on many biologically important compounds. During the first decades of the present century, respiration calorimetry or "whole body calorimetry" was rather extensively employed for measurements on animals and humans at rest and when performing mechanical work.

A rather special branch of physiological calorimetry was initiated around 1910 by A. V. Hill, who used very sensitive thermopile instruments for measurements of heat production in nerves and in muscles, see e.g. (3).

During the first part of this century, numerous combustion calorimetric measurements were performed on food products, biological waste products and on purified, simple organic compounds known to take part in metabolic processes.

A few reaction calorimetric investigations on biochemical systems were performed prior to 1956, in particular by Meyerhof and his group, see (4). Sturtevant and coworkers and a few other investigators reported results of some reaction calorimetric measurements during the 50's and early 60's (4), but it was not until modern microcalorimeters became available that the present era of biocalorimetry began. As the middle and late sixties approached, the field developed very rapidly, mainly due to the fact that commercially produced instruments became available.

Bio-calorimetry: thermodynamic and analytical investigations

Most processes, be they physical, chemical or biological, are accompanied by heat effects. Calorimetric measurements of thermal power thus provides a method for the characterization of many types of processes, e.g. in biology. The heat quantity evolved during a process is quantitatively related to the extent of the process and the power to its rate. It is thus clear that calorimetry, in addition to its importance in thermodynamics, also can serve as a general analytical tool.

Calorimetric methods are not specific, which is a serious limitation for many types of analytical problems. However, in biology the inherent specificity of the reaction systems often allows the use of an unspecific analytical method. Furthermore, for a complex reaction system like that of living cells, the use of an unspecific method is often advantageous. It is then more likely that unknown phenomena will be discovered than if a narrowly specific analytical technique is employed. Naturally, in such cases it is usually desirable to identify the process or sequence of processes by more specific methods than calorimetry.

All living systems produce heat. If a calorimeter has the adequate sensitivity and stability, it can thus form a monitor for the cellular process without any need for addition of reagents, which might interfere with the biological processes. On the other hand, the use of reagents in calorimetric analytical experiments can be of significant advantage, as it may lead to a higher specificity of the experiment. A feature of particular importance in work on biological systems is that calorimetric methods, in contrast to spectrophotometric methods, do not require optically clear objects but can be used on non-transparent systems having any aggregation state.

Thermodynamic studies on well-defined biochemical systems form a central part of bio-calorimetry. Results are often difficult to interpret on the molecular level and there is also a need for studies of model systems of differing complexity. Within a series of model compounds, the structure can be varied systematically and correlations can be made between thermodynamic data and structural features; this is usually not possible for the more complex biochemical compounds. A range of non-thermodynamic calorimetric determinations is also carried out on biochemical compounds, in particular relating to enzyme assays and determinations of substrate concentrations, see e.g. (5), (6).

In calorimetric work on living systems, the aim is sometimes to compare experimental power-time curves with power values calculated from results of rate determinations and known or estimated enthalpy values for corresponding reaction steps. Studies are also undertaken to provide total heat balances for, for instance, animals or for eco-systems. But often the calorimetric instruments are primarily used as monitors for "biological activity" of cellular systems, without any serious attempt to interpret the results in terms of thermodynamic quantities.

Calorimetry applied to living systems

Calorimetry has been applied to a wide variety of living systems and the nature of the experimental work ranges from micro-calorimetric measurements involving thermal powers on the microwatt level to whole-body calorimetry, where the power evolution is on the order of 10^8 times larger. Table 1 summarizes the most important fields for current calorimetric work on living systems.

TABLE 1. CALORIMETRIC WORK ON LIVING SYSTEMS

    MICROORGANISMS
        Bacteria (7), (8), Mycoplasma (9), Yeast (10)

    ANIMAL CELLS
        Blod cells (11), (12), Macrophages (13), Liver Cells (14), Fat Cells (15), Tumour Cells (16), Cultured Tissue Cells (17)

    TISSUES AND INTACT ORGANS FROM ANIMALS
        Muscle (Striated, Smooth, Heart), Nerve, Electric Organs, Fat Tissue, Brain, Kidney, Liver (7)

    PLANT MATERIALS
        Seed Germination (18)
        Tissues (19)

    ANIMALS
        Small Animals (< 1 cm^3) (20)
        Large Animals ("Whole Body Calorimetry") (21)

The present paper will concentrate on work conducted on micro-organisms and animal cells, for which micro-calorimeters

(5) (cf. also chapter by I. Wadsö in this volume) normally are used. In such work thermopile heat conduction instruments are those which are most commonly used. A brief description of this calorimetric principle will therefore be given.

Thermopile heat conduction calorimeters. In the ideal heat conduction calorimeter, heat evolved in the reaction vessel is quantitatively transferred to a heat sink, usually a metal block, which surrounds the reaction vessel. With these calorimeters, a property proportional to the heat flow is measured. Usually the heat flow is monitored by positioning a "thermopile wall" between the calorimetric vessel and the heat sink.

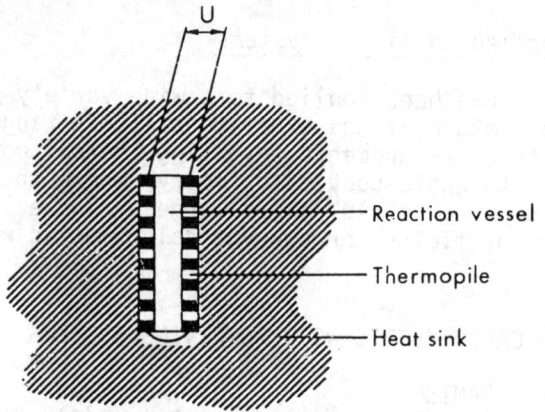

Fig. 1 The principle of a thermopile heat conduction calorimeter. U is the thermopile voltage.

The temperature gradient over the thermopile will give rise to a voltage signal, U, which is proportional to the heat flow, $\frac{dq}{dt}$.

$$\frac{dq}{dt} = \varepsilon \cdot U \tag{1}$$

Under steady-state conditions, the heat flow is equal to the thermal power evolved, P.

$$P = \varepsilon \cdot U \tag{2}$$

Integration of (1) leads to

$$q = \varepsilon \int U dt \tag{3}$$

The heat evolved is thus proportional to the surface area under the voltage-time curve:

$$q = \varepsilon A \tag{4}$$

With a good approximation, eqn. (2) also holds for processes where U changes slowly compared to the time constant of the instrument. For processes where rates change rapidly, the recorded voltage-time curve will be significantly different from the true kinetic curve (the power-time curve). To obtain accurate kinetic information it is then necessary to apply a "dynamic correction", see e.g. (22), and the chapter by H. Tachoir in this volume. It may be observed that eqn. (4) is not dependent on the thermal inertia of the instrument.

The (first) time constant (τ_1) is equal to the quotient between the "effective" heat capacity of the reaction vessel and the heat conductivity between the vessel and the surroundings. For a heat conduction calorimeter using semi-conducting thermocouple plates ("Peltier effect coolers") and a small reaction volume (1 ml), the time constant is typically in the order of 10^2 s. This means that for many biological processes with nearly constant thermal power, the dynamic correction can be neglected. In other cases the simple equation

$$P = \varepsilon \cdot U + \varepsilon \cdot \tau \cdot \frac{dU}{dt} \tag{5}$$

often leads to adequate power values from the recorded values for the thermopile voltage.

The properties of several thermopile heat conduction calorimeters and their use in biology was reviewed in (5). Most microcalorimeters of this kind are arranged as twin instruments. The thermopile of the reaction vessel is then connected in opposition to that of a reference vessel and it is thus the differential voltage which is recorded.

In the following section some examples will be given of the analytical use of microcalorimetry on cellular systems. In figures where power-time curves are presented, results were obtained by use of thermopile heat conduction calorimeters. No dynamic corrections were applied.

A few examples of the analytical use of microcalorimetry on cellular systems.

Microbial growth curves. Fig. (2) shows two power-time curves obtained from experiments with a strain of E. coli grown under aerobic conditions in a complex medium (5). A flow microcalorimeter was used. In experiment A, 10 mg of ampicillin was added to the growth vessel, which then contained about 0.5 l of bacterial suspension. After about 1 hr the power-time curve was significantly affected and the power decreased to near zero. Comparison of the earlier parts of the two power curves shows that even minor characteristics are reproducible. The highly profiled power-time growth curves represent a much more detailed record of the processes than, for example, corresponding turbidimetric curves. In a few simple cases the different phases of such growth curves have been correlated with results of biochemical analysis of the growth medium and the biomass, see e.g. (7). The shape of the growth curves depends on the strain and on the composition of the growth medium, including the availability of oxygen, but also on other factors such as the temperature and the history of the innoculum. Provided that the growth medium is well-defined and the experimental parameters are carefully controlled, the finger-print calorimetric growth curves can be used for identification purposes (23).

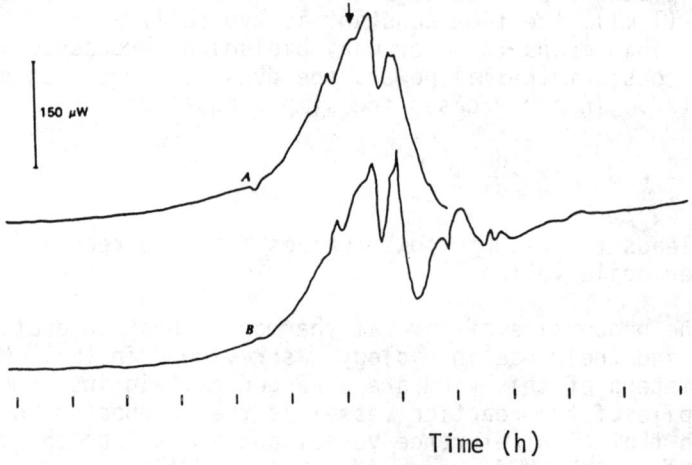

Fig. 2 Calorimetric growth experiments with E. coli grown on complex media. Ampicillin was added at time indicated by arrow. Curves A and B are two consecutive experiments.

Fig. (3) shows growth curves for some strains of lactic acid bacteria used in the dairy industry. A complex synthetic medium was used. It is seen that for each strain, the replicate curves are reproducible, whereas there are very marked differences in curves for different strains.

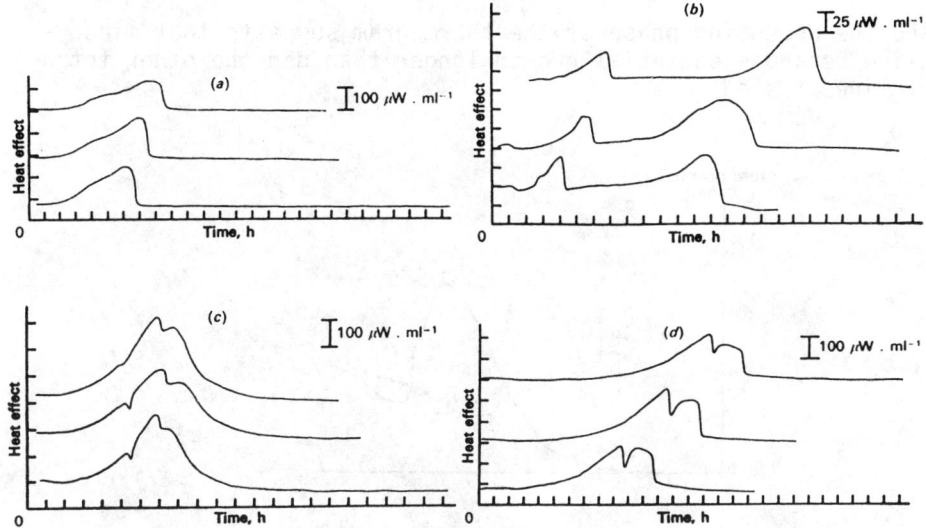

Fig. 3 Growth of thermophilic strains in defined medium at 37 °C. Streptococcus thermopilus CNRZ 302 (a), Lactobacillus bulgaricus CNRZ 36 (b), L. helveticus CNRZ 303 (c), and L. acidophilus 1748 (d). Three replicate thermograms are shown for each strain. From (25).

Different strains of yeast (23) and mycoplasma (9) can also show significant differences in their calorimetric growth patterns, which may be used for identification.

Beezer and coworkers (24) have investigated the heat effects produced during growth by many bacterial strains isolated from urine; these workers concluded that flow microcalorimetry could be used for their numeration down to a level of 10^5 cells ml^{-1}. Ljungholm et al. (9) have demonstrated that mycoplasmas can be numerated calorimetrically.

Testing of antibiotic activity. In several studies, static and flow microcalorimetric methods have been used to assess antibiotic sensitivity and the kinetics of antibiotic action on microorganisms (23). Fig. (4) summarizes some results from one of these studies. E. coli was grown in the presence of minocycline, doxycycline, oxytetracycline, or tetracycline, 0.4 µg/ml (half the minimum inhibitory concentration). The antibiotic was present from the start of the experiment, i.e. introduced during the lag phase of the culture. The figure also shows the power-time curve from a control experiment where no antibiotic was present. The time interval between the start of the experiment and the ascending phase of the thermogram suggests that minocycline retarded bacterial growth longer than did the other tetracyclines tested.

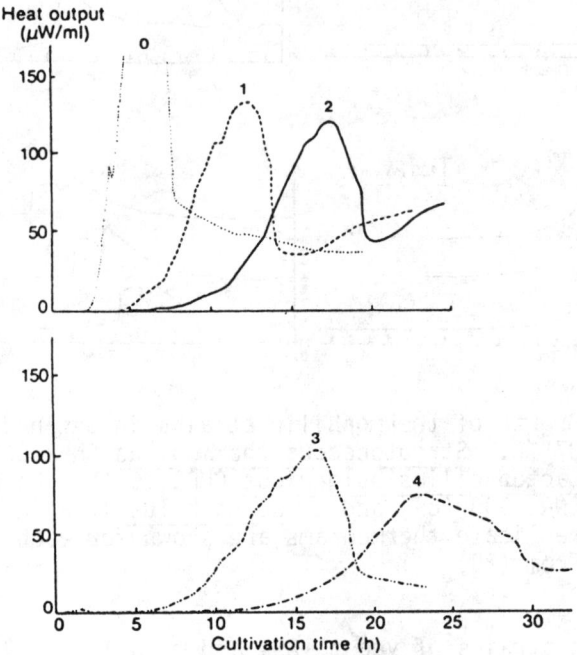

Fig. 4 Upper panel. Heat output of E. coli cultured in the presence of tetracycline (1), doxycycline (2), or in the absence of antibiotic (0). Lower panel. Heat output of E. coli in the presence of oxytetracycline (3) or minocycline (4). Each antibiotic was present at a concentration of 0.4 µg/ml (0.5xMIC). All experiments were carried out in an ampoule microcalorimeter. From (26).

Microbial activity in soil. Methods for microcalorimetric assessment of microbial activity in soil have recently been developed, see (27). Homogenized samples (1-10 ml) are enclosed in ampoules. The thermal power produced varies markedly with the soil type but is normally in the range of 5-50 µW g^{-1}, i.e. of a magnitude which can be measured conveniently by a modern microcalorimeter.

Figure 5 summarizes results from a series of experiments where a soil was treated with different fertilizers. It is seen that addition of a salt mixture does not affect the heat production significantly. In the case of cellulose addition, a small increase in heat production is found after a lag phase of about 10 days. However, after the addition of salts and cellulise, the heat effect increased dramatically after a lag phase of 3 days. It is believed that model experiments of this type can be developed into practically useful techniques in e.g. agriculture and forestry and for pollution control.

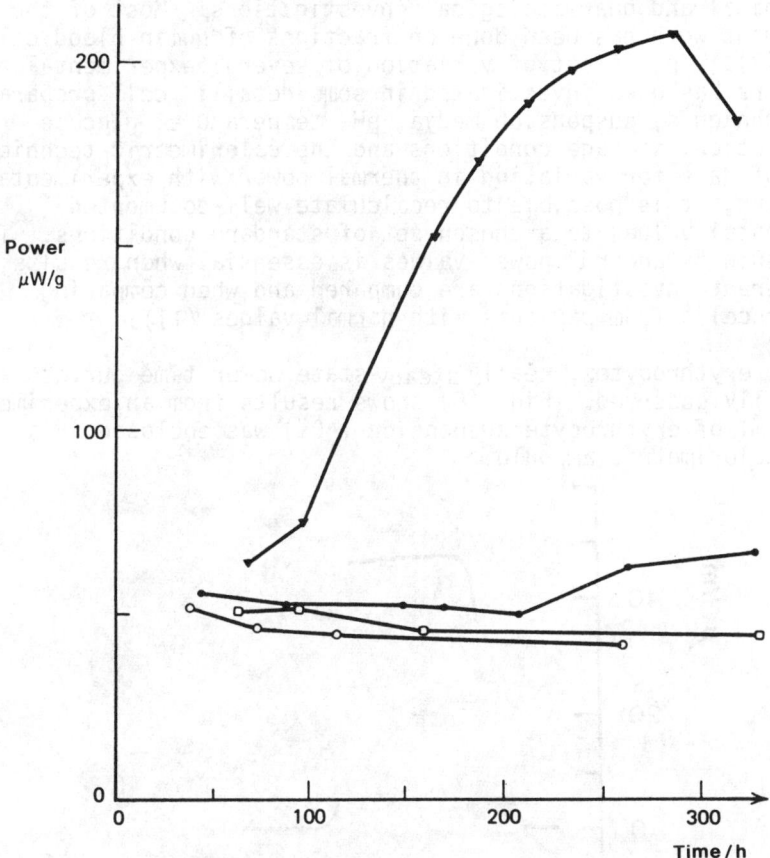

Fig. 5 Power evolution from a soil treated with fertilizers. Nutrients added: cellulose powder (2%), salt mixture: $NaNO_3$ (0.1%), $MgSO_4 \cdot 7H_2O$ (0.1%), K_2HPO_4 (0.2%); (o) untreated; (□) salt mixture added; (●) cellulose powder added; (▼) salts and cellulose added. In all samples the water content was the same (34%) and the temperature was 25 °C. From (28).

In the experiments summarized in Fig. 5, calorimetric ampoules were charged with soil immediately before each calorimetric measurement. It is often desirable to perform repeated measurements on individual samples during long periods of time (months). However, if a sample is hermetically enclosed in a metal ampoule, there will be poisoning effects after a few hours, caused by high concentrations of CO_2, cf. (29).

Human cells. Calorimetric studies on blood cells and other animal tissue cells have received significant attention during recent years, not least because of their potential importance for clinical and pharmacological investigations. Most of the development work has been done on fractions of human blood cells (11), (12). The effect of variation of several experimental parameters has been investigated in some detail: cell preparation techniques, suspension media, pH, temperature, glucose concentration, storage conditions and the calorimetric technique. By use of data for variation in thermal power with experimental conditions, it is possible to recalculate well-documented experimental values to a chosen set of standard conditions. The use of such "standard" power values is essential when results of different investigations are compared and when comparing data for cells from patients with normal values (11).

For erythrocytes, nearly steady-state power-time curves are usually observed. Fig. (6) shows results from an experiment where 1 ml of erythrocyte suspension (40%) was enclosed in a static calorimetric ampoule.

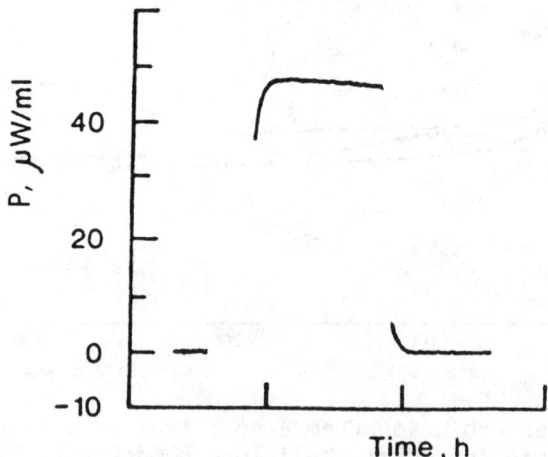

Fig. 6 Typical power-time curve for erythrocytes obtained under static conditions. Sample volume was 1 ml and the cells were suspended in plasma, 40% hematocrit (Monti and Wadsö).

Good results are also obtained with flow calorimetry. There are more problems connected with the other main fractions of blood cells: thrombocytes, lymphocytes and granulocytes. Granulocytes in particular are difficult to work with, due to their tendency to adhere to reaction vessels and flow lines and to the fact that the power produced by these cells can be greatly influenced by the experimental conditions.

Calorimetric studies with some tissue cells are preferably made with the cells attached as monolayers to a solid support, such as the walls of the reaction vessel or to glass beads contained in the vessel, see (17). It is believed that calorimetric methods applied to systems of this sort can be of a very significant importance for cell biology. However, the calorimetric techniques and working procedures are as yet in an early stage of development.

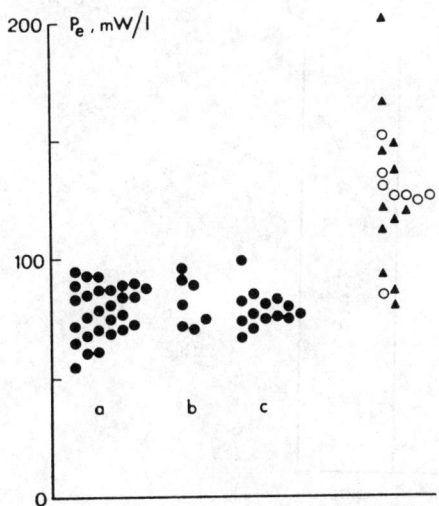

Fig. 7   Heat effects of erythrocytes from normal subjects and from anemic patients: (●) normal subjects, (O) sideropenic anemia, (▲) other anemias. P_e is the heat effect produced per liter of erythrocytes. a, b, and c denote different methods of preparation of cells from normal subjects. From (30).

Figures (7) and (8) summarize results from two clinical studies on human cells. In Fig. (7) the heat production of erythrocytes from normal subjects and from patients with different types of anemia are compared. It is seen that cells from the patients have significantly increased power values. Fig. (8) shows the heat production of tumour cells from non-Hodgkins lymphoma patients. A significant difference was observed with cells obtained from patients who responded to medical treatment and those who did not. Results of the type shown in Figs. (7) and (8) suggest that calorimetry can be of practical diagnostic value in clinical analysis. It is believed that the specificity of the analytical experiments often can be much increased by addition of reagent known to have a specific interaction (stimulation or inhibition on the cell metabolism, see e.g. (11), (12), and the chapter on cellular thermochemistry by G. Rialdi in this volume.

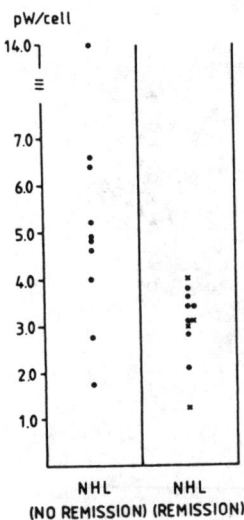

Fig. 8   Heat production of tumour cells from NHL patients.

● Patients with no remission and with complete remission

× Patients with partial remission.

From (37).

Some notes about errors and artefacts

It was pointed out in the introduction that all kinds of processes are accompanied by heat effects; this makes calorimetry such a universal tool. However, it will also make calorimetric measurements very susceptible to systematic errors and misinterpretations due to different kinds of artefacts (29). Such risks are naturally greater in microcalorimetry than in cases where larger heat effects are studied. Table 2 summarizes some of the more common causes of errors and artefacts in micro-reaction calorimetry. In the following, such problems are discussed with particular reference to work on living cells.

TABLE 2. SOME TYPICAL CAUSES OF ERRORS IN MICRO-REACTION CALORIMETRY

CALIBRATION
Microcalorimeters are sometimes difficult to calibrate electrically. Calibration values should be checked by suitable test reactions.

MECHANICAL EFFECTS
Ampoule breaking, opening of valves and stirring effects often give rise to large heat effects, which are not sufficiently reproducible to allow for adequate correction. Frictional effects in flow experiments can vary during a process.

EVAPORATION AND CONDENSATION. GASEOUS REACTION COMPONENTS.
Enthalpy of vaporization of water is very high and virtually no uncontrolled evaporation or condensation can be tolerated. Gas phase composition can change as a result of a reaction. It must be carefully established to what extent a gaseous reaction component is dissolved in the liquid medium before and after the reaction.

ABSORPTIONS
Adsorption of reaction components (or components of a buffer) into the walls of a microcalorimetric vessel can lead to very significant heat effects. Adsorption of living cells to calorimetric vessels or flow lines can lead to large systematic errors.

INCOMPLETE MIXING, SLOW REACTIONS
In flow-mixing calorimetry it is very important to establish that the mixing is adequate and that the process takes place within the calorimetric vessel

(that part of the vessel for which the calibration value holds).

CHANGE OF INSTRUMENT DESIGN
Even small design changes can give rise to significant systematic errors. The instrument should be tested by a suitable test reaction.

Calibration and test experiments. It may seem unnecessary to carry out careful calibrations in terms of well-defined energy units for calorimeters used for analytical experiments rather than thermodynamic work. However, it is believed that there is a general and a lasting value in well-documented energy data for biological systems. Further, in order to make a precise comparison between results from different studies, it is essential that the results are expressed in well-defined units. Thus arbitrary units like "mm recorder deflection" should be avoided. Thermal power is correctly expressed in terms of watts ($W = Js^{-1}$). The Interunion Commission on Biothermodynamics has prepared recommendations for use of units and nomenclature and for the presentation of results in connection with calorimetric measurements on cellular systems (31).

Calorimeters are normally calibrated electrically, which is very convenient, but the experimenter should be aware of the risks associated with systematic errors. The actual measurement of electrical energy, or power, is today a trivial procedure, which can easily be made with an accuracy far exceeding the needs in biological experiments. The problem is to make sure that the electrical energy is released in a manner that is closely comparable with that of the process studied. For practical reasons microcalorimeters used in biological work are not always well-suited for strict comparisons; for instance, it is sometimes difficult to produce a calibration heater of an ideal design, or to place it in the best position. It is suggested that, when a new calorimetric design is tested or when modifications have been made on an existing instrument, calibration experiments should be performed with different types of heaters and/or with heaters placed in different positions. It is usually possible to incorporate a heater which can be judged to be nearly ideal, although it may not be realistic to use it in routine work. Comparison of calibration values from such heaters with those obtained with the regular heater can give the experimenter a realistic feeling for the magnitude of possible errors in the calibration value.

In particular, results obtained with flow-through vessels in current use should be judged with some caution. Provided that kinetic corrections are not needed, the shape of the thermograms is presumably correct, but the power values may be seriously in error. Flow vessels which consist of flat gold tube spirals placed between copper plates (like the regular flow vessels used with the LKB flow calorimeters, see e.g. (5), (29)) have proved not to give very serious errors. The copper plates will distribute the electrical calibration heat evenly along the flow path and the thermopiles will thus not sense the difference between a given electrical power and one released in the biological system contained in the flow-spiral. However, in flow vessels where the liquid is not in as good thermal contact with the thermopile and with the calibration heater, the errors can be serious. This is believed to be the case for some insertion vessels, in particular if they are used with a high flow rate, see (29).

Some flow vessels (5), (29) are designed to allow a mixed flow of liquid and gas. It is then very important to know the liquid volume (containing the power-producing cells) which is contained in the calorimetric vessel.

There is also another, more general, problem with the assessment of the "practical" value for the volume of flow-through vessels in heat-conduction calorimeters. In an experiment with a cell suspension, some heat will be produced in the flow line between the heat exchanger and the flow vessel. Part of this heat will be recorded by the instrument, but in an electrical calibration experiment there will be no corresponding power produced. It therefore appears that many flow calorimetric vessels which are used for power measurements are best calibrated by use of a chemical calibration procedure (29). Very recently a new test and calibration process was developed for this purpose (32). This process consists of the hydrolysis of triacetin in imidazole/acetic acid buffer. The power can be regulated by changing the buffer composition. Power values can be calculated accurately ($\pm$ 0.5%) as a function of time during extended reaction periods, about 20 h or more.

In flow calorimetric experiments it is necessary to account for the power produced by the viscous heating in the flow vessel. With a heat conduction instrument this is incorporated in the "base-line value". For a cell suspension the position of the base line is usually established by pumping pure suspension medium (e.g. a buffer, nutrient broth, plasma) through the calorimeter. The base-line value thus accounts for the viscous heating, but in addition other effects, possibly due to insufficient temperature equilibration of the incoming flow, pressure effects on the thermopile and on the liquid flow, will also be included. Often the base line is close to the value at zero flow rate but for viscous liquids, and at high flow rates, the difference can be substantial. It is then important that the liquid used in the base-line experiments has flow properties which are closely related to those of the cell suspension. At low cell concentrations there is normally no problem in this respect but for dense suspensions it is important to watch out for errors. In cases where the viscosity changes during the experiment, such as in microbial growth experiments, it is possible to obtain a false calorimetric growth pattern for this reason. There appears to be no good working procedure by which such effects can be overcome, except by employing low flow rates and using flow vessels where the frictional effects are minimized. Such conditions are frequently difficult to arrange in practice because of problems with supply of oxygen or due to sedimentation effects.

Unexpected processes. It is often an advantage that calorimetric methods are so unspecific - one feature is that unexpected phenomena are likely to be detected. But there is also a risk that such effects can be misinterpreted. For instance, thrombocytes have been shown to produce significant bursts of heat when undergoing a mild agitation during the rotation of an LKB batch microcalorimeter (33) during a mixing process. If proper control experiments had not been conducted, results of the mixing process could have been gravely misinterpreted. Heat effects presumably due to aggregation (thrombocytes) and adhesion (granulocytes) are other examples of effects which can easily be misinterpreted, see (29).

Aerobic processes. In many types of calorimetric experiments with aerobic cells, it is difficult to arrange a sufficient supply of oxygen. The enthalpy of vaporization of water is very high, 43.9 kJ mol^{-1} at 25 ^{o}C. This means that more than 2 mJ of heat is absorbed per µg of water evaporated. If measurements are made at a high sensitivity, it is therefore very difficult to avoid significant heat effects from evaporation or condensation processes when a flow of gas is bubbled through an aqueous medium in a calorimetric vessel. A direct and efficient aeration of the medium in the calorimetric vessel is therefore not possible.

Let us first consider the amount of oxygen which may be contained in a given volume of an aqueous medium. Pure water in equilibrium with air dissolves about 0.2 µmol of oxygen per ml at 37 ^{o}C. If saturated with air, 1 ml of water or aqueous solution will thus contain enough oxygen for the complete oxidation of about 33 nmol of glucose. For a steady-state process this would correspond, for example, to a power of the order of 13 µW during 2 h. Thus, with dilute cell suspensions and with very sensitive calorimeters, it is possible to make rather long experiments (hours) with aerobic cells without other provisions than air or oxygen saturation of the medium prior to the experiment.

In cases where the liquid medium is shallow or where the cellular material is floating (fat cells), significant amounts of oxygen can be transferred to the biological material from a gas phase above the liquid medium even in a static system.

If the liquid medium is stirred or agitated the rate of oxygen uptake from a gas phase will increase. Further, stirring of the liquid will enable the cells to utilize the dissolved oxygen more efficiently in cases where they do not form a homogeneous suspension, e.g. when they are adsorbed to a solid support or when they tend to sediment. Examples of suitable vessels and stirring/mixing procedures are discussed in, for instance, (5), (29) and (34).

During a recent methodological study ((27), (29)), it was shown that the thermal power produced by microorganisms in soil is drastically influenced by the gas phase composition in the calorimetric vessel. When soil is hermetically enclosed in a calorimetric ampoule, after some time there will be a drastic reduction in the power level, in particular if the soil has been enriched with glucose. Results indicated that the decreased heat effect was due to high levels of carbon dioxide rather than to depletion of oxygen. In order to avoid significant changes in the gas phase during long-term experiments, the following ampoule technique was developed. Samples were enclosed in polythene ampoules with top and bottom consisting of membranes made from 1 mm silicone rubber. This latter material has a high permeability both for oxygen and for carbon dioxide. The plastic ampoule fits snugly into the calorimetric steel ampoules. During an extended experimental period (weeks or months), the plastic ampoule can be exposed to a controlled atmosphere outside the calorimeter, except for the brief calorimetric observation periods during which the insert ampoule is enclosed by the steel ampoule.

Sedimentation. Partial or complete sedimentation of cellular material during a calorimetric experiment can in some cases lead to serious errors and in any case to poorly defined power values. For some types of cells "crowding effects", i.e. inhibition of the metabolic rates with increasing cell concentration, have been demonstrated. In other cases, where one cannot talk about "crowding effects", the metabolic rate can still be affected by local differences in concentrations of, for example, oxygen or of pH, which may develop as a result of sedimentation.

In a flow calorimetric experiment, sedimentation of cells in the flow line prior to the reaction vessel can lead to systematically low power values. Probably more common, the cells may preferentially sediment in the flow vessel and power values recorded can then be increasingly larger than the value representative for the nominal cell concetration.

For many types of cells like bacteria, thrombocytes or mycoplasmas, there are normally no problems with sedimentation, even in static vessels, during short calorimetric experiments. For long experiments, sedimentation can easily be prevented by a very gentle stirring or agitation or by using a flow method. However, for large and heavy cells, like yeast cells and erythrocytes, and with aggregates of cells, it can be very difficult to prevent sedimentation during a calorimetric experiment (29).

Adhesion. Many types of cells have a tendency to adhere to the walls of the calorimetric vessel or to the tubes in a flow calorimeter. As for sedimentation, there are two different problems which can be caused by such effects. First, the metabolic activity for a cell which is free in suspension may be different to one which has been attached to a wall. For flow systems, we have the additional problem of poorly defined quantities of cells contained in the calorimetric vessel.

For experiments of short duration, wall growth by bacteria can usually be neglected. But it is a common experience that in extended bacterial growth experiments, such as with continuous cultures, the performance of analytical sensors can be seriously impaired by cell attachments. It is then advisable to clean the calorimetric flow lines at regular intervals by brief flushings with a suitable cleaning liquid, e.g. dilute NaOH solution.

Among the major fractions of blood cells, it is the granulocytes which are most difficult to handle in a calorimetric experiment. Some figures may illustrate the problem. In a typical calorimetric experiment run under static conditions, 5×10^6 cells were contained in a 1 ml calorimetric ampoule. This was made from stainless steel but was coated with teflon (35). After about 1 h at 37 $^{\circ}$C the average cell count in the suspension had decreased by 10 per cent if a phosphate buffer was used as suspension medium. In plasma suspension up to 50 per cent of the granulocytes had become attached to the wall of the ampoule. If a perfusion ampoule was used with an air flow of 20 ml h^{-1}, about 50 per cent of the cells disappeared from a phosphate suspension during 1 h. If no air flow was used the decrease was the same as for the standard ampoules, about 10 per cent. For cells with these properties, the most suitable conditions for measurements would probably be obtained if the cells were allowed to adhere quantitatively to the walls of the reaction vessel or to some support contained therein.

Levin (11), (36), has reported flow calorimetric experiments on mixtures with thrombocytes and leucocytes suspended in plasma. An LKB calorimeter equipped with a standard flow-through vessel made from gold tubing was used. The peristaltic pump sucked the cells through the calorimeter. Cell counts performed on the suspension leaving the calorimeter showed a substantial reduction in concentration of granulocytes, at times 50 per cent of the original count. On no occasion did Levin observe any significant reduction in numbers of platelets or monomorphonuclear leucocytes (i.e. mainly lymphocytes). The instrument base-line was established with plasma before and after passage of the cell suspension. Sometimes very substantial base-line shifts were noted, typically of the order 5 µW, which corresponds to approximately 25% of the total power recorded for the leucocyte mixture. The results clearly show that the granulocytes adhered to a significant extent to the flow vessel and that such effects can lead to very large errors for which it is difficult to correct.

Conclusions. Calorimetric investigations on living cells represent a vast experimental area where many specialized instrumental properties and working procedures are needed. It is felt that further attention must be given to the design of specialized reaction vessels where the following conditions for the cells can be verified during a calorimetric measurement: supply of oxygen, sedimentation, adhesion, cell concentration in a flow vessel, pH, etc. More attention should also be given to the use of processes suitable for tests and calibrations of microcalorimeters for power measurements. For many cellular systems it is important that values be determined for the influences of various experimental parameters on the heat effect values. In cases where such relationships are known - in particular for non-growing cells - it will be possible to recalculate well-documented results to a chosen set of standard conditions. The use of such "standard" power values is believed to be essential when results of different investigations are to be compared.

REFERENCES

1. Kleiber, M., "The Fire of Life", pp. 4, 116. Wiley, New York, 1961.

2. Armstrong, G. T., J. Chem. Educ. (1964)) 41, 297-307.

3. Wooledge, R. C. in "Biological Microcalorimetry" (A. E. Beezer, ed.), pp. 145-162. Academic Press, London, 1980.

4. Sturtevant, J. M. in "Experimental Thermochemistry", vol. 2 (H. A. Skinner, ed.) pp. 427-442. Interscience, London, 1962.

5. Spink, C. and Wadsö, I., in "Methods of Biochemical Analysis", vol. 23 (D. Glick, ed.) pp. 1-159. Wiley, New York, 1976.

6. Danielsson, B., Mattiasson, B. and Mosbach, K., in "Appl. Biochem. Bioeng.", vol. 3 (L. B. Wingard, Jr., E. Kabzir-Katchalski, L. Goldstein, ed), pp. 97-141, Academic Press, New York, 1981.

7. Belaich, J. P., in "Biological Microcalorimetry", (A.E. Beezer, ed.) pp. 1-42, Academic Press, London, 1980.

8. Kresheck, G. C. in "Biochemical Thermodynamics", (M. N. Jones, ed.) pp 281-307. Elsevier, Amsterdam, 1979.

9. Ljungholm, K., Wadsö, I. and Mårdh, P. A., J. Gen. Microbiology (1976) 96, pp. 283-288.

10. Lamprecht, I. in "Biological Microcalorimetry", (A. E. Beezer, ed.), pp. 43-112. Academic Press, London, 1980.

11. Monti, M. and Wadsö, I., in "Biochemical Thermodynamics" (M. N. Jones, ed.), pp. 256-280. Elsevier, Amsterdam, 1979.

12. Levin, K. in "Biological Microcalorimetry" (A. E. Beezer, ed), pp. 131-144. Academic Press, London, 1980.

13. Loike, J. D., Silverstein, S. C. and Sturtevant, J. M. Proc. Natl. Acad. Sci. USA (1981) 78, pp. 5958-5962.

14. Jarrett, J. G., Clark, D. G., Filsell, O. H., Harvey, J. W. and Clark, M. G. (1979), Biochem. J. 180, 631-638.

15. Monti, M., Nilsson-Ehle, P., Sörbris, R. and Wadsö, I. Scand. J. Clin. Invest (1980) 40, pp. 581-587.

16. Monti, M., Brandt, L., Ikomi-Kumm, J., Olsson, H., and Wadsö, I. (1981). Scand. J. Haematol. 27, pp. 305-310.

17. Kemp, R. B. in "Biological Microcalorimetry" (A. E. Beezer, ed.) pp. 113-130. Academic Press, London, 1980.

18. Calvet, E. and Prat, H. Recent Progress in Microcalorimetry. Pergamon Press, London, 1963.

19. Bogie, H. E., Kresheck, G. C. and Hormat, K. H., Plant Physiol. (1976), 57. 842-845.

20. See e.g. Gnaiger, E., Thermochem Acta (1981), 49, pp. 75-85 and Kurtti, T. J., Brooks, M. A., Wensman, C. and Lovrien, R., J. Therm. Biol. (1979) 4, 129-136.

21. See e.g. Irsigler, K., Veitl, V., Sigmund, A., Tschegg, E. and Kunz, K., Metabolism (1979) 28, pp. 1127-1132.

22. Randzio, S. L. and Suurkuusk, J. in "Biological Microcalorimetry" (A. E. Beezer, ed), pp. 311-342. Academic Press, London, 1980.

22. Delin, S., Monk, P., and Wadsö, I. Science Tools (1969) 16, 22-24.

23. Newell, R. D. in "Biological Microcalorimetry" (A. E. Beezer, ed), pp. 163-186. Academic Press, London, 1980.

24. Bettelheim, K. A. and Shaw, E. J. in "Biological Microcalorimetry" (A. E. Beezer), pp 187-194. Academic Press, London, 1980.

25. Fujita, T., Monk, P. and Wadsö, I. J. Dairy Sciences (1978) 45, pp. 457-463.

26. Mårdh, P. A., Ripa, T., Andersson, K. E. and Wadsö, I. (1976), Antimicrob. Ag. and Chemother. 10, pp. 604-609.

27. Ljungholm, K., Norén, B., Sköld, R. and Wadsö, I. (1979) OIKOS, 33, pp. 15-23.

28. Mortensen, U., Norén, B. and Wadsö, I. Bull. Ecol. Res. Comm. (Stockholm)(1973), 17, pp. 189-196.

29. Wadsö, I. in "Biological Microcalorimetry" (A. E. Beezer, ed), pp. 247-274. Academic Press, London, 1980.

30. Monti, M. and Wadsö, I., 1973, Scand. J. Clin. Lab. Invest. 32, pp. 47-54

31. Belaich, J. P., Beezer, A. E., Prosen, E. and Wadsö, I. Calorimetric Measurements on Cellular Systems: Recommendations for Measurements and Presentation of Results. See CODATA Bulletin 44 (1981) or Pure & Appl Chem. (1982), 54, pp. 671-679.

32. Chen, A.-t. and Wadsö, I. (1982), J. Biochem. Biophys. Methods, in press.

33. Ross, P. D., Fletcher, A. P. and Jamieson, G. A. (1973). Biochem. Biophys. Acta, 313, pp. 106-118.

34. Suurkuusk, J. and Wadsö, I., (1982), Chemica Scripta, in press.

35. Bandmann, U. and Wadsö, I. (1977), unpublished results.
36. Levin, K., (1971), Clin. Chim. Acta, pp. 87-94.
37. Monti, M., Brandt, L., Ikomi-Kumm, J., Olsson, H., and Wadsö, I. (1981), Scand. J. Haematol. 27, 305-310.

APPLICATION OF CALORIMETRY TO THE LIFE SCIENCES

H.Klump

Institut für Physikalische Chemie der
Universität Freiburg, West Germany

ABSTRACT

Unlike the frequently used physico-chemical methods like NMR or CD or IR spectroscopy e.g. calorimetry does not center directly on the object under research but on the object and its environment, since nearly all chemical and biochemical processes are combined with a production or consumption of heat, and therefore with a flow of heat between the object and its surrounding.

It is beyond the scope of this review to describe the different instruments and methods in detail. We will rather try to give an impression of the wide area of applications of calorimetry in life sciences as an analytical tool. Concerning instrumentation we will only briefly describe calorimeters which are commerciablely available and will meet certain demands for research work. Most experimental work presented here deals with model compounds. There are forthcoming an increasing number of papers on calorimetry of microorganisms and the occasional attempt to follow the generation of heat in complete biological objects. This last field of application is the most recent and the examples serve to show how far the applicability of calorimetry in life sciences can reach today.

1. INTRODUCTION

In the conclusion of the thermodynamic analysis of the universe Rudolf Clausius formulated (1) the following statements.

"The energy of the universe is constant", which is the famous energy conservation low in his words and: "The entropy of the universe increases until it reaches a maximum". To understand these statements we have to bear in mind that Clausius viewed the universe as a closed system.

Thermodynamics would not have developed into such a marvelous, logic intellectual structure, if it concentrated on the speculation, whether Clausius' assumptions about the universe were correct. The power of his theory lies in the fact, that it was limited to closed systems, where the starting conditions and the boundary conditions are reproducibly controlled. This theory is fit to describe the conditions of a laboratory rather than the whole universe. Under these limiting conditions it allows to throw some light on the closed systems in the inanimated and the animated world. Keeping this in mind we can use thermodynamics to gain some insight into living systems. The results we wish to obtain are valid for biopolymers as well as for synthetic polymers. The talk is not limited to special systems but to the kind of examples which are exclusively selected from life sciences.

We will present experimental results obtained for the three main classes of biopolymers, namely lipids, proteins and nucleic acids. Prior to a detailed discussion of the individual systems we have to introduce some formalism to facilitate the understanding of the results (2).

2. FORMALISM TO EVALUATE THE TRANSITION ENTHALPY

This scetch of the formalism will be useful for those which don't deal with biopolymers routinely to bridge the gap between the experimental results and the conclusions drawn for a molecular system.

As a "structural change" we will lable those changes within a macromolecule, where the degree of order is changed,

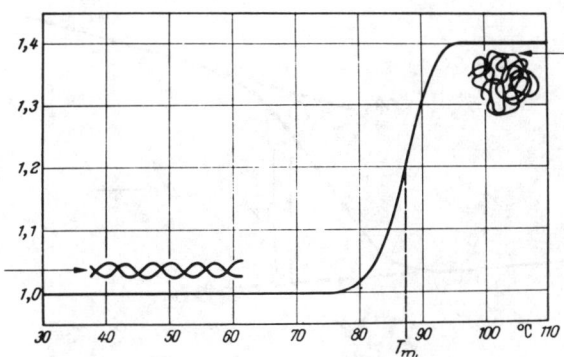

i.e. we will look at the transition from an ordered helical state to a disordered random coil. As initial state we start either from a α-helix as in proteins or from a double-helix

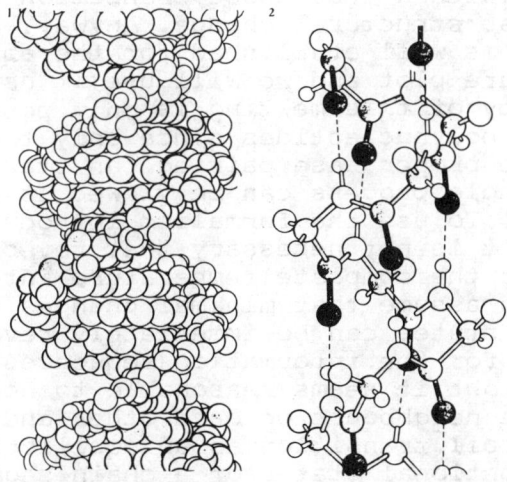

as in a couple of nucleic acids. The conformational change is reflected by a number of changes of characteristic physical properties. It is convenient and extremely sensitive to monitor the absorbance at a fixed wave length as a function of the temperature. The slide demonstrates schematically the relationship between the UV-absorption and the order disorder transition of a double helix.

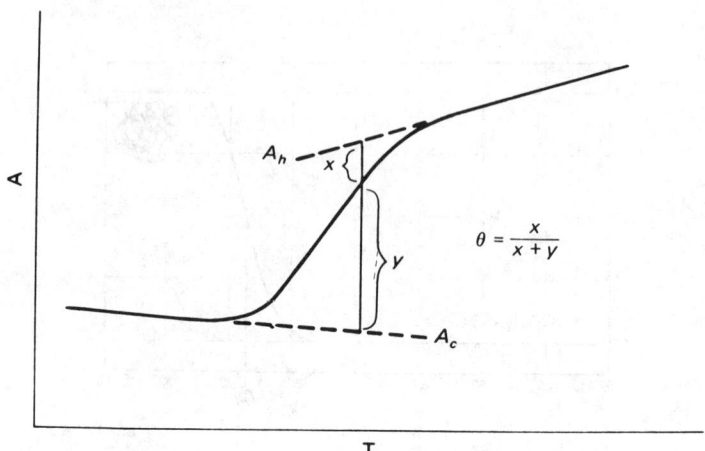

Noticeable is the small temperature interval where all the absorbance change occurs. It is convenient to call the temperature at the half conversion change the "melting temperature" Tm. This is not correct since we are not dealing with a true phase transition but with an intramolecular structural change. Anyway, the term "melting curve" is well etablished for the absorbance versus temperature plot and we will use it usually. The reversibility of this melting is only possible in the case of the polynucleotides since only there the formation of the proper base pairing can start at any position. The whole process can be viewed as an equilibrium process. To use the formalism introduced for the equilibria it is not necessary that reversibility is possible over the complete temperature interval, it is sufficient to assume that minimal changes between closely related states can be immediately reversed. This holds also for the information sequences of the DNA. At first sight it seems reasonable to neglect the influence of the neighbours on each other and to deal with the helix coil transition as an equilibrium between two conformational states of a chain segment with conserved primary structure of the polymer. We can apply the mass law in the following way.
The equilibrium constance is defined as

$$K = \frac{\text{concentration of segments in the helical state}}{\text{concentration of segments in the coiled state}}$$

It will be apt to formulate a degree of formation of the ordered state by the following relation.

$$f = \frac{\text{concentration of segments in the helical state}}{\text{concentration of all segments}}$$

and to express the equilibrium constance with the help of f.

$$K = \frac{f}{1-f}$$

The temperature course of the degree of formation f(T) for biopolymers shows some characteristic behaviour. It follows directly from the temperature dependence of the equilibrium constant which is given by text book thermodynamics by the relation

$$\frac{1}{K}\frac{dK}{dT} = \frac{\Delta H}{RT^2}$$

We will replace

$$\frac{dK}{dT} \quad \text{by} \quad \frac{dK}{df} \cdot \frac{df}{dT}$$

and write according to the definition given above

$$\frac{dK}{df} = \frac{1}{(1-f)^2}$$

So we will get for T=Tm with K=1 and f=$\frac{1}{2}$

$$\frac{df}{dT} = \frac{\Delta H}{4RT_m^2}$$

Based upon this model of independent segments it should be easy to determine the transition energy by following a signal which is directly proportional to the temperature course of the transition. We would only need the slope of this plot at the half conversion temperature. An evaluation of the transition curve for a DNA would result in a transition enthalpy of 250 kcal/mole segment. All of you know that this exceeds the energy of breaking a covalent bond. For biopolymers whose structure is stabilized exclusively by weak secondary bonds this will be much to high if the primary structure will be unchanged. We would expect an amount of about 10 kcal/mole for the conformational change of a segment. The results of this deduction is that the independence of the segments is not valid, that the formalism is insufficient and has to be improved to take the influence of the neighbours on the conformational state of a given segment into account. The transition from one state into the other has to be cooperative.

There are a couple of theoretical approaches to calculate the influence of the kind and the number of neighbours on the thermal behaviour of a segment (3). The quantity we would like to evaluate by this approach is the mean cooperative length $\langle m \rangle$. The slope of the experimental curve is determined by this quantity. This aspect of cooperative leads to the modification of the formalism in such a way that $\Delta H_{v.H.}$ has to be replaced by $\Delta H_{cal} \cdot \langle m \rangle$. But to solve this equation we need either a method to determine $\langle m \rangle$ or ΔH_{cal} independently. We will denote this product $\Delta H \cdot \langle m \rangle$ the "apparent transition enthalpy". The cooperative length is not experimentally accessible, since in the case of polynucleotides we are not dealing with "Avogadro-type-molecules" of equal molecular weight. So to solve this problem we have to introduce a method, which enables us to determine the "true" transition enthalpy. This quantity is fit to compare the stability of secundary structures under different experimental conditions. The only model free method to reach this goal is microcalorimetry. (4)

Before we turn to the experimental details we still have to complete our formalism.

Under the assumption that within the transition interval the transition enthalpy is independent from the temperature (this should show up as a finite ΔCp as we will see later) the differential change of the enthalpy Δh should be proportional to the differential change to the conformational state $\langle n \rangle$. The definition of ΔH is given by the following equation

$$\frac{\Delta h}{\Delta n} = \Delta H$$

APPLICATION OF CALORIMETRY TO THE LIFE SCIENCES

Since we can determine the concentration of the subunits by a chemical analysis we can write

$$\frac{df}{dT} = \frac{1}{N}\frac{dn}{dT}$$

Combining the two equations we get

$$\frac{1}{N}\frac{dh}{dT} = \Delta H \cdot \frac{df}{dT}$$

The left hand side of this equation represents the first derivative of the transition enthalpy to the temperature for one segment. Refered to a mole this represents the molar heat capacity change due to the helix coil transition. This additional heat capacity is proportional to the differential quotient $\frac{df}{dT}$. A plot of this function is shown in the lower part of the next slide.

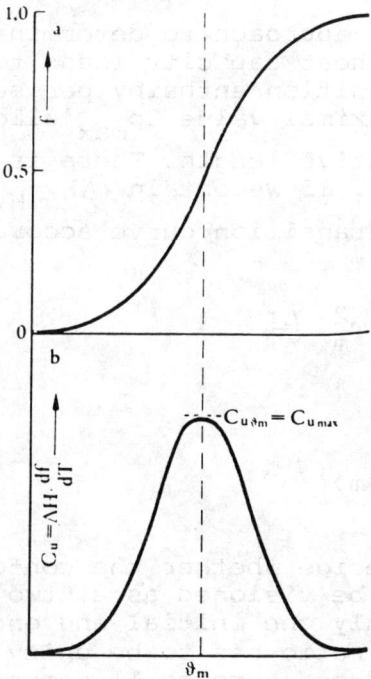

As we have seen from the optical transition curve, the slope is steepest at Tm, therefore the heat capacity function must have a maximum here. The total area is represented by the integral

$$\int_{T_2}^{T_1} Cp \, dT = \Delta H_1 \,.$$

The determination of the area and comparison with a calibration peak allows to calculate the transition enthalpy. We will combine the two equations

$$\left(\frac{df}{dT}\right) = \frac{\langle m \rangle \Delta H}{4RT_m^2} \quad \text{and} \quad \frac{df}{dT} \cdot \Delta H = Cp_{trans}.$$

to the equation

$$Cp_{max} = \frac{\langle m \rangle \Delta H^2}{4RT_m^2}$$

The experimental approach to determine the temperature function of the heat capacity leads to the determination of the transition enthalpy per segment as a molar basis and the maximal value Cp_{max} allows to determine the mean cooperative length. There is a second method to determine $\langle m \rangle$, if we obtain $\Delta H_{v.H.}$ from the slope of the optical transition curve according to the equation

$$\Delta H_{v.H.} = 6RT_m^2 \frac{\Delta f}{\Delta T}$$

Then the ratio

$$\frac{\Delta H_{v.H.}}{\Delta H_{cal}} = \langle m \rangle$$

enables us to decide whether the conformational change under study can be visioned as a "two state process", i.e. there is only one initial and one final state, since than this ratio has to be unity. This simple formalism introduced here shall demonstrate that the thermodynamic analysis can only be successful if we can relay on experimentally determined thermodynamic functions. The method to chose, calorimetry, belongs to the traditional techniques from the early days of physical chemistry. Bunsen has designed an apparatus of this kind, Eucken and Eigen successfully improved the set and as I will demonstrate here today there is a rapid developement of this technique.

3. DIFFERENTIAL SCANNING CALORIMETER

To give you an opportunity to picture yourself the experimental possibilities of adiabatic scanning calorimetry (5) we will open up one of these gray boxes.

This particular one was designed by Prof. P. Privalov from the institute of protein research of the Soviet Academy of Sciences. It is a good example for the state of the art. There are a couple of these instruments around also in western countries and almost all results published on dilute aqueous solutions of biopolymers were obtained with the help of this instrument. It is called a differential adiabatic scanning calorimeter. The scan monitors the supplement heat capacity due to the transition. Adiabatic conditions are necessary to avoid heat leakage to the surrounding. The differential setup is chosen as to cancel out all influences due to heat capacity changes of the solvent. This is very important since the solvent represents more than 99.9% of the total heat capacity. To exemplify the difficulties: In case of naked viruses (6), I will come back to this later in more detail, the polymer concentration was 0.15 g/ltr (a tenth of a permil). To determine the transition enthalpy of this quantity to an accuracy of 5% is a tough job.

Experimentally we approach our aim in the following way. We fill the cell and the reference cell (1 ml) with the appropriate degassed solutions and apply an external pressure of 1 atm on the surface of both liquid phases to avoid the developement of air boubbles during the heating process. After both cells have equilibrated within 10^{-5} °C the constant heating rate is set to the predetermined value. As long as there is no change of the conformation of the dissolved biopolymer the difference of the heat capacity of the sample and the reference solution will be small and constant so that the temperature difference between the two cells is also kept constant. At the same time the adiabatic shields are continuously adjusted to the cell temperature to minimize the heat leakage to the surrounding. When in the course of the conformational change of the biopolymer an additional amount of energy is required the heating voltage in the cell is automatically increased to keep the differential thermal signal very close to zero. We record this additional heating power $\frac{U^2}{R}$ as function of time (6).

This is equivalent to the energy input as a function of temperature

$$\int_{t_1}^{t_2} \frac{U^2}{R} dt = \int_{T_1}^{T_2} C_p \, dT$$

The result of this integration is the transition enthalpy h. The knowledge of the analytical concentracions, in case of the nucleic acids the knowledge of the phosphorus concentration, enables us to calculate the transition enthalpy per mole of segments of base pairs. The area of a peak represents an energy of a few mcal, i.e. the extra energy input is sufficient to rise the temperature of the solution for a couple of 10^{-3} °C. As you can see we really save energy.

After completing the formalism of the thermodynamic analysis and after scetching the experimental prerequisites to measure the extreme low heat effects we will concentrate on the molecular systems and the conclusions we will draw for elementary steps in molecular biology.

4. THERMAL DENATURATION OF A DNA

Our first example is the temperature induced conformational transition of the DNA-double helix (7) to a random coil. The opening of the ordered double helix is the provision for the controlled replication, the production of a set of complementary strands. It is also the provision for the transcription, the production of a given m-RNA, which carries all the information for a certain protein sequence. The detailed analysis of the transition of the different DNAs will give us the opportunity to subdivide the individual of stabilizing forces of the nucleic acid double helix. To remind those of you, which are not dayly dealing with this material, all nucleic acids

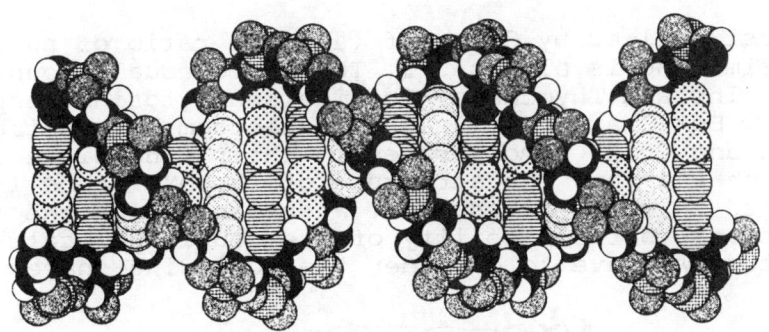

consist of the same polymer backbone namely a strictly alternating sequence of ribose or desoxiribose and a phosphate group (8). This phosphate group is bound by a 3'ester linkage to the previous and a 5'ester bond to the following sugar ring. Under physiological conditions each phosphate group carries negative charge. To the C-1-position of the sugar ring one out of the four heterocyclic bases, Adenine, Thymine (or Uracil), Guanine or Cytosine , is linked via a N-glycosidic bond. The complete segment hence consists of phosphate, sugar and a heterocycle. This unit is called a nucleotide (9). The bases are either derived from the pyrimidine as Thymine and Cytosine or from purine as Adenine and Guanine.

As first noticed by Chargaff (10) the ratio of purine to pyrimidine is always 1:1. There are equal amounts of Adenine and Thymine or Guanine and Cytosine respectively. But Chargaff missed to call them base pairs. Watson and Crick demonstrated in their fine piece of research (11) how the structure of the DNA fibre is arranged in three dimensions. When discussing the contributions to the stability of the double helical structure we have to consider the following shares (12).

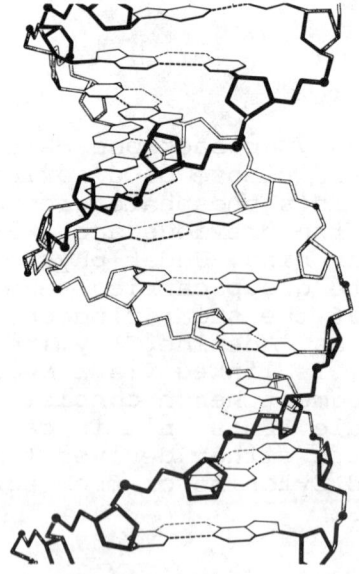

1. Electrostatic energy of dipole/dipole interactions and 2. dissociation energy of hydrogen bonds. AT base pairs form two, GC base pairs three H-bonds. These contributions can be labelled "inplane" interactions. Dispersion forces give a stabilizing share along the helix axis. Usually they are called stacking interactions (13). There are also contributions from changes of the entropy of the system, encluding changes from the solvent solute interaction and from conformational entropy of the polymer. We will not suppress the destabilizing contributions of the Coulomb repulsion of the phosphate groups, which space only 1.7 Å along the helix axis.

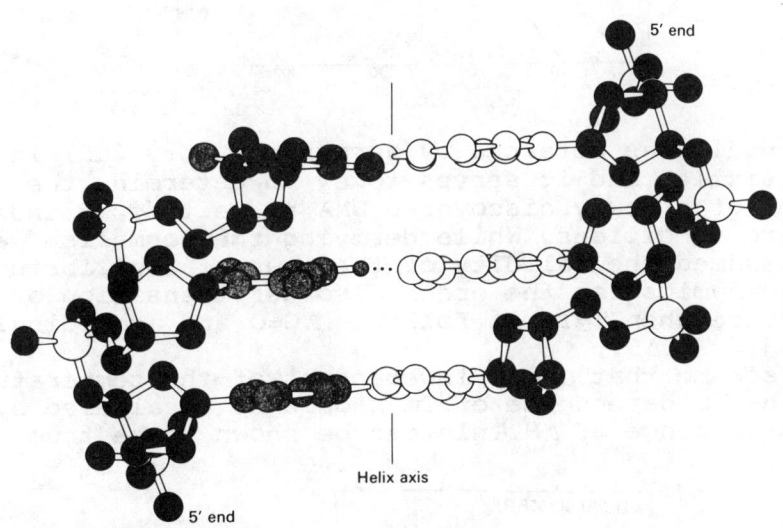

Helix axis
5' end

A decrease of the total ion concentration decreases the transition temperature according to a logarithmic function (14). Therefor the total ion activity as well as the neutral pH value have to be known for each experiment. The results demonstrate that there is a linear dependence of the transition temperature from the GC content of the nucleic acid.

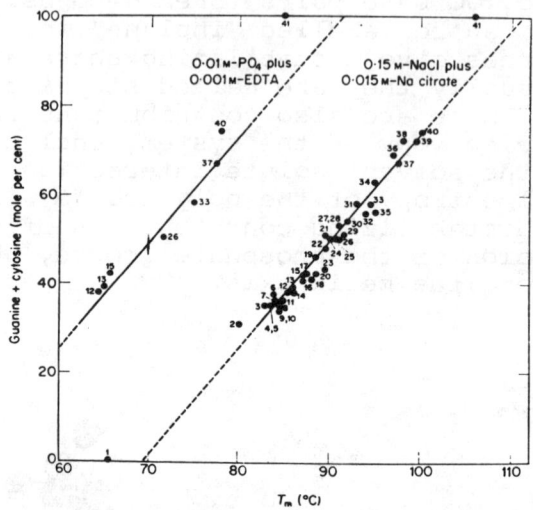

This result was obtained by Marmur and Doty (15) in the early sixties and it serves today to determine the GC content of a newly discovered DNA by melting it under standard conditions. While deriving the formalism we have assumed the validity of the laws of equilibrium thermodynamics for the order disorder transition of a DNA. Hence when T=Tm it follows $\Delta G=0$ and accordingly
$\Delta H = Tm \Delta S$.
If we assume that ΔS is independent of the temperature than the GC dependence of Tm should be paralleled by the same dependence of ΔH. This can be shown to be true (16).

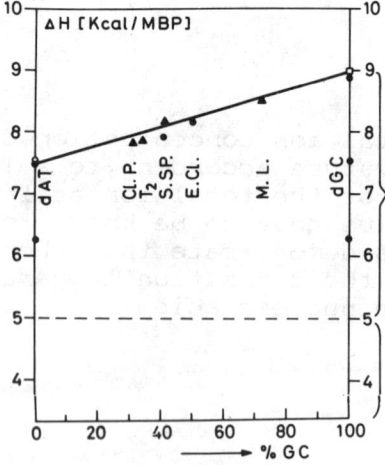

The transition enthalpy per mole of base pairs calculated for a hypothetical polymer of 100% GC amounts to 9 kcal and for 0% GC, i.e. 100% AT amounts to 7.5 kcal. The difference is due to one additional H-bond in the GC pair compaired to an AT pair. This result enables us to separate the contribution of the H-bond to the over all stability of the double helix. If we subtract the value for two H-bonds from the transition enthalpy of an AT pair and for three H-bonds from the enthalpy of a GC pair we remain with the basic contribution of 4.5 kcal from all sources. The message is that the H-bonds which serve exclusively to discriminate the complementary bases, contribute only half the amount of the stabilizing forces.

5. EVALUATION OF THE STACKING ENTHALPY OF POLY A

There is an independent method to determine the stacking enthalpy. The helix formation of polyadenylic acid and polyuridylic acid can be monitored in an isoperibolic calorimeter at different temperatures to give the temperature dependence of the energy of formation of an AU base pair (17). To calculate the stacking enthalpy of adenine from these results we have to argue as follows:

The final state under all temperature conditions chosen for these experiments is the complete double helix, i.e. the final state is always identical. The initial state is a function of the temperature in so far as the conformational state of adenine, the degree of single strand stacking is more pronounced at lower temperatures. Consequently the degree of order in the adenine strand is changed less when the helix formation occurs at lower temperature. This is reflected by a lower heat of formation at lower temperatures. Extrapolation to a value of complete order also in the single stranded structure allows to calculate the stakking enthalpy per adenine ring. The amount is 2.5 kcal per residue or about 5 kcal per base pair. This is in good agreement with our extrapolated base value for the DNA. In the mean time Breslauer and Sturtevant (18) on one hand looking for oligonucleotides and Privalov and coworkers (19) on the other hand looking for polyadenylic acid have confirmed the stacking enthalpy os adenine. So we can be pretty sure about the reliability of our energy separation scheme. In the mean time we have completed our investigations on the different base pairing so that we have at our disposal a set of data to calculate and compare DNA sequence stabilities in advance. We will demonstrate this for a sequence of 359 bases of a viroid.

6. THERMAL DENATURATION OF VIROIDS

Viroids are extremely small, covalently closed single stranded RNA rings,

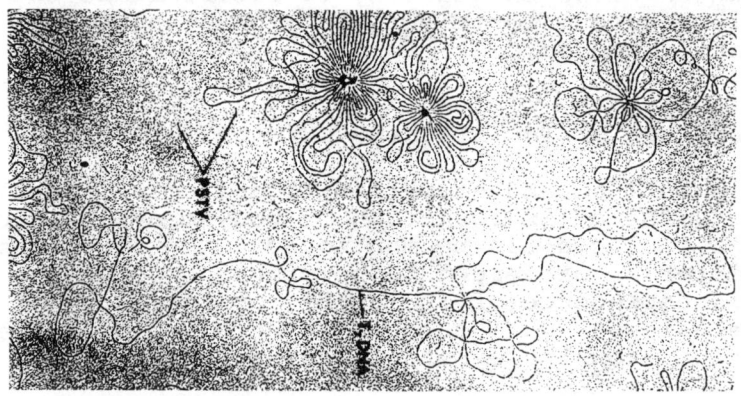

an unique arrangement for heredity material (20), which in the native state are almost completely doublestranded. From the UV melting it was known that the helix coil transition occurs at a lower temperature as should result from GC content of the species and the temperature interval is extremely narrow -2°- unknown for any information bearing sequence. From the absorbance versus temperature function the apparent transition enthalpy was calculated and it proved that the description of the transition of the viroid by a two state model was insufficient. An elaborated statistical mechanical approach was performed taking some special assumptions about the shape of the virion into account. The picture shows the model of the structure.

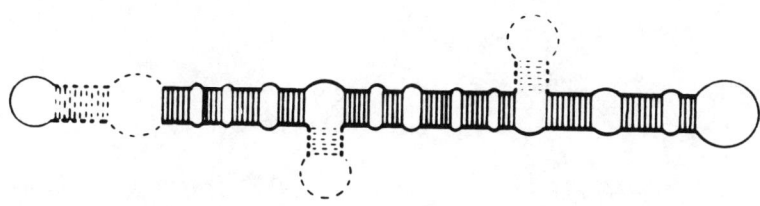

We deal with a dumbbell structure with two loops at
the ends and an imperfect helix in between. This particular structure was exclusively derived from physical
chemical studies and it was confirmed by the sequence
work (21) published almost two years later.

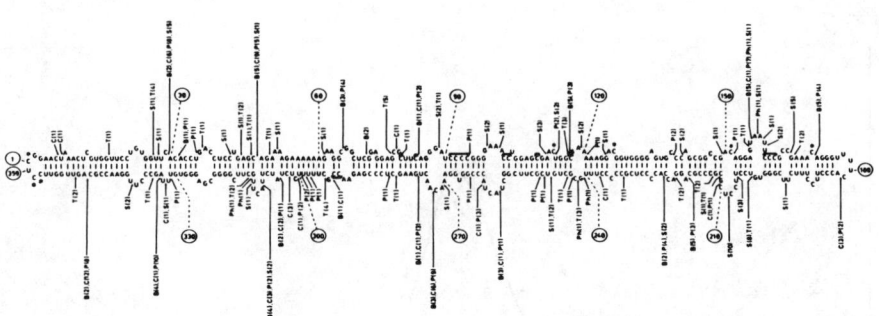

To complete the energy calculations it was necessary to
determine the true transition enthalpy of the complete
viroid by calorimetry. The question then was whether
the sensitivity of any calorimeter would be sufficient
to cope with 150 ug of the purified material, the maximum amount available for physico chemical measurements (22). This meant at that time to measure the

smallest amount of transition enthalpy up to now, some kind of world championship for calorimeters. The slide shows the original heat capacity curve,

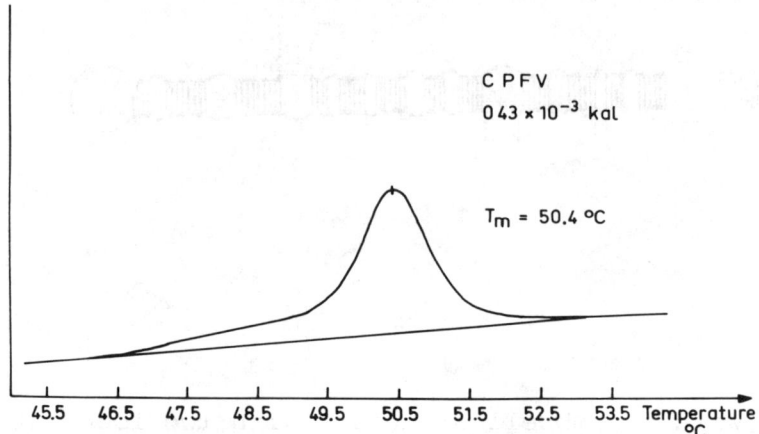

registred by the Privalov calorimeter of the University of Oregon in Eugene. It shows the experimental data for citrus exocortis viroid (CEV) and for potato tuber spindle viroid (PTSV).

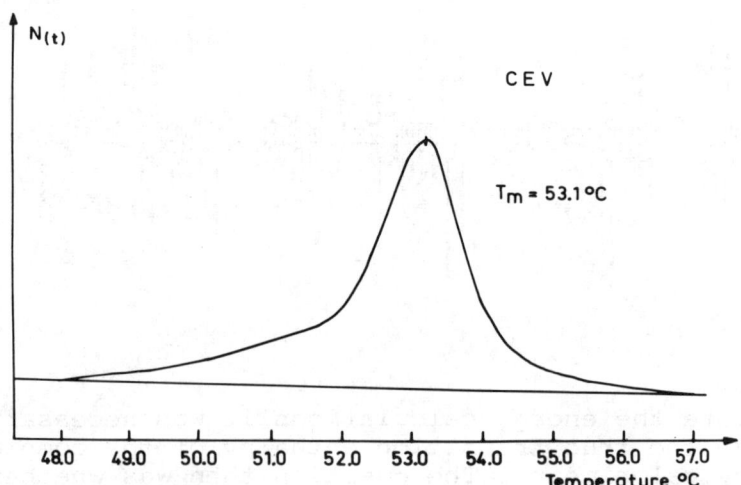

The peaks are almost symmetric with a slight tail at the low temperature side. Repeated heating and cooling of the same sample resulted in a shift of the peak area in favor of the tail.

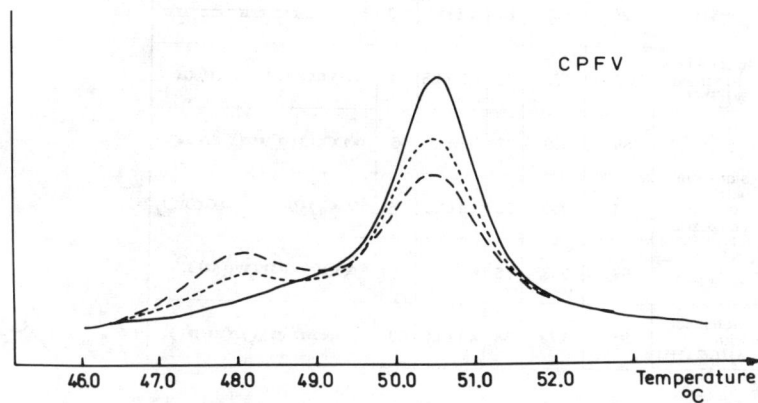

The gel chromatography could show that this resulted from turning part of the covalently closed circles into open chains with no change in molecular weight. There is only one single strand break within the molecule. The calorimetric results confirm the optical measurements. from comparison to a calibration area the transition enthalpy for a viroid molecule can be determined. The partial results for the two viroid species are very similar. The calorimetric determined transition enthalpy of 3930 kJ/mole reflects the amount for the complete molecule. It is independent from any assumption of the shape of the molecule. Division by the transition enthalpy of a base pair demonstrates that in the initial state 112 base pairs are formed, arranged in a helix with small single stranded regions interdispersed. The calculation of the stability of the secondary structure by utilizing the increments (23)

Sequence	Number of base pairs	$\Delta G^{25°C}$ [kJ/mol]	T_m [°C]	Average helical length	$\Delta T_{1/2}$ [°C]	Secondary structure
PSTV	124	641	84	4.8±1.8	1.0	
hypothetical (−) strand of PSTV	117	552	72	4.3±1.5	>1	
statistical with nucleotide composition of PSTV	99	460	77	5.0±1.9	15	
	101	360	69	4.1±1.0	5	
	99	353	68	4.3±1.2	5	
statistical with A:U:G:C = 1:1:1:1	95	272	58	4.1±1.0	30	

derived from polynucleotide sequence with neglect of all structural pecularities only taking the computerized base pairing scheme, results in a total value which matches the calorimetric value within 2%. This assures us to rely on the increment values for further calculations.

7. UNFOLDING OF NUCLEOSOMES

Stabilities of naked nucleic acid are irrelevant with only a few exceptions. Far more important are investigations of functioning nucleic acid protein complexes. As an example we will discuss the analysis of the temperature induced unfolding of the smallest ordered structure within the chromosomes,

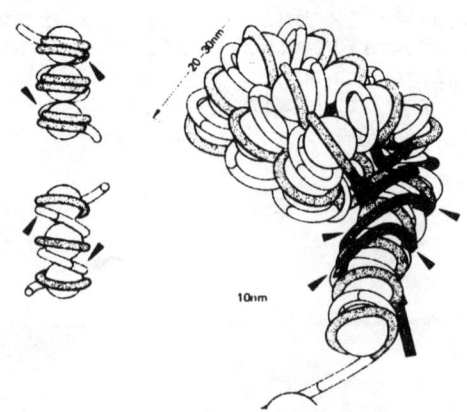

the nucleosome (24). The calorimetric results were also obtained with the Privalov calorimeter of the institute of molecular biology in Eugene.

The native isolated nucleosomes were prepared from chicken erythrocytes to get extremely homogeneous protein/nucleic acid complexes. They consist of a sequence of 140 bp of DNA and an octamer of the following four histons (H2A, H2B, H3, H4)$_2$. This structure is the basic building block of chromatin and the overwhelming part of all DNA is arranged in this manner. Therefore the illucidation of the stabilizing energies which contribute to structure are of prime importance. Which is the character of forces? Are all parts in the complex arranged in the same fashion? Which are the conditions for a conformational change of a core particle? The thermal denaturation of a nucleosome is depicted in the next slide and shows some typical pecularities which can also be shown by following the UV absorbance or the optical rotatory dispersion at two different wave lengths at 273 nm for the DNA and 223 nm for the protein as a function of the temperature. The intensity of the signal begins to change at about 50°C and the first maximum is approached at 60°C. The signal starts to increase again at 70°C and reaches a new and higher maximum at 74°C. The course of the optical rotatory dispersion shows that both peaks contain contributions from the DNA but only the main transition includes a contribution from the protein. The analysis of the UV-melting curve shows that 25% of the conformational change of the DNA occurs within the temperature interval of the pretransition and 75% occurs within the interval of the main transition. From these results we can conclude that out of 140 bp about 20 bp from both ends can change their conformation without breaking the whole structure apart. This rise of the temperature beyond the main transition maximum causes a thermal denaturation which cannot be reversed any more. Ultracentrifugal analysis shows that the molecular weight of the particles stays the same at all temperatures. I.e. the histons stay at the single strands and prevent the double helix from reformation. Which are the physical forces to keep this structure in its spacial arrangement? Nucleosomal arrangement is found in all eucariotic cells. Even the DNA of bacteria and of virions can be arranged in this manner (25). The universal distribution within the living world is only rasonable if we assume that an universal structural element of the DNA, namely the sugar-phosphate backbone, is evolutionary adapted to the surface of the histon core, and is highly conserved throughout the developement of species. The rearrangement starting at high salt and low temperature and proceeding stepwise to low ionic strength is mainly caused by electrostatic forces. Predominantly the phosphate groups of the backbone and the frequent lysine residues (25%) within the histons are involved. If this is true we should be able to prepare nucleosomes with no information bearing sequences involved and with ribose instead of desoxyribose as the sugar constituent.

Here I will demonstrate that this can be done (26) by reassociation of poly uridylic acid and classical histone cores under conditions, where poly U is shown to form a double helix. From melting curves

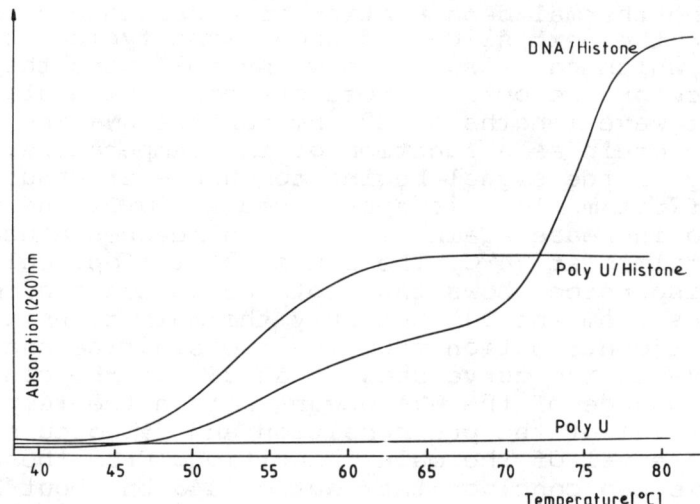

and from electron micrographs it is possible to demonstrate that the reconstituent nucleosomes are identical with native structures from chromatin. It is neither the bases nor the sugar which contributes to the special structure, it is the charge pattern on the surface of the double helix and on the surface of the histone core which have to match properly to come up with the right complex.

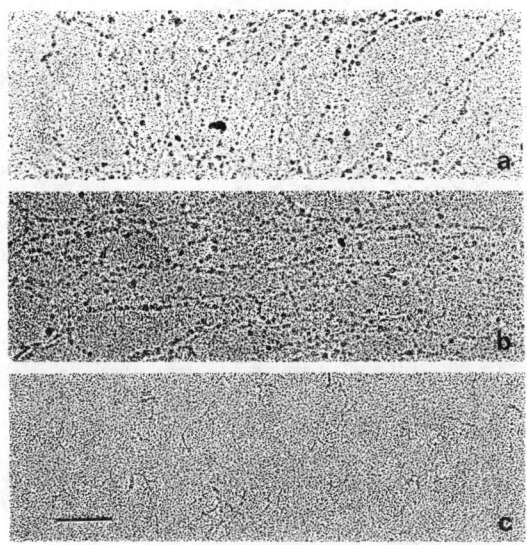

8. CONFORMATIONAL CHANGES OF LIPIDS

Finally I will demonstrate that also in the case of two dimensional structures calorimetry, in addition with Raman spectroscopy (27) can be used for a thermodynamic analysis of the conformational change of lipids within a bilayer. The aim of this research is to establish that selectively deuterated lipids

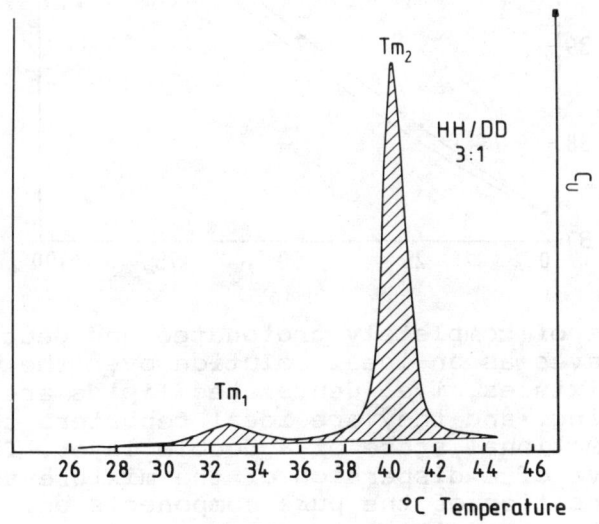

can serve as a nondisturbing reporter molecule. In principal the conformational change of Dipalmitoyl-phosphatidylcholin (DPPC) can be followed by Deuterium Resonance Spectroscopy or by investigations of the acoustical longitudinal skeletal modes of the hydrocarbon backbone vibrations. Since this two dimensional structural change of lipids is also a cooperative effect, we have to use calorimetry to determine the transition enthalpy per lipid molecule (28). To cut a long story short I will only sum up the results.

TABLE I

Thermodynamic Parameters for Aqueous Dispersions of Deuterated and Non-Deuterated Phospholipids

Sample	T	ΔH^*_{cal}	ΔH_{VH}	ΔS	Cooperative‡ Unit	T	ΔH_{cal}	Δ_{VH}	ΔS	Cooperative Unit
DPPC	41.4	8.8	839	28.0	95	35.4	1.2	338	3.8	323
DPPC-d_{62}	37.3	10.9	712	35.3	65	28.8	1.2	226	3.8	192
DPPC/DPPC-d_{62}										
.75	40.2	9.0	648	28.8	72	32.8	1.2	236	4.1	189
.50	39.0	8.9	361	28.5	41	31.6	1.2	190	4.0	154
.25	38.2	9.8	639	31.5	65	29.0	1.2	226	3.6	187
DPPC-d_{31}	39.1	9.6	827	30.7	86	30.9	1.3	281	4.2	216

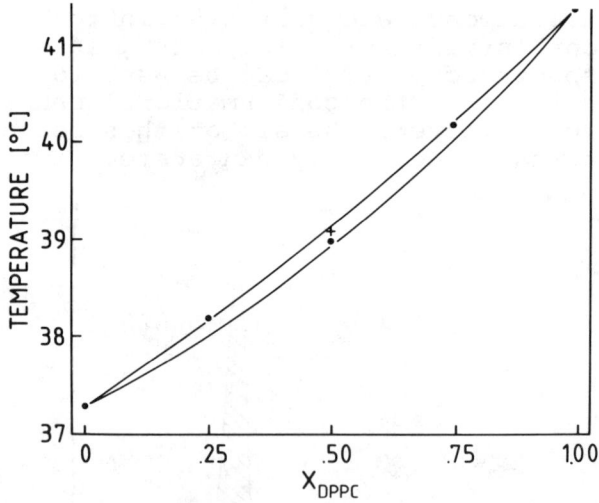

The mixture of completely protonated and deuterated lipids behaves as an ideal solution over the whole range of mixtures, i.e. deuterated lipids are truly nonperturbing, and they are ideal reporters to monitor the conformational state of a double layer. The transition curve of a dispersion of the mixture shows equal to the transition of the pure components only one pre- and one main- transition.

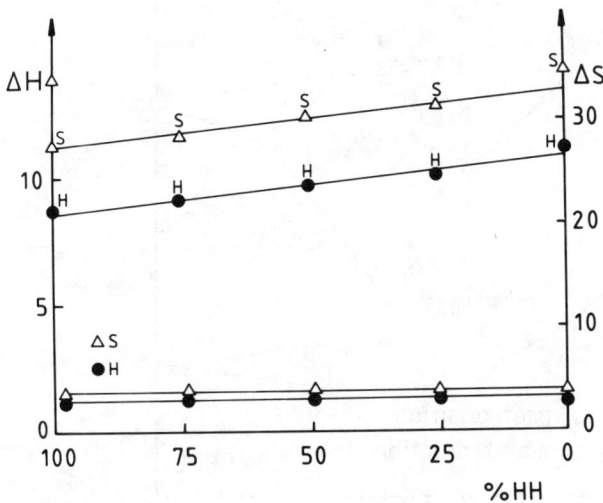

There is a linear dependence of the transition enthalpy on the mole fraction. There is almost no detectable phase separation during the transition. If we form dispersions of the components prior to the mixing,

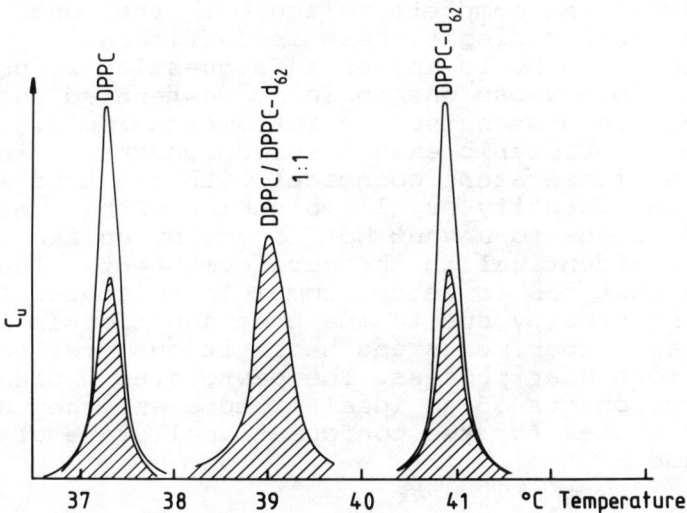

however, we see the separate transitions of the components. (Here clearly occurs a phase separation.) The calorimetric curves can also serve to calculate the $\Delta H_{v.Hoff}$. The plot of these enthalpies or entropies vs. mole fraction shows a distinct minimum at 50% DPPC-^{2}H.

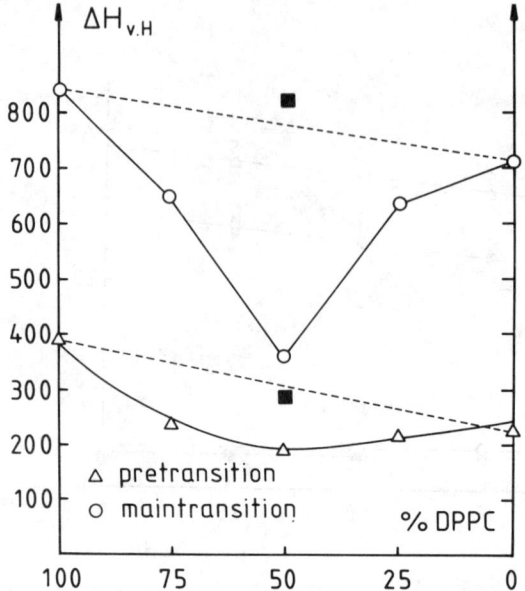

There is no melting temperature depression according to Raoul's law. The plot of the ratio $\Delta H_{v.H.}/\Delta H_{cal}$ vs. mole fraction demonstrates that the number of cooperative molecules drops to a minimum, when the fraction of deuterated chains reaches 50%. There is still to check whether the complete molecule is the fundamental unit or whether a single chain is sufficient to be considered as such. To answer this question we used a synthetic lipid whose one chain is deuterated and the other chain is protonated, an intramolecular 1:1 mixture. The calorimetric experiment demonstrates that the melting temperature coincides with the true mixture, that the identity of all molecules within the dispersion leads to a van't Hoff enthalpy unlike the mixture and identical to the pure components. The cooperative unit has to be the complete molecule. The transition enthalpy due to the pretransition is identical in all experiments and hence it must reflect the same molecular process. The deuterated lipids behave as components of an ideal mixture and they are excellent probes for the conformational state of a lipid phase.

9. CONCLUSION

This short review was compiled to demonstrate, that the thermodynamic analysis of conformational changes of biopolymers can tell us a lot about the molecular mechanism and that for these cooperative processes calorimetry is the only method to come up with quantitative description of the systems in terms of enthalpy and entropy changes.

REFERENCES

1. Clausius,R.: 1850, Annalen der Physik 155, pp.368ff.
2. Ackermann, Th. and Rüterjans, H.: 1964, Z. Physik. Chemie 41, pp. 116ff.
3. Ising, E.: 1925, Z. Physik 31, pp. 253ff.
4. Sturtevant, J.M.: 1969, Proc. 1st. Int. Conf. Calorimetry Warsaw.
5. Privalov, P.L., Plotnikov, V.V. and Filimonov, V.V.: 1975, J. Chem. Thermodynamics 7, pp. 41ff.
6. Diener, T.O.: 1974, Ann. Rev. Microbiol. 28, pp 23 - 39.
7. Marmur, J. and Doty, P.: 1962, J. Mol. Biol. 5, pp. 109ff.
8. Steiner, R.F. and Beers, R.F.: 1961, "Polynucleotides", pp. 21-63, Elsevier Publ. Comp., Amsterdam.
9. Watson, J.D.: 1977, "Molecular Biology of the Gene", pp. 73ff, W.A.Benjamin Inc., Menlo Park, Cal..
10. Chargaff, E.: 1950, Experimentia (Basel) 6, pp. 201ff.
11. Watson, J.D. and Crick, F.H.: 1953, Nature 171, pp. 737ff.
12. Klump, H. and Ackermann, Th.: 1971, Biopolymers 10, pp. 513ff.
13. Van Holde, K.E., Brahms, J. and Michelson, A.M.: 1965, J. Mol. Biol. 12, pp. 726ff.
14. Gruenwedel, D.W.: 1974, Biochem. Biophys. Acta 340, pp. 16-30.
15. Schildkraut, C.L., Marmur, J. and Doty, P.: 1962, J. Mol. Biol. 4, pp. 430ff.
16. Klump, H.: 1968, Thesis, Münster.
17. Krakauer, H. and Sturtevant, J.M.: 1968, Biopolymers 6, pp. 491ff.
18. Breslauer, K.J. and Sturtevant, J.M.: 1975, J. Mol. Biol. 99, pp. 552ff.
19. Privalov, P.L.: 1974, FEBS Letters 40, pp. 140ff.
20. Langowski, J., Henco, K., Riesner, D. and Sänger, H.L.: 1978, Nucl. Acid Res. 5, pp. 1589ff.
21. Gross, H.J., Domday, H. et al.: 1978, Nature 273, pp. 203-208.
22. Klump, H., Riesner, D. and Sänger, H.L.: 1978, Nucl. Acid Res. 5, pp. 1581ff.
23. Klump, H. and Breslauer, K.J.: to be published in Biopolymers.

24. Weischet, W., Tatchell, K., Van Holde, K.E. and Klump, H.: 1978, Nucl. Acid Res. 5, pp. 139-151.
25. Olins, D.E. and Olins, A.L.: 1974, Science 183, pp. 330ff.
26. Klump, H. and Hütig, H.: 1980, Ber. Bunsenges. Physik. Chemie 84, pp. 250-253.
27. Klump, H., Schmid, E.D. and Moschallski, M.: 1982, VIIIth. Intern. Conf. on Raman Spectroscopy, Bordeaux.
28. Klump, H., Gaber, B.P., Peticolas, W.L. and Yager, P.: 1981, Thermochim. Acta 48, pp. 361-366.

SUPERHELICITY AND ENERGETICS OF STRUCTURAL TRANSITIONS OF NUCLEIC ACIDS

H. Klump

Institut für Physikalische Chemie der
Universität Freiburg, West Germany

ABSTRACT

The model of the DNA double helix, first proposed by Watson and Crick in 1953, has been unchallenged for almost twentyfive years. Advances in polynucleotide synthesis and more recently in oligonucleotide chemistry ,however, have given access to a large variety of sequences with particular structural features. In this paper we will review the different helical structures and outline the experimental methods to evaluate the energetics of structural transitions of nucleic acids.

Besides the traditional B-DNA structure we will first discuss proposals of alternative arrangements of the two single strands. We will further demonstrate that single strands, triple strands and even quadruple strands can contribute to ordered secundary structures, which are stabilized by the same group of interactions as the Watson Crick structure, namely H-bonds within the plane of the bases ,and stacking interactions along the the axis of the helical rods.We can vary the number and the direction of the H-bonds ,we can omit the covalently linked backbone and still form a linear helix.

The next higher level of complexity is reached when both ends of a linear helix are linked to form a closed circel. To compare circular closed DNAs with linear sequences the concept of superhelicity is introduced to describe the order disorder transitions of DNA rings.

1. INTRODUCTION

Since the double helix was discovered almost 30 years

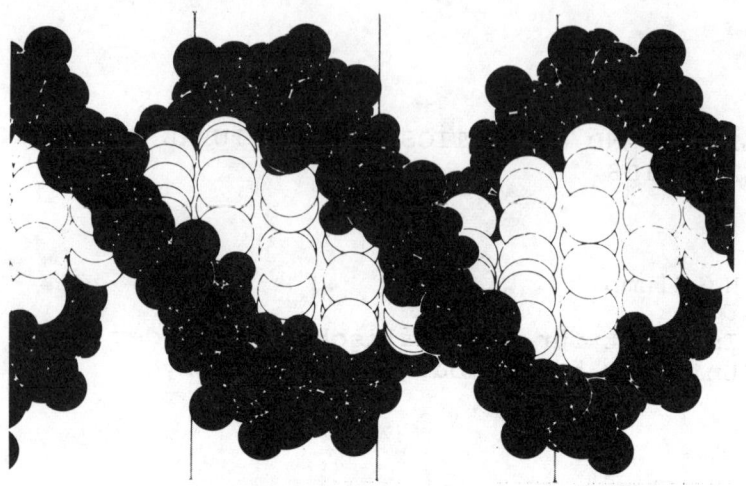

ago, most of its general features of the structure have stood up very well to experimental tests (1). It could be shown that DNA is usually double stranded with the exception of some viruses and phages.

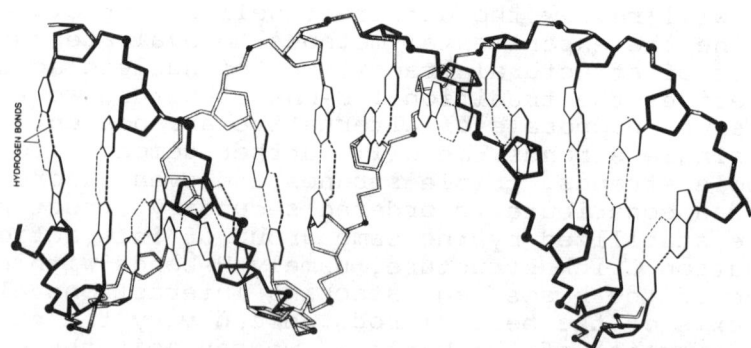

The sugar-phosphate backbones run antiparallel rather than parallel. The bases, one from each chain, are paired in the classical way, that is adenine exclusively pairs with thymine and guanine does so with cytosine. These statements are supported by a large body of experiments. On the other hand, there are other features of the structure which are highly plausible so that we are willing to accept them intuitively but which are not supported by experiment to anything like the same extent. In a number of recent papers it is suggested in contrast to the classical picture that the two strands of DNA do not coil round one another but lie side-by-side (2).

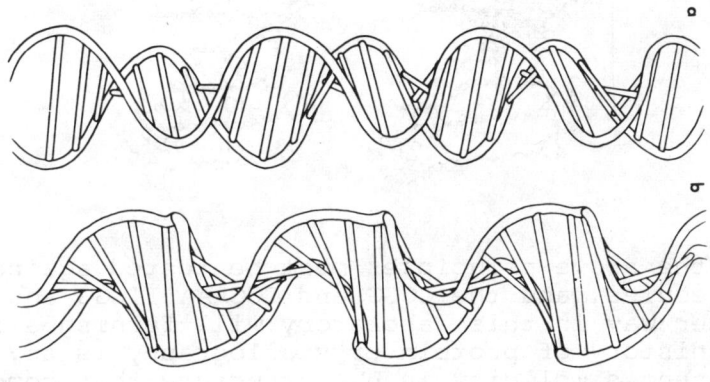

In this outline we will try to answer the question: Is the structure really a proper double helix, with the two chains wound plectonemically round a common axis? Is the helix predominantly right-handed (as originally claimed) or left-handed? But we will also demonstrate that there are a lot of different structures possible for the arrangement of nucleic acids from single stranded, open helices to four stranded helices. Aditionally the first structural analysis of a DNA single crystal by Dickerson and Drews (3) suggests that the easy times of a homogeneous continuing double helix are gone in favor of a much microheterogeneous picture of the DNA structure. To use Dickerson own words in the last paragraph of their paper: "After this first B-DNA single -crystal structure, we are in roughly the same position that protein crystallographers were in 1959 after the first globular protein structure analysis. We know much about one molecule, but are unsure as to how much this can safely be generalized to similar molecules with different sequences.

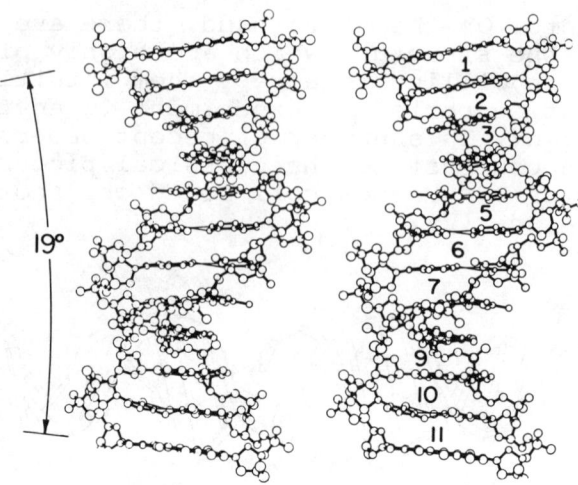

All of the above principles must be tested against other
DNA molecules, and the CCGG and CCCAAATTTGGG analysis
now under way in this laboratory will furnish a start.
If the history of protein crystallography is any guide,
the present simplicity in DNA structure that comes
mainly from ignorance will shortly be replaced by a
bewildering complexity of new data. before it ultimately settles down again to the simplicity that means
that we truely understand matters. The CGCGAATTCGCG
structure is a beginning.

2. THE Z- DNA STRUCTURE

After twenty years of slow motion there is a lot of
whirling in the field of nucleic acids. The first
major shock occured in 1972 when Pohl and Jovin (4)
published their paper on the CD measurements of poly
d (G–C) in 4M NaCl. The CD signal was completely
reversed as compared to the signal at 0.5M NaCl, suggesting that the secondary structure changed from the
classical right handed form into the left handed form.
For some time there was no more experimental evidence
for this besides the inversion of the Cotton effect.
Finally Rich and coworkers came up with the X-ray
defraction analysis of this structure (5) derived from
hexanucleotides of the same sequence which demonstrated
the existence of the zig-zag DNA-structure. The mere
existence of the Z-DNA as this structure was called
besides the allready well known B-DNA and the less
important A-DNA was noteworthy.

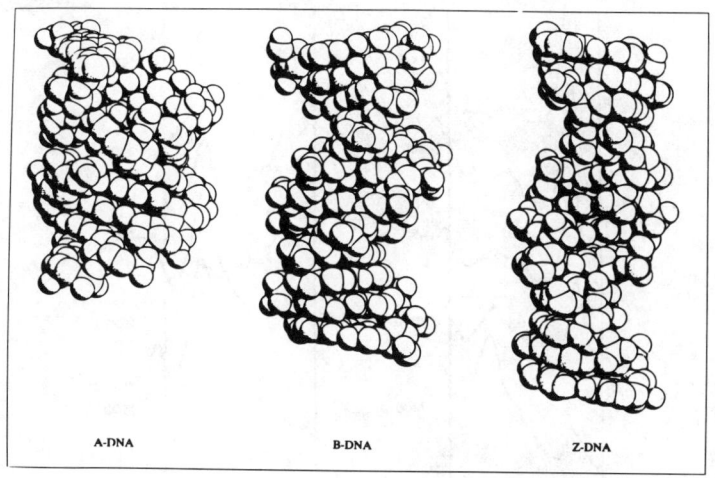

A-DNA B-DNA Z-DNA

Since then it was possible to induce antibody reaction against this Z-DNA (6). These antibodies were used to find out wether Z-DNA exists in chromatin e.g. There are kinds of streches of this structure in the silent genes and it was suggested that B-DNA - Z-DNA switches are used to switch off genes and the reverse reaction to turn them on again. It is still very speculative if this is all true but it is a suggestive possibility to answer the fundamental question how genes are activated.

The calorimetry can serve as a tool to test wether this inversion of the secundary structure is accompanied by any energetic effect or if this is an entropy driven reaction. The switch from B to Z is accompanied from a destinct change in the Raman spectrum (7). When the ionic strength of a poly d(GC) solution is drastically above 3M Na^+ the occurence of a new Raman band at 625 cm^{-1} is found which can be utilized to evaluate the kinetics of this reaction at different temperatures.

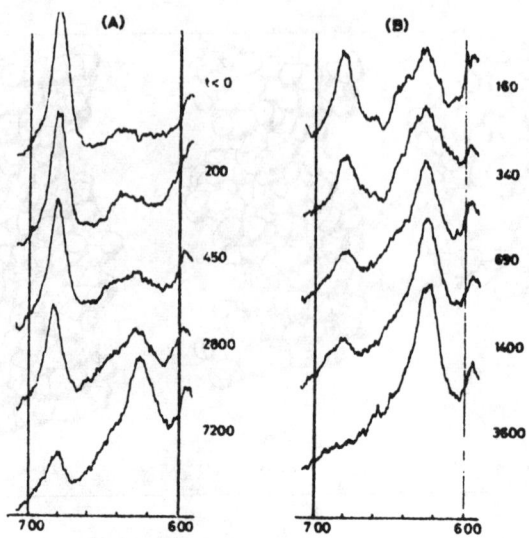

The process is very slow at 5°C, so that the poly d(GC) can be mixed with a certain amount of cold saturated NaCl-solution to yield a final cation concentration of 3.5M. This solution can be filled into the calorimeter cell and the scan to follow the conversion can start within 15 min from the mixing. The rate of reaction is increased by increasing the temperature of the cells, and the polynucleotide is pushed into its new conformational state. It was somehow disappointing that we could not detect any enthalpy change during this process. This was also controlled by performing the backward reaction by dilution of a high salt solution in the instrument. We have to conclude from this that B Z conversion is an isenthalpic or entropy driven process. This is a result of energetics of structural transition of nucleic acids as it stands totay (8).

3. THE WOBBLE PAIR

But let us turn a little to some earlier investigations in this field. As already mentioned before the B-DNA or "Watson-Crick Helix", consists of two polynucleotide strands which coil plectonemically round one another with the sugar-phosphate backbone running antiparallel. The spacing between two adjacent base pairs is 3.4 Å and the helical turn contains 10.3 base pairs. The base pairs, one from each chain, are paired in the classical way (9). Hoogsteen (10) could demonstrate from X-ray crystallography of mononucleotides that under special conditions there are quite different pairing schemes possible and we will come back to this when we argue on polynucleotide structures. Already Crick pointed out in the elaboration of the genetic code that a GU-base pair (11), the so-called "wobble base pair" should exist. It is also realized in parts of the t-RNA structure. How can the existence of wobble base pairs be approached experimentally? This question can be tackled as follows (12). Since the wobble pairs are assumed to be less stable than the classical base pairs one has to construct a double helix, which consists of a major fraction of standard base pairs such as AU pairs with occasionally inserted GU base pairs. This can be achieved by pairing poly uridylic acid in one strand with a statistical copolymer of adenylic acid and guanylic acid (poly A,G) with an input ratio of 4:1 (A:G). This ratio assures that there are at least short AU-helices with small loops when the GU pair does not form. The mean cooperative length should be around four. The existence of wobble pairs within the AU host helix will lead to a drastic increase of the cooperative length. A calorimetric investigation, accompanied by spectroscopic methods could demonstrate, that GU pairs are formed and the formation enthalpy of a wobble pair is about 5 Kcal/mole (13). The cooperative length is about 25 residues and the transition temperature is about 15°C lower than the T_m of pure AU double helix under the same experimental conditions. It is crucial for the success of this investigation to keep the guanine fraction low to avoid other G_s as nearest neighbours, for guanine can form more stable bonds to guanine than to uracil. A higher guanine content would exclusively lead to these GG-pairs leaving out all possible GU pairs. Thus it is possible to demonstrate that wobble pairs are possible but they can only occur at separate positions. A stretch of wobble pairs within a sequence is excluded by these results.

4. THE VARIETY OF POLYNUCLEOTIDE STRUCTURES

But double helices are not the exclusive secondary structures for polynucleotides. We will go through a couple of examples to demonstrate that besides the classical double helix with its antiparallel arrangements of the strands and Watson and Crick base pairing scheme there are structures with parallel strands and there are structures with non-Watson Crick base pairs and we find single strand structures, triple strand helices and even four stranded helices. Finally we will come to the supercoil which is essentially a four stranded helix with no particular interactions between the two double helices.

4.1. The poly G structure

When heating highly polymerized poly G at ph7 in 10nM NaCl we will obtain the following results (14). In the lower part of the figure the calorimetric transition curve is shown.

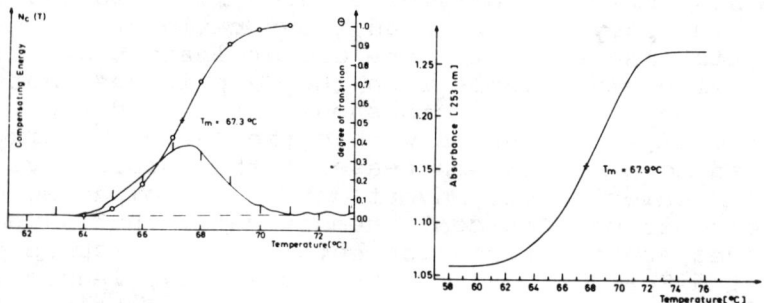

The sigmoid curve (with open circles) represents the degree of transition calculated from the experimental curve. The area under the heat absorption is proportional to the enthalpy change accompanying the thermal transition of poly G. The heat denaturation profile, obtained from the absorbance at 253 nm versus the temperature is almost identical with the integrated calorimetric curve. It is evident that the temperature course of the two parameters (additional heat capacity and UV absorption) follow the same transition of poly G from an ordered single stranded structure to the randomly coiled state. How can we be sure that we are looking at a single strand to coil transition? The evidence comes from two independent sources. First, the ionic strngth dependence of this transition temperature is in contrast to a double helix coil transition almost negligible. Second, the experimental transition enthalpy H is only 2.2 ± 0.3 Kcal per mole and the transition entropy is about 7 e.u. This low value is due to the fact that poly G is exclusively stabilized by stacking interactions. The transition enthalpy is equivalent to the stacking enthalpy. The value is in good agreement with the result we obtained for the stacking of adenine residues (15). The cooperative length can be calculated from the ratio of $\Delta H_{v.Hoff}$ and ΔH_{cal}. $\Delta H_{v.H.}$ may be calculated according to the following equation.

$$\Delta H_{v.H.} = 4RT_m^2 \frac{\delta \theta}{\delta T}$$

$\Delta H_{v.H.}$ yields 136 Kcal, leading to a cooperative length of 62 bases. Privalov and coworkers measured the thermal transition of poly A under comparable experimental conditions. They obtained 2.5 Kcal for the stacking enthalpy of adenine. In this case the cooperativity is much smaller but in both cases calorimetry is the only method to determine this quantity unambiguously.

4.2. Single stranded structures

Guanine residues can form a different "single stranded helix" even without any backbone. It is a well known fact that concentrated solitions of guanylic acid can produce highly viscous gels in aqueous solutions of ph 5.0 . For 5'- GMP the structure actually formed is a continuous helix with each nucleotide joined to the next by two hydrogen bonds. The helix consists of 15 nucleotides in four turns with a pitch of 3.33 Å.

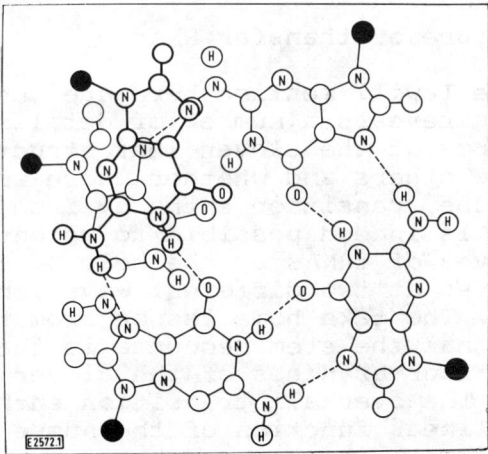

In contrast to the "normal" Watson Crick helix it is formed by these monomers excluding any backbone support and it is stabilized by hydrogen bonding and stacking interactions. Within the range of concentra-

tion investigated here the dependence of H and S on the monomer concentration is linear. The extrapolation to the reference state of ideal dilution gives
$\Delta H = 2.3 \pm 0.5$ Kcal/mole and $\Delta S = 9.0$ cal per mole and degree.

4.3. The viroid structure

Let us turn from the single stranded structures to the non classical double stranded helices. The viroids clearly belong into this class. To remind you we will briefly recall that viroids are single stranded covalently closed RNA-rings with a high fraction of base pairing (16). A detailed analysis of the secondary structure shows that we find 16 helical sequences interrupted by small loops of three to five unpaired residues. The loop is sufficiently small that it does not entirely interrupt the interaction between adjacent helices but it leads to a decrease of the transition temperature. This is an important result for the discovery of the viroid replication. It is generally noticable that RNA sequences are more stable than the homologous DNA sequences. In the case of viroids this defective RNA helix was accepted by the DNA-dependent RNA polymerase as if it was a DNA. The thermodynamic stability of the secondary structure serves on the molecular basis to discriminate between the different nucleic acids.

4.4. The structure of transfer RNAs

Another example I will mention here are the transfer RNAs. The tRNAs reveals a number of details, e.g. which of the arms of the clover leaf structure is more stable than the others and whether there is a linear dependency of the transition enthalpy from the number of H-bonds. It is indeed possible to deconvolute the transition curves of tRNAs by the help of differential UV-melting curves at two different wave lengths (260 nm and 280 nm). The take home lesson from these investigations is that the stem sequence is the most stable and the four branches of the clover leaf melt independently. The over all transition enthalpy per molecule is a linear function of the number of H-bonds per molecule.

4.5. The structure of methylated poly A

To finish up with the odd double helices I will present you a structure with exclusively Hoogsteen-base pairs (18). When poly A is treated with dimethylsulfate under basic conditions the first attack of the methyl group centers on the N-6 position.

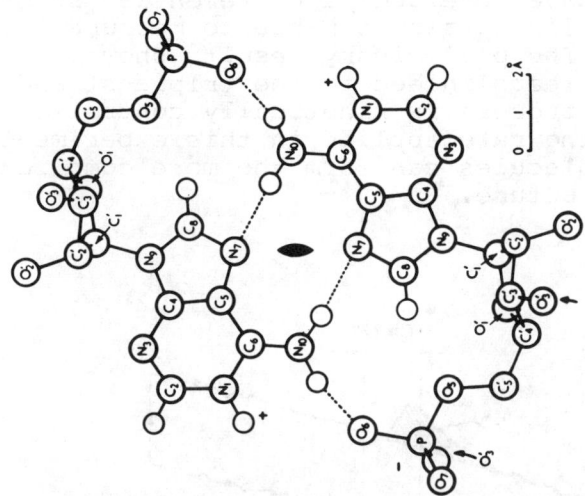

The consecutive reaction is a Dimroth-rearrangement which brings the methyl group into N-1 position so that this position is excluded from the possibility to function as an hydrogen acceptor in a Watson Crick base pair. The pairing scheme of the new arrangement is shown on the slide. The H-bonds bridge the N-6 and N-7 position of each adenine in a symmetrical fashion. The same structure is obtained when poly A is protonated below pH 5.9.

4.6. The triple helix formation

Helix formation does not stop with a double stranded structure. It can be demonstrated that triple stranded helices can exist when the concentration of monovalent or divalent cations in the solution exceed a threshhold value. This was shown very early in the sixties for the system poly A + poly U when a polynucleotide solution of stoichiometric amounts of both polymers was heated in the presence of 0.30M NaCl (19). The UV absorbance monitored at 280 nm starts to drop at about 50°C, due to a new structure, and it starts to rise again above 60°C. This new structure results from a rearrangement of two double helices into a triple helix of poly A 2 poly U and a partly ordered poly A single strand. It is interesting to speculate whether the reversed process, the cooling of a solution which contains only random coils of all species present, will lead immediately to the double helices or whether the triple helix formation is an indistensable intermediate step in both directions. In the mean time Grubert and Messner (20) have developed a differential scanning calorimeter which is also capable to measure the cooling process. The preliminary results show that the reverse process leads indeed to the triple strand formation but this process is kinetically controlled and with the cooling rate applied in this experiment only part of the molecules can form the more complicate intermediate structure.

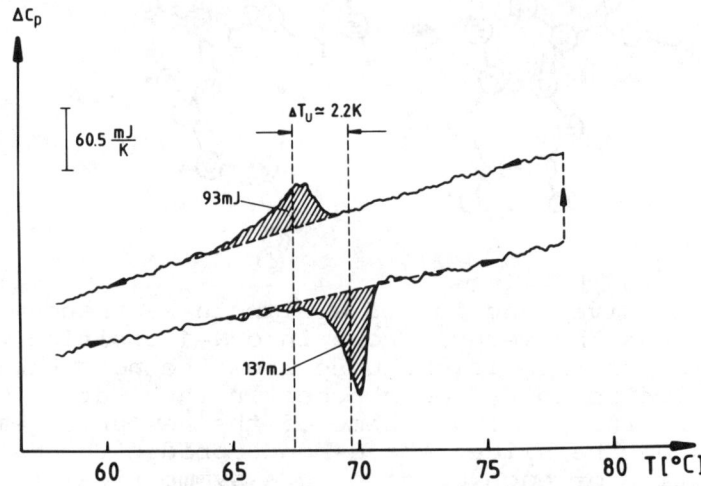

Finally all strands end up in the double helices again. But this rearrangement to a triple helix is an exception. The usual way is either mixing of the constituents of the ratio 1 : 1.1 under proper conditions or stepwise formation of a double helix followed by the addition of a third polymer strand.

The general composition is of the type : $(pyrimidine)_n \cdot (purine)_n \cdot (pyrimidine)_n$, with the $(purine)_n$-strand consisting either of monomers or oligomers or polymers. The two homopolymer strands of a purine and a pyrimidine polymer form a "Watson Crick" double helix and the second pyrimidine strand binds via "Hoogsteen" H-bonds to the purine. From the X-ray analysis of the tertiary structure of t-RNAs it is shown that the base triplets serve to stabilize this conformation. The only significant triple helix, however, might be the transcription complex of the DNA double helix and the messenger RNA. Since this complex is very weak it is impossible to isolate it for biophysical measurements. Poly dA · poly dT · poly rU (21) represents a suitable model system for investigation of DNA-mRNA like interactions. Mixing of equal amounts of poly dA and poly dT in 1 mM phosphate, 10 mM NaCl at neutral pH leads to a double helix and adding stoichiometric amounts of poly U under the same buffer conditions with 1 mM $MgCl_2$ present leads to the proper triple helix formation, which can be followed in the UV as function of the Mg^{2+} concentration. Band velocity sedimentation measurements decide clearly in favour of the triple helix formation. The thermal transition curves indicate that the poly U strand dissociates first from the complex at 50°C, followed by the helix coil transition of poly dA · poly dT at about 70°C. The enthalpy of formation of the double helix poly dA · poly dT was measured directly by the help of d LKB batch calorimeter to yield 8.2 Kcal/MBP. This value is slightly higher than the enthalpy of formation of the poly A·poly U helix. The difference results from the different starting conditions. Poly dA is less ordered than poly A so that the formation of the former helix gains from the increased stacking of poly dT in the ordered secondary structure. The addition of the poly U strand yields 2.3 Kcal per mole uracil. As compared to the poly A · 2 poly U structure the second poly pyrimidine strand in the hybrid triple helix is less tightly bound, leading to a consecutive dissociation of the mRNA like strand while the DNA like double helix remains stable under these conditions. This behavior is assumed to reflect the conditions in the native transcriptional complex.

4.7. The four stranded helices

Let us now leave the triple helical structures and turn to the next higher structure, the four stranded helices. Model building of helical structures shows that crossing over of DNA double helices within chromosomes can lead to intermediates with a four stranded secondary structure. This might be plausible but it is by no means proven by experiments. There are indeed true four stranded helices available. They all result from the fact that guanine residues are capable to form simultaneously Watson Crick and Hoogsteen base pairs which lead to an "in plane" arrangement of four bases, connected by eight H-bonds. The ease with guanine derivatives can form molecular aggregates probably explains why guanylyl-(3'-5') guanosine (GpG) e.g. has been avoided in many studies involving the other dinucleotides (22). Nevertheless, a reasonable model has been proposed for GpG by Chantot (23). In this model the guanosine residues form a pseudooctamer from four GpG molecules in a four stranded helix arrangement, where one layer of bases is rotated 45° with respect to the other layer. The Raman investigations can be used to demonstrate that a cooperative melting transition takes place in GpG, resulting in the breakdown of the polymeric superstructure. The calorimetric measurements were again made on a Privalov differntial scanning calorimeter. The calorimetric heat capacity curves obtained were differentiated in order to obtain the vant Hoff enthalpy change associated with the transition. For a simple two state transition involving the equilibrium $M_x \rightleftharpoons x\,M$ the enthalpy is calculated according to

$$\Delta H_{v.H.} = 2(X+1)\,R\,T_m^2 \left(\frac{\delta\theta}{\delta T}\right)_{T_m}$$

This simple model is indeed adequate to describe the melting of GpG at very low concentrations where the aggregation does not go beyond the tetramer stage $(GpG)_4$.

TABLE
Thermodynamic Data for the Melting of GpG in Aqueous Solution at pH 7

Method	Molarity	t_m (°C)	ΔH_{VH} (kcal)[a]	ΔH (kcal/mol) GpG	ΔS (e.u.)	$\Delta H_{VH}/\Delta H$
Raman	3.8×10^{-2}	60	130	—	—	—
Calorimetry	6.8×10^{-3}	54	117	9.20	28.1	12.8
	4.4×10^{-3}	54	123	9.24	28.3	13.3
	2.0×10^{-3}	51.5	135	8.64	26.6	15.6
	1.0×10^{-3}	49	100	8.95	27.8	11.1
	5.3×10^{-4}	46	97	8.99	28.2	10.8
uv	5.3×10^{-5}	44	37	(9.2)[a]	(28)	(4.1)[a]
CD	5.8×10^{-4}	47	125	9.00	28.1	13.8

The desintegration of these particularly stable species was probably th only cooperative step involved in the UV-melting curve determined at 5×10^{-5}M concentration. If we use a value of 4 for X in the formula given above the $\Delta H_{v.H.}$ calculated for process is 36.8 Kcal. Deviding this value by the number of subunits (for monomers) involved in the cooperative transition gives for the enthalpy of melting (ΔH) of GpG a value of 9.2 Kcal/mole which is almost exactly equal to the average (9.0 Kcal/mole) of the value obtained from the direct calorimetric measurements. With increasing concentration $\Delta H_{v.H.}$ increases. Even though in this concentration range the simple model adopted for the melting transition is not valid any more, it is clearly shown that the average number of particles involved in the cooperative step, as given by the ratio $\Delta H_{v.H.}/\Delta H$ becomes larger as the concentration is increased, indicating that octamers, dodecamers and probably higher aggregates are present. The calorimetric value of enthalpy of formation of the GpG aggregates remains approximate constant 9.0 Kcal per mole over a large concentration range. The average enthropy change upon melting GpG as calculated from the relation $\Delta S = \Delta H/T_m$ is 13 cal per mole and degree of residue, comparable to the values obtained for larger polynucleotides. It is noteworthy to recall that the entropy of transition per nucleotide increases stepwise from 10 to 20 to 30 to 40 in the row of single, double, triple and four-stranded helices. This can also be used as a hind for the particular secondary structure present in an experiment.

5. SUPERHELICITY

Finally we will come to superhelicity and the influence of supercoiling on the thermal stability of secondary structures. Initially there was evidence, largely from autoradiography and electron microscopy, suggesting that all DNA molecules were linear and had two free ends. However, as it became possible to look more easily at undegrated DNA, it was surprising to see that many DNA molecules have circular forms. This was first shown for small viruses like SV40 (24). But not only small DNAs can be circular. Most, if not all bacterial chromosomes turned out to have a circular configuration.

An appreciation of this structural feature and its consequences is therefore essential to a complete understanding of the biology of DNA. Supercoiling can take a variety of forms, as the superhelical solenoid of DNA in the chromosomes, but we will restrict ourselves to a type of supercoiling, in which no protein core is needed, it is the supercoiling of closed circular DNA. A review of this problem was recently published in Scientific American (25). In the following we will give a compressed excerpt from this paper.

The existence of these twisted rings of DNA was first postulated to explain a surprising discovery about the DNA of the polyoma virus, a small virus that causes tumors in the mouse. When this DNA is suspended in a solvent and spun in a centrifuge, it resolves into three components with different sedimentation velocity (24). As it turned out, these components, labelled I, II and III in order of their decreasing velocity, do not differ in molecular weight, in other words, the molecules had to have different shapes. The difference between component I and II of polyoma DNA is that the circular molecules of component II are "nicked", i.e. there is at least one nick or break in one of the polynucleotide chain. In contrast, the circular molecules of component I are closed and underwound. The closed circular molecules resist such underwinding and they compensate by forming supercoils, leading to a very compact structure. Virtually every physical, chemical and biological property of DNA are affected by closed circularity and the particular deformations associated with supercoiling.

5.1. The mathematical model

We will first outline briefly a mathematical model of this closed circular DNA and then discuss the implication for real DNA. To study supercoiling mathematically it is most convenient to construct a model in which the structure is represented as a twisted ribbon of infinitesimal thickness (25). The axis of the ribbon coincides with that of the double helix. In addition we specify that the ribbon must always lie perpendicular to the pseudodyades or twofold axes of rotation, that are distributed along the helix. In addition to enclude the sequence of the two chains in the model the edges of the ribbon will be assigned opposite orientations. When the ends of the ribbon are joined, each edge describes a closed curve in three dimensional space. This also implies that a number of 360-degree twists are introduced before the ends are joined. The two curves described by the ribbon edges are linked.

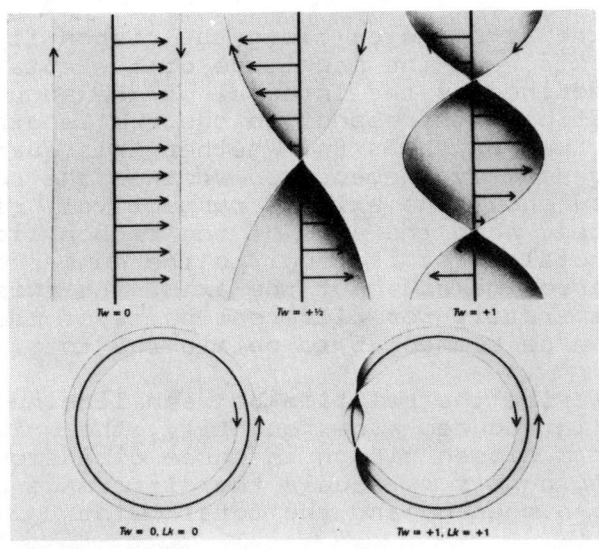

It is now impossible to separate the curves without cutting one of them. They are joined by a topological bond. No part of one molecule is covalently joined to any part of the other, it is nonetheless necessary to break a covalent bond in order to separate the two. This relation can be described mathematically by a linking number LK whose magnitude expresses the number of times one curve is linked through the loup of the other and whose sign depends on the way the curves are labeled. One way to compute this value quantitatively is to examine a projection, a two-dimensional representation of the two curves and to each point where one curve crosses over the other assign an index number according to the following rule: If a clockwise rotation is required to move the top piece to coincide with the bottom piece, then +1 is assigned to the crossing point, and -1 for the opposite movement. LK is than calculated by adding up the index numbers and dividing the number of curves (here: 2). The linking number obtained in this way is a signed integer equal to 0 if the two curves are unlinked, to +1 or -1 if one curve links through the other just once, to +2 or -2 if one curve links through the other twice. The sign of the number will change if the orientation of either one of the curves is changed. But the number remains the same no matter how the two curves are deformed. Another way to analyze the ribbon model of DNA is by not looking at the relation between its edges but at the way the ribbon twists. Twisting of a ribbon can be assigned a numerical value by placing a small arrow on the ribbon perpendicular to its axis and pointing to one of its

edges. As the arrow moves along the ribbon it rotates
about its axis, and the magnitude of the total twist
T_W can be defined as the integral of the angular rate
of its rotation with respect to the arc length of the
curve described by the axis. Whether this quantity is
positive or negative depends on whether the rotation
of the arrow about the axis is respectively righthanded
or lefthanded. When the axis of the ribbon lies in a
plane the total twist is equal to the number of rota-
tions the arrow makes about the axis. The twist can be
computed separately for different parts of the ribbon
and can then be summed up to obtain the total value.

To describe the relation between linking and twist,
Fuller has introduced a new quantity, the writhing
number. For a closed ribbon in three dimensional space
the writhing number W_r equals the difference between
the linking number L_k and the total twist T_w.

$$W_r = L_k - T_w$$

Linking number and total twist of a ribbon are equal
when the axis of the ribbon lies entirely in a plane.
These two quantities do not necessarily have the same
value, as can be shown by an example where the linking
number is positive, yet the total twist is close to
zero. When a ribbon is wound around a cylinder as shown
at the bottom left, L_k, T_w and W_r are all zero.

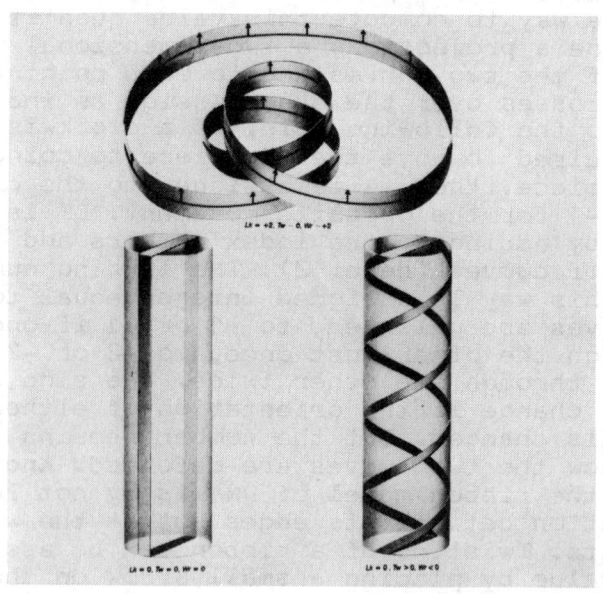

Rotating the top of the cylinder about its central axis results in a interwound right handed helix, in which the linking number is clearly unchanged but the twist is $+4 \sin\alpha$ and W_r therefore equals $-4 \sin\alpha$. The final configuration is a simplified model of supercoiled DNA. There is a difference between the linking number in the native closed circular state and the L_k of the same molecule in the relaxed closed circular state. This deficit in the linking number causes the molecule to supercoil.

5.2. Experimental results

All this mathematical reasoning seems to be an academic game, but in fact all the plasmids investigated so for are supercoiled. The question is how does this affect the energy of denaturation? There are very few experimental results but it can clearly be shown that superhelical molecules are much more stables than relaxed circular species. Sonication for 10 seconds turns them completely into relaxed molecules. The transition enthalpy of the relaxed DNA is in complete agreement with the results for linear DNA namely 8.5 Kcal per mole of base pairs (27). We have some preliminary results of supercoiled DNA.

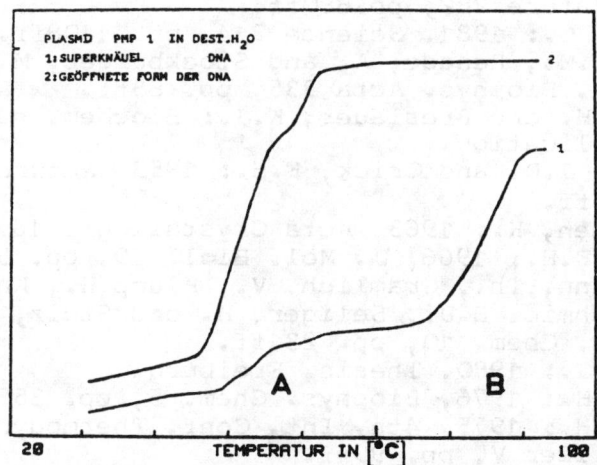

The transition enthalpy is about 50% higher, resembling the results for DNA which forms a different type of supercoils around the nucleosomal core.

CONCLUSION

The impression I wanted to give You in this talk was that there are much more helices around than the DNA and in the DNA case that the B-form is the least interesting one

ACKNOWLEDGEMENT

The author acknowledges the support of the Deutsche Forschungsgemeinschaft throughout the work published from this institute.

REFERENCES

1. Crick, F.H., Wang, J.C. and Bauer, W.R.: 1979, J. Mol. Biol. 129, pp. 449ff.
2. Pohl, W.F. and Roberts, G.W.: 1978, J. Math. Biol. 6, pp. 383ff.
3. Dickerson, R.E. and Drews, H.R.: 1981, J. Mol. Biol. 149, pp. 761ff.
4. Pohl, F. and Jovin, T.: 1972, J. Mol. Biol. 67, pp. 375ff.
5. Wang, A.H., Quigley, G.J., Kolpak, F.J., Crawford, J.L., Boom, H.J., van der Marel, G. and Rich, A.: 1979, Nature 282, pp.680ff.
6. Kolata, G.: 1981, Science 214, pp. 1108ff.
7. Pohl, F.M., Renade, A. and Stockburger, M.: 1973, Biochim. Biophys. Acta 335, pp. 85ff.
8. Klump, H. and Breslauer, K.J.: Biochem. submitted for publication.
9. Watson, J.D. and Crick, F.H.: 1953, Nature 171, pp. 737ff.
10. Hoogsteen, K.: 1963, Acta Crystallogr. 16, pp. 907ff.
11. Crick, F.H.: 1966, J. Mol. Biol. 19, pp. 548ff.
12. Ackermann, Th., Gramlich, V., Klump,H., Knäble, Th., Schmid, E.D., Seliger, H. and Stulz, J.: 1979, Biophys. Chem. 10, pp. 231ff.
13. Stulz, J.: 1980, Thesis, Freiburg.
14. Klump, H.: 1976, Biophys. Chem. 5, pp. 359ff.
15. Klump, H.: 1975, 4th. Int. Conf. Thermodyn. Montpellier V, pp. 99ff.
16. Klump, H., Riesner, D. and Sänger, M.L.: 1978, Nucl. Acid Res. 5, pp. 1581ff.
17. Schott, F.J., Grubert, M., Wangler, W. and Ackermann, Th.: Biophysical. Chem. (in press)

18. Klump, H., Sturm, J. and Peticolas, W.L.: 1981, Ber. Bunsenges. Phys. Chem. 85, pp. 661ff.
19. Felsenfeld, G.: 1958, Biochim. Biophys. Acta 29, pp. 133ff.
20. Messner, G. and Grubert, M.: 1982, J. Sci. Instr., submitted for publication.
21. Klump, H.: 1978, Ber. Bunsenges. Phys. Chem. 82, pp. 805ff.
22. Savoie, R., Klump, H. and Peticolas, W.: 1978, Biopolymers 17, pp. 1335ff.
23. Chantot, J.-F., Haertle, T. and Guschlbauer, W.: 1974, Biochimie 56, pp. 501ff.
24. Vinograd, J., Lebowitz, J. and Watson, R. : 1968, J. Mol. Biol. 33, pp. 173ff.
25. Bauer, W.R., Crick, F.H. and White, J.H.: 1980, Scientific American 284, pp. 100ff.
26. Fuller, F.B.: 1971, Proc. Natl. Acad. Sci. 68, pp. 814ff.
27. Klump, H.: 1982, Biochemistry, submitted for publication.

CALORIMETRIC STUDIES IN BINDING REACTIONS IN BIOCHEMISTRY

G. Rialdi and S. Raffanti

Centro Studi Chimico Fisici Macromolecole
CNR - Corso Europa 30, 16132 GENOVA-ITALY

Binding between ions, molecules and macromolecules is an important, often pivotal step in many biological processes. The simple initial event of binding - two chemical entities coming together in a state of mutual attraction - is central to the vast and complex occurences that are studied under the heading of biochemistry.
Binding is usually an initial step which leads to subsequent more complicated effects. Transport of molecules across a membrane or in the serum of an organism is dependent upon the binding between the carrier and the compound being transported. The many hormonal effects so important in the delicate control of the cellular environment are all related to the function of binding as expressed by extracellular and intracellular receptors. Binding-induced conformational change is an important factor in the aggregation effects seen in immunology. All these phenomena are based on the central event of binding which can be represented by the following equation:

$$M + X \rightleftarrows MX \qquad (1)$$

We shall consider this equation reversible in order to apply classical thermodynamics. Of course the equilibrium can be shifted completely to the right, as in the case of precipitation or aggregation reactions.
For the reversible reaction described in equation (1):

$$K' = \frac{[MX]}{[M][X]} = \frac{MX}{[M_o-MX][X_o-MX]}$$

Where, K' is the association constant, $[M_o]$ and $[X_o]$ are the initial concentrations of M and X species.

We must also limit our discussion to simple monovalent binding taking place between two molecules. Polyvalent interactions, though involving a myriad of theoretical complications, can be analysed using the same approach. Major considerations would include independent versus dependent cooperative binding, which is beyond the scope of this discussion.

In equation (1) and (2), M could stand for a protein or nucleic acid while X could be any ligand, - a proton, ion, small molecule or large protein.

Though the binding _per se_ may be very simple as described in equations (1) and (2), the situation may be more complicated if the ligand is also involved in a change in pH or protonation of some buffer. In that case the situation would be represented as follows:

$$MH + X \rightleftarrows M + H^+ + X \rightleftarrows MX + H^+ \qquad (3)$$

$$H^+ + Buffer \rightleftarrows Buffer-H^+ \qquad (4)$$

Equation (2) would be modified to include the total of the two effects involved, namely binding and protonation.

Another complication of our simple binding situation could occur when X is affected by a conformational change. In the event that the conformationally transformed X_B has a higher affinity for M than X_A, the reaction is represented by:

$$M + X_A \rightleftarrows M + X_B \rightleftarrows MX_B \qquad (5)$$

The more common or possibly more biochemically significant situation occurs when the macromolecule, M, undergoes a conformational change upon binding the ligand:

$$M_A + X \rightleftarrows M_B + X$$
$$\updownarrow \qquad\qquad \updownarrow$$
$$M_A X \rightleftarrows M_B X$$

The most likely route taken for the formation of the ligand bound transformed molecule ($M_B X$) will be $M_A + X \rightleftarrows M_A X \rightleftarrows M_B X$ or, shortened, $M_A + X \rightleftarrows M_B X$. Calculation of K' depends on the respective affinities of $M_A X$ and $M_B X$.

Aggregation effects which occur following simple binding reactions offer another example of possible associated sequel to binding reactions. This reaction can be somewhat simply described by the equation:

$$M + X \rightleftarrows MX \rightleftarrows [MX]_n \qquad (6)$$

More often, all the above-mentioned phenomena – protonations, conformational changes, aggregation – are involved to some degree. A good example of this is the binding of Manganese to Concanavalin A (see below).

The calorimetric study of these binding reactions requires that all thermodynamic parameters of the reaction be recognized and related to the empirically obtained values. Measurement of thermodynamic parameters with standard methods relies upon the use of the two equations relating $\Delta G°'$, the apparent standard Gibbs energy change, to the binding constant (K') and the thermodynamic values for change in enthalpy ($\Delta H°'$) and entropy ($\Delta S°'$).

$$\Delta G°' = \Delta H°' - T\Delta S°' = -RT\ln K' \qquad (7)$$

The binding constant is usually determined by measuring the concentration of the product (MX, for example) as a function of one of the reactants using equilibrium dialysis, spectroscopic changes etc. $\Delta H°'$ can then be estimated from temperature dependent measurements of K' using the Van't Hoff equation:

$$-\frac{d(\ln K')}{d(1/T)} = \frac{\Delta H°'}{R} \quad (8)$$

$\Delta S°'$ can be calculated using the equation:

$$\Delta S°' = \frac{\Delta H°' - \Delta G°'}{T} \quad (9)$$

and finally the change in heat capacity ($\Delta C_p°'$) can be calculated from different $\Delta H°'$ values:

$$\Delta C_p°' = \frac{d \Delta H°'}{dT} \quad (10)$$

Microcalorimetry allows us to obtain both the binding constant and $\Delta H°'$ of a reaction simultaneously with good accuracy (1). The precise $\Delta H°'$ values obtained also permit calculation of $\Delta G°'$ from only a few experiments. Comparison of these values, which represent the overall heat effects, to those obtained with the Van't Hoff equation, especially when another "marker" reaction is involved (i.e. spectroscopic changes), allows us to separate the two effects and have a better idea of the reaction under study. The $\Delta H°'$ we record in the calorimeter is the sum of all heat contributions of each single event taking place in the reaction vessel.

At the moment we are not interested in going into details about the heat of dilution and other thermal effects which can be easily subtracted from the system. Nor are we interested in discussing the various types of calorimeters and accessory equipment available today (for a thorough discussion of these topics see Dr. Wadso's lecture).

An example of simple binding of a ligand to a nucleotide is provided by the reaction of Mg ions with various mononucleotides (2). The binding constant is as reported in equation (2) and the heat produced during the reaction will be proportional to the amount of MX formed according to the equation:

$$Q = \frac{[MX]}{[M]_o} \Delta H^{\circ\prime} \qquad (11)$$

where $\Delta H^{\circ\prime}$ is the molar enthalpic contribution of the reaction. Combining equations (2) and (11) and rearranging to produce a linear mathematical equation gives us:

$$\frac{1}{Q} = \frac{1}{\Delta H^{\circ\prime}} + \frac{1}{\Delta H^{\circ\prime} \cdot K^\prime} \cdot \frac{1}{[X]_o - [MX]} \qquad (12)$$

Repeating the experiments at several different concentrations, inserting the obtained value and solving by the iterative least square method can give the best-fitting values for $\Delta H^{\circ\prime}$ and the $K^\prime$. In fig. 1 the curve obtained experimentally for the binding of Mg ion with some mononucleotides is shown.

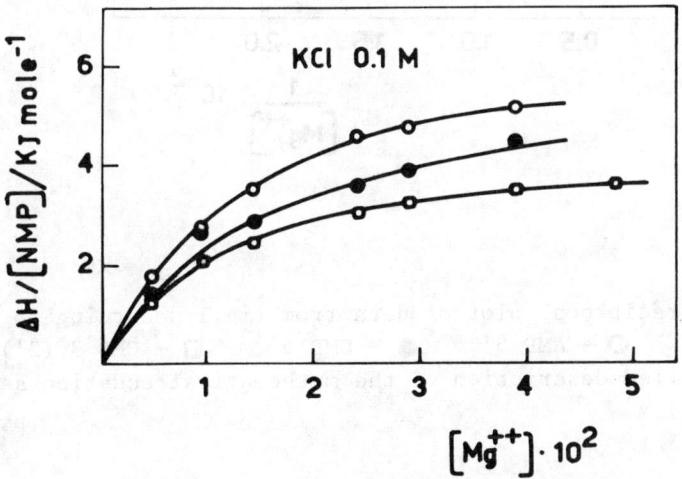

Fig.1 - Heats of binding of magnesium ion to mononucleotides.
○ -CMP 3'(2'), ● - UMP 5', □ - AMP 5'. Temp.25 °C, pH 7.3, KCl 0.1 M.

Solving as described above gives the curve displayed in fig. 2.

The experimental conditions in which the reaction takes place can have a significant effect on the K'. Variations in the ionic strength of the solution due to different concentrations of KCl have no effect on the ΔH°' (fig. 3).

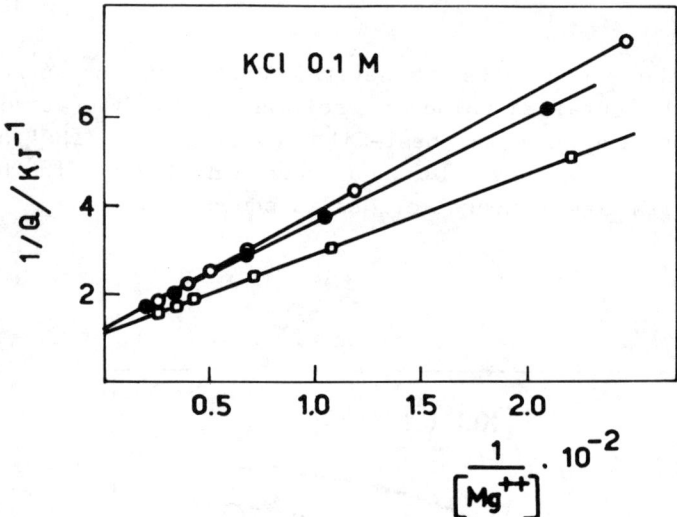

Fig.2 - Double reciprocal plot of data from fig.1 according to equation (12).   O - AMP 5',   ● - UMP 5',   □ - CMP 3'(2').
For a more detailed description of the mathematical equation see Bolen et al, (1).

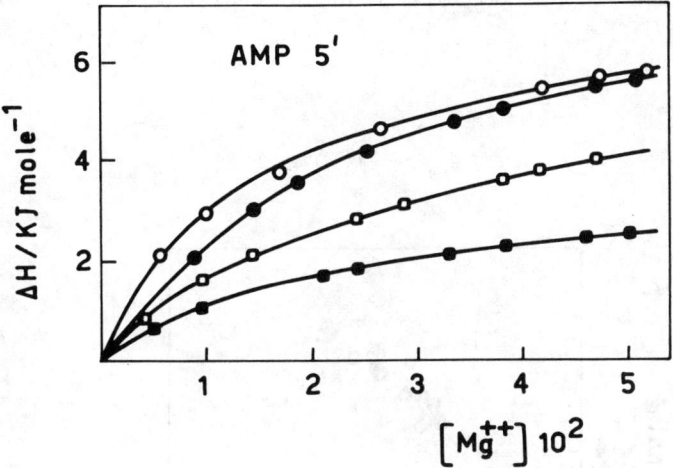

Fig.3 - Heats observed upon mixing of AMP 5' solution with magnesium ion solution at different concentrations of KCl. Temp. 25 °C, pH 7.3. Double reciprocal plot is not reported.

The presence of KCl has, of course, modified the activity coefficient of the magnesium ion. Therefore if we want to take a correct thermodynamic approach, all parameters of the reaction should be calculated according to the activity of magnesium and not the concentration of the ionic species. Fig. 4 shows that all the curves are superimposable when corrected for the activity coefficient.

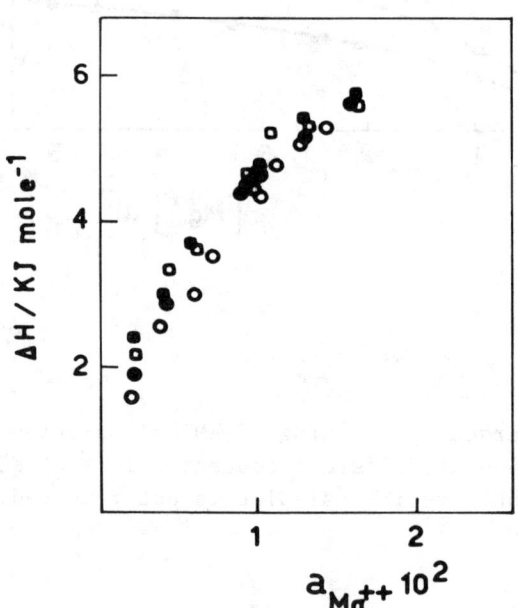

Fig.4 - Heats of binding of magnesium ion to AMP solution at different concentrations of KCl plotted versus the activity of magnesium ion.   O - KCl 0.1 M,   ● - KCl 0.07 M,   ■ - KCl 0.05 M,   □ - KCl 0.03 M.

We recommend that this correction be made when possible, and if the complexity of the system renders calculation of the activity coefficient impossible (i.e. with large proteins etc.), this should be taken into account when interpreting results. Many authors have spoken of the "protection" given by some ions against the carcinogenic effect of compounds like Ni, Be, Cd, Cr binding to DNA. The changes in binding equilibrium should be considered in terms of changes in the activity coefficient of the carcinogenic ion involved.

In the simple binding reaction described above, the thermodynamic parameters can be calculated at different temperatures to determine the $\Delta C_p^{o'}$ according to equation (10).
Other examples of calorimetric studies using AMP, ADP, ATP have been reported. (3).

Table I

T	$\overline{\Delta H^\circ}$	$\overline{\Delta G^\circ}$	$\overline{\Delta S^\circ}$	$\overline{K}$	ΔC_p
37°	2.13 ± 0.03	−3.62 ± 0.3	18.6	365 ± 18	34.0
32°	1.96 ± 0.05	−3.48 ± 0.5	17.8	317 ± 27	32.9 33.3
25°	1.73 ± 0.03	−3.40 ± 0.3	17.2	316 ± 15	

Table 1 - The thermodynamic parameters of magnesium ion binding to AMP at different temperatures.

As we mentioned earlier, binding reactions are usually not as simple as that represented by equation (1), but carefully designed experiment can limit the number of events taking place. An early study by Rialdi and Hermans (4) provides a good example of a binding reaction limited to two events: the binding of a proton to poly-L-glutamate and the subsequent coil to helix conformational change of the poly-L-glutamic acid. In this simple titration experiment, acidification of solutions containing poly-L-glutamate causes the coil to helix transition of the polypeptide, which is detected by a change in optical rotation. Calorimetric measurement of the pH changes gave fig. 5.

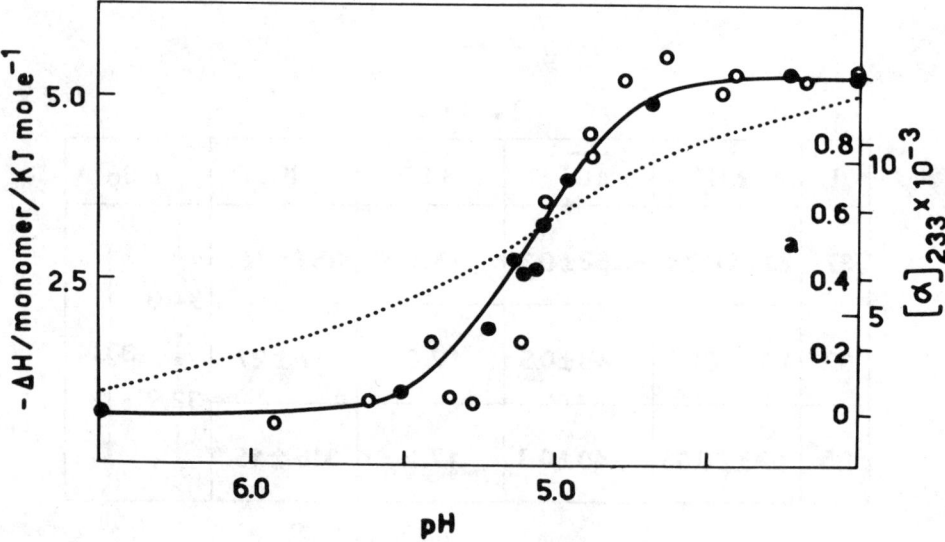

Fig.5 - Molar heat content, $\Delta H°$ (open circles), specific rotation at 233 nm $[\alpha]_{233}$ (closed circles), and degree of ionization, (dashed curve), of poly-L-glutamic acid in 0.1 N KCl as a function of pH at 30 °C.

It is obvious that the enthaply does not follow the degree of ionization but changes abruptly in the pH range where the coil to helix transition occurs. The enthalpic contribution due to protonation of the carboxyl group is very low while the contribution from the helix to coil transition is easily detected.

A similar reaction has been studied using an ion as a ligand. (5) Both poly-L-ornithine and poly-L-lysine react with KSCN at 25 °C, pH 3. Poly-L-lysine undergoes a conformational change upon binding whereas poly-L-ornithine does not. The binding involved in both reactions is very similar hence the additional heat evolved in the poly-L-lysine reaction should be related to the conformational change. Fig. 6 shows the heat effect of the two reactions measured by calorimetry.

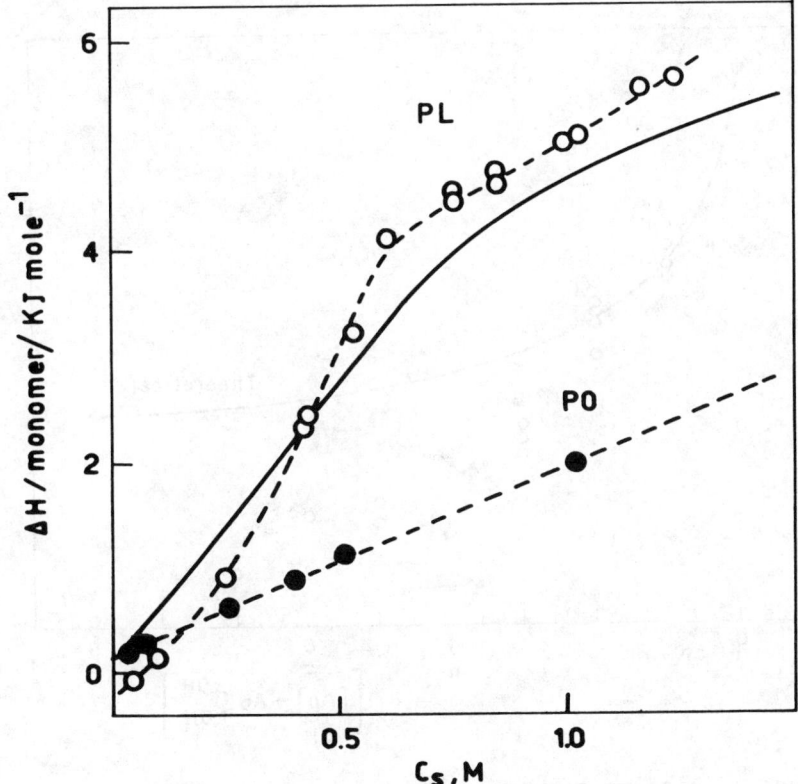

Fig.6 - Corrected heat of reaction for the systems Lys - KSCN and Orn. - KSCN.

If there is more than one binding site involved in the reaction, the total enthalpic contribution will be the sum of all binding contributions. For example, the enthalpic contribution of the binding of magnesium ion to a dinucleotide will be the sum of the binding of the terminal and intrachain phosphate group to the ion. Lengthening of the nucleotide chain diminishes the contribution of the enthalpic effect due to binding of the terminal phosphate group. It has been reported the average binding enthalpies for poly U and Mg^{++} $\Delta H°' = +5.4$ KJ per nucleotide (6). In this case we have a simple binding but similar experiments done with poly A of different lengths demonstrate a coil to helix transition induced by the presence of Mg ions (Fig. 7) (7). Krakauer has reported a $\Delta H°'$ of binding for poly A equal to -8.3 KJ/ mole monomer (6). Magnesium binding to DNA has been measured by Ross and in this case the binding contribution plus the modification of the double helix can be observed. (8)

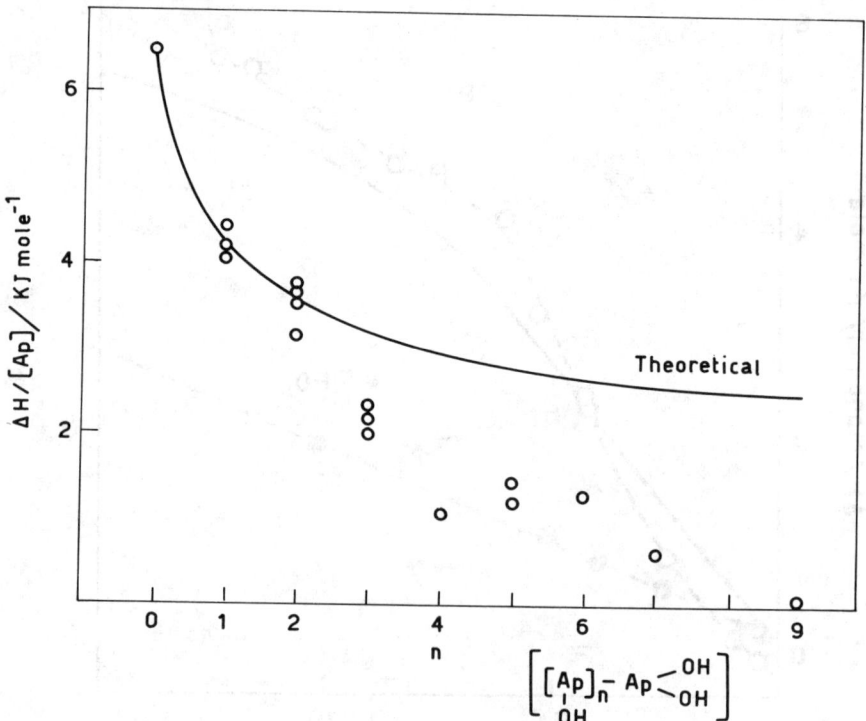

Fig.7 - Heat of binding of magnesium ion to oligonucleotides. Solid curve represents the theoretical contribution expected in the absence of induced helix transition.

All the reactions mentioned so far have fulfilled a rather important requirement for calorimetric study; they all produce measurable heat effects. In some cases $\Delta H = 0$, making calorimetric measurements of these reactions impossible. One way of circumventing this problem is to follow a non heat-producing reaction by relating it in some way, to a measurable reaction. Rialdi et al. (9) have shown that phenylalanine specific RNA from yeast is able to bind Mg ion with an associated enthalpy change of virtually zero. Calorimetric techniques were then used to measure the extent of binding to RNA as a function of Mg concentration. t-RNA was dialyzed against differing amounts of $MgCl_2$. The amount of magnesium ion in the solution and dialysate was measured calorimetrically by reaction with excess ethylenediaminetetra-acetate (EDTA). In this study the situation can be represented by the following equation:

$$RNA-Mg + EDTA \rightleftarrows RNA + Mg^{++} + EDTA \rightleftarrows RNA + MgEDTA$$

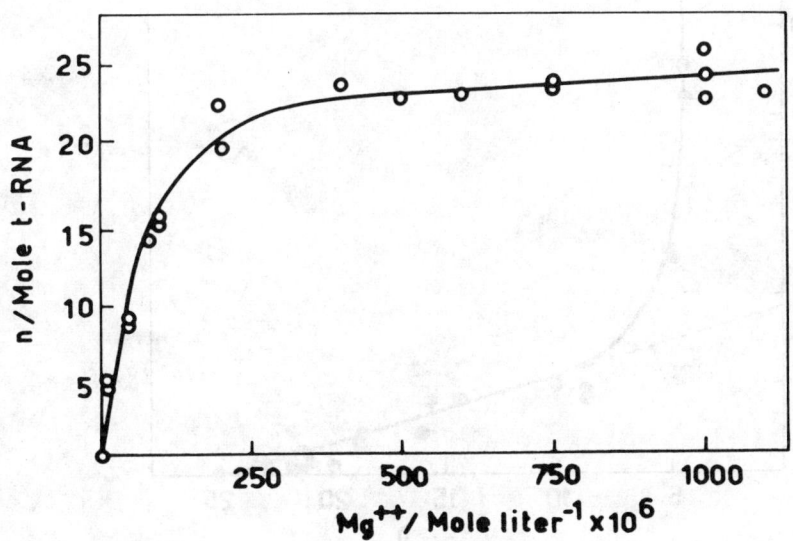

Fig.8 - Average number of Mg^{2+} / t-RNA^{phe}, ra as a function of free Mg^{2+} concentration at 25 °C, pH 7.2, 0.01 M salt.

With this experimental set-up, the relationship between the amount of free magnesium and the degree of binding to tRNA was determined over a concentration range of 0.1-2.5 mM. Fig. 8 shows the results of this study, with the extent of binding plotted versus free Mg ion concentration. Two different binding sites were detected.

During the binding reaction a conformational change takes place in the tRNA molecule with a very low enthalpic change, which we were unable to detect.

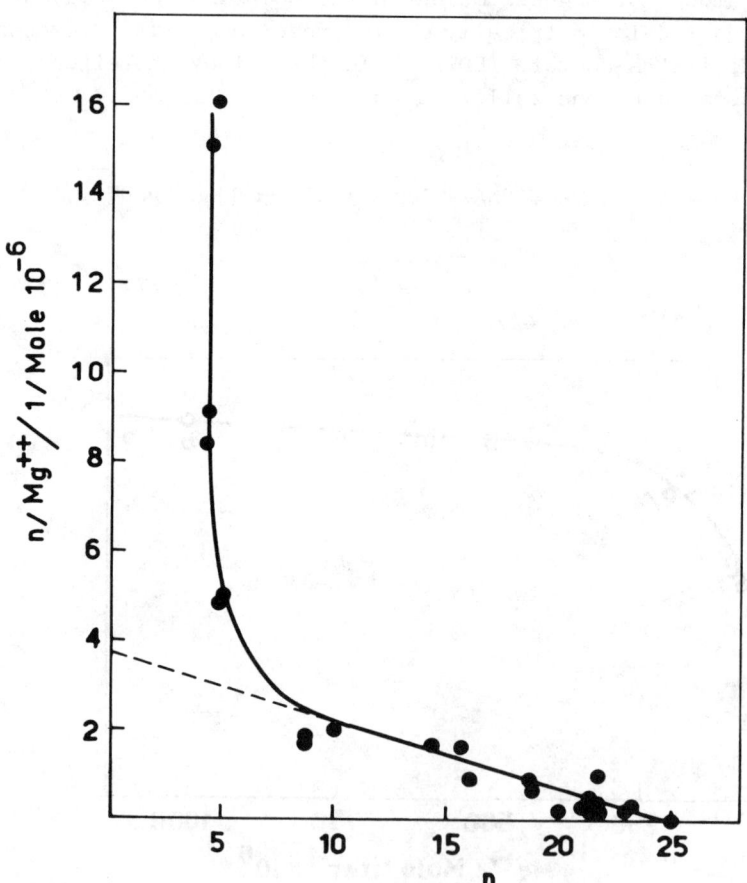

Fig.9 - Scatchard representation of the data in Fig. 8.

An example of a more complex conformational change induced by binding of an ion is provided by the reaction of manganese to Concanavalin A. Concanavalin A is a protein isolated from the jack-bean, which reacts with polysaccharides. Divalent ions are required for its interaction with sugars and two distinct binding sites, S1 and S2, have been reported (10). Manganese binds the first site, forming the native state with a high binding constant, and induces a conformational change, whereupon a calcium ion binds to the second site. At pH 4 - 5.5 Con A exists as a dimer, made up of two identical subunits each with one calcium, one manganese and a single sugar binding site. We have studied the binding of Mn^{++} to Con A in this pH range and our results show that the two identical sites, S1, of the dimer are independent. The binding enthalpy we measured ($\Delta H^{o'} = 95$ KJ/mole and $\Delta H^{o'} = 65$ KJ/mole) is high for the heat of complexation for a simple ion ligand joined to a carboxyl group. It has been shown that the ion is not only bound to the carboxyl group on the protein but enters into the structure of the molecule, interacting with other residues of the metal binding region and inducing a conformational change (11).

When the Con A molecule attains the correct configuration, i.e. the binding of manganese and calcium has transformed the molecule into the biologically active compound, the two subunits can react with sugar derivatives.
Sugar derivatives in solution exist in predominately the chair structure, so a significant heat contribution would not be expected from the boat to chair transition - see equation (5) -. The sugar binding site on the Con A subunit is a complex site with well defined dimensions. For each hydroxyl group on the sugar molecule there must be a corresponding hydrophilic group on the protein molecule. Hydrophobic portions of the two molecules must be properly aligned. For these reasons the hydroxyl group on the number 3 carbon (C_3) of the sugar residue must be in an axial position, as it is in mannose and glucose. An equatorially positioned hydroxyl group (galactose) does not permit binding.
Calorimetric studies on the binding of Con A to saccharides have borne out these considerations. Table III provides a summary of the binding enthalpies of Con A reacting with various saccharides. Binding of the glucose and mannose derivatives results in the largest $\Delta H^{o'}$ (12).
The position of the hydroxyl group on C2 determines a different amount of heat produced, with the axial position (mannose)

Table II

	ΔH	ΔG	ΔS	K x 10⁻³
25°C	93.6 ± 1.1	-22.7 ± 0.3	399 ± 4	9.5 ± 0.8
30°C	63.9 ± 0.7	-24.2 ± 0.3	291 ± 7	15.2 ± 0.2

Table 2 - Thermodynamic parameters of the binding of manganese ion to concanavalin A at pH 4, in acetate buffer 10^{-3} M, temp. 25 ° and 30 °C. ΔH calculated from Van't Hoff equation = 70.4 = 18 KJ/M.

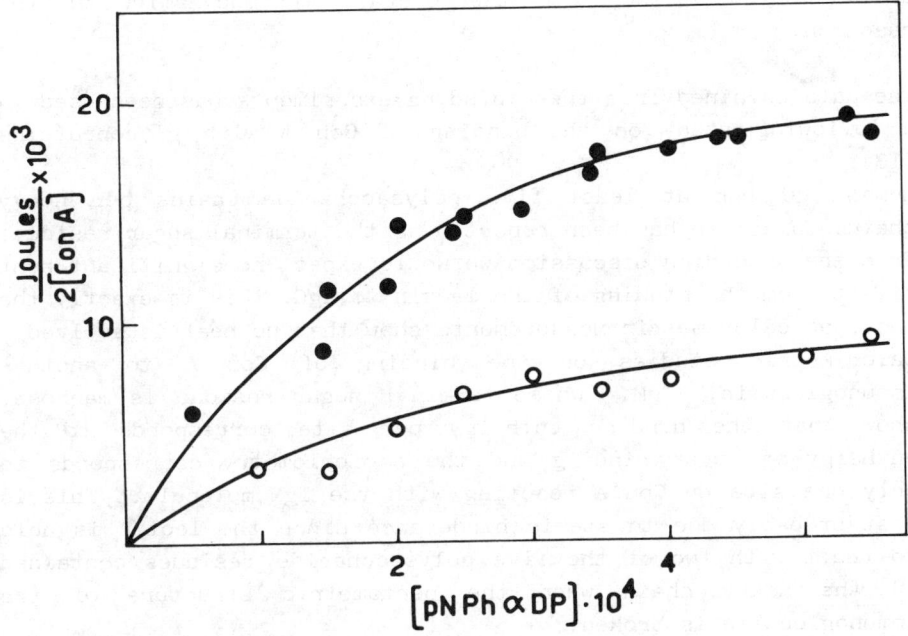

Fig.10 - Thermal titration of the heat of reaction between concanavalin A and p-nitrophenyl α-D-manoside (●), p-nitrophenyl α-D-glucoside (○), at 30 °C in acetate buffer 10^{-3} M, pH 4.5

TABLE 3

Pyranoside	$-\Delta H°'$ KJ mol^{-1}	$-\Delta G°'$ KJ mol^{-1}	$\Delta S°'$ J/mol per K
p-nitrophenyl-α-D-manno	+15.6 ± 0.2	25.6 ± 0.3	33.5 ± 0.3
p-nitrophenyl-α-D-gluco	14.6 ± 0.2	21.7 ± 0.3	23.8 ± 0.3
p-nitrophenyl-β-D-gluco	3.2	-	-
o-nitrophenyl-β-D-gluco	2.2	-	-
o-nitrophenyl-α-D-galacto	0	-	-

Thermodynamic parameters of the binding of p-nitrophenyl pyranosides with concanavalin A. Temp. 25 °C, pH 4.5 in acetate buffer 10^{-3} M. in the presence of calcium and manganese.

resulting in a greater heat contribution than the equatorial position (glucose). Of course the other carbon atoms of the sugar derivatives contribute to the overall thermodynamics of the reaction.

The data obtained from these binding experiments has been used in a following study on the binding of Con A with glycoproteins (13).
Human IgG has at least five polysaccharide chains per heavy chain. Galactose has been reported as the terminal sugar residue; from the preceding discussion we would expect no significant heat effect from the binding of the lectin to IgG. This is exactly the case, as calorimetric measurements show that no heat is evolved.
Calorimetric studies on the binding of Con A to another immunoglobulin, IgM, whose terminal sugar residue is mannose, show that the binding enthalpy per site corresponds to the enthalpy of sugar binding and the stoichiometry corresponds to only one site on Con A reacting with the IgM molecule. This is most probably due to steric hinderance since the lectin is able to react with two of the five polysaccharide residues contained in the heavy chain when the pentametric structure of the immunoglobulin is broken.
Binding of IgM with an antigen induces a conformational change in the molecule which allows Con A to bind with the antigen-bound IgM in a ratio somewhere between the minimum (1 ConA molec./heavy chain) seen with the intact molecule and the maximum (2 ConA molec./heavy chain) observed with the free subunits.

Immunology is a field based on specific binding interactions which can be studied with great precision using calorimetric techniques. Biltonen (14) and Barisas (15,16,17) have shown that antigen-antibody and hapten-antibody binding is similar in thermodynamic terms. The $\Delta H°'$ of the binding reaction is due to binding of a specific site to 3-6 amino acids. The great differences in affinity seen between antibodies is most probably due to extra-site interactions between large binding surfaces and the resultant entropy change.
These reactions are normally studied using light scattering techniques or precipitation phenomena, all of which are based on the formation of large aggregates.

In this short presentation we have given some examples of the different effects induced by ligand binding reactions as studied by calorimetry. We have seen that the binding reaction cannot be considered a simple reaction per se, but often involves changes in conformation with subsequent changes in hydration, biological activations, etc. The complexity of the reaction often renders any sort of analysis difficult even in the most simple binding situations. Clear experimental design becomes even more important when more complex systems are involved. Calorimetry can often provide insight into the thermodynamic changes occurring in even the most complex biological systems as long as basic thermodynamic considerations are respected.

This work was supported by research grants C.N.R. - Progetti Finanziati per la Chimica Fine e Secondaria -. Nos. 8205835/6

Reviews:

1) Brown, H.D.: 1969 *Biochemical Microcalorimetry* Academic Press (New York, London)
2) Sturtevant, J.M.: 1971 *Calorimetry, Physical Methods of Chemistry* Vol.1., Part V Ed. A. Weissberger and B.W. Rossiter, John Wiley and sons (New York), pp. 347-425.
3) Rialdi, G. and Biltonen, R.L.: 1975 *Thermodynamics and Thermochemistry of Biologically Important Systems, Internat. Rev. Sci., Phys. Chem.* (2) Butterworths, London 10, pp. 147-189.
4) Spink,C. and Wadso, I.: 1976 *Calorimetry as an Analytical Tool in Biochemistry and Biology, Methods of Biochemical Analysis* Vol. 23 Ed. D. Glick, John Wiley (New York), pp. 1-159.
5) Barisas, B.G. and Gill, S.J.: 1978 *Microcalorimetry of Biological Systems*, Ann. Rev. Phys. Chem. 29, pp. 141-166.
6) Langerman, N. and Biltonen, R.L.: 1979 *Meth. Enzym.* pp. 61, 261.
7) Martin, C.J. and Marini, M.A.: 1979 *Microcalorimetry in Biochemical Analysis, Critical Reviews in Analytical Chemistry* Vol. 8 CRC Press, Inc., pp. 221-286.
8) Eftink, M. and Biltonen, R.L.: 1980 *Thermodynamics of Interacting Biological Systems* in *Biological Microcalorimetry* Ed. A.E. Beezer, Academic Press. (London, New York), pp. 343-412

References

1) Bolen, D.W., Flogel, M. and Biltonen, R.L.: 1971 *Biochemistry* 10, pp. 4136-4140.
2) Rialdi, G., Schiano, I. and Braggio, R.: 1975 *Thermodynamics of Magnesium Ion Binding to Mononucleotides* in *L'Eau et les Systemes Biologiques* Ed. A. Alfsen et A.J. Berteaud, Colloques internationaux du CNRS n. 246, pp. 271-274.
3) Belaich, J.P. and Sari, J.C.: 1969 *Proc. Nat. Acad. Sci.* 64, pp. 763.
4) Rialdi, G. and Hermans, J.: 1966 *J. Am. Chem. Soc.* 88, pp. 5719-5720.
5) Conio, G., Patrone, E., Rialdi, G. and Ciferri, A.: 1974 *Macromolecules* 7, pp. 654-659.
6) Krakauer, H.: 1972 *Biopolymers* 11, pp. 811.
7) Rialdi, G.: (unpublished results).
8) Ross, P. and Shapiro, J.: 1974 *Biopolymers* 13, pp. 415-416.
9) Rialdi, G., Levy, J. and Biltonen, R.L.: 1972 *Biochemistry* 11, pp. 2472-2479.
10) Kalb, A.J. and Levitzki, A.: 1968 *Biochem. J.* 109, pp. 669 - 672.
11) Toselli, M., Battistel, E., Manca, F. and Rialdi, G.: 1981 *Biochim. Biophys. Acta* 667, pp. 99.
12) Dani, M., Manca, F. and Rialdi, G.: 1981 *Biochim. Biophys. Acta* 667, pp. 108-117.
13) Dani, M., Manca, F. and Rialdi, G.: 1982 *Molecular. Imm.* 19, pp. 907-911.
14) Halsey, J.F. and Biltonen, R.L.: 1975 *Biochemistry* 14, pp. 800-804.
15) Barisas, B.G., Sturtevant, J.M. and Singer, S.J.: 1971 *Biochemistry* 10, pp. 2816.
16) Barisas, B.G., Singer, S.J. and Sturtevant, J.M.: 1972 *Biochemistry* 11, pp. 2741-2745.
17) Johnston, M.F., Pecht, I., Sturtevant, J.M. and Barisas, B.G.: 1979 *Molecular Immunology* 16, pp. 681-689.

CELLULAR THERMOCHEMISTRY: RESTING STATE AND STIMULATION

G. Rialdi and S. Raffanti

Centro Studi Chimico Fisici Macromolecole
CNR - Corso Europa 30, 16132 GENOVA-ITALY

Living organisms, though extremely complex in the quantity and diversity of biochemical processes involved, follow the same thermodynamic laws that govern any reaction. Energy is neither created nor destroyed but trasnformed into chemical products, entropy changes and heat. For centuries man has known that living systems produce heat and measurement of this phenomemum dates back to the earliest biological experiments. In recent years many researchers have shown great interest in the heat developed by cultured cells (1, 2, 3, 4, 5).
Although living organisms are open systems, it is useful to look upon as if they were ideal, isolated, closed systems. This allows us to apply, as a first approximation, classical thermodynamic laws to the energetics of the biochemical processes taking place. Calorimetric study of cell systems requires precise measurement of the overall heat effect of the system under study plus a plausible and experimentally sound correlation with some biochemical or biological event.
Early calorimetric studies of biological systems dealt with the most basic situation possible: the cell in the resting state. Several authors were able to measure the heat evolved in cells in the resting state and correlate the "Q" measured with the number of cells present (6, 7, 8, 9). Not only was there good agreement between heat produced and cell number, but subsequent studies were able to demonstrate characteristic thermograms for different species of bacteria and yeasts, allowing identification of up to 200 different species (10, 11, 12).

The relative ease of working with human erythrocytes led to their use in the initial calorimetric studies on eucaryotic cells in vitro (13, 14, 15, 16). A value of 62 ±7 nW/L of erythrocytes was calculated using the batch calorimeter while flow techniques gave a value of 115 nW/L of packed cells. Subsequent studies by Monti and Wadso (17, 18) compared the effect of different suspension media, calorimetric techniques, pH, temperature and other factors on heat production in human erythrocytes.

Heat production in cells from healthy and diseased individuals has been investigated by various groups and includes studies on erythrocytes from normal and anemic subjects (19), erythrocytes from hyperthyroid patients before, during and after treatment (20), and adipocytes from obese and normal weight individuals (21).

Quantitative correlation of heat production with a biological event, in this case cell multiplication, was achived in studies on heat production in growing bacterial cell cultures (22). Measurement of heat effect in E.coli cultures not only showed good agreement with the increase in dry weight of bacteria during the exponential growth period, but also revealed a second peak that could not be explained by bacterial growth (Fig. 1).

Krakauer (23) reported that the heat produced in stimulated lymphocytes correlated quite closely with the increased uptake of radioactive thymidine, concluding that the increased heat effect was directly proportional to cell number.

Loike et al have used differential scanning microcalorimetry in the study of heat production and glucose oxidation in murine macrophages (24). Measurement of lactic acid levels and heat production showed good agreement between the rates of the two processes (Figure 2).

The major difficulty involved in any attempt to correlate biochemical processes taking place within the cell with the total heat effect measured in the calorimeter lies in the incredible complexity of the reactions taking place at any given moment. The only correct way to overcome this problem would be to measure the reactants and products of all the reactions taking place in the cell and then analyse the observed heat effect according to these data.

The total heat effect that we measure in the calorimeter is equal to the sum of all these anabolic and catabolic reactions as expressed in the following equation reported by Belaich (2):

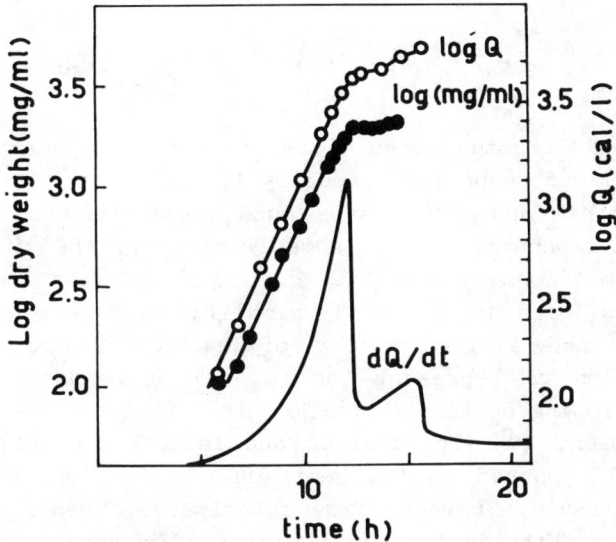

Fig.1 - E. coli grown aerobically on glucose (4 g/l at 27 °C):
——, thermogram, ▲ log dry weight, -O- log total heat.

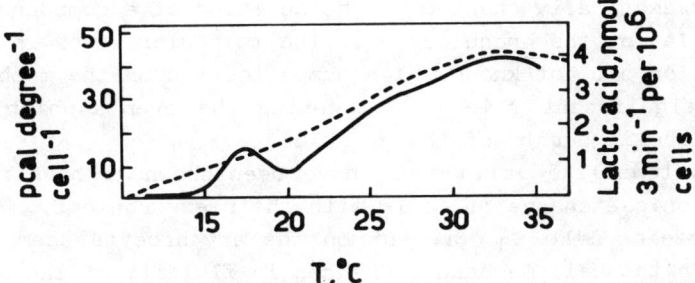

Fig.2 - Comparison of the rates of lactic acid production (———)
and heat evolution (- - -) by thioglycollate - elicited perito-
neal cells (TEP) vs temperature.

$$Q = \Delta H_{met} = \Sigma_i M_i \Delta H_{cat,i} + \Sigma_j N_j \Delta H_{an,j}$$

where M_i and N_j are the moles of reaction for components i and j of catabolism and anabolism respectively.

Obviously, this approach leaves much to be desired. The only reasonable alternbative is to somehow simplify the system under study and limit ourselves to the more important reactions. Human neutrophils offer a significantly simplified system since the few mitochondria and limited number of anabolic reactions taking place allow us to represent the major heat producing reaction with the following outline: (see Fig.3)

CO_2 production, O_2 consumption and lactate production can be measured and compared to the heat effects seen under different conditions. Levin, using a flow calorimeter, reported results obtained with platelet leucocyte mixtures in heparinized plasma suspensions. He calculated a value of $Q = 5.9 \pm 1.6$ pW/cell for leukocytes alone when using a gold reaction vessel and 6.2 ± 1.4 pW/cell when the entire flow line was made of Teflon (25, 26). Leukocytes from defibrinated blood gave a value of 2.2 pW/cell in the same experimental set-up, while Bandman et al have reported a value of 3.5 ± 1.0 pW/cell for granulocytes prepared with the Ficoll-Hypaque technique and resuspended in plasma (27).

Once a suitable design has been established to measure the heat evolved by neutrophils in the resting state, the entire situation can be dramatically changed by the addition of a compound capable of stimulating the granulocytes. The molecular steps involved in stimulation are not known. Ion pumps located on the membrane may have a significant role in triggering the events which lead to the "metabolic burst" of the granulocyte.

Crown systems like valinomycin have been shown to have an effect on the ionic exchange of cells with their environment.

Some proteins like C_3 present on the erythrocyte seem to "open holes" in the cell membrane. Changes in fluidity of the membrane, clustering of surface molecules and externally - induced internal conformational protein changes have all been cited as possible mechanisms involved in external stimulation phenomena. Brandts has proposed a large entropic contribution in multivalent binding due to rearrangement of the membrane proteins (28). Regardless of the mechanism involved, stimulation of neutrophils with compounds

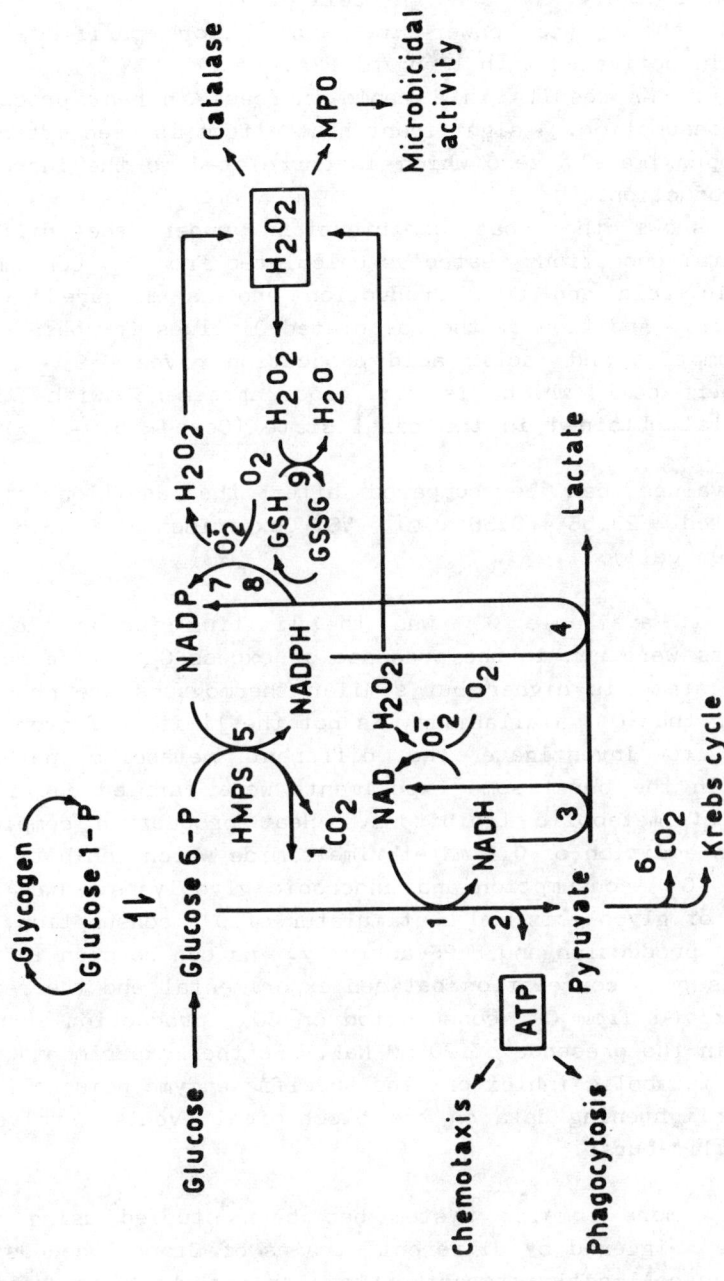

Fig.3 - Schematic representation of carbohydrate metabolism in neutrophils. Taken from The Neutrophil: Function and Clinical Disorders, S.J.Klebanoff and R.A.Clark, North-Holland Publishing Co.1978

such as Concanavalin A and Phorbol-myristate-acetate (PMA) results in a highly "switched on" cell (29).
Figure 4 shows the power-time curve for purified human neutrophils activated with 10 ng/ml PMA.
Addition of PMA results in a rapid increase in heat production and O_2 consumption. A significant heat effect is seen after pO_2 equals approximately zero which is correlated to the increasing 1-CO_2 production.
Table 1 shows the heat contributions under the different experimental conditions tested as calculated from O_2 consumption and lactic acid and CO_2 production and as measured in the calorimeter. Addition of the calculated Q values for basal state O_2 consumption and lactic acid production gives 6.9 $\pm$ 2.29 x 10^{-9} J per cell which is in good agreement with the Q experimental obtained in the basal state (Qexp = 6.0$\pm$ x10^{-9} J per cell).
Similar values can be compared after the addition of PMA (Qcalculated = 23.55 $\pm$ 0.36 x 10^{-9} VS. Qexperimental = 23.5 $\pm$ 1.2 x 10^{-9} J per cell).

To check if available O_2 was the limiting factor the same experiments were run in the presence of excess O_2. The resting metabolic state is higher but similar thermograms are produced, indicating that O_2 availability is not the limiting factor.
In order to investigate the different metabolic pathaways involved in the burst some experiments were carried out in the presence of metabolic inhibitors. Heat production completely stops upon addition of 0.1 mM ethylmaleimide which inhibits shunt activity, O_2 consumption and anaerobic glycolysis. NaF is an inhibitor of glycolysis and a stimulator of O_2 consumption, O_2 and H_2O_2 production and HMPS activity. As can be seen in table 1 there is good correlation between experimental end theoretical values derived from O_2 consumption or CO_2 production when PMA is added in the presence of 20 mM NaF. Further experiments using different metabolic inhibitors and specific enzyme poisons should provide enlightening data on the biochemical events involved in the metabolic burst.

A slightly more complex system has been studied using human phagocytes triggered by different strains of Staphylococcus. In this case a cell-cell interaction takes place which triggers the complex chain of events leading the phagocytosis of the bacteria. Preliminary results using 3 different strains of Staphylococcus

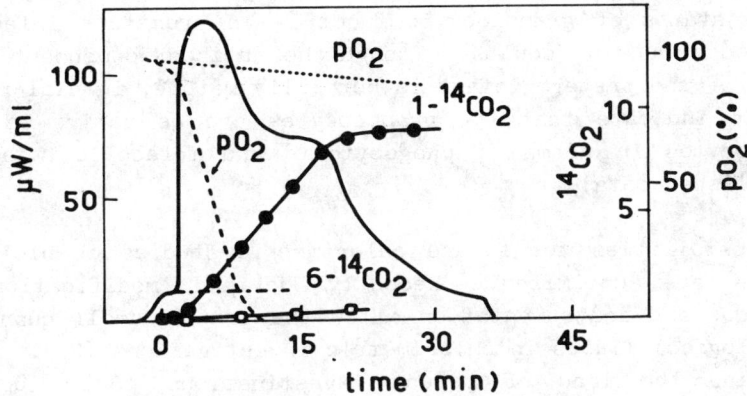

Fig.4 - Power-time curves of human neutrophils activated with 10 ng/ml PMA.

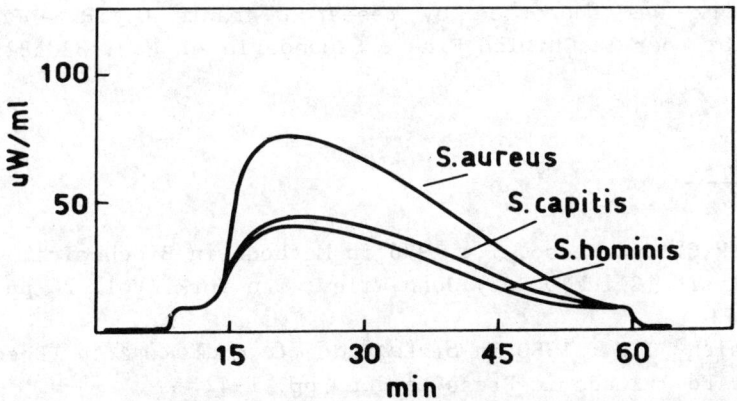

Fig.5 - Power-time curve during phagocytosis of Staphylococcus aureus, Staphylococcus hominis and Staphylococcus capitis.

are displayed in Figure 5 (30).
There seems to be a correlation between pathogenecity and increased heat production. Experiments on purified granulocytes from different patients have also given some interesting results on the status of granulocytic function in premature infants as compared to normal adults. Though the antibody-dependent immune system of the preterm infant is not fully mature, initial results seem to indicate that the granulocytes are as active as adult granulocytes in terms of phagocytosis and metabolic activation (unpublished data).

The possibilities for future calorimetric studies on biological systems are unlimited. Recently devised modifications in technique and instrumentation allow the use of small quantities of biological fluids in calorimetric investigations (31).
Following the lead of recent investigations (32), upcoming studies could include whole blood experiments aimed at detecting possible cellular defects in diseased donors.
Interaction between the different components of blood could be analysed calorimetrically and pharmaceutical or storage effects could be investigated.

This work was supported by research grants C.N.R. - Progetti Finanziati per la Chimica Fine e Secondaria -. Nos. 8205835/6

References

1) Spink,C. and Wadso, I.: 1976 in Methods in Biochemical Analysis (Glick D.Ed.) John Wiley, New York. Vol. 24 pp. 1 - 159.
2) Belaich, J.P.: 1980 in *Biological Microcalorimetry* (Beezer A.E. Ed.) Academic Press London. pp. 1-42.
3) Lamprecht, I.: 1980 in *Biological Microcalorimetry* (Beezer A.E. Ed.) Academic Press London. pp. 43-112.
4) Levin , K.: 1980 in *Biological Microcalorimetry* (Beezer A.E. Ed.) Academic Press London. pp. 131-144
5) Monti, M. and Wadso, I.: 1979 pp. 256-280
   Kresheck, G.C.: 1979 pp. 281-307
   Crabtree, B. and Taylor,D.J.: 1979 pp. 333-337

in *Biochemical Thermodynamics* (Jones M.N. Ed.) Elsevier Scientific Publishing Co., Amsterdam.
6) Beezer, A.E., Bettelheim, K.A., Newell, R.D. and Stevens, J.: 1974 *Science Tools* 21,13.
7) Beezer, A.E., Bettelheim, K.A., Al-Salihi, S. and Shaw, E.J.: 1978 *Science Tools* 25,1 pp. 6-8.
8) Cooney, C.L., Wang, D.I.C. and Mateks, R.I.: 1968 *Biotech. Bioeng.* 11, pp. 269-281.
9) Monti, M. and Wadso, I.: 1977, *Scand. J. Haematol.* 19, pp. 111-115.
10) Russell,W.J., Farling, S.R., Blanchard, G.C. and Boling, E.A. 1975 in *Microbiology - 1975* (Schlessinger, D. Ed.) American Society for Microbiology, Washington D.C. pp. 22-31.
11) Boling, E.A., Blanchard, G.C. and Russel, W.J.: 1973 *Nature* 241, pp. 472-473.
12) Lungholm, K., Wadso, I. and Mardh, P. A.: 1976 *J. Gen. Microbiol.* 96, pp.283-288.
13) Levin, K. and Boyo, A.E.: 1971 *Scand. J. Clin. Lab. Invest.* Suppl. 118 pp. 55.
14) Boyo, A.E. and Ikomi-Kumm, J.A.: 1972 *Lancet* 1, pp. 1912.
15) Monti, M. and Wadso, I.: 1973 *Scand. J. Clin. Lab. Invest.* 32, pp. 47.
16) Bandmann, U., Monti, M. and Wadso, I.: 1975 *Scand. J. Clin. Lab. Invest.* 35, pp. 121-127.
17) Monti, M. and Wadso, I.: 1976 *Scand. J. Clin. Lab. Invest.* 36, pp. 565.
18) Monti, M. and Wadso, I.: 1976 *Scand. J. Clin. Lab. Invest.* 36, pp. 573-580.
19) Monti, M. and Wadso, I.: 1973 *Scand. J. Clin. Lab. Invest.* 32, pp. 47-54.
20) Monti, M. and Wadso, I.: 1976 *Acta Med. Scand.* 200, pp. 301-308.
21) Sorbis, R., Nilsson-Ehle, P., Monti, M. and Wadso, I.: *FEBS Letter* 101,2, pp. 411-414.
22) Eriksson, R. and Wadso, I.: 1969 in *First European Biophysics Congress* (Broda, Locker, Springer-Lederer Ed.) Part IV Wiener Med. Acad., pp. 319.
23) Krakauer, T. and Krakauer, H.: 1976 *Cellular Immunol.* 26, pp. 242-253
24) Loike, J.D., Silberstein, S.C. and Sturtevant J.M.:1981 *Proc. Natl. Acad. Sci. USA* 78, pp. 5958-5962.
25) Levin, K.: 1971 *Clin. Chim. Acta* 32, pp. 87

26) Levin,K.: 1973 *Scand. J. Clin. Lab. Invest.* 32, pp. 67
27) Bandmann, U., Monti, M. and Wadso, I.: 1975 *Scand. J. Clin. Lab. Invest.* 35, pp. 121.
28) Brandts, J.F. and Jacobson, B.S.: 1982 (personal communi= cation).
29) Eftimiadi, C. and Rialdi, G.:1982 *Cell Biophysics* (in press)
30) Eftimiadi, C. and Rialdi, G.:1980 *Atti XIX Congresso Naz. Soc. It. Microbiologia* - Catania.
31) Suurkuusk, I. and Wadso, I.: 1982 in *A Multichannel Micro= calorimetry System, Chemica Scripta, Sweden* (in press).
32) Monti, M. and Fagher, B.: 1980 *Clin. Chim. Acta* 102, pp. 83-89

DIFFERENTIAL SCANNING CALORIMETRY APPLICATIONS TO PROTEIN SOLUTIONS

Pedro L. Mateo

Department of Physical Chemistry, Faculty of Sciences,
University of Granada, Granada, Spain.

This report surveys the recent advances in differential scanning calorimetry as applied to the thermodynamic study of protein solutions.

A brief comment is made upon high-sensitivity instrumentation and the basic data analysis of the technique.

The application of this microcalorimetric method to the investigation of simple compact globular proteins is extended to more complex examples including a newly developed theoretical analysis of the excess heat capacity function.

Final comments are made on some special problems such as the proposed non-denaturational conformational transitions in proteins.

The particular advantages of scanning calorimetry are stressed both in the several systems and the problems here described.

1. INTRODUCTION

Differential Scanning Calorimetry (DSC), a method to measure the relative heat capacity of a system as a function of temperature, has proved to be a fundamental technique for studying the temperature-induced conformational transitions in molecular biological systems such as the unfolding of proteins (1,2) and nucleic acids (3), and phase transitions in biomembranes and related model systems (4,5).

The problem of stability of the native structure of proteins has received great attention during the last years (6,7,8). Since the stability of a given system can be obtained by measuring the work required for its disruption, it follows that DSC is very suitable for the investigation of protein thermal denaturation, thus providing new insight into the stability of these biological macromolecules. The study of the transition from a state defined as *native* to the denaturated one, approximated by the random coil model, is especially important in the understanding of the organization of the three-dimensional structure of proteins, and the nature of the forces responsible for determining and maintaining such a structure. A specific advantage of DSC towards this goal is, for example, the simultaneous determination of the calorimetric and van't Hoff enthalpies of the unfolding process. The comparison of these two enthalpy values can elucidate unambiguously whether we are dealing with a two-state process (9).

The DSC application to a group of simple compact globular proteins has shown that their denaturation is very close to being a two-state process (7). When dealing with somewhat more complex proteins, however, this situation no longer applies and its interpretation requires a more detailed analysis of the calorimetric results. This information can now be obtained by using high-sensitivity DSC instruments able to determine partial heat capacities of proteins in diluted solution and their heat capacity increase on denaturation, as well as by means of the theoretical deconvolution of the excess heat capacity function into individual transitions. Thus, the overall enthalpy of the process can be calculated from DSC data by direct integration of the C_p function. Once this enthalpy is known as a continuous function of the temperature, the partition function of the system, and consequently the statistical thermodynamic description of the system, can be obtained. The analysis of that enthalpy-temperature dependence can now be carried out following the algorithm of Biltonen and Freire (10,11) which permits the deconvolution of the overall thermal process into its constituent two-state transitions without any *a priori* assumptions on their possible interdependence.

It is then possible to obtain some information about the structural domains which undergo denaturation more or less independently. A recent example is provided by the DSC study of pepsin and pepsinogen denaturation (12) where four structural domains have been assigned to the pepsin molecule, while only two co-operative blocks seem to pertain in the case of pepsinogen, both including a pair of domains merged together. The example of this study, as well as others currently being carried out, displays new possibilities for the use of DSC in the investigation of protein stability and their thermodynamic study at the level of co-operative structural domains.

2. INSTRUMENTATION

Briefly described a differential scanning calorimeter consists of two cells, the reference cell and the sample cell, which can be heated at a constant rate in such a way as to maintain an equal temperature in both cells throughout the scan. When a temperature-induced process takes place in the sample cell, the DSC control system supplies the adecuate excess heat to this cell through appropriate heaters to maintain the same temperature as the other cell. Usually the output from the DSC is the excess electric power, which is, in its turn, proportional to the difference in heat capacity of the two cells and their contents as a function of temperature.

A major difficulty in the DSC applications to protein solutions is the high dilution which is necessary to avoid intermolecular interactions during the scan, particularly when the macromolecule is unfolded. The problem which then arises is that the heat effects are comparatively very small, especially in comparison with the heat flow continuously supplied by the system to carry out the temperature scan. Therefore, highly sensitive instruments, which have only become accesible during the last few years, are necessary to overcome this difficulty.

A very good critical description of scanning microcalorimeters currently available has recently been published by Privalov (13).

The DSC developed by Privalov and co-workers (14), DASM-1M, is at present, in the author's opinion, the most suitable for biological studies, mainly due to its high sensitivity and good baseline stability. This stability is always a serious problem in DSC experiments and has a dramatic influence when trying to calculate the partial heat capacity of proteins.

The DASM-1M has two coin-shaped gold cells with equal volumes of 1 cm^3 each, and a diameter of about 20 mm (Figure 1). Both cells are enclosed in two concentric silver adiabatic shields with heaters. Thermopile systems detect any temperature difference between the cells themselves, and also between the two adiabatic shields and the cells. The voltage produced in these thermopiles feeds servomechanisms which control the heating of the cells and the shields so that at no time throughout the scan is there any temperature difference inside the overall adiabatic system. The loading of the sample is made through platinum capillary filling tubes, and to avoid the production of bubbles during the scan a constant extra N_2 pressure of 1 to 2 atmospheres is kept on the liquid in the cells. The adiabatic condition of the whole cell unit and the fact that it is a non-dismountable block are the basic reasons for the outstanding sensitivity of this sophisticated model.

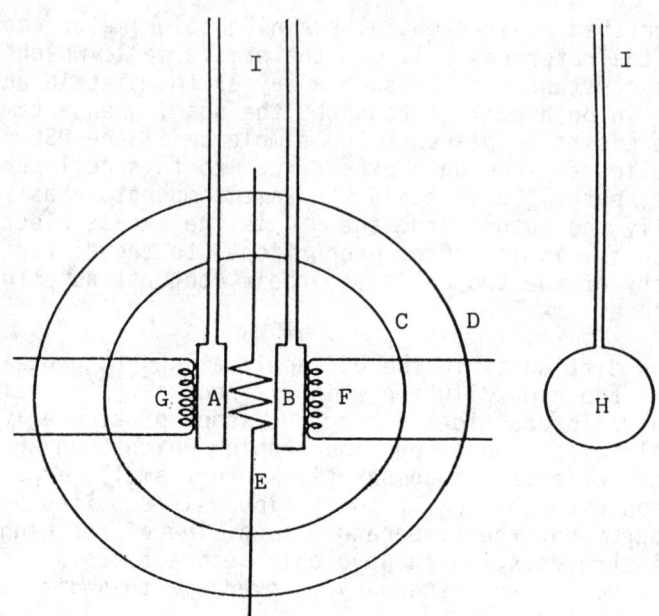

Figure 1. Schematic diagram of the DASM-1M. A and B, cells; C and D, internal and external adiabatic shields; E, thermopile system between the two cells; F and G, main heaters; H, lateral view of the cells; I, capillary filling tubes. See the text for details. Taken from reference (4).

Most of the work described below has been carried out with the DASM-1M.

3. DATA ANALYSIS OF DSC

I intend to present in this section, firstly, a brief analysis of an ideal system undergoing a two-state process without any change in the heat capacities of both states and secondly, we will see how to obtain the thermodynamic parameters of a given transition from the DSC profile of the process.

3.1. Two-state transition

Let us consider the simple two-state system

$$A \rightleftarrows B \atop 1-\alpha \quad \alpha \qquad (1)$$

which is heated through the temperature range where the endothermic process takes place. The absorption heat curve is that shown in Figure 2A, which can also be taken as the reaction extent, α, simply assuming that the heat evolved is proportional to this extent. The equilibrium constant, $K = a_B/a_A$ where a_A and a_B are activities, changes with temperature according to the van't Hoff equation

$$\left(\frac{\partial \ln K}{\partial T}\right)_P = \frac{\Delta H^{v.H}}{RT^2} \qquad (2)$$

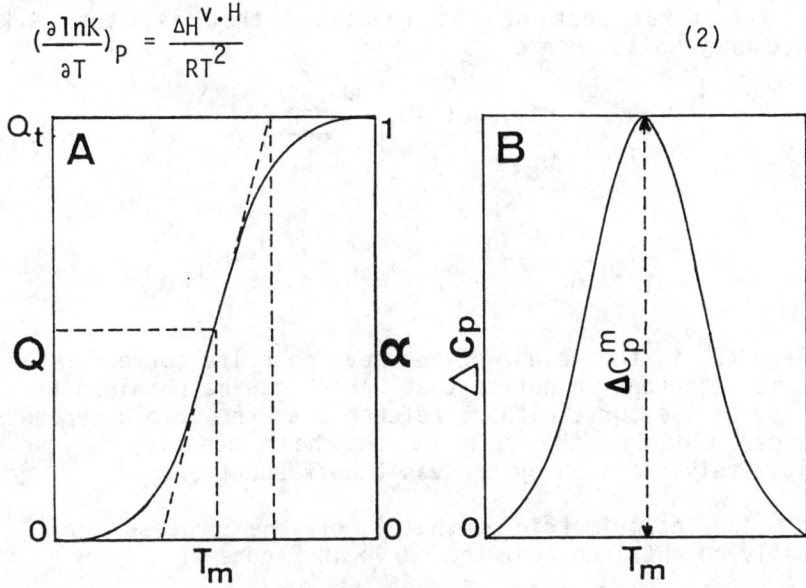

Figure 2. (A) Heat evolution, Q, and reaction extent, α, as a function of temperature during a two-state process, $A \rightleftharpoons B$, without overall change in the heat capacity; (B) Temperature dependence of the excess heat capacity for the same process. The excess heat capacity reaches a maximum, ΔC_p^m, at the transition temperature, $T = T_m$.

From now on, we will assume activities to be equal to concentrations, as we usually do in biochemical studies. Therefore all our thermodynamic properties should be considered as the apparent ones. Then, the equilibrium constant is $K = \alpha/(1-\alpha)$ from which results

$$\frac{\Delta H^{v.H}}{RT^2} = \left(\frac{\partial \ln \frac{\alpha}{1-\alpha}}{\partial T}\right)_P = \frac{1}{\alpha(1-\alpha)} \left(\frac{d\alpha}{dT}\right)_P \qquad (3)$$

Given that $\alpha = 1/2$ at the transition temperature, $T = T_m$, the temperature at which A is half converted into B, we have

$$\Delta H^{v.H} = 4RT_m^2 \left(\frac{\partial \alpha}{\partial T}\right)_{T=T_m} \qquad (4)$$

Figure 2B shows the excess heat capacity curve corresponding to this process, which is obviously the derivative of the former curve, since $\Delta C_p = (\partial Q/\partial T)_p$. In this case $\alpha = Q(T)/Q_t$ where $Q(T)$ is the heat absorbed at a given temperature and Q_t the total heat of the reaction, that is, the peak area in energy units. Hence

$$\left(\frac{\partial \alpha}{\partial T}\right) = \frac{1}{Q_t}\left(\frac{\partial Q(T)}{\partial T}\right)_p = \frac{\Delta C_p}{Q_t}$$

$$\left(\frac{\partial \alpha}{\partial T}\right)_{T=T_m} = \frac{\Delta C_p^m}{Q_t} \qquad \therefore \qquad \Delta H^{v.H} = 4RT_m^2 \frac{\Delta C_p^m}{Q_t} \qquad (5)$$

where ΔC_p^m is the height of the heat capacity curve for $T = T_m$. It is important to notice that $\Delta H^{v.H}$ can be obtained from the *shape* of the curve without reference to the sample concentration since, clearly, the form of the heat capacity curve is ultimately dictated by the van't Hoff equation.

The calorimetric enthalpy of the process, ΔH^{cal}, can easily be obtained from the curve in Figure 2B

$$\Delta H^{cal} = M \int_{T_1}^{T_2} \Delta C_p \, dT = M \cdot \Delta h \qquad (6)$$

where M is the molecular weight, ΔC_p the excess of specific heat capacity, and Δh the specific enthalpy of the transition. $\Delta H^{v.H} = \Delta H^{cal}$ for a two-state process.

Although $\Delta H^{v.H}$ has a thermodynamic interpretation in the two-state case only, it is useful to employ its value as an indication of the sharpness (co-operativity) of a transition. For the sake of clarification we can explore two different situations:

a) For a system made up of identical and independent molecules $\Delta H^{cal} \geq \Delta H^{v.H}$ where equality holds for a two-state process. This equation can also be expressed as $\Delta H^{v.H}/\Delta h \leq M$. Only large enough molecules showing intramolecular co-operativity will produce transitions which can be observed calorimetrically.

Globular proteins, or polypeptides and polynucleotides fall into this category, and in the last two cases $\Delta H^{V.H}/\Delta h$ is much smaller than the molecular weight. In fact, the value $\Delta H^{V.H}/\Delta h$ gives the size in grams of the average co-operative unit of the transition, which has been defined properly by Sturtevant (4) as *that quantity of substance such that if the sample were composed of independent units of that size undergoing a two-state transition, the transition curve would have the observed shape.*

b) For a process with intermolecular co-operation, such as a phase transition, $\Delta H^{V.H}/\Delta h >> M$, and the value $\Delta H^{V.H}/\Delta H^{cal}$ provides an estimation of the number of molecules included in the co-operative unit. An example of this case is the thermal transitions in phospholipids bilayers (15). For a first order phase transition the last ratio approaches infinity, which seems to be the case for a phospholipid bilayer of sufficient purity (16).

3.2. Thermodynamic parameters and protein stability

An original DSC recording for lysozyme is shown in Figure 3, where it can be seen that the C_p of the protein is not the same before and after the denaturation process having an apparent increase, $\Delta_d C_p^{ap} > 0$, as is the case for all the proteins so far investigated. It is possible to obtain the partial specific heat capacity of the protein, $C_{p,p}(T)$, as a function of temperature from the distance between the DSC recording of the protein and the base line, $-\Delta C_p^{ap}(T)$ (see reference (1) for details):

$$-\Delta C_p^{ap}(T) = C_{p,p}(T) \cdot m_p - C_{p,s}(T) \cdot \Delta m_s \qquad (7)$$

where $C_{p,p}(T)$ and $C_{p,s}(T)$ are the partial specific heat capacities of the protein solution and solvent respectively, m_p is the amount of protein in grams within the sample cell, and Δm_s the amount of solvent displaced by m_p.

Since

$$\Delta m_s = m_p \frac{v_p(T)}{v_s(T)} \qquad (8)$$

where $v_p(T)$ and $v_s(T)$ are the corresponding partial specific volumes of protein solution and solvent, it follows

$$C_{p,p}(T) = C_{p,s}(T) \frac{v_p(T)}{v_s(T)} - \frac{\Delta C_p^{ap}(T)}{m_p} \qquad (9)$$

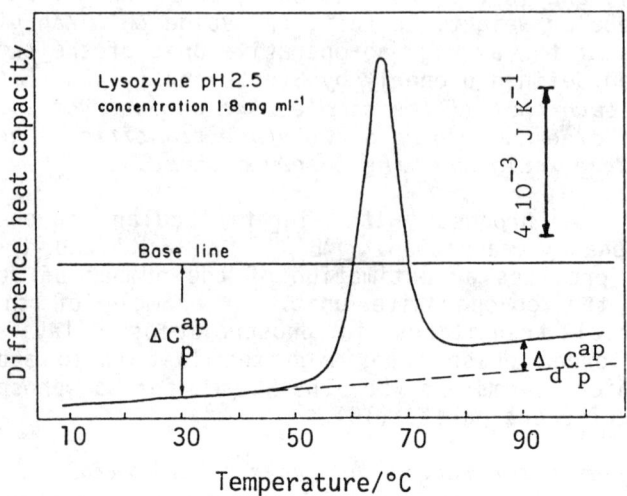

Figure 3. An example of DSC recording of a diluted protein solution. Taken from reference (7).

With regard to some of the approximations and assumptions currently made in studies of this type, the author's opinion is that the main source of error in the quantitative results of these experiments is the one coming from the uncertainty in measuring the protein concentration, this being determined by using either a known extinction coefficient for a certain wavelength or by any other usual method to calculate the concentration.

Thus, $C_{p,p}$ values for lysozyme at different pH's are shown in Figure 4. The $\Delta H^{v \cdot H}$ can be obtained by equation (5), and the specific thermal effect of the denaturation $\Delta_d h$ is calculated from the area under the peak hatched area in Figure 4, using an appropriate calibration signal and expressed per gram of protein. Here we have to extrapolate the $C_{p,p}$ of the native and denaturated states to the denaturation temperature (T_d), thus obtaining $\Delta_d C_p$ at that temperature. On the other hand, to use a step instead of the sigmoid does not imply any appreciable error given the symmetry of the curve (1).

The change in Gibbs energy for the process N ⇌ D (native⇌denaturated states) gives a measure of the protein stability as long as it is a reversible process. The question of the thermodynamic characterization of the initial and final states of the process, e.g., the folded and unfolded states,

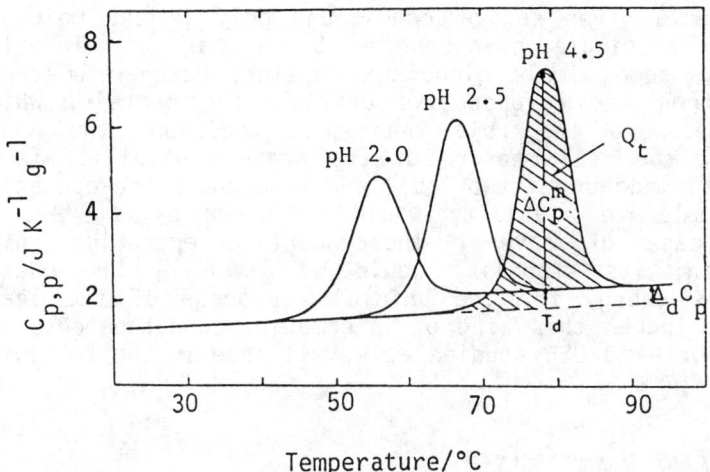

Figure 4. Temperature dependence of the partial specific heat capacity of lysozyme solution at different pH values. Taken from reference (7).

has been recently discussed by Privalov (7) and Pfeil (8). If we know the temperature dependence of the protein heat capacity it is possible to obtain the changes in enthalpy and entropy at any temperature, according to the equations (10) and (11):

$$\Delta_d H(T) = \Delta_d H(T_d) - \int_T^{T_d} \Delta_d C_p \, dT \qquad (10)$$

since $\Delta_d G(T_d) = 0$, then $\Delta_d S(T_d) = \Delta_d H(T_d)/T_d$, and

$$\Delta_d S(T) = \frac{\Delta_d H(T_d)}{T_d} - \int_T^{T_d} \frac{\Delta_d C_p}{T} \, dT \qquad (11)$$

where $\Delta_d X(T)$ is the change in X due to the unfolding process of the protein at any temperature (1,7). It follows then that

$$\Delta_d G(T) = \Delta_d H(T) - T \cdot \Delta_d S(T) =$$

$$= \Delta_d H(T_d) \frac{T_d - T}{T_d} - \int_T^{T_d} \Delta_d C_p \, dT + T \int_T^{T_d} \Delta_d C_p \, d\ln T \qquad (12)$$

The value of $\Delta_d G$ will provide a direct measure of the protein stability for a given set of conditions only if the molecule consists of a single co-operative block, as is virtually the case for some simple globular proteins. However if there were more than one co-operative unit in the protein, which seems to be more the rule than the exception, the above value is not the real measure of the protein stability (e.g. for a protein made up of two equal and independent co-operative units the molecule stability would be given by $\Delta_d G/2$). In a general case of several independent co-operative units in a protein, its stability would be given by the change in the Gibbs energy for the unfolding process of the least stable unit. Indeed this kind of information cannot be obtained without recourse to DSC studies as we will see in the following sections.

4. SIMPLE GLOBULAR PROTEINS

Privalov and co-workers have carried out a very complete DSC study of several simple compact globular proteins showing some interesting common thermodynamic features in their denaturation (7).

For each of these proteins the change in partial molar heat capacity on denaturation is independent of the pH conditions, being a characteristic value for each of these proteins. This value also coincides with the slope of $\Delta_d H$ versus denaturation temperature, T_d, in agreement with Kirchoff's law, $\Delta C_p = d \Delta H/dT$. This fact justifies the method followed to calculate the calorimetric enthalpy, and also indicates that the denaturation enthalpy is a direct function of the transition temperature but not of pH (the transition temperature is not a lineal function of pH).

A remarkable situation appears if we consider the temperature dependence of the specific enthalpy values as in Figure 5. For some proteins (Figure 5A) their enthalpy values merge at about 110°C with a $\Delta_d h \approx 54$ J g^{-1}, while for a second group of proteins (Figure 5B) there is not that coincidence with lower $\Delta_d h$ values at around 110°C. The present-day interpretation of this fact is that those proteins which reach 54 J g^{-1} for the denaturation enthalpy at about 110°C are very compact having a high internal saturation of hydrogen bonds, and the lower the values of $\Delta_d h$ at around 110°C the less compact and the fewer are the internal hydrogen bonds, as is the case for those proteins included in Figure 5B.

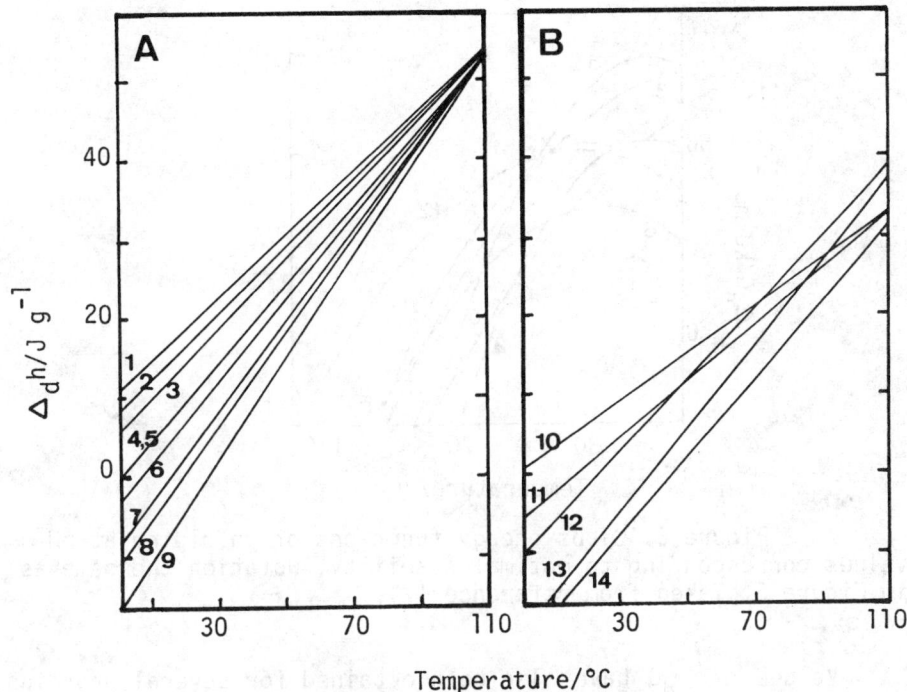

Figure 5. Temperature dependence of specific enthalpies of unfolding for several globular proteins. A: 1, ribonuclease; 2, parvalbumin; 3, lysozyme; 4, α-chymotripsin; 5, β-trypsin; 6, cytochrome c; 7, carbonic anhydrase B; 8, metmyoglobin; 9, papain. B: 10, α-lactalbumin; 11, ribosomal protein L7; 12, pancreatic trypsin inhibitor; 13, serum albumin; 14, histone H1. Taken from reference (7).

The slope of these lines, which as has been pointed out coincides with their respective $\Delta_d C_p$, is in its turn mainly related to the quantity and intensity of well developed hydrophobic cores within the native structure of the protein, which are water-exposed after denaturation (17). A distribution of the $\Delta_d C_p$ values between hydrophobic and vibrational contributions has been proposed by Sturtevant (18).

For some simple globular proteins such as ribonuclease A, lysozyme, α-chimotrypsin, cytochrome C and metmyoglobin the ratio $\Delta H^{cal} / \Delta H^{v.H} = 1.05 \pm 0.3$ which can be interpreted, as a first aproximation, as being due to the two-state character of the transition (denaturation), being the possible intermediate states of a very low relative concentration with a very low thermodynamic stability (1,7).

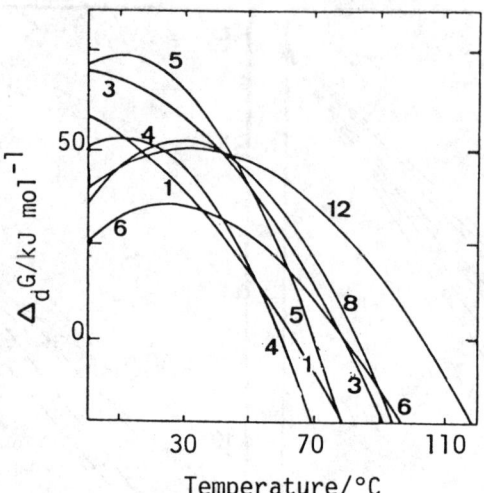

Figure 6. Gibbs energy functions of unfolding at pH values corresponding to optimal stability. Notation the same as in Figure 5. Taken from reference (7).

Values of $\Delta_d G$ have also been obtained for several proteins and their variations with temperature are shown in Figure 6. From the figure it is clear that the stability of these proteins is very similar without correlation with their molecular weight, and also that stability is rather low around physiological temperatures (50 ± 20 kJ mol^{-1}) (7). A final comment about these curves is the possibility of *cold denaturation*, as first pointed out by Brandts (19).

5. MORE COMPLEX PROTEINS

Not all proteins so far investigated by DSC provide a $\Delta H^{cal} / \Delta H^{v.H.}$ ratio close to one, since for many of them this value is clearly higher. An example would be papain whose melting profile does not qualitatively distinguish from those of simple proteins (7). The ratio for papain, however, is 1.8 (20) thus indicating that its denaturation is not a two-state transition. In cases like this, the interpretation could be either a molecule composed of more than one co-operative region, or that the whole molecule undergoes an overall transition with intermediate states, and it is the task of the investigator to find the appropriate ways to elucidate the correct explanation. Papain is known to have a deep cleft which divides the macromolecule into two halves or structural domains (21), so the most probable interpretation of the above ratio value is that there are two transitions corresponding to these two independent and co-operative regions. From this example the danger of using equilibrium data of denaturation processes without knowledge about their possible mechanisms and the ambiguity of the thermodynamic parameters thus obtained becomes clear (7,22). Other examples would be the Bence-Jones protein where the ratio $\Delta H^{cal}/\Delta H^{v.H}$ is 1.90 (23) or pepsinogen with that ratio being also very close to two (22). The Bence-Jones protein however consists of four domains and then it would have to be assumed that pairs of domains for a unique co-operative block. The consequence here is that the knowledge of the structural domains is not sufficient to establish the number of co-operative regions in the macromolecule. Pepsin, which is commented upon later is another example supporting this last statement. Other proteins which show rather complex DSC melting profiles are Troponin C (24), myosin and some of its fragments (25,26), parvalbumin in the presence of Ca^{+2} at pH <6.0 (27), fibrinogen (28), and some others currently being carried out.

5.1. Pepsin and Pepsinogen

A recent study carried out with porcine pepsin and its zymogen, pepsinogen, shows new possibilities of DSC investigation on the thermodynamic aspects of protein structure (12,22).

The pepsinogen molecule consists of 371 amino acid residues and after acidic activation releases a peptide made up of 44 residues from its N-terminal end thus becoming active pepsin. The sequence of both proteins is known although the three-dimensional structure is only known for pepsin at 3 Å resolution (29), which consists of two distinct lobes separated by a deep cleft as is the case of all acid proteases studied up until now (30,31).

Figure 7 shows $C_{p,p}$ values of pepsinogen solution as a function of temperature at three different pH values. Under all conditions investigated there was only one sharp and reversible denaturation peak with a $\Delta H^{cal}/\Delta H^{v.H}$ ratio verging on two (12,22).

The temperature dependence of $C_{p,p}$ for pepsin is indicated in Figure 8, where the dotted line refers to the second heating of the same solution. The thermodynamic parameters for the two main transitions (peak A and B in Figure 8) were obtained as shown in this Figure (see reference (12) for more details) and for both transitions, under different conditions, the ratio $\Delta H^{cal}/\Delta H^{v.H}$ was always close to two.

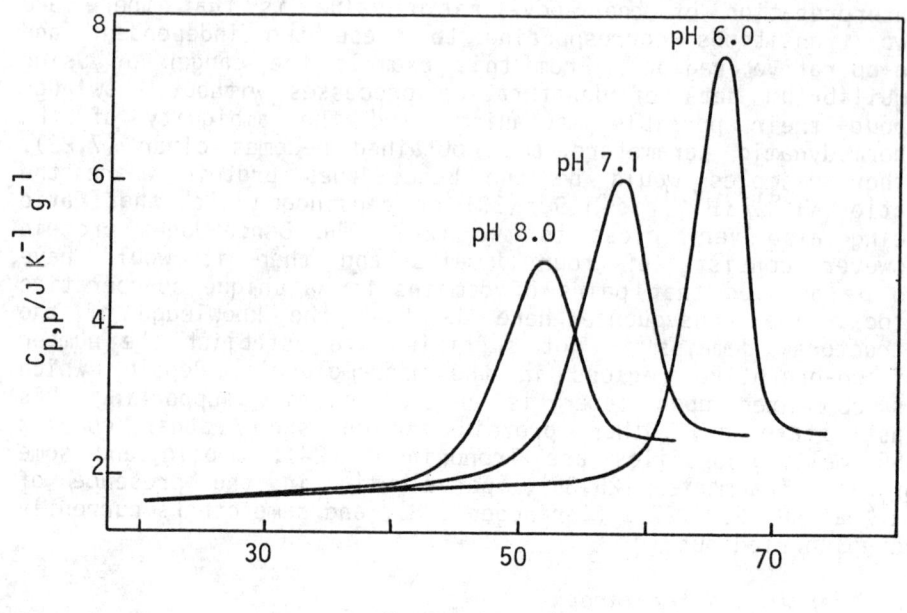

Figure 7. Partial specific heat capacity functions of pepsinogen at various pH values of solution. Taken from reference (12).

The assignment of the two pepsin transitions to specific regions of the molecule was carried out by DSC investigation of an inhibited pepsin fragment (residues 180-327 with an inhibitor attached to the aspartic acid residue 215), DDE-F, and the whole inhibited pepsin, DDE-P, as a direct reference (12).

Calorimetric recordings of DDE-P were under all conditions similar to those of active pepsin, showing the split of the curve into two well-defined peaks, except the higher temperature of the second peak in DDE-P. On reheating DDE-P, however, that transition temperature becomes equal to that of pepsin at corresponding conditions (Figure 9). A similar situation is obtained with DDE-F, where only this second peak appears with the total absence of the first one, peak A.

The change in temperature of the second peak on reheating has been proved to be due to the release of the label (inhibitor) at high temperatures during the first scan (12).

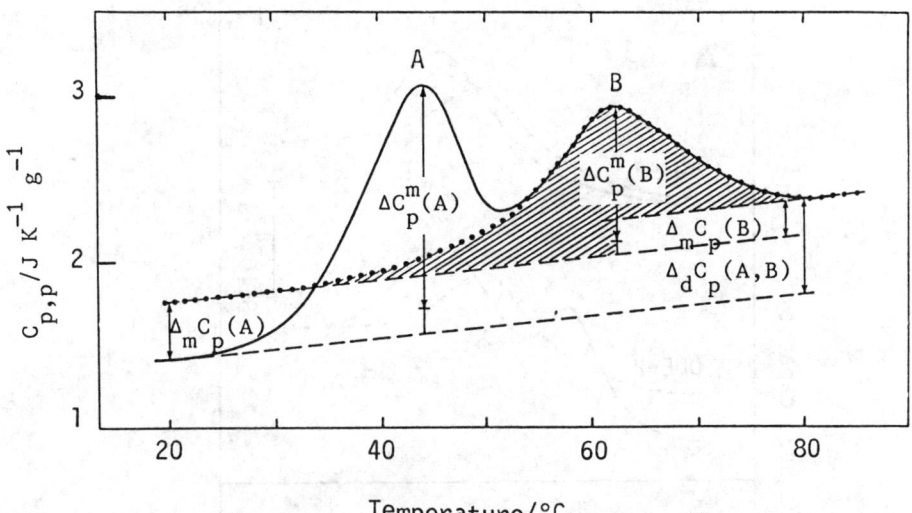

Figure 8. Partial specific heat capacity function of pepsin solution, pH 6.5, 100 mM-NaCl. Solid line, first heating; dotted line, second heating of the same sample. Dashed lines show extrapolation of heat capacity of native, partially denatured and denatured states. Hatched area corresponds to the heat effect of the second stage at denaturation. Taken from reference (12).

Under all conditions the ratio $\Delta H^{cal}/\Delta H^{v.H}$ for DDE-F was close to 2, both during the first and second heatings.

From the DDE-F results it is then possible to assign the first transition of pepsin to the N-terminal lobe of the molecule and the second transition to the C-terminal one (Figure 10). From an appropriate choice of the weight of each pepsin lobe according to Andreeva's scheme of the molecule (12,31) it is then possible to calculate the specific enthalpy values of melting for both lobes. When these enthalpy values, as well as those of pepsinogen, are plotted versus their respective unfolding temperature the slope of each line compares well with the respective increase of heat capacity on melting, and all of them extrapolate to the value 54 ± 4 J g^{-1} at 110°C, which indicates a high saturation of internal hydrogen bonds (about 0,7 bond/residue). From the slope of the lines it is possible to obtain 1,3 and 1,2 non-polar contacts/residue for pepsinogen and each of the pepsin lobes respectively (12).

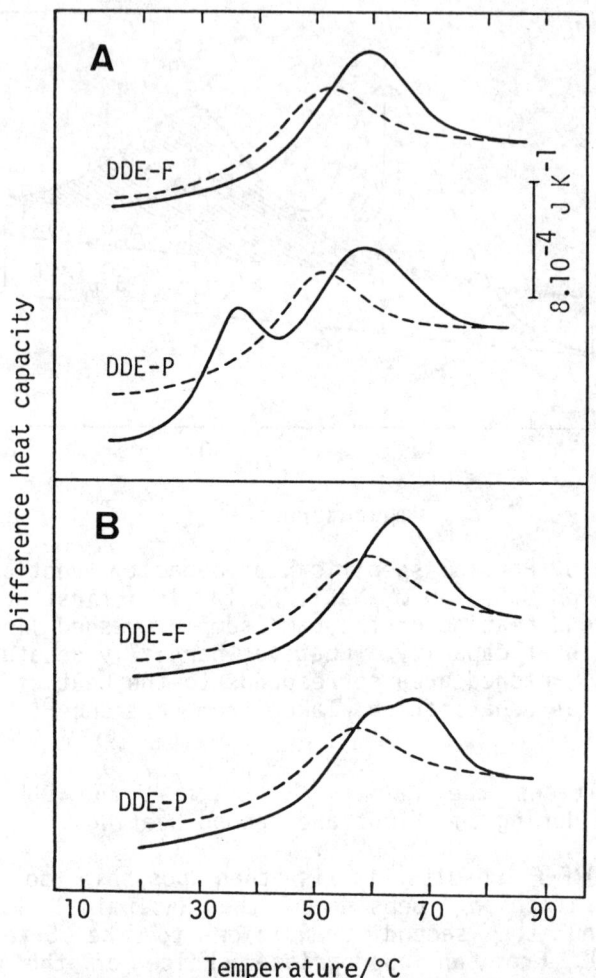

Figure 9. Calorimetric recordings of heat absorption on heating inhibited pepsin (DDE-P) and inhibited pepsin fragment (DDE-F) in solutions at pH 7.1 and (A) 10 mM-NaCl and (B) 100 mM-NaCl. Solid line, first heating; dashed line, second heating. Taken from reference (12).

As has been mentioned, the $\Delta H^{cal}/\Delta H^{v.H}$ ratio is higher than one for both pepsin transitions, pepsinogen and DDE-F, reaching the value of 2,0 under some conditions. It follows then that none of these transitions represents a two-state process. These results can be interpreted as an indication that each lobe of pepsin, the pepsin fragment and pepsinogen

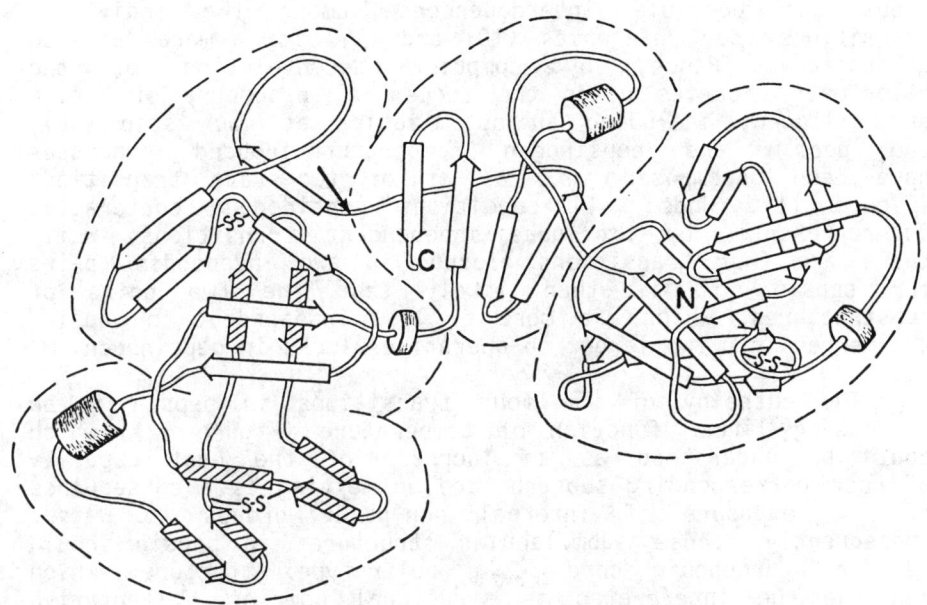

Figure 10. Scheme of pepsin structure according to Andreeva and Gustchina (31). Dashed lines encircle the compact regions with hydrophobic cores. The arrow indicates the place of cleavage of the polypeptide chain at fragmentation (residues 179-180). Taken from reference (12).

consists of two more or less independent co-operative substructures. At this point it is possible to carry out a more detailed analysis of these calorimetric curves by using the method proposed by Freire and Biltonen (10,11).

Experimentally, the average excess enthalpy function of the system, $<\Delta H>$ during a thermally induced transition, is obtained from the DSC data. Once $<\Delta H>$ is known as a continuous function of the temperature, the partition function of the system, $Z(T)$, can be calculated according to the equation

$$Z(T) = \exp \int_{T_o}^{T} \frac{<\Delta H>}{RT^2} dT \qquad (13)$$

From the partition function and using a recursive form the number of discrete macroscopic energy states and the enthalpy changes between them can be calculated without any *a priori* assumptions about the possible interdependence among the individual transitions (see references (10) and (11) for a more detailed discussion). Hence, by computer deconvolution of the calorimetric curves using the sequential procedure of Freire and Biltonen, including an optimization at each step (32), the pepsin and pepsinogen temperature-induced processes have been decomposed into a set of two-state transitions (Figure 11). Under all conditions pepsinogen denaturation is represented by two quasi-independent transitions, while there are four transitions grouped in two independent pairs for pepsin. It is then likely that the two pairs of substructures, which are more or less independent in pepsin, are merged into two larger co-operative blocks in pepsinogen.

The enthalpy of the four transitions in pepsin is an increasing linear function of temperature (Figure 12), which could be understood as an increase of the heat capacity of each corresponding substructure on melting as a consequence of the exposure of internal non-polar groups to water. Consequently, those submolecular structures in pepsin should have a hydrophobic core of globular-type structure, which can then be interpreted as structural domains. A tentative assignment of these four co-operative domains in the pepsin molecule is shown in Figure 10 by broken lines. This finding reminds one of the structural observations of Andreeva and Gustchina (31). These authors have proposed that, in considering the supersecondary structure of this protease, the polypeptide chain of the porcine pepsin molecule consists of four topologically equivalent structural units, each pair of them forming a domain. The calorimetric results bring up the question of the real definition of *domain* in a protein molecule. Since a geometrically defined domain can consist of several co-operative regions which fold independently of the rest of the molecule the term *domain* should perhaps be given more properly to those co-operative subunits.

Pepsinogen should be expected to have the same domains as pepsin, but here the pairs of domain in the lobes are much more connected with each other forming one co-operative block. This notable difference with pepsin might be the result of the loss of the 44 amino acid residues at the N-terminal end on activation, which are filling the gap between the lobes mainly connected to them by hydrophobic interactions (12). In this context, it is interesting to note that the presence of pepstatin, a well known inhibitor of acid proteases, also leads to the co-operation of two domains in the N-terminal lobe of pepsin, presenting a single co-operative block as it is found in pepsinogen. But in contrast to the removable 44 residues fragment in pepsinogen,

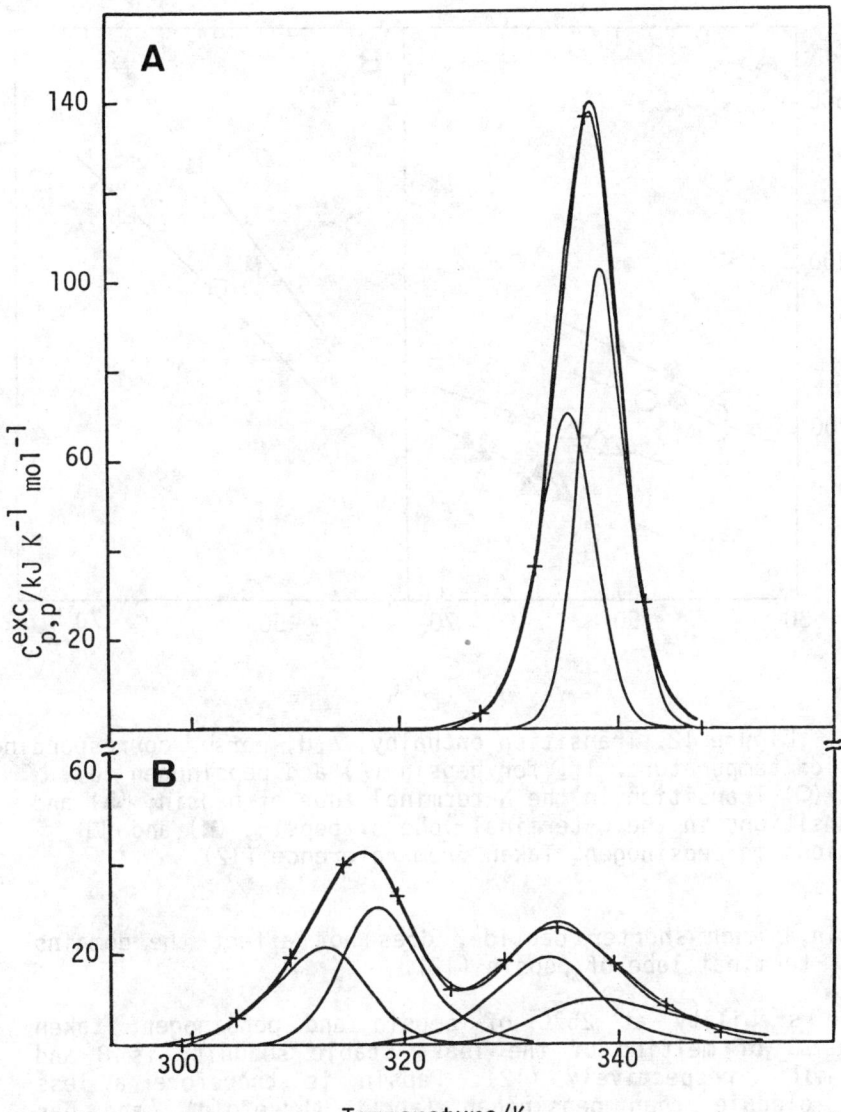

Figure 11. Computer deconvolution of the excess heat capacity function, $C_{p,p}^{exc}$, of pepsinogen (A) and pepsin (B) at the same solvent condition: pH 6.5, 100 mM-NaCl. Crosses indicate calculated excess heat capacity function, which virtually coincides with the experimental one. Taken from reference (12).

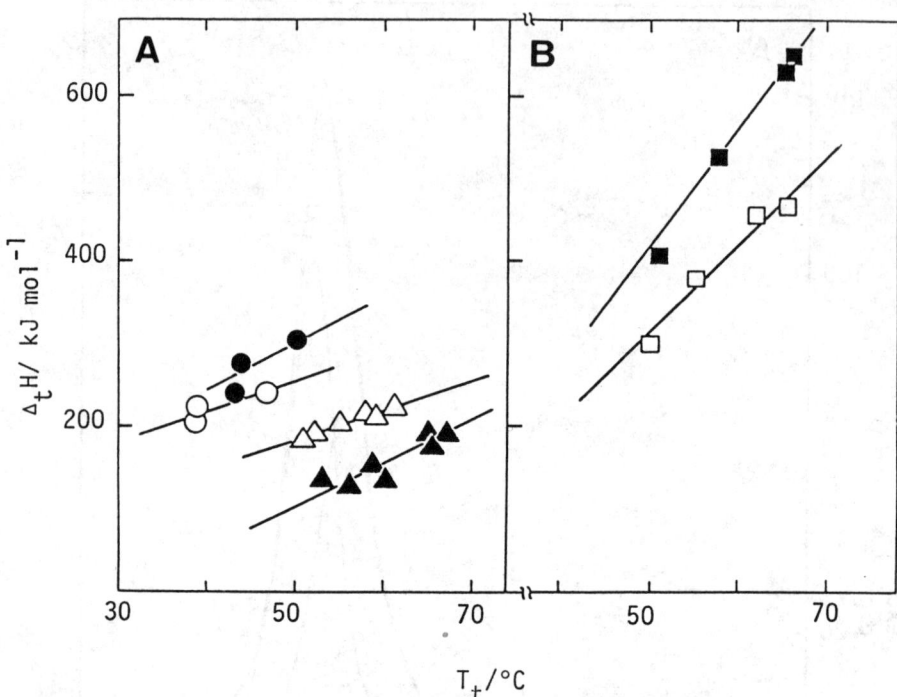

Figure 12. Transition enthalpy, $\Delta_t H$, versus corresponding transition temperature, T_t, for pepsin (A) and pepsinogen (B). (●) and (○) Transition in the N-terminal lobe of pepsin; (▲) and (△) transitions in the C-terminal lobe of pepsin; (■) and (□) transitions in pepsinogen. Taken from reference (12).

pepstatin, a much shorter peptide, does not affect the domains in the C-terminal lobe of pepsin (12).

The stability at 25°C of pepsin and pepsinogen, taken as the ΔG of melting of the least stable subunit, is 8 and 26 kJ mol^{-1} respectively (12). Pepsin is therefore a less stable molecule than pepsinogen (about threefold), and has a looser domain structure being thus a less tight molecule. This internal motility of pepsin might be related to its function as an enzyme.

5.2. Pepsin A from Aspergillus Awamori

The bilobal structure is a common feature of all the acid proteases studied by X-ray crystalography (30,31). A DSC calorimetric study has been recently carried out with pepsin

A from *Aspergillus Awamori* (PAA) and its results compared with those of porcine pepsin and pepsinogen (33). The thermal denaturation of PAA is also a complex process where at least two main transitions are observed (Figure 13). There are only two transitions, however, for the overall PAA melting curve, which is in contrast to what is found for porcine pepsin, and rather resembles the pepsinogen results (33). This similarity between pepsinogen and PAA at the level of co-operative domains, together with the closer $\Delta_d C_p$ values for both proteins than for PAA and porcine pepsin (33), reminds one of the fact that there are no zymogens known for acid proteases of fungous origin. It is then tempting

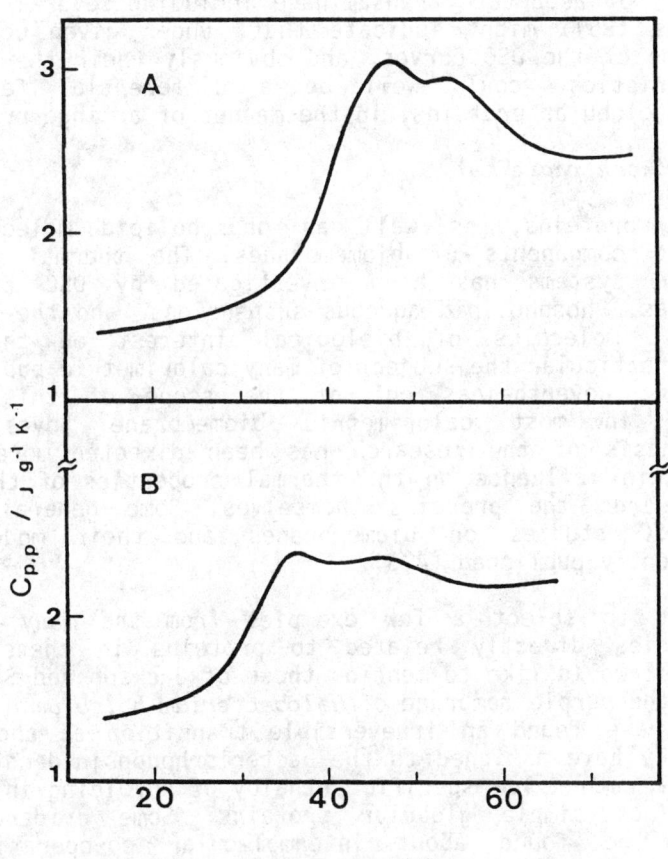

Figure 13. Partial specific heat capacity of pepsin from *Aspergillus Awamori* in solution as a function of temperature: A, pH 6.5; B, pH 6.9, 100 mM-NaCl.

to speculate that these fungous proteases could possess some structural characteristics that through molecular evolution manifest themselves in the corresponding zymogens of more developed species.

It is finally interesting to note the distinct difference found between the DSC curves for two proteins (porcine pepsin and PAA) which are so similar *a priori*. In this respect, the replacement of only one amino acid residue, Gly-211 by Arg or Glu, in the α-subunit of tryptophan synthase of *E. Coli* gives rise to clearly distinguishable DSC curves with significant differences between their respective thermodynamic parameters of denaturation (34). These results and some others (e.g. DSC study of aspartate transaminase including several substrate analogues (35)) might indicate that, under given conditions, the shape of the DSC curves, and obviously their thermodynamic characteristics, could well be a differential feature of at least globular proteins, in the manner of a fingerprint.

5.3. *Membrane proteins*

Membrane proteins, as well as phospholipid molecules are important components of biomembranes. The thermal behaviour of these systems has been investigated by DSC and other techniques. Phospholipid aqueous suspensions, and the influence of small molecules of biological interest on them, have been in particular the subject of many calorimetric publications which are nevertheless out of the scope of this report. Actually in most calorimetric biomembrane investigations the emphasis of the research has been directed more towards the protein influence on the thermal properties of the lipids than towards the proteins themselves. Some general reviews about DSC studies on biomembranes and their models have been recently published (4,5).

Just to select a few examples from the many of those DSC studies directly related to proteins in these complex systems, I would like to mention those of Jackson and Sturtevant (36) on the purple membrane of *Halobacterium halobium*. These authors have found an irreversible transition at about 100°C which they have assigned to the bacteriorhodopsin denaturation, showing a rather low specific enthalpy of unfolding in relation to that of simple globular proteins. Some evidence seems also to be found about intermolecular co-operativity in bacteriorhodopsin denaturation.

Small and co-workers using closely related systems, lipoproteins, have carried out an extensive DSC work of the human high density lipoproteins (HDL) and their main component, apolipoprotein A-1, including experiments both in the presence and absence of pure additional phospholipids (37-40).

Recently Pfeil and Bendzko have carried out a rigorous DSC study of cytochrome b_5, a typical membrane protein with two well defined regions, a hydrophobic one buried in the membrane and a haeme-containing hydrophilic one in contact with the solvent (41,42). They have shown that the thermal melting of the tryptic (sequence 1-90) or chymotryptic (12-97) fragments corresponds in both cases to a two-state transition with the same ΔG of stabilization.

The difficulty in the DSC studies of membrane proteins is related to the problem of purification and particularly the difficult solubility of these proteins; they usually need detergents, non-polar solvents or phospholipids to avoid agregation in aqueous media. This ultimately makes their study a more challenging one and there are reasons to believe that the near future will see more quantitative calorimetric studies of these proteins as well as of membrane systems in general.

6. NON-DENATURATIONAL CONFORMATIONAL TRANSITIONS

Non-denaturational conformational transitions have been often reported in the Literature, being in some cases temperature-induced processes (43-46). DSC would be in these latter cases a suitable method to study the thermodynamics of such transitions. Nevertheless not much calorimetric work has been published on the subject. Here I would like to comment briefly on some of these results and the consequent fact that some proposed transitions, as studied using other techniques than DSC, cannot in fact be considered as *real* conformational transitions in macromolecules.

An appropriate example would be D-amino acid oxidase (DAAO) for which Massey et al. (46) found a break in the Arrhenius plot around 12°C, and also a decrease in the fluorescence quantum yield of the protein by about 30% with increasing temperature along a sigmoidal curve centered at 12°C, with a van't Hoff enthalpy of 326 kJ mol^{-1}. The interpretation given was that of a temperature-dependent equilibrium between two forms showing different properties. Other authors found a somewhat lower $\Delta H^{v.H}$ for that temperature-dependent fluorescence change, about 213 kJ mol^{-1}, also centered at 12-13°C (47). Accordingly, a $\Delta H^{cal} \geq \Delta H^{v.H}$ should be expected and therefore easily observed in a high-sensitivity calorimeter like the DASM-1M. Figure 14 shows such a DSC experiment without any trace of the transition, where the

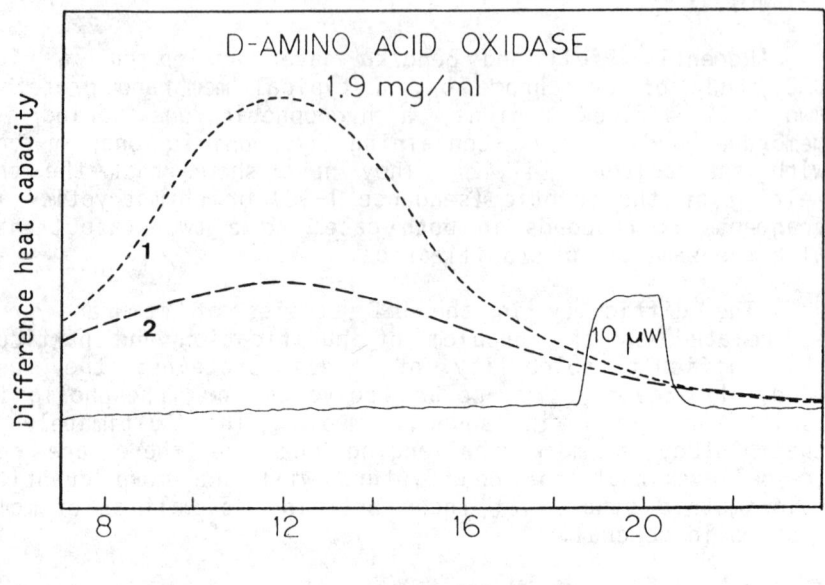

Figure 14. Solid line: original DSC recording of a DAAO solution, the peak at about 20°C is due to calibration heat supplied to the reference cell. Dashed lines (1) and (2): expected curves for a conformational transition centered at 12°C with enthalpy absorption of 326 and 213 kJ mol^{-1} respectively. Scan rate, 1 K min^{-1}.

dashed curves indicate the expected transition curves for a $\Delta H^{cal} = \Delta H^{v.H} = 326$ and 213 kJ mol^{-1} (47). This apparently puzzling result has to be explained by rejecting the idea of a conformational transition since DSC is a technique certain to detect such a transition. There is little difficulty in finding an alternative explanation for the kinetic evidence. A non-linear Arrhenius plot can be fitted to a curved one considering a non-zero $\Delta C_p^{\neq}$. In the data of Massey et al. (46), the required value of $\Delta C_p^{\neq}$ would be about $-0,84$ kJ K^{-1} mol^{-1}, which is a very modest quantity as it is less than half the heat capacity change, $-2,22$ kJ K^{-1} mol^{-1}, of the coenzyme binding, FAD, to the apoenzyme (48). It is, moreover, reasonable to expect $\Delta C_p^{\neq}$ for macromolecular systems far from the special value $\Delta C_p^{\neq} = 0$, thus implying that non-linear Arrhenius plots should normally be expected within experimental uncertainty. On the other hand, Silvius et al. (49) have shown that an enzyme with a temperature-dependent K_m can yield a variety of Arrhenius plot artifacts, most notably erroneus *breaks* if activity is assayed at a fixed substrate concentration. In other words, serious errors in the interpretation of Arrhenius plots can arise if temperature variations in substrate-binding affinity are not considered when determining the temperature effect on the catalyzed reaction rate.

The explanation for the fluorescence results is not so evident. Recently, however, Cooper (50) has given a simple and elegant possible explanation for this apparent paradox. According to Cooper the temperature-dependent fluorescence decrease could be due to a dynamic quenching process of large activation energy rather than a change in the conformational state of the protein. Thus, if the fluorescence transition were due to a thermal quenching process an apparent equilibrium constant of the form

$$K_{app} = \frac{BT}{k_f + k_o} \exp(-\Delta H^{\neq}/RT) \qquad (14)$$

would be obtained, where $B = (k_B/h) \exp(\Delta S^{\neq}/R)$ (k_B and h are the Boltzmann and Planck constants respectively), k_f is the rate constant for the radiative decay process and k_o is the temperature-independent term of the rate constant for the nonradiative decay process. The fluorescence experimental results are well fitted by a theoretical curve using an activation enthalpy, $\Delta H^{\neq}$, of 326 kJ mol^{-1} and assuming a typical value of 1 nsec for the fluorescence lifetime, $(k_f + k_o)^{-1}$. The case of DAAO does not have to be a particular one, neither fluorescence intensity the only property potentially affected by dynamic processes. Actually, properties such as fluorescence depolarization, NMR and EPR, enzyme kinetics, and others whose magnitude depends on stochastic relaxation processes, could show similar changes when having relaxation mechanisms with different thermal coefficients (50). Therefore, much care should be taken in considering experimental evidence for temperature-induced conformational transitions in biological macromolecules, and DSC is in that sense a definite check for the proposed structural changes and, in case of their existing, for their direct thermodynamic characterization.

7. CONCLUDING REMARKS

I have tried to give a general view of the present possibilities of DSC and some of its particular and unique advantages in studying protein unfolding and any structural change which is dependent on temperature. Thus, the simultaneous determination of the ΔH^{cal} and $\Delta H^{v.H}$ values is possible and hence the conclusions one can make from their ratio, as well as obtaining the unfolding heat capacity change from a single calorimetric recording. This approach allows us to investigate the co-operative nature of structural transitions, a property achievement which is not yet clearly understood, as well as their direct thermodynamic characterization.

On the other hand, although thermodynamics is not obviously concerned with mechanism and molecular arrangements, we have also seen how structural information, at the level of co-operative structural domains, can be obtained through a detailed quantitative analysis of high-sensitivity DSC curves.

Of course, new refinements in instrument design, that is, higher sensitivity and the requirement of smaller quantities of sample as seems to be the case of the new model DASM-4M (13), as well as improvements in the theoretical analysis of the DSC results, will make for a deeper understanding of the self-organization of proteins and related thermodynamic properties.

Finally, it is clear that I have just chosen a few recent examples out of many other important advances in this field. I have tried, however, to draw particular atention to the specific physico-chemical approach to the thermodynamic studies of the present selection, as developed during the last few years, without dwelling on their physiological or even biochemical aspects, which are also, of course, important facets of these studies.

NOTE

Those figures and data taken from the Literature and expressed in calories have been rearranged here to have them in joules in agreement with the recommendations of the Interunion Commission on Biothermodynamics.

ACKNOWLEDGEMENTS

I am particularly indebted to Prof. J.M. Sturtevant and Prof. P.L. Privalov, who, through direct collaboration originally aroused my interest in this field. I also thank Prof. M. Cortijo for reading and commenting upon this manuscript, and more generally for his help in setting up a Microcalorimetric section at Granada University. I would finally like to acknowledge the financial support from the *Comisión Asesora*.

REFERENCES

1.- Privalov, P.L. and Khechinashvili, N.N.: 1974, J.Mol.Biol. 86, pp. 665-684.
2.- Pfeil, W. and Privalov, P.L.: 1979, *Biochemical Thermodynamics* (ed. M.N. Jones) Vol. 1, Elsevier Scientific Publishing Co., Amsterdam, pp. 75-115.
3.- Privalov, P.L. and Filimonov, V.V.: 1978, J.Mol.Biol. 122, pp. 447-464.
4.- Mabrey, S. and Sturtevant, J.M.: 1978, Methods Membr.Biol. 9, pp. 237-274.
5.- Bach, D. and Chapman, D.: 1980, *Biological Microcalorimetry* (ed. A.E. Beezer) Academic Press, London, pp. 275-311.
6.- Pace, C.N.: 1975, Crit.Rev.Biochem. 3, pp. 1-43.
7.- Privalov, P.L.: 1979, Adv.Protein Chem. 33, pp. 167-241.
8.- Pfeil, W.: 1981, Mol.&Cell.Biochem. 40, pp. 3-28.
9.- Lumry, R., Biltonen, R. and Brandts, J.F.: 1966, Biopolymers 4, pp. 917-944.
10.- Freire, E. and Biltonen, R.L.: 1978, Biopolymers 17, pp. 463-479.
11.- Biltonen, R.L. and Freire, E.: 1978, CRC Crit.Rev.Biochem. 5, pp. 85-124.
12.- Privalov, P.L., Mateo, P.L., Khechinashvili, N.N., Stepanov, V. and Revina, L.: 1981, J.Mol.Biol. 152, pp. 445-464.
13.- Privalov, P.L.: 1980, Pure & Appl.Chem. 52, pp. 479-497.
14.- Privalov, P.L., Plotnikov, V.V. and Filimonov, V.V.: 1975, J.Chem.Thermodynamics 7, pp. 41-47.
15.- Mabrey, S. and Sturtevant, J.M.: 1976, Proc.Natl.Acad.Sci. USA 73, pp. 3862-3866.
16.- Albon, N. and Sturtevant, J.M.: 1978, Proc.Natl.Acad.Sci. USA 75, pp. 2258-2260.
17.- Kauzmann, W.: 1959, Adv.Protein Chem. 14, pp. 1-63.
18.- Sturtevant, J.M.: 1977, Proc.Natl.Acad.Sci. USA 74, pp. 2236-2240.
19.- Brandts, J.F.: 1964, J.Am.Chem.Soc. 86, pp. 4291-4301.
20.- Tiktopulo, E.I. and Privalov, P.L.: 1978, FEBS Lett. 91, pp. 57-58.
21.- Drenth, J., Jansonins, J.N., Krekoek, R., Swem, H.M. and Wolthers, B.G.: 1968, Nature 218, pp. 929-932.
22.- Mateo, P.L. and Privalov, P.L.: 1981, FEBS Lett. 123, pp. 189-192.
23.- Zav'yalov, V.P., Troitsky, G.V., Khechinashvili, N.N. and Privalov, P.L.: 1977, Biochim.Biophys.Acta 492, pp. 102-111.
24.- Tsalkova, T.M. and Privalov, P.L.: 1980, Biochim.Biophys.Acta 624, pp. 196-204.
25.- Potekhin, S.A. and Privalov, P.L.: 1978, Biofizika (USSR) 23, pp. 219-223.

26.- Potekhin, S.A. and Privalov, P.L.: 1979, Biofizika (USSR) 24, pp. 46-50.
27.- Filimonov, V.V., Pfeil, W., Tsalkova, T.N. and Privalov, P.L.: 1978, Biophys.Chem. 8, pp. 117-122.
28.- Privalov, P.L. To be published.
29.- Andreeva, N.S., Fedorov, A.A., Gustchina, A.E., Riskulov, R.R., Shutzkever, N.E. and Safro, M.G.: 1978, Mol.Biol. USSR 12, pp. 922-935.
30.- Tang, J., James, M.N.G., Hsu, I.N., Jenkins, J.A. and Blundell, T.L.: 1978, Nature 271, pp. 618-621.
31.- Andreeva, N.S. and Gustchina, A.E.: 1979, Biochem.Biophys. Res.Commun. 87, pp. 32-42.
32.- Matveev, S.V., Potekhin, S.A., Filimonov, V.V. and Privalov, P.L.: 1980, *Konformatsioniye Izmenenija Biopolymerov v Rastvorakh*, Proc. 5th Conf. Telavi, Izdatelstvo *Metsniereba*, Tbilisi (USSR) pp. 74.
33.- Mateo, P.L.: 1982, An.Quim., in the press.
34.- Matthews, C.R., Crisanti, M.M., Gepner, G.L., Veliçelebi, G. and Sturtevant, J.M.: 1980, Biochemistry 19, pp. 1290-1293.
35.- Relimpio, A., Iriarte, A., Chlebowski, J.F. and Martínez-Carrión, M.: 1981, J.Biol.Chem. 256, pp. 4478-4488.
36.- Jackson, M.B. and Sturtevant, J.M.: 1978, Biochemistry 17, pp. 911-915.
37.- Tall, A.R., Small, D.M., Shipley, G.G. and Lees, R.S.: 1975, Proc.Natl.Acad.Sci. USA 72, pp. 4940-4942.
38.- Tall, A.R., Shipley, G.G. and Small, D.M.: 1976, J.Biol.Chem. 251, pp. 3749-3755.
39.- Tall, A.R. and Small, D.M.: 1977, Nature 265, pp. 163-164.
40.- Tall, A.R., Small, D.M., Deckelbaum, R.J. and Shipley, G.G.: 1977, J.Biol.Chem. 252, pp. 4701-4711.
41.- Pfeil, W. and Bendzko, P.: 1980, Biochim.Biophys.Acta 626, pp. 73-78.
42.- Bendzko, P. and Pfeil, W.: 1980, Acta Biol.Med.Germ. 39, pp. 47-53.
43.- Sizer, I.W.: 1943, Adv.Enzymology 3, pp. 35-62.
44.- Levy, H.M., Sharon, N. and Koshland, D.E.: 1959, Biochim. Biophys.Acta 33, pp. 288-289.
45.- Kayne, F.J. and Snelter, C.H.: 1965, J.Am.Chem.Soc. 87, pp. 897-900.
46.- Massey, V., Curti, B. and Ganther, H.: 1966, J.Biol.Chem. 241, pp. 2347-2357.
47.- Sturtevant, J.M. and Mateo, P.L.: 1978, Proc.Natl.Acad.Sci. USA 75, pp. 2584-2587.
48.- Mateo, P.L. and Sturtevant, J.M.: 1977, BioSystems 8, pp. 247-253.
49.- Silvius, J.R., Read, B.D. and McElhaney, R.N.: 1978, Science 199, pp. 902-904.
50.- Cooper, A.: 1981, Proc.Natl.Acad.Sci. USA 78, pp. 3551-3553.

THERMOCHEMISTRY OF MOLECULES, MOLECULE-IONS, AND FREE RADICALS. FACTORS INFLUENCING BOND ENERGIES IN THESE SPECIES

Henry A. Skinner

Department of Chemistry,
University of Manchester,
Manchester, M13 9PL, U.K.

The thermochemist is entrusted with the experimental task of measuring heat changes in chemical processes, and establishing reliable values for the enthalpies of formation of chemical compounds in general. Precision calorimetric procedures have by now provided a wealth of data, particularly for organic compounds, sufficient to allow reliable predications of the enthalpies of formation of compounds for which measurements have not been made. Essentially, these empirical schemes attempt to correlate the total binding energy of a molecule with its molecular structure.

In this paper, the Allen scheme [1,2] is examined and applied to organic molecules. This scheme starts from the premise that the energy of atomization (ΔE_a^o) of a molecule in the gas-phase is given by the sum of the bond contributions in the molecule, and the sum of the bond-bond interactions between neighbouring bonds in the molecule. It resembles the "group-additivity" scheme successfully developed by Benson and co-workers [3].

The detailed Allen-scheme, by suitable algebraic manipulation [2], is rendered simple to apply in organic molecules, based on three bonding parameters, namely:-

(1) the bond-energy contributions, B(C-X)
(2) the bond-bond interaction terms, Γ_{CCX}

and

(3) 'trio-interactions, Δ, between three bonds attached to the same central atom.

To apply the simple Allen scheme, consider isopropyl chloride, C_3H_7Cl, as an example. The bond-energy contributions obtainable from the _full_ structure, are

$$[7B(CH) + 2B(CC) + B(CCl)];$$

the bond-bond interactions however are obtainable from the _skeleton_ structure ((C-H) bonds eliminated)

$$\begin{array}{c} C \\ \diagdown \\ C \\ \diagup \diagdown \\ C Cl \end{array}$$

and comprise _one_ C-C-C interaction (Γ_{CCC}), _two_ C-C-Cl interactions (Γ_{CCCl}), and _one_ trio interaction (CC; CC; CCl). Accordingly, the scheme predicts

$$\Delta E_a^o(\underline{iso}\text{-}C_3H_7Cl) = 7B(CH) + 2B(CC) + 2B(CC) + B(CCl)$$
$$+ \Gamma_{CCC} + 2\Gamma_{CCCl} + \Delta_{CCCl} \quad (1)$$

There are many molecules for which the Allen-scheme prediction is inadequate, and for good reasons. These include:-

(i) Sterically hindered molecules;
(ii) Ring structures under "strain";

and

(iii) resonance stabilized molecules, for which the chosen structure (e.g. _one_ of the Kekulé forms of benzene) is clearly inadequate as a representation of the true structure.

Where prediction is at fault the scheme nevertheless provides a measure of the disturbing factor, be this "strain" or "resonance" in character.

Application of the Allen scheme requires _empirical_ evaluation of the bond parameters, and values of these for carbon bonds in saturated organic molecules are given in Table 1. The application to various groups of compounds (hydrocarbons, alcohols, ethers, amines, thiols, halides) is summarized in Tables 2-15. The experimental ΔE_a^o values are obtained from enthalpies of formation, $\Delta H_f^o(g)$, taken mainly from recent critical evaluations of the experimental data (e.g. by Pedley and Rylance (4)), and from enthalpies of formation of ground-state atoms recommended by CODATA (5).

1. Bond Contributions (kJ mol^{-1})

$B(CH) = 413.31$; $B(NH) = 388.4$
$B(OH) = 461.0$; $B(SH) = 364.31$

$B(CC) = 328.1$; $B(NN) = 152.7$
$B(OO) = 140.78$; $B(SS) = 243.3$

$B(CN) = 270.75$; $B(CO) = 326.1$
$B(CF) = 432.1$; $B(CS) = 271.9$
$B(CCl) = 324.2$; $B(CBr) = 269.9$
$B(CI) = 212.2$;

2. Bond-Bond Interactions (kJ mol^{-1})

$\Gamma_{CCC} = 11.21$;
$\Gamma_{CCN} = 17.5$; $\Gamma_{CNC} = 18.5$
$\Gamma_{CCO} = 23.8$; $\Gamma_{COC} = 22.65$
$\Gamma_{CCF} = 23.7$;
$\Gamma_{CCCl} = 20.7$;
$\Gamma_{CCBr} = 18.0$;
$\Gamma_{CCI} = 14.9$;
$\Gamma_{CCS} = 14.2$; $\Gamma_{CSC} = 11.8$
$\Gamma_{CNN} = 22.5$; $\Gamma_{NCN} = 42$
$\Gamma_{COO} = 35.2$; $\Gamma_{OCO} = 54.5$
$\Gamma_{CSS} = 16.5$; $\Gamma_{FCF} = 63.0$
$\Gamma_{ClCCl} = 6.1$

$\Gamma_{NCO} = 52.6$;

3. Trio Interactions (Δ-values/kJ mol^{-1})

$\Delta_{CCC} = -2.7$ $\Delta_{CCF} = (-8.5)$
$\Delta_{CCN} = -4.7$ $\Delta_{CCCl} = -7.5$
$\Delta_{CCO} = -6.0$ $\Delta_{CCBr} = -6.0$
$\Delta_{CCS} = -5.3$ $\Delta_{CCI} = -5.5$
$\Delta_{COO} \sim (-14)$ $\Delta_{CFF} = -12.1$

$$\Delta_{CClCl} = -11.3$$
$$\Delta_{OOO} \sim (-30) \qquad \Delta_{FFF} = -36.9$$
$$\Delta_{ClClCl} = -11.0$$
$$\Delta^N_{CCC} = -8.9 \qquad \Delta^N_{CCN} = -8.4$$

TABLE 1. ALLEN SCHEME PARAMETER VALUES.

The application to <u>paraffins</u> (Tables 2,3) is well-tested and is rendered very satisfactory provided that small steric corrections arising from <u>gauche</u> 1,4 CH:CH interactions in branched chains are made [2]. <u>Alcohols</u> (Tables 4,5) fit the scheme satisfactorily, provided that a slightly smaller value (10.7 kJ mol^{-1}) for Γ_{CCC} is used than in hydrocarbons [6]. Ethers and polyethers (Tables 6,7) are also accurately predicted by the scheme; the deviations (δ) in dialkylethers containing bulky alkyl groups (e.g. di-<u>t</u>-butyl ether) provide a measure of steric interference in these molecules. Straight-chain polyethers have been considered in terms of the Allen scheme by Månsson [7]. In the case of 3,6-dioxaoctane, the deviation (δ = 9 kJ mol^{-1}) is outside experimental error limits, and indicates a destabilizing factor in this molecule overlooked in the scheme. The molecule has neighboring CO dipoles, <u>aligned in opposition</u>, contrasting with most of the other examples in Table 7. This unfavorable dipole alignment is also present in 3,7-dioxanonane, but here the opposing dipoles are separated from each other, and the deviation is much reduced (δ = 1.8 kJ mol^{-1}). The application to amines (Table 8) to thiols and alkyl sulphides (Tables 9,10), and to chloro-, bromo-, and iodo-alkanes (Tables 11-14) shows that the deviations are generally negligible, other than where steric interactions or strain (e.g. Table 10 : ring strain in cyclo-alkane sulphides) are to be expected. The destabilization effect of adjacent dipolar groups is revealed in polychloro-alkanes (Table 12) and dibromo-alkanes (Table 13), but in these cases may also be attributable to steric repulsion of the adjacent halogen atoms on neighboring carbon atoms.

	$\Delta H_f^o(g)$ kJ mol^{-1}	ΔE_a^o kJ mol^{-1}	ΔE_a^o(calc) kJ mol^{-1}	δ kJ mol^{-1}
CH_4	$-(74.5 \pm 0.4)$	1653.24	1653.24	–
C_2H_6	$-(84.0 \pm 0.4)$	2807.97	2807.97	–
C_3H_8	$-(104.7 \pm 0.7)$	3973.9	3973.9	+0.0
C_4H_{10}	$-(125.7 \pm 1.0)$	5140.2	5139.8	-0.4
C_5H_{12}	$-(146.5 \pm 1)$	6306.1	6305.8	-0.3
C_6H_{14}	$-(167.2 \pm 1)$	7472.1	7471.7	-0.4
C_7H_{16}	$-(187.5 \pm 1.3)$	8637.6	8637.6	0.0
C_8H_{18}	$-(208.5 \pm 1.5)$	9803.8	9803.6	-0.2
C_9H_{20}	$-(229.0 \pm 1.5)$	10969.6	10969.5	-0.1
$C_{10}H_{22}$	$-(249.5 \pm 2)$	12135.2	12135.4	+0.2
$C_{12}H_{26}$	$-(290.1 \pm 2.4)$	14466.4	14467.3	+0.9

$\delta = \Delta E_a^o(\text{calc.}) - \Delta E_a^o(\text{expt.})$
$B_{CH} = 413.31$ kJ mol^{-1}
$B_{CC} = 328.10$ kJ mol^{-1}
$\Gamma_{CCC} = 11.21$ kJ mol^{-1}

TABLE 2. ENTHALPIES OF FORMATION AND ATOMIZATION OF n-ALKANES.

Alkane	$\Delta H_f^o(g)$	ΔE_a^o	ΔE_a^o(calc.)	δ
2 Me propane	-(134.2 ± 0.7)	5148.6	5148.3	-0.3
2 Me butane	-(153.6 ± 1.0)	6313.2	6313.0	-0.2
22 diMe propane	-(168.5 ± 1.0)	6328.1	6328.6	+0.5
2 Me pentane	-(174.8 ± 1.2)	7479.7	7478.9	-0.8
3 Me pentane	-(172.1 ± 1.0)	7477.0	7477.0	0.0
22 diMe butane	-(186.4 ± 1.0)	7491.3	7490.9	-0.4
23 diMe butane	-(179.4 ± 1.0)	7484.3	7484.3	0.0
2 Me hexane	-(194.6 ± 1.4)	8644.7	8644.8	+0.1
3 Me hexane	-(191.3 ± 1.9)	8641.4	8642.9	+1.5
3 Et pentane	-(189.6 ± 0.8)	8639.7	8639.7	0.0
22 diMe pentane	-(206.6 ± 0.7)	8656.7	8656.7	0.1
23 diMe pentane	-(197.7 ± 0.8)	8647.8	8648.0	0.2
24 diMe pentane	-(201.7 ± 1.1)	8651.8	8652.0	+0.2
33 diMe pentane	-(201.2 ± 1.1)	8651.3	8651.2	-0.1

$\Delta_{CCC} = -2.7$ kJ mol^{-1}
$S_{11}(12) = -1.3$ kJ mol^{-1}
$S_{12}(12^2) = -1.6$ kJ mol^{-1} ; $S_{12}(12^3) = -1.7$ kJ mol^{-1}
$S_{12}(13^2) = -1.8$ kJ mol^{-1} ; $S_{12}(13^4) = -2.3$ kJ mol^{-1}
$S_{11}(2^22^2) = -2.2$ kJ mol^{-1} ; $S_{11}(2^22^3) = -2.3$ kJ mol^{-1}
$S_{12}(2^22^3) = -2.6$ kJ mol^{-1}

TABLE 3. BRANCHED-CHAIN ALKANES.

Alcohol	$\Delta H_f^\circ (g)$ kJ mol^{-1}	ΔE_a° kJ mol^{-1}	ΔE_a°(calc) kJ mol^{-1}	δ kJ mol^{-1}
CH_3OH	$-(201.6 \pm 0.5)$	2027.0	2027.0	–
C_2H_5OH	$-(234.8 \pm 0.5)$	3205.5	3205.5	–
C_3H_7OH	$-(255.5 \pm 0.5)$	4371.0	4371.0	0.0
C_4H_9OH	$-(275.0 \pm 0.8)$	5536.1	5536.4	+0.3
$C_5H_{11}OH$	$-(295.3 \pm 1.2)$	6701.6	6701.8	+0.2
$C_6H_{13}OH$	$-(315.8 \pm 1.0)$	7867.4	7867.2	-0.2
$C_7H_{15}OH$	$-(336.5 \pm 1.1)$	9035.3	9032.7	-0.5
$C_8H_{17}OH$	$-(356.3 \pm 2.1)$	10198.1	10198.1	-0.2
$C_9H_{19}OH$	$-(376.5 \pm 1.8)$	11363.7	11363.5	-0.2
$C_{10}H_{21}OH$	$-(396.6 \pm 1.9)$	12529.1	12528.9	-0.2
$C_{12}H_{23}OH$	$-(436.7 \pm 1.9)$	14859.6	14859.8	+0.2

ΔE_a° (calc) $= (2n+1)B_{CH} + (n+1)B_{CC} + B_{CO} + B_{OH} + \Gamma_{CCO} + (n-2)\Gamma_{CCC}$

$[B_{CO} + B_{OH}] = 787.1$ kJ mol^{-1} ; $B_{OH} = 461.0$ kJ mol^{-1}

$B_{CO} = 326.1$ kJ mol^{-1}

$[B_{CC} + \Gamma_{CCC}] = 338.8$ kJ mol^{-1} ; $B_{CC} = 328.1$ kJ mol^{-1}

$\Gamma_{CCC} = 10.7$ kJ mol^{-1}

(alcohols)

$[B_{CC} + \Gamma_{CCO}] = 351.9$ kJ mol^{-1} ; $\Gamma_{CCO} = 23.8$ kJ mol^{-1}

TABLE 4. ENTHALPIES OF FORMATION AND ATOMIZATION OF n-ALKANOLS.

Compound	ΔH_f^o(g)	ΔE_a^o kJ mol^{-1}	ΔE_a^o (calc)	δ
2-propanol	−(272.5 ± 0.5)	4388.4	4388.8	+0.4
2-Me-1-propanol	−(283.6 ± 0.6)	5544.7	5544.4	−0.3
2-butanol	−(292.8 ± 0.6)	5555.9	5554.2	−0.7
2-Me-2-propanol	−(312.4 ± 1.4)	5573.5	5574.0	+0.5
2-Pentanol	−(313.8 ± 1.2)	6720.1	6719.6	−0.5
3-Pentanol	−(314.7 ± 1.2)	6721.0	6719.6	−1.4
2-Me-1-butanol	−(302.0 ± 1.4)	6708.3	6708.5*	+0.2
3-Me-1-butanol	−(301.3 ± 1.4)	6707.6	6708.5*	+0.9
2-Me-2-butanol	−(330.8 ± 1.3)	6737.1	6738.1*	+1.0
3-Me-2-butanol	−(315.7 ± 1.1)	6722.0	6726.3*	+4.3

$\Delta_{CCO} = -6.0$ kJ mol^{-1}

* Includes the steric 1,4(CH:CH) correction, $S_{11}(12) \sim -1.3$ kJ mol^{-1}.

TABLE 5. BRANCHED-CHAIN ALCOHOLS; C_3-C_5

Compound	$\Delta H_f^\circ(g)$ kJ mol^{-1}	ΔE_a° kJ mol^{-1}	ΔE_a°(calc) kJ mol^{-1}	δ
Me$_2$O	−(184.0 ± 0.5)	3154.7	3154.7	−
MeOEt	−(216.4 ± 0.7)	4332.3	4333.2	+0.9
MeOPr	−(237.9 ± 1.1)	5499.0	5499.2	+0.2
Et$_2$O	−(251.7 ± 1.0)	5512.8	5511.7	−1.1
MeOBu	−(258.1 ± 1.1)	6664.4	6665.1	+0.7
EtOPr	−(272.2 ± 1)	6678.5	6677.7	−0.8
Pr$_2$O	−(292.3 ± 1.7)	7843.9	7843.6	−0.3
Bu$_2$O	−(333.9 ± 1)	10176.0	10175.5	−0.5
MeO^{i}Pr	−(252.0 ± 0.9)	5513.1	5517.0	+3.9
MeO^{t}Bu	−(291.6 ± 5)	6697.9	6703.2	+5.3
^{i}Pro^{t}Bu	−(357.7 ± 5)	9054.5	9065.5	+11.0
^{i}Pr$_2$O	−(318.8 ± 2.3)	7870.4	7870.4	+8.8
^{s}Bu$_2$O	−(360.9 ± 1.6)	10203.0	10211.1	+8.1
^{t}Bu$_2$O	−(362.0 ± 1.1)	10204.1	10251.7	+47.4

$B_{CO} = 326.1$ kJ mol^{-1} ; $\Gamma_{CCO} = 23.8$ kJ mol^{-1}

$B_{CC} = 328.1$ kJ mol^{-1} ; $\Gamma_{CCC} = 11.21$ kJ mol^{-1}

$\Gamma_{COC} = 22.65$ kJ mol^{-1} ; $\Delta_{CCO} = -6.0$ kJ mol^{-1}

TABLE 6. DI-ALKYL ETHERS, ΔH_f° AND ΔE_a° VALUES.

Compound	$\Delta H_f^o(g)$ kJ mol^{-1}	ΔE_a^o kJ mol^{-1}	ΔE_a^o (calc)	δ
Me(OCH$_2$)$_2$H	-(348.2 ± 0.7)	4710.7	4710.7	-
Et(OCH$_2$)$_2$Me	-(414.8 ± 0.8)	7067.8	7067.8	0.0
Et(OCH$_2$)$_3$Me	-(581.1 ± 1.0)	8626.1	8623.7	-2.4
Et(OCH$_2$)$_4$Me	-(741.0 ± 1.4)	10177.7	10179.7	+2.0
Et(OCH$_2$)$_5$Me	-(905.9 ± 1.9)	11734.7	11734.7	+1.0

$\Gamma_{COC} = 54.4$ kJ mol^{-1}

Compound	$\Delta H_f^o(g)$	ΔE_a^o	ΔE_a^o(calc)	δ
EtOCH$_2$CH$_2$OEt	-(408.2 ± 1)	8206.5	8215.54	+9.0
EtOCH$_2$CH$_2$CH$_2$OEt	-(436.2 ± 1.5)	9379.7	9381.5	+1.8

TABLE 7. STRAIGHT-CHAIN POLYETHERS, R(OCH$_2$)$_n$R'.

Compound	$\Delta H_f^o(g)$ kJ mol^{-1}	ΔE_a^o kJ mol^{-1}	ΔE_a^o (calc)	δ kJ mol^{-1}
NH_3	$-(45.94 \pm 0.35)$	1165.17	1165.17	-
CH_3NH_2	$-(23.0 \pm 0.4)$	2287.46	2287.46	-
$C_2H_5NH_2$	$-(47.5 \pm 0.7)$	3457.2	3459.7	+2.5
$C_3H_7NH_2$	$-(70.2 \pm 0.5)$	4625.1	4625.6	+0.5
$(CH_3)_2CHNH_2$	$-(83.8 \pm 0.5)$	4638.7	4638.4	-0.3
$C_4H_9NH_2$	$-(92.0 \pm 2.7)$	5792.1	5791.5	-0.6
$C_2H_5CH(NH_2)CH_3$	$-(104.8 \pm 0.9)$	5804.9	5804.3	-0.6
$(CH_3)_3CNH_2$	$-(120.9 \pm 0.5)$	5821.0	5821.0	-
$(CH_3)_2NH$	$-(18.5 \pm 0.4)$	3428.2	3428.2	-
$(C_2H_5)_2NH$	$-(72.6 \pm 1.9)$	5772.7	5772.7	0.0
$(CH_3)_3N$	$-(23.7 \pm 0.6)$	4578.6	4578.6	-
$(C_2H_5)_3N$	$-(92.8 \pm 0.6)$	8083.4	8095.3	+11.9
$H_2NCH_2CH_2NH_2$	$-(17.8 \pm 2.1)$	4113.2	4111.4	-1.8
$CH_3CH(NH_2)CH_2NH_2$	$-(48.5 \pm 2)$	5289.1	5290.1	+1.0
$C_2H_5CH(NH_2)CH_2NH_2$	$-(74.0 \pm 0.8)$	6459.9	6456.1	+3.8
$(CH_3)_2C(NH_2)CH_2NH_2$	$-(90.3 \pm 0.6)$	6476.2	6472.7	+3.5

$B_{NH} = 388.39$ kJ mol^{-1}
$B_{CN} = 270.75$ kJ mol^{-1}
$\Gamma_{CCN} = 17.5$ kJ mol^{-1} : $\Delta_{CCN} = -4.7$ kJ mol^{-1}
$\Gamma_{CNC} = 18.5$ kJ mol^{-1} : $\Delta_{CCC}^N = -8.9$ kJ mol^{-1}
$\delta = \Delta E_a^o(calc) - \Delta E_a^o(expt)$

TABLE 8. AMINES AND DIAMINES.

Compound	$\Delta H_f^o(g)$ * kJ mol^{-1}	ΔE_a^o kJ mol^{-1}	ΔE_a^o (calc)	δ
H_2S	$-(20.6 \pm 0.4)$	728.62	728.62	–
CH_3SH	$-(22.9 \pm 0.6)$	1876.14	1876.14	–
C_2H_5SH	$-(46.3 \pm 0.6)$	3044.8	3045.1	+0.3
C_3H_7SH	$-(67.9 \pm 0.8)$	4211.6	4211.0	-0.6
C_4H_9SH	$-(88.1 \pm 1.3)$	5377.0	5376.9	-0.1
$C_5H_{11}SH$	$-(110.1 \pm 1.3)$	6544.3	6542.8	-1.5
$C_{10}H_{21}SH$	$-(212.2 \pm 2)$	12372.5	12372.5	0.0
$HS(CH_2)_2SH$	$-(9.7 \pm 1.2)$	3282.7	3282.2	-0.5
$HS(CH_2)_3SH$	$-(29.7 \pm 1.3)$	4447.9	4448.1	+0.2
$HS(CH_2)_4SH$	$-(50.4 \pm 1.9)$	5613.8	5614.0	+0.2
$HS(CH_2)_5SH$	$-(71.0 \pm 1.5)$	6779.7	6779.9	+0.2
$(CH_3)_2CHSH$	$-(76.2 \pm 0.6)$	4219.9	4219.9	0.0
$(CH_3)_2CHCH_2SH$	$-(97.3 \pm 0.8)$	5386.2	5385.4	-0.8
$(C_2H_5)CH(CH_3)SH$	$-(96.9 \pm 0.8)$	5385.8	5385.8	0.0
$(CH_3)_3CSH$	$-(109.6 \pm 0.8)$	5398.5	5397.9	-0.6
$(CH_3)_2CHCH_2CH_2SH$	$-(114.9 \pm 1.2)$	6549.1	6550.0	+0.9
$(CH_3)_2CHCH(CH_3)SH$	$-(121.3 \pm 0.9)$	6555.5	6558.3	+2.8
$(C_2H_5)(CH_3)CHCH_2SH$	$-(115.1 \pm 0.9)$	6549.3	6549.1	-0.2
$(CH_3)_3CCH_2SH$	$-(129.0 \pm 0.9)$	6563.2	6563.7	+0.5
$(CH_3)_2(C_2H_5)CSH$	$-(127.1 \pm 0.9)$	6561.3	6561.4	+0.1

B_{SH} = 364.31 kJ mol^{-1} ; B_{CS} = 271.9 kJ mol^{-1}
Γ_{CCS} = 14.2 kJ mol^{-1} ; Δ_{CCS} = -5.3 kJ mol^{-1}

* $\Delta H_g^o(g)$ values from Pedley (4)

TABLE 9. ALKYL THIOLS AND DI-THIOLS.

Compound	$\Delta H_f^o(g)$ kJ mol^{-1}	ΔE_a^o kJ mol^{-1}	ΔE_a^o (calc)	δ kJ mol^{-1}
Me$_2$S	-(37.5 ± 0.5)	3036.0	3035.5	-0.5
MeSEt	-(59.6 ± 1.1)	4203.3	4204.4	+1.1
Et$_2$S	-(83.5 ± 0.8)	5372.4	5373.3	+0.9
MeSPr	-(82.2 ± 0.9)	5371.1	5370.3	-0.8
MeS^{i}Pr	-(90.5 ± 0.7)	5379.4	5379.2	-0.2
MeSBu	-(102.2 ± 0.7)	6536.4	6536.2	-0.2
EtSPr	-(104.7 ± 0.7)	6538.9	6539.2	+0.3
EtS^{i}Pr	-(117.2 ± 2.4)	6551.4	6548.1	-3.3
MeS^{t}Bu	-(121.3 ± 0.7)	6555.5	6557.3	+2.1
(CH$_2$)$_2$S	(82.1 ± 1.2)	2485.3	2565.3	+80.0
(CH$_2$)$_3$S	(60.7 ± 1.0)	3652.0	3731.2	+79.2
(CH$_2$)$_4$S	-(34.1 ± 0.9)	4892.0	4897.1	+5.1
(CH$_2$)$_5$S	-(63.5 ± 0.7)	6066.6	6063.1	-3.5
(CH$_2$)$_6$S	-(65.8 ± 2.0)	7214.1	7229.0	+14.9

$\Gamma_{CSC} = 11.8$ kJ mol^{-1}

TABLE 10. DIALKYL SULPHIDES.

Compound	$\Delta H_f^o(g)$ kJ mol^{-1}	ΔE_a^o kJ mol^{-1}	ΔE_a^o (calc)	δ kJ mol^{-1}
CH_3Cl	$-(82.0 \pm 0.7)$	1564.1	1564.1	–
C_2H_5Cl	$-(112.3 \pm 0.7)$	2739.6	2739.6	–
C_3H_7Cl	$-(132.5 \pm 0.9)$	3905.0	3905.5	+0.5
$(CH_3)_2CHCl$	$-(145.0 \pm 0.9)$	3917.5	3918.7	+1.2
C_4H_9Cl	$-(154.6 \pm 1.3)$	5072.3	5071.4	-0.9
$(CH_3)_2CHCH_2Cl$	$-(159.4 \pm 8.4)$	5077.1	5079.9	+2.8
$(CH_3)(C_2H_5)CHCl$	$-(161.2 \pm 8.4)$	5078.9	5084.6	+5.7
$(CH_3)_3CCl$	$-(182.1 \pm 1.2)$	5099.8	5098.8	-1.0
$C_5H_{11}Cl$	$-(175.2 \pm 1.4)$	6238.2	6237.3	-0.9
$(CH_3)_2CHCH_2CH_2Cl$	$-(179.5 \pm 8.5)$	6242.5	6244.5*	+2.0
$(CH_3)_2(C_2H_5)CCl$	$-(202.2 \pm 8.4)$	6265.2	6263.4*	+1.8
$C_8H_{17}Cl$	$-(238.9 \pm 2.1)$	9737.5	9735.1	-2.4
$C_{12}H_{25}Cl$	$-(320.4 \pm 2.9)$	14399.9	14398.9	-1.0
cyclo-$C_6H_{11}Cl$	$-(163.6 \pm 3.6)$	6940.7	6940.4	-0.3

$\beta(C-Cl) = 324.2$ kJ mol^{-1}

$\Gamma_{CCCl} = 20.7$ kJ mol^{-1} ; $\Delta_{CCCl} = -7.5$ kJ mol^{-1}

* ΔE_a^o(calc) includes $S_{11}(12) = -1.3$ kJ mol^{-1}

TABLE 11. CHLOROALKANES.

Compound	$\Delta H_f^o(g)$ kJ mol^{-1}	ΔE_a^o kJ mol^{-1}	ΔE_a^o (calc)	δ kJ mol^{-1}
CH_2Cl_2	$-(95.7 \pm 0.8)$	1481.1	1481.1	–
CH_3CHCl_2	$-(127.8 \pm 1.1)$	2658.4	2665.9	+7.5
$(CH_3)_2CCl_2$	$-(173.2 \pm 8.5)$	3849.0	3847.0	-2.0
CH_2ClCH_2Cl	$-(129.7 \pm 1.7)$	2660.3	2671.1	+10.8
$CH_3CHCl \cdot CH_2Cl$	$-(162.6 \pm 8.4)$	3838.6	3850.3	+11.7
$CH_2Cl \cdot CH_2 \cdot CH_2Cl$	$-(159.5 \pm 8.4)$	2835.5	3836.7	+1.2
$CHCl_3$	$-(104.8 \pm 2)$	1393.5	1393.2	-0.3
CH_3CCl_3	$-(142.3 \pm 1.4)$	2576.2	2576.1	-0.1
CH_2ClCCl_3	$-(152.4 \pm 1.4)$	2489.6	2507.7	+18.1
$CH_2ClCHCl_2$	$-(148.9 \pm 2.4)$	2582.8	2597.5	+14.7
$CH_2Cl \cdot CHCl \cdot CH_2Cl$	$-(182.9 \pm 1.8)$	3762.1	3781.9	+19.8
CCl_4	$-(97.2 \pm 2.9)$	1289.2	1289.4	+0.2
$CHCl_2CHCl_2$	$-(149.7 \pm 5)$	2486.9	2523.9	+37.9
$CHCl_2CCl_3$	$-(141.9 \pm 5.9)$	2286.4	2344.3	+57.9

$\Gamma_{ClCCl} = 6.1$ kJ mol^{-1}

$\Delta_{CCl_2} = -11.3$ kJ mol^{-1} ; $\Delta_{Cl_3} = -11.0$ kJ mol^{-1}

TABLE 12. POLYCHLORO-ALKANES.

Compound	$\Delta H_f^o(g)$ kJ mol^{-1}	ΔE_a^o kJ mol^{-1}	ΔE_a^o (calc)	δ kJ mol^{-1}
CH_3Br	$-(37.2 \pm 0.9)$	1509.8	1509.8	–
C_2H_5Br	$-(63.6 \pm 2.0)$	2681.4	2682.5	+1.1
C_3H_7Br	$-(84.5 \pm 2.0)$	3847.5	3848.6	+1.1
$(CH_3)_2CHBr$	$-(98.3 \pm 2.5)$	3861.3	3860.6	-0.7
C_4H_9Br	$-(107.3 \pm 1.4)$	5015.6	5014.4	-1.2
$C_2H_5(CH_3)CHBr$	$-(120.9 \pm 0.5)$	5029.2	5026.4	-2.8
$(CH_3)_3CBr$	$-(132.4 \pm 1.3)$	5040.7	5040.9	+0.2
$C_5H_{11}Br$	$-(128.9 \pm 1.6)$	6182.4	6180.3	-2.1
$C_6H_{13}Br$	$-(148.3 \pm 1.8)$	7347.0	7346.3	-0.7
$C_7H_{15}Br$	$-(167.8 \pm 1.9)$	8511.8	8512.2	+0.4
$C_8H_{17}Br$	$-(189.4 \pm 2.5)$	9678.6	9678.1	-0.5
$C_{12}H_{25}Br$	$-(269.9 \pm 2.6)$	14340.0	14341.9	+1.9
$C_{16}H_{33}Br$	$-(350.1 \pm 3.2)$	19001.1	19005.6	+4.5
CH_2BrCH_2Br	$-(37.8 \pm 1.5)$	2549.5	2557.1	+7.6
$CH_3CHBrCH_2Br$	$-(71.4 \pm 1.1)$	3728.3	3735.1	+6.8
$C_2H_5CHBrCH_2Br$	$-(95.1 \pm 5)$	4897.3	4901.0	+3.7
$CH_3CHBrCHBrCH_3$	$-(102.8 \pm 0.8)$	4904.9	4913.0	+8.1
$(CH_3)_2CBrCH_2Br$	$-(113.3 \pm 0.8)$	4915.5	4915.5	0.0

$B(C-Br) = 269.9$ kJ mol^{-1}
$\Gamma_{CCBr} = 18.0$ kJ mol^{-1}
$\Delta_{CCBr} = -6.0$ kJ mol^{-1}

TABLE 13. BROMOALKANES AND DIBROMOALKANES.

Compound	$\Delta H_f^o(g)$ kJ mol^{-1}	ΔE_a^o kJ mol^{-1}	ΔE_a^o (calc)	δ kJ mol^{-1}
CH_3I	(15.4 ± 0.9)	1452.1	1452.1	–
C_2H_5I	$-(9.0 \pm 0.9)$	2621.7	2621.7	–
C_3H_7I	$-(32.5 \pm 1.7)$	3790.5	3787.7	-2.8
$(CH_3)_2CHI$	$-(41.6 \pm 1.7)$	3799.6	3797.3	-2.3
$(CH_3)_3CI$	$-(72.0 \pm 2.2)$	4975.2	4975.4	+0.2
ICH_2CH_2I	(66.3 ± 1.4)	2435.2	2435.5	+0.3
CH_3CHICH_2I	(35.7 ± 3.4)	3611.0	3610.9	-0.1

$B(C-I) = 212.2$ kJ mol^{-1}
$\Gamma_{CCI} = 14.9$ kJ mol^{-1}
$\Delta_{CCI} = -5.5$ kJ mol^{-1}

TABLE 14. IODOALKANES AND DI-IODOALKANES.

The most pronounced evidence for destabilization from adjacent dipolar bonds is shown in Table 15, dealing with fluoralkanes and fluorocarbons. Unfortunately, there are gaps in the experimental data, particularly on 'key compounds' including CH_3F and C_2H_5F, which make for difficulties in establishing reliable values for the Allen parameters for C-F bonds; these gaps need to be closed, and present an experimental challenge which hopefully will be met in the near future.

The deviations δ are large for all the fully fluorinated hydrocarbons, each grouping $-CF_2-CF_2-$ reducing the stability by ca. 40-45 kJ mol^{-1}. When separated, as in $CF_3CH_2CF_3$, the destabilizing effect is markedly reduced.

Compound	$\Delta H_f^\circ(g)$ kJ mol^{-1}	ΔE_a° kJ mol^{-1}	ΔE_a° (calc)	δ kJ mol^{-1}
CH_2F_2	$-(452.3 \pm 0.9)$	1753.8	1753.8	-
CHF_3	$-(698.8 \pm 1.4)$	1861.7	1861.7	-
CF_4	$-(934.5 \pm 0.4)$	1958.8	1958.8	-
CH_3CHF_2	$-(497.0 \pm 8)$	2943.8	2943.8	-
CH_2FCHF_2	$-(682.0 \pm 17)$	2990	2986.3	-3.7
CH_3CF_3	$-(744.6 \pm 1.7)$	3052.7	3051.2	-1.5
C_2F_6	$-(1350.2 \pm 2.3)$	3242.5	3294.5	+52.0
C_3H_7F	$-(285.8 \pm 2.2)$	4016.4	4016.4	0.0
$(CH_3)_2CHF$	$-(293.3 \pm 1.5)$	4023.9	4031.7	+7.7
$CF_3CH_2CF_3$	$-(1406.2 \pm 7)$	4443.8	4449.2	+5.4
C_3F_8	$-(1783.2 \pm 7.3)$	4543.5	4614.6	+71.1
cyclo-C_4F_8	$-(1542.6 \pm 10.6)$	5017.1	5280.4	+263.3
cyclo-$C_6F_{11}CF_3$	$-(2899.7 \pm 1.4)$	8978.3	9248.6	+270.3
C_7F_{16}	$-(3385.8 \pm 1.6)$	9618.2	9895.1	+276.9
cyclo-$C_6F_{11}C_2F_5$	$-(3304.9 \pm 2.3)$	10251.5	10568.7	+317.2

$B(C-F) = 432.1$ kJ mol^{-1}
$\Gamma_{CCF} = 23.7$ kJ mol^{-1} ; $\Gamma_{FCF} = 63.0$ kJ mol^{-1}
$\Delta_{CCF} \sim -8.5$ kJ mol^{-1} ; $\Delta_{CCF} = -12.1$ kJ mol^{-1}
$\Delta_{FFF} = -36.9$ kJ mol^{-1}

TABLE 15. FLUOROALKANES AND FLUOROCARBONS.

Olefines and Unsaturated Compounds

The application of an Allen-type scheme to olefines can be made in more than one way, depending on the model chosen to describe the -C=C- double-bond. The normal description of this as a σ-π combination allows the total binding energy to be divided in σ- and π-bond contributions. In practice, this scheme requires evaluation of bond-energy contributions (B(C*X) and interactions $\Gamma_{XC^*C^*}$ (where C* indicates trigonal carbon), which cannot be presumed equal to their unstarred conterparts in saturated organic molecules. Moreover, whereas values B(CH), B(CC), B(CX) can be evaluated progressively from $CH_4 \rightarrow C_2H_6 \rightarrow CH_3X$, it is not possible to isolate B(C*H) from other contributions in olefins.

An alternative approach treats the -C=C- bond as a pair of 'strained' or 'banana' C-C single bonds, and has the initial advantage of allowing transfer of Allen parameters already evaluated from saturated compounds. The interaction of a substituent X in $XCH=CH_2$ with the double bond is equated to twice Γ_{XCC}, requiring that the double bond contribution B(C=C) = [2B(CC) + $2\Gamma_{CCC}$ - S], where S is the "strain" in the double-bond.

This scheme gives, as typical examples,

$$\Delta E_a^o(C_2H_4) = 4B(CH) + B(C=C)$$

$$\Delta E_a^o(CH_3CH=CH) = 6B(CH) + B(CC) + B(C=C) + 2\Gamma_{CCC} + \Delta_{CCC}$$

$$\Delta E_a^o((CH_3)_2C=CH_2) = 8B(CH) + 2B(CC) + B(C=C) + 5\Gamma_{CCC} + 4\Delta_{CCC}$$

and

$$\Delta E_a^o(\text{butadiene}) = 6B(CH) + B(CC) + 2B(C=C) + 4\Gamma_{CCC} + 2\Delta_{CCC}$$

$$\dots\dots \quad (2)$$

but makes no specific allowance for "conjugation" energy between double-bonds (as in 1,3-butadiene), nor for "hyperconjugation" of alkyl groups adjacent to a formal -C=C- bond in olefines. Table 5, 16 and 17 present the application of this scheme to olefines and polyenes. The small deviations for olefines virtually disappear if one allows a hyperconjugative stabilizing increment of ca. 2.5 - 3 kJ mol^{-1} per alkyl-C=C attachment. In non-conjugated dienes (Table 17), the deviations are again small, consistent with slight stabilization from hyperconjugation : in conjugated dienes (e.g. 1,3 butadiene), the stabilization is much larger, and a 'resonance energy' contribution of ca. 20 kJ mol^{-1} for the C=C-C=C π-conjugation is implied.

The large positive deviations in cyclopropene and cyclobutene provide a measure of the ring-strain which indicate this to be larger in these small rings than in the corresponding cycloalkanes. In the larger rings, however, the ring-strain is possibly less in cycloalkenes than in cycloalkanes.

The application to selected aromatic hydrocarbons is shown

in Table 18. The deviations, δ, are large, and measure the energy gap between a typical Kekulé structure (fixed double-bonds) and the actual resonance-hybrid molecule. The values correspond closely with accepted 'resonance' energies in these aromatic systems.

Compound		$\Delta H_f^o(g)$ kJ mol^{-1}	ΔE_a^o kJ mol^{-1}	ΔE_a^o(calc) kJ mol^{-1}	δ kJ mol^{-1}
C_2H_4		(52.2 ± 1.2)	2240.7	(2240.7)	–
$CH_3CH=CH_2$		(20.2 ± 0.6)	3418.0	3415.1	-2.9
$C_2H_5CH=CH_2$		-(0.4 ± 0.9)	4583.8	4580.9	-2.7
$C_3H_7CH=CH_2$		-(21.9 ± 1.2)	5750.5	5747.0	-3.5
$C_4H_9CH=CH_2$		-(41.4 ± 0.9)	6915.2	6912.9	-2.3
$C_5H_{11}CH=CH_2$		-(62.5 ± 1.7)	8081.6	8078.9	-2.7
$CH_3CH=CHCH_3$	(t)	-(12.2 ± 0.5)	4595.6	4589.6	-6.0
	(c)	-(7.8 ± 0.5)	4591.2	4585.2	-6.0
$C_2H_5CH=CHCH_3$	(t)	-(32.7 ± 0.9)	5761.3	5755.5	-5.8
	(c)	-(29.1 ± 0.9)	5757.7	5751.1	-6.6
$C_3H_7CH=CHCH_3$	(t)	-(53.9 ± 1.6)	6927.7	6921.4	-6.3
EtCH=CHEt	(t)	-(54.4 ± 1.3)	6928.2	6921.4	-6.8
$(CH_3)_2C=CH_2$		-(16.9 ± 0.6)	4600.3	4594.2	-6.1
$(C_2H_5)(CH_3)C=CH_2$		-(35.6 ± 0.7)	5764.2	5758.8	-5.4
$(CH_3)_2C=CHCH_3$		-(42.1 ± 0.6)	5770.7	5764.2	-6.5
$(CH_3)_2C=C(CH_3)_2$		-(69.3 ± 0.8)	6943.1	6938.9	-4.2
(t) = trans	(c) = cis				
$B(C=C) = 587.46$ kJ mol^{-1} $\Gamma_{C-C=C} = 2\Gamma_{CCC} = 22.4$ kJ mol^{-1} $(S_{cis} \sim -4.4$ kJ mol$^{-1})$					

TABLE 16. OLEFINES.

Compound	$\Delta H_f^o(g)$ kJ mol^{-1}	ΔE_a^o kJ mol^{-1}	ΔE_a^o (calc)	δ kJ mol^{-1}
Buta-1,3-diene	(109.9 ± 0.8)	4042.5	4022.3	-20.2
Penta-1,3-diene (c)	(81.1 ± 1.2)	5216.5	5195.1*	-21.4
Penta-1,3-diene (t)	(76.3 ± 0.6)	5221.3	5199.5	-21.8
Penta-1,4-diene	(105.7 ± 0.6)	5191.9	5188.2	-3.7
2Me-1,3-butadiene	(75.3 ± 0.8)	5222.3	5198.2*	-24.1
Hexa-1,5-diene	(84.1 ± 0.6)	6358.7	6354.2	-4.5
2,3 diMe-1,3-butadiene	(43.9 ± 1.0)	6398.9	6374.0*	-24.9
Cyclopropene	(277.1 ± 2.5)	2730.0	2947.6	+217.6
Cyclobutene	(156.7 ± 1.5)	3995.7	4113.4	+117.7
Cyclopentene	(32.7 ± 1.7)	5264.9	5279.4	+14.5
Cyclohexene	-(4.6 ± 0.5)	6447.4	6445.3	-2.1
Cycloheptene	-(9.4 ± 0.9)	7597.4	7611.3	+13.9
Cyclooctene	-(27.0 ± 1.1)	8760.3	8777.2	+16.9
Cyclopentadiene	(130.8 ± 3.8)	4735.7	4729.2	-6.2
1,3-cyclohexadiene	(106.3 ± 1.0)	5905.5	5895.1	-10.4
1,3-cycloheptadiene	(94.2 ± 0.9)	7062.8	7061.0	-1.8
1,3,5-cycloheptatriene	(182.8 ± 1.2)	6543.2	6510.8	-32.4
1,3,5,7-cyclooctatetraene	(297.6 ± 1.3)	7142.8	7126.5	-16.3

* 1,4(C-H)[CCC=C] $\sim S_{cis} \sim$ -4.4 kJ mol^{-1}

TABLE 17. DI-ENES, CYCLOALKENES AND DIENES.

Compound	$\Delta H_f^\circ(g)$ kJ mol^{-1}	$\Delta E_a^\circ(g)$ kJ mol^{-1}	ΔE_a° (calc)	δ kJ mol^{-1}
C_6H_6	(82.9 ± 0.3)	5497.8	5344.9	-152.9
$C_6H_5CH_3$	(50.1 ± 0.3)	6675.9	6525.1	-150.8
1,3 diMe benzene	(17.3 ± 0.6)	7853.9	7705.3	-148.6
1,3,5 triMe benzene	-(15.9 ± 1.3)	9032.3	8885.5	-146.8
Styrene	(147.7 ± 0.7)	7287.5	7132.3	-155.2
Naphthalene	(150.4 ± 1.4)	8718.1	8460.6	-257.5
Anthracene	(230.3 ± 2.5)	11926.0	11576.4	-349.6
Diphenyl	(182.3 ± 1.4)	10545.7	10237.9+	-307.8

+ Steric correction, = -4.4 kJ mol^{-1} included

TABLE 18. AROMATIC HYDROCARBONS.

The application of the scheme to aromatic compounds and to pyridine derivatives is presented in Tables 19 and 20. The deviations, δ, provide a measure of the π-conjugation or ring-resonance energies in these compounds, and show that in the pyridines the ring-resonance is ca. 30 kJ mol^{-1} less than in benzene derivatives. The aromatic compounds C_6H_5X, yield resonance energies of similar order to that in benzene for $X = NH_2$, OH, SH and SCH_3, but are less for $X = OCH_3$ and halogen. The resonance in C_6H_5X is expected to involve the contribution from structures of the type

    (1)   (11)   (111)

when the group X is potentially capable of back-coordination, and the smaller values for δ for X = halogen are not in accord with this expectation.

The deviations δ for methyl- and hydroxy-pyridines are generally similar to that for pyridine. It is particularly interesting to note that in the hydroxy-pyridines (recently investigated thermochemically by Suradi et al [8]) the marked increase in stability of the 2-hydroxy derivatives with respect to the 3-, or 4-hydroxy compounds is accounted for not in terms of a larger conjugation or resonance energy effect, but because 2-hydroxy substitution involves the strong interaction Γ_{NCO} (∼ 52.6 kJ mol^{-1}) where 3- or 4-hydroxy substitution involves Γ_{CCO} (∼ 23.8 kJ mol^{-1}). In like manner, 2-Me pyridine involves Γ_{NCC} (∼ 17.5 kJ mol^{-1}) where 3-Me and 4-Me involve the slightly weaker Γ_{CCC} (∼ 11.2 kJ mol^{-1}).

Compound	$\Delta H_f^o(g)$ kJ mol^{-1}	ΔE_a^o kJ mol^{-1}	ΔE_a^o (calc)	δ kJ mol^{-1}
Phenol	−(96.3 ± 0.8)	5923.7	5772.1	−151.6
2Me-phenol	−(128.6 ± 1.3)	7101.2	6952.4*	>(−148.8)*
3Me-phenol	−(132.3 ± 1.2)	7104.9	6952.4	−152.5
4Me-phenol	−(125.3 ± 1.6)	7097.9	6952.4	−145.5
3-OH phenol	−(274.7 ± 2.1)	6348.8	6199.3	−149.5
$C_6H_5OCH_3$	−(68.0 ± 1.1)	7040.7	6899.7	−141.0
$C_6H_5OC_2H_5$	−(101.7 ± 0.5)	8219.6	8078.2	−141.4
$C_6H_5NH_2$	−(87.1 ± 0.8)	6179.4	6017.6	−161.8
C_6H_5F	−(116.0 ± 1.4)	5558.1	5409.3	−148.8
1,2 $C_6H_4F_2$	−(293.8 ± 1.0)	5597.3	5473.7*	>(−123.6)*
1,3 $C_6H_4F_2$	−(309.2 ± 1.1)	5612.7	5473.7	−139.0
1,4 $C_6H_4F_2$	−(306.6 ± 1.1)	5610.1	5473.7	−136.4
C_6H_5Cl	(51.3 ± 0.7)	5432.7	5295.4	−137.3
1,2 $C_6H_4Cl_2$	(30.2 ± 1.8)	5357.1	5245.9*	>(−111.2)*
1,3 $C_6H_4Cl_2$	(25.7 ± 2.1)	5361.6	5245.9	−115.7
1,4 $C_6H_4Cl_2$	(22.5 ± 1.6)	5364.9	5245.9	−119.0
C_6H_5Br	(104.3 ± 3.1)	5370.3	5237.5	−132.8
C_6H_5I	(162.2 ± 4.6)	5307.3	5172.0	−135.3
C_6H_5SH	(112.4 ± 0.8)	5742.8	5594.5	−148.3
$C_6H_5SCH_3$	(97.8 ± 1.2)	6902.7	6753.8	−148.9

* No correction made for cis-steric interaction

TABLE 19. AROMATIC COMPOUNDS.

Compound	$\Delta H_f^o(g)$ kJ mol^{-1}	ΔE_a^o kJ mol^{-1}	ΔE_a^o (calc)	δ kJ mol^{-1}
Pyridine	(140.2 ± 1.2)	4981.0	4857.4	-123.6
2 Me-pyridine	(99.2 ± 0.7)	6167.3	6042.0	-125.3
3 Me-pyridine	(106.4 ± 0.5)	6160.1	6037.7	-122.5
4 Me-pyridine	(103.8 ± 1.2)	6162.7	6037.7	-125.0
23-diMe-pyridine	(68.3 ± 1.3)	7343.4	7222.2	-121.2*
24-diMe-pyridine	(63.9 ± 0.9)	7347.8	7222.2	-125.6
25-diMe-pyridine	(66.5 ± 1.0)	7345.2	7222.2	-123.0
26-diMe-pyridine	(58.7 ± 1.6)	7353.0	7226.7	-126.3
34-diMe-pyridine	(70.1 ± 1.1)	7341.6	7217.9	-123.7
35-diMe-pyridine	(72.8 ± 0.9)	7338.9	7217.9	-121.0
2-OH pyridine	-(79.7 ± 1.5)	5447.6	5322.2	-125.4
3-OH pyridine	-(43.7 ± 1.7)	5411.6	5284.6	-127.0
4-OH pyridine	-(40.8 ± 2.1)	5408.7	5284.6	-124.1
2Me, 3-OH pyridine	-(84.5 ± 1.8)	6597.6	6469.2	-128.4
2Me, 4-OH pyridine	-(71.7 ± 1.7)	6584.8	6469.2	-115.5
2Me, 5-OH pyridine	-(69.8 ± 2.6)	6582.9	6469.2	-113.7
2Me, 6-OH pyridine	-(120.3 ± 2.5)	6633.4	6504.7	-128.7

* No correction for cis-diMe interaction.

$B(C=N) + \Delta_{NNC} + \Delta_{CCC}^{N} = 527.5$ kJ mol^{-1} (from BuN=CH^{i}Pr)

$\Delta_{CNN} = -6.7$ kJ mol^{-1} ; $\Delta_{OCN} \sim -9$ kJ mol^{-1}

$\Delta_{ONN} \sim -20$ kJ mol^{-1}.

TABLE 20. PYRIDINE DERIVATIVES.

Alkyl Radicals

The alkyl radicals (CH_3, C_2H_5) offer scope for the evaluation of the trigonal C bond parameters, $B(C^*H)$, $B(C^*C)$, (where C^* is trigonal), but there are some difficulties. The structural evidence from I.R. spectral analysis of matrix-isolated t-butyl favours C_{3v} symmetry (9), implying that the central carbon atom is not trigonal in this free radical. The evidence for planarity in CH_3 is strong, but it remains to be established that alkyl radicals in general adopt a trigonal planar conformation in the gaseous-state (10).

The enthalpies of formation of the simpler alkyl radicals are reasonably well-established (within ± 5 kJ mol^{-1}) following extensive kinetic studies (shock-tube alkane decompositions; very low pressure pyrolyses; alkyl radical reactions), and measurements of ionization (photoelectron spectra) of alkyl radicals and appearance potentials of alkyl positive ions (9,10,11,12,13,14). The relevant thermochemical data are given in Table 21.

The σ-bond contributions may be written as follows, where $B(C*C) = E(C*H) - 3/2P_1 + 3P_2 - P_1^* + 2P_2^* - 2T_1 + 3T_2 - 2/3T_1^* + T_2^*$, using the symbolism of ref.[2].

$$CH_3 = 3B(C*H)$$
$$C_2H_5 = 2B(C*H) + B(C*C) + 3B(CH)$$
$$iC_3H_7 = B(C*H) + 2B(C*C) + 6B(CH) + \Gamma_{CC*C}$$
$$tC_4H_9 = 3B(C*C) + 9B(CH) + 3\Gamma_{CC*C} + \Delta^*_{CCC}$$

and the total binding energy, ΔE_a^o, requires the addition of π-hyperconjugation effects. In Table 21, the derived B(C*C) values <u>include</u> the hyperconjugation contribution from the alkyl group attached to C*.

Radical	$\Delta H_f^o(g)$	ΔE_a^o	
CH_3	(146 ± 2)	1217	$B(C*H) = 405.7$ kJ mol^{-1}
C_2H_5	(115 ± 5)	2393	$B(C*C) = 342$ kJ mol^{-1}
$n-C_3H_7$	(93 ± 8)	3561	$B(C*C) = 343$ kJ mol^{-1}
$i-C_3H_7$	(80 ± 5)	3574	$B(C*C) + ½\Gamma_{CC*C} = 344$ kJ mol^{-1}
$t-C_4H_9$	(40 ± 5)	4759	$B(C*C) + \Gamma_{CC*C} + 1/3\Delta^*_{CCC} = 346$ kJ mol^{-1}

TABLE 21. ALKYL RADICALS.

If we assume that the bond-bond interaction terms Γ_{CCC*} and Γ_{CC*C} are of similar magnitude ($\Gamma \sim 11$ kJ mol^{-1}), and that $\Delta^*_{CCC} \sim \Delta_{CCC} \sim -3$ kJ mol^{-1}, the <u>effective</u> B(C*C) values in ^{i}Pr and ^{t}Bu are ~ 336 kJ mol^{-1} respectively - slightly less than in the

Et radical. This is consistent with a reduced hyperconjugation contribution and with distortion from the planar trigonal arrangement.

The $B(C^*H)$ value appropriate to the CH_3 radical is not necessarily transferable to olefines, in that the effect of the π-electron on the adjacent C^*H bonds in CH_3 is replaced in olefines by that of the π^2-bond. We may fit an Allen-type scheme to olefines, by writing:-

$$B(C^*H) = (405.7 + x) \text{ kJ mol}^{-1}$$

$$B(C=C) = (617.9 + 4x) \text{ kJ mol}^{-1}$$

where x is undetermined, and taken as zero in free radicals. The Allen scheme then leads to "effective" $B(CC^*)$ values in olefines, e.g. in propene,

$$B(CC^* + \Gamma_{CC^*C^*}) = (343.4 + x); \quad B(CC^*) \sim (332 + x) \text{ kJ mol}^{-1}$$

The difference $[B(CH^*) - B(CC^*)] \sim 74$ kJ mol^{-1}, is independent of x, and by comparison with the difference in paraffins (~ 85 kJ mol^{-1}), a value of ca. 11 kJ mol^{-1} is to be attributed to the overall interaction in the $CH_3-C=C$ grouping, compared with that in $CH_3CH_2CH_3$.

These values give insight into the successful application of the "bent bond" Allen scheme model for unsaturated hydrocarbons. In effect this model adds 2Γ to ΔE_a^o for each $C-C=C$ grouping, as the contribution from adjacent $C-C-C$ bond interactions. The trigonal C model adds $(\Gamma_{CC^*C^*} + h_\pi)$, where h_π is the overall "hyperconjugation" stabilization. The above analysis shows $h_\pi \sim 11$ kJ mol$^{-1} \sim \Gamma$, provided that $\Gamma_{CCC} \sim \Gamma_{CC^*C^*}$, so that both schemes should coincide. The 'bent bond' scheme may well be at fault, however, in $X-C=C$ systems should Γ_{XCC} (from saturated molecules) differ significantly from the π-interaction of X with the $\pi\pi$ component of the $-C=C-$ bond.

Alkyl Cations

The enthalpies of formation of lower molecular weight alkyl cations have been determined by photoelectron-photoion coincidence studies on the fragmentation of alkyl halides [14], from proton affinity measurements with olefines, and by combining measured adiabatic ionization potentials of the free alkyl radicals with enthalpies of formation of the neutral radicals [10]. There are some discrepancies, and it is doubtful that the 'adiabatic' radical ionization potentials obtained from photoelectron spectroscopic studies are always reliable. The vertical and adiabatic ionization potentials reported by Houle and Beauchamp [10] are the same for the $.CH_3$, alkyl and benzyl radicals, but there are significant differences between the vertical and adiabatic values for C_2H_5, $\underline{i}$-C_3H_7 and $\underline{t}$-C_4H_9. These may well reflect geometric changes in passing from the neutral radicals to the positive ions, involving displacements of the -CH bonds in the fashion

Cation	ΔH_f^o kJ mol^{-1}	ΔE_a^o kJ mol^{-1}	
CH_3^+	1094	1362	$B(CH) = 454$
$MeCH_2^+$	904	2697	$B(CC) + h(Et^+) \sim 549$
Me_2CH^+	799	3947	$2B(CC) + h(^iPr^+) \sim 1002$
Me_3C^+	694	5198	$3B(CC) + \Delta^+ + h(^tBu^+) \sim 1444$
$EtMeCH^+$	771	5121	$2B(CC) + h(^sBu^+) \sim 1010$
Me_2CEt^+	661	6376	$3B(CC) + \Delta^+ + h(C_5)^+ \sim 1457$
$Me_2C\underline{i}Pr^+$	627	7555	$3B(CC) + 2\Delta^+ + h(C_6)^+ \sim 1459$
(C_5^+ = isoamyl; C_6^+ = 2-3 dimethylbutyl)			

TABLE 22. ALKYL CATIONS.

The experimental ΔH_f^o values in Table 22 are mainly from Rosenstock et al [14], Traeger [13,15] and Solomon and Field [16]. Error limits are of the order ± 4 kJ mol^{-1}, but may be larger in some cases. The ΔE_a^o values relate to disruption into ground-state atoms and one C^+ion (ΔH_f^o = 1809.4 kJ mol^{-1}).

For the formal structures, based on trigonal C^+,

the Allen scheme gives:-

$$\Delta E_a^o(\overset{+}{C}H_3) = 3B(\overset{+}{C}H)$$

$$(C_2H_5^+) = 2B(\overset{+}{C}H) + B(\overset{+}{C}C) + 3B(CH) + h(Et^+)$$

$$(C_3H_7^+) = B(\overset{+}{C}H) + 2B(\overset{+}{C}C) + 6B(CH) + \Gamma_{\overset{+}{C}CC} + h(^iPr^+)$$

$$(C_4H_9^+) = 3B(\overset{+}{C}C) + 9B(CH) + 3\Gamma_{\overset{+}{C}CC} + \Delta_{\overset{+}{C}CC} + h(^tBu^+)$$

where the terms in h allow for the hyperconjugation of CH_3 groups adjacent to the $\overset{+}{C}$ vacant π-orbital. In fitting these 'formal' structures to the Allen scheme (Table 22), the interaction terms $\Gamma_{\overset{+}{C}CC}$, $\Gamma_{C\overset{+}{C}C}$ have been assumed unchanged from $\Gamma_{CCC} \sim 11.2$ kJ mol^{-1} for neutral C atoms. It is clear that the fitting requires unusually large values for the hyperconjugation energy terms, h, and indeed if the difference [B(CH) - B(CC)] is of similar magnitude (85 kJ) to [B($\overset{+}{C}$H) - B($\overset{+}{C}$C)], the required values are $h(Et^+) \sim 180$ kJ mol^{-1}; $h(^iPr^+) \sim 264$ kJ mol^{-1}; $h(^sBu^+) \sim 272$ kJ mol^{-1}; $h(^t_rBu^+) \sim (327-\Delta^+)$ kJ mol^{-1}; $h(C_5)^+ \sim (350-\Delta^+)$ kJ mol^{-1}, and $h(C_6)^+ \sim (352-2\Delta^+)$ kJ mol^{-1}. The size of these terms points to "protonated olefines" as a more realistic formulation of the cation structures, and it is of interest to examine the fitting of structures of this type.

1. **Ethyl cation.**

$$E_a^o = [4\bar{B}(CH) + \bar{B}(C_2H)^+]$$

where $\bar{B}$ indicates bonds within or directly attached to the protonated bridge

2. **i-Propyl cation.**

Two equivalent bridge structures

can be written, and the actual cation is a hybrid of the two:

$$\Delta E_a^o \sim [5\bar{B}(CH) + \bar{B}[C_3H_2]^+]$$

3. s-Butyl cation.

Two bridge structures,

[Structure: CH₃CH₂ and H groups bridged via C—⊕—C with H substituents] [Structure: CH₃ and H on C—⊕—H, bonded to C with CH₃ and H]

leading to a hybrid

[Hybrid structure with CH₃, H on one C, central C⊕ bridging with H, and C with H, H]

and
$$\Delta E_a^o = 3B(CH) + [4\bar{B}(CH) + \bar{B}(CC) + \bar{B}(C_3H_2)^+]$$

4. t-Butyl cation.

Three equivalent bridge structures, of the type

[Structure: (CH₃)₂C—⊕—CH with H substituents]   giving a hybrid

[Hybrid structure showing three bridged CH groups with central C⊕]

and $\Delta E_a^o = [6\bar{B}(CH) + \bar{B}(C_4H_3)^+]$

5. i-Amyl cation.

Three bridge structures

[Structure 1: (CH₃)₂C—⊕—C(CH₃)H]   [Structure 2: H-C(H)—⊕—H with CH₃, C—CH₂, CH₃]   [Structure 3: CH₃, CH₃ on C—CH₂ with H⊕ bridged to CH with H, H] forming a hybrid,

for which

$$\Delta E_a^o = 3B(CH) + \Gamma_{CCC} + [5\bar{B}(CH) + \bar{B}(CC) + \bar{B}(C_4H_3)^+]$$

6. **2,3 diMe butyl cation.**

Three bridge structures, to form a hybrid

with

$$\Delta E_a^o = 6B(CH) + 3\Gamma_{CCC} + \Delta + [4\bar{B}(CH) + 2\bar{B}(CC) + \bar{B}(C_4H_3)^+]$$

These structures lead to the following set of equations: (values in kJ mol^{-1})

1. $4\bar{B}(CH) + \bar{B}(C_2H)^+ = 2697$
2. $5\bar{B}(CH) + \bar{B}(C_3H_2)^+ = 3947$
3. $4\bar{B}(CH) + \bar{B}(CC) + \bar{B}(C_3H_2)^+ + \Gamma_{CCC} = 3881$
4. $6\bar{B}(CH) + \bar{B}(C_4H_3)^+ = 5198$
5. $5\bar{B}(CH) + \bar{B}(CC) + \bar{B}(C_4H_3)^+ + \Gamma_{CCC} = 5136$
6. $4\bar{B}(CH) + 2\bar{B}(CC) + \bar{B}(C_4H_3)^+ + 3\Gamma_{CCC} + \Delta = 5075$

from which, with $\Gamma = 11.2$ kJ mol^{-1}, (and Δ given a large value to allow for steric crowding (-11 kJ mol^{-1}), we may obtain:

(2)-(3) $\bar{B}(CH) - \bar{B}(CC) \sim 77$
(4)-(5) $\bar{B}(CH) - \bar{B}(CC) \sim 73$
(5)-(6) $\bar{B}(CH) - \bar{B}(CC) \sim 72$

and

(2)-(1) $\bar{B}(CH) + \bar{B}(C_3H_2)^+ - \bar{B}(C_2H)^+ \sim 1250$
(4)-(2) $\bar{B}(CH) + \bar{B}(C_4H_3)^+ - \bar{B}(C_3H_2)^+ \sim 1251$

indicating a degree of internal consistency within this scheme.

The equations are insufficient to isolate $\bar{B}(CH)$, but it is reasonable to expect that this will be in between the extremes of $B(C^\pm H) = 454$ and $B(C-H) \sim 413$ kJ mol^{-1} : accepting $\bar{B}(C-H) \sim 434$ kJ mol^{-1}, we then have

$$B(C_2H)^+ \sim 961; \quad \bar{B}(C_3H_2)^+ \sim 1777; \text{ and } B(C_4H_3)^+ \sim 2594;$$

These figures imply that the bridge contribution per CC bond diminishes from 961 to 889 to 865 kJ mol^{-1} in passing from the $C_2 \to C_3 \to C_4$ bridges.

The wealth of experimental data that has accumulated on organic cations in recent years, mainly from proton affinity measurements (see, eg. Kebarle [17]), is open for examination in the manner outlined above, and is not attempted here.

In conclusion, it is our view that the Allen scheme provides a good empirical approach to molecular structural stability, and is capable of more extensive application. It has tended to be discarded in favour of more sophisticated computations under the general heading of 'molecular mechanics'. There remains the entire area of organometallic compounds (wherein the tetrahedral or trigonal carbon atom is no longer the starting-point) virtually unexplored by other than the crudest empirical schemes, and now in need of more detailed examination.

References

1. T.L. Allen, J.Chem.Phys., 1959, 31, p. 1039.

2. H.A. Skinner, J.Chem.Soc., 1962, p. 4396.

3. S.W. Benson, 'Thermochemical Kinetics' 2nd Edition, Wiley, New York, 1976.

4. J.B. Pedley and J. Rylence, N.P.L. Computer Analysed Thermochemical Data, University Sussex, 1977.

5. CODATA, J.Chem.Thermodynamics, 1978, 10, p. 903.

6. P. Sellers, G. Stridh and S. Sunner, J.Chem.Eng.Data, 1978, 23, p.250

7. M. Månsson, J.Chem.Thermodynamics, 1969, 1, p. 141.

8. S. Suradi, N. El Saiad, G. Pilcher and H.A. Skinner, J.Chem. Thermodynamics, 1982, 14, p. 45.

9. J. Pecansky and J.S. Chang, J.Chem.Phys., 1981, 74, p. 5539.

10. F.A. Houle and J.L. Beauchamp, J.Amer.Chem.Soc., 1979, 101, p. 4067.

11. W. Tsang, Int.J.Chem.Kinetics, 1978, 10, p. 821.

12. A.L. Castelhano, P.R. Marriott, D.R. Griller, J.Amer.Chem. Soc., 1981, 103, p. 4262.

13. J.C. Traeger and R.G. McLoughlin, J.Amer.Chem.Soc., 1981, 103, p. 3647.

14. H.M. Rosenstock, R. Buff, M.A.A. Ferreira, S.G. Lias, A.C. Parr, R.L. Stockbauer and J.L. Holmes, J.Amer.Chem. Soc., 1982, 104, p. 2337.

15. J.C. Traeger, Org.Mass Spectrom., 1981, 16, p. 193.

16. J.J. Solomon and F.H. Field, J.Amer.Chem.Soc., 1975, 97, p. 2625.

17. P. Kebarle, Ann.Rep.Phys.Chem., 1977, 28, p. 445.

EXPERIMENTAL METHODS TO STUDY THE THERMOCHEMISTRY OF GAS PHASE
IONS AND ION-MOLECULE REACTIONS

Rod S. Mason

Department of Chemistry and Molecular Sciences
University of Warwick
Coventry, CV4 7AL.

INTRODUCTION

Ionic species and ion-molecule reactions provide, in one way or another, a major contribution to the Earth's chemistry. Although the most familiar reactions occur in solution (where they are often significantly moulded by solvent interaction) there are a number of important areas where ions exist and react in the gas phase. The upper atmosphere is an obvious and important example[1][2]. An even more hospitable environment for free ions lies outside the Earth, in outer space. It is only just being revealed how important such reactions are in the creation and propagation of interstellar clouds[3]. Gas phase ion-molecule reactions also occur in flames[4], electrical discharges[50], and are sometimes a corrosive nuisance in the coolant gases of nuclear reactors[5]. Finally, they have become an important tool in analytical mass spectrometry[6].

Inevitably, the study of gas phase ion thermochemistry has extended far beyond the classical measurements of ionisation and appearance potentials, and has become an important aid to the understanding and prediction of their reactions. Indeed, it is now possible to study many of the reactions of solution chemistry in the gas phase, thus avoiding the effect of solvation energy, the data so obtained being more reliable and consistent than solution measurements[7]. Frequently, heats of formation of neutral species derived via appropriate thermodynamic cycles from ion thermochemical data[8] are now more reliable than those derived from neutral chemistry.

In thermochemical terms, the fundamental property of an ionic, as for a neutral species, is its heat of formation ΔH_f^0, which except for elements in their standard states (when ΔH_f^o = ionisation potential) is not directly measurable. However, the properties of free ions that can be measured directly are the ionisation potential (I_p), electron affinity (EA) and appearance energy (A_e). These are defined for processes (1) to (3) respectively, in which the reactants and products are at zero kelvin and hence in their ground electronic, vibrational and rotational states.

$$M \to M^+ + e^- \qquad I_p(M) \qquad (1)$$

$$M^- \to M + e^- \qquad EA(M) \qquad (2)$$

$$ABC \to AB^+ + C + e^- \qquad A_e(AB^+, ABC) \qquad (3)$$

When defined in terms of the heats of formation of reactants and products, e.g.

$$I_p(M) = \Delta H_{f_o}^o (M+) + \Delta H_{f_o}^o (e^-) - \Delta H_{f_o}^o (M) \qquad (4)$$

it is seen that the derivation of reliable values for $\Delta H_{f_o}^o (M+)$ depends not only on the measured ionic property but also on the availability of accurate information on the neutral species involved. Usually the free electron is treated as an element in its standard form, hence $\Delta H_{f_o}^o (e^-) = 0$ (however, see the following chapter.)

In addition to the above, a new group of measurements has become available over the last 16 years. This group comprises measurements of the free energies and enthalpies of ion-molecule reactions, quantities which had hitherto been measurable only in solution. The impetus to measurements of this nature came with the development of High Pressure Source Mass Spectrometry[9] and Ion Cyclotron Resonance techniques[10]. Proton transfer, as represented by (5) and (6) is an important class of reactions in this group[10][11].

$$AH^+ + B \rightleftarrows A + BH^+ \qquad (5)$$

$$A^- + BH \rightleftarrows AH + B^- \qquad (6)$$

The respective free energy changes for these reactions are a measure of the relative basicity and acidity of the reactant molecules (see Table 1). Other reactions provide such data as relative ionisation energies[9][12], hydride affinities[9][11], electron affinities[9][11], halide affinities[9], ligand binding energies[13] and heats of association[9], all defined, as in Table 1, at temperatures above zero kelvin. From these data it is possible to derive $\Delta H_{f_T}^o$ values. The use of the data to forecast molecular and ionic reactivity as a function of structure and bonding is discussed in the later chapter, "The Thermochemistry of Gas Phase Ion-Molecule Reactions".

TABLE 1
Definition of Some Thermochemical Terms in Gas Phase Ion Chemistry

SYSTEM at TdegK		ΔH_T^o
$AH^+ \rightarrow A + H^+$		Proton affinity (basicity). PA(A).
$HA \rightarrow H^+ + A^-$		Heterolytic bond dissociation energy (acidity)
$A \rightarrow A^+ + e^-$		Ionisation energy
$RH \rightarrow R^+ + H^-$		Hydride (H^-) affinity
$RX \rightarrow R^+ + X^-$		Halide (X^-) affinity
$ML^+ \rightarrow M^+ + L$		Ligand (L) affinity (M = metal atom)
$A^+A \rightleftarrows A^+ + A$		Heat of association

For general examples see Ref. 9.

The present chapter is devoted to a discussion of experimental techniques. In the N.B.S. compilation "Gaseous Ion Energetics"[14], 19 distinct methods are listed for investigating cations and 31 for anions, representing developments up to 1971. The methods discussed here are restricted to studies of polyatomic ions and represent the Author's view of the best methods currently available for the measurement of each of the thermochemical quantities described above. The range of experiments covered is listed in Table 2, and only those which have gained prominence in the last 10 years are described in any detail. Thermochemical measurements of polyatomic ions in the gas phase offer two distinct advantages over other systems: first, they are free of solvent interaction, and second, they readily lend themselves to selection or detection by mass spectrometry, the use of which occurs in most of the experiments reported. The principles of mass spectrometry are well known[15], although the application of one particular technique, Ion Cyclotron Resonance (ICR) Spectrometry which has contributed significantly to the advance in gas phase ion thermochemistry is probably not well known outside the field. ICR is important in three of the methods to be discussed here, so the basic principles involved are described briefly before a discussion of the whole range of techniques in Table 2.

ION CYCLOTRON RESONANCE SPECTROMETRY[16]

When an ion moves through a magnetic field it is deflected in a plane perpendicular to the field direction (Fig. 1) according to the "right" (cations) or "left hand rule" (anions.) If the field is sufficiently large, circular motion results. Thus, for an ion of mass m, charge e and velocity v perpendicular to the field B, the force $F = ma = evB$, where a is the acceleration imparted to the ion. For circular motion, $a = v^2/r$, where r is the radius of the ion path, so that $ma = mv^2/r = evB$. The angular frequency of the ion ω_c, is then given by $\omega_c = v/r = eB/m$ rad s^{-1} and is independent of v. The frequency ν_c is given by

$$\nu_c = eB/2\Pi m \quad \text{cycles s}^{-1} \quad (7)$$

TABLE 2
Experimental Techniques Reviewed

Method		Quantity Measured
OPTICAL		
Optical Spectroscopy		I_p, EA of atoms/small molecules
Photo-electron Spectroscopy	PES	I_p, EA of polyatomics
Laser Photodetachment Spec.	LPS	EA of atoms/small molecules
THRESHOLD		
ICR Photodetachment		EA of polyatomics
(electron impact mass spec.		I_p, A_e, general)
Photoionisation Mass Spec.	PIMS	I_p, A_e, general
Threshold Photo-Ion Photo-Electron Coincidence	TPIPECO	A_e
ION-MOLECULE EQUILIBRIA		
K^o vs. T		ΔG^o, ΔH^o, ΔS^o
Bracketing		(ΔG^o)
High Pressure Pulses Source Mass Spec.	HPMS	relative acidities
Ion Cyclotron Resonance	ICR	etc, See Table 1.
Flowing Afterglow	FA	
Selected Ion Flow Tube	SIFT	

For example, the frequency of CO_2^+ in a magnetic flux density of 1.0 Tesla ($\equiv 10^4$ Gauss) is 350kHz which is in the radio frequency (rf) range. Although this frequency is independent of the ionic velocity, the diameter of the ion path is given by $d = v/\pi\nu$, so that for the CO_2^+ ion above, at 300K, $d = 0.4$mm. Since thermal velocity decreases with mass, so does d for ions at the same temperature.

If the ions are trapped in the magnetic field, but also between the upper and lower metal plates of a capacitor to which an rf signal ω_o is applied, such that $\omega_o = \omega_c$, then the ionic motion is in resonance and absorbs power, proportional to the intensity of the ions. As the ion absorbs energy, its velocity increases and hence the diameter of its orbit. If sufficient power is absorbed the ion is ejected to the sides of the cell. It is the measurement of this power consumption that allows a detection of the ions. The relation (7) indicates that scanning of the spectrometer for different masses can be achieved either by a scan of B or of ω, there being advantages and disadvantages to both depending upon the application. Both methods are used[16].

The first ICR experiments employed a "drift cell"[16] (see Fig. 2). This consists of a set of parallel metal plates, here shown as a three section cell suspended in a magnetic field. Ions

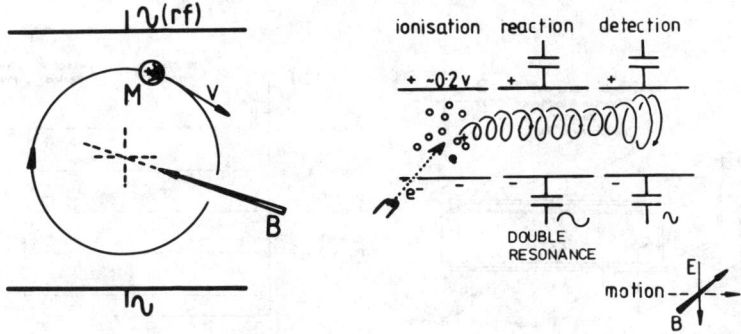

Fig. 1. The motion of an ion in a strong magnetic field.
Fig. 2. Schematic of ICR Drift Cell.

are produced usually by electron impact in the source region. They are made to drift to the reaction region by the presence of a potential difference across the plates above and below the electron beam. The electric field intensity vector between the plates is perpendicular to the magnetic field vector. These crossed fields cause the ions to move in a direction perpendicular to both. The ions eventually drift into the analyser region where they are detected by the application of an appropriate rf signal. Side plates carrying a small positive (negative) charge prevent cations (anions) from drifting to the sides of the cell. The motion of the ions down the cell is described as cycloidal.

This is an essentially low pressure instrument since collisions tend to destroy the cycloidal motion of the ions. However, collisions are clearly necessary in order to study ion-molecule reactions. Therefore typical operating pressures are in the region of $10^{-5}-10^{-6}$ torr, although up to 10^{-3} torr can be tolerated[10]. Ion residence times are in the region of 2 ms still allowing time for several hundred collisions.

An unwanted ion can be ejected from the reaction region by the application of a suitable resonance signal to the middle plates. This is the basis of the "double resonance" experiment, in which ionic intermediates in a sequence of reactions can be directly determined. Thus if in the reaction $A^+ + B \rightarrow C^+ + D$, the product ion C^+ is monitored in the analyser region and a signal resonant with A^+ is applied to the reaction region sufficient to remove A^+, the signal for C^+ will be seen to decrease accordingly, and a direct relationship is established between the two. The Pulsed Trapped Ion Cell[16] was a slightly later development (Fig. 3). This has only one top and bottom plate and operates in a pulsed

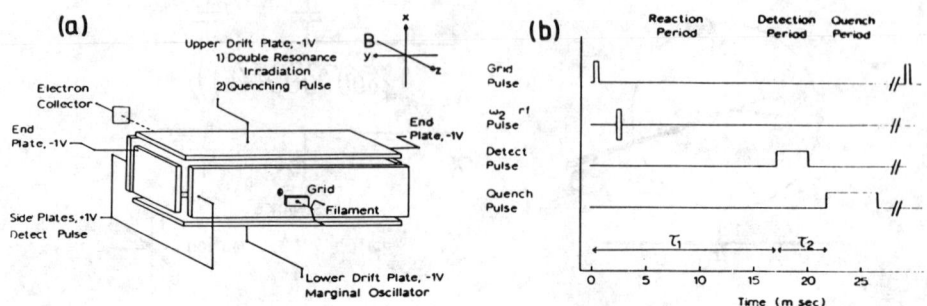

Fig. 3. (a) Pulsed Trapped Ion ICR cell, (b) Sequence of pulses; (adapted from ref. 16).

mode. A pulse of electrons causes ionisation, and an rf pulse applied to the top and bottom plates sets the ions in cycloidal motion. Appropriate dc potentials (see Fig. 3) on the top, bottom, side and end plates keep the ion trapped for up to several seconds, provided the pressure is low enough ($\leq 10^{-6}$ torr.) A detection rf pulse applied at various times after the initial electron pulse allows ion intensities to be followed as a function of time. Again provided the residence time is sufficiently long, a significant number of collisions can take place. In between times, a set of rf pulses or even a continuous signal can be used to eject unwanted ions.

Developments of ICR are used in three applications to be described later: anion photodetachment, ionic equilibria and kinetic studies.

SPECTROSCOPIC MEASUREMENTS OF IONISATION POTENTIALS AND ELECTRON AFFINITIES

Photoelectron Spectroscopy

The most precise I_p's are obtained for atoms and diatomic molecules by classical optical spectroscopy[18]. However, this is not usually suitable for polyatomic molecules because the spectra are complex, consisting of many overlapping vibronic transitions complicated by autoionisation transitions which are difficult to interpret. In addition, symmetry (selection-rules) and Franck-Condon like considerations may prevent population of the ground state ion. The best method for studying polyatomics has therefore been photoelectron spectroscopy because it avoids many of these complicating factors[19].

This well established technique investigates the process

$$M + h\nu_{fixed} \to M_a^+ + e^- \quad (8)$$

If the molecule M in its ground state is singly ionised by radiation with a fixed energy, $h\nu$ (usually $\equiv$ 21.21eV, the He 58.4nM line), leaving M^+ in a state a, then the ejected photoelectron has a translational energy given by

$$e_{trans} = h\nu - I_p(M \to M_a^+) \quad (9)$$

The spectrum is generated by plotting the change in the current of electrons collected via an energy analyser, as a function of the electron translational energy. The limitations on accuracy arise from contact potentials in the analyser and for this reason calibration of the energy scale against a known compound is usually required. The resulting precision is in the order of 0.01eV ($<$1 kJ mol^{-1}).

In order to measure the first I_p, the ground state transition must be identified, but again symmetry factors may give this process a very small probability, making it not suitable for all molecules. Similar considerations also apply to the study of negative polyatomic ions[20]. In this case interaction with the light causes photodetachment of the electron from various states of the negative ion, the energy for the ground state process being equal to the electron affinity of the molecule. An additional difficulty here lies in obtaining a high enough concentration of the negative ions.

Laser Photodetachment Spectroscopy[20][21]: Electron Affinities

A way of overcoming the anion intensity problem is to use a mass spectrometer to select a beam of the required ions and to cross this with a Laser beam (Fig. 4.) The spectrum[22] generated from a beam of O_2^- ions interacting with an Argon laser (488nM$\equiv$2.54leV) is shown in Fig. 5. Provided the transitions can be identified, an accurate EA measurement can be made to within 0.01eV($<$1 kJ mol^{-1}). Here the ground state transition occurs at a photodetached electron translational energy of 2.1eV yielding an EA(O_2^-) of 0.44eV$\equiv$42.5 kJ mol^{-1}. This type of experiment has been performed in a number of laboratories[20][22] but relatively few precise measurements have been reported for large polyatomic ions ($>$3 atoms) by these methods.

THRESHOLD MEASUREMENTS OF ELECTRON AFFINITY: ICR Photodetachment Spectroscopy[20][23]

A schematic diagram of this technique, developed by Braumann and co-workers[23] is shown in Fig. 6. The specially designed

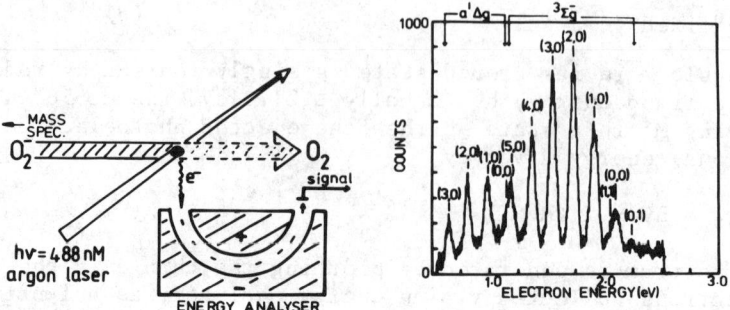

Fig. 4. Crossed Ion Beam Laser Beam Photodetachment.
Fig. 5. LPD spectrum of O_2^- (adapted from ref. 22).

drift cell enclosed in a vacuum housing sits between the poles of a powerful electromagnet. The negative ions of interest are generated in the source region by a suitable sequence of ion-molecule reactions initiated by electron impact. In the example[24] chosen here, NF_3 is a rich source of F^- ions formed by dissociative electron attachment

$$e^- + NF_3 \rightarrow F^- + NF_2 \tag{10}$$

In the presence of a molecule containing an acidic hydrogen atom, proton transfer to F^- occurs rapidly to form the anion of interest (here the enolate anoin of acetaldehyde) and the neutral HF.

$$F^- + CH_3COH \rightarrow CH_2=C\bar{O}H + HF \tag{11}$$

The anions drift slowly into the detector region (lengthened to increase the path length) where they can be monitored in a suitable rf field. Light from a tunable dye laser passes through a window at the opposite end of the cell. The highly polished source plates are bent at right angles to form a high reflection mirror, reflecting the incident light back through the cell and the ions are constrained by suitable dc voltages to remain trapped for periods of up to 1 sec. A particular advantage of ICR is that, using a special detection circuit (a marginal oscillator), small changes in ion current can be detected without removing ions from the system. Thus adequate time is allowed for the interaction of the laser beam with the ions. The relative decrease in ion intensity (caused by photodetachment) is monitored as a function of wavelength (see Fig. 7.) Provided the pressure is kept low (commensurate with sufficient ion signal) and with the low translational energy of the ions (close to thermal) collisional detachment of the ions is not a problem, unlike the beam

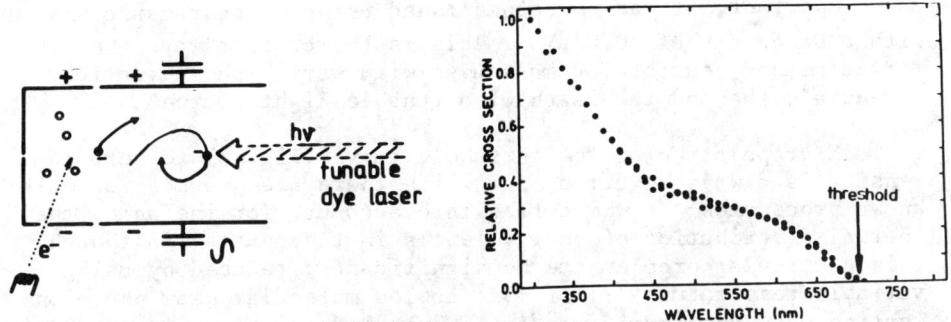

Fig. 6. ICR photodetachment experiment.
Fig. 7. Relative photodetachment cross sections for acetaldehyde enolate anion (●) generated using F^- from NF_3, and (o) generated using F^- from SO_2F_2 (adapted from ref. 7).

methods mentioned above. The threshold, which can be determined usually to within ±0.03eV (<3 kJ mol^{-1}), is equated with the EA. In general, however, since transitions are vertical, it gives an upper limit to the real value when Franck-Condon or symmetry factors are unfavourable for ground state transitions. It is also possible, if the anions are formed in "hot" vibrational levels for the figure to be underestimated. Care must also be taken that there is no interference from other ions or molecules by absorption of light in the region of the threshold. Despite these reservations, this method has provided a dramatic increase in "precise" EA values available for polyatomics[20].

THRESHOLD MEASUREMENTS OF APPEARANCE ENERGIES

In general it is more difficult to obtain reliable A_e data than it is to obtain either I_p or EA values because the process (3), involving unimolecular decomposition frequently preceded by rearrangement of the parent ion, is considerably more complex. All methods require a threshold type of measurement (Fig. 8) in which the intensity of the fragment ion AB^+ is monitored by mass spectrometry as a function of the decreasing ionising energy transferred to the parent molecule ABC in the mass spectrometer ion source, until the limit of detection is reached.

Electron impact has been the most widely used but least reliable method of ionisation[25]. The main difficulties stem from the low energy resolution attainable for an electron beam and the gradual nature of the threshold curve at onset, both of which limit the precision at best to <0.1eV[26]. Photoionisation, another well known technique[26], in many cases exhibits a step function at threshold giving (in principle) a sharper onset and, more important, it can be accomplished using monochromatic photons with a bandwidth of <0.01eV. This is therefore always the preferred method, except for molecules with very high ionisation potentials (beyond the reach of a tunable light source).

Extrapolation of the threshold curve (Fig. 8) to zero intensity is always required, for which there are a number of well known procedures[26] which take into account, for instance, the thermal distribution of energy levels in the neutral (although this particular problem can be significantly reduced by using variable temperature[27] or even cooled molecular beam photoionisation mass spectrometry[28]). Then, under ideal circumstances (represented in Fig. 9a) the energy at threshold for the process $ABC \rightarrow AB^+ + C$ is just enough to produce the ionic and neutral fragments in their ground states. Provided the dissociation is a simple cleavage, the process is thermodynamically equivalent to

$$A_e(AB^+, ABC) = BDE(AB-C) + I_p(AB) \qquad (12)$$

It is unlikely, however, that this situation is ever realised in practice, for the reasons described below:

(a) Excitation Energy Distribution of Parent Ion

Even though the ionising photons may be monochromatic, the parent ion ABC^+ is formed with an initial distribution of energy states determined by Franck-Condon and symmetry factors, just as in photoelectron spectroscopy, so that the excitation energy E^* for any one photon absorbed (energy = $h\nu$) is given by

$$E^* = h\nu - I_p(ABC) - e_{trans} \qquad (13)$$

where e_{trans} is the translational energy of the ejected electron. The true energy transferred to the parent molecule is only known if the photoelectron has zero kinetic energy, or if e_{trans} is actually measured. Of course the thermal distribution of energies in the parent neutral causes a further broadening of the initial ionic state population.

(b) Vertical Ionisation

The ionisation process occurs vertically, and symmetry or Franck-Condon factors may not allow parent ions (ABC^+) to be populated at the appropriate bond dissociation limit.

(c) Reverse Energy of Activation, Parent Ion Rearrangement

There may be an energy barrier, E_a, for the reverse reaction to the formation of products in their ground state (Fig. 9c). This would

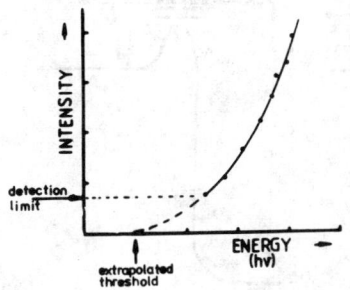

Fig. 8. A_e measurement by extrapolation of threshold curve.

occur particularly if internal rearrangement of the parent ion is necessary prior to dissociation. The observed products would, therefore, be formed with excess internal energy, E_{int}. Such reactions require time for the rearrangement to occur, often leading to "metastable" decompositions, where the half life of the parent ion is $>10^{-6}$s, and it occurs outside a conventional mass spectrometer source leading to further problems (see (e) below). In addition some of the excess energy may be dissipated as kinetic energy of the products (see (d) below).

(d) Kinetic Energy Release

If, because of (b) and (c) above, or (e) below, the parent ion is excited at threshold above its dissociation limit, the excess energy is dissipated among the various bonds until sufficient has accumulated in the one to be broken (if there are sufficient degrees of freedom this process is described by statistical theory[29]). If, during this dissipation down the vibronic ladder, a repulsive potential surface correlating with the fragmentation products is encountered (Fig. 9b), then predissociation can occur causing ion and neutral to fly apart with excess kinetic energy, E_{trans}. Such kinetic energy release is usually associated with metastable ions for which accurate measurements can be made[30] outside the ion source; however, it may also occur for the faster unimolecular fragmentations ($\tau_{\frac{1}{2}}<10^{-6}$s) which take place in the mass spectrometer ion source.

(e) The Kinetic Shift[31]

The rate of unimolecular decomposition depends on the excess energy above threshold. At very low excitation energies therefore the

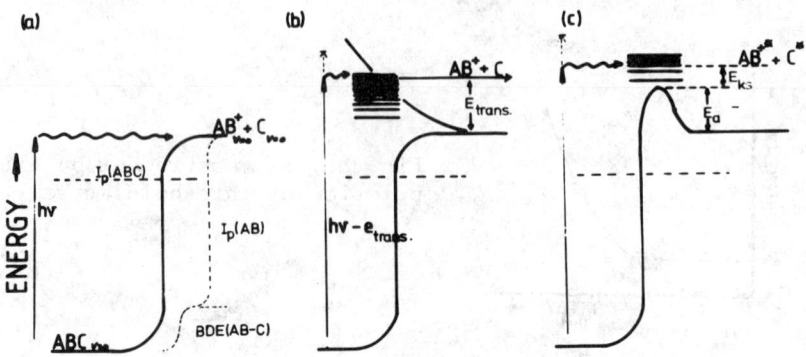

Fig. 9. Appearance energy potential diagram, (a) The ideal, (b) Vertical ionisation above threshold plus kinetic energy release, (c) Reverse energy of activation and kinetic shift.

lifetime of metastable parent ions may become so long that the number of fragment ions produced prior to mass analysis would be too small to be detected. This shifts the observed onset to an energy higher than the thermodynamic one by an amount, E_{ks}, known as the "kinetic shift" (see Fig. 9c). The effect can be measured by Trapped Ion Mass Spectrometry (TIMS)[32], in which ions are kept trapped in a specially designed ion source for up to 1s after the initial ionisation process and prior to mass analysis, allowing time for dissociations to occur. Recently, for example, Lifshitz[32] measured by TIMS an A_e of 12.1eV for the process $C_5H_5N \rightarrow C_4H_4^+ + HCN$ compared with a previous estimate of 12.5eV, a kinetic shift of 0.4eV. A 1.5eV shift has been observed for H atom loss from benzene[33].

In the very worst circumstances, the observed threshold value is equivalent to the expression given in (14) where $\Delta E(ABC^{+*} \rightarrow AB^+ + C)$ is the ionic fragmentation energy and includes any reverse

$$\text{Threshold}_{obs} = h\nu - e_{trans} = I_p(ABC) + \Delta E(ABC^{+*} \rightarrow AB^+ + C) + E_{int} + E_{trans} + E_{ks} \quad (14)$$

activation energy which might be present should prior rearrangement occur. A method which attempts to account for most of the above considerations is described below.

Threshold Photoion Photoelectron Coincidence (TPIPECO)[34][35]

A schematic diagram of this method[35] is shown in Fig. 10. Photoionisation by monochromated light (band width 10nM) yields fragment ions and photoelectrons. The latter are accelerated towards an electron analyser-detector system by a weak static electrostatic field. The electrons are energy selected such that principally only electrons with near-zero kinetic energy reach the

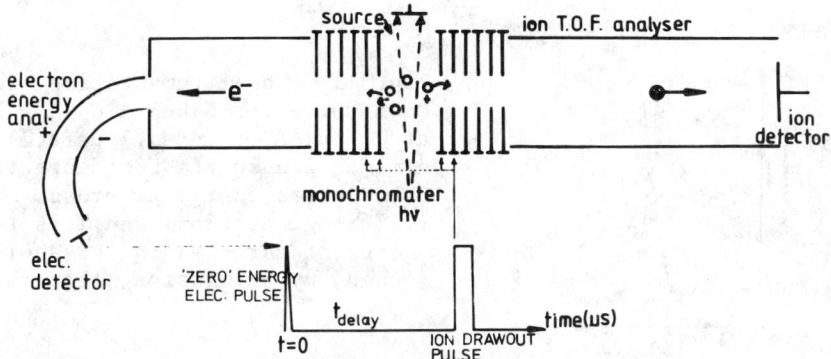

Fig. 10. Schematic of Threshold Photoion Photoelectric Coincidence apparatus, based on ref. 35.

detector. Ions formed in coincidence with these electrons have a known internal energy distribution equal to the photon energy. The electron transit time before detection is ca. 400 nS during which time the much heavier mass ion remains essentially stationary in the ionisation region. The detection of the threshold photoelectron can therefore be used to trigger an ion drawout pulse across the plates which define the ionisation region, thus ejecting the ions into a drift space where they separate in time according to mass and initial kinetic energy. The beauty of this method is that the signal averaged spectrum which emerges is due to parent ions whose internal excitation energy is exactly specified by the ionising photon. Time of flight mass analysis is performed and from the spread of each bunch of ions of the same mass arriving at the ion detector, further information about any excess kinetic energy release can be obtained. The triggering of the drawout pulse can be delayed to allow extra time (presently only up to approximately 7μs) for fragmentation and hence to investigate the "kinetic shift". The relative parent and fragment ion abundances can thus be measured as a function of photon energy, and the data assembled to give the breakdown curve at one or another ion source residence time.

The breakdown curve of ethyl iodide shown in Fig. 11 is an example taken from a recent study[36] of alkyl halides. No shift is detectable for the two residence times indicated, and therefore it is argued that within the time scale and energy resolution of this experiment ionisation is a step function. This means that if the photon energy sampled was truly single valued, and the neutral molecule had no thermal energy, then the relative abundance of the fragment ion $C_2H_5^+$ would rise vertically at the zero kelvin threshold from 0 to 100% and the parent ion decrease from 100 to 0%. The

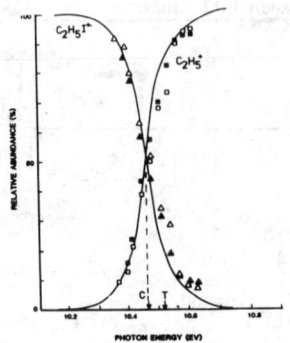

Fig. 11. The fragmentation breakdown curve for $C_2H_5I \rightarrow C_2H_5^+ + I$; at 0.7 μs($\triangle$,$\square$) and 5.7 μs($\blacktriangle$,$\blacksquare$) nominal ion source residence times. C indicates energy at crossover (50% fragmentation) and T is the energy of zero-kelvin threshold. (taken, by permission, from ref.36.)

spreading of these curves to their observed sigma shapes is then assumed to be the product of convolution by the "apparatus function" (i.e. the finite bandwidth of the incident photon beam + the distribution of photoelectron energies actually sampled) and the thermal energy distribution in the neutral, both of which are known. It is on this basis that the curves (indicated by the solid lines in Fig. 11) to fit the experimental points are calculated, the only "unknown" being the zero-kelvin threshold value which was thus determined here to be 10.53 ±0.01eV. Presumably the cross-over point which is accurately determined experimentally is shifted below this value by an amount equal to the average thermal energy in the neutral, plus a contribution from the "apparatus function". This approach, as with any other photoionisation method, has to make some assumptions about the amount of internal energy from the neutral actually available to the fragmentation of the parent ion.

If a kinetic shift is observed[31][36], then two sets of curves are obtained for two different ion residence times and a step function is no longer appropriate. In order to fit these, a calculation based on the statistical theory of unimolecular decomposition (the "QET" theory) must be performed, into which a second unknown, the reverse activation energy (E_a in Fig 9c) must be introduced. Any additional kinetic energy release must also be corrected for. Again good fits can be made and threshold values are still quoted to within ±0.01eV.

Unfortunately, although there are a number of laboratories with this type of apparatus[34][35][37], only one[34] is so far apparently capable of producing precise data. Therefore A_e measurements will continue to rely heavily on photoionisation mass spectrometry data, even though it would appear from the above considerations that many values will only be upper limits.

MEASUREMENTS OF ION-MOLECULE EQUILIBRIA

Most of these measurements are based on the simple idea of establishing, in the gas phase, an ion-molecule equilibrium system of the type represented by (4), and monitoring reactant and product

$$AH^+ + B \underset{k_r}{\overset{k_f}{\rightleftarrows}} A + BH^+ \qquad (4)$$

ion intensities by mass spectrometry. The equilibrium constant K_T, given by $K_T = k_f/k_r = [A][BH^+]/[B][AH^+]$ can be measured provided the relative partial pressures of A and B are known. K_T is related by the usual thermodynamic equations (15) and (16) to ΔG^o, ΔH^o and ΔS^o, all of which can be determined from measurements as a function of temperature T.

$$-RT \ln K_T^o = \Delta G^o \qquad (15)$$

$$\Delta G^o = \Delta H^o - T\Delta S^o \qquad (16)$$

Bracketing(39)

In the early days of High Pressure Mass Spectrometry it appeared that ion-molecule reaction rates were always very fast, proceeding at close to the collision rate ($k_f \approx 10^{-9}$ cm^3 molecules-1 s^{-1}) which, because of the large ion-induced dipole attractive force, is larger by two orders of magnitude than for neutral collisions. Hence if an ion-molecule reaction was observed it was assumed to be exothermic and to proceed without an energy of activation barrier. Consequently no attempt was made to measure K values. Instead a series of such reactions as (4) and (17) were examined(38). If (4) was observed to occur (i.e. BH^+

$$AH^+ + B \rightarrow A + BH^+ \qquad (4)$$

$$CH^+ + B \nrightarrow C + BH^+ \qquad (17)$$

was formed at the expense of AH^+) but (17) was not, then (4) is assumed to be exothermic and (5) not. Since a proton will always transfer to a substance of higher basicity, the value for $\Delta G_b^o(B)$ was bracketed between that of A and C: $\Delta G_b^o(C) > (B) > (A)$.

In actual fact endothermic reactions do occur, their rate constants having an Arrhenius form ($k_r \approx 10^{-9} e^{-\Delta H_r/RT}$, where ΔH_r is the endothermicity) and the method is therefore a qualitative version of later ones. Early ICR measurements(39) also used this method, and it is still used when it is experimentally difficult to achieve equilibrium. In fact, by judicious choice of reaction systems, in which the acidity or basicity values of A and C are known, values of ΔG^o can be pinned down to within $\pm$ 10 kJ mol^{-1}.

Measurements of K_T^o

In order to make an accurate measurement of K_T there are three general requirements. (i) The reactant ions must be in thermal equilibrium with their surroundings. This condition is aided by higher reaction pressures and/or longer ion residence times. (ii) Sufficient time must be allowed for the system to reach equilibrium before making the measurement, particularly where the values of K are large. (iii) There should be no interfering reactions. The most common problem in proton transfer studies is the formation of proton bound dimers[10]:

$$AH^+ + B \rightleftharpoons BH^+ + A \qquad (6)$$

$$\begin{array}{ccc} A\downarrow \searrow B & A\downarrow \searrow B & \\ AH^+A & AH^+B & BHB^+ \end{array}$$

If one or another of these reactions is fast enough to compete with the proton transfer reactions they can prevent the establishment of equilibrium, or lead to a false steady state condition. Since they are formed in "3-body" reactions (with a negative temperature coefficient) they can be discouraged by use of lower pressures and/or higher temperatures.

Three quite different experimental techniques have been applied to gas phase ion-molecule thermochemistry: High Pressure Pulsed Source Mass Spectrometry (HPMS), Ion Cyclotron Resonance (ICR), and Flowing Afterglow (FA) (including the Selected Ion Flow Tube (SIFT) method).

High Pressure Pulsed Source Mass Spectrometry[9][40]

This technique was developed by Kebarle and co-workers. Reactions take place in a mass spectrometer ion source, which has been specially modified to operate at pressures up to >5 torr and whose electron beam is pulsed. A schematic diagram of such an experiment[41] is shown in Fig. 12. Gas, usually containing the premixed reaction mixture, is introduced into the source, escaping into the evacuated mass spectrometer source housing only through very narrow (25-50μ) electron entrance and ion exit slits. The housing is rapidly pumped to minimise collision induced decomposition of ions emerging from the ion exit slit. The electron beam, which is gated on only for a few micro-seconds, enters the ion source with an energy of ~400eV causing the initial ionisation from which ion-molecule reactions ensue. Reactant and product ions are eventually lost by diffusion to the walls of the reaction chamber, but can be detected up to several milliseconds after the initial electron pulse, depending on the pressure. Those ions which happen to drift or diffuse out through the ion exit slit are accelerated into the mass spectrometer for analysis. The reaction chamber itself

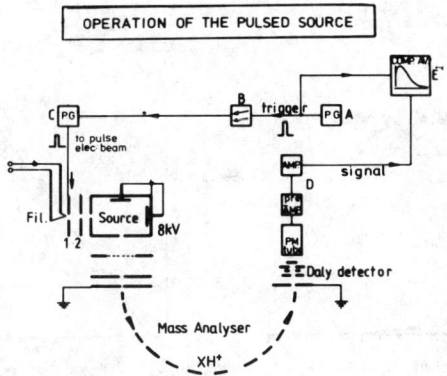

Fig. 12. Schematic diagram of the HPMS pulsed-source experiment.

is kept field free so that the energy of the reacting ions is as close as possible to thermal.

The reaction mixture is carefully chosen in order to generate the ions of interest. Thus, in order to study proton transfer between fluoro- and chlorobenzenes[41], the neutrals are premixed as vapours in a known ratio of partial pressures to <1% in a bath gas of methane. Initial ionisation by electron impact leads to the following sequence of reactions:

$$e^- + CH_4 \rightarrow CH_4^+ + 2e^- \qquad (18)$$

$$CH_4^+ + CH_4 \rightarrow CH_5^+ + CH_3^\cdot \qquad (19)$$

$$CH_5^+ + \begin{array}{c} phF \\ phCl \end{array} \rightarrow CH_4 + \begin{array}{c} phFH^+ \\ phClH^+ \end{array} \qquad (20)$$

$$phClH^+ + phF \rightleftarrows phCl + phFH^+ \qquad (21)$$

(18) is the most likely electron impact reaction because CH_4 is present in very much larger quantities than other reactant gases. (19) is a very rapid reaction, however CH_5^+ does not react further with methane (except by symmetrical proton transfer) but will transfer its proton very quickly to any other species present with a higher proton affinity (PA), a process which is to all practical purposes irreversible if the negative enthalpy of the reaction is >30kJ mol^{-1}. On the other hand if the two reactants of interest are chosen such that their relative proton affinities are close (≤13kJ mol^{-1}) they will happily swap a proton back and forth much more rapidly than their rate of diffusion to the walls.

In a typical experiment the mass spectrometer is focussed on each ion in turn and a profile of ion intensity as a function of residence time is recorded on a multichannel analyser triggered by the same signal which pulses on the electron beam. An example[41]

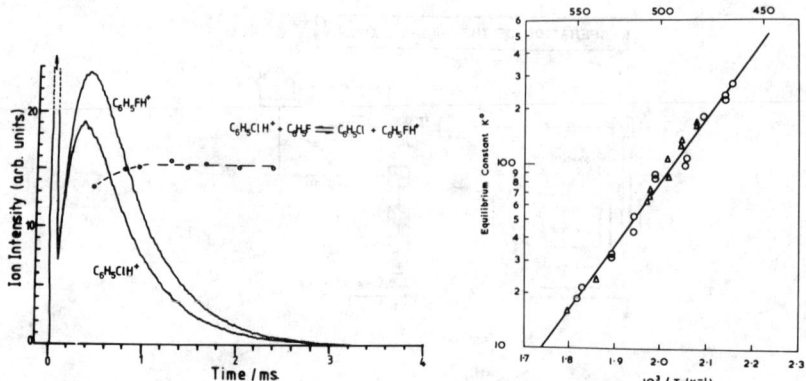

Fig. 13. Example of signal averaged ion intensity versus time profiles, from a mixture of C_6H_5Cl and C_6H_5F in CH_4 at 382 K and total pressure 1.2 torr. Elec. pulse width = 40 μs.

Fig. 14. Van't Hoff plot for the reaction $CO_2^+ + 2CO_2 \rightleftharpoons (CO_2)_2^+ + CO_2$.

is shown in Fig. 13. Equilibrium is assumed to be reached when the ratio of ion intensities ($phClH^+/phFH^+$ in Fig. 14) reaches a constant, from which the value of K is calculated. The early part of the profile is complicated by high charge density processes, but these are rapidly dissipated (for a full discussion see ref. 41 and references quoted therein). K values are usually checked for consistency over a range of pressures and ratios of the neutral reactants.

The ion chamber is heated to a constant measured temperature which in the Author's laboratory[41] can be varied between ~200-600K. The quality of data for a typical van't Hoff plot is shown in Fig. 14. Possible mass discrimination due to the manner in which the ions flow through the exit slit is only important for ions which have a very large mass difference. When the interfering reactions discussed above are a problem, the experiments are performed at high temperatures. For this reason much of the data gathered by Kebarle and co-workers[10],[40] is quoted at 600K. In order to relate this to thermochemical data at different temperatures, the experimental ΔG^o values are adjusted using appropriate entropy corrections (see next Chapter).

Ion Cyclotron Resonance Spectrometry[10],[16],[39]

The principles of this technique were described earlier. Its application to equilibrium constant measurements was developed in various laboratories (Beauchamp[39], Bowers[10], Brauman[43], and McIver[10]). Both drift and pulsed cells have been used[10],[11]. As for the HPMS method, an appropriate reaction mixture must be introduced into the cell in order to generate the equilibrium system of interest, but at considerably lower pressure (see above). The approach to equilibrium (signified by a constant value of K) is followed by measuring the ratio of ion densities as a function of pressure ($10^{-6} \rightarrow 10^{-3}$ Torr). The lower pressures limit the range of K values which can be measured to <50 ($\equiv \Delta G^0$ 10kJ mol^{-1}).

This method is less prone to interference from proton bound dimer formation because of the lower pressures employed. However, if it does become a problem the pulsed trapped ion cell can be used since it operates at even lower pressures (<10^{-6} Torr). In this case, ion intensities are measured as a function of residence time (see above). In cases where reactions are too slow to reach equilibrium, the product ion formed in either direction can be ejected and the rate of decay of the reactant ion measured. Hence rate constants for both the forward and reverse directions can be obtained and combined to give a value for K.

Relatively few variable temperature measurements[44] have been performed by ICR, most data being quoted at 300K. There is, in any case, some doubt as to whether the ions formed and trapped by ICR are actually thermal. Evidence from both selected ion flow tube experiments[45] and ICR kinetic energy release measurements[46] indicate that they are "suprathermal" by ca. 0.1eV. Nevertheless there is, in general, very good agreement between the different methods, especially for relative acidity and basicity measurements[10][11], but this is most probably because many gas phase ion-molecule reactions, particularly proton transfer, are relatively insensitive to temperature (see Chapter following). In any case an enormous amount of useful data has been collected by this technique.

Flowing Afterglow (FA) and the Selected Ion Flow Tube (SIFT)[47]

The FA technique was developed by Ferguson and co-workers[48] primarily to measure thermal ion-molecule reaction rates. The ions are created in a carrier gas (usually Helium) flowing fast (~10^4 cm s^{-1}) down a long cylindrical tube. The reactant neutral gas is introduced at fixed points downstream, hence obtaining a separation of primary ion formation and subsequent reaction. The basic idea here was to avoid exposing the neutral reactant to the ionising medium, and hence to avoid any possibility of the neutral becoming excited. This is always a potential problem in "stationary afterglow" or high pressure mass spectrometer sources, and indeed ICR machines. In these methods it is necessary to rely on any excited neutral species having either very short lifetimes or being rapidly collisionally deactivated. This FA arrangement also allows a little more flexibility in systems that can be studied.

Reactant and product ions are detected by sampling them through a small orifice at the end of the flow tube, leading to a differentially pumped quadrupole mass spectrometer system. Their intensities are followed as a function of time usually by varying

the flow rate of the carrier gas and as a function of neutral
partial pressure to obtain decay rate constants and product ion
distributions. The pressure range attainable (usually around
~0.5 Torr) is more restricted than the HPMS, and temperature variation is experimentally less easy to achieve although a temperature range of 80-900K is possible[47]. There are other
potential problems associated with this technique, and these are
discussed at length in refs. 48 and 47. Equilibrium constants
are generally obtained by combining separate measurements of
forward and backward rate constants, although equilibrium systems
can sometimes be set up[58][59].

The most significant advance in thermal gas phase ion
chemistry in recent years has been Smith and Adam's development
of the Selected Ion Flow Tube[47]. A simplified version of
their apparatus is shown in Fig. 15. The reactant ions are
produced in, and selected by, a quadrupole mass spectrometer.
The mass analysed ion beam is then injected at energies as low as
5eV (but up to 30eV) into the fast flowing helium carrier. This
method of ion selection in a low pressure instrument, but injection
into a high pressure reactor, is achieved by a cunning design.
The carrier emerges at high velocity through small holes positioned
annularly around the ion injector orifice. This prevents backflow
of gas by directing its momentum away from the orifice, in addition
carrying the emerging ions forward as they are caught by the forward motion of the gas. Detection and analysis of the ions is
similar to the FA method.

The initial very high kinetic energy of the ions is rapidly
dissipated by collision before the ions meet the reactant gas.
It is not always clear, however, exactly how much of this energy
may be converted to internal energy of the reacting ion. Nevertheless, because ion densities are low, there are no free electrons
in the reactor, there is only one reactant ion, and the range of
potential reactant ions is almost unlimited, this is the "cleanest"
and most versatile method for kinetic studies, and hence for
critical thermochemical investigations in the future.

Internal Consistency Checks

Because of the potential sources of error in any method it
has become the practice to measure relative free energies, such
as acidities and basicities, of a new compound against several
other compounds. A ladder of overlapping values is obtained
such as that shown in Fig. 16 for relative proton affinity values
of the halogenotoluenes[49]. Checks for internal consistency
can then be made by comparing the sum of values between two compounds by different routes. In this example the relative
basicities of toluene and p-fluorotoluene may be obtained from
the sum of ΔG^o values for (a) the linked systems o-fluorotoluene
+ toluene, p-chlorotoluene + o-fluorotoluene and p-fluorotoluene
+p-chlorotoluene giving a value of 19.8kJ mol^{-1}, or (b) m-fluoro-

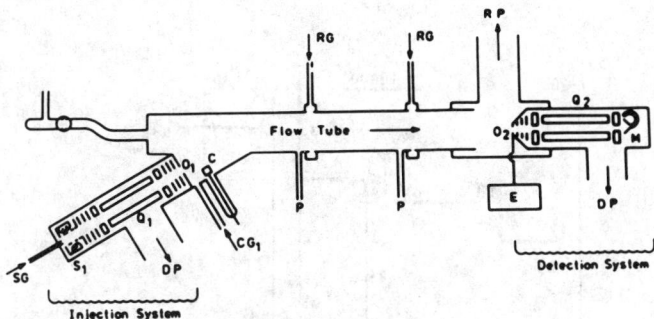

Fig. 15 Schematic diagram of the Selected Ion Flow Tube apparatus S_1-ion source, Q_1,Q_2-quadrupole mass filters; O_1,O_2-orifices; C-ion collector; SG-ion source gas; CG-carrier gas; RG-reactant gas; P-pressure measurement ports; E-electrometer; M-channel multiplier; RP-roots pump. [taken, by permission, from ref.47]

toluene + toluene, m-chlorotoluene + m-fluorotoluene, o-fluorotoluene + m-chlorotoluene, o-chlorotoluene + o-fluorotoluene and p-fluorotoluene + o-chlorotoluene giving a total of 19.7kJ mol^{-1}. The self consistency is a test of the reliability of the measurements(10)(11) The agreement between all three experimental methods is often to within 1 kJ mol^{-1}.

CONCLUSIONS AND SUMMARY

Thermochemical data on gas phase ions falls into four categories: ionisation potentials; electron affinities; appearance energies and reaction enthalpies or free energies. The first three are defined at zero kelvin, whereas the fourth is measured at temperatures above zero. It is found, however, that corrections for temperature are often small and easily calculated (this is discussed in the next Chapter). The fourth category can itself be divided up into many more, the most important of which, so far, have been the relative acidity and basicity scales obtained from proton transfer studies on anions and cations respectively.

This paper has described the most reliable and prolific methods presently available to each type of measurement for polyatomic ions. Thus, for choice, PES is used to determine I_p's, ICR Threshold Photodetachment for EA's, Threshold PIPECO for A_e's and either HPS, ICR or SIFT for free energy and enthalpy measurements. Each is capable of a precision to within <1kJ mol^{-1}.

The equilibria measurements have given rise to a significant expansion in the range of ionic systems studied and hence the amount

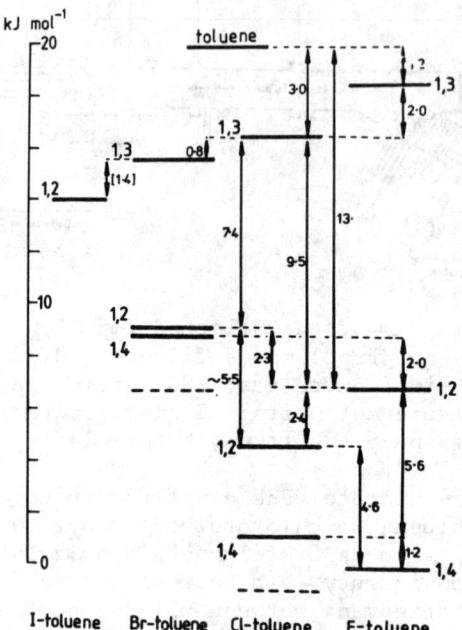

Fig.16. Relative basicities of toluene and halogenotoluenes [$\Delta G°_{369}$ kJ mol^{-1} values for reactions XH$^+$ + Y $\rightleftarrows$ X + YH$^+$ (taken from ref. 49)].

of thermochemical data now available. By far the bulk of this data has been obtained by the HPMS or ICR methods. The former offers probably the simplest and fastest method of investigation over a range of temperatures and pressures, but is severely restricted in the systems with which it can cope. ICR, particularly the pulsed trapped ion cell, appears to be more versatile. This is enhanced by its accessibility to different ionisation techniques such as Laser Desorbtion which, for example, facilitates the study of metal containing compounds. Temperature control and measurement is not so straightforward, however, and it is less suited to the study of "cluster" ions. SIFT, which has not so far addressed itself with any conviction to thermochemistry as such, points the way to an even greater expansion of systems available to the investigator, and will no doubt provide standards against which other methods may be judged. In principle, any ion which can be created in the gas phase could be studied in the SIFT (mass resolution permitting). Ionisation techniques have made great advances in recent years such that it is conceivable that systems of very large molecule ions, perhaps even of biochemical interest, may be studied in the gas phase.

In many cases where it may not be possible to make an accurate absolute thermochemical measurement of, say, an I_p, accurate values relative to another compound whose absolute value is known can be determined from an equilibrium measurement[12]. As will be seen in the following Chapter, the extensive scales of data generated from equilibria measurements need to be anchored to absolute values only at one or two points. This data can be combined, using appropriate thermodynamic cycles, with neutral data compilations[50] to provide good estimates of heats of formation, not only of ionic, but also of neutral polyatomic species. Obviously some systems will not be amenable to any of the techniques discussed here, and for these other less precise (such as high energy ion beam collision methods[51]) or more indirect thermodynamic cycle methods are called for.

A number of assorted compilations of gas phase ion thermochemical data are available. Of the four NBS publications, refs. 14, 52, 53 give compilations of I_p, EA and A_e up to 1971, whilst the most recent (54) compiles similar data published in the period 1971-1981. EA data up to 1978 is to be found in ref. 20. Gas Phase basicities (or proton affinities) can be obtained from a number of sources (refs. 10, 40 and 55) the most recent of which lists over 400 compounds (up to 1980). Acidities up to 1978 appear in ref 11. A critical evaluation of all gas phase ion thermochemical data is presently being compiled[56].

Finally, by analogy with Benson's "Additivity Rules" (see chapter by S. Benson in this volume) for predicting the thermodynamic properties of unknown neutral compounds, similar rules are being investigated for application to ions[57].

REFERENCES

1. Ferguson E.E., Fehsenfeld F.C., and Albritton D.L.: 1979, in Gas Phase Ion Chemistry, Vol. 1, (ed. M.T. Bowers), Academic Press, Chap. 2, pp 45-82.
2. Smith D., and Adams N.G.: 1980, Topics in Current Chemistry 89, pp 1-43.
3. Viggiano A.A., Haworka F., Albritton D.L., Fehsenfeld F.C., Adams N.G., and Smith D.: 1980, The Astrophys.J. 236, pp 492.
4. Bohme D.K.: 1979, in Kinetics of Ion-Molecule Reactions, (ed. P. Ausloos), Plenum, New York, pp 323-343.
5. Wickham A.J., Best J.V., and Wood C.J.: 1977, Radiat. Phys. Chem. 10, pp 107.
6. Jennings K.R.: 1979, in Gas Phase Ion Chemistry, Vol. 2 (ed. M.T. Bowers), Academic Press, Chap. 12, pp 123-151.
7. Taft, R.W.: 1979, in Kinetics of Ion Molecule Reactions, (ed. P. Ausloos), Plenum, New York, pp 323-343.
8. For example: Traeger J.C., and McLoughlin R.G.: 1981, JACS 103, pp 3647-3652.
9. Kebarle, P.: 1975, in Interaction Between Ions and Molecules, (ed. P. Ausloos), Plenum, New York, pp 459-487.
10. Aue D.H., and Bowers M.T.: 1979, in Gas Phase Ion Chemistry, Vol. 2, (ed. M.T. Bowers), Academic Press, pp 1-51.
11. Bartmess J.E., and McIver R.T.: 1979, in Gas Phase Ion Chemistry Vol. 2, (ed. M.T. Bowers), Academic Press, pp 87-121.
12. For example: Moet-Ner M., Sieck L.W., and Ausloos P.: 1981, JACS 103, pp 5342-5348.
13. For example: Uppal J.S., and Staley R.H.: 1982, JACS 104, pp 1229-1238.
14. Rosenstock H.M., Draxl K., Steiner B.W., and Herron J.T.: 1977, J. Phys. and Chem. Ref. Data 6, Suppl. 1.
15. Kiser R.W.: "Introduction to Mass Spectrometry and Its Applications" 1965, Prentice Hall Inc.
16. Lehman T.A., Bursey M.M.: Ion Cyclotron Resonance Spectrometry, 1976, Wiley-Interscience; McIver R.T., and Dunbar R.C.: 1971, Int.J.Mass.Spectrom.Ion. Phys. 7, pp 471; McIver R.T., Rev.Sci. Instrum. 41, pp 555.
17. Bowers M.T., Aue D.H., Webb H.M., and McIver R.T.: 1971, JACS 93, pp 4314.
18. Herzberg G.: 1944, Atomic Spectra and Atomic Structure, Dover; 1950, Molecular Spectra and Molecular Structure, Vol. 1, Spectra of Diatomic Molecules, Van Nostrand.
19. Turner D.W., Baker A.D., Baker C., and Brundle C.R.: 1970, High Resolution Molecular Photoelectron Spectroscopy, New York, J. Wiley.
20. Janousek B.K., and Brauman J.I.: 1979, in Gas Phase Ion Chemistry, Vol. 2., (ed. M.T. Bowers), pp 53-86.
21. Siegel N.W., Celotta R.J., Hall J.L., Levin J., and Bennett R.A.: 1972, Phys. Rev. A6, pp 607-630.

22. Celotta R.J., Bennett R.A., Hall J.L., Siegel M.W. and Levine J.: 1972, Phys. Rev. A6, pp 631.
23. Smyth K.C., and Brauman J.I.: 1972, J.Chem.Phys. 56, pp 1132-1142.
24. Zimmerman A.H., Reed K.J., and Brauman J.I.: 1977, JACS 99, pp 7203.
25. Beynon J.H., Cooks R.G., Jennings K.R., and Ferrer-Correia A.J. 1975, Int.J.Mass Spectrom. Ion Phys. 18, pp 87-99.
26. Morrison J.D.: 1972, Org. Phys. Chemistry Series One, Vol. 5, (ed. A. Maccoll), pp 25-54; Maccoll A.: 1982, Org. Mass Spectrom. 17, pp 1-9.
27. See for example: McCulloh K.E.: 1976, Int.J. Mass Spectrom. Ion Phys. 64, pp 333-342.
28. Parr G.R., and Taylor J.W.: 1973, Rev. Sci. Instrum. 44. pp 1578.
29. See: Chesnavich W.J., and Bowers M.T.: 1979, in Gas Phase Ion Chemistry, Vol. 1, (ed. M.T. Bowers), pp 119-151, and refs. therein.
30. See, for instance: Cooks R.G., Beynon J.H., Caprioli R.M., and Lester G.R.: 1973, Metastable Ions, Elsevier Scientific Pub. Co.; Burgers P.C., and Holmes J.L.: 1982, Org. Mass Spectrom. 17, pp 123-130.
31. See: Stockbauer R., and Rosenstock H.M.: 1978, Int.J.Mass Spectrom. Ion Phys. 27, pp 185-195, and refs. therein.
32. Lifshitz C.: 1982, J.Phys.Chem 86, pp 606.
33. Lifshitz C., Peers A.M., Weiss M., and Weiss M.J.: 1973, Adv. Mass Spectrom. 6, pp 871.
34. Werner A.S., and Baer T.: 1975, J. Chem. Phys. 62, pp 2900-2910.
35. Stockbauer R.: 1977, Int.J. Mass Spectrom. Ion Phys. 25, pp 89.
36. Rosenstock H.M., Buff R., Ferreira M.A.A., Lias S.G., Parr A.C., Stockbauer R.L., and Holmes J.L.: 1982, JACS 104, pp 2337-2345.
37. Baer T.: 1979, Gas Phase Ion Chemistry, Vol. 1 (ed. M.T.Bowers), Academic Press, pp 153-196, and refs. therein.
38. Tal'roze V.L., and Frankevich E.L.: 1956, Dokl. Akad. Nauk. SSR 111, pp 376.
39. Beauchamp J.L: 1971, Annu. Rev. Phys. Chem. 22, pp 527.
40. Kebarle, P.: 1977, Annu. Rev. Phys. Chem., pp 445-476.
41. Bohme D.K., Stone J.A., Mason R.S., Stradling R.S., and Jennings K.R.: 1981, Int. J. Mass Spectrom Ion Phys. 37, pp 283-296.
42. Headley J.V., Mason R.S., and Jennings K.R.: 1982, J. Chem. Soc. Faraday Trans. 1,V78, pp 933-945.
43. Brauman J.I., and Blair L.K.: 1968, JACS 90, pp 5636.
44. See, for example: Hartman K., and Lias S.G.: 1978, Int. J. Mass Spectrom. Ion Phys. 28, pp 213-223.
45. See, for instance: Smith D., Adams N.G., and Lindinger W.: 1982, J. Chem. Phys. 75, pp 3365-3370.

46. Marx R., "Energetics and Dynamics of Positive Ion-Molecule Reactions in Thermal Energy Collisions", NATO Advanced Study Institute: Chemistry of Ions in the Gas Phase, Portugal, 1982.
47. Smith D., and Adams N.G.: 1979, in Gas Phase Ion Chemistry, Vol. 1, (ed. M.T. Bowers), Academic Press, pp 1-44.
48. Ferguson E.E., Fehsenfeld F.C., and Schmeltekopf A.L.: 1969, Adv. At. Mol. Phys. 5, pp 1-56.
49. Mason R.S., Böhme D.K., and Jennings K.R.: 1982, J. Chem. Soc Faraday Trans. 1, 78, pp 1943-1952.
50. Pedley J.B., and Rylance J.: 1977, Sussex-NPL Computer Analysed Thermochemical Data, Organic and Organometallic Compounds, University of Sussex Press.
51. Cooks R.G. (ed.): 1978, Collision Spectroscopy, Plenum Press.
52. Wagman D.D., Evans W.H., Parker V.B., Halow I., Bailey S.M., and Schumm R.H.: 1968, Selected Values of Chemical Thermodynamic Properties, NBS Technical Note 270-3: 1971, JANAF Thermochemical Tables, Nat. Stand. Ref. Data Ser., N.B.S. (US) 37.
53. Franklin J.L., Dillard J.G., Rosenstock H.M., Herron J.T., Draxl K., and Field F.H.: 1969, Nat. Stand. Ref. Data Series, N.B.S. (US) 26.
54. Levin R.D., Lias S.G., Ionisation Potential and Appearance Potential Measurements, 1971-1981, N.B.S. Washington, 1983, In Press.
55. Walder R., and Franklin J.L.: 1980, Int. J. Mass Spectrom. Ion Phys. 36, pp 85-112.
56. Lias S.G., Holmes J.L., Bartmess J., and others, to be published by National Bureau of Standards (US) in 1983.
57. Holmes J.L., Fingas M., and Lossing F.P.: 1981, Can. J. Chem. 59, pp 80-93.
58. Schiff H.I., and Bohme D.K.: 1975, Int. J. Mass Spectrom. Ion Phys. 16, pp 167-189.
59. Pack J.L., and Phelps A.V.: 1966, J. Chem. Phys. 44, pp 1870.
60. Studniarz S.A: 1972, in Ion-Molecule Reactions Vol. 2 (ed. J.L. Franklin), Butterworths, London, Chap. 14, pp 647-671.

THE THERMOCHEMISTRY OF GAS PHASE ION-MOLECULE REACTIONS

Rod S. Mason

Department of Chemistry and Molecular Sciences,
University of Warwick, Coventry, CV4 7AL, U.K.

INTRODUCTION

In the previous chapter[1] a description was given of how thermochemical measurements on ions have been extended to gas phase ion-molecule reactions. One of the major advantages of making these measurements is that intrinsic effects on the reactivity of molecules can be studied in the absence of the solvent effects usually present in solution chemistry. This has led to the collation of a large amount of self consistent thermochemical data, by many different laboratories, from which it is possible to build a self-consistent picture of molecular reactivity in relation to molecular structure. It also provides a bank of reliable data from which to compute ionic heats of formation and against which to test increasingly sophisticated theoretical calculations. In addition, much effort has gone into studying the solvation process itself, not only by direct comparison of gas and solution phase data, but also through the investigation of ion "clustering" or "solvation" reactions in the gas phase, perhaps even giving an insight into nucleation phenomena and the phase transition process itself. It is these aspects of gas phase ion thermochemistry to which this chapter is devoted.

This is not intended to be an exhaustive review of the subject, but an attempt to span, in a general way, across the many areas of study which have arisen in the past 10 years or so. The many different views of the subject are covered in three of the NATO Advanced Study Institutes Series devoted to gas phase ion chemistry[2-4], as well as excellent review articles in the book[5], "Gas Phase Ion Chemistry" Vols. 1 and 2 (ed. M.T. Bowers) and the references quoted therein. Important papers in the field up to 1977 are covered by one of the "Benchmark Papers in Physical Chemistry" series[6].

PROTON TRANSFER

One of the oldest concepts of molecular reactivity in solution is the acid-base idea. The Brönsted definition of an acid is that of a substance which is a proton-donor, whilst a base is a proton acceptor, and it is this concept which is one of the most convenient to study in the gas phase. The reactivity of a substance towards a proton is formally measured by its proton affinity (PA) defined as the enthalpy change in reaction (1). It is a measure, therefore,

$$AH^+ \rightarrow A + H^+ \tag{1}$$

of the basicity of the neutral molecule A. Since this process usually requires much more energy than is available under thermal conditions, proton affinity has to be measured in a proton transfer equilibrium reaction where the enthalpy (3) is equal to a difference

$$AH^+ + B \rightleftharpoons BH^+ + A \tag{2}$$

$$PA(A) \downarrow \qquad \downarrow PA(B)$$

$$A + H^+ + B$$

$$\Delta H^\circ = PA(A) - PA(B) \tag{3}$$

in proton affinities, a measure of the relative basicities of A and B.

Alternatively, reaction (4) can be studied, where a proton is

$$A^- + HB \xrightleftharpoons{\Delta H^\circ} HA + B^- \tag{4}$$

$$\Delta H^\circ_{acid}(HB) \downarrow \qquad \downarrow \Delta H^\circ_{acid}(HA)$$

$$B^- + H^+ + A^-$$

$$\Delta H^\circ = \Delta H^\circ_{acid}(HB) - \Delta H^\circ_{acid}(HA) \tag{5}$$

being transferred to a negative ion. $\Delta H^\circ_{acid}(HA)$ defined as the heterolytic bond dissociation energy in reaction (6) is the absolute

$$HA \rightarrow H^+ + A^- \tag{6}$$

measure of the acidity of HA. Again, in the gas phase, this process does not occur at normal temperatures, whereas equilibria of the type (4) occur readily, ΔH° giving a measure of relative acidities of HA and HB. (Reaction (6) may well occur in solution because much of the energy required is offset by the heat of solvation for each ion.) The terms basicity and acidity are usually reserved for ΔG° values, by analogy with solution chemistry, and therefore contain entropy contributions, the magnitude of which is discussed later. The extension of these investigations to Lewis acids and bases is also discussed in a later section.

Extensive scales of relative PA and ΔH°_{acid} values are already available (references were quoted and experimental details were discussed previously[1]). They can be anchored to an absolute scale by reference to a few compounds whose absolute value is known, or can be calculated.

Absolute Proton Affinities

Rosenstock has pointed out[7] that although proton affinity is a term used for room temperature measurements, the term "affinity" really implies, thermodynamically, the zero kelvin or "threshold" value. The most accurate absolute threshold values are determined from heats of formation. At zero kelvin $\Delta E_o = \Delta H_o$ which for reaction (3) is given by (7). $\Delta H_f(A^+)_o$ must be taken from neutral

$$\Delta H_o = \Delta H_f(A)_o + \Delta H_f(H^+)_o - \Delta H_f(AH^+)_o = PA(A) \quad (7)$$

data compilations (and in many cases is the major source of uncertainty in $\Delta H_f^o(M^+)$ calculations). $\Delta H_f(H^+)_o$ is well known[8] from spectroscopy measurements to be 1530 kJ mol^{-1}, whilst $\Delta H_f(AH^+)$ must be measured by either an appearance energy (A_e) or an ionisation potential (I_p) measurement. For the sake of example and simplicity, let us assume that AH^+ can be generated by fragmentation of the parent ion AHB^+ and its appearance energy (9) accurately measured (a real example is given later). By convention the electron is

$$AHB \rightarrow AH^+ + B + e^- \quad (8)$$

$$A_e(AH^+, AHB) = \Delta H_f(AH^+)_o + \Delta H_f(B)_o + \Delta H_f(e^-)_o - \Delta H_f(AHB)_o \quad (9)$$

treated as an element in its natural state so that $\Delta H_f(e^-) = 0$. To determine $\Delta H_f(AH^+)_o$ we must again rely upon the accuracy of neutral data for AHB and B. For most substances it is improbable that accurate values for all this data will be available. Absolute estimates of PA are, therefore, restricted to a few compounds which can act as anchor points for the scales of experimentally measured relative values.

First, however, the threshold values determined above must be corrected to the temperature, T, of the experimental values. According to the familiar thermodynamic relation (10), the correction is given by the difference in heat capacities, C_p, at constant

$$\Delta H_T = \Delta H_o + \int_o^T C_p(\text{products}) - C_p(\text{reactants}) \cdot dT \quad (10)$$

pressure, between reactants and products, which can often be estimated using statistical mechanics[9]. At this stage, a point of confusion may arise by the convention adopted for the heat capacity of an electron. For example, consider the ionisation process (11) at

$$M_T \xrightarrow{I_e} M_T^+ + e_T^- \quad (11)$$

a temperature T. The heat of reaction is called the ionisation energy, I_e, and is related to the heats of formation by (12) and to its threshold value, I_p, by (13) which combine to give

$$I_e = \Delta H_f(M^+)_T + \Delta H_f(e^-)_T - \Delta H_f(M)_T \quad (12)$$

$$I_e = I_p + \int_0^T C_p(\text{products}) - C_p(\text{reactants}) \cdot dT \quad (13)$$

(expanding the integral) (14). Since M and M^+ are usually very

$$\Delta H_f(M^+)_T = I_p + \Delta H_f(M)_T - \Delta H_f(e^-)_T + \int_0^T C_p(M^+) - C_p(M) \cdot dT$$
$$+ \int_0^T C_p(e^-) \cdot dT \quad (14)$$

similar in structure the term $C_p(M^+) - C_p(M)$ is frequently close to zero (although this assumption is discussed later). Again, by convention $\Delta H_f(e^-)_T = 0$ (ΔH_f [element in its standard state] = 0 at all temperatures). However, there are two conventions for estimating the heat capacity of an electron[7,10]. The "Thermal Electron" Convention: The electron is considered to be an ideal gas for which $C_p = 5/2RT$, where R is the universal gas constant. This convention was adopted in two thermochemical data compilations[11] and in ref. 22. The "Stationary Electron" Convention: The electron is considered to be stationary, in which case $C_p = 0$, a convention adopted in later compilations[8] and now used by most people. The difference in ionic heats of formation at 298K is given by

$$\Delta H_{f_{298}}(M^+) \text{ 'thermal'} = \Delta H_{f_{298}}(M^+) \text{ 'stationary'}$$
$$+ 6.197 \text{kJ mol}^{-1} \quad (15)$$

Difference in Heat Capacity

Although the molecular ion will usually be very similar in structure to the neutral, let us examine, briefly, the justification for the frequent assumption that $\Delta C_p(T) = \int_0^T C_p(\text{ion}) - \int_0^T C_p$ (neutral) $\simeq 0$. The calculation of C_p can be divided up in the usual way[9] into electronic, vibrational, rotational and translational components (16). Most stable neutral molecules (M) are even

$$\Delta C_p = \Delta C_p(\text{elec}) + \Delta C_p(\text{trans}) + \Delta C_p(\text{rot}) + \Delta C_p(\text{vib}) \quad (16)$$

electron species which therefore have a singlet ground state, consequently their ions have an odd electron (conventionally represented by $M^{+\cdot}$) with a doublet ground state. A splitting in the degeneracy (e.g. Jahn-Teller distortion) could result in a maximum contribution to ΔC_p (elec) at 400K $\simeq$ 0.8kJ mol^{-1} (e.g. benzene $\to$ benzene$^+$, ref. 10). Since there is no effective change in mass ΔC_p (trans) = 0, and, except where there is a change in rotational symmetry (such as $CF_3 \to CF_3^-$ [12]), ΔC_p (rot) $\simeq$ 0. The largest contribution is, therefore, likely to come from ΔC_p(vib) due to changes in frequency of the bond or bonds with which the ejected electron is most closely associated. Obviously removal of a bonding electron reduces the vibrational frequency, removal of an antibonding electron causes an increase, and a non bonding electron has little effect. In any case these effects can be calculated using the standard statistical thermodynamic formulae[9]. The biggest variations come from molecules such as ethylene[10] when an electron is removed from the C=C π-bond resulting in a lowering of symmetric C=C stretch (1623cm^{-1} $\to$ 1230cm^{-1}) and the twisting around the C=C bond (1027cm^{-1} $\to$ 430cm^{-1}). Using equation (12) gives an estimate of I_e (350K) = I_p + 0.92 = 1014.7 $\pm$0.4kJ mol^{-1}. For very many molecules, however, there is no measurable difference[10].

Reference Standards for Proton Affinity Scales

The sort of data, outlined above, required for a precise absolute determination of proton affinity is available to very few compounds. However, since there is now an extensive scale of relative gas phase proton affinities, their absolute values could be placed by reference to only a few, or perhaps one, compound whose absolute value is known with certainty. Kebarle[13] found it convenient, as an approximation, to delineate low, medium and high scales of PA values by reference to H_2, H_2O and NH_3 respectively as indicated in Fig. 1.

On this scale a protonated molecule will preferentially transfer a proton to any of the compounds above it (basicity increases with increasing proton affinity). The nitrogen bases lie above NH_3, whilst the inert gases fall below H_2. Many of the interesting organic molecules have values between water and ammonia[14]. Since the position of isobutene is well established within this region, and the information required for an absolute determination was also available, the PA (isobutene) was chosen as a reference standard[14]. It is determined from the following sequence of reactions:

$$(CH_3)_2C=CH_2 + H^+ \to (CH_3)_3C^+ \qquad (17)$$

$$PA(\text{isobutene}) = \Delta H_f(\text{isobutene}) + \Delta H_f(H^+) - H_f(t-C_4H_9^+) \quad (18)$$

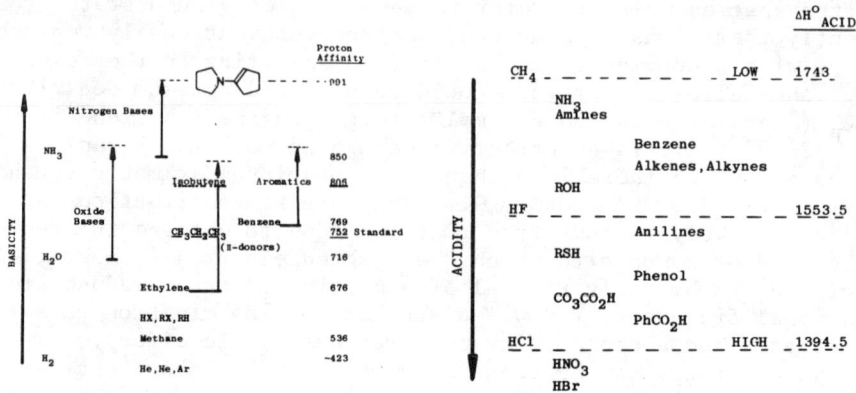

Fig.1. Schematic diagram of the Proton Affinity Scale for ranges of common compounds. Absolute values (kJ mol^{-1}) based on isopropylene as reference standard.

Fig. 2. Acidities. Absolute values taken from ref. 17.

$$(CH_3)_3C^{\cdot} \rightarrow (CH_3)_3C^+ + e^- \qquad (19)$$

$$\underline{I_p(t-C_4H_9^{\cdot})} = \Delta H_f(t-C_4H_9^+) - \underline{\Delta H_f(t-C_4H_9^{\cdot})} \qquad (20)$$

where values for the underlined quantities are available. Rosenstock et al have recently[15] pointed out, however, that there are now a number of disparate values for both $I_p(t-C_4H_9^{\cdot})$ and $\Delta H_f(t-C_4H_9^{\cdot})$ and consequently varying estimates of $\Delta H_f(t-C_3H_9^+)$ and PA(isoC$_3$H$_8$) absolute. These authors report[15] accurate A_e measurement by Threshold Photoion Photoelectron Coincidence for the processes (21) where X$^{\cdot}$ is a halogen atom, giving a value for

$$(CH_3)_2CH-X \rightarrow (CH_3)_2CH^+ + X^{\cdot} + e^- \qquad (21)$$

$\Delta H_f(i-C_3H_7^+)_{298} = 798.8 \pm 2 kJ\ mol^{-1}$. The PA (propylene) is thus determined:

$$CH_3CH=CH_2 + H^+ \rightarrow (CH_3)_2CH^+ \qquad (22)$$

$$PA(propylene) = \underline{\Delta H_f(propylene)} + \Delta H_f(H^+) - \underline{\Delta H_f(i-C_3H_7^+)} \qquad (23)$$

Since propylene is now also well placed[16] in the PA scale it is suggested that this is a more reliable reference standard, because there are consistent values in the literature for the quantities underlined in (21) and (23). A range of absolute values determined from the literature is shown in Table 1 by comparison with the data based on the above value for $\Delta H_f(i-C_3H_7^+)_{298}$. Clearly, the absolute scale still cannot be relied upon to better than 8kJ/mol.

Table 1: The Assignment of Absolute Values (kJ/mol) to the Proton Affinity Scale. Data taken from ref. 15.

M	Relative Proton Affinity	absolute values based on $\Delta H_f(i\text{-}C_3H_7^+)_{298}$ = 798.8 ± 2 kJ/mol	Other lit values*
	98.3	850.1	860.2
NH_3	94.6	846.4	
	93.3	845.1	
$C_6H_5CH_2$	81.2 ± 6.3	833.0 ± 7.9	832.6 ± 5.9
			828.9 ± 5.9
$i\text{-}C_4H_8$	54.3 ± 2.1	806.1 ± 3.8	816.7 ± 3.3
			812.1 ± 5.4
			815.5 ± 7.9
CH_3CHCH_2	0.0	751.8 ± 2.9	748.0 ± 2
			755.6 ± 8
			752.3 ± 4
trans-2-C_4H_8	-0.4 ± 5.0	751.4 ± 6.7	746.8 ± 3

* See ref. 15 and refs. quoted therein.

Absolute Acidities

As for proton affinities there are extensive scales of relative gas phase acidities[17] and the positions of a range of organic compounds are indicated schematically in Fig. 2. By analogy with Fig. 1 proton transfer from compounds towards the bottom of the scale takes place to the conjugate bases of compounds above. The acidity therefore <u>decreases</u> in going from bottom to top. Again, these values must <u>be related</u> to standard reference compounds whose absolute values are well known. In this case, however, there are a number of acids, such as HCl, whose absolute gas phase acidity is well defined by thermodynamic cycles of the type (24)

$$
\begin{array}{ccc}
 & \Delta H^\circ_{acid} & \\
HCl & \rightleftarrows & H^+ + Cl^- \\
D(H-Cl) \downarrow & & \downarrow EA(Cl) \\
 I_p(H\cdot) & & \\
H\cdot + Cl\cdot & \rightarrow H^+ + e^- + Cl\cdot &
\end{array}
\qquad (24)
$$

$$\Delta H^\circ_{acid}(HCl) = D(H-Cl) + I_p(H\cdot) - EA(Cl) \qquad (25)$$

$$431.2 \pm 1 \quad 1312.1 \quad 348.9 \pm 1 \text{ kJmol}^{-1}$$
(at 298K)

From (25) the heterolytic bond dissociation energy of HCl, ΔH°_{acid} (HCl), is governed by the bond dissociation energy D(H-Cl), the ionisation potential of the hydrogen atom $I_p(H\cdot)$ and the electron affinity of atomic chlorine $EA(Cl\cdot)$. All these quantities have been accurately determined from spectroscopic data. Values of

$D(A-H)$ are reported at 298K; there is no real structural difference between H^+ and $H^\cdot$, or between $Cl^\cdot$ and Cl^- so that $I_p(H^\cdot)_T \simeq I_p(H^\cdot)_o$ and $EA(Cl^\cdot)_T \simeq EA(Cl^\cdot)_o$, hence $\Delta H^o_{acid}(HCl)_T = 1394.5$ kJ/mol. Bartmess and McIver list 36 absolute acidities calculated in this way[17].

Entropy Effects

Enthalpy (ΔH) and Entropy (ΔS) terms can only be measured by studies of equilibria as a function of temperature ($-RT\ln K^o = \Delta G^o = \Delta H^o - T\Delta S^o$). This is not always possible for the reasons outlined previously[1] and, in fact, much of the tabulated data is derived from free energy changes, ΔG^o, determined at only one temperature, for which the $T\Delta S$ term has been calculated from statistical mechanical formulae. The entropy change in reaction (1), given by (26), is likely to be very small because of the similarity between reactants and products. ΔS can be partitioned

$$AH^+ + B \rightleftarrows A + BH^+ \qquad (1)$$

$$\Delta S = S(A) - S(AH^+) + S(BH^+) - S(B) \qquad (26)$$

in the usual way (27). Since the masses differ only by one a.m.u.

$$\Delta S = \Delta S_{trans} + \Delta S_{elec} + \Delta S_{vib} + \Delta S_{rot} \qquad (27)$$

$\Delta S_{trans} < 0.5$ J mol^{-1} K^{-1} (when m.wt >30)[10]. Assuming the ions are in their ground state $\Delta S_{el} \simeq \Delta S_{vib} \simeq 0$, and for polyatomic molecules whose moments of inertia are also likely to be similar, the rotational term reduces to (28) where σ represents rotational

$$\Delta S_{rot} = R\ln(\sigma_{AH^+} \cdot \sigma_B / \sigma_{BH^+} \cdot \sigma_A) \qquad (28)$$

symmetry numbers. This is in fact the major contribution in many proton transfer reactions. The evaluation of ΔS_{rot} therefore requires knowledge of the symmetry of the protonated species which often may depend upon the site of protonation. Take, as an example, proton transfer between benzene and fluorobenzene[19]. The symmetry changes indicated in Fig. 3 assume[19] that a proton is attached to a specific carbon atom of the benzene ring, which therefore exhibits a tetrahedral sp^3 valence structure in agreement with indirect experimental evidence, and that fluorobenzene is protonated exclusively in the para position. The calculated value of ΔS_{rot} is -14.19 J mol^{-1}k^{-1}. An experimental van't Hoff plot for this reaction is shown in Fig. 4, and indicates exact agreement (see Table 2). The systems C_6H_6/C_6H_5Cl and C_6H_5Cl/C_6H_5F were also examined and fall in between what would be expected if C_6H_5Cl was protonated in the para or the ortho position (see Table 2), the indication being that chlorobenzene probably protonates at both sites. This points up the fact that although the protonation site can often be predicted

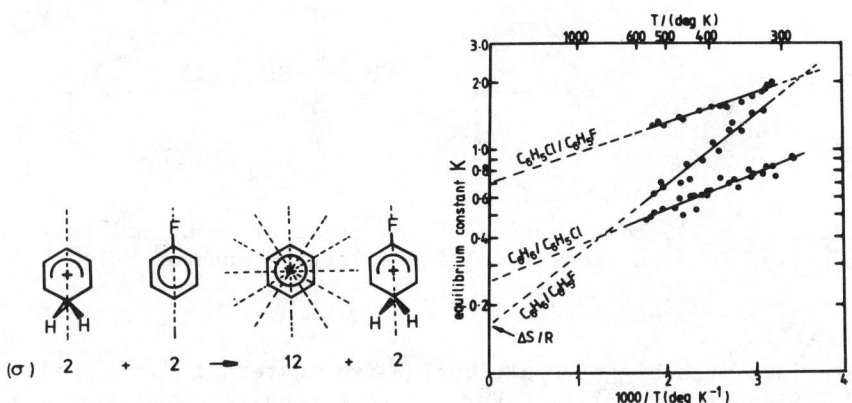

Fig. 3.

Fig. 4. Van't Hoff plots for reactions $C_6H_5XH^+ + C_6H_5Y \rightleftarrows C_6H_5YH^+ + C_6H_5X$; (——) l.m.s. fit

Table 2: Thermodynamic data for process: $XH^+ + Y \rightarrow YH^+ + X$

System		$-\Delta H^\circ$/kJ mol^{-1}	$-\Delta S^\circ$/J mol^{-1}K^{-1}	$-\Delta S_{calc}$	
X	Y			(a)	(b)
C_6H_6	C_6H_5F	5.9±1.5	15±2	14.9	9.13
C_6H_6	C_6H_5Cl	3.0±0.8	11±2	14.9	9.1
C_6H_5Cl	C_6H_5F	2.5±0.6	2.8±1.0	0	5.8*

(a) assuming para protonation of halogenobenzene, (b) assuming ortho protonation.
* assumes para protonation of C_6H_5F and ortho protonation of C_6H_5Cl

and rationalised in mechanistic terms, and hence ΔS_{rot} calculated, it is not always obvious, and experimental verification is required. In this instance the TΔS term is sufficiently large to change the order of relative basicities at different temperatures as indicated by the crossing of the van't Hoff plot lines in Fig. 4. Where there is no overall change in symmetry between reactants and products, TΔS values are usually <0.5kJ mol^{-1} at 300K[10].

<u>Anions</u> The situation can be more complex for negative ions because it is sometimes possible to get a significant change in internal rotation[20]. This is best illustrated by reference to the acetyl enolate anion and its conjugate acid, acetone, and involves the conversion of the near free internal rotation of a methyl group (I) to the rotation of a CH_2 group in the anion which is very hindered because of the partial double bond formation (II) and (III). Again these quantities can be calculated from statistical mechanics. There are also small changes in the vibrational frequencies. On the average the ΔS contribution is of the order of 8-13 J mol^{-1}K^{-1}

$$A^- + \underset{(I)}{CH_3\text{-}\underset{\|}{\overset{O}{C}}\text{-}CH_3} \longrightarrow \underset{(II)}{CH_3\text{-}\underset{\|}{\overset{O}{C}}\text{-}\bar{C}H_2} + HA$$

$$\updownarrow$$

$$CH_3\text{-}\underset{\|}{\overset{O^-}{C}}\text{=}CH_2 \text{ (III)}$$

leading to a TΔS contribution of the order of 2-4kJ mol^{-1} at 300K. Naturally, for reactions involving molecules such as HCl:

$$Cl^- + AH = HCl + A^- \tag{29}$$

there is a much bigger contribution from external rotational changes and T$\Delta S_{300K} \simeq$ 6kJ mol^{-1}. For a more extensive discussion see refs. 17 and 20.

<u>Hydrogen Bonding</u> If hydrogen bonding takes place in the cation, ΔS is affected dramatically. A famous example was provided by Kebarle's study of α,ω diamines[21]. ΔS was measured and compared for the series of reactions involving proton transfer to alkyl amines (30) and to the α,ω diamines (31). The results showed (Table 3) the ΔS changes for (30) to be insignificant, but that

$$BH^+ + CH_3\text{-}(CH_2)_n\text{-}NH_2 \rightarrow B + CH_3(CH_2)_nNH_3 \tag{30}$$

$$BH^+ + NH_2\text{-}(CH_2)_n\text{-}NH_2 \rightarrow B + \underset{(IV)}{\underset{(CH_2)_{n-2}}{H_2N\diagdown\underset{H_2C}{|}\diagup\overset{H^+}{\cdots}\diagdown\underset{CH_2}{|}\diagup NH_2}} \tag{31}$$

they were very large for protonation of the diamines when n>1. This was interpreted as being the result of hydrogen bonding of the proton to the second amine group creating a cyclic cation, ΔS reaching a constant once the possibility of a six or larger membered ring is possible. Comparison of the enthalpy change for reaction (31), in which BH$^+$ $\equiv$ CH$_3$-(CH$_2$)$_n$NH$_3^+$, with the larger values for the analogous proton bound dimer reaction (32) gives an estimate of ring strain in the cyclic product of (31). Naturally there

$$CH_3(CH_2)_nNH_3^+ + CH_3(CH_2)_nNH_2 \underset{\rightarrow}{\overset{M}{\leftarrow}} (CH_3(CH_2)_nNH_2)_2H^+ \tag{32}$$

are many other types of compound for which intramolecular hydrogen bonding can occur in the protonated species without necessarily incurring an overall symmetry change. However, the extra stability gain is reflected in an enhanced value for their proton affinity[22].

Table 3: Thermodynamic data for reactions (30) and (31); PA' ≡ proton affinity relative to PA(Ammonia).

n	Monoamines PA' (kJ/mol)	diamines PA'	ΔS J mol^{-1}K^{-1}
2	64	117	53
3	66	152	86
5	68	152	84
7	68	152	84

PROTON TRANSFER - STRUCTURAL EFFECTS

In the gas phase most exothermic proton transfer reactions[23] appear to proceed, without an energy of activation, at or close to the collision rate via an "orbiting collision complex" (however, see ref. 24 for significant exceptions) and this is very rapid because of the long range ion-dipole or ion-induced dipole forces involved. In a cationic equilibrium system the degree to which the proton sticks to either one or the other of the neutral bases, on emerging from the complex, therefore depends on the relative stabilities of the protonated products. This in turn can be regarded as the ability of the neutral to accommodate a positive charge. It will be seen that in the case of proton transfer from a neutral acid to leave behind a negative ion, the reactivity depends not only on the ability of the anion to accommodate its negative charge but also on the strength of the hydrogen bond in the acid. These factors which determine basicity and acidity naturally depend upon the composition and structure of the reacting species and, based on the large body of data already available[17,22], a self-consistent picture of intrinsic structural effects on the stability of organic cations and anions is emerging. The major effects, of which only a few examples are discussed here, are well documented. In many cases there is agreement with those earlier qualitative results of solution chemistry which gave rise to the classical ideas of physical organic chemistry invoking qualitative concepts such as induction, resonance and polarisation. Occasionally, however, there are a few dramatic reversals.

The Polarisability of Alkyl Groups

One of the first, and now famous, effects studied was the stabilising influence of alkyl groups. According to solution chemistry it appeared that alkyl groups were "electron releasing" and hence stable towards positive charge (by the inductive effect) but destabilising towards anions. This was supported by studies of carbonium ion stability, which showed tBu$^+$>C$_2$H$_5$$^+$·>CH$_3$$^+$·, and the solution order of alcohol acidities: MeOH>EtOH>t-BuOH. The stability of carbonium ions in the gas phase is indeed in the order expected[22], and the basicity of compounds such as the alkyl amines and alcohols is also in the order Propyl- > Ethyl-> Methyl- etc. On the other hand, gas phase acidities of the alcohols and, indeed, other anionic systems is[17]: tBu-OH > Propyl- > Ethyl > H_2O. This is the reverse of solution results, and indicates that alkyl groups stabilise both positive and negative charge. It is now accepted that the effect arises because of the polarisability of the alkyl group[25], such that any charge, positive or negative, induces an appropriate dipole (Fig. 5a) within the alkyl groups and that the stabilisation energy, E, would have the form of the ion-induced dipole interaction: $E = -\alpha q^2/2 \varepsilon r^4$, where α is the group polarisability, ε the dielectric constant, q the charge and r is the effective separation of charges. The $1/r^4$ term should cause a rapid tailing off of the effect as the alkyl group gets bigger. The PA of the long chain amines which have been studied up to n-decylamine, do in fact continue to increase. This is shown in Fig. 5b where relative PA's of R-OH and $R-NH_2$, and relative acidities of R-OH are plotted as a function of the number of carbon atoms, n, in the alkyl group, along with a curve proportional to $1/r^4$. It has been explained as being the result of the flexibility of long chain alkyl groups which allow coiling of the chain to bring it back into close proximity to the charged site, thereby effecting a continuing, albeit small, stabilising effect[22]. Recent ionisation energy measurements on alkanes[27] appear to concur with this finding. Fig 6a plots the difference between adjacent members of the homologous series in both ionisation energy and entropy as a function of the number of carbon atoms. The negative values of $\Delta(\Delta H)_{n-1,n}$ indicate a continuing stabilisation of the charge. In addition, on going from the 6 to the 7 membered homologue there is a discontinuity arising, presumably from a significant (although small) additional stabilising contribution because of the possibility of forming an unstrained ring. It should be remembered that the charge on an ionised alkane is unlikely to reside upon the terminal carbon atom, but rather on a secondary carbon because of the relative stability of secondary carbonium ions over primary. The additional stabilisation therefor occurs on going from hexane to heptane. By analogy with the previous discussion on diamines we would also expect the relative

(V)

entropy changes to reflect this change, as indeed they do (Fig. 6b). Such effects have been described as "intramolecular solvation"[25].

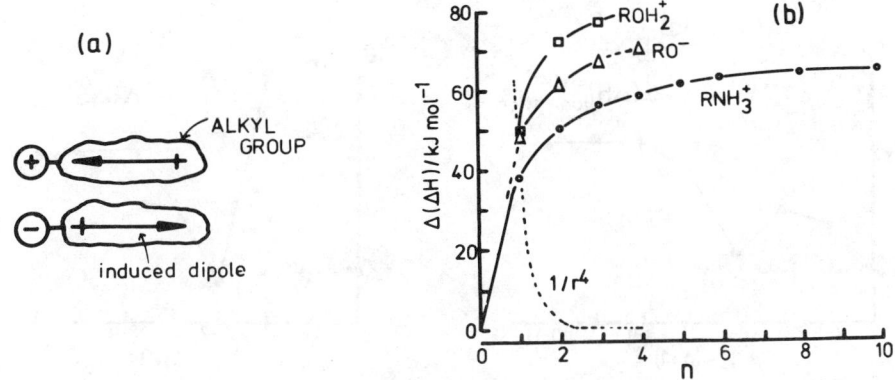

Fig. 5a. Ion-induced dipole stabilisation by alkyl groups;

Fig. 5b. $\Delta(\Delta H) = PA(C_nH_{2n-1}X) - PA(HX)$, where $X = OH$ (□) and NH_2 (○), and $\Delta(\Delta H) = \Delta H^o_{acid}(H_2O) - \Delta H^o_{acid}(ROH)$ for (△).

The Nature of the Acidic Site[17]

$$\begin{array}{ccc} & \Delta H_{acid} & \\ AH & \longrightarrow & H^+ + A^- \\ D(A-H) \downarrow & I_p(H^\cdot) \downarrow & \downarrow EA(A^\cdot) \\ H^\cdot + A^\cdot & \longrightarrow & H^+ + e^- + A^\cdot \end{array}$$

$$\Delta H_{acid} = D(A-H) + I_p(H^\cdot) - EA(A^\cdot) \qquad (32)$$

The general thermodynamic cycle (32) above indicates that since the $I_p(H^\cdot)$ is a constant, the factors contributing to the acidity of a compound, HA, can be considered to be a combination of the H-A bond strength and the electron affinity of A. Strong acids could therefore result either from a weak (A-H) bond, or a high value of $EA(A^\cdot)$. It is well known that the electron attracting power of atoms (electronegativity) increases towards the upper right in the periodic table. This is reflected in their electron affinities, except that the electron configuration of some atoms can strongly effect their electron-attracting power[28]. For instance, the group VA elements have half-filled p shells. The addition of a further electron means doubly occupying one of these p orbitals resulting in a strong electron repulsion and a consequent diminishment of the electron affinity (e.g. EA(C)>EA(N), EA(Si)>EA(P) etc.) Similar arguments are used to explain variations in ionisation potentials. Therefore, in order to compare the electron attracting power of the elements within small molecules, it is important to compare isoelectronic species. So a demonstration of the manner in which (32) operates to determine

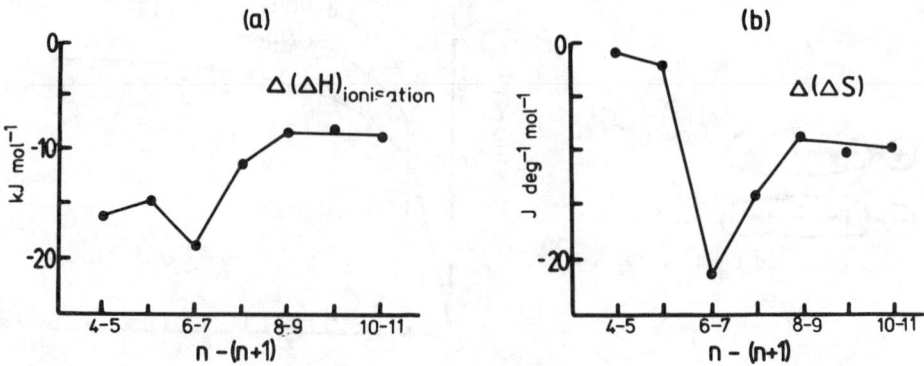

Fig. 6. Changes in ionisation energy, ΔH, and entropy, ΔS, between adjacent members of the homologous series of normal alkanes[27] at 300K, (a) $\Delta(\Delta H) = \Delta H_n - \Delta H_{n+1}$, (b) $\Delta(\Delta S) = \Delta S_n - \Delta S_{n+1}$

Table 4: Acidity of the simple hydrides. Values are $\Delta H^o_{acid}/kJ\ mol^{-1}$

Group	IV(A)		V(A)		VI		VII
	CH_4 1743 ± 4	<	NH_3 1689 ± 4	<	H_2O 1635 ± 2	<	HF 1554 ± 3
	∧		∧		∧		∧
	SiH_4 1554 ± 8	<	PH_3 1550 ± 8	<	H_2S 1479 ± 8	<	HCl 1395 ± 1
	∧		∧		∧		∧
	GeH_4 1509 ± 12	≈	AsH_3 1502 ± 29	<	H_2Se 1418 ± 21	<	HBr 1354 ± 1
							∧
							HI 1315 ± 1

relative acidities is provided by a comparison of the values for isoelectronic hydrides as shown in Table 4 (taken from ref. 17).

It should be remembered that a <u>high</u> value for ΔH^o_{acid} indicates a <u>low</u> acidity. Thus we see acidity increasing (but ΔH^o_{acid} decreasing) across the rows as might be expected from classical ideas of the increasing electronegativity of the elements. The vertical order is opposite to this, even though EA values increase as expected. For the columns, therefore, changes in D(H-A) dominate the change of relative acidities.

Comparison with Theoretical Calculations

The study of substituent effects on protonation of the aromatic ring or the acidities of substituted aromatic compounds has, naturally, received particular attention because of the large number of solution data which have been used to parameterise substituent effects in the form of semi-quantitative linear free-energy relations such as the Hammett acidity functions. The effects are usually qualitatively discussed in terms of charge distribution within the ring by invoking[29] the qualitative concepts of induction, resonance and polarisation, and more recently perturbation molecular orbital theory[30]. In recent years great advances have been made in quantitative quantum mechanical methods of calculation, particularly the ab-initio methods of Pople, Hehre and others[31]. Because they are free of solvent effects, the gas phase basicities and acidities serve as a good data base against which to test such calculations. Proton transfer equilibria are particularly appropriate because the reactions are isodesmic, i.e. the total number and type of bonds remains the same in both reactants and products such that many potential errors arising in the calculation are cancelled out. This is well illustrated by a study of reactions between various substituted phenol

(33)

molecules and the phenoxide anion (33). A comparison of experiment and theory appears in Fig. 7. The ab initio calculations were performed[32] using the "STO-3G" basis set and are values of $\Delta E = \Delta H_0$, the experimental results are those of Kebarle et al[33] and are acidities relative to phenol corrected for the $T\Delta S$ contribution where appropriate. On the whole there is remarkably good agreement, except for o-Nitrophenol. It appears to have an acidity which is lower than predicted by theory, but this would be explained at least in part by the likelihood of intramolecular hydrogen bonding in acid (VI) which would add to its stability

(VI)

relative to the anion. However the energy of 58 kJ mol^{-1} which would be required to bring theory and experiment into line is rather high by comparison with conventional H-bonding values[34].

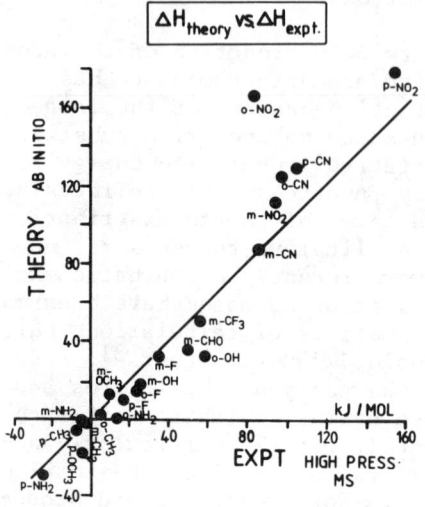

Fig. 7.

Correlation of theoretical[32] and experimental[33] relative acidities (at 300K) of various substituted phenols in reaction (33)

In general the success of these calculations provides a rigorous explanation of structural effects on reactivity in terms of the movement of charge within the molecule. They do not, of course, preclude the very useful qualitative predictions or rationalisations of mechanistic organic chemistry. However, the matching up of theoretical calculation with experimental gas phase ion data is proving a powerful tool in the prediction of ionic structures, and particularly the relative stabilities of isomeric forms[36].

ORGANOMETALLIC CHEMISTRY, METAL$^+$-LIGAND BINDING ENERGIES

The general involatility of inorganic compounds has, in the past, been one of the major limitations to the study of their ion chemistry in the gas phase. However, a technique which is directly analogous to the measurement of proton affinities has been developed for the measurement, in an ICR spectrometer, of metal cation affinities[37]. Initially the generation of metal atom ions was either by electron impact on volatile organometallics such as $C_pNi(NO)$ ($C_p \equiv \eta^5 - C_5H_5$, cyclopentadienyl radical) to generate C_pNi^+, or the thermionic ionisation of a compound of the metal (e.g. an oxide) coated onto a special filament inside the ion source. The latter method is suitable only for metals with low ionisation potentials (the alkali metals). Recently a Laser (YAG) Desorbtion source method has been developed[38] which is capable of volatising and ionising very many different metal targets. Studies on B, C, Mg, Al, Ti, V, G, Mn, Fe, Co, Ni, Cu, Zn, Zr, Nb, Rh and Ag cations have been reported[39] by this method.

The idea is to set up an equilibrium of the type (34) where

$$ML_1^+ + L_2 \overset{\Delta H}{\rightleftarrows} ML_2^+ + L_1 \qquad (34)$$

$$\downarrow D(M^+-L_1) \qquad \downarrow D(M^+-L_2)$$

$$M^+ + L_1 \qquad\qquad M^+ + L_2$$

$$\Delta H = D(M^+-L_1) - D(M^+-L_2) \qquad (35)$$

$M \equiv$ metal atom, $L \equiv$ ligand. $D(M^+-L)$, the metal–ligand binding energy, is a measure of the cation affinity of the ligand and is exactly analogous to proton affinity. It is found[37] that if ML^+ is generated in the presence of another ligand type, the exchange reaction (31) is very rapid in comparison with any other possible reactions, thus allowing the equilibrium to be established and an equilibrium constant determined by measurement of $[ML_2^+]/[ML_1^+]$ and knowing the partial pressures of L_1 and L_2. The secret, as in proton transfer, is to find an ion-molecule reaction sequence which will generate the ions of interest. An example is taken from the recent literature in which ligand exchange reactions of Al^+ were reported[40]. Al^+ ions produced in the presence of ethanol

$$Al^+ + ROH \rightarrow Al(H_2O)^+ + Alkene \qquad (36)$$

$$Al(H_2O)^+ + L \rightarrow Al(L)^+ + H_2O \qquad (37)$$

undergo reactions (36) and (37) without further reaction. The sequence can be followed in a pulsed ICR as a function of time as shown in Fig. 8a. The Al^+ is formed at time zero by a pulse from the laser, as the Al^+ decays the intermediate $Al(H_2O)^+$ rises and decays as the final product emerges. In a binary mixture of alcohols (37) is followed by reaction (34), eventually reaching equilibrium as shown in Fig. 8b. From this the relative binding energies can be determined. From the values of ΔG_{300K} for many different organic ligands, a ladder of relative ligand binding energies was obtained. The results can be compared with proton affinities (represented as $D(B-H^+)$ here) as in Fig. 9.

Where it is possible to compare, absolute metal ion affinities are a factor of five lower than proton affinities (e.g. $D(Li^+-NH_3) \simeq 171$ kJ mol^{-1} [41], cf $PA(NH_3) \simeq 846$ kJ mol^{-1}). This is to be expected since H^+ does not have any core electrons. The large repulsions associated with the Li^+ core result in a bond of largely ionic character. The absence of such repulsion in the protonated complex results in a more intimate reaction which is of a largely covalent nature.

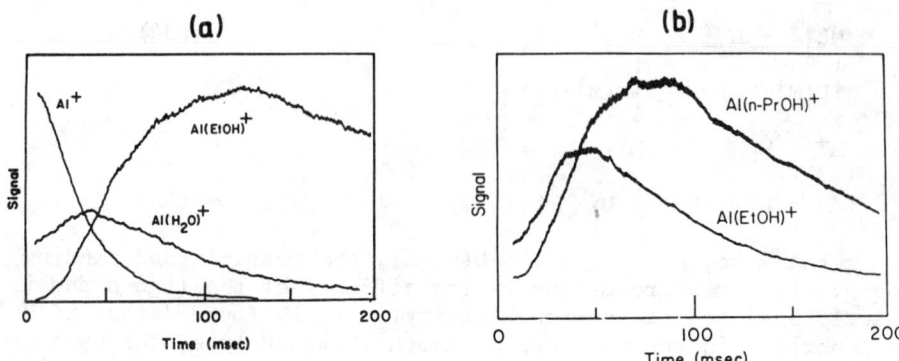

Fig. 8. Pulsed ICR signals showing sequence of reactions of Al^+ in presence of (a) EtOH, and (b) EtOH + n-PrOH; taken, by permission, from ref. 40.

There is, even so, very good correlation here between proton and metal cation affinities. The correlation is even better if the data is restricted to molecules having the same functional group, such as the six esters in Fig. 9. If it were perfect the slope, S, of the line in Fig. 9 would be 1.0. Overall the line leans quite heavily (S = 1.43) towards the proton affinity axis indicating that functional group effects are stronger in H^+ species. It is argued that this is expected since in the Al-(ligand)$^+$ complex the charge centre is further away from the centre of the ligand base - hence electrostatic stabilising effects are bound to be smaller. The slopes for linear correlations in similar plots for various other Lewis acids have also been obtained[42] and are listed in Table 5. They are, in fact, correlations for binding only to Oxygen bases and the slopes follow a systematic pattern, indicating that the relative binding energies are in the order of $H^+ > Al^+ > Mn^+ > Li^+ > C_pNi^+$. On a simplistic level this is probably related to the effective metal-ligand bond distance[42], however, these data still await a sound theoretical treatment, which may provide a deeper theoretical understanding of organometallic bonding. It is also probable that many more aspects of inorganic chemistry will become amenable to these methods in the future.

SOLVATION

How does gas phase data relate to the solution phase ? In this final section three aspects of the transition from gas to solution are briefly considered: (1) the effect of solvent on the intrinsic thermochemical properties of the solute, (2) the nature of the solvation process, and (3) the phase transition.

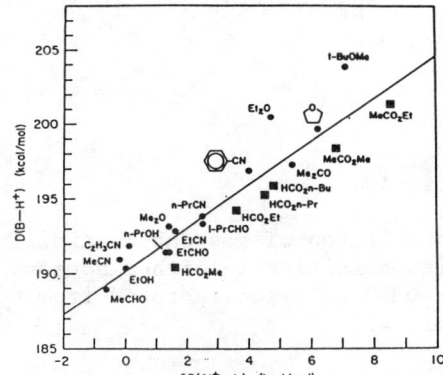

TABLE 5: Slopes for Linear Correlations in Plots of Ligand Binding Energies to One Metal vs. Another, for Oxygen Bases

metal cation	versus				
	H^+	Al^+	Mn^+	Li^+	$CpNi^+$
H^+	1.00	1.53	2.08	2.71	3.38
Al^+	0.65	1.00	1.22		1.68
Mn^+	0.48	0.82	1.00	1.08	1.15
Li^+	0.37		0.93	1.00	
$CpNi^+$	0.30	0.60	0.87		1.00

Taken from ref. 42.

Fig. 9. Correlation of Al^+-ligand binding energies with H^+-ligand binding energies $[\equiv D(B-H^+)$, 1 kcal $\equiv$ 4.183 kJ] at 300K; taken, by permission, from ref. 40.

Solvent Effects

Solvent effects on gas phase thermochemistry are approached simply by considering a thermodynamic cycle similar to that of (38). This represents proton transfer between various acids HA and the conjugate base $C_6H_5CH_2^-$ to give acidity values ΔH_{acid} relative to that for toluene.

$$HA_g + C_6H_5CH_2^-{}_g \xrightleftharpoons{\Delta H_{acid,g}} A^-_g + C_6H_5CH_3{}_g$$

$$\Delta H_{sol} \downarrow \quad \downarrow \quad \quad \downarrow \quad \downarrow$$

$$HA_s + C_6H_5CH_2^-{}_s \xrightleftharpoons{\Delta H_{acid,s}} A^-_s + C_6H_5CH_3{}_s$$

$$\Delta H_{acid,g} = \Delta H_{acid,s} + \delta\Delta H_{sol}(AH-A^-) + \underline{\delta\Delta H_{sol}(C_6H_5CH_2^- - C_6H_5CH_3)} \qquad (38)$$

Gas phase (g) acidity is modified in solution (s) by the difference in heats of solution (ΔH_{sol}) between reactants and products. Fig. 10 shows a plot of gas phase acidities against solution phase acidities relative to toluene for a variety of acids HA, and in two different solvents: water and dimethyl sulfoxide (DMSO)[43]. In this case the term underlined in (38) remains constant. The straight line drawn with unit slope therefore represents $\Delta H_{acid,g} = \Delta H_{acid,s} + \delta\Delta H_{sol}(C_6H_5CH_2^- - C_6H_5CH_3)$ and any deviations from the line are due to the $\delta\Delta H_{sol}(AH-A^-)$ term. As you might expect the biggest deviations occur for aqueous solution data,

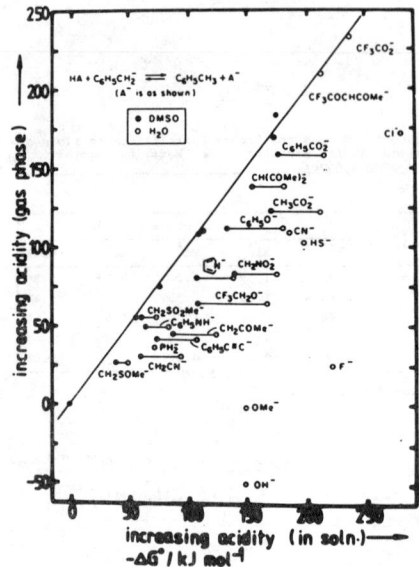

Fig. 10.

Correlation of gas phase acidities with acidities in either aqueous or DMSO solution, adapted from ref. 43.

and particularly for acids which are not very water soluble. The data is therefore pulled over to much higher acidity values in aqueous solution compared to DMSO because of the greatly enhanced stabilisation of the anions. There is in fact quite good correlation for much of the DMSO data. In general solvent effects serve merely to weaken intrinsic effects on reactivity, but they are occasionally enough, as we have seen earlier, to completely reverse trends seen in the gas phase, the acidity of alkyl alcohols being a case in point.

The Nature of Ion Solvation: Ion Clustering

In high pressure gas phase experiments the effect of the solvation process can be followed directly by studying reactions of ionic clusters as a function of the number of solvent molecules, n, attached to the ion. Bohme and co-workers have made a number of measurements of this type using the flowing afterglow and SIFT technique[45]. An example[46] is shown in Fig.11 in which experimental equilibrium constants for proton transfer between OH^- and CH_3OH are displayed as a function of the extent of solvation in both water and methanol. In all cases so far examined the biggest change in K from the solvent free reaction comes with addition of only one solvent molecule. Although, in water, the process shown in Fig.11 rapidly approaches thermo-neutrality (K=1) it does tail off exponentially to remain

THE THERMOCHEMISTRY OF GAS PHASE ION-MOLECULE REACTIONS

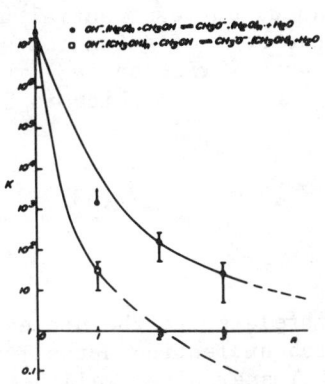
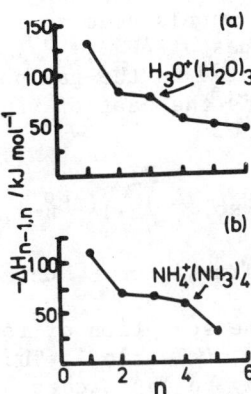

Fig. 11. Variation of equilibrium constant, K (at 300 deg.K), with increasing solvation number, n; taken by permission, from ref.46.

Fig. 12. Variation of the heat of association[48] for increasing cluster sizes of (a) $H_3O^+(H_2O)_n$, and (b) $NH_4^+(NH_3)_n$.

exothermic as n increases. In methanol as solvent the extrapolation indicates that proton transfer from CH_3OH becomes rapidly endothermic after the addition to the anion OH^- of only two solvent molecules or so.

Similar investigations have been conducted into the actual clustering process itself. Indeed, the first high pressure gas equilibrium measurements were of the clustering of water molecules around H_3O^+ and ammonia around NH_4^+ ions[47]. The processes (40) - (42) were studied, where S is the solvent molecule. At equilibrium

$$A^+ + S \rightleftarrows A^+S \qquad (40)$$

$$A^+S + S \rightleftarrows A^+2S \qquad (41)$$

$$A^+(n-1)S + S \rightleftarrows A^+nS \qquad (42)$$

$\Delta H^0_{0,n} = \sum_1^n \Delta H^0_{n-1,n}$. As before measurements as a function of temperature reveal ΔH^0 and ΔS^0 values without the interference of the bulk solvent. Recent measurements[48] on the water and ammonia clusters reveal the shell structure of the solvated cluster. This is seen by plotting $\Delta H_{n-1,n}$ (which are energies of association) versus n as in Fig. 12a and 12b when a break in continuity is seen corresponding to $H_3O^+(H_2O)_3$ and $NH_4^+(NH_3)_4$ respectively, showing the extra stabilisation on completion of the first shell. The first shell structure of solvated water had already been predicted by an **ab initio** calculation[49]. Similar discontinuities are found for ammonia clustering around Li^+ and Na^{+50}. However the larger alkali metals seem to follow a continuous but exponential decrease in values of $\Delta H_{n-1,n}{}^{50}$. Ultimately the heat of solvation should be related to the gas phase heats of association by (43) where ΔH^o_{vap} is the heat of vapourisation of the system which is independent of n.

$$\Delta H^o_{sol} = \sum_{n-1}^{\infty}(\Delta H^o_{n-1,n} + \Delta H^o_{vap}) \qquad (43)$$

Phase Transition: Nucleation

The solvation of ions is closely related to the process of ion induced nucleation. The point at which nucleation actually occurs can only be approached at very large cluster sizes well beyond those amenable to gas phase ion equilibria studies because such large clusters tend to be stable only at very low temperatures or high pressures. Fig.13 shows $\Delta H_{n,n+1}$ values for $H_3O^+(H_2O)_n$ up to n = 26 obtained by a free jet expansion (and therefore cooled) experiment[50]. Although it is possible to get much higher clusters by supersonic cooling in a nozzle beam experiment, it is not possible so far to make equilibrium or rate constant measurements. This is not necessarily a handicap since $\Delta H^o_{n,n+1}$ values tend to fall rapidly during the first few solvations (ref.45,50) quickly approaching a nearly constant value, or at least a slowly declining value thereafter, as displayed in Fig.13 for $n \geq 10$. This trend, which is general, is not surprising since the distance over which the central charge has an appreciable effect is rather short. The observed ability of different ions to induce nucleation also appears[50] to follow their relative binding energies reached at these low values of n (although not the $\Delta H_{0,1}$ value) since the initial rates of decrease in ΔH, although large, vary considerably.

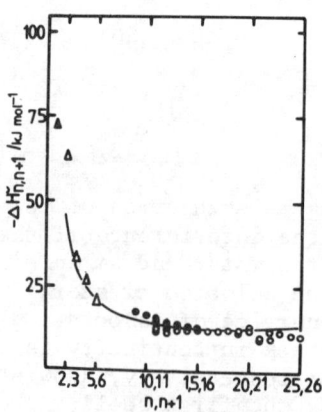

Fig.13. Heats of association for $H_3O^+(H_2O)_n \rightarrow H_3O^+(H_2O)_{n+1}$; ($\Delta$) data taken from ref.48; (o,•) data taken from Fig.8. of ref.50, as is the solid curve which represents calculated values from the Thomson equation.

The solid line drawn in Fig.13 represents the predictions of the classical Thomson equation[51]. This was developed to calculate the energy of interaction of a point charge in a liquid droplet with a well defined surface tension. It is surprising, therefore, to find that there is remarkably good agreement here down to the smallest cluster sizes. Unfortunately the agreement is not so good for other systems[50]. Nevertheless, this result demonstrates that a successful theory of nucleation should extrapolate even to small cluster sizes, for which reliable experimental thermochemical data are now obtainable from gas phase ion-equilibria studies.

REFERENCES

1. Mason, R.S., "Experimental Methods to Study the Thermochemistry of Gas Phase Ions and Ion-Molecule Reactions" Chapter in this Volume.
2. Interaction Between Ions and Molecules, ed. P. Ausloos, 1975, Nato Advanced Study Institute, Series B, Vol. 6, Plenum Press pp 1-690.
3. Kinetics of Ion-Molecule Reactions, ed. P. Ausloos, 1979, Nato Advanced Study Institute, Series B, Vol. 40, Plenum Press pp 1-508.
4. Chemistry of Ions in the Gas Phase, ed. P. Ausloos, 1983, Nato Advanced Study Institute, Series B, Plenum Press, to be published.
5. Gas Phase Ion Chemistry, ed. M.T. Bowers, 1979, Vols. 1 and 2. Academic Press, pp 1-354 and 1-346.
6. Ion-Molecule Reactions, Parts I and II, ed. J.L. Franklin, Benchmark Papers in Physical Chemistry and Chemical Physics, V3, 1979, Dowden, Hutchinson and Ross Inc., pp 1-399 and 1-378.
7. Rosenstock H.M., 1979, ref. 3, pp 246-249.
8. Rosenstock H.M., Draxl K., Steiner B.W., and Herron J.T., 1977, J. Phys. and Chem. Ref. Data 6, Suppl. 1.
9. See for instance, Benson S.W., 1976, Thermochemical Kinetics, 2nd Edition, J. Wiley & Sons, pp 19-77.
10. Lias S.G., 1979, ref. 3, pp 223-245.
11. Wagman D.D., Evans W.H., Parker V.B., Halow I., Bailey S.M. and Schumm R.H., 1968, Selected Values of Chemical Thermodynamic Properties, NBS Tehcnical Note 270-3; JANAF Thermochemical Tables, Nat. Stand. Ref. Data Ser. NBS (US) 37.
12. Richardson J.H., Stephenson L.M., and Brauman J.I., 1975, Chem. Phys. Lett. 30, p 17.
13. Kebarle P., 1977, Annu. Rev. Phys. Chem. 28, pp 445-476.
14. Yamdagni R., and Kebarle P., 1976, J. Am. Chem. Soc. 98, pp 1320-1324.
15. Rosenstock H.M., Buff R., Ferreira M.A.A., Lias S.G., Parr A.C., Stockbauer R.L., and Holmes J.L., 1982, J. Am. Chem. Soc. 104, pp 2337-2345.

16. Lias S.G., Shold D.M., and Ausloos P., 1980, J. Am. Chem. Soc. 102, p 2540.
17. Bartmess J.E., and McIver R.T., 1979, ref. 5, Vol. 2, pp 87-121.
18. For example, Pedley J.B., and Rylance J., 1977, "Sussex-NPL Computer Analysed Thermochemical Data: Organic and Organometallic Compounds". University of Sussex Press.
19. Bohme D.K., Stone J.A., Mason R.S., Stradling R.S., and Jennings K.R., 1981, Int. J. Mass Spectrom. Ion Phys. 37, pp 283.
20. Cumming J.B., and Kebarle P., 1978, Can. J. Chem. 56, pp 1-9.
21. Yamdagni R., and Kebarle P., 1973, J. Am. Chem. Soc. 95, pp 3504-3510.
22. Aue D.H., and Bowers M.T., 1979, ref. 5, pp 2-51.
23. Bohme D.K., 1975, ref. 2., pp 489-504.
24. Farneth W.E., and Brauman J.I., 1976, J. Am. Chem. Soc. 98, pp 7891-7898.
25. Brauman J.I., and Blair L.K., 1970, J. Am. Chem. Soc. 92, pp 5986-5992.
27. Meot-Ner M., Sieck L.W., and Ausloos P., 1981, J. Am. Chem. Soc. 103, pp 5342-5348.
28. Janousek B.K., and Brauman, J.I., 1979, ref. 5, pp 53-86.
29. Hine J., 1975, Structural Effects on Equilibria in Organic Chemistry, Wiley and Sons, New York.
30. Dewar M.J.S., and Dougherty R.C., 1975, The PMO Theory of Organic Chemistry, Plenum Press, New York.
31. Hehre W.J., Radom L., and Pople J.A., 1972, J. Am. Chem. Soc. 94, pp 1496-1504.
32. Pross A., Radom L., and Taft R.W., 1980, J. Org. Chem 45, pp 818-826.
33. McMahon T.B., and Kebarle P., 1977, J. Am. Chem. Soc. 99, pp 2222-2230.
34. See for instance: Pimentel G.C., and McClellan A.L., 1960, The Hydrogen Bond, W.H. Freeman and Co.
36. See for example: Bouma W.J., Ross R.H., and Radom L., 1982, J. Am. Chem. Soc. 104, p 2929.
37. Corderman R.R., and Beauchamp J.L., 1976, J. Am. Chem. Soc. 98, pp 3998-4000.
38. Cody R.B., Burnier R.C., Reents W.D. Jr., Carlin T.J., McCrery D.A., Lengel R.K., and Freiser B.S., 1980, Int. J. Mass Spectrom. Ion Phys. 33, pp 37-43.
39. Uppal J.S., and Staley R.H., 1982, J. Am. Chem. Soc. 104, pp 1229-1234.
40. Uppal J.S., and Staley R.H., 1982, J. Am. Chem. Soc. 104, pp 1235-1238.
41. Woodin J.L., and Beauchamp J.L., 1978, J. Am. Chem. Soc. 100, pp 501-508.
42. Uppal J.S., and Staley R.H., 1982, J. Am. Chem. Soc. 104, pp 1238-1243.
43. Taft R.W., 1979, in ref. 3, pp 271-293.
44. Arnett E.M., Jones F.M., Taagepera M., Henderson W.G., Beauchamp J.L., Holtz D., and Taft R.W., 1972, J. Am. Chem. Soc. 94, pp 4724-4726.

45. Bohme D.K., 1983, "Effect of Solvation in Ion-Molecule Reaction Rates" in ref. 4, to be published.
46. Mackay G.I., Bohme D.K., 1978, J. Am. Chem. Soc. 100, p 327.
47. Hogg A.M., Haynes R.M., and Kebarle P., 1966, J. Am. Chem. Soc. 88, pp 28-31; Kebarle P., Searles S.K., Zolla A., Scarborough J., and Arshadi M., 1967, J. Am. Chem. Soc. 89, pp 6393-6399.
48. Lau Y.K., Ikuta S., and Kebarle P., 1982, J. Am. Chem. Soc, 104, pp 1462-1469.
49. Newton M.D., 1977, J. Chem. Phys. 67, p 5535.
50. Castleman A.W., 1979, in ref. 3, pp 295-321.
51. See Castleman A.W., Holland P.M., and Keesee R.G., 1978, J. Chem. Phys. 68, pp 1760-1767, and refs. therein.

THE APPLICATION OF LOW-TEMPERATURE CALORIMETRY TO SOME CONTEMPORARY PROBLEMS IN SOLID STATE CHEMISTRY

L.A.K. Staveley

The Inorganic Chemistry Laboratory, Oxford University,
South Parks Road, Oxford OX1 3QR, U.K.

The contribution which low-temperature calorimetry can make to contemporary problems in solid state chemistry is illustrated, first by showing how the heat capacity-temperature curve of $RbAg_4I_5$ reflects the growing disorder of the silver ions, which gives this substance its remarkably high electrical conductivity. The use of C_p data to obtain information about movement in a crystal is then considered for the particular case of the restricted rotation of the ammonia ligands in the complex cations in the salt $Co(NH_3)_6Cl_3$. There follows a review of the polymorphism and magnetic ordering in the layer compounds $(RNH_3)_2MX_4$, where R = an alkyl group, M = metal, X = halogen. Finally, the entropies of transition in orientational order-disorder transitions are considered, with special reference to recent work on metallic nitrites.

INTRODUCTION

One important application of low-temperature calorimetry may be mentioned at the outset. This is the use of the heat capacity (C_p) of a substance and of the enthalpies of any phase changes to evaluate the calorimetric entropy of a mole at some arbitrary standard conditions (usually at one atmosphere pressure and 298.15K), so that such entropy data can be combined with standard enthalpies of formation to calculate the standard Gibbs energy change, and hence the equilibrium constant, for a particular chemical reaction. For convenience, it has now become more or less general practice, when publishing new heat capacity data, to give smoothed values at rounded temperatures not only of C_p and $S^o(T) - S^o(0)$, but also of the functions $H^o(T) - H^o(0)$ and $-[G^o(T) - H^o(0)]T^{-1}$. We shall not discuss further this important application of heat capacity measurements. It is not, of course, novel, but it is certainly not obsolete, and one can predict with confidence that such measurements will be made on hitherto unstudied compounds, as occasion requires, for many years to come.

A second well-established use of heat capacity data may also be mentioned, namely given a value of ΔH for a chemical reaction at one temperature, to estimate its value at other temperatures, using the Kirchoff equation

$$(\partial(\Delta H)/\partial T)_p = \Sigma C_p \tag{1}$$

where ΣC_p is the heat capacity of the products less that of the reactants. Usually, this equation is applied to estimate ΔH above room temperature, starting with a ΔH value for, say, 298.15K. But with increasing use of high-temperature cells to provide thermodynamic information which in the first instance necessarily relates to the high-temperature used, C_p data and the Kirchoff equation are also applied to derive standard thermodynamic parameters for 298.15K. In both applications, high-temperature rather than low-temperature C_p results are required. In this review we are primarily concerned with the latter, but it should be noted that in classifying temperature regions as 'high' and 'low', it is quite arbitrary to take room temperature as marking the change from one to the other. It is true that for a long time calorimeters for low-temperature C_p work differed in construction from those used for high-temperature measurements, but 'low-temperature' calorimeters have now been modified to enable measurements to be made with them up to $\sim$550K (1). By using two calorimeters, one operating from $\sim$2K to $\sim$100K (2) and the other from $\sim$80K to $\sim$550K, the range 2K to 550K can be covered. We shall therefore not hesitate in this review to refer to C_p results for temperatures well above room temperature which were obtained with what would still be conventionally described as a 'low-temperature' calorimeter.

Before considering low-temperature studies of some selected systems, a general observation may be made. It is the author's firm conviction that the thermodynamicist should wherever possible consider his findings, not in isolation, but in relation to information about the system he has studied which has been derived from quite different techniques. In studies of disorder in crystals, for example, usually a battery of different experimental methods will be brought to bear on a particular solid, of which experimental thermodynamics is one, generally neither superior to, nor inferior to, any of the others. There is certainly no evidence that the need for experimental thermodynamic data is declining. Indeed, the signs point rather in the other direction, in that the demand for such data seems to exceed the present rate of supply.

SUPERIONIC CONDUCTORS

A type of solid in which there is considerable interest at the present time is that in which ions are relatively free to move, and so give the crystal an unusually high electrical (ionic) conductivity. Such solids are known as 'superionic' or 'fast ion' conductors. Besides their own intrinsic interest, they have important uses, for example as the electrolyte in solid state batteries. Perhaps the most famous is silver iodide in the cubic form stable above 421K, where the iodide ions form the lattice and the silver ions move with remarkable freedom between them. A similar superionic conductor is the compound $RbAg_4I_5$, which even at room temperature conducts about as efficiently as a saturated aqueous solution of sodium chloride, due to the facile movement of the silver ions. The heat capacity of this salt was measured from 5K to 340K by Johnston et al. (3). The C_v-temperature curve is shown in Fig.1.

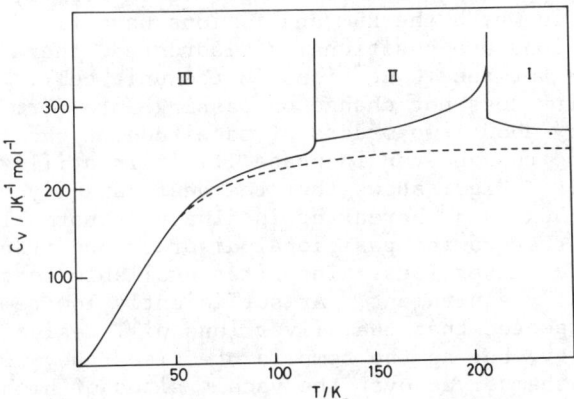

Fig.1. C_v vs. T for $RbAg_4I_5$, after Johnston et al. (3). Upper curve, experimental results. Lower curve, baseline.

Before commenting on particular features of this, we may note that this example illustrates an important aspect of the interpretation of the heat capacity of solids when there are grounds for thinking that 'abnormal' or 'unusual' factors are contributing to the overall heat capacity, namely that it is very helpful if a reasonably reliable estimate of the 'normal' or 'baseline' heat capacity can be made. Essentially, this means assessing the contribution from the progressive excitation of the lattice vibrations as the temperature is raised - the vibrations here including the torsional oscillations and internal vibrational modes of diatomic or polyatomic molecules or ions if they are present. In addition, since theories of the heat capacity of solids primarily deal with C_v, whereas the measurements are carried out at constant pressure, an estimate has to be made of the quantity C_p-C_v, though usually this is fairly small. A whole review could easily be devoted to ways and means of estimating the baseline heat capacity of a solid, but here our observations on this matter must necessarily be brief. Sometimes comparison with a structurally similar solid which, however, lacks the 'abnormality' of the solid under examination can be helpful. And sometimes it is apparent from the C_p-temperature curve for the compound being studied that over a range of low temperatures its heat capacity is 'normal'. C_p in this region can then be empirically fitted to a suitable function, or combination of such functions. Thus, in the case of $RbAg_4I_5$, Johnston et al., having corrected their experimental measurements to give C_v, fitted their C_v values below 50K to a combination of Debye and Einstein functions, and obtained a baseline curve up to 340K by extrapolation. This baseline curve is shown in Fig.1.

At room temperature, where $RbAg_4I_5$ is in form I, it has a cubic lattice in which the Rb^+ and I^- ions have fixed positions, while the Ag^+ ions are positionally disordered, there being 56 possible sites for the 16 Ag^+ ions in the unit cell. The electrical conductivity does not change on passing into form II on cooling, but it drops by about two orders of magnitude at the lower II → III transition, though even in phase III it is still remarkably high for a salt. Fig.1 shows that the heat capacity begins to rise at about 50K, and thereafter is always 'abnormally' high. This reflects the growing positional disorder and freedom of movement of the silver ions. The sites available to these are not all energetically equivalent. At sufficiently low temperatures, it is to be expected that the silver ions will reside in the sites of lowest energy, but as the temperature rises they will begin to distribute themselves over the vacant sites of higher energy, which of course requires an extra energy intake. The randomisation of the silver ions among the possible sites therefore begins at about 50K, causing the appearance of a configurational contribution to C_p and the development of electrical conductivity.

Given a reliable baseline heat capacity curve, the total configurational entropy gain at any temperature can be estimated. For $RbAg_4I_5$ at 300K, this amounts to the considerable figure of 45.2 $JK^{-1}mol^{-1}$, or 5.4R, implying a high degree of disorder of the silver ions. The entropy gains at the two transitions, however, are relatively small - 0.79R at the III → II transition, and 0.5R at the II → I transition. So most of the positional disorder prevailing at 300K has been acquired gradually rather than abruptly.

There is another unusual feature displayed by $RbAg_4I_5$, namely that in phase I C_v appears to be decreasing with rising temperature. (Admittedly in this case the effect is not very pronounced, but the same behaviour can be seen more markedly in the high-temperature phase of other super-ionic conductors). A mole of monatomic ions undergoing simple harmonic vibration at a temperature sufficiently high for the equipartition of energy principle to be applicable would contribute $3R$ to the heat capacity. The corresponding contribution for the same particles undergoing translational movement in the dilute gaseous state would, of course, be less, namely $3R/2$. Accordingly, the downward trend in C_v in phase I of $RbAg_4I_5$ may be an indication that the motion of the silver ions is tending to change to translation from vibration.

Finally, there is an interesting chemical point connected with the stability of $RbAg_4I_5$. Strictly, this substance is thermodynamically unstable at room temperature, though it can be kept indefinitely, it seems, in the absence of moisture. However, the considerable configurational entropy is a useful addition to the total entropy of the solid, which acts in the direction of reducing the Gibbs energy, thereby helping the compound to be less unstable than it would otherwise be.

LATTICE DYNAMICS

The subject of the movement of atoms, ions or molecules in a crystal lattice is one of very long standing. It is scarcely necessary to remind the reader, for example, of the attempts made to interpret the heat capacity of monatomic solids down to low temperatures in terms of the vibrational movement of the atoms, of which attempts that of Debye is still the most famous. Nevertheless, there is more still to be learnt, and in this Section we will illustrate the use which can be made of low-temperature heat capacity results to provide information about the parameters governing the torsional movement of ligands within a complex ion in a crystal. The two salts we shall consider are $Co(NH_3)_6Cl_3$ and $Ni(NH_3)_6I_2$, which contain the octahedral complex ions $Co(NH_3)_6^{3+}$ and $Ni(NH_3)_6^{2+}$. It could scarcely be expected that the NH_3 ligands in these ions carry out literally free one-dimensional rotation about the metal-nitrogen axes, either because of the intraionic ligand-ligand interaction, or due to the interaction of the ligands with surrounding ions, or to both of these causes simultaneously. It might be that at higher temperatures something approaching free rotation is attained, but at sufficiently low temperatures each NH_3 unit must rock about the metal-nitrogen bond in a potential well. Owing to the well-known phenomenon of quantum-mechanical tunnelling (here on the part of the protons) the torsional oscillation energy levels are split, the splitting being greater the less the height of the potential energy barrier. The torsional oscillational levels are eightfold degenerate, and to a first approximation split into two groups of four (4,5). Since the energy difference between the split levels is constant (i.e. independent of the relative extent to which they are occupied), the system shows a Schottky effect in its C_p-temperature curve. At a sufficiently low temperature only the lower levels are occupied. At sufficiently high temperatures, the particles distribute themselves impartially between the two sets of levels. If the splitting is between two levels of the same degeneracy, at sufficiently high temperatures both levels are equally populated. The passage with rising temperature from the low-temperature to the high-temperature condition, requiring as it does the intake of extra energy, gives rise to an extra contribution to the heat capacity somewhat resembling, in its temperature dependence, a hump-backed bridge. In the case of $Ni(NH_3)_6I_2$, the maximum in this anomaly occurs at $\sim$0.3K. In the deuterated form of this salt, the deuteron, owing to its higher mass, has a much smaller tunnelling propensity than the proton, and the energy level splitting is considerably less. Accordingly, the heat capacity maximum for the Schottky anomaly in the deuterated salt falls at a much lower temperature (5).

Figure 2 shows the C_p-temperature curve for the salt $Co(NH_3)Cl_3$ from $\sim$2K to $\sim$309K (2). There is nothing very remarkable about it at first sight, but there appears to be a 'knee' in the curve between $\sim$165 and 185K. Useful information, however, lies concealed in the C_p values at the lowest temperatures. In extrapolating heat capacity results to 0K, use is often made of the Debye T^3-law, which is really the first term of an expansion

$$C_p = aT^3 + bT^5 + \ldots \qquad (2)$$

When this equation was applied to the results for the cobalt salt it revealed that the C_p values below $\sim$4K, so far from obeying eqn. (2) more closely with falling temperature, were showing an increasing divergence from it. A feature of a Schottky anomaly is that the 'extra' heat capacity as it declines asymptotically to zero on the high-temperature side - the high-temperature 'tail' of the effect - is proportional to T^{-2}, and on the assumption that the 'abnormality' below 4K was a manifestation of the participation of this tail, C_p at the lowest temperatures was represented by the equation

$$C_p = aT^3 + bT^5 + cT^{-2} \qquad (3)$$

To test the adequacy of this equation and to evaluate the coefficient c, eqn.(3) was rearranged to give

$$T^2(C_p - bT^5) = c + aT^5 \qquad (4)$$

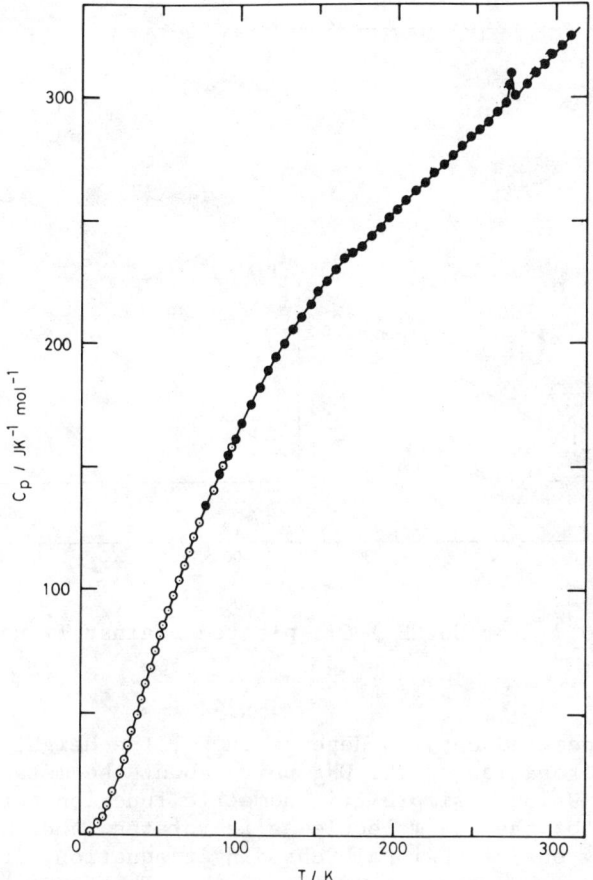

Fig.2. Molar heat capacity, C_p, of $Co(NH_3)_6Cl_3$. The two sets of points were obtained with two calorimeters (2).

and $T^2(C_p - bT^5)$ plotted against T^5. This gave a good straight line (Fig.3), which did <u>not</u> pass through the origin. The intercept on the vertical axis, though small, is not zero, and it gives a reasonably reliable estimate of the coefficient c. From this, the splitting of the energy levels ΔE which produces the Schottky effect can be estimated, the connection being made by the equation

$$c = \frac{6g_1 g_2 R}{(g_1+g_2)^2} \left(\frac{\Delta E}{k}\right)^2 \tag{5}$$

where g_1 and g_2 are the degeneracies of the two split levels.

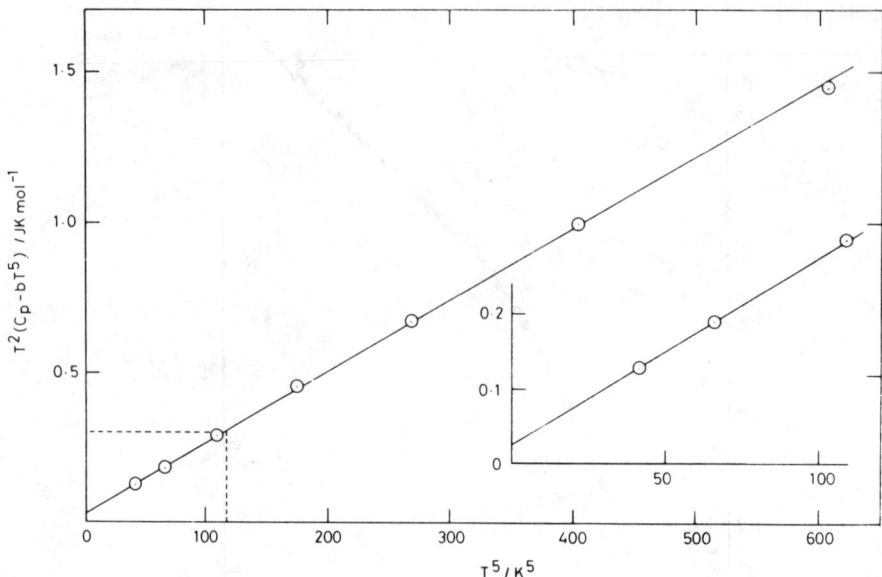

Fig.3. $T^2(C_p - bT^5)$ for $Co(NH_3)_6Cl_3$ plotted against T^5 up to 3.6K.

As already pointed out, ΔE depends on V_o, the height of the barrier opposing rotation of the NH_3 units about the metal-nitrogen bonds. Using a simple trigonometric function for the potential energy of the NH_3 molecule as it rotates, and solving the corresponding one-dimensional Schrödinger equation, it is possible to evaluate V_o from the value of ΔE. This gave $V_o/k = 560K$, or 4.7 kJ mol^{-1}.

The torsional movement of the NH_3 molecules in the $Co(NH_3)_6^{3+}$ ions in the lattice therefore presents an example of a well-known problem in intramolecular dynamics, namely the restricted rotation of a group about a single bond. The statistical mechanics of this were first dealt with by Pitzer (6) who applied his results to molecules such as ethane. He showed how the contribution to the heat capacity over a range of temperature from such rotation can be used to estimate the barrier height V_o. In our case, knowing V_o, we can proceed in the other direction, and calculate the contribution to the heat capacity of $Co(NH_3)_6Cl_3$ from the restricted one-dimensional rotation of the NH_3 molecules. This contribution is shown in Fig.4. (The logarithmic scale should be noted). The maximum at very low temperatures is the Schottky effect (of which only a very small portion was actually observed and used), while the second maximum shows the highest value reached from the restricted rotation before the heat capa-

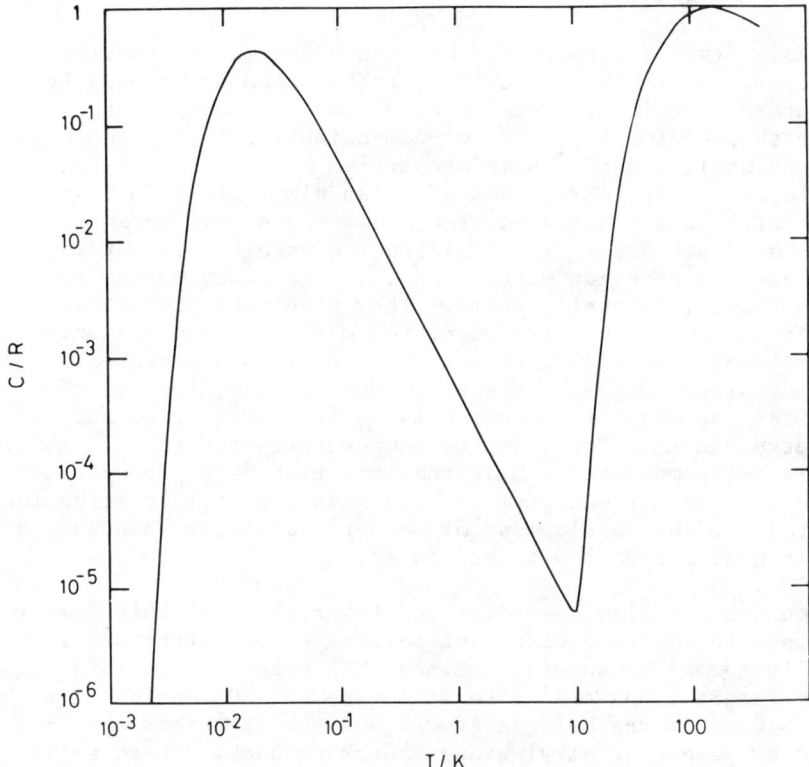

Fig.4. Logarithmic plot of C/R against T, where C is the calculated contribution to the heat capacity of $Co(NH_3)_6Cl_3$ per mole of NH_3 molecules from the Schottky effect and from the restricted rotational movement about the Co-N axis.

city starts to decline towards the high-temperature limit for unrestricted free rotation. There seems little doubt from the temperature at which this second maximum is reached that it is the cause of the 'knee' in the overall C_p- temperature curve to which we have already referred.

Accordingly, we conclude that the barrier to one-dimensional rotation of the NH_3 groups in the cations in crystalline $Co(NH_3)_6Cl_3$ is relatively low, such that at room temperature a not negligible fraction of the NH_3 molecules will at any instant be 'above the barrier', that is in a state of rotation, even though rather an uneven one. There are many solid coordination compounds which contain NH_3 ligands in a complex ion, and it would be interesting to know how the restriction on their rotation varies from one such solid compound to another.

LOW-DIMENSIONAL SOLIDS

In some crystals, there can be much stronger interaction in one direction, or in two directions, rather than about equally strong interaction in all three dimensions, and such crystals are often spoken of as one- and two-dimensional solids. In this Section, we shall consider some properties of crystals of the general formula $(RNH_3)_2MX_4$, where R is an alkyl group, M a divalent metal, and X a halogen. These compounds have layer lattices, with relatively weak interaction between the layers and much stronger interaction within them, so that they can be regarded as two-dimensional systems. They have recently attracted a considerable amount of attention, for more than one reason. If the ion M^{2+} has unpaired electrons, the crystals approach closely to two-dimensional magnetic systems. When cooled, in some of these solids the magnetic ordering is antiferromagnetic, and in others ferromagnetic. Thus, at low temperatures the salt $(CH_3NH_3)_2 CrCl_4$ is a ferrromagnet which at the same time is a transparent solid. Heat capacity measurements can make a useful contribution to the study of the development of magnetic disorder when such a crystal is heated from low temperatures.

Attention has also been directed to crystals of this family of compounds for quite a different reason. Their structure is essentially formed by sheets of linked MX_6 octahedra, within which the interactions are quite strong. The RNH_3^+ cations are hydrogen-bonded to the halogen atoms, and the MX sheets are separated by layers of alkyl groups, back to back. These solids have therefore been proposed as models for lipid bilayer systems (7).

Finally, these compounds can be remarkably polymorphic. Thus, the salt $(CH_3NH_3)_2MnCl_4$ exists at atmospheric pressure in four forms, and the compound $(CH_3CH_2CH_2NH_3)_2MnCl_4$ in no less than seven. Some at least of the transitions encountered in progressing from the lowest to the highest-temperature form are of the order-disorder kind, the disorder being, in effect, orientational disorder of the alkyl chains. In general, configurational problems in two dimensions are more amenable to theoretical treatment than those in three dimensions, and it is not surprising that there has been a considerable amount of theoretical work on the transitions in these solids, notably by Blinc and Kind and their collaborators (8,9,10). An essential feature of their treatment is that the order-inducing factor is taken to be the hydrogen-bonding between the NH_3 groups and the halogen atoms of the infinite two-dimensional anions.

The heat capacity of a number of salts of this kind has been measured in the author's laboratory, primarily by Dr.M.A. White, mostly between 10K and 300K. Some of the information so obtained on the transitions in these solids is summarised in Table 1. (Other transitions exist for some of these compounds above room temperature).

R	M	T_t/K	$\Delta S_t/JK^{-1}mol^{-1}$		Ref
CH_3	Cd	164.2	10.8	M → T	(11)
		275-285	0.24	T → O	
CH_3	Mn	94.37	7.76	M → T	(12)
		257.02	0.45	T → O	
CH_3	Cu	229.33	0.50	M → O	(13)
$CH_3CH_2CH_2$	Cd	105.5	13.96	?M → T	(14)
		156.8	3.85	?T → O	
		178.7	6.16	O → O	
$CH_3CH_2CH_2$	Mn	112.8	5.47	M → T	(14)
		164.3	3.29	T → O	
$CH_2 = CHCH_2$	Cd	206.9	10.5		(15)
		266.7	7.87		

Table 1. Results of calorimetric studies of transitions in salts of the type $(RNH_3)_2Cl_4$. T_t= transition temperature, ΔS_t= entropy of transition. Where there is a known crystal structure change, this is indicated (M = monoclinic, T = tetragonal, O = orthorhombic).

Before discussing the implications of the results in Table 1, we may briefly refer again to the use of a baseline heat capacity. With the more complicated salts with which we are now dealing, it is less easy to construct a satisfactory baseline heat capacity-temperature curve from a combination of functions which were originally derived with monatomic solids in mind. A useful alternative is then to use a structurally similar compound as a reference. This approach is not new. Thus, it was used by Stout and his coworkers to separate from the total C_p of a salt such as $CuCl_2$ the magnetic contribution, by using as a reference a similar but diamagnetic solid (e.g. $ZnCl_2$), or else a salt such as $MnCl_2$ in which the magnetic effects occur at much lower temperatures than for the copper compound (16,17). The same kind of method was used in analysing the heat capacity of the layer compounds under discussion, to aid in evaluating the contribution to ΔS_t from the gradual part of a transition, and also in assessing the magnetic contributions to the entropy which are discussed below.

The theory elaborated by Blinc, Kind et al. had as one of its objectives an explanation of the interesting sequence of phases namely (starting from low temperatures): monoclinic → tetragonal → orthorhombic → tetragonal. Four possible orientations were assigned to each cation. The monoclinic phase was considered to be completely orientationally ordered, and the high-temperature teragonal form as being completely disordered, in that each cation makes full independent use of its four possible orientations. Since there are two cations per 'molecule', the expected total configurational entropy gain is 2Rln4, or 23.05 $JK^{-1}mol^{-1}$. A prediction of this theory was that the minimum possible entropy gain at the lowest transition would be Rln4 (= 11.5 $JK^{-1}mol^{-1}$), and that the actual value would be 14.8 $JK^{-1}mol^{-1}$. It will be seen (Table 1) that ΔS_t for the lowest transition in $(CH_3NH_3)_2CdCl_4$ is somewhat less even than Rln4, and that ΔS_t for the corresponding transition in the manganese compound is still smaller. Perhaps there is persistence of local order above the transition. As for $\Sigma(\Delta S_t)$ for all three transitions, the only values available for ΔS_t at the highest transition (which brings the solid back to a tetragonal structure) have been obtained using the DSC technique (18), and are surprisingly small. Thus, for the transition at 484K in the cadmium compound, $\Delta S_t = 0.17$ $JK^{-1}mol^{-1}$, and use of this gives $\Sigma(\Delta S_t) = 11.2$ $JK^{-1}mol^{-1}$, or less than half the predicted 2Rln4. However, it is well-known that many order-disorder transitions, while having an isothermal completion, are partly gradual, and often extend over a considerable temperature range. The entropy gained in the final isothermal stage may only be quite a small fraction of the total. This isothermal component will be measured in the DSC experiment, but the assessment of the contribution acquired in the gradual approach to the completion of the transition demands good C_p values over this region of gradual change, and comparison of this with a reliable baseline. There is no doubt that in this kind of situation there is considerable scope for accurate C_p measurements above room temperature made by adiabatic calorimetry.

In the compounds with R = $CH_3CH_2CH_2$, one would expect that increased flexibility associated with the longer alkyl group would increase the possibilities of disorder, and indeed the manganese compound with R = $CH_3CH_2CH_2$ undergoes no less than six transitions at atmospheric pressure, It therefore does not seem surprising that the entropy gained by the corresponding propyl-ammonium cadmium compound at the first two transitions exceeds that acquired by the methylammonium cadmium salt. On the other hand, $\Sigma(\Delta S_t)$ for the first two transitions in the propylammonium manganese compound is surprisingly low. Also, although the replacement of the $CH_3CH_2CH_2$ group in the cadmium salt by the more rigid $CH_2 = CHCH_2$ group causes the transition temperatures to rise considerably, it does not lead to a drop in the combined entropy increase at the first two transitions.

The copper compound $(CH_3NH_3)_2CuCl_4$ might have been expected to be different in view of the Jahn-Teller distortion, which gives rise to a less regular arrangement of the six chlorine atoms surrounding a copper atom. In fact, between 10K and 300K it only has one transition, with a trivial entropy change. Rahman et al. (19) studied the corresponding chromium compound, in which the chromium atom likewise has a Jahn-Teller distorted environment, and no obvious transition was found. There is, however, a broad anomaly in the heat capacity curve, to which we shall return shortly.

The heat capacity work on these layer compounds has therefore disclosed some interesting features which invite further theoretical consideration and suggest further experimental work (especially structural studies). It is clear from the differences between one compound and another to which we have drawn attention that the metal atom has a considerable influence on the characteristics of the transitions - for example, the temperatures at which they occur, whether they are more gradual than isothermal or vice versa, the values of ΔS_t, and indeed whether a transition will occur at all. No doubt the hydrogen bonding is of fundamental importance, and presumably the effect of the metal atom operates through this - by changes, for instance, in the metal-halogen bond distance and the N-H--Cl distance, and by altering the effective charge on the halogen atoms.

As regards the change from magnetic order to disorder, in the manganese and chromium methylammonium compounds, use of the diamagnetic cadmium compound as a reference showed that the extra C_p, C(mag), for the manganese salt extends from $\sim$20K to $\sim$130K with a maximum at about 60K, the dependence of C(mag) on temperature being that expected for a two-dimensional Heisenberg antiferromagnet. The propylammonium compound is very much the same in these respects. The methylammonium chromium compound is particularly interesting in that it orders ferromagnetically. Here, the analysis of the C_p results suggested that the magnetic entropy of Rln5 is acquired in two stages, the first being the breakdown in three-dimensional order, and the second the loss of the remaining two-dimensional order (19). This latter gives a maximum in C(mag) at $\sim$170K, and is responsible for the 'shelf' in the overall C_p-temperature curve.

ORIENTATIONAL DISORDER IN CRYSTALS

Part of the interest in the layer compounds discussed in the previous Section derives from the possibility of orientational disorder of the cations. The subject of orientational disorder in crystals is one which has attracted considerable interest in the last two or three decades, sufficient in fact to have been assigned a Gordon Conference, held in alternate years. It is a large subject which deals with a wide variety of substances and has many aspects, both experimental and theoretical. Here we shall limit ourselves to just one matter, namely the interpretation of entropies of transition on an order-disorder basis. Many, indeed probably most, of the entropy values at present available for interpretation are the product of low-temperature calorimetry.

If a diatomic or polyatomic molecule or ion passes from a condition in which it has $\underline{n}_1$ distinguishable but energetically equivalent orientations to one in which it has a larger number $\underline{n}_2$ of distinguishable and equivalent orientations, then the entropy gain on this account (the configurational entropy gain) will be $R\ln \underline{n}_2/\underline{n}_1$. Often, of course, $\underline{n}_1$ will be unity. If there were no other contributions to the entropy difference between the two conditions, ΔS_t would then be $R\ln \underline{n}$, where $\underline{n}$ = an integer or a relatively simple fraction. In general, however, the transition will probably involve a structural change, though this may be slight, and an associated volume change, and the consequential change in the phonon spectrum will also contribute something to ΔS_t. Nevertheless, an analysis of the ΔS_t values of transitions which a priori appeared to involve passage from orientational order to disorder showed a clear grouping of the experimental values round $R\ln 2$, $R\ln 3$, and to a lesser extent round $R\ln 4$ [20]. Sometimes, the interpretation of a ΔS_t approximately equal to an $R\ln \underline{n}$ of integral $\underline{n}$ seems straightforward. Probably the most famous case is that of the λ-transition in ammonium chloride. Each ion has two possible orientations in the cubic unit cell. At low temperatures all the ions adopt the same orientation (i.e. are parallel) while above the transition both orientations are used on an equal footing [21].

In general, one might say that the chance of interpreting ΔS_t on an $R\ln \underline{n}$ basis is at its best if the transition only involves a very slight change in structure and a small volume change, and occurs at a low rather than a high temperature. Even if these conditions are reasonably well fulfilled and if ΔS_t is found to be alomst exactly $R\ln \underline{n}$ of integral $\underline{n}$, its physical interpretation need not necesarily be self-evident. Thus, the nitrates of rubidium, caesium and thallium(I) all undergo a transition from an orthorhombic to a cubic, CsCl-like form, passage through which can be accomplished with a single crystal without it shattering. For all three compounds, ΔS_t for this transitions is within two or three per cent of $R\ln 3$. Newns and Staveley [20] suggested that this should be interpreted in terms of positional disorder rather than orientational disorder, while Strømme [22] treated it as a case of orientational disorder, but as a non-simple one in that it was necessary to consider the limitations imposed on the orientation of any one ion by its neighbours - in which case, the very close approach of ΔS_t to Rln3 is, in a sense, fortuitous. It should also be pointed out that the larger $\underline{n}$, the more unreliable may be the interpretation of ΔS_t on the simple basis we have indicated. For example, the difference between Rln24 and Rln32 is only 2.39 $JK^{-1}mol^{-1}$.

Accordingly, attempts to explain observed ΔS_t values in order-disorder transitions on an Rln$\underline{n}$ basis should be viewed with circumspection. But they should not be dismissed as being quite misleading or useless. Another interesting group of salts from the standpoint of orientational disorder are the alkali nitrites, some of which have been studied calorimetrically in the author's laboratory, using a calorimeter covering the range ~80K to ~500K. The triangular NO_2^- ion is, of course, a less symmetrical entity than the NO_3^- ion. A priori, therefore, the configurational entropy gain on passing from a low-temperature form of a nitrite with ordered anions to a high-temperature form in which these ions are disordered would be expected to be larger than for a nitrate. Thus, KNO_2 exists in at least three forms at atmospheric pressure. The low-temperature form III is orientationally ordered, but in phase II the NO_2^- ions are disordered, and in the high-temperature phase I still more so. This was shown by the X-ray diffraction work of Solbakk and Strømme (23), who concluded that in phase II each anion has six possible orientations, while in I there are two non-equivalent positions for each anion, associated with each of which there are 16 orientations. This gives 32 positions altogether for each nitrite ion in I. Rubidium nitrite is in form I at room temperature, and appears to have the same structure as KNO_2-I.

The calorimetric study of KNO_2 (24) showed that the III → II and II → I transitions are both partly gradual (Fig.5), the lower having ΔS_t = 23.54 $JK^{-1}mol^{-1}$, and the upper ΔS_t = 7.01 $JK^{-1}mol^{-1}$. The transitions clearly overlap to some extent. Whatever process is occurring at the II → I transition has, it seems, already started in the lower transition. In fact, ΔS_t for the lower transition is considerably larger than Rln6 (= 14.9 $JK^{-1}mol^{-1}$), while the sum of the two ΔS_t values is about 6 per cent larger than Rln32. The total entropy gain is therefore certainly consistent with the possibility that the anions in phase I make effectively full use of 32 orientations. The single transition found in $RbNO_2$, from II to I, proved to take place predominantly isothermally at 263K with ΔS_t = 32.3 $JK^{-1}mol^{-1}$, which is also larger than Rln32(= 28.8 $JK^{-1}mol^{-1}$)(25). With $CsNO_2$, on the other hand, a careful calorimetric study from 83K to 479K showed only one major transition, which begins gradually and reaches completion isothermally at 208.8K, with ΔS_t having the much lower value of 13.3 $JK^{-1}mol^{-1}$, or Rln7 (26). (There was a minor thermal anomaly at 408K with

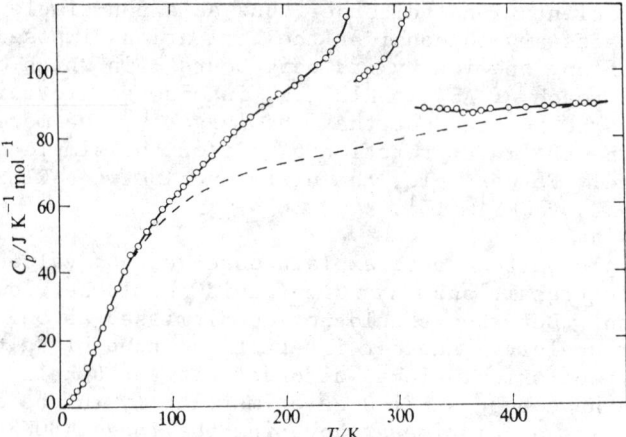
Fig.5. Molar heat capacity C_p of KNO_2 plotted against temperature. The dashed line is the estimated baseline heat capacity curve (24).

$\Delta S_t \approx 0.05$ $JK^{-1}mol^{-1}$). This behaviour was different from that expected from earlier reports, which recorded a transition at 179K and another between 353K and 393K. Perhaps this is a case (others are known) where the polymorphism of the solid is qualitatively affected by the presence of impurities.

A fourth nitrite studied was thallium(I) nitrite, $TlNO_2$, (25). This is an interesting solid - it is bright orange, whereas the ions Tl^+ and NO_2^- do not absorb in the visible. It proved to have one transition between 90K and 444K, which occurred at 282K with $\Delta S_t = 18.76$ $JK^{-1}mol^{-1}$, or Rln9.6. But what is chiefly remarkable here is the high and almost constant value of C_p in phase I (which is a cubic structure). Between the transition and 400K, this is about 105 $JK^{-1}mol^{-1}$, as compared with 91.5 $JK^{-1}mol^{-1}$ for $CsNO_2$, and 85 to 89 $JK^{-1}mol^{-1}$ for $RbNO_2$. The reason for this high heat capacity of $TlNO_2$ has not yet been elucidated, but it is a curious property which merits further investigation and consideration, as does the more general matter of the effect which changing the cation can have on the polymorphism of nitrites.

The author once heard the late Professor Simon (Sir Francis Simon, one of the great pioneers of low-temperature physics) remark that any experiment has been worthwhile if it suggests another worthwhile experiment. From the above brief survey of the heat capacity studies so far carried out on metallic nitrites, it is clear that these studies readily suggest further investigations which might prove to be illuminating and informative - not only thermodynamic investigations, but others based on diffraction or on some of the spectroscopic techniques now available (67). Other systems discussed in this review could also provide examples of such cross-fertilisation, to serve as a reminder that the thermodynamicist should always keep in touch with his non-thermodynamic colleagues.

REFERENCES

1. Andrews, J.T.S., Norton, P.A., and Westrum, E.F.: 1978, J. Chem. Thermodynamics 10, pp.949-958.
2. Clayton, P.R., Staveley, L.A.K., and Weir, R.D.: 1981, J. Chem. Phys. 75, pp.5464-5473.
3. Johnston, W.V., Wiedersich, H., and Lindberg, G.W.: 1969, J. Chem. Phys. 51, pp.3739-3747.
4. Van Kempen, H., Duffy, W.T., Miedema, A.R., and Huiskamp, W.J.: 1964, Physica 30, pp.1131-1140.
5. Van Kempen, H., Garofano, T., Miedema, A.R. and Huiskamp, W.J.: 1965, Physica 31, pp.1096-1106.
6. Pitzer, K.S.: 1937, J. Chem. Phys. 5, 469-472.
7. Blinc, R., Kozelj, M., Rutar, V., Zupancic, I., Zecs, B., Arend, H., Kind, R., and Chapuis, G.: 1980, Faraday Discussions of The Chemical Society No.69, Phase Transitions in Molecular Solids, pp.58-65.
8. Blinc, R., Burgar, M., Lozar, B., Seliger, J., Slak, J., and Rutar, V.: 1977, J. Chem. Phys. 66, pp.278-287.
9. Blinc, R., Zeks, B., and Kind, R.: 1978, Phys. Rev. B17, pp. 3409-3420.
10. Kind, R., Blinc, R., Zeks, B.: 1979, Phys. Rev. B19, pp. 3743-3754.
11. Rahman, A., Clayton, P.R., and Staveley, L.A.K.: 1981, J. Chem. Thermodynamics 13, pp.735-744.
12. White, M.A., Granville, N.W., and Staveley, L.A.K.: 1982, J. Phys. Chem. Solids 43, pp.341-349.
13. White, M.A., and Staveley, L.A.K.: J. Phys. Chem. Solids, in press.
14. White, M.A., Granville, N.W., Davies, N.J. and Staveley, L.A.K.: 1981, J. Phys. Chem. Solids 42, pp.953-965.
15. White, M.A. and Staveley, L.A.K.: J. Chem. Thermodynamics, in press.
16. Stout, J.W., and Catalano, E.: 1955, J. Chem. Phys. 23, pp. 2013-2022.
17. Stout, J.W., and Chisholm, R.C.: 1962, J. Chem. Phys. 36, pp.979-991.
18. Tello, M.J., Arriandiaga, M.A., Fernandez, J.: 1977, Solid State Commun., 24, pp.299-302.
19. Rahman, A., Staveley, L.A.K., Bellitto, C., and Day, P.: J. Chem. Soc., Faraday Transactions, in press.
20. Newns, D.M., and Staveley, L.A.K.: 1966, Chem. Rev., pp. 267-278.
21. Callanan, J.E., Weir, R.D., and Staveley, L.A.K.: 1980, Proc. R. Soc. Lond. A372, pp.497-516.

22. Strømme, K.O.: 1971, Acta Chem. Scand. 25, pp.211-218.
23. Solbakk, J.K., and Strømme, K.O.: 1969, Acta Chem. Scand. 23, pp.300-313.
24. Mraw, S.C., Boak, R.J., and Staveley, L.A.K.: 1978, J. Chem. Thermodynamics 10, pp.359-368.
25. Boak, R.J., and Staveley, L.A.K. Unpublished results.
26. Mraw, S.C., and Staveley, L.A.K.: 1976, J. Chem. Thermodynamics 8, pp.1001-1007.
27. Parsonage, N.G., and Staveley, L.A.K. Disorder in Crystals, The Clarendon Press, Oxford, 1979, Chap.4.

THERMODYNAMICS OF CRYSTALS - 1982[1]

Edgar F. Westrum, Jr.

Department of Chemistry, University of Michigan, Ann Arbor, Michigan 48109, U.S.A.

After a terse review of highlights in recent calorimetric developments in solid state calorimetry and the crises in Gibbs energy at phase and especially melting transitions, as well as the currently much pursued study of orientational disorder in crystals ("ODIC"), several systems of quite diverse nature will be considered in depth. These will include:
- *the plastically crystalline state*
- *the π - π molecular complexes — typical molecular crystals with sandwich-stacked structures and interesting vibrational orientational disorder.*
- *the uranium intermetallic compounds UM_3 where elucidation of electronic structure and delocalization of electrons are in question, and*
- *the salts of organic acids with much metastability and many surprises.*

INTRODUCTION

Both the molecularly-crystalline state and the ionic-crystalline state are among the relatively neglected areas of modern solid-state science. The abrupt changes which occur in the orientational disorder and/or in the molecular freedom of a molecule or an ion in such crystalline solids are the most fascinating and still the least understood of physical problems. The transitions which often accompany these events are dramatic changes in the physical characteristics of the substances which are disproportionate to the subtle structural metamorphoses which occur at the molecular level. Although many experimental techniques do provide insight into the nature and mechanism of molecular freedom in crystalline substances, calorimetric studies of heat capacity and of related phenomenon have contributed much to the understanding of molecular motion and of transitions as well as in the revalation of the energetic spectrum itself of condensed matter. An important by-product for technology of such studies are the derived quantities such as the transitional entropy increments and the thermodynamic functions. However, heat capacity determination very often profits by adjuvant measurements, either equilibrium, neutron diffraction, vapor pressure, crystallographic structures, etc., etc., in completing the appreciation and understanding of the crises in the Gibbs energy which occur at transitions.

Interest in the orientational disorder in crystals is held by a fraternity of scientific workers known as ODIC (Orientational Disorder in Crystals). At least some 400 people claim a predominant interest in this subject and receive regularly the ODIC Newsletter. Heat capacity proves to be a particularly revealing parameter of energetics at the molecular and crystalline level as the temperature drops well below the thermal chaos. Yet, even at high temperatures—as shown in another paper (1)— the Schottky contribution and heat capacity can be identified and resolved. At temperatures where statistical effects cloud the issue, the entropy derived from the heat capacity—for example, that of an order-disorder transition — supplants it as the significant and useful criterion. Moreover, in the higher-temperature regions, characteristic of much current scientific and technological endeavors, the entropy may even become the dominant factor in the Gibbs energy equation since its contribution is multiplied by the absolute temperature.

To reduce the magnitude of the hundreds of publications which might be cited and which bear on the title topic to manageable proportions, this review does tend to emphasize the special interests of this author, and by a thoroughly arbitrary decision, minimizes the coverage of recent theoretical developments, correlations, and calculations. It excludes any attempt at being truly encyclopedic in scope. Further papers showing the development of the topic may be mentioned as (2-6). The excellent book by Wallace entitled "Thermodynamics of Crystals" deals with other aspects of this problem (7).

Samples of four recent rather diverse crystalline state problems are cited to show the diversity of structures that can be elucidated by heat capacities and adjuvant measurements. One of these—the plastically-crystalline state—has been an area of considerable current interest from the ODIC viewpoint.

THE PLASTICALLY-CRYSTALLINE STATE

The nature of the so-called plastically-crystalline state can be best appreciated if we contrast it with certain other phases, some of which may be confused with it. In fig. 1 we see a simple phase diagram of the temperature dependence of the properties of a normal hydrocarbon, n-pentane, and the transitions identified as T_m (temperature of melting), T_b (the temperature of boiling at normal pressure) and in the second part of the figures, the value, T_r, (temperature of transition—usually solid-solid).

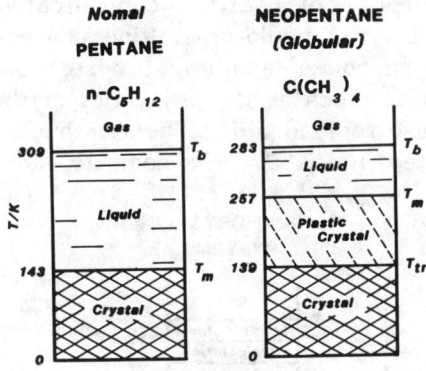

Fig. 1. Schematic phase diagram for n-C_5H_{12} and tetramethylmethane $C(CH_3)_4$.

Although the molecular weight of both isomers is the same, phase behavior is complicated in the second compound— neo-pentane because at the approximate temperature which the n-pentane melts there here exists a transition from the crystalline phase to the plastically-crystalline phase and this phase does not melt until a temperature nearly twice as high is reached. The very characteristic thermodynamic properties of plastically-crystalline phases may be summarized as follows. The macroscopic properties are:
- Large ΔH_t and ΔS_t for transitions
- Small ΔH_m and ΔS_m for melting
  $\Delta S_m/R$ 2.5
- "Tacky", isotropic crystals
- Volatile solid
- Plastic (i.e., extrudable)
- Diffusion in solid

and the microscopic properties are:
- Close-packed crystal structure
- Highly symmetrical structure
- Molecules reorient ("flip") ca. 10^{13} sec^{-1}

The plastically crystalline phase should not be confused with the vitreous phase—which is readily produced by rapid cooling—of many molecular and ionic crystals. Here the liquid phase assumes a certain rigidity at T_g (the glass temperature) without significant isothermal desorption of heat and the resultant <u>vitreous</u> phase is characterized by an essentially amorphous structure and the absence of long-range order. It also remelts without significant isothermal absorption of heat. The plastically crystal phase should not be confused with the liquid-crystalline phases (the name of which is somewhat of a misnomer and might better be described as <u>crystalline-liquid</u> phases in as much as the properties of these phases are essentially 90 per cent anisotropic liquid). The vitreous and liquid-crystalline behavior are depicted in fig. 2. The nematic, smectic, and cho-

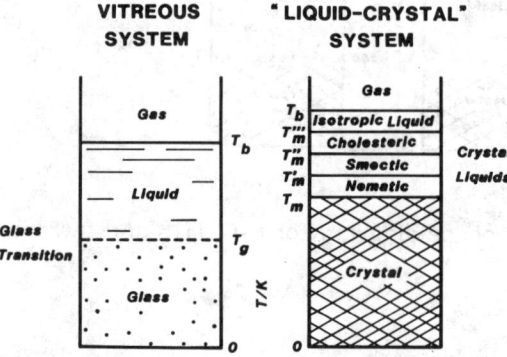

Fig. 2. Schematic phase diagram of a glass-forming substance and a liquid-crystal. The cholesteric, nematic, and smectic liquids are anisotropic.

lestric liquid phases are all anisotropic; not until some such sequence of phases has appeared does the truly liquid isotropic phase occur. The major heat effect, however, is involved in the transition from the crystalline to the first anisotropic liquid; the higher temperature transitions involve but a few per cent of the enthalpy increments of the former.

From an historical point of view the plastically crystalline phase is one identified originally by Prof. Timmermans in the 1930's (8) .Concern about this phase was largely lost until after World War II when it was revived in England with a 1961 symposium at Oxford University initiated by Prof. Staveley (9). Timmermans enunciated a rule of thumb definition that a plastic crystal was composed of globular molecules for which the entropy of melting ($\Delta S_m/R$) was less than 2.5. If we adopt the Planck expression

$$\Delta S = R \ln (N''/N'''),$$

the magnitude of the entropy of the plastically-crystalline transition is typically such that it would correspond to about 10 distinct orientations in the plastically-crystalline phase. Unlike ordinary molecular crystals for which the entropy of melting is related to the number of atoms in the molecule, a great many of these globular molecules have much smaller entropies of melting than might be expected and indeed the communal entropy of melting is very often approached. These crystals do lack long range order in the solid state, flow easily under stress, and tend to show diffusion rates about 10 000 times as high as those in ordinary molecular crystals. This is ascribed to the occurrence of relaxation around the vacancies in the plastically crystalline phase yielding a small heterophase region in which the molecules move cooperatively.

Instrumental Interlude

Two instruments which have best enabled us to study the transitional behavior in molecular crystals are the cryogenic adiabatic calorimeter shown in fig. 3 and the superambient-temperature calorimeter pictured in fig. 1 in another paper in this volume (10). The former calorimeter provides data from about 5 to 350 K and the latter overlaps this range somewhat and is typically utilized between 300 and about 650 K so that it can conveniently provide data on the heat capacity and thermophysical functions well into the liquid phase. Measurements made by adiabatic calorimetry provide equilibrium data as distinct from a differential thermoanalysis (DTA) and differential scanning calorimetry (DSC) which in many situations involving hysteresis do not provide equilibrium data. The achievement of true thermal equilibration after an energy input in a transition region may be a matter of hours or even a day's wait for equilibrium, and the adiabatic calorimeter enables one to make measurements even in the presence of such hysteresis.

Succinonitrile: a Typical Globular Molecule

An impression of the thermodynamics of a plastic crystalline molecule may be received by considering the thermodynamic data obtained by adiabatic calorimetry on this substance (11). The heat-capacity curve is char-

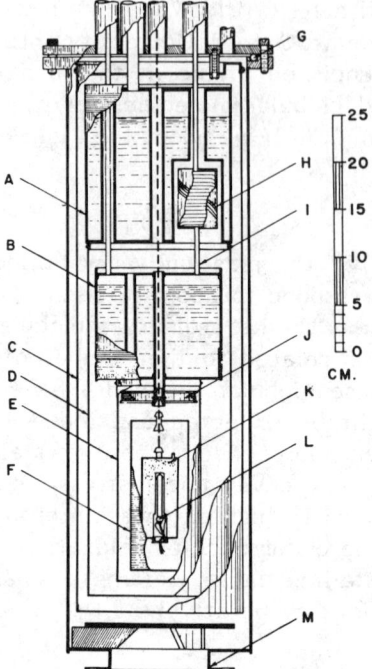

Fig. 3. Schematic cross section of modern cryostat for low temperature adiabatic calorimetry: A, liquid nitrogen tank; B, liquid helium tank; C, D, and E, radiation shields; F, adiabatic shield; G, O-ring gasket sealing brass vacuum jacket; H, effluent helium vapor exchanger ("economizer"); I, helium exit tube; J, ring for adjusting temperature of leads; K, calorimeter assembly; L, platinum-resistance thermometer; and M, connection to vacuum diffusion pump.

acterized by the presence of two first-order transitions (1), the lower one which involves an order-disorder transition at 223.31 K and the onset of the plastically crystalline phase and is followed by melting at a temperature of 331.30 K as shown in fig. 4.

The melting data on this compound in table 1 are of considerable interest. The small magnitude of the entropy increment is particularly noticeable, and it is interesting to observe that the data in best accord with the recent adiabatic calorimetric value are the oldest data on record; subsequent values are much poorer. The solid phase is clearly a plastic crystal.

The third-law entropy of gaseous succinonitrile—based primarily upon the experimental data shown in fig. 4, but admittedly incorporating adjuvant data—is given in table 2 and represents an important by-product of our

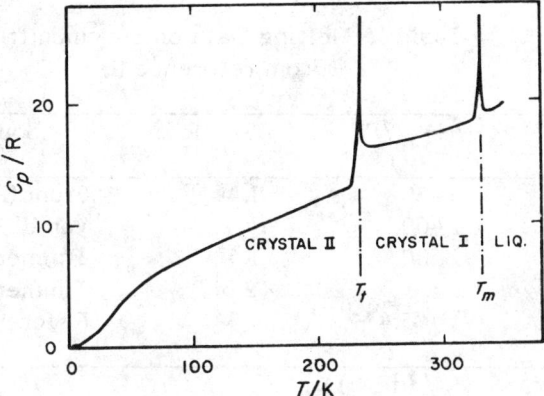

Fig. 4. The heat capacity of succinonitrile showing the plastic-crystal transition and melting.

study of transitions. The comparison with the calculated spectroscopic entropy of gaseous succinonitrile presented in table 3 is in excellent accord with the third-law value of table 2 and thereby evinces the absence of zero-point entropy in this substance (12).

Even more germane is the analysis of the entropy increment of the plastically crystalline transition presented in table 4. Because at and above the transition temperature the gauche molecules are converted to trans-and gauche-isomers with the structure shown in fig. 5, the entropy of mixing of these isomers must be taken into account. In other respects the calculation of the entropy of the order/disorder transition is admittedly a somewhat naive one, but in spite of this, it does result in a calculated value which is in excellent accord with that observed in the calorimetry. A more sophisticated analysis of some aspects of this transition is to be found elsewhere (13). The excellence of the accord obtained here is typically that shown in the case of other plastically-crystalline substances and shows at once both the precision and the power of the method. Molecules which are even more symmetrical show an enhanced behavior.

Unlike liquid crystals which are industrially important, plastic crystalline behavior seems not to have been found useful in practical or commercial applications — despite their scientific interest.

Almost-symmetrical-top Molecules

Carboranes. These molecules with ten boron atoms and two carbon atoms on an essentially spherical surface are superb plastic crystals and are particularly interesting because of the increase of the dipole moment and

Table 1. Melting Data on Succinonitrile
(From reference 11)

T_m/K	$\Delta H_m/R$	$\Delta S_m/R$	Source
327.5	470	1.44	Bruni (1909)
330	1 400	4.25	Van de Vloed (1939)
330.0	500	1.51	Timmermans
330.0		2.31 [a]	Timmermans
331.30	445.4	1.34	Reference 11

[a] From $S = V(dP/dT)$.

Table 2. The Third-law Entropy of Gaseous Succinonitrile at 298.15 K
(From reference 11)

(T or T_{range})K	Contribution	S_1/R
0-5	Debye extrapolation	0.01
5-233.31	Crystal II	15.72
233.31	Transition	3.19
233.31-298.15	Crystal I	4.12
298.15	S, Crystal I	23.04 ± 0.02
298.15	Sublimation	28.24
298.15	Compression to 1 atm.	−11.50
298.15	Ideal gas correction	0.00
298.15	S, Real gas (IIIrd Law)	39.78 ± 0.05

Table 3. Spectroscopic Entropy
of Gaseous Succinonitrile at 298.15 K [a]
(From reference 11)

Contribution	$\Delta S_t/R$
Translation and rotation	32.19
Vibration	4.29
Internal rotation	2.55
Entropy of mixing trans and gauche isomers	0.77
S, Real gas (spectroscopic)	39.80 ± 0.05

[a] The values are for the equilibrium mixing of isomers.

Table 4. Analysis of ΔS_t for Plastic-Crystal
Transition in Succinonitrile
(From reference 11)

Factors	ΔS_t
Alignment of axes	1.386
Rigid rotation of molecule	0.693
Entropy of mixing isomers	1.045 ± 0.02
Volume change on transition	0.07 ± 0.03
Calculated: ΔS^o_t	3.19 ± 0.05
Observed: ΔS^o_t (calorimetric)	3.19 ± 0.02

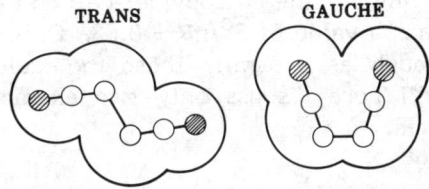

Fig. 5. The molecular structure of succinonitrile showing gauche and trans isomers.

the symmetry changes as one goes from the para- to the meta- and from thence to the ortho-compound. Their thermodynamics reveal very interesting transitional aspects (14).

Bicycloalkanes, etc. (15,16) The bicycloheptanes, bicyclo-octanes, and the nitrogen, phosphorus, and sulphur derivatives of these molecules present an interesting series in which the presence of one or more double bonds can be used to perturb the symmetry in the achievement of tests of the effects of theories on thermal behavior. Since there is as yet no completely unifying theory which accounts quantitatively for the transitions—systematic investigations of carefully selected systems, quantitative establishment of the role of molecular crystal symmetry, of the nature of the motion of other internal groups or of whole molecules at lattice sites, and the effect of volume are certainly desiderata. Summaries of these studies may be found in other references (17,18).

CONDENSED POLYNUCLEAR AROMATIC HYDROCARBONS

Among the interesting species of molecules provided by nature, the hydrocarbon derivatives of benzene provide a very interesting series. No doubt, other derivatives would prove equally or more interesting, but very little data presently exists. They do exemplify very nicely the Planck relation noted earlier, the theory devised by Guthrie, and McCullough (19) and a more recent interpretation provided by a group in Belfast (20). Their packing, reorientation, and thermophysical properties have been widely discussed (e.g., 17, 21-23). These "disc", "pancake", or "flat-stacker" molecules have been extensively studied and time permits a discussion of only a limited number of these.

However, let us note that if the spatial extension represented by these planar molecules is extended to infinity in two dimensions, the ultimate result is a molecular layer of graphite itself. Insofar as the series has been studied, the asymptotic limit of the entropy per carbon atom might well be that of graphite which has a value of $S^{\circ}/nR = 0.68$. Only benzene seems to be deviant (24). Although as a family these molecules are worthy of further interest, we will here discuss only two examples—azulene, and charge-transfer complexes.

Azulene

The controversy among early crystallographers as to whether or not the azulene crystal was disordered at 300 K was resolved in that the disorder seemingly arose from a random distribution of azulene molecules at 300 K among two possible orientations at each lattice site differing from each other by 180° rotation about an axis perpendicular to the plane of the molecule (fig. 6). Gunthard (25) proposed this model on the basis of x-ray diffraction studies and supported it by vapor-pressure measurements and

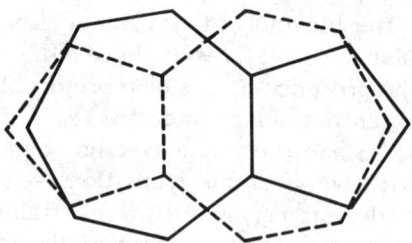

Fig. 6.   The azulene molecule with a second molecule rotated 180° and superimposed.

gas-phase entropy values from infra-red spectral data. He suggested the probable existence of an R ln 2 contribution to the entropy of the azulene crystal probably resulting from a transition from a low-temperature, ordered phase to a high-temperature disordered one. Such a transition should manifest itself in the heat capacity versus temperature curve. We measured the heat capacity of azulene over the temperature range 5 to 400 K by means of adiabatic calorimetry (26) and found no evidence for a cooperative transition other than that of melting. Our tests for residual entropy at zero Kelvin using vapor-phase entropies of Kovats et. al. (27) and the vapor-pressure data of Bauder and Gunthard (28) seemed to indicate zero entropy at zero Kelvin within the reliability of the data. Hence it appeared that the disorder present at 300 K is absent at very low temperatures.

The comparison of the heat-capacity values for azulene with those of its isomer, naphthylene (which has two, six-membered aromatic rings instead of the five/seven combination of azulene) reveals that the curves are very similar with azulene slightly higher than that of naphthylene. The difference is never more than that 0.41 ($\Delta C_p/R$). Most of the excess entropy of azulene over that of naphthylene arises from a gradual onset of orientational disorder in azulene beginning at very low temperature and reaching a value of R ln 2 ($C_2/R = 0.70$) near the triple point of naphthylene and is consistent with both the crystallographic and thermodynamic results. The implicit assumption made is that the lattice contribution to the total entropy is essentially the same for azulene as for naphthylene (in view of their close similarity in molecular and crystal structures). Recent nuclear magnetic resonance studies on solid azulene by Fyfe and Kuppferschmidt (29) were claimed to show azulene molecules in dynamic equilibrium between two alternative orientations (as shown in fig. 6) the motion being a 180° flip about an axis perpendicular to the plane of the molecule. They claimed, however, to see evidence for a transition centered at 310 K and predicted a structure determination below 270 K would show a single molecular orientation rather than a distribution among two. No evidence of such a transition is found in the study of the infra-red spectrum by Van Tets and Gunthard (30) nor in the heat-capacity measurements just discussed.

Serious errors do exist in the literature relative to the entropy of melting of azulene based on vapor-pressure studies and differential scanning calorimetry, and, moreover, considerable uncertainty still exists in the normal coordinate analysis of the spectrum of azulene and in the vapor-pressure measurements. We hope, however, to resolve this problem eventually by a third-law analysis and to understand more fully the nature of the disorder present in this molecule.

Charge-Transfer Complexes

The ambient temperature structure of more than fifty crystalline π-π molecular donor-acceptor compounds involving aromatic hydrocarbons and various electron acceptors have been made. They have been the subject of many structural, spectroscopic, and theoretical studies, and recently a few of them have been studied also by calorimetry (fig. 7). The fourth and fifth items in the table have been studied elsewhere (31,32), and all except the fourth have been studied in our laboratory (33-36). These very interesting molecules have a characteristic "sandwich" arrangement in mixed stacks with an alternating donor-acceptor sequence along the stack axis which is parallel to the C-axis of the unit cell, and the planes of the donor and acceptor molecules are nearly parallel but inclined slightly to the C-axis.

Many of these compounds have been studied by nmr spectra (37,38). Their spin-lattice relaxation times (39) as a function of temperature, show that the disorder indicated by diffraction studies may be of a dynamic rather than a static nature. In any event, two distinct kinds of motion are observed for the naphthylene compounds created at different temperatures. Naphthylene molecules are considered to occupy two orientations related by an in-plane rotation about the C-axis. At low temperatures, the molecules flip over an acute angle from one orientation to the other. As the temperature increases, the molecules begin to flip over the supplementary angle as well. This results in complete in-plane reorientation of the naphthylene molecules. However, as the angle between the two orientations becomes smaller it becomes more difficult for the nmr technique to distinguish between the occurrence of the small-angle flipping between the two sites and the motion in which the molecule sits at one orientation and undergoes a large in-plane vibration. For example, for N-TCNB it appears certain that the molecules occupy two sites separated by about $36°$ and that the molecules switch from one orientation to the other by jumping over the small angle at 77 K. In N-TCNE where the motion must be over about a $17°$ angle to account for the observed second moment, a large-amplitude vibration is also a possible explanation. The heat capacity measurements can be utilized in making a decision between the two modes of motion in charge-transfer complexes and in determining the temperature at which the motion begins. If the molecular motion arises from small-angle flipping between two distinguishable orientations, then disorder is introduced into the crystal. The onset of a disordering process is usually observed in a heat-capacity curve as a transition with an entropy of the order of the now familiar $R \ln N$. If, on the other hand, the naphthylene molecules occupy a single orientation and the observed motion arises from large amplitude vibrations, no disorder is present and the heat capacity will show no anomaly.

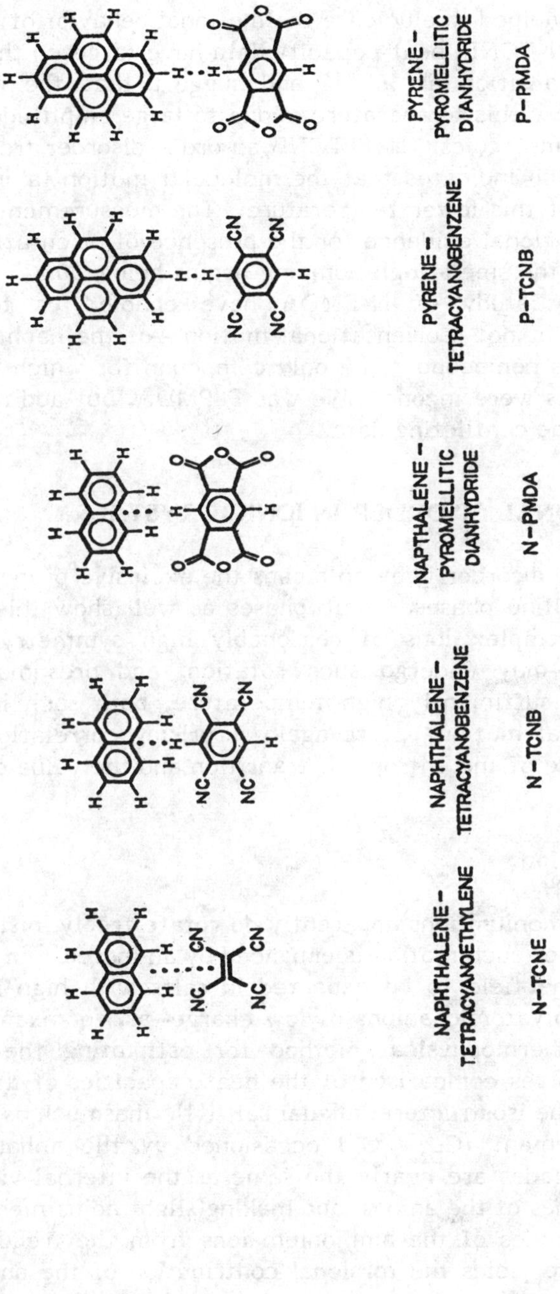

Fig. 7. Structures, names, and codes for the five charge transfer compounds discussed.

Although the calorimetric data are very different for each of the five compounds and the interpretation of it is reasonably complex, it is fair to say that the addition of such thermophysical data to that derived by other methods has helped to elucidate the motional behavior of several molecular crystals. In N-TCNE, heat-capacity data have targeted the occurrence of a structural transition at 160 K and suggest that the molecular motion observed below this temperature is due to large amplitude vibration of the naphthylene molecules. In P-TCNB an order-disorder transition was found at 232 K which indicates that the molecular motion taking place at 300 K must begin at this lower temperature. The measurements on the N-TCNB provided additional evidence for the presence of orientational disorder and suggest that the small-angle jumping begins below 50 K. The results of the heat-capacity study of N-PMDA have enabled us to conclude that vibrational—but not reorientational motion —of the naphthylene molecules occurs in this compound. The only compound for which the heat-capacity measurements were inconclusive was P-PMDA, but additional studies may yet resolve the conflicting data.

ORIENTATIONAL DISORDER IN IONIC CRYSTALS

Orientational disorder is by no means the exclusive property of the molecularly crystalline phases. Ionic phases as well show this property particularly for complex ions of reasonably high symmetry. Either ion—if symmetrical—may undergo such rotation, and presumably in favorable instances at sufficiently high temperature, both such ions may undergo reorientational motion. Here again surprising correlation exists between the magnitude of the entropy of transition and the value of the Planck R ln (N"/N') (40).

Symmetrical Ions

Although ammonium ions apparently do rotate freely in simple halide salts, the prospect of such motion is enhanced by an increase in the uniformity of the crystalline field to be expected in salts with high lattice symmetry containing polyatomic anions of low charge—as, for example, in NH_4ClO_4 (41). The thermophysical method for estimating the extent of ionic freedom involves comparison of the heat capacities of the ammonium salt and that of the isostructural alkalai salt (41). Inasmuch as the contributions to the increment $(C_p - C_v)$ occasioned by the anharmonicity of the vibrational modes are nearly the same as the internal vibrational and the torsional modes of the anions, and making slight adjustment for the internal vibrational modes of the ammonium ions from the frequency assignment, the difference yields the torsional contribution of the ammonium ions. In the instance mentioned, the measured heat capacity of ammonium perchlorate compared with that of potassium perchlorate lead to the conclusion that the rotation of the ammonium ion in this salt is opposed by a potential energy barrier $\Delta E/R \sim 5000$. The combination of the thermophysical values with ammonium perchlorate aqueous thermochemical data indicates the absence of both zero Kelvin entropy and disorder.

Alkali Hydrogen Difluorides and Azides

The alkali hydrogen difluorides have a very interesting pattern of orientational disorder involving reorientation of the <u>linear</u> (FHF)⁻ ions for ordered perpendicular positions at right angles in the plane perpendicular to the C-axis through a random orientation of these ions along the four cube diagonals of the cells. The crystals of rubidium hydrogen difluorides are similar in their behavior; although the cesium compound has the same low-temperature structure, its high-temperature structure is one in which these ions line up along the three cartesian axes of the cubic crystal.

The potassium compound at ambient temperatures is a normal inorganic white crystal. At the temperatures even slightly above the 413 K transition temperature it is a waxy, translucent solid behaving very much like a block of paraffin from which curliques can be cut with a scalpel.

In the case of the cesium compound, the anions in adjacent cells along a particular cartesian direction may not similarly orient along this line without interference and so the entropy increment is slightly smaller than would be calculated for random order. Just below melting a transition to yet another unknown phase occurs in this compound. These phenomenon have been studied by a variety of techniques including x-ray diffraction, heat-capacity measurements (42), as well as by high-pressure studies (43), and excellent confirmation of the behavior has been obtained by the adjuvant techniques. Moreover, the corresponding azides (possessing linear, essentially comparable in mass and size (NNN)⁻ ions) show exceedingly similar behavior both as a function of temperature and pressure throughout the several phases (44). Details of the transitions and of the corresponding entropies of both sets of isostructural compounds at 300 K show very close parallels under constraints of both temperature and pressure.

URANIUM INTERMETALLIC COMPOUNDS

The thermodynamic properties of actinide-intermetallic compounds are technologically useful since it has been found that metallic inclusions in nuclear fuel formed during fission in the fast breeder reaction consist of such high-stability alloys. The uranium intermetallics are also of interest because of their thermophysical properties. The 5f-electrons of the actinides exhibit behavior intermediate between that of the four itinerant d-electrons of the transition metal and the localized 4f-electrons of the lanthanides. Moreover, the 5f-electrons are of considerable interest in the study of magnetism and of crystalline electric-field (CEF) levels by optical and neutron spectroscopy. For these reasons we studied compounds of the formula UMe_3 in which Me = Ru, Rh, and Pd. The results of measurements on the first two compounds were of interest primarily in establishing the nature of the Fermi level of electrons and the broad 6d-5f hybridized band which overlaps the Fermi level (45). For UPd_3 the large spatial extent of the 5f-electrons gives them a more important role in bonding than that of the 4f-electrons in the much better understood lanthanide systems. Moreover, the evaluation of band-electronic structures in thermophysical data is a topic urgently needing further elucidation. The heat capacity of UPd_3 supplements that on the other two compounds (46). Resistivity studies have been interpreted in terms of enhanced spin fluctuations and a narrow 5f-band, and these conclusions are not inconsistent with ESR measurements. Neutron inelastic-scattering experiments on UPd_3 revealed several peaks identified as crystalline-field transitions suggesting that 5f-electrons might be localized on the uranium sites and that the $5f^2$ ($^3H_4$) configuration of the U^{4+} is split in the hexagonal field to give a singlet ground state. Measurements of the heat capacity of UPd_3 are shown to exhibit a jump near 7 K flanked by two distinct but small maxima nearby. From neutron inelastic scattering, resistivity, magnetic, and other data the electronic structure emerges as one which sharply contrasts those of URu_3 and URh_3 involving discrete energy levels and localized electrons. From our calorimetric measurements, the confirmation of the above statement arises both from the existence of experimental Schottky levels and from the relatively low value of the conduction electron coefficient near T = 0. Moreover, the essentially linear separation of C_p/R for both URu_3 and URh_3 from the value for UPd_3 above ambient temperature shown in fig. 8 is consistent with the conduction electron coefficient and the localization of electrons in UPd_3.

In this substance uranium atoms occur in both hexagonal and cubic sites and each possesses its own set of crystalline electric field levels, the lower ones of which have been identified by neutron elastic scattering and the higher ones have been delineated by theoretical means (46). Resolution of the heat capacity data by procedures outlined elsewhere (12) and confirmatory methods (46) give a calorimetric Schottky which is in good accord with the Schottky function derived from energy levels as shown in fig. 9. The agreement provides an excellent confirmation of the reliability of the lower energy levels.

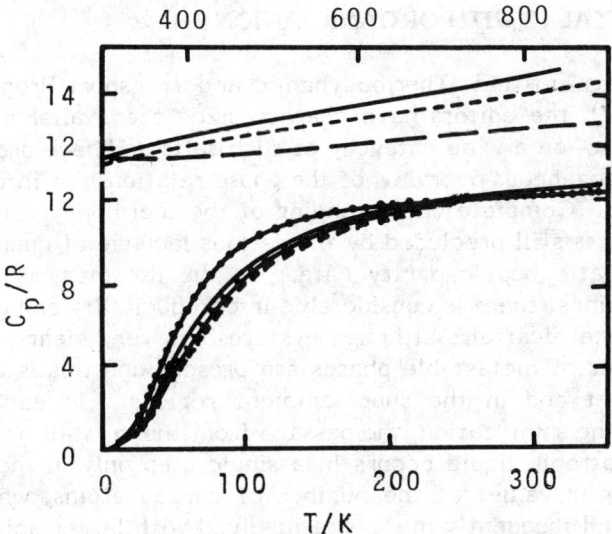

Fig. 8. The heat capacities of the intermetallic compounds UPd_3 (•, – – – –), URh_3 (————), and URu_3 (△, – - – -).

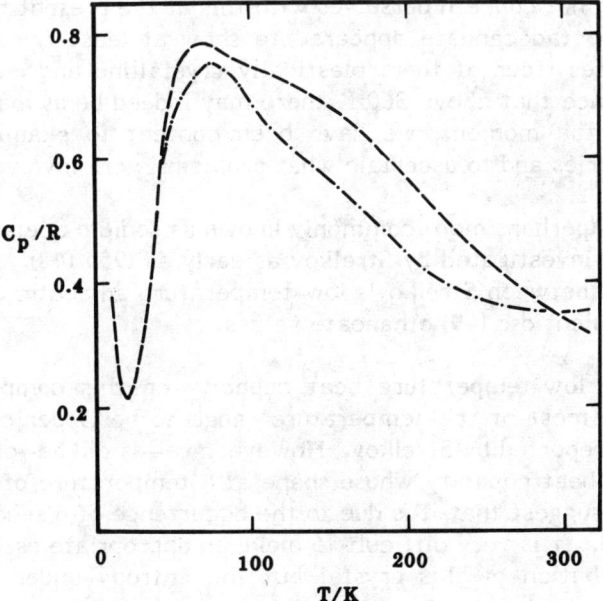

Fig. 9. The Schottky contribution in UPd_3. The curve calculated from spectroscopic and theoretical levels — . — . — . . The curve based on heat capacity data — — — .

TRANSITIONS IN SALTS WITH ORGANIC ANIONS

In the opus magnus entitled "Thermodynamic and Transport Properties of Organic Salts" (47), the editors have characterized the available thermal data—mostly by dsc—on a wide category of such salts and have provided an extensive and homogeneous overview of the phase relationships in the alkali alkanoates to 1979. Complete understanding of the thermophysics of these materials however is still precluded by the serious lacunae of quantitative, equilibrium, adiabatic heat-capacity data. In the dsc measurements on super-ambient regions, there is considerable irreproducibility and confusion as to the true state of affairs although hysteresis is very clearly present. Moreover a number of metastable phases are present and this is true both in the sub-ambient and in the super-ambient regions. In each family characterized by the same cation, the passage from the crystalline solids to a conventional isotropic liquid occurs in a single step only in the case of homologs with lower values of the number of carbon atoms, whereas at least two steps—and frequently more—are involved with larger anions. The additional—largely plastic phases—impose themselves between the crystalline solids, and the anisotropic liquids and mesomorphic liquid phases are also formed. A complicated and to some extent a confused terminology regarding both the phase changes and the designation of the phases has arisen but we need not concern ourselves with this at the present time. For example, sodium octadecanoate appeared to show at least five different mesomorphic phases (four of them plastically crystalline and one liquid), but there is evidence that above 360 K there may indeed be as many as ten transitions. For the moment, we have been content to examine lower members of the series and to ascertain what problems were involved there.

Sodium ethanoate (perhaps more commonly known as sodium acetate) has in fact already been investigated by Strelkov as early as 1955 (48). A serious discrepancy exists between Strelkov's low-temperature adiabatic calorimetric and super-ambient dsc (49) ethanoate values.

We examined the low-temperature heat capacity on this compound and found values over most of the temperature range to be 15 per cent or so higher than those reported by Strelkov. However, we—as did he—observed a sharp peak in the heat capacity whose shape at a temperature of 21 K and easy reversibility suggest that it is due to the occurrence of a second-order transition (fig. 10). It is very difficult to make an appropriate estimate for the lattice contribution of this crystal but the entropy under our peak curve—as seen in part by the actual heat-capacity values— must be approximately an order of magnitude larger than that which he reported much earlier. Inasmuch as he was certainly one of the best Soviet calorimetrists it is difficult to understand the discrepancy between his work and ours.

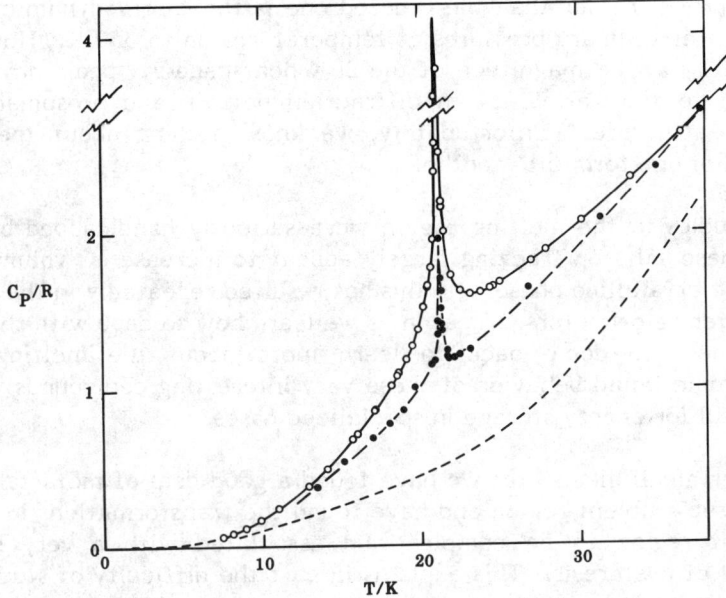

Fig. 10. The 21 K transition in CH_3COONa. Values from our laboratory ———o———; those from Strelkov et al. (48) ———•———. The curve - - - - - represents an estimate of the probable lattice heat-capacity contribution but is not overly reliable. (No explanation for the origin of this transition is known.)

In an attempt to have at least some idea of what might be involved at so low a transition temperature, we persuaded colleagues to undertake a crystal structure of this garden variety molecule on which shockingly little structural information was available. The two phase structures which they found (50) were both prepared by crystallization from aqueous solutions above the peritectic temperature of the trihydrate (i.e., about 68 °C). These two phases—designated Form I and Form II—are both orthorhombic with relatively complex structures. Both phases have sodium ion-carboxylate layers alternating with layers of methyl group contacts, with the C-C bonds lying along the b-axis. Form I with 12 molecules per unit cell has eight molecules in general positions and four on two-fold axes. Form II was refined assuming a partial disorder in all of the oxygen atom positions, but the nature of the disorder was not fully established.

We were able to ascertain from powder patterns that our heat-capacity measurements had indeed been run on Form II. This implies, paradoxically, that Form II rather than the ordered one is the thermodynamically stable form under ordinary pressures at temperatures up to 350 K. However, the variability in the magnitude of the 21 K heat-capacity peak may indeed be related to the variability in diffraction pattern and presumably to the domain structure. Unfortunately, we know nothing about the transformation of one form into another.

Our studies in the melting region were seriously handicapped by the fact that these salts on freezing largely appear to increase in volume in going into the crystalline phase, and this has resulted repeatedly in the rupture of our silver calorimeters. As soon as we learn how to cope with this problem we expect to come back to learn more about the melting and the anisotropic liquid behavior of these very interesting compounds which may be useful for energy storage in specialized cases.

In other alkali alkanoates we have found a good deal of metastability even in the sub-ambient region and have found the transformations in the super-ambient region to be complex and associated with a very significant amount of hysteresis. This explains in part the difficulty of studying these materials by dsc. It is a challenging area and one that we hope to examine further in the near future.

NOTE

1. The author, like virtually all other speakers, took the opportunity at his first presentation to thank the host organization for their kindness and hospitality and to note the excellence of the arrangements. He also expresses his gratitude to the Thermodynamics and Structural Chemistry Section of the Chemistry Division of the National Science Foundation for its continuing support of the research endeavors on which this presentation is based under Grant CHE-8007977, and G EAR-8009538.

REFERENCES

1. Westrum, E.F. Jr.: 1984, this volume, paper entitled "Calorimetry of Phase, Schottky and Other Transitions."

2. Westrum, E.F. Jr., McCullough, J.P.: 1963, in "Physics & Chemistry of the Organic Solid State," Vol. 1, ed. Fox, D. (Interscience) pp. 1-178.

3. Gopal, E.S. Raja: 1966, "Specific Heats at Low Temperatures" (Plenum).

4. Westrum, E.F. Jr.: 1974, Pure Appl. Chem. 38, pp. 539-555.

5. Westrum, E.F. Jr.: 1978, J. Thermal Analysis 14, pp. 5-13.

6. Westrum, E.F. Jr.: 1980, in "Proceedings of 75 Aniversario de la Real Sociedad Espanola de Fisica y Quimica" pp. 239-257.

7. Wallace, D.C. : 1972, "Thermodynamics of Crystals" (John Wiley & Sons).

8. Timmermans, J.: 1935, Bull. Soc. Chem. 44, pp. 17-40, 178-191.

9. Staveley, L.A.K., et. al.: 1961, J. Phys. Chem. Solids.18, pp. 1-92.

10. Westrum, E.F. Jr.: 1984, this volume, paper entitled "Computerized Adiabatic Thermophysical Calorimetry."

11. Wulf, C.A., Westrum, E.F. Jr.: 1963, J. Phys. Chem. 67, pp. 2376-81.

12. Westrum, E.F. Jr.: 1984, this volume, paper entitled "Thermophysics of Mineral and Rock Systems."

13. Descamps, M., et al., personal communication..

14. Westrum, E.F. Jr. and Henriques, S.: unpublished data.

15. Westrum, E.F. Jr., Wong, S.: 1967, in "Thermodynamik-Symposium," Section II, No. 10, ed. Schafer, Kl. (Werbund and Weber).

16. Westrum, E.F. Jr.: 1964, J. Pure and Applied Chemistry 8, pp. 187-214.

17. For summary, see also Reference 6.

18. For details, see also Reference 2.

19. Guthrie, G.B., McCullough, J.P.: 1961, J. Phys. Chem. Solids 18, pp. 53-61.

20. Westrum, E.F. Jr., Wulff, C.A.: 1971, in "Proceedings of the First International Conference on Calorimetry and Thermodynamics" (Polish Scientific Publishers).

21. Ubbelohde, A.R.: 1965, "Melting and Crystal Structure" (Clarendon).

22. Bondi, A.: 1963, "Physical Properties of Molecular Crystals, Liquids, and Glasses" (Wiley).

23. Stull, D.R., Westrum, E.F., Jr., Sinke, G.C.: 1969, "The Chemical Thermodynamics of Organic Compounds" (John Wiley & Sons).

24. Westrum, E.F. Jr. and coworkers, unpublished studies.

25. Gunthard, H.H., Plattner, P.A., Brandenberger, E.: 1948, Experimentis 4, pp. 425-531.

26. Westrum, E.F. Jr., Brink, I., unpublished data.

27. Kovats, E., Gunthard, H.H., Plattner, P. A.: 1955, Helv. Chim. Acta 38, pp. 1912-1919.

28. Kovats, E., Gunthard, H.H., Plattner,P.A.: 1955, Helv. Chim. Acta 38, p. 1912; ibid 1957, 40, p. 2008.

29. Fyfe, C.A., Kupferschmidt, C.J.: 1973, Can. J. Chemistry 51, pp. 3774-3780.

30. Van Tets, A., Gunthard, H.H.: 1963, Spectrochimica Acta 19, pp. 1495-1530.

31. Clayton, P.R., Worswick, R.D., Staveley, L.A.: 1976, Mol. Cryst. Liq. Cryst. 36, pp. 153-163.

32. Dunn, A.G., Rahman, A., Staveley, L.A.K.: 1978, J. Chem. Thermodyn., pp. 787-796.

33. Boerio-Goates, J., Westrum, E.F., Jr., Fyfe, C.A.: 1978, Mol. Cryst. Liq. Cryst. 48, pp. 209-218.

34. Boerio-Goates, J., Westrum, E.F. Jr.: 1979, Mol. Cryst. Liq. Cryst. 50, pp. 249-267.

35. Boerio-Goates, J., Westrum, E.F. Jr.: 1980, Mol. Cryst. Liq. Cryst. 60, pp. 237-248.

36. Boerio-Goates, J., Westrum, E.F. Jr.: 1980, Mol. Cryst. Liq. Cryst. 60, pp. 249-266.

37. Fyfe, C.A.: 1974, J. Chem. Soc. Faraday Trans., II, 70, pp. 1633-1641.

38. Fyfe, C.A.: 1974, J. Chem. Soc. Faraday Trans., II, 70, pp. 1642-1649.

39. Fyfe, C.A., Harold-Smith, D., Ripmeester, J.: 1976, J. Chem. Soc. Faraday Trans., II, 72, pp. 2269-2882.

40. Newns, D.M., Staveley, L.A.K.: 1966, Chemical Reviews 66, pp. 267-278.

41. Westrum, E.F. Jr., Justice, B.H.: 1969, J. Chem. Phys. 50, pp. 5083-5087.

42. Westrum, E.F. Jr., Landee, C.P., Takahashi, Y., Chavret, M.: 1978, J. Chem. Thermodynamics 10, pp. 835-846, and earlier papers in this series.

43. Pistorius, C.W.F.T.: 1976, Progress in Solid State Chem. 11, pp. 1-151.

44. Carling, R.W., Westrum, E.F. Jr.: 1978, J. Chem. Thermodynamics 10, pp. 1181-1200, and earlier papers in this series.

45. Cordfunke, E.H.P. et al. : J. Chem. Thermo. (submitted).

46. Burriel, R. et al.: J. Chem. Thermo. (submitted).

47. Sanesi, M., Cingolani, A., Tonelli, P.L., Franzosini, P.: 1980, in "Thermodynamic and Transport Properties of Organic Salts," IUPAC Chemical Series, No. 28, eds. Franzosini, P., Sanesi, M. (Pergamon) pp. 29-117.

48. Strelkov, I.I.: 1955, Ukr. Khim. Zh. 21, p. 551.

49. Ferloni, P., Sanesi, M., Franzosini, P.: 1975, Z. Naturforsch 30a, p. 1447.

50. Hsu, L., Nordman, C.E.: 1982 submitted to Acta Cryst.

CALORIMETRY OF PHASE, SCHOTTKY AND OTHER TRANSITIONS[1]

Edgar F. Westrum, Jr.

Department of Chemistry, University of Michigan, Ann Arbor, Michigan 48109 U.S.A.

With an eye toward direct estimation of thermophysical properties we consider the morphology of heat-capacity curves and the major contributions thereto (i.e., lattice-vibration, magnetic, and electronic state contibutions) as well as first and second order transitions, order-disorder transitions, etc. Then we note that the besetting problem of resolution of <u>excess</u> heat capacities of many types of transitions--whether structural or electronic, or otherwise--requires a reliable estimate of the lattice contribution. We see that corresponding state theories have been less than satisfactory; but Lindemann schemes are suggestive. The Latimer scheme has been widely used. For chemical thermodynamic purposes use of a volume-weighted scheme over the range where entropy development largely occurs is demonstrated to provide resolution of Schottky contributions for $Ln(OH)_3$ and $LnCl_3$ systems and has been shown to be superior to other approaches. The lanthanide sesquioxides, sesquisulfides, halides, and especially the trihydroxides provide exemplary models in which the Schottky contributions (occasioned by the crystal-field splitting of the ground state manifold) provide an excellent basis for testing the resolution of the lattice heat-capacity contribution, since the Schottky contribution can be evaluated both by spectroscopy and by thermophysical data. Agreement of the excess contribution then provides basis for judging the quality of the lattice contribution resolution.

INTRODUCTION

One of the most exciting events in the measurement of solid-state heat capacities is the occurence of a "transition"—be it phase or otherwise.

Whether described thermodynamically as a crisis in the Gibbs energy—or microscopically in terms of molecular or ionic freedom, the resultant phenomena seems totally disproportionate to the subtle "causes" involved. Although time does not permit a thorough treatment we will gallop through "observing flowers from horseback" and focus mainly on the Schottky contribution to heat capacity—for its own interest—and then use it as a tool to deduce the reliability of a scheme for evaluation of the lattice contribution to the heat capacity—an essential step in analyzing the morphology of a heat-capacity curve.

FIRST-ORDER TRANSITIONS

A first-order or "phase" transition is characterized thermodynamically by a discontinuity in heat capacity (i.e., a heat capacity approaching infinity) or an isothermal adsorption of energy—which used to be described as a latent heat (1). The calorimetric observations deviate significantly from this model in showing premonitory effects (2), a somewhat less dramatic rise, and often unexplained post-transition effects (3). The solid-solid transition in pure NH_4Br shown in fig. 1 and table 1 is an example par excellence of a first-order transition with very high heat capacity and an almost isothermal energy increment (99% over a 0.01 K wide region) (4). That of solid-solid methanol in fig. 2 is perhaps more characteristic (5) but does reveal not only the possibility of undercooling, but that of superheating the transition and the catalytic effect of the impurity (water) on the transition leading to the elimination of both above phenomena. Work by other investigators (who because of impure samples, poor calorimetry, etc., deduced this to be a second-order transition) shows clearly the effect of impurities on pre- and post-monitory heat capacities.

Since even a non-prejudicial single lecture cannot hope to encompass the entire spectrum of topics associated with transition, three monographs are particularly recommended as supplementary sources (6,7,8).

SECOND-ORDER TRANSITIONS

The second order λ-transition earliest known from work on the superfluid (II)/fluid transition in liquid helium, has come of "calorimetric" age in the work of Suga, Seki, and Matsuo (9) on the stannous chloride dihydrate. Using an auxiliary thermistor thermometer and a high quality shield control, they produced some of the most convincing and precise data yet shown on scaling theory—an important aspect of modern day transition theory.

CALORIMETRY OF PHASE, SCHOTTKY AND OTHER TRANSITIONS

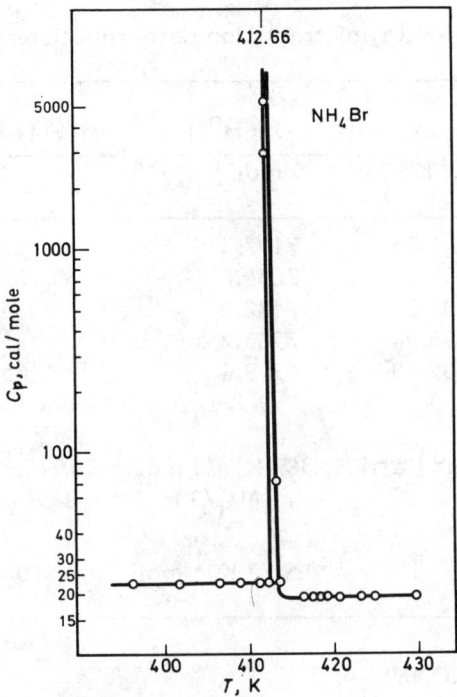

Fig. 1. The NH_4Br solid-solid phase transition—a typical first order transition.

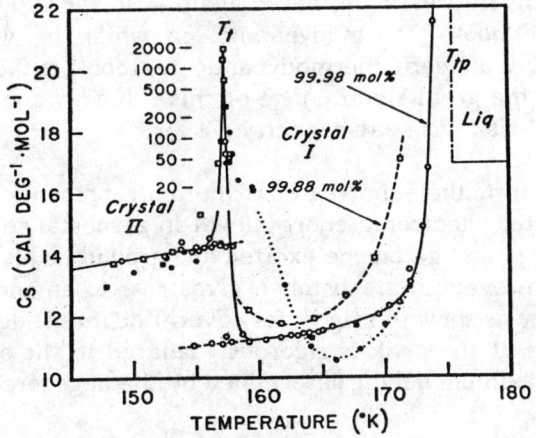

Fig. 2. The heat capacity of methanol in the region of the solid-solid phase transition and melting. Undercooled determinations are shown. Other data on less pure samples are also depicted.

Table 1. Molar enthalpy of transition determinations for NH_4Br

Determination	T_1/K	T_2/K	$H°(T_2) - H°(T_1)$ / J mol^{-1}	$H°(444\ K) - H°(398\ K)$ / J mol^{-1}
A	399.12	441.65	7197.3	8538.3
B	387.92	434.07	7654.6	8542.1
C	389.46	435.01	7582.2	8539.5
D	397.05	443.06	7503.2	8533.3
E	396.58	435.60	6949.6	8535.8
				Mean: 8537.8±3.4

$\{H°(\text{lattice}, 444\ K) - H°(\text{lattice}, 387\ K)\}$ / J mol^{-1} = 4943.4±3.4

ΔH_t / J mol^{-1} = 3594±4

T_t = 412.66 K

ΔS_t / J K^{-1} mol^{-1} = 8.710±0.008

SCHOTTKY "TRANSITIONS"

Werner Schottky, who had received world-wide renown for the Schottky effect and the Schottky defect, also had a third brainchild called the Schottky anomaly which he discovered more than half a century ago, published (10), but, interestingly, never again — so far as I know — referred to even in the book "Thermodynamik" on which he was a co-author. Unfortunately, few modern thermodynamic textbook authors either recognize or mention the great significance of this Schottky anomaly, transition, or "contribution" vis a vis heat-capacity curves.

In its simplest form, the Schottky contribution represents simply the presence of an excited electronic energy level in a substance with respective degeneracies of g_1 and g_0 on the excited and ground levels (cf. fig. 3). The presence of such levels in a substance gives rise to an anomolous shape in the heat capacity as shown in fig. 4 for several different degeneracy ratios. The temperature of the peak is rigorously related to the separation of the levels and the maximum height determined by the degeneracy ratio.

Moreover, the total entropy under the curve is related to the degeneracies involved. At sufficiently low temperatures, the Schottky contribution represents an obvious anomaly on a heat capacity curve. At higher temperatures, the maximum may be so spread out as to be unrecognized unless

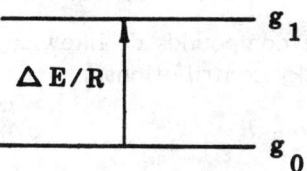

Fig. 3.   The simplest two-level electronic Schottky function.

heat capacities of isostructural substances with nearly identical masses are compared but, nonetheless, it does represent an important aspect of the morphology of heat capacity curves.

Interestingly, Schottky heat-capacity functions, Sch (u), are rigorously related to Einstein heat-capacity functions, Ein (u), by both an additive and a multiplicative identity:

and
$$Ein (u) - Ein (2u) = Sch (u)$$
$$E(u) \, Sch (u) = u^2 \, Ein (2u).$$

A great many other mathematical relationships with Einstein and even with Debye functions can be generated. More complex Schottky contributions with many excited levels are, of course, typically found in compounds. Transition-element compounds characteristically reveal Schottky transi-

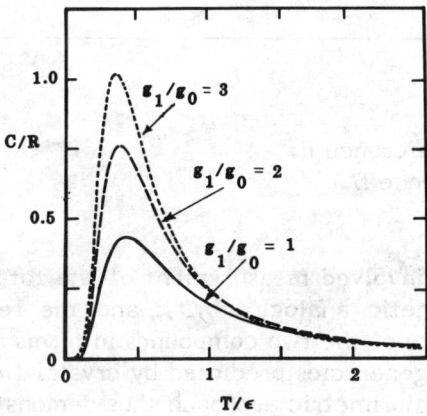

Fig. 4.   Schottky contributions for various degeneracy ratios.

tions, lanthanide and actinide compounds do likewise, and mineral samples also are known to have Schottky contributions.

Lanthanide Sesquioxides

Recognition of the importance of the Schottky thermophysical contribution came with a series of papers by Justice and Westrum (11-15) who studied the lanthanide sesquioxides and obtained an unusually rich yield of data concerning the energetics of the trivalent ions in these compounds. Although the Schottky contributions may also be studied spectroscopically, the general unavailability of single crystal samples for absorption spectroscopy or for paramagnetic resonance experiments had tended to favor the calorimetric approach. Our initiatory measurements (11) on neodymium sesquioxide yielded results in levels more than an order of magnitude smaller than those estimated by Penny (16) from crystal-field splittings. But any discredit was of short duration, since spectroscopy (17) confirmed, in this instance, the values that had been obtained by calorimetry. The quantitative comparison is made in table 2.

Table 2. Stark levels/cm^{-1} for Nd_2O_3

Level	g_i[a]	Calc.[b]	Calorimetry[c]	Spectroscopy[d]
0	2	0	0	0
1	2	492	21	22
2	2	1476	81	83
3	2	2952	400	390
4	2	4920	---	---

[a] degeneracy of level.
[b] Penny, Reference 16.
[c] Justice and Westrum, Reference 11.
[d] Henderson et al., Reference 17.

The method of approach involved measurement of the total heat capacity of Nd_2O_3, of a diamagnetic analog, La_2O_3, and the resolution of the difference in heat capacity of the two compounds in terms of a sequence of Schottky levels of the degeneracies predicted by crystal-field theory. The power of the cryogenic calorimetric approach thus demonstrated was later extended by the same authors to include most other lanthanide sesquioxides (even those containing C_2 and C_{3i} sets of levels) by adjustment of the lattice heat capacity as described later. It was demonstrated that the levels were valid not only for the cryogenic heat-capacity contribution, but as well for temperatures in excess of 1000 K. Unfortunately, the sesquioxides crystallized in A-, B-, and C-forms thus making it harder to recognize the underlying trends. A recent paper extends this fitting and correlation to new electronic Raman scattering data (18) and is discussed later.

Lanthanide Trichlorides

The desirability of extending such studies to other systems led Sommers and Westrum to examine the lighter lanthanide trichlorides (19,20). Their study further heightened the understanding of the trend and regularities involved and showed the importance of the Schottky contribution to the thermophysical functions. The latter were in excellent accord with those predicted by the scheme of Westrum (21) based upon the treatment of Grønvold and Westrum (22). The quality of the accord can be seen in table 3.

Table 3. Comparison of some trichloride entropy estimation schemes

Compound	S (298.15K) - S (0) /cal			
	Latimer	Latimer augmented [a]	Westrum augmented [b]	Experimental Refs. 19, 20
$LaCl_3$	34.5	34.5	33.1	32.88
$CeCl_3$	34.5	38.1	36.1	(36.0) [c]
$PrCl_3$	34.5	38.9	36.8	36.64 [d]
$NdCl_3$	34.6	39.2	36.8	36.67
$PmCl_3$	34.7	39.5	36.8	(37.0) [c]
$SmCl_3$	34.8	38.4	35.7	35.88
$EuCl_3$	34.8	37.3	34.5	34.43
$GdCl_3$	35.0	39.1	36.0	36.19

[a] By R ln(2J+1); the $(Cl_3)^{3-}$ ion contribution is taken as 20.7 cal K^{-1} mol^{-1}.
[b] By R ln(2J+1); the $(Cl_3)^{3-}$ ion contribution is taken as 17.9 cal K^{-1} mol^{-1}.
[c] Parentheses denote interpolated lattice and calculated Schottky contributions.
[d] Based on 0.294 K.

Lanthanide Pnictides

The heat capacities of a set of lanthanide mononitrides has been achieved by Stuttius (23) on materials less well characterized than desirable. We are presently undertaking a reinterpretation of these data. His work is interesting in that he did utilize crystal parameters in interpolating lattice heat capacities across the lanthanide series.

Lanthanide Hexaborides

Several of these compounds have been studied over the cryogenic range by Westrum et al. (24); these studies are interesting and will be examined.

In some instances Schottky levels can be deduced from spectroscopic observations (or from theoretically evaluated crystal-field parameters) and a comparison made between such a "spectroscopic" Schottky contribution and a "calorimetrically" deduced Schottky contribution--but to do so requires resolution of the Schottky contribution from the observed total heat capacity. This requires that we have a suitable scheme to deduce the lattice heat-capacity contribution. A suitable scheme has not been available so in the next section we discuss the utilization of such a comparison--first to test the reliability of a particular scheme and second the resolution of Schottky functions by this so-called "volumetric scheme."

TEST OF LATTICE HEAT-CAPACITY CONTRIBUTIONS VIA SCHOTTKY CONTRIBUTIONS

Because the heat capacity of a system is determined by the energy levels of that system, heat-capacity measurements themselves are often used to deduce those energy levels. Very often one has a model of a system and can calculate the resultant excess heat capacity associated with the model. Unfortunately, the heat capacity of interest is often superimposed on a background contribution and the separation of the excess contribution must be made from that of the background. All too often, this resolution is made by good judgment only. The relevant background contribution is usually the so-called lattice contribution and inasmuch as it typically represents 80 to 100% of the total measured heat capacity, if it is going to be estimated, the procedure must be one of relatively high precision. Consequently, whenever resolution of transitions (be they magnetic, structural, Schottky, order-disorder) is to be done, one is confronted with the necessity of an evaluation of the lattice contribution. When estimates are made of the heat capacity and/or the thermodynamic functions of substances, the lattice represents the bulk of the quantity. Moreover, when adjustment of thermophysical properties, e.g. of minerals or of petrological materials is involved, the same problem arises. Consequently, an urgent need exists for evaluation of the lattice contribution.

We consider that we have recently made significant progress in this matter and that the volume of the crystal over an important portion of the temperature scale is clearly involved. Nonetheless, mass, coordination number, and structure-type are also relevant parameters.

The Grønvold-Westrum Scheme

Historically, the story begins with the 1961 Grønvold/Westrum paper on the correlation of entropies for transition-element chalcogenides (22) and the graphical representation of the surprising result at a 1977 Moscow Plenary lecture (25,26) in which this endeavor was depicted against the Latimer (27,28) scheme for entropy contributions for cations in fig. 5. Latimer considered that these contributions were proportional to the logarithm of the mass of the cation. Three aspects of our findings surprising to me were

- The main stream 3d, 4d, and 5d-transition-element-cations all show the same contribution to the cationic entropy contribution rather than a Latimer-like mass dependence (27,28). Borderline transition-element chalcogenides are possibly intermediate.

- The entropy values for 4f (lanthanide) cations are essentially diametrically perpendicular to the mass dependence predicted by the Latimer scheme suggesting that although masses and volumes change by roughly the same fractional magnitude through the system, the heat-capacity trend here clearly favored volume rather than mass dependence.

- Although the first surprise hides much of interest, the second is the one of immediate concern and shows clearly the consequences of the lanthanide contraction.

On the basis of these abstracts, we have chosen to study lanthanide systems, first the trichlorides, $LnCl_3$ (19,20,29) and then the trihydroxides, $Ln(OH)_3$ (30-35), which have the great advantage of being isostructural throughout the series.

In the discussion which follows, frequent reference is made to Schottky contributions—interesting in their own right—but used here initially and primarily as a means of testing the success of the evaluation of the lattice contribution.

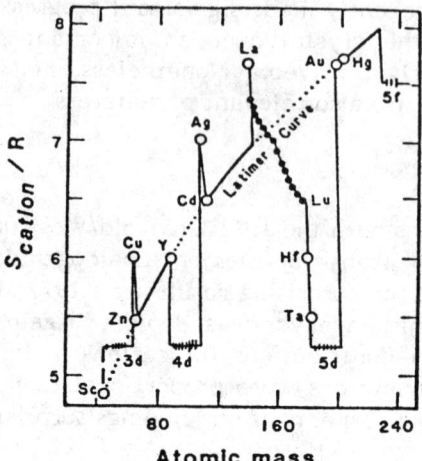

Fig. 5. The Grønvold-Westrum scheme (22) superimposed on the d-f electron portion of the Latimer (25) scheme.

Other Interpolation Schemes

In principle, the best way to evaluate C_{latt} is from the phonon dispersion relation, $k(\omega)$, determined by inelastic neutron scattering.

However, occasionally Einstein functions (in which ω_E = constant) or Debye functions (in which $\omega_D = \gamma_0 k$) can be used to get results reliable within about 10 per cent.

Attempts to get an experimental estimate of C_{latt} in X compound by measuring the heat capacity of an isostructural diamagnetic (ID) compound are frequent. Here the corresponding states assumption

$$C_{latt}(X\ cpd)(T) = C_{latt}(ID) \times (k*T)$$

in which k is experimentally deduced is often employed.

Alternatively, the Debye theta approximation may be couched in the mass (M) of the molecules

$$\Theta_D(X)/\Theta_D(ID) = [M(ID)/M(X)]^{1/2}$$

The more refined Lindemann's relationship using melting points, T_m, and molal volumes, V, is also used.

$$\Theta_D^2 = k'T_m/MV^{2/3}$$

Corresponding states approaches are often used; but for nearly half a century the Latimer scheme (27,28) has been a favorite way of taking into account the differences between compounds in iso-anionic series. This time-honored scheme--devised primarily for entropy estimates--is not without its flaws, despite the several times it has been adjusted by Latimer himself.

The Volumetric Scheme

The scheme that we have advocated involves linear interpolation on the basis of the molal volumes of the compounds in question. In particular, the formula by which the lattice heat capacity of the praseodymium trihydroxide may be calculated is indicated below:

$$C_p [Pr(OH)_3, \text{lattice}] = xC_p [La(OH)_3] + (1-x)C_p [Gd(OH)_3]$$

and in which x is the fractional molal volume increment, i.e.,

$$x = \{V [Pr(OH)_3] - V [La(OH)_3]\} / \{V [Gd(OH)_3] - V [La(OH)_3]\}$$

It should be noted that utilization of other than linear interpolations would have involved differences in only second- order effects. The importance of volume was appreciated (25, 26) on recognition of the fact that for the lanthanide chalcogenides the lattice contribution decreased with increasing atomic number and was, therefore, diametrically opposed to the trend in the Latimer scheme based on mass. The lanthanide contraction provided the clue (fig. 5). Other authors (36, 37) have been engaged in a polemic as to the relevance of volume versus mass in providing interpolation schemes for lattice contributions. Moreover, Kieffer (38-42) has undertaken a theoretical and experimental correlation of the lattice vibrations of minerals. This takes into account the many factors involved and discusses particularly the analysis of the vibrational contribution, which has been discussed also by Sommers and Westrum (20).

But does the volumetric scheme really work? Perhaps the best way of testing the validity of a lattice contribution scheme is in the calculation of the calorimetric Schottky contribution and the comparison of this excess heat capacity with that calculated from spectroscopic data on the samples itself. However, this comparison can only be made when one utilizes the

Stark levels of the <u>concentrated</u> compounds. Measurements made on <u>doped</u> lanthanide halides, for example, need to be extrapolated by some technique—discussed elsewhere (30)—or by calculations based on crystal-field parameters.

Testing with Lanthanide Trihydroxides and Trichlorides

Since resolution of Schottky contributions from the generally much larger vibrational (lattice) heat capacities of lanthanide compounds has been limited by the uncertainty in the magnitude of the lattice contribution, such subtle effects as dependence of the Stark levels on temperature and host lattice have been heretofore undetected calorimetrically. Since the lanthanide trihydroxides are an iso-anionic series having relatively small lattice contributions and their lower-lying Stark levels have been spectroscopically deduced for many of the concentrated compounds, this series is the most nearly ideal system yet studied in an attempt to resolve Schottky contributions in the 5 to 350 K range. Three examples illustrate the success of the scheme described on $Ln(OH)_3$ systems; moreover, two examples from $LnCl_3$ systems demonstrate that excellent agreement obtains here as well.

<u>$Eu(OH)_3$</u>. The Schottky contribution to the heat capacity of the Eu(III) analog is unique in that it arises entirely from thermal populations of excited [SL] J-manifolds. This invariably results in the lowest excited Stark levels being much higher in energy for the Eu(III) analog than for any other series member. The calculated Schottky heat capacity is consequently relatively insensitive to small shifts in the Stark level energies and, therefore, is expected to be the most accurate approximation to the true Schottky heat capacity within any lanthanide series.

The energy levels of concentrated $Eu(OH)_3$ were determined by Cone and Faulhaber (43) from absorption and fluorescence spectra at 4.2 and 7.7 K. Stark levels arising from the $^7F_0$, $^7F_1$, $^7F_2$, and $^7F_3$ manifolds all contribute to the Schottky heat capacity below 350 K. The derived <u>calorimetric</u> Schottky contribution shown in fig. 6 is seen to be in excellent accord with that calculated from the <u>spectroscopic</u> data (32).

<u>$Pr(OH)_3$</u>. The crystal-field splitting of the $^3H_4$ manifold of $Pr(OH)_3$ has been determined from the absorption spectra of mulls at 95 K (44). The observed spectra were not as highly resolved as one might obtain from measurements on single crystals. This lack of resolution is reflected in a ± 3 cm^{-1} uncertainty in the Stark level energies. As seen in fig. 6 the <u>calorimetric</u> and <u>spectroscopic</u> Schottky curves are in very good agreement

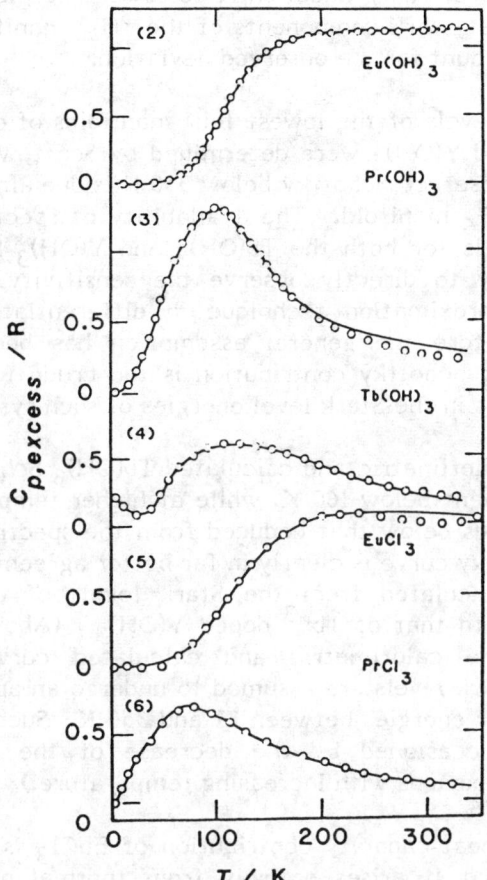

Fig. 6. Calorimetrically (——) and spectroscopically (o) determined Schottky contributions. (a) .... for Eu(OH)$_3$; (b) .... for Pr(OH)$_3$ (31); (c) .... for Tb(OH)$_3$ (33); (d) .... for EuCl$_3$ (30); and (e) .... for PrCl$_3$. (The successive curves are displaced by one unit of C/R).

between 15 and 230 K. Below 25 K a cooperative magnetic contribution of unknown magnitude plus the uncertainty in the energy of the lowest excited Stark level preclude any attempt to accurately determine the Schottky contribution in this temperature region. Above 260 K the calorimetric curve trends below the calculated band. Such a decrease in the high-temperature calorimetric Schottky curve could be due to a gradually decreasing crystal-field intensity within the Pr(OH)$_3$ crystals as the lattice expands with temperature. (A gradual shift to lower energies of 5 to 10 cm^{-1} by the four highest Stark components of the $^3H_{4-}$ manifold between 100 and 350 K would account for the observed deviation.)

Tb(OH)$_3$. The energy levels of the lowest four manifolds of concentrated Tb(OH)$_3$ and Tb^{+3} doped Y(OH)$_3$ were determined by Scott, Meissner, and Crosswhite (45). The observed Schottky below 350 K is due almost entirely to population of the $^7F_6$ manifold. The availability of spectroscopically determined energy levels for both the Tb(OH)$_3$ and Y(OH)$_3$ host lattices provides an opportunity to directly observe the sensitivity of the new lattice-contribution approximation technique in differentiating between such systems. Heretofore the general assumption has been that any calorimetrically derived Schottky contribution is too crude to detect the effect of any differences in the Stark level energies of such systems.

As seen in fig. 6 the calorimetric and calculated Tb(OH)$_3$ Schottky curves are in excellent agreement below 160 K, while at higher temperatures the calorimetric curve trends below that deduced from the spectral data (33). The calorimetric Schottky curve is clearly in far better agreement with the spectroscopic curve calculated from the Stark levels of concentrated Tb(OH)$_3$ rather than with that of Tb^{+3} doped Y(OH)$_3$. (Above 160 K the difference between the calorimetric and calculated curves may be accounted for if the Stark levels are assumed to undergo an approximately 6 per cent shift to lower energies between 77 and 350 K. Such a shift may be postulated to be occasioned by the decrease of the crystal-field intensity as the lattice expands with increasing temperature.)

EuCl$_3$. The Schottky heat-capacity contribution of EuCl$_3$ is unique—like that of Eu(OH)$_3$—in that it arises entirely from thermal population of excited [SL]J-manifolds. The first excited states are near 355 and 405 cm^{-1}. The Schottky contribution was calculated from energy levels of (1 and 4 per cent) Eu^{+3} doped LaCl$_3$ determined from the absorption spectrum at 4 K and the fluorescence spectrum at 4 and 77 K studied by Deshazer and Dieke. The energy levels of concentrated EuCl$_3$ are not expected to be identical to those of Eu^{+3} doped LaCl$_3$. The stronger crystal field in concentrated EuCl$_3$—compared to that in the LaCl$_3$ host—is expected to increase the Stark splitting and simultaneously to lower the center of gravity of the $^7F_1$-manifold, i.e., to lower the energy of the $\mu = 1$ doublet and to leave the $\mu = 0$ level essentially unchanged. The effect of the stronger crystal field will be countered to some extent by expansion of the EuCl$_3$ lattice at higher temperatures (i.e., in the region of the Schottky maximum); however, this is anticipated to be insufficient to fully nullify the effect. Because the energies of the Stark levels contributing to the Schottky heat capacity are relatively high, a shift of the $\mu = 1$ doublet by as much as 10 to 15 cm^{-1} will have but a small effect on the calculated Schottky contribution. The derived <u>calorimetric</u> Schottky heat capacity shown in fig. 6 is seen to be in excellent accord with that derived from the <u>spectroscopic</u> data (30).

PrCl$_3$ The analysis of the Schottky contribution to the heat capacity of PrCl$_3$ is complicated by unusual shifts in the Stark level energies as the intensity of the crystalline field is varied. However, the basic arguments remain essentially unchanged from those applied to the preceding compounds. The energy levels of Pr^{+3} doped LaCl$_3$ were determined from absorption and fluorescence spectra by Sarup and Crozier. Although the wealth of spectroscopic data may be best deduced by reference to the definitive paper (30), both fig. 6 and table 4 attest to the good agreement between the two Schottky contributions (30).

Table 4. Energy levels/cm^{-1} for ^{3}H$_4$-state of PrCl$_3$

State Technique	Pr^{+3}: LaCl$_3$ Abs. Fluorescence	Calorimetry	Concentrated PrCl$_3$			Absorption
			CEF	ERS'	ERS''	
Levels	0^a	0^b	0^c	0^d	0^e	0^f
	33.1	---	29	30.5	32	31.8
	96.4	---	99	99	100	99.6
	130.2	155	152	145	139	---
	137.0	168	176	160	---	---
	199.1	235	230	(228)	---	---

^{a} Energy levels by absorption and/or fluorescence spectroscopic data for Pr^{3+} doped into LaCl$_3$. (Sarup and Crozier) (63).
^{b} Levels deduced from heat-capacities by volumetric lattice contribution method (30).
^{c} Calculated from estimated crystal-field parameters (30).
^{d} Observed in electronic Raman scattering (Chirico, et al.) (30).
^{e} Observed in electronic Raman scattering (Hougen and Singh) (64).
^{f} Observed in absorption spectra data (Dorman) (65).

The Role of Mass

The dashed and continuous curves of fig. 7 represent $\{C_p[Gd(OH)_3] - C_p[La(OH)_3]\}$ and $\{C_p[Y(OH)_3] - C_p[La(OH)_3]\}$ with the cooperative magnetic contribution to the Gd(OH)$_3$ heat capacity deleted. The dotted curve is an estimate of the quantity $\{C_p[Lu(OH)_3] - C_p[La(OH)_3]\}$ derived by extrapolation of the experimentally observed lattice heat-capacity variation between La(OH)$_3$ and Gd(OH)$_3$ (e.g., see reference (33)). If the lattice heat-capacity contribution for the lanthanide trihydroxides were exclusively a linear function of the molar volume, then the $C_p[Y(OH)_3] - C_p[La(OH)_3]$ curve would lie almost exactly midway between the dashed and dotted curves from 5 to 350 K. In contrast, if the trend in the lattice contributions was determined principally by the molar-mass variation, it would lie entirely below the dotted curve instead of only for temperatures below about 120 K.

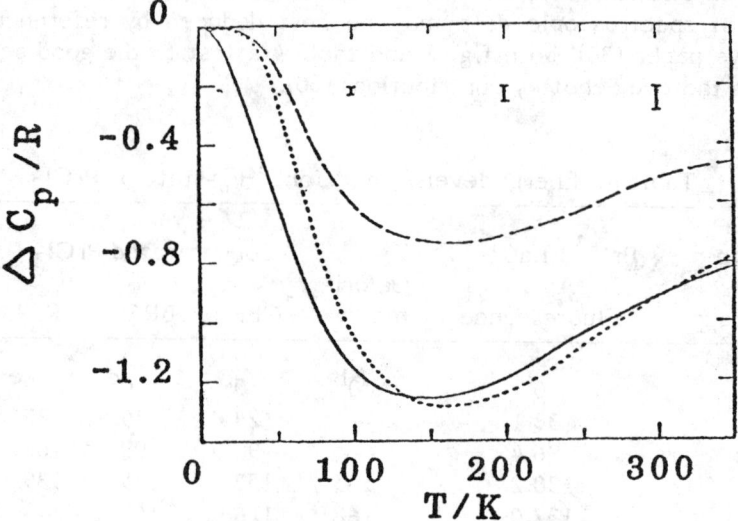

Fig. 7. Temperature dependence of C_p/R for the trihydroxide pairs:
C_p Gd(OH)$_3$ - C_p La(OH)$_3$, (— — —);
C_p Y(OH)$_3$ - C_p La(OH)$_3$, (———); and the estimated
C_p Lu(OH)$_3$ - C_p La(OH)$_3$ (-----).
The error bars represent 0.5 per cent of C_p La(OH)$_3$.

From the results obtained for Y(OH)$_3$, a shift in the relative importance of the cationic mass and volume in determining the trend in the lattice contributions across the series occurs near 100 K. The experimental observations may be rationalized by considering the type and nature of the lattice vibrational modes being activated at each temperature. At very low temperatures, low-frequency modes—roughly characterized as unit-cell vibrations—are those primarily activated. The lanthanide contraction which is an intramolecular contraction, has little effect upon these vibrations. In essence the force constants between the unit cells are unchanged, while the cell masses increase across the series. This occasions a decrease in the vibrational frequencies of the unit cells with increasing atomic number and, therefore, a corresponding increase in lattice contribution at a given temperature.

At higher temperatures an increasing proportion of the observed heat capacity is due to thermal activation of optical vibrational modes. The effect of the lanthanide contraction upon these modes is to increase their frequency by increasing the intramolecular force constants to such an extent that the counteracting effect of the increased cationic mass is largely overshadowed. An analogous effect is routinely observed in temperature dependence of vibrational spectra. As the temperature is decreased (i.e., as the molecule contracts) the vibrational mode frequencies are generally seen to increase. In the case of the lanthanide trihydroxides the molar-mass variation between the lanthanum and gadolinium iso-anionic compounds is small enough to be insignificant in determining the trend in lattice contribution across the series between 10 and 350 K. Even when considering $Y(OH)_3$, which has a molar mass approximately 2/3 that of the lanthanide compounds, the effect of molar mass is clearly dominant only below 80 K. The difference between the heat capacities of lutetium and yttrium ethylsulfates (46,47) clearly exhibits the same type of behavior as that of the corresponding trihydroxides.

The apparent differences with temperature of the functional dependence of the lattice contribution upon molar mass and volume makes it imperative that lattice approximations (e.g., the method of "corresponding states") which employ observations made at high temperatures to imply low-temperature properties or vice versa be applied with caution. Related problems have been discussed by Saxena (36), by Cantor (37), and by Kieffer (38-41).

Thus the trend in lattice heat capacities of iso-anionic series of lanthanide compounds may be rationalized in terms of two contributing factors: molar mass and molar volume. At low temperatures, the lattice contribution is due primarily to thermal activation of acoustic lattice modes and molar mass is the dominant factor. At higher temperatures increasing thermal activation of optical lattice modes, which are strongly affected by the lanthanide contraction, results in lattice heat-capacities which are related predominantly to the trend in molar volume. For the light lanthanide trihydroxides the molar-mass variation is dominated by the molar-volume effect at least above 50 K. Only for much lighter $Y(OH)_3$ is the mass effect clearly visible and then only below 100 K.

Hence, in emphasizing the importance of volume, we do not mean to slight mass—especially not at lower temperatures. Data on $U(OH)_3$, which is isostructural with the $Ln(OH)_3$'s should help to clarify and to test the roll of mass. Although we have demonstrated the great utility of the volumetric scheme as an interpolation device for $C_p(T)$ or S^o (298 K) for a system of isostructural compounds, what about the broader implications? How generally does it supplant the Latimer rule even when "augmented" to provide magnetic contributions, etc.?

We have examined isostructural series on which sufficient data exist to make a judgment. Many interesting trends are observed. For example, as seen in fig. 8, extrapolation by the Latimer scheme from the entropies (298 K) of MoS_2, WS_2, PtS_2 to that of TiS_2 would lead to a Latimer-scheme value of $S^o/R = 7$; on the other hand the volumetric approach would lead to $S^o/R = 9.5$. Experiment ($S^o/R = 9.4$) confirms the latter: In other instances the general trend of cation mass with molar volume in iso-anionic series often tend to make choice between the two systems difficult inasmuch as molar mass and molar volume usually go hand in hand. Identification of key compounds on which to test the scheme and to develop more reliable correlations is underway.

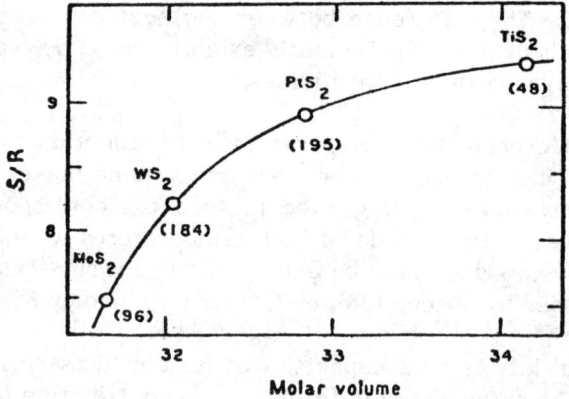

Fig. 8. Entropy versus molar volume for the MS_2 compounds at 298.15 K. The cation masses appear in parentheses below the points.

Lanthanide Sesquisulfides

Exceedingly interesting spectroscopic studies, together with heat-capacity measurements at very low temperatures, are beginning to probe lanthanide sesquisulfides, Ln_2S_3, which are being prepared as stoichiometric single crystals, as well as in the hyper-and hypo-stoichiometric forms. Since most of this work is as yet unpublished in definitive form, one can only herald these endeavors to explore Schottky functions at the cryogenic temperatures in a collaborative endeavor with the Ames Laboratory.

Revisiting the Bixbyite Lanthanide Sesquioxides

As has already been noted, the cubic lanthanide sesquioxides possess two inequivalent cation sites, C_2 and C_{3i}, the latter with inversion symmetry. They have been of scientific and technological interest for many years in part because the lattice itself is an excellent host material for some of the most powerful lasers built (48-50). Moreover, since the early sixties, lasers have made possible the study of many low-probability non-linear optical phenomena, (51) including vibrational and electronic Raman scattering (52-53). Coherent, polarized, and tuneable excitation sources permit identification through electronic Raman scattering and double-photon absorption of crystal-field states of lanthanide ions in sites having inversion symmetry (53-54). Moreover, energy transfer between inequivalent cation sites in the cubic lanthanide sesquioxides has been investigated quantitatively using laser excitation sources (55).

Although now already two decades old, the heat-capacity data of Justice and Westrum for the lanthanide sesquioxides (11-15) provide an important check on the assignments made to the crystal-field split [SL]J-levels for lanthanide ions in both C_2 and C_{3i} cation sites as deduced from Raman and optical spectra (24). Such a check is particularly valuable in the interpretation of the Raman scattering experiments, which in certain cases exhibit vibrational and electronic spectral peaks of similar magnitude (56). Moreover, a recent comprehensive study of the Ln^{+3} crystal-field splitting of the 4 $\underline{f}^n$ [SL]J-manifolds in both C_2 and C_{3i} sites in Y_2O_3 provides an additional independent check of the experimentally deduced Stark splittings (57). In addition, these crystal-field calculations make possible further interpretation or reinterpretation of EPR, Mossbauer, and heat-capacity data (58-62).

A self-consistent interpretation of the Raman and optical spectra with the heat-capacity measurements of concentrated (i.e., 100%) Dy_2O_3, Er_2O_3, and Yb_2O_3 were presented (10) by analyzing unpublished, infrared and electronic Raman scattering data as well as extant data (56-57). The deduced Stark splitting for both sites were then compared with the crystal-field calculations of Chang, Gruber, Leavitt and Morrison (57).

Only the results on Er_2O_3 are selected here as a typical example and are shown in fig. 9.

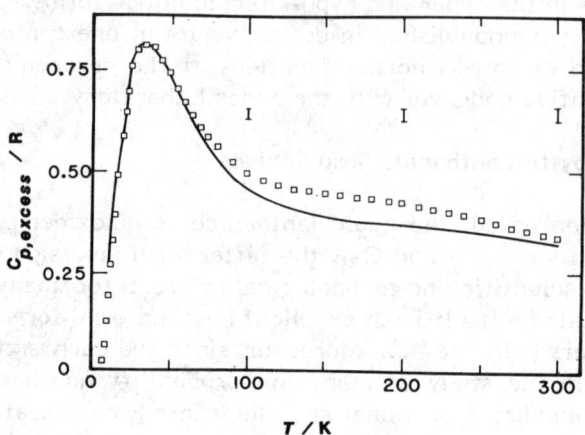

Fig. 9. Calorimetric Schottky contribution, i.e., $C_p(Er_2O_3) - 0.64\ C_p(Lu_2O_3) + 0.36\ C_p(Gd_2O_3)$ ─────.
Spectroscopic Schottky contribution calculated from Stark levels □ □ □ □ .

C_2 = 0, 38, 75, 88, 159, 265, 490, 505
C_{3i} = 0, 41, 80, 168, 359 (or 328), 416, 485, 580.

NOTE

1. The author, like virtually all other speakers, took the opportunity at his first presentation to thank the host organization for their kindness and hospitality and to note the excellence of the arrangements. He also expresses his gratitude to the Thermodynamics and Structural Chemistry Section of the Chemistry Division of the National Science Foundation for its continuing support of the research endeavors on which this presentation is based under Grant CHE-8007977, and G EAR-8009538.

REFERENCES

1. Stull, D.R., Westrum, E.F. Jr., Sinke, G.C.: 1969, "The Chemical Thermodynamics of Organic Compounds" (John Wiley and Sons).

2. Westrum, E.F. Jr., McCullough, J.P.: 1963, in "Physics and Chemistry of the Organic Solid State," Vol. 1, ed. Fox, D. (Interscience), pp. 1-178.

3. Gopal, E. S. Raja: 1966, "Specific Heats at Low Temperatures" (Plenum).

4. Bartel, J.J., Callanan, J.E., Westrum, E.F. Jr.: 1980, J. Chem. Thermodyn. 12, p. 753.

5. Carlson, H.G., Westrum, E.F. Jr.: 1971, J. Chem. Phy. 54, p. 1464.

6. Rao, C.N.R., Rao, K.J.: 1978, "Phase Transitions in Solids" (McGraw-Hill).

7. Harrison, W.R.: 1970, "Solid State Theory" (McGraw-Hill).

8. Samuelsen, E.J., Andersen, E., Feder, J. (eds.): 1971, "Structural Phase Transitions and Soft Modes" (Universitets forlaget).

9. Kishimoto, K., Suga, H., Seki, S.: 1980, Bull. Chem. Soc. Jpn., 53, p. 2748.

10. Schottky, W.: 1927, Zeit. Phy. 21, p. 465.

11. Justice, B.H., Westrum, E.F. Jr.: 1963, J. Phys. Chem. 67, p. 339.

12. Justice, B.H., Westrum, E.F. Jr.: 1963, J. Phys. Chem. 67, p. 345.

13. Justice, B.H., Westrum, E.F. Jr.: 1963, J. Phys. Chem. 67, p. 659.

14. Justice, B.H., Westrum, E.F. Jr., Chang, E., Radebaugh, R.: 1969, J. Phys. Chem. 73, p. 333.

15. Justice, B.H., Westrum, E.F. Jr.: 1969, J. Phys. Chem. 73, p. 1959.

16. Penny, W.G.: 1933, Phys. Rev. 43, p. 2515.

17. Henderson, J.R., Muramoto, M., Gruber, J.B.: 1967, J. Chem. Phys. 46, p. 2515.

18. Gruber, J.B., Chirico, R.D., Westrum, E.F. Jr.: 1982, J. Chem. Phys. 76, p. 4600.

19. Sommers, J. A., Westrum, E.F. Jr.: 1977, J. Chem. Thermodyn. 9, p. 1.

20. Westrum, E.F. Jr., Chirico, R.D., Gruber, J.B: 1980, J. Chem. Thermodyn. 12, p. 717.

21. Westrum, E.F. Jr., Clever, H.L., Andrews, J.T.S., Feick, G: 1966, in "Rare Earth Research III." ed. Eyring, L. (Gordon and Breach), plus unpublished work.

22. Grønvold, F., Westrum, E.F. Jr.: 1962, Inorg. Chem. 1, p. 36.

23. Stull, Daniel R., Westrum, E. F., Jr., Sinke, Gerard C.: 1969, "The Chemical Thermodynamics of Organic Compounds," (John Wiley and Sons).

24. Stuttius, W.G.: 1975, personal communication. See Stuttius, W.G.: 1969, Phys. Kondens. Mater. 10, p. 152.

25. Kovats, E., Gunthard, H.H., Plattner,P.A.: 1955, Helv. Chim. Acta 38, p. 1912; ibid 1957, 40, p. 2008.

26. Brink, I., Westrum, E.F. Jr.: unpublished.

27. Latimer, W.M.: 1921, J. Am. Chem. Soc. 43, p. 818.

28. Latimer, W.M.: 1951, J. Am. Chem. Soc. 73, p. 1480.

29. Chirico, R.D., Westrum, E.F. Jr.: 1980, J. Chem. Thermodyn. 12, p. 311.

30. Westrum, E.F. Jr., Chirico, R.D., Gruber, J.B: 1980, J. Chem. Thermodyn. 12, p. 717.

31. Chirico, R.D., Westrum E.F. Jr., Gruber, J.B., Warmkessel, J.: 1979, J. Chem. Thermodyn. 11, p. 835.

32. Chirico, R.D., Westrum,E.F. Jr.: 1980, J. Chem. Thermodyn. 12, p. 71.

33. Chirico, R.D., Westrum, E.F. Jr.: 1981, J. Chem. Thermodyn. 13, p. 519.

34. Chirico, R.D., Westrum, E.F. Jr., Boerio-Goates, J.: 1981, J. Chem. Thermodyn. 13, p. 1087.

35. Westrum, E. F. Jr.: 1981, in Proceedings of the 6th International Conference on Thermodynamics (Merseburg), p. 1.

36. Saxena, S.K.: 1976, Science 193, p. 1241.

37. Cantor, S.: 1977, Science 198, p. 206.

38. Kieffer, S.W.: 1979, Rev. of Geophys. & Space Phys. 17, p. 1.

39. Kieffer, S.W.: 1979, Rev. of Geophys. & Space Phys. 17, p. 20.

40. Kieffer, S.W.: 1979, Rev. of Geophys. & Space Phys. 17, p. 35.

41. Kieffer, S.W.: 1980, Rev. of Geophys. & Space Phys. 18, p. 862.

42. Kieffer, S.W.: in press.

43. Cone, R.L., Faulhaber, R.: 1971, J. Chem. Phys. 55, p. 5198.

44. Cited in Reference 31.

45. Scott, P.D., Meissner, H.E., Crosswhite, H.M.: 1969, Phys. Lett. 28A, p. 489.

46. Gerstein, B.C., Jennings, L.D., Spedding, F.H.: 1962, J. Chem. Phys. 37, p. 1496.

47. Gerstein, B.C., Penny, C.J., Spedding, F.H.: 1962, J. Chem. Phys. 37, p. 2610.

48. Dieke, G.H.: 1968, "Spectra and Energy Levels of Rare Earth Ions in Crystals" (Interscience).

49. Hufner, S: 1978, "Optical Spectra of Transparent Rare Earth Compounds" (Academic).

50. Reisfeld, R., Jørgensen, C.K.: 1977, "Lasers and Excited States of Rare Earths" (Springer).

51. DiBartolo, B.: 1968, "Optical Interactions in Solids" (Wiley).

52. Yariv, A.: 1967, "Quantum Electronics" (Wiley).

53. Axe, J.D.: 1964, Phys. Rev. A 42, p. 136. See also Elliott, R.J., Loudon, R.: 1963, Phys. Lett. 3, p. 189.

54. Gruber, J.B.: 1971, in "Proceedings of the 9th Rare Earth Research Conference, Blacksburg, Virginia" (NTIS, U.S. Dept. of Commerce) p. 465.

55. Heber, J., Hellwege, K.H., Kobler, U., Murmann, H.: 1970, Z. Phys. 237, p. 189, and references therein.

56. Lejus, A.M., Michel, D.: 1977, Phys. Status Solidi 84, p. K105.

57. Gruber, J.B., Chang, N.C., Leavitt, R.P., Morrison, C.A. (to be published).

58. Mandel, M.: 1963, Appl. Phys. Lett. 2, p. 197.

59. Schafer, G.: 1969, Phys. Kondens. Materie 9, p. 359, and references therein.

60. Vivien, D., Lejus, A.M., Collongues, R.: 1978, Nov. J. de Chem. 2, p. 569.

61. Forester, D.W., Ferrando, W.A.: 1976, Phys. Rev. B 14, p. 4769.

62. Moon, R.M., Koehler, W.C., Child, H.R., Raubenheimer, L.J.: 1968, Phys. Rev. 176, p. 722.

63. Sarup, R., Crozier, M.H.: 1965, J. Chem. Phys. 42, p. 371.

64. Hougen, J.J., Singh, S.: 1964, Proc. Royal Soc. (London) A277, p. 193.

65. Dorman, E.: 1966, J. Chem. Phys. 44, p. 2910.

THERMOPHYSICS OF MINERAL AND ROCK SYSTEMS[1]

Edgar F. Westrum, Jr.

Department of Chemistry, University of Michigan, Ann Arbor, Michigan 48109, U.S.A.

The present decade may almost be described as that in which geoscientists "discovered" thermodynamics and the application thereof to the genesis and stability of mineralogical and petrological structures.

As in other areas the first lacunae is that of accurate and reliable data for "key substances", although there have been endeavors to fill the gaps by estimation schemes and by theoretical analysis. The former has been done by compilers generally but especially by Hendrickson; the latter by Kieffer. Others are endeavoring to provide direct data.

The theoretical approach has involved first an examination based on Debye theory and secondly a recognition of observed thermodynamic deviations in minerals. A generalized lattice model has evolved from these studies; however, the reliability is—in many instances—inadequate.

On the experimental side considerable progress is being made both on carefully picked natural mineral specimens and on synthesized specimens. Since the time scale in nature is much greater than that in the laboratory—ordering is often higher in natural samples—but they are unlikely to have end member composition.

Particular examples will be examined—especially the bifurcation of the Verwey transition in magnetite, the stability of members of the marcasite/pyrite system, the CASH system, and others with emphasis not only on the thermophysics involved in the calorimetry, but also on the adjuvant types of equilibrium and phase diagram data involved.

INTRODUCTION

Asking the question of "why are we interested in thermodynamics and geoscience?" is a little bit like asking the question of "why do people climb mountains?". In a sense, because they are there. In another sense which is a little bit facetious, one notes that geoscientists have discovered thermodynamics and what it can do for them. Indeed they have discovered the utility of a great many modern tools in recent years, but in a real sense they have awakened to the reality of the implications of a classical discipline in their scientific endeavors. This is particularly evident from an abundance of texts (e.g., 1-3), from a number of excellent monographs (4-11), and indeed of several new compilations of thermodynamic data for geoscience supplementing extant ones which have not seen revision in several decades (14). To a thermodynamicist this "honeymoon" is to be warmly applauded, and I think that one can safely predict the next decade is going to reveal some very exciting developments in the application of thermodynamics to mineralogy and petrology and the development of computer programs to deduce the stability and the genesis of ore and rock deposits.

For most problems in chemistry the application of thermodynamics is direct and rewarding. Although large numbers of geoscientists do apply thermodynamic treatment to a variety of problems in earth and planetary sciences, the complex nature of the inorganic and organic mineralogical and petrological systems involved as a consequence of their involved crystal structure and their multicomponent character, is such that the results often seem hard to interpret. For these reasons, thermodynamic approaches to geo-/planetological problems should be attempted only after a clear understanding of the crystallochemical and thermochemical characterization of each mineral or rock.

Interestingly enough we see history repeating itself in the initial stages of the encounter; that is to say, first the discovery of the projected utility, recognition that most of the data that one wants are simply not at hand, the subsequent attempt to estimate such data, followed in due course by the recognition of the need for measurement of key compounds, the development of approaches, and—finally—the actual data determination itself.

In the United States, for example, we have had for many years an excellent Bureau of Mines Experimental Station at Bartlesville, Oklahoma. This group has provided leadership not only in the determination of important quantities in petroleum and petrochemical refining and development, but indeed have contributed mightily over the years to the establishment of high standards of calorimetric and thermodynamic endeavor as well as in the development of instrumentation and techniques. Without these data it is improbable that this important industry would have flourished as quickly and rapidly as it has. It is with sadness that we learn of the probable demise of this laboratory in early 1983. The experiment station at the University of California at Berkley under the direction of K. K. Kelley did indeed concern itself with minerals but much of the work has had to be repeated since the investigators failed to take the measurements of heat capacity below 50 K. It has been moved to Albany, Oregon and is presently under the direction of Nev Gokcen and is indeed producing data on selected mineral systems but has not been permitted to undertake continuing production of the tables. The present interest in thermodynamics, however, stems largely from academia and other government laboratories, including that of the Geological Survey at Reston, Virginia. I am not so well acquainted with the situation in other countries of the world, but am fully aware of the need for the generation of important thermodynamic data to permit calculations, data banking, etc.

For many years a close collaboration with F. Grønvold of the University of Oslo has featured compounds like the transition-element chalcogenides and pnictides. In the course of this endeavor we had occasion to study compositions of matter which exist in nature in the form of minerals. At the present time, however, our mineralogic and petrographic endeavor is somewhat more focussed and directed to the provision of data on key substances. Table 1 provides a quick summary of the mineral substances we have examined and puts them in a broad time frame perspective. In this presentation we can hope to look at only a few of these systems and to show some of the interesting aspects of them.

SAMPLES FOR CALORIMETRY

Recognising that minerals are chemical substances and can be expected to show at least as many—if not more—interesting phenomenon than do garden variety inorganic chemicals, one recognizes the probability of finding magnetic, ferro-electric, order-disorder, and a great many other transitions such as those which have already been enumerated (15).

The procurement of reliable thermodynamic data requires, of course, a high-purity, well-characterized sample. Purists will not call such a sample a mineral unless it was produced in nature, however much its composition might resemble that of a naturally-occurring substance.

Table 1. Partial List of Mineral Species Studied at Ann Arbor

(Mineral names in parentheses represent samples prepared synthetically; others are selected natural materials.)

Chalcogenide Minerals

(Pyrrhotites) $Fe_{1-x}S$
(Troilite)   FeS
(Millerite)   NiS
(Cooperite)   PtS
(Acanthite)   AgS
(Argentite)   AgS

Molybdenite MoS_2
Hauerite MnS_2
Marcasite FeS_2
Pyrite FeS_2

(Chalcocite) Cu_2S
(Djuerleite) $Cu_{1.95}S$
(Anilite) $Cu_{1.75}S$
(Digenite) $Cu_{1.95}S$
(Covelite)   CuS

(Frobergite) $FeTe_2$
(Melonite)   NiTe

Oxide Minerals

(Hematite) Fe_2O_3
(Magnetite) Fe_3O_4
Magnetite Fe_3O_4

α-Quartz SiO_2
β-Cristobalite SiO_2
Coesite SiO_2
Stishovite SiO_2

Silicate Minerals

Fayalite $FeSiO_4$
Ferrosilite $FeSiO_3$
Hercynite $FeAl_2O_4$
Almandine $Fe_3Al_2Si_3O_{12}$

C-A-S-H System Minerals

Lawsonite $Al_2O_3, 2SiO_2 \cdot 2H_2O$
Margarite $CaAl_4H_2Si_2O_{12}$
Prehnite $H_2Ca_2Al_2(SiO_4)_3$

Miscellaneous Minerals

Monticellite $MgO \cdot CaO \cdot SiO_2$
Ilmenite $FeTiO_3$
Chlinochlore
   $(Mg_{4.25}Fe_{0.376}Al_{1.08}Cr_{0.29}Mn_{0.07})$
   $(Si_{3.05}Al_{0.95})O_{10}O_{0.42}(OH)_{7.58}$
Topaz $Al_2SiO_4Fe_2$

Grossular $Ca_3Al_2Si_3O_{12}$
Pyrope $Mg_3Al_2Si_3O_{12}$
Pyrope/Grossular

Although nature has produced much material of gem quality, it does not always have the desired composition. If one wishes to have terminal-member composition, departure from this is clearly not a desiderata. A great many minerals owe even their characteristic coloration to impurities which are present in significant amounts. However, minerals can usually be expected to have been around long enough to have achieved equilibrium—even though in some instances this may be a meta-stability. Often naturally occurring samples have reasonably large crystals the composition of which can be studied by microprobe on the surface. Care needs to be taken occasionally to be sure that one is not confronted with a vitreous rather than a crystalline phase.

The other alternative is the preparation of synthesized samples corresponding in composition and crystal form to the desired mineral substance. The preparation technique is often quite exotic, requiring extreme conditions, careful control, etc. Preparation of minerals by hydrothermal means is a typical example, and the research endeavor which has gone into the preparation of a single-crystal magnetite is also a very impressive one. In making minerals by hydrothermal means one is often obliged to use gold capsules and a reasonably small scale of preparation. Calorimetrists are perhaps best known for their "greediness" in desiring to have samples massing between 50 and 200 grams. This is often totally out of the question if the samples are to be synthesized. It may also be difficult if the samples have to be hand picked from a mineralogical occurrence.

Although one can certainly control the composition in close approximation to terminal member composition and hopefully can avoid in most instances the introduction of impurities, this does not automatically insure that the sample will be of high purity. It does not, moreover, insure that the crystal structure will be highly ordered and free of defects, and it does not insure that for mineral substances which may be disordered, the corresponding synthetic preparation will not actually show zero-point entropy. This will be discussed in more detail later.

Often the procurement of a high-purity, terminal-member composition sample of well characterized crystal structure and zero-point disorder is the most difficult aspect of the experimental process.

EXPERIMENTAL METHODS OF CALORIMETRY APPLIED TO MINERALOGY AND PETROLOGY

Fundamentally, calorimetry does provide two types of data. The first of these is essentially thermophysical, and measurement of heat capacities of non-reacting systems is a good example of it. The second is essentially thermochemical and involves measurements of heats of chemical reaction (including relative stability of competing phase assemblages, solution calorimetry, or mixing in solids and liquid solutions).

Thermophysical Techniques

The importance of heat capacity measurements at sub-ambient temperatures is of course occasioned by the fact that (apart from a zero-point disorder and imperfections) the entropy is given by the integral of C_p/T from zero to the temperature of interest. Below the lowest temperature of measurement the heat capacity is typically given by the Debye limiting law (C_p proportional to T^3). Recent progress by Kieffer (7-11) in estimating low-temperature heat capacities in minerals using a more sophisticated semi-empirical model for lattice vibrations provides useful correlations but does not eliminate the need for measurements. Magnetic transitions at low temperatures do occur and lead to λ-type anomalies in the heat capacity and can have a very significant effect upon the entropy (16).

Low-temperature heat capacities are perhaps best measured in adiabatic calorimeters and a still largely valid summary of the state of the art in 1968 has been made (17). Such measurements were a demanding and relatively time-consuming operation. In recent years due to advances in computer-aided operation and data acquisition, the task has been relatively lightened and the ease of making repetitive measurements greatly enhanced (18).

The relatively large samples that have been needed in the past have tended to restrict measurements of minerals to natural samples with the concomitant problems of impurities, but the development of miniaturized heat capacity calorimeters operating on approximately one gram of sample should facilitate utilization of synthetically prepared samples.

The development of differential scanning calorimeters (dsc) provides a possibility for survey heat-capacity measurement from about 250 to 1000 K. The ease and rapidity of operation and the ability to use a very small (5-25 mg) sample are counterbalanced in part at least by the fact that low quality data are very often produced—even though they do not look to be obviously in error. Typical dsc heat capacities have an accuracy of about 2 per cent and hence are less accurate than those by alternative methods. The upper limit however is unfortunately nearer the threshold of a temperature range where interesting phenomenon (melting, glass transformations, and phase transformations) frequently occur in silicates, etc.

Differential thermal analysis (dta) is useful in the detection of phase transitions although it does not often provide quantitative data. Utilization however in piston-cylinder apparatus at high pressures is a significant convenience.

Calorimetry by the-method-of-mixtures or "drop" calorimetry often provides quite accurate data at high temperatures. The measured enthalpy increments may be differentiated to yield values of C_p. Relatively large samples, long equilibration times and very careful operation are required to provide suitable data. Transposed temperature drop calorimetry in which a sample at the ambient (room) temperature is dropped into a calorimeter near 1000 K often requires smaller samples (about 100 mg) than the ten to 100 g samples required for conventional drop calorimetry and avoids the frozen-in equilibria sometimes encountered in usual drop calorimetry (19).

Thermochemical Techniques

Enthalpies of reaction are usually determined by solution calorimetry, that is on measuring the heat of solution of reactants in a suitable solvent. Silicates are notoriously unreactive and hydrofluoric acid (HF) solution calorimetry either at the ambient temperature or at temperatures as high as 90 °C is often used. Moreover, oxide melt solution calorimetry handles refractory oxides with ease at temperature ranges from 600 to 1000 °C. This is usually done in a Calvet-Tian type calorimeter which measures and integrates the heat flow as a function of time. It utilizes a sample of approximately 20 to 50 mg in size and has an attainable precision in heat of solution of approximately one per cent corresponding to an attainable precision of enthalpies of formation from oxide components of about 0.4 kj/mol. The utility of the technique has been enhanced by recent developments (20).

DISORDER IN ALUMINOSILICATE MINERALS

The calorimetric data for cordierite, sapphirine, and anorthite in table 2 show differences between the enthalpies of formation of natural and synthetic samples which cannot be ascribed to compositional variations. These differences are perhaps better attributed to differences in structural state with greater disorder in the synthetic than in the natural samples. These values should constitute a warning to the petrologist against uncritical application of experimental data based on synthetic systems to natural situations even when the natural phases closely approach the composition of the synthetics. The warning applies also to the utilization of calorim-

Table 2. Third-Law Disorder in Samples

Substance	Formula	Provenance	ΔH_f
Cordierite	$Mg_2Al_4Si_5O_{18}$	Natural	-15.9 ± 0.3
		Synthetic	-16.3 ± 0.5
Sapphirine	$Mg_7Al_{18}Si_3O_{40}$	Natural	-38.7 ± 1.8
		Synthetic	-46.3 ± 2.0
Anorthite	$CaAl_2Si_2O_8$	Natural	-23.1 ± 0.4
		Synthetic	-23.9 ± 0.5

etric data, phase equilibria determined in relatively short runs on synthetic systems, and indeed to the generation and utilization of self-consistent data banks. However, calorimetry when combined with structural characterization of the samples used does provide the framework for the understanding of this structural complexity and even for the prediction of its thermodynamic consequences.

An excellent discussion of the contributions of this sort to the third-law entropies of silicates has been presented by Ulbrich and Waldbaum (16). These authors note that the discrepancy can usually be attributed to neglected residual or unextracted entropies related either to site-mixing and molecular disorder or to the lack of a significant magnetic ordering at the lowest temperatures reached by heat-capacity measurements. This effect in entropy is well known in feldspars but there is reason to believe that residual entropies may also be present in many other classes of minerals including zeolites and silicates. The paper focuses on the causes resulting in these discrepancies and indicates the uncertainties involved in deriving third-law entropies from either equilibrium data or from calorimetric investigation especially in dealing with systems with vacancies, containing transition metals, or the possibility of magnetic ordering.

An excellent discussion of other order-disorder transformations, the lambda transitions in minerals, and the extrapolation of thermodynamic data as a function of temperature and pressure has been presented by Thompson and Perkins (21). The thermodynamic analysis of simple mineral systems by Holland (22) also provides much useful advice.

SCHOTTKY CONTRIBUTIONS FROM CRYSTAL FIELD EFFECTS

Crystal-field theory is a simple electrostatic model which deals with the energetic effects of coordinating anions about a central cation which has unfilled d- or f- atomic orbitals. In a succinct review of the first row transition ions in the sites present in silicate minerals, Wood (23) concludes that crystal-field stabilization energies make significant contribution to the Gibbs energies of these ions in silicate minerals. Although it has been difficult to accurately predict the Gibbs energies of silicate minerals, their heat capacities and entropies do obey simple relationships provided the appreciation of the special relationships involved in the distorted cation sites in, for example, olivine structures.

Crystal-field effects do play an important role in many minerals and need to be taken into account whenever transition elements are dealt with.

MAGNETITE—THE BIFURCATED VERWEY TRANSITION

Magnetite has long been known to undergo a phase transition near 120 K characterized by a lattice distortion and a sharp change in the electrical conductivity. The unusually high conductivity above 120 K has been explained by various schemes some of them somewhat more complex than that proposed by Verwey who first studied the transitions. But the transition itself has been found to be complex, and in 1969 Westrum and Grønvold provided the first evidence for the step-wise nature of the Verwey transition in their observation of the bifurcation of the heat-capacity anomaly in pure Fe_3O_4 (24). Additional studies in the same laboratory on pure and doped Fe_3O_4 (25,26) confirmed the bifurcation of the heat capacity anomaly and the association of the lattice distortion with the lower-temperature anomaly, note fig. 1. Samples were prepared by carefully doping pure Fe_3O_4 with a variety of dopants. A sample of composition $Zn_{0.005}Fe_{2.995}O_4$—in fact the identical sample on which heat-capacity and Mossbauer measurements (27) were previously performed—was subjected to neutron diffraction analysis (28). The results of this experiment proved that the heat-capacity anomaly occurring in this sample at 110.6 K is associated with the lattice transformation as concluded previously but less definitively on the basis of Mossbauer measurements, although no lattice distortion was found to be associated with the higher-temperature anomaly. These findings, taken together with the results of redistributing measurements, support the idea that the higher-temperature anomaly is a semi-metal to semi-conductor transition.

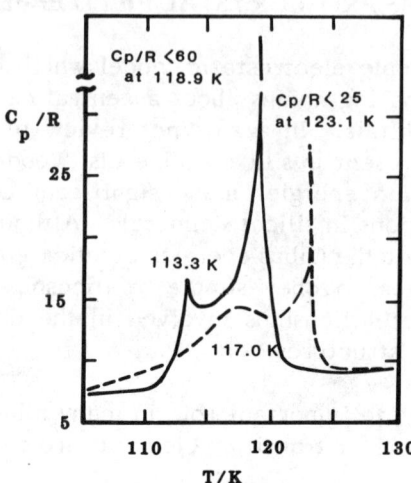

Fig. 1. The heat capacity of magnetite. Two samples showing different bifurcation of the Verwey transition.

Even at present, the furcation of the heat-capacity anomaly at the Verwey transition in Fe_3O_4 continues to be somewhat of a conundrum—albeit a well-established one. Perhaps, nature in the raw is seldom simple!

IRON DISULFIDES—MARCASITE/PYRITE

Iron disulfide occurs in nature and can be prepared also in the laboratory in two polymorphic forms, marcasite and pyrite. Pyrite is formed under widely varying conditions while marcasite only under strict limitations. Hydrothermal laboratory experiments have shown that the formation of pyrite is favored in elevated temperatures in neutral or slightly basic solutions, while marcasite forms in colder, more acidic solutions. Pyrite and marcasite both occur together under laboratory conditions and in nature as is commonly the case for monotropic polymorphic forms. Some natural crystals of the marcasite habit have, in fact, already undergone transformation to pyrite or do so upon crushing. Although the irreversibility of the marcasite to pyrite transformation had been noted by many investigators, quantitative information about its instability from thermodynamic measurements and the possible delineation of a limited stability range at lower temperatures as a consequence of subtle differences in vibrational properties needed consideration. It seemed a priori more reasonable that the heat capacity of marcasite would be higher than that of structurally related pyrite due to the larger molar volume of the former. We, therefore, measured the heat capacities of both substances from 5 to 700 K and slightly above (29,30), and the results are depicted in figs. 2 and

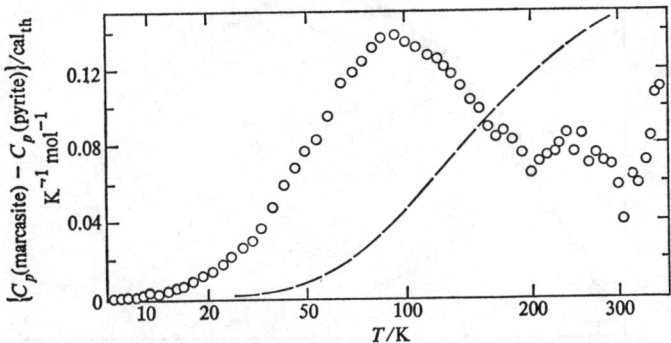

Fig. 2. Deviation of the low-temperature heat capacity of marcasite (o) from the smoothed values for pyrite. The dashed line indicates 0.1 per cent deviation.

3. As will be seen the low-temperature heat capacities are very nearly the same with marcasite slightly higher. The same is true for the higher-temperature data except that marcasite at a temperature of about 706 K began evolution of heat which then took place over a period of 98 hours and the heat capacity at higher temperatures was found comparable to that of pyrite. Even in the absence of zero-point entropy in the system, marcasite—even in the vicinity of the transformation temperature—is thermodynamically metastable with respect to pyrite; hence the transformation is monotropic and reversible. Utilizing an enthalpy cycle we deduced from the enthalpy increment for the marcasite-to-pyrite transformation at 706 K that the zero-point enthalpy difference of marcasite over pyrite is (0.99 ± 0.05) kcal/K mol^{-1}. Marcasite is clearly seen to be metastable with respect to pyrite at all temperatures above $T = 0$. Hence, the geological occurrence of pyrite pseudomorphs after marcasite is consistent with the above interpretation. Marcasite occurs in low-temperature hydrothermal deposits and apparently occasionally transforms to pyrite at temperatures well below 700 K.

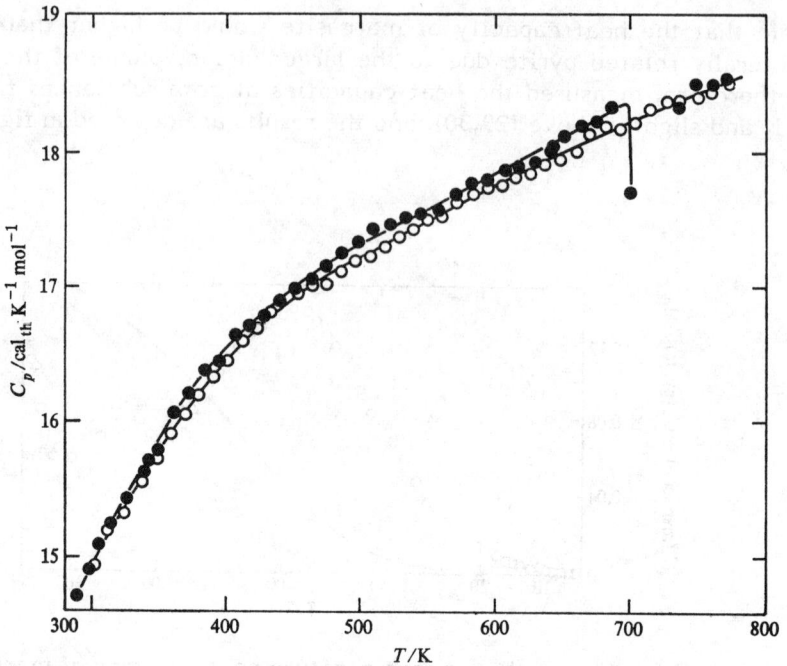

Fig. 3. Heat capacities of marcasite (●) and pyrite (o) at higher temperatures.

IRON AND TRANSITION-ELEMENT-BEARING SYSTEMS—AN EXAMPLE

As geochemical modelling of natural processes becomes more quantitative the need for accurate thermodynamic and phase equilibrium data becomes more acute. During the past decade there has been a marked increase in both the quality and quantity of thermodynamic data for rock-forming materials (31,12). But since thermochemical and phase-equilibrium investigations have focused mainly on iron-free systems, there are few accurate thermodynamic data for iron- and other transition-element-bearing geological materials.

Transition-element-bearing minerals (silicates, oxides, carbonates, etc.) are significant consituents of many crustal rocks including metasediments and metaigneous materials, as well as of some igneous and hydrothermal products. Equilibria in iron-bearing systems include a number of potentially useful geothermometers and geobarometers. These systems also provide sensors of the fugacities of volatile components (e.g., O_2 and S_2) involved in redox reactions and also in decarbonation and dehydration reactions. Ferrous and ferric phases form end-members of extensive crystalline solution series and thermodynamic data are needed in the evaluation of activity-composition relations in—and element partitioning between—crystalline solutions. A firm base of thermodynamic and phase equilibrium data for such minerals will thus contribute to a more quantitative understanding of a wide array of geochemical processes and products.

The calorimetric determination of the low-temperature heat capacities of synthesized, well-characterized, transition-element minerals are again the prime requisite. The data are of fundamental interest in the evaluation of lattice, magnetic, and electronic configuration contributions to the entropies of these minerals. Heat capacities are required in the evaluation of thermodynamic properties including the Gibbs energy (G), the entropy (S), and the enthalpy increment (H). The entropy data, in conjunction with the molar volumes of solid phases and the Gibbs energies of volatiles can be used with critically-evaluated, experimental, phase-equilibrium data (see next section), and, where possible, solution of calorimetric enthalpies of formation, to derive internally consistent sets of thermodynamic data for such minerals. This information will be utilized to develop phase diagrams for these systems useful in interpreting a range of petrological and geochemical processes. Heat-capacity measurements are also important in clarifying the thermal energy spectrum and the electronic behavior of materials. Experimental measurements are made down to some low temperature [unfortunately, this is about 50 K for older measurements on minerals (14), but more often 5 K for some modern experiments (32-33)] and the data extrapolated to lower temperatures by appropriate procedures.

In thermophysical studies the third-law entropy (S) is generally utilized and incorporates contributions such as S (lattice), S (electronic), S (magnetic; ΔS_{mag}), ΔS_{trs} (the entropy associated with first-order phase transition), and ΔS_{res} (that associated with residual contributions at zero K) (16, 34).

Thermally activated vibrations of atoms in a structure account for the "lattice" heat capacity, the single major entropy contribution (35-36). For temperatures above 300 K the heat capacities of many silicates, complex oxides, etc., approximate the sum of that of their constituent oxides (37-39, 13). Vibrational modes which are not fully activated at T ⩾ 300 K are largely dependent on the nature of individual cation-oxygen bonds rather than on the atomic arrangement of the complex solid. On the other hand low-temperature heat capacities are not additive (despite Kopp's Law), implying that the atomic vibrations are strongly structure dependent. At present, for most complex minerals, low-temperature "lattice" heat capacity contributions cannot be accurately calculated or estimated—although some progress is being made in this respect (36, 7-11). The sparsity of heat capacity data available for ferruginous, rock-forming minerals, together with known "complications" in the heat-capacities (see below) require accurate low-temperature calorimetric measurements for entropy evaluations. Although "lattice" heat capacity contributions above 300 K can usually be estimated with some confidence, super-ambient temperature measurements for certain phases are also desiderata.

In addition to their lattice-contribution terms, transition-element compounds often exhibit significant magnetic-skin configurational entropy contributions occasioned by ordering of their unpaired electrons. This effect has been detailed in a number of studies (40-42). Below their Curie temperatures ferrimagnetic materials such as magnetite have adjacent, unequal electron spin states arranged in antiparallel arrays. The transition to the paramagnetic state (disordered electron spins) at higher temperatures involves a lambda-shaped maximum in the heat capacity versus temperature curve—with concomitant magnetic transition entropy (43). Other transition element compounds (e.g. $MnCO_3$; $CoCl_2$ and $MnCl_2$; FeF_2; $FeWO_4$; and Fe_2SiO_4 which are paramagnetic at high temperatures transform to antiferromagnetic at low temperatures. (Similar magnetic transitions may be expected in other transition-element minerals at low temperature.) The molal entropy increment accompanying such magnetic transitions approaches the spin only value

$$\Delta S_{mag}/R = \ln(2s + 1)$$

where s represents the spin state of the transition ion. For Fe^{2+} (s = 2), $\Delta S_{mag}/R = 1.6$, for Fe^{3+} (s = 5/2), $\Delta S_{mag}/R = 1.78$, and for Mn^{2+} (s = 5/2), $\Delta S_{mag}/R = 1.78$. Even for the available low temperature heat capacity measurements on rock-forming minerals this contribution of the magnetic transition has not generally been adequately assessed. For example Kelley's work (44) on fayalite (the only iron silicate in the literature so far) involved C_p measurements only down to 55 K—in the region of the magnetic transition for this phase. Kelley made a highly uncertain extrapolation through this region to zero K—resulting in a serious error shown by our measurements on fayalite. For fayalite and other minerals such errors in the entropies could lead to discrepancies on the order of several kilocalories in enthalpies.

Although ΔS_{mag} may be predicted in simple instances, even in fayalite, a more subtle magnetic ordering is observed below the Curie temperature

(45). There is no adequate substitute for direct equilibria heat-capacity measurement today (46). Moreover, heat-capacity data enhance our understanding of magnetic/electronic behavior of minerals.

At low temperatures transition-element bearing compounds may have α-electron configurations reflecting orbital splitting due to crystal-field effects (47). With increasing temperature the d-electron distribution may become partially or completely disordered, contributing an electronic entropy term (35,46,48). Wood and Strens (48) have calculated this electronic entropy contribution for Fe^{2+} in octahedral coordination in orthopyroxene and note that the full entropy contribution may not be attained below temperatures >1500 K. The electronic entropy contribution can be evaluated from low-temperature adiabatic calorimetric measurements by subtraction of lattice and magnetic terms, estimated by corresponding state methods utilizing isostructural diamagnetic materials and the Schottky heat-capacity contribution thus inferred. The volumetric estimation of the lattice contribution described in another paper of this series (15) is essential here. Super-ambient temperature heat-capacity measurements (by adiabatic calorimetry or by D.S.C.) are also desiderata for evaluation of the total heat capacity and contributing electronic terms within this stability field of the minerals involved. Such measurements are useful in examining the accuracy of Wood and Strens' models for electronic contributions over a range of iron-bearing mineral structures and assist in developing models for the estimation of the heat capacities of other transition-element minerals.

The main cause for the existence of residual entropies is the persistence at low temperatures of site disorder instead of the ordering or unmixing that should take place under equilibrium conditions. Substitutions and vacancies also contribute to residual terms. Detailed reviews of these effects in minerals are available in the literature (34,13,39). Residual entropy terms may be evaluated from some combination of third-law analysis of phase-equilibrium data using accurate heat capacity functions (43) and theoretical models of configurational entropy terms (34,13). Third-law analyses of tightly reversed phase-equilibrium data (see next section) using our heat-capacity measurements may also reveal unsuspected residual entropy terms in the experimental materials.

In summary, calorimetric measurements contribute significantly to the understanding of the systematics of heat-capacity and entropy contributions in transition-element bearing minerals and are valuable in the development of more adequate theories of the thermal behavior of crystalline materials. The calorimetric data are essential in the determination of accurate thermodynamic properties for these minerals. As discussed in the following section, the derived heat-capacity and entropy functions are used in the retrieval of thermodynamic properties from phase equilibrium data and for development of geochemically useful phase diagrams for these systems.

To facilitate the thermodynamic calculations, a computer program package such as EQUILI, originally developed by Wall and Essene (1971) (50) and modified by Slaughter, Wall, and Kerrick (51), and subsequently greatly amplified by Wall (52) is appropriate. The programs can handle reactions involving crystalline materials and common volatile species (H_2O, CO_2, H_2S, O_2, F_2, H_2, etc.). Data bases include complete thermodynamic functions for the major volatiles to 30 kb, provision for non-ideal mixing of volatiles (53) as well as molar volumes, available Gibbs energy data, heat-capacity functions, thermal expansion and compression functions for over two hundred crystalline materials.

The prime source of data for phase diagram calculation and the generation of thermodynamic data are critically-evaluated, well-reversed experimental phase relations. There are few reliable thermochemical measurements on Fe-bearing minerals and the uncertainties in the measured enthalpies of formation commonly introduce larger errors than good experimental data (for extensive discussion see Helgeson et al. 13). Where a number of independent reactions involving a set of phases is available the Gibbs energy functions will be simultaneously regressed to ensure consistency in the data set.

Such thermodynamic and phase equilibrium calculations provide a markedly improved thermodynamic data base for transition-element minerals and phase diagrams valuable in the quantitative evaluation of physiochemical conditions for a range of rock-forming environments. For example, from published experimental and thermochemical data and new C_p measurements, phase diagrams for much of the system Ca-Fe-Si-C-O-H-S (applicable to skarn development and iron formation metamorphism) can be developed. Equilibria among iron-rich aluminosilicates, oxide, and sulphide phases, many of which are difficult to determine experimentally, can also be derived for a wide range of conditions. Moreover apart from mineralogy and petrology, such thermodynamic investigation form a basis for further key experimental and thermochemical studies of transition-element compounds.

Quite apart from uses of entropy data in calculations of phase equilibria, entropy measurements of a large number of iron phases will give us a much greater appreciation of the effects of local bonding on S(lattice),

S(electronic), and S(magnetic). While Fe^{2+} is present in octahedral sites in many ferrous iron compounds, the difference in entropy between the equivalent Mg compound and the Fe compound per mole of Mg-Fe^{++} yields the sum differences among all these terms for each structure as a function of temperature. This will enable better estimates of entropies in related compounds than are at present feasible. For instance, tests can be made as to whether differences in pyroxene entropies are better models for estimates of ΔS(Fe-Mg) in amphiboles than differences in oxide entropies (13). Similar tests for Al-Fe^{3+} in octahedral sites by comparing jadeite vs. aegirine, grossular vs. andradite and the available data for corundum vs. hematite enables extension of the entropy estimation schemes of Latimer (37), of Fyfe (38), of Helgeson et al. (13), and of Westrum and Chirico (54) to complex solid solutions in pyroxene, amphiboles, micas, and oxides with critical tests attesting to the accuracy of such estimates.

CORRELATIONS OF ENTROPY AND ENTHALPY VALUES

The great utility of the thermodynamic approach to the study of reactions in mineralogical and petrological systems as opposed to direct experimentation is that provided the thermodynamic data are available it is possible to calculate the position of equilibrium in reactions which cannot be studied experimentally because of the time required for equilibration. This applies particularly to reactions in which only solids are involved and especially when they take place at relatively low temperatures. Moreover, provided activity-composition relationships are available for solution phases it is possible to calculate the effect of variable composition in any phase in the reaction of interest. To attempt to do this experimentally on all conceivable bulk compositions and pressure and temperature ranges is clearly not feasible. However, the validity of the thermodynamic approach depends upon the availability of entropy, enthalpy, and volume increment data for pure phases and knowledge of the mixing properties of multicomponent phases.

Enthalpy of formation data relative to the oxides—or to the elements—may be derived calorimetrically by accurate measurement of the heat evolved upon dissolution of the compounds in a suitable solvent. At relatively low temperatures hydrofluoric acid in aqueous solution is utilized; at higher temperatures molten salts have proven to be a convenient solvent for many classes of minerals. By taking advantage of the fact that both enthalpy and entropy increments are functions of state, one may obtain reaction data if tabulated values are not conveniently accessible.

Extraction of Thermodynamic Data from Phase Equilibrium Experiments

Let us suppose that a simple reaction occurs between four mineralogical species designated as A, B, C, and D; i.e., A + B = C + D, and utilize the hypothetical data which define the stability fields for the two simple assemblages in that system. Let us further assume that the $\Delta V^o_{298,\,1\,bar}$ is known (and is positive). As shown in fig. 4, C and D are then the phases stable together at high temperatures and low pressures.

At equilibrium, the Gibbs energy increment will be that given by the following equation provided the standard states of all components are the pure phase at the pressure and temperature of interest:

$$\Delta G^o_{P,T} = 0 = \Delta H^o_{1\,bar,T} - T\Delta S^o_T + \int_1^P \Delta V^o dP$$

Provided that ΔC_p is zero and that ΔV^o is constant, two points on the equilibrium boundary would be adequate to obtain $\Delta H^o_{1\,bar}$ and ΔS^o. However, as suggested in figure 4 phase "equilibrium" experiments are performed in such a way that equilibration between reactants and products is rarely obtained. Product and reactant assemblages are mixed together and held at a prescribed pressure and temperature for an appropriate length of time (usually days or weeks). Some experiments result in the growth of A and B, others in the growth of C and D. It is rare for no reaction to be observed. The actual position of the equilibrium boundary is, therefore, bracketed by the experiments in which growth of either reactants or products occurs. From these bracketing values thermodynamic data must be derived.

Obviously, if the pressure and temperature of the experiment is close to—or even on—the equilibrium boundary, no reaction should take place. It is therefore tempting to assume that "no reaction" corresponds to equilibrium between products and reactants and to use the results to obtain enthalpy and entropy data. The inadequacy of this conclusion is revealed in that although two experiments which yielded "no reaction" lie close to the equilibrium boundary, two others are nearly 100 K distant from it. The reason for the errors in the latter points is probably that reaction rates are slow at low temperatures and little reaction may be observed in long experiments even under conditions far removed from equilibrium. Hence, "no reaction" is not an adequate criterion of equilibrium. Only those experimental runs in which reaction did occur may be used to provide data.

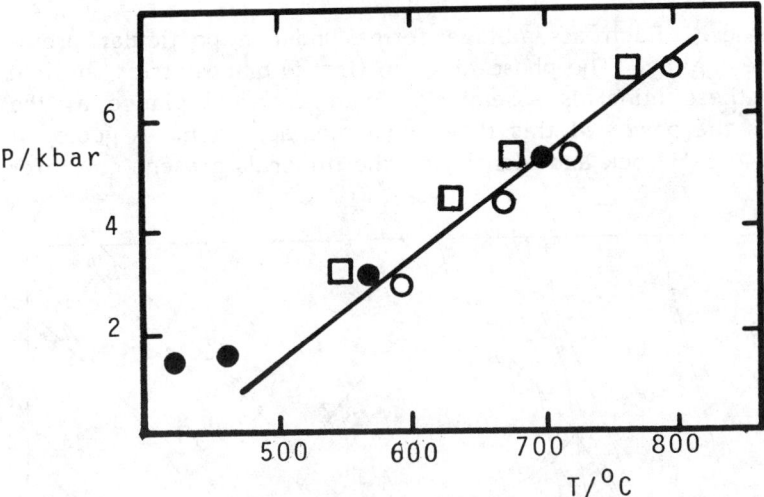

Fig. 4. Stability field for a real mineral reaction designated as A + B = C + D; $\Delta V^o > 0$. Results of bracketing experiments: A + B found stable = □ ; C + D found stable = o; and no reaction = ●.

It will be seen that only certain ranges of values of $\Delta H^o_{1\,bar}$ and ΔS^o are consistent with the data. The values of ΔS^o and ΔH^o are interdependent, and it is customarily assumed that the best estimate of ΔS^o is that value which gives a maximum range of internally consistent values of $\Delta H^o_{1\,bar}$ (equivalent to maximum uncertainty). This is equivalent to fitting a best slope to the experimental bracketing values, i.e., a slope which most closely bisects all of the bracketing points shown in the figure. Usually this is close to half way between the two extreme values.

Different procedures are needed for extracting thermodynamic data from experiments involving volatile products or reactants and indeed an extended discussion of this topic has been provided elsewhere (3,55).

ANOTHER EXAMPLE—THE C-A-S-H SYSTEM

The most important phase diagram that we have so far assembled is that of the so-called C-A-S-H System. Although it may sound like a get-rich-quick scheme, it actually abbreviates an assemblage of chemical components found in many rocks: calcium oxide (C), aluminum oxide (A), silicon dioxide (S), and water (H). These four compounds may be combined into a myriad of mineral phases. Each assemblage forms under a particular pressure/temperature regime. The phase diagram (fig. 5) places strict physical limits on how these minerals assemble. Even a casual glance at the diagram reveals the power of this thermodynamic approach. A geologist picking up a C-A-S-H rock and identifying the minerals present can infer

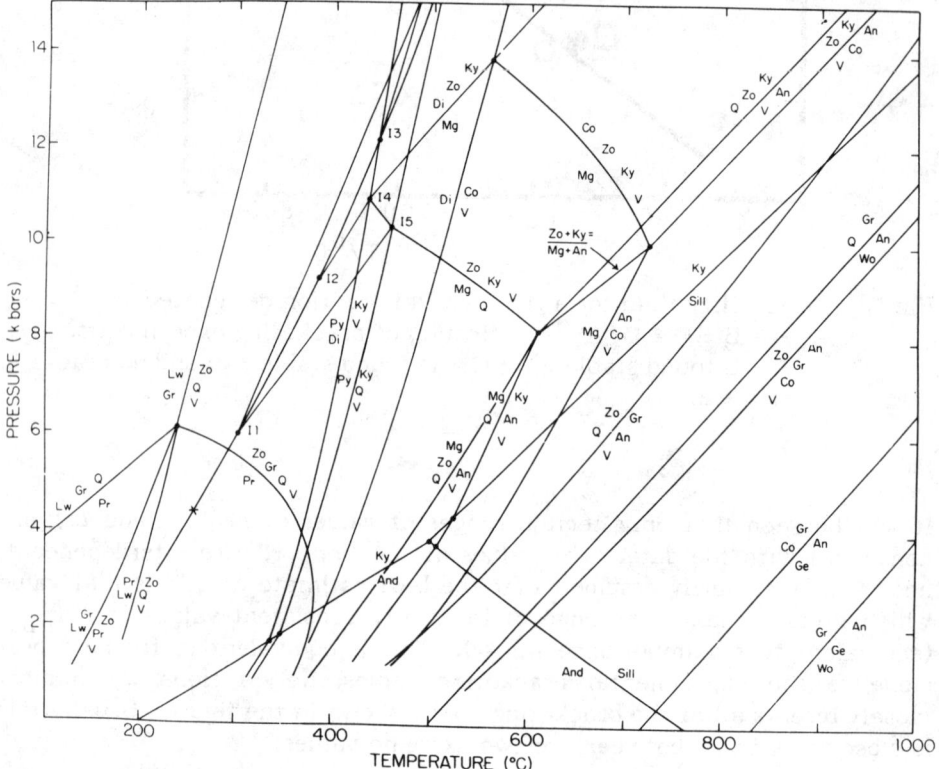

Fig. 5. The phase diagram (simplified by excluding the reactions involving zeolites or kaolinite) for the C-A-S-H system. (See text for identification of abbreviations and other interpretations.)

the temperature/pressure regime which created the rock. Since such information is necessary before other conditions (such as the vapor pressures of gases) can be deduced, the C-A-S-H diagram makes a whole new array of information accessible to geologists. The diagram tends to illustrate the complexity of mineralogical systems. Each line represents a chemical reaction which can yield mineral products on either side. The minerals represented here are: And = andorsite, An = anorthite, Co = corundum, Di = diaspore, Gr = grossular, Ky = kyanite, Lw = lawsonite, Mg = margarite, Pr = prehenite, Py = pyrophillite, Q = quartz, Sill = sillimanite, V = water. It should be noted that this is a simplified version of the diagram.

An interesting aspect of work on this system concerns eclogite, a garnet-rich mineral which has stirred considerable controversy among geologists interested in the evolution of the earth's crust. Eclogite is the high-pressure equivalent of basalt but is much denser and consists of garnet and pyroxine minerals.

Another aspect concerns understanding the metamorphosed sulfide deposits near Ducktown, Tennessee using the phase diagram of fig. 5. Bruce Nesbitt showed that high water pressures were maintained during the metamorphic event some million years ago (56). He mapped a halo in fluid compositions around the formation. Such discoveries provide geologists with yet another tool in exploring for valuable minerals in metamorphosed rocks. Although further examples might be considered a desiderata, only one extraterrestrial application will be permitted by the available time.

THE ROCKS AND MINERALS OF THE VENUSIAN TROPOSPHERE

Based on the data of the Soviet Venera 11 and 12 missions and the American Pioneer Venus missions the thermodynamic calculation of a 16-component multisystem has been performed for modelling the Venusian troposphere surface-rock interactions. The mineral assemblages in surface rocks are considered to be the result of chemical interaction of basalts and rhyolites with the ascending and descending gas fluxes of the global convective tropospheric cells. These fluxes are assumed to differ in water vapor content as well as in microcomponent concentrations. Hydroxyl, chloride, fluoride, and sulfur incorporation into surface rock mineral phases has been predicted (57).

An even more recent thermodynamic study treats only to the first approximation the complicated processes in the system troposphere-surface rocks but has lead to the conclusion that the formation of solid antimony oxychlorides is plausible in the cloud cover, but the calculations are hindered by lack of adequate data on these substances. The condensation of metastable, monoclinic arsenic oxide crystals within the main cloud layer or their identification as the mode 3 particles by the probes is not excluded by the thermodynamic calculations. Elemental sulfur particles associated with the dominant sulfuric acid droplets are considered to be the most likely candidates. The stability field of orthorhombic sulfur coincides with the pressure/temperature conditions of the main cloud layer. Small undercooled sulfur droplets might also be good candidates. Iron and mercury chlorides occur only in negligible mass concentrations and the existence of ammonium and aluminum chlorides is excluded by the thermodynamics. It seems unlikely that liquid hydrochloric acid and sulfuric acid aerosols coexist within the cloud cover. These facts lead to the theoretical implication of the important role of sulfur, chlorine, arsenic, and possible antimony in Venusian meteorology playing a role similar to that of water in the terrestrial atmosphere (58).

ROCKS: A PERSPECTIVE ON THERMODYNAMIC DATA NEEDS

Recent research emphasis on geological isolation of radioactive nuclear wastes and geothermal energy resource development has created a renewed demand for thermophysical property data on rocks and other geological materials at various conditions including elevated pressures and temperatures. In contrast to fabricated engineering materials with specified properties, rocks are complex, naturally-occurring materials having properties which must be characterized—rather than specified—for technological utilization. However, since a fair amount of data are available on the constituent minerals of the rocks the values can usually be synthesized adequately for technological purposes (59). There is every indication that during the next decade the thermophysical and thermodynamic properties of minerals and rocks will be a prime subject of investigation.

NOTE

1. The author, like virtually all other speakers, took the opportunity at his first presentation to thank the host organization for their kindness and hospitality and to note the excellence of the arrangements. He also expresses his gratitude to the Experimental and Theoretical Geochemistry Program of the Division of Earth Science and to the Thermodynamics and Structural Chemistry Section in the Chemistry Division of the National Science Foundation for continuing support of the research endeavors on which this presentation is based under Grant G EAR-8009538 and CHE-8007977.

REFERENCES

It should be particularly noted that the references given here are suggestive rather than exhaustive in order to maximize space for the text itself.

1. Powell, R.: 1978, "Equilibrium Thermodynamics in Petrology: An Introduction" (Harper & Row).

2. Greenwood, H.J. (ed.): 1981, "M.A.C. Short Course in Application of Thermodynamics to Petrology and Ore Deposits" (Evergreen).

3. Wood, B.J., Fraser, D.G.: 1977, "Elementary Thermodynamics for Geologists" (Oxford).

4. Saxena, S.K.: 1973, "Minerals, Rocks and Inorganic Materials, no. 8: Thermodynamics of Rock-Forming Crystalline Solutions," eds. von Engelhardt, W., Hahn, T., Roy, R. (Springer-Verlag).

5. Newton, R.C., Navrotsky, A., Wood, B.J. (eds.): 1981, "Advances in Physical Geochemistry, Vol. 1: Thermodynamics of Minerals and Melts" (Springer-Verlag).

6. Saxena, S.K. (ed.): 1982, "Advances in Physical Geochemistry," Vol. 2 (Springer-Verlag).

7. Kieffer, S.W.: 1979, Rev. of Geophys. & Space Phys. 17, pp. 1-19.

8. Kieffer, S.W.: 1979, Rev. of Geophys. & Space Phys. 17, pp. 20-34.

9. Kieffer, S.W.: 1979, Rev. of Geophys. & Space Phys. 17, pp. 35-59.

10. Kieffer, S.W.: 1980, Rev. of Geophys. & Space Phys. 18, pp. 862-886.

11. Kieffer, S.W.: 1980, Rev. of Geophys. & Space Phys.

12. Robie, R.A., Hemingway, B.S., Fisher, J.R.: 1978, "Thermodynamic Properties of Minerals and Related Substances at 298.15 K and 1 Bar (10^5 Pascals) Pressure and at Higher Temperatures," Geological Survey Bulletin 1452 (U.S. Gov't. Printing Office).

13. Helgeson, H.C., Delany, J.M., Nesbitt, H.W., Bird, D.K.: 1978-A, Amer. Jour. Science 278-A.

14. Kelley, K.K., King, E.G.: 1961, "Entropies of the Elements and Inorganic Compounds" Bulletin 592 (U.S. Bureau of Mines, U.S. Gov't. Print. Off.). and related documents.

15. Westrum, E.F., Jr.: 1982, this volume, paper entitled "Calorimetry of Phase, Schottky and Other Transitions."

16. Ulbrich, H.H., Waldbaum, D.R.: 1976, Geochimica et Cosmochimica Acta. 40, pp. 1-24.

17. Westrum, E.F., Jr., Furukawa, G.T., McCullough, J.P.: 1968, in "Experimental Thermodynamics," Vol. 1, eds. McCullough, J.P., Scott, D.W. (Butterworth's & Co.) pp. 133-214.

18. Westrum, E.F., Jr.: 1982, this volume, paper entitled "Computerized Adiabatic Thermophysical Calorimetry."

19. Holm, J.L., Kleppa, O.J., Westrum, E.F.,Jr.: 1967, Geochim. Cosmochim. Acta. 37, pp. 2289-2307.

20. Navrotsky, A.: 1977, Phys. Chem. Min. 2, pp. 89-104, and Navrotsky, A.: 1979, Ann. Rev. Earth Planet. Sci. 7, pp. 93-115.

21. Thompson, A.B., Perkins, E.H.: 1981, in "Advances in Physical Chemistry," Vol. 1, eds. Newton, R.C., Navrotsky, A., Wood, B.J. (Springer-Verlag) pp. 35-62.

22. Holland, T.J.B.: 1981, in "Advances in Physical Chemistry," Vol. 1, eds. Newton, R.C., Navrotsky, A., Wood, B.J. (Springer-Verlag) pp. 19-34.

23. Wood, B.J.: 1981, in "Advances in Physical Chemistry," Vol. 1, eds. Newton, R.C., Navrotsky, A., Wood, B.J. (Springer-Verlag) pp. 63-84.

24. Westrum, E.F., Jr., Grønvold, F.: 1969, J. Chem. Thermodyn. 1, p. 543.

25. Bartel. J.J., Westrum, E.F., Jr.: 1975, AIP Conf. Proc. 24, p. 86; and Bartel, J.J., Westrum, E.F., Jr.: 1976, J. Chem. Thermodyn. 8, p. 583.

26. Evans, B.J., Westrum, E.F., Jr.: 1972, Phys. Rev. B5, p. 3791.

27. Evans, B.J.: 1975, AIP Conf. Proc. 24, p. 73.

28. Kelber, J., Shirane, G., Evans, B.J.. Westrum, E.F., Jr.: 1977, Solid State Communications 21, pp. 551-554.

29. Grønvold, F., Westrum, E.F., Jr.: 1962, Inorg. Chem. 1, p. 36.

30. Grønvold., F., Westrum, E.F., Jr.: 1976, J. Chem. Thermodyn. 8, pp. 1039-1048.

31. Robie, R. A., Waldbaum, D.R.: 1968, "Thermodynamic Properties of Minerals and Related Substances at 298 15° K (25.0° C) and One Atmosphere (1.013 bars) Pressure and Higher Temperatures," Geological Survey Bulletin 1259 (U.S. Gov't. Printing Office).

32. Westrum, E.F., Jr., Essene, E.J., Perkins, D.: 1979, Jour. Chem. Thermodyn. 11, pp. 5766.

33. Perkins, D. Essene, E.J., Westrum, E.F., Jr.: 1977, Contr. Miner. Petrol. 64, pp. 137-147.

34. Swalin, R.A.: 1972, "Thermodynamics of Solids," 2nd Ed. (John Wiley).

35. Gopal, E.S.R.: 1966, "Specific Heats at Low Temperatures" (Plenum).

36. Chirico, R.D., Westrum, E.F.,Jr., Gruber, J.B., Warmkessel, J.: 1979, J. Chem. Thermodyn. 11, pp. 835-850.

37. Latimer, W.M.: 1952, "The Oxidation States of the Elements and Their Potentials in Aqueous Solutions" (Prentice Hall).

38. Fyfe, W.S., Turner, F.J., Verhoogen, J.: 1958, Geol. Soc. Amer. Mem. 73, p. 259.

39. Helgeson, H.C.: 1969, Amer. J. Sci. 267, pp. 729-804.

40. Stout, J.W., Giauque, W.F.: 1941, Jour. Amer. Chem. Soc. 63, p. 714.

41. Lyon, D.H., Giauque, W.F.: 1949, Jour. Amer. Chem. Soc. 71, p. 1647.

42. Lyon, W.G., Westrum, E.F., Jr.: 1974, J. Chem. Thermodyn. 6, pp. 781-786.

43. Grønvold, F., Sveen, H.: 1974, Jour. Chem. Thermodyn. 6, pp. 859-872.

44. Kelley, K.K.: 1941, Jour. Am. Chem. Soc. 63, p. 2750.

45. Santoro, R.P., Newnham, R.E., Nomura, S.: 1966, Jour. Phys. Chem. Solids. 27, pp. 655-666.

46. Lyon, W.G., Westrum, E.F., Jr.: 1974, J. Chem. Thermodyn. 6, pp. 763-780.

47. Burns, R.G.: 1970, "Mineralogical Applications of Crystal Field Theory" (Cambridge Univ. Press).

48. Wood, B.J., Strens, R.G.J.: 1972, Mineral. Mag. 38, pp. 910-917.

49. Thompson, J.B., Jr.: 1969, Amer. Mineral. 54, pp. 341-375.

50. Wall, V.J., Essene, E.J.: 1972, Geol. Soc. Amer. (Abstr. with Program) 4:7, p. 400.

51. Slaughter, J., Wall, V.J., Kerrick, D.M.: 1976, Contr. Mineral. Petrol. 54, pp. 157-171.

52. Wall, V.J.: unpublished.

53. Wall, V.J., Burnham, C.W.: 1981, personal communication; and MRK II by Holloway, M.J., personal communication.

54. See, for example, Westrum, E.F., Jr.: 1983, J. Chem. Thermodyn. (Rossini Lecture) (in press).

55. See reference #2, chapter 13.

56. Nesbitt, B.: 1982, personal communication.

57. Barsukov, V.L., Volkov, V.P., Khodakovsky, I.L.: 1980, in "Proc. Lunar Sci. Conf. 11th," (Pergamon) pp. 765-773.

58. Barkusov, V.L., Khodakovsky, I.L., Volkov, V.P., Sidorov, Yu. I., Dorofeeva, V.A., Andreeva, N.E.: 1981, in "Proc. Lunar Planet. Sci. Conf., 12 B," (Pergamon) pp. 1517-1532.

59. Hendricks, R.C.: 1980, in "The Technological Importance of Accurate Thermophysical Property Information," eds. Sengers, J.V., Klein, M. Nat'l. Bur. of Standards Bulletin 590 (U.S. Gov't. Printing Office).

COMPUTERIZED ADIABATIC THERMOPHYSICAL CALORIMETRY[1]

Edgar F. Westrum, Jr.

Department of Chemistry, University of Michigan, Ann Arbor, Michigan 48109 U.S.A.

The dedicated minicomputer is lightening the load and enhancing the productivity of the adiabatic calorimetrist over all regions of temperature.

What approaches are favored and how are such systems really working out? What hazards and risks are introduced and what problems are eliminated? Can automated-adiabatic calorimetry really compete with dsc in terms of demands on the scientists' time? What advantages does it possess?

A number of different operating systems will be examined and their characteristic features studied. The probable effect on trends in calorimetry will be noted.

INTRODUCTION

The capabilities of both mini- and micro computers, self-balancing bridges and DVM's, as well as programmable amplifiers and precision power supplies facilitate the automation of calorimetric gear in virtually every temperature range. Moreover the very repetitiousness of the measurements is well suited to automation. The measurements and preliminary calculations can proceed without operator intervention for significant periods of time.

Obviously many levels of automation exist from simple data recording for off- or on-line processing, computer control of the experiment itself with self-balancing instrumentation, on-line data processing with feed-back loops to insure proper design of subsequent experiments, and even automated computer-controlled adiabatic shield control.

The consequences of automation include:

- the removal of tedium and fatigue of extended experiments for the operator freeing him for interpretation of data and devising better experiments,

- patience in awaiting thermal equilibration (especially in regions of hysteresis or slow equilibrium)

- elimination of many human errors in recording data

- enhanced performance of calorimetric system

- immediate plotting of heat capacity curves (CRT screen)

- possibility also of design of digital processes for adiabatic shield control, equal to that of analog controllers.

- provision of better populated (more points/more precise points) heat-capacity curves

- improved experimentally derived values of heat capacity—and of deduced thermophysical properties

THE ANN ARBOR SYSTEM

Let us first look at a modern system to get a feeling for and an impression of the modus operandi and then proceed to consider other systems in use, and finally examine in greater detail some aspects and constraints. Most of my remarks will pertain to the chemical thermodynamic and the super-ambient ranges—but most of them will also be applicable—with occasional modifications, caveats, etc. — to lower- and higher-temperature ranges.

The System Itself

A low-temperature adiabatic cryostat (1,2) described previously and a super-ambient thermostat (1,3) (cf. fig. 1) represent without significant modification the actual calorimeter sites and are both serviced by analog shield control units, an ASL 7-digit inductance bridge with modified switching unit, and high-precision dc constant-current supply. These units are interfaced with a dedicated DEC PDP-11-V03 computer with dual floppy disks and connections to the Michigan Terminal System. The controls of the apparatus are shown in fig. 2. The computer often operates one or the

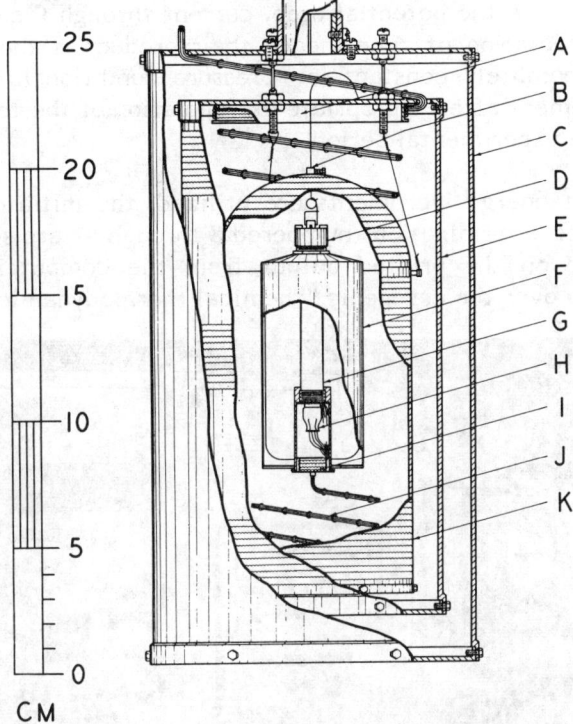

Fig. 1. Cross section of superambient temperature adiabatic calorimeter thermostat. The symbols represent:
A. Guard shield ring. B. Calorimeter suspension collar, C. Primary radiation shield, D. Calorimeter closure assembly, E. Guard shield, F. Calorimeter assembly, G. Thermometer-heater assembly, H. Thermometer, I. Thermal equilibration spool, J. Lead Bundle, K. Adiabatic shield.
(In more recent models an additional guard shield has been added.)

other of the calorimeters for continuous periods of 24 hours and records and displays the relevant data.

Data Acquisition and Calorimeter Control

The heat capacity is characteristically obtained from the equation

$$C_p = \lim_{\Delta T \to 0} \Delta H/\Delta T \cong \Delta H/\Delta T = EI\Delta t/\Delta T = "C_p" \quad (1)$$

in which E,I and t are the potential drop, current through the calorimeter "heater" and the duration of time (hence their product is the energy increment at approximately constant low-pressure conditions). ΔT is the temperature increment after appropriate extrapolation of the temperature-time drift into the experimental period.

Since intermittent energy increments are utilized, the initiatory instructions for a series of determinations numbered 8 through 47 are shown in fig. 3 as incorporated on the printed output from the computer using the ADCAL 2 program over the cryogenic "chemical thermodynamic range."

Fig. 2 The control unit for the Ann Arbor cryogenic and superambient sytems. The computer is in the foreground. The left bank of the cabinet has the 5-channel analog shield control system with, (from top to bottom) Keithley microvolt amplifiers, 3-pen recorders, Leeds and Northrup CAT 80 controllers, Kepco power supplies, shield/drift adjusters. The central bank has a 3-pen recorder, the ASL bridge and switching units, and a programmable power supply. The right bank has a digital quartz clock, the microcomputer interface, the digital nanovolt meter, the scanner (multiplexer), timers, programmable constant voltage unit to drive the constant current power supply, 5-channels of improved shield control unit, and a constant temperature bath for the standard resistors.

Typical initial global constraints and preprograming for the runs presented in fig. 3 are discussed in table 1.

The computer then commences with a presentation of drift (cf. fig. 4) together with the first and second time derivations. After criteria to permit the computer to determine the lead resistance, the full display of the enhanced sensitivity of the ASL automated bridge assessed from (repeated) linear integration of the imbalance of the inductance occurs and least-squares fitting of the linear equation to the drift occurs, (cf. fig. 5) and continues until the preassigned criteria are satisfied.

The energy input mode is then undertaken by the computer and recorded. The potential and current readings are taken repeatedly as seen in fig. 6. Here the constant current source is controlled by a computer programmable constant voltage source and the integrated power used to ascertain the energy supplied.

It will be noted that the heater resistance in fig. 6 varies by several per cent. This was occasioned by a faulty solder joint between the Karma wire of the heater and the copper leads. Despite the real variation in the resistance, the integration procedures for the energy insured reproducibility of the data to within several hundreths of a per cent.

At this point the computer extrapolates the drift to the midpoints of adjacent experimental periods (or only forward if the drift is the first of a series) and if adequate data have been recorded calculates the <u>apparent</u> heat capacity at constant pressure (fig. 7)

$$"C_p" = \Delta H / \Delta T \qquad (2)$$

Subsequent adjustment for curvature correction (4) is made in the MICHI-THERM program (6) for all except points in transition regions. Eventually the FITAB program (7) provides molal values, smoothed heat-capacity values, and derived thermodynamic functions at selected rounded temperatures in form suitable for publication.

Alternate drift and energy runs follow, and in the end a summary of the data series is printed out as shown in fig. 8. At the present time, adiabatic shield control is provided either by 5-channels of commercial approach, proportional, rate, and reset control, or 5-channels of locally devised and fabricated proportional, integral, differential controllers, both monitored by 3-track recording systems.

```
ADIABATIC CALORIMETER - DATA AQUISITION

08-JUN-82
22:31:43

O/P FILE
*HO2S3O.DA2

PUNCH FILE
*HO2S3P.DA2

RESISTOR NO., SENSITIVITY, TIME UNIT, MAX DRIFT
                                  -> 3,.002,15,2.E-05

TITLE-> HO2S3 HELIUM REGION ABOVE 10 K G.W.M.

DATA (#, SEC, DELT, C/H/Z, NAVG, NSTD, TMAX, DMAX, NO)
>8,150,-2.69,
>9,150,0,0,0,0,0,0,6
>15,240,0,0,0,0,0,0,12
>27,360,0,0,0,0,0,0,10
>37,420,0,0,0,0,0,0,10
>47,480,3.00,0,0,0,0,0,3
>
```

#	SEC	DT		AV	+/-	TM	DM
8	150	-2.6900	0	5	1	0	0.00E+00
9	150	0.0000	0	5	1	0	0.00E+00
10	150	0.0000	0	5	1	0	0.00E+00
11	150	0.0000	0	5	1	0	0.00E+00
12	150	0.0000	0	5	1	0	0.00E+00
13	150	0.0000	0	5	1	0	0.00E+00
14	150	0.0000	0	5	1	0	0.00E+00
15	240	0.0000	0	5	1	0	0.00E+00
16	240	0.0000	0	5	1	0	0.00E+00
17	240	0.0000	0	5	1	0	0.00E+00
18	240	0.0000	0	5	1	0	0.00E+00
19	240	0.0000	0	5	1	0	0.00E+00
20	240	0.0000	0	5	1	0	0.00E+00
21	240	0.0000	0	5	1	0	0.00E+00
22	240	0.0000	0	5	1	0	0.00E+00
23	240	0.0000	0	5	1	0	0.0 +00
24	240	0.0000	0	5		0	
25	240	0	0	5			

Fig. 3 Initiatory global constraints and preprogramming of runs 8 through 47 on Ho_2S_3.

Table 1. ADCAL 2 Run Control Parameters

Global Control Parameters
(These refer to entire series of determinations and must be specified)

Resistor no.:	Identification of standard resistor by number 2 = 10 ohms resistor used for $T \geq 70$ K 3 = 1 ohm resistor $\quad\quad T \leq 70$ K
Sensitivity:	Sensitivity of bridge to small changes. (value-dependent on standard resistor, temperature, and bridge settings—is computed using a special program called ADSEN)
Time unit:	Time interval (in seconds) between measurements
Max drift:	Value of first time derivative of temperature that must be reached before lead resistance is to be read

Run Parameters
(These can be varied for each run, but some need not be specified; i.e., zero indicates the default value)

#	Run number
Sec:	Length of heating run (in seconds). It must be divisible by 2 (time units in seconds)
DEL T:	Temperature increment desired DEL T < 0, a negative value is read as heater current. (The current must be specified for the first run of a series). DEL T = 0, default value if T < 100 K $\Delta T = T/20$ $\quad\quad\quad\quad\quad\quad\quad\quad$ T > 100 $\Delta T = 5$ K DEL T > 0, specifies desired temperature-increment in Kelvins.

C/H/Z	Options "0" = normal heat capacity run "1" = enthalpy run "2" = zero drift extrapolation; T_1 & T_2 computed as if drift were zero
NAVG:	Number of drift determinations which must agree to "fix" the drift.
NSTD:	Number of standard deviation units required before the program accepts a drift value as an "equilibrium" value, (Default = 1). The criterion is that the last NAVG points must lie within NSTD (n standard deviation units) of the mean of the last NAVG points.

The next two parameters provide exceptions to the above rule:

TMAX:	Maximum time allowed for measurement of drift (Default = infinity). If time >TMAX then program will accept the current drift value, or if a value hasn't been computed--it will accept the first value subsequently computed.
DMAX:	Maximum allowable drift. If drift < DMAX then program will accept this value; otherwise normal criteria apply. (Default = zero drift)
NO:	Number of runs to be performed using these parameters.

The system routinely provides about twice as many data points

i.e., $\Delta T = 0.05\ T$ but < 5K.

as we usually took manually and they are of equal or better precision.

The total (computer) time involved with C_p measurements from 5 to 350 when no transitions are involved is about 40 hours. Apart from loading and unloading operations, four or five hours of operator time are usually involved. This cost in terms of operator time is thus not greatly in excess of that used for DSC measurements which can cover only the upper half of the range.

COMPUTERIZED ADIABATIC THERMOPHYSICAL CALORIMETRY

An entirely similar program operates the super-ambient calorimeter.

A quartz-digital clock and internal time will soon provide higher accuracy confirmatory energy input time duration records than are presently available. (cf also references 7 & 8)

	T	T'	T"
4: 9:30	30.35773	2.024D+00	1.349D-01
4: 9:45	30.35622	-1.007D-04	-1.349D-01
4:10: 0	30.35600	-1.457D-05	5.742D-06
4:10:15	30.35582	-1.221D-05	1.569D-07

. R 0.3113514423   R* 0.3113446506   RL 7.75

4:10:45	30.3555793	-1.607D-05	-2.372D-07
4:11: 0	30.3554481	-8.747D-06	4.882D-07
4:11:15	30.3553774	-4.716D-06	2.688D-07
4:11:30	30.3552869	-6.035D-06	-8.794D-08
4:11:45	30.3551112	-1.171D-05	-3.786D-07
4:12: 0	30.3550479	-4.220D-06	4.995D-07
4:12:15	30.3549609	-5.797D-06	-1.051D-07
4:12:30	30.3548912	-4.651D-06	7.642D-08
4:12:45	30.3547977	-6.231D-06	-1.053D-07
4:13: 0	30.3548183	1.374D-06	5.070D-07
4:13:15	30.3547096	-7.247D-06	-5.747D-07
4:13:30	30.3546414	-4.550D-06	1.798D-07
4:13:45	30.3546247	-1.111D-06	2.293D-07
4:14: 0	30.3545637	-4.069D-06	-1.972D-07
4:14:15	30.3544996	-4.274D-06	-1.363D-08
4:14:30	30.3544802	-1.288D-06	1.991D-07
4:14:45	30.3544432	-2.471D-06	-7.885D-08
4:15: 0	30.3543798	-4.223D-06	-1.168D-07
4:15:15	30.3543686	-7.486D-07	2.317D-07
4:15:30	30.3543155	-3.542D-06	-1.862D-07
4:15:45	30.3542791	-2.428D-06	7.427D-08
4:16: 0	30.3542807	1.101D-07	1.692D-07
4:16:15	30.3540943	-1.243D-05	-8.358D-07
4:16:30	30.3541224	1.875D-06	9.535D-07
4:16:45	30.3541918	4.627D-06	1.835D-07
4:17: 0	30.3541178	-4.937D-06	-6.376D-07

Join to figure 5

Fig. 4.   Drift (K/sec). RL represents the resistance of the leads (in ohms).

A	+/-	B	+/-	SDT	DEV	
30.35716026	1.208E-04	-6.009926396D-06	3.551E-07	4.838E-05	9.209E-05	
30.35700372	1.253E-04	-5.556879954D-06	3.528E-07	4.806E-05	3.978E-05	
30.35688711	1.312E-04	-5.232286761D-06	3.546E-07	4.831E-05	3.030E-05	
30.35666022	1.292E-04	-4.630716817D-06	3.358E-07	4.575E-05	4.830E-05	
30.35642461	8.190E-05	-4.045171600D-06	2.048E-07	2.791E-05	2.006E-05	A 1
30.35635978	8.048E-05	-3.889196692D-06	1.941E-07	2.644E-05	6.591E-06	A 1
30.35623369	7.962E-05	-3.590574154D-06	1.853E-07	2.525E-05	2.388E-05	A 1
30.35613131	8.386E-05	-3.354486534D-06	1.886E-07	2.570E-05	2.266E-05	A 1
30.35598117	4.813E-05	-3.043397978D-06	1.035E-07	1.288E-05	-3.592E-06	A 1
30.35605045	9.311E-05	-3.169405549D-06	1.962E-07	2.674E-05	2.963E-05	A 1
30.35589357	5.262E-05	-2.853222232D-06	1.075E-07	1.465E-05	5.443E-06	A 1
30.35583955	4.960E-05	-2.746232240D-06	9.835E-08	1.340E-05	4.860E-06	A 1
30.35576312	7.350E-05	-2.588648030D-06	1.415E-07	1.929E-05	3.194E-05	A 1
30.35568299	7.475E-05	-2.441616193D-06	1.420E-07	1.650E-05	-1.237E-04	A 1
30.35578287	1.028E-04	-2.641386012D-06	1.892E-07	2.355E-05	-3.599E-05	A 2
30.35566053	1.373E-04	-2.404652729D-06	2.452E-07	3.197E-05	4.624E-05	A 5
30.35561629	1.361E-04	-2.328066733D-06	2.360E-07	3.166E-05	3.083E-06	OK

Fig. 5 Least squares fitting of drift. These are contiguous with the last data points of fig. 4 as indicated by the dashed lines. The constraint imposed was satisfied at minute 4:17 (A.M.) and the computer shifted from drift to energy increment mode.

```
4: 3:15  *ON*  RUN   30   360 SEC
     15    0.1117339E+01 E        30    0.7042405E-02 I
     45    0.1117328E+01 E        60    0.7042695E-02 I
     75    0.1117328E+01 E        90    0.7042685E-02 I
    105    0.1117310E+01 E       120    0.7042385E-02 I
    135    0.1117301E+01 E       150    0.7042785E-02 I
    165    0.1117274E+01 E       180    0.7042825E-02 I
    195    0.1117308E+01 E       210    0.7042435E-02 I
    225    0.1117335E+01 E       240    0.7042375E-02 I
    255    0.1117390E+01 E       270    0.7042835E-02 I
    285    0.1117422E+01 E       300    0.7042595E-02 I
    315    0.1117441E+01 E       330    0.7042385E-02 I
    345    0.1117516E+01 E                              HTR
4: 9:15  *OFF* RUN   30      2.832854E+00  JOULES    158.64256
```

Fig. 6. Energy input for determination number 30. E = potential drop across resistor (in volts), I = current through heat (in milliamperes).

```
          4: 6:15     30.3556163   T2( 30)
          4:20:30     30.3536258   T1( 31)


       DT( 30) =   1.3506973    T( 30) =   29.6802676


                    CP( 30) = 0.2097327E+01
```

Fig. 7. Calculation of apparent heat capacity.

An additional convenience of the utilization of automated control systems includes the possibility of monitoring cooling operations and learning at once about the presence of transitions, undercooling hysteresis phenomena, etc. This makes it conveniently possible to begin measurement near 4 K in the absence of transitions and to continue through to the highest temperature with suitable overlaps, enthalpy-type runs, etc. for checking reproducibility, etc.

Not only can a "free" cooling curve be made but one with a constant-head (in terms of thermocouple e.m.f.) can be made automatically. By the use of a suitable program to make radiation corrections and "invert" the data, heat-capacity measurement may be made in the cooling direction by the continuous method.

It is interesting to note that a new adiabatic calorimeter for heat-capacity measurements in the cooling direction (9) has been designed to remove heat quantitatively—on an intermittent basis—by means of thermal contact with a cold, copper block of known heat capacity. This was tested on the heat capacity of an empty calorimeter.

The research personnel involved most in the creation and operation of the automated adiabatic calorimeters are Prof. J.T.S. Andrews (Consultant), Prof. Ramon Burriel (Post Doctoral), William Plautz (Programmer), and J. Scott Westrum (Computer Technician).

HO2S3 HELIUM REGION BELOW 10 K G.W.M.

RUN	HTR. RES.	ENERGY	T1
1	164.37053	0.38559485D-01	7.162018
2	162.83144	0.15713443D+00	7.316102
3	163.00203	0.16077231D+00	7.820183
4	163.25729	0.16465656D+00	8.322971
5	165.58214	0.16920435D+00	8.829523
6	164.21785	0.17178202D+00	9.332433
7	163.07356	0.17639458D+00	9.832621

T*	CP	T2	DELTA T
7.225234	0.3049819	7.288450	0.126432
7.570623	0.3086871	7.825143	0.509041
8.074515	0.3160674	8.328847	0.508665
8.579702	0.3206794	8.836433	0.513462
9.083672	0.3328843	9.337821	0.508298
9.583896	0.3415653	9.835359	0.502926
10.083717	0.3512492	10.334813	0.502192

Fig. 8. The summary of a set of data determinations. Note that heater resistance, energy, (extrapolated) initial and final temperatures, ΔT, T, and C_p are summarized for each run.

OTHER AUTOMATED CALORIMETERS

Before enquiring into several aspects of the computerization problem a brief summary of several other automated systems will be made. This section will be restricted to presentation of selected more or less fully automated systems, i.e., those which involve automatic digital experimental control and data logging and either automatic analog or digital shield control. No attempt will be made to be exhaustive either in scope or detail and representative systems will be explained in essentially tabular format. Not all of these represent the latest developments, some are included for historical reasons, others because of their innovation.

The first group, table 2, covers the "chemical-thermodynamic range" (from 5-350 K) as described elsewhere in this volume (10) and deals first with historical development and concludes with one in which no computer is required.

Those automated calorimeters which operate over the chemical-thermodynamic range (5 - 350 K) are listed with salient points in table 2. Those in the superambient range are in table 3, and those going to higher temperature in table 4. A few computerized calorimeters for use below the usual chemical-thermodynamic range are listed in table 5. Readers desiring further information are urged to consult the cited references. Many other calorimeters in various ranges are known to be in various stages of automation.

It is to be noted once again that complete automation offers many advantages in that the operator is freed from the tedious repetitive task of measurement, human errors in measurement and in data logging are eliminated. More consistency and, therefore, more precise data are obtained. Moreover, the price of duplicate (repetitive) series of measurements is so low that multiple runs may be made to obtain an enhanced confidence level. Automatic checks may be programmed to ensure that the software, bridges, and apparatus are performing well.

THE QUESTION OF SAMPLE SIZE

The size of the sample is a matter which can be and is discussed apart from automation, but the two do go hand-in-hand in modern calorimetry. When Professor Giauque, the recently deceased Nobel Laureat of low-temperature calorimetry, measured the heat-capacity of gold he enjoyed the luxury of having a 2.5 kilogram sample. The rest of us have a reputation for largish samples and whether we are dealing with materials which are

Table 2. Other Computerized Adiabatic Low-Temperature Calorimeters

---

Stull, D. R. (11) late 1960's
    (Dow Chemical, Midland, MI, U.S.A.)
        Semi-automated, precomputer-era automation, a forerunner
        Commercial (industrial) components
        Early recognition of importance of thermophysical data in technological milieu
        Range: 50-300 K
        Accuracy: ~ 0.3 per cent

Furukawa, G. T. (12) late 1960's
    (National Bureau of Standards, Washington D.C., U.S.A.)
        Extensive experimentation with self-balancing thermometer bridge. Megadollar units.
        Paper tape logging of data

Kishimoto, K.S.; Suga, H.; Seki,S. (13) (1980)
    (University of Osaka, Japan)
        Automation
        Analog shield control
        Auxilliary thermistor thermometer
        Digital data recording
        Intermittent energy input
        Inorganic compunds
        Range: 13-300 K

Chang, S.S. (14,15) (1976, 1977)
    (National Bureau of Standards, Washington, D.C., U.S.A.)
        Mechanization of Rubicon Potentiometer for temperature measurement
        Controlled guard shield around adiabatic shield
        See fig. 9 for block diagram
        Analog shield-control
        Digital data acquisition
        Intermittent energy input
        Range: 2-380 K
        Glasses, polymers, plastics, organic compounds
        Precision 0.02 per cent ($\Delta T > 1$ K)

Beyer, R. P. (16) (1979)
 (Bureau of Mines, Albany Research Center, Albany, Oregon, U.S.A.)
  Automatic analog shield controls
  Digitally controlled data acquisition
  Programmable dc voltage source
  Temperature measurements made potentiometrically
  Intermittent energy input
  Range: 5-300 K
  Minerals, rocks, for mineral resource technology

Arvidson, K.: Falk, B.; Sunner, S. (17) (1976)
 (University of Lund, Sweden)
  Automatic analog shield control
  Digital data acquisition
  ASL automated inductance temperature bridge
  Minco thermometer (not bifilar winding, large inductive component at He temperatures)
  Intermittent energy input
  Range: 5-330 K
  Organic and inorganic compounds

Junod, A. (18) (1979)
 (University of Geneva, Switzerland)
  <u>NO</u> computer
  Graphical data output $(C_p)^{-1}$ vs T
  Continuous adiabatic
  Range: 80-300 K
  Samples: ca. 1 gram
  Accuracy between ± 1 to ± 3 per cent

Moses, D.; Ben-Aroya, O.; Lupu, N. (19) (1977)
 (Technion-Israel Institute of Technology, Haifa, Israel)
  Discrete or continuous energy input
  Samples attached with varnish to thin copper plates
  Range: 3-300 K
  Samples: single crystals inorganic oxide (2 cm^3)
  Accuracy: 0.3 per cent

Robie, et al. (20) (1980)
    (U.S. Geological Survey, Reston, Virginia, U.S.A.)
        Analog shield control
        Platinum thermometry
        Digital data acquisition
        Intermittent heating
        Range: 5-300
        Minerals

Schaake, R.C.F. et al. (21) (1979)
    (State University of Utrecht, Belgium)
        Analog shield control
        Intermittent energy input
        ASL Model A6 AC Bridge
        Range: 90-300 K
        Samples: Organic compound
        Accuracy 0.1 per cent

---

prohibitively expensive, dangerous chemically, even explosive, radioactive, or hard to synthesize, or purify, there is a need to scale down our sample requirements.

A partial list of achievements in that direction is made in table 6 but further standards will need to be measured in order to ensure the true accuracy and precision of gram scale operation.

COMPUTERIZED ADIABATIC THERMOPHYSICAL CALORIMETRY

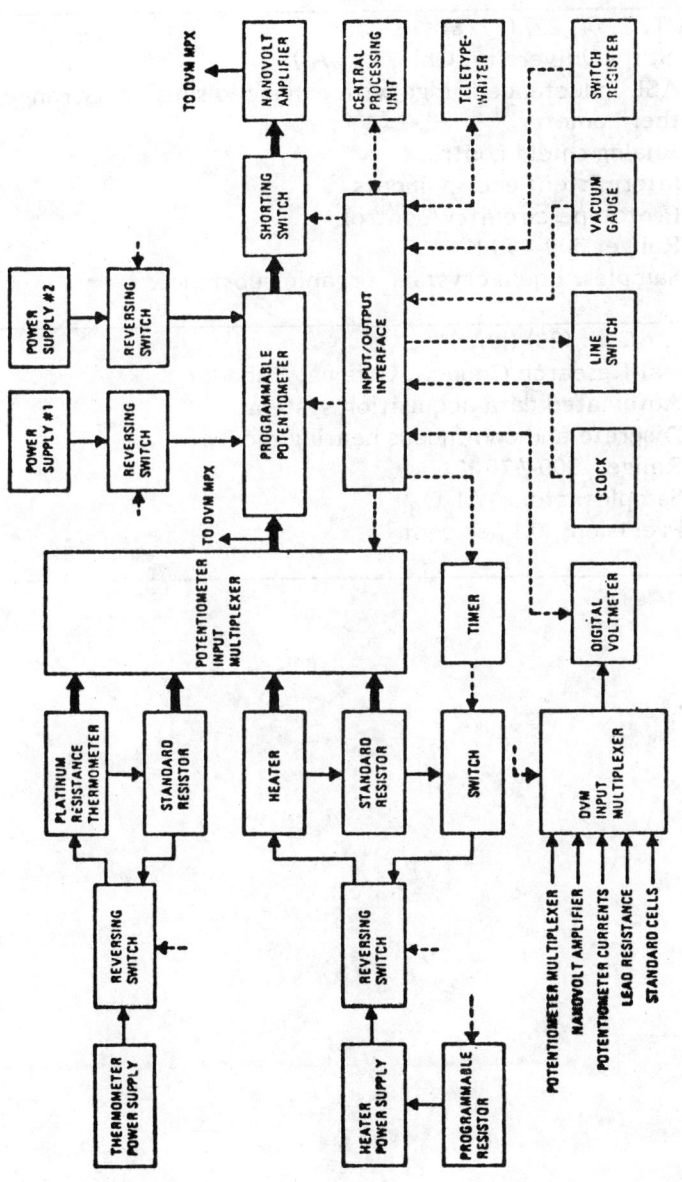

Fig. 9. Block Diagram of Automated Calorimetric Measurement System. Chang (14) ⎯⎯⎯ high-level signal, ⎯⎯ low-level signal, --- digital signal. DVM = digital voltmeter, MPX = multiplexer.

Table 3. Other Computerized Adiabatic, Super-ambient Temperature Calorimeters

---

Andrews, J. T. S. (4, 22) (1978)
    (Kent State University, Ohio, U.S.A.)
        ASL inductance bridge with enhanced sensitivity range. Platinum thermometry
        Analog shield control
        Intermittent energy inputs
        Prototype circuitry/control
        Range: 300-650 K
        Samples: liquid crystals, organic substances

Martin, D.L., et. al. (23) (1970)
    (National Research Council, Ottawa, Canada)
        Automated data acquisition system
        Discrete and continuous heating modes
        Range: 300-470 K
        Samples: std. $\alpha\text{-}Al_2O_3$
        Precision: 0.1 per cent

Table 4. Computerized, Adiabatic, High-Temperature Calorimeters

---

Grønvold, F. (24) (1967-82)
    (University of Oslo, Blindern, Norway)
        Forerunner in computer automation, modernized computer
        ca. 1980, vitreous silica sample container
        Mechanized self-balancing potentiometer, platinum thermometry
        Adiabatic calorimetry in surrounding furnace
        Intermittent heating
        Range: 298-1048 K
        Samples: transition-element prictides, chalcogenides, metals

---

Cash, W.M.; Stansbury, E.E.; Moore, C.F.; Brooks, C.R. (25) (1981)
    (University of Knoxville, Tennessee, U.S.A.)
        Potentiometric temperature measurement using Pt, Pt-Rh
        thermocouples (Programmed voltage suppression)
        Real time, time-sharing computer
        Digital shield control ($\pm$ 0.1 K)
        Continuous heating at constant heating rate
        Range: 300-1300 K
        Samples: metals
        Accuracy: 0.6 to 1.0 %

Table 5. Computerized, Adiabatic Very-low-temperature Calorimeters

---

Martin, D.L., et. al. (26) (1973)
    (National Research Council, Ottawa, Canada)
        On-line minicomputer
        Manual adjustment of heater current
        Copper "tray" technique calorimeter
        Analog shield control
        Range: 3-30 K
        Samples: metals
        Accuracy: 0.3 per cent (limited by thermometer calibration)

Table 6.   The Trend to Smaller Samples in Adiabatic Thermophysical
Calorimetry

Staveley, L. A. K. et al. (27) (1970)
(Oxford University, Oxford, U.K.)
Isotopically substituted methanol 3-4 g

Arvidson, K. (28) (1974)
(University of Michigan, Ann Arbor, Michigan, U.S.A.)
Minco Pt thermometer
Samples: 1-2 g

Arvidson, K. et al. (17) (1976)
(University of Lund, Sweden)
(See table 2)
Minco Pt. thermometer
Samples: several grams, miscellaneous substances

Paukov, I. E. et al. (29) (1968)
(Institute of Inorganic Chemistry, Akademgorodok, U.S.S.R.)
Thermometer and heater on copper tube calorimeter
Samples: standard α-Al_2O_3, <1 g

Gorbunov, V.E. et al. (30) (1980)
(Institute of General Chemistry, Moscow, U.S.S.R.)
Calorimeter mounts in detachable platinum containers
Analog shield control
Range: 10-350 K
Samples: 1-3 g inorganic compounds
Accuracy: $\pm$ 0.2 per cent

Khodakovsky, V. et al. (31) (1978)
(Institute of High Temperatures, Moscow, and Institute of Experimental Mineralogy, Chernogolovka, U.S.S.R.)
Similar in many respects to above instrument
Samples: 1 g mineral samples

Ann Arbor (unpublished)
(University of Michigan, Ann Arbor, Michigan, U.S.A.)
L&N calibrated Pt-resistance thermometer on shield. 4-junction AuFe, AgCo, chromel-P, constantan thermel to record T difference if any
Samples: 1-2 g minerals, inorganic, organic compounds

REFERENCES

1. Westrum, E.F. Jr.: 1984, this volume, Paper entitled "Calorimetry of Phase, Schottky and Other Transitions.

2. Westrum, E.F. Jr., Furukawa, G.T., McCullough, J.P.: 1968, in "Experimental Thermodynamics," Vol. I, eds. McCullough, J.P., Scott, D.W. (Butterworths) pp. 133-214.

3. West, E.D., Westrum, E.F. Jr: 1968, in "Experimental Thermodynamics," Vol. 1, eds. McCullough, J.P., Scott, D.W., (Butterworths) pp. 333-367.

4. Andrews, J.T.S., Norton, P.A., Westrum, E.F. Jr.: 1978, J. Chem. Thermodyn. 10, pp. 949-958.

5. Justice, B.H.: 1961, "Thermodynamic Properties and Electronic Energy Levels of Eight Rare Earth Sesquioxides" unpublished doctoral dissertation, University of Michigan.

6. Justice, B.H.: 1969, "Thermal Data Fitting with Orthogonal Functions and Combined Table Generation. The FITAB Program" (Univ. of Michigan, Ann Arbor).

7. Martin, D.L., Bradley, L.L.T., Cazemier, W.J., Snowdon, R.L.: 1973, Rev. Scie. Inst. 44, pp. 675-84.

8. Martin, D.L., Snowdon, R.L.: 1970, Rev. Sci. Inst. 41, pp. 1869-76.

9. Yoshimoto, Y., Atake, T. Chihara, H.: 1982, Netsusokutei 9, pp. 57-60.

10. Westrum, E.F. Jr.: 1984, this volume, paper entitled "Thermodynamics of Crystals."

11. Stull, D.R.: 1957, Analytica Chemica Acta 17, pp. 133-143.

12. Furukawa, G.T., unpublished research.

13. Kishimoto, K., Suga, H., Seki, S.: 1980, Bull, Chem. Soc, Jpn. 53, pp. 2748-2754.

14. Chang, S.S.: 1977, in "Proceedings of the 7th Symposium on Thermophysical Properties" (ASME) pp. 75-82.

15. Chang, S.S.: 1977, in "Proceedings of the 7th Symposium on Thermophysical Properties" (ASME) pp. 83-90.

16. Beyer, R.P.: 1981, in "Proceedings of the Workshop on Techniques for Measurement of Thermodynamic Properties, Albany, Oreg., 1979" (Bureau of Mines I.C. 8853) pp. 113-119.

17. Arvidson, K., Falk, B., Sunner, S.: 1976, Chemica Scripta 10, pp. 193-200.

18. Junod, A.: 1979, J. Phys. E.: Sci. Instrum. 12, pp. 945-952.

19. Moses, D., Ben-Aroya, O., Lupu, N.: 1977, Rev. Scie. Instrum. 48, pp. 1098-1103.

20. Robie, R., unpublished work.

21. Schaake, R.C.F., Offringa, J.C.A., van der Berg, G.J.K., van Miltenburg, J.C.:1979, Recueil, Jorunal Royal Neth. Chem. Soc. 98, pp. 408-412.

22. Andrews, J.T.S.: unpublished work.

23. Martin, D.L., Snowdon, R.L.: 1970, Rev. Sci. Instrum. 41, pp. 1869-1876.

24. Grønvold, F.C.: 1967, Acta. Chem. Scand. 21, pp. 2-1713. See also Grønvold, F.C.: 1966, in "Proceedings of IUPAC Symposium on Thermodynamica, Vienna 1965" (Vienna) pp. 35-52.

25. Cash, W.M., Stansbury, E.E., Moore, C.F., Brooks, C.R.: 1981, Rev. Sci. Instrum. 52, pp. 895-901.

26. Martin, D.L., Bradley, L.L.T., Cazemier, W.J., Snowdon, R.L.: 1973, Rev. Sci. Instrum 44, pp. 675-684.

27. Staveley, L.A.K., Gupta, A.K.: 1959, Trans. Faraday Soc. 45, p. 50.

28. Arvidson, K.: unpublished work.

29. Paukov, I.E., Anishin, V.F., Anishimov, M.P.: 1972, Zh. Fiz. Khim. 46, p. 778. see also Sukhovei,K.S., Anishin, V.F., Paukov, Gurevich, V.M. and Khlyustov, V.G.: 1978, Geokhimiya, p. 829.

30. Gorbunov, V.E., Gavrichev, K.S., Gurevich, V.M., Sharpataya, G.A.: 1980, in "Abstracts of Poster Papers, 6th International Conference on Thermodynamics, Merseburg, 1980." see also prototype in Gorbunov, V.E., Palkin, V.A.: 1972, J. Fiz. Khim. 46, p. 1625.

31. Gorbunov, V.E., Gurevich, V.M., Gavrichev, K.S.: 1979, in "Proceeding of the VIII All-Union Conference on Calorimetry and Chemical Thermodynamics 1979 Ivanova, Izdatel" (Ivanovski Chemical Technological Inst., Ivanov) p. 113.

LECTURES ON THERMOCHEMISTRY AND KINETICS

Sidney W. Benson

University of Southern California

INTRODUCTION

Two of the major goals of physical chemistry is the prediction of equilibrium constants for chemical equilibria and the prediction of rate constants for elementary reactions. With the basic knowledge available to us today of molecular structure and geometry together with the techniques of quantum statistical mechanics it is now possible to estimate absolute entropies and heat capacities of molecules and radicals with good accuracy. Combining this information then allows us to estimate equilibrium constants. Where data on heats of formation are not available it is possible to use empirical methods to estimate them usually to within ±1 kcal/mole.

In the following two lectures we will examine in detail the methods available to make these estimates, their accuracy and limitations. Most of this material is summarized in the first two chapters of my book on "Thermochemical Kinetics" [Ref. 1] and in a popular article [Ref. 2] which are currently available. In consequence the accompanying notes are simply that, notes on the material to be covered and highlights of the topics.

    I. Additivity Relations for Physical Properties—
Thermochemistry

        1. Relation of Thermochemistry to Chemical Reactions

            Basic thermochemical data used in predicting states of chemical equilibria

$$aA + bB \rightleftarrows rR + qQ,$$

$$K_P = \frac{(R)^r(Q)^q}{(A)^a(B)^b}$$

and

$$\Delta G_T^\circ = -RT \ln K_P$$
$$= r \Delta G_{fT}^\circ(R) + q \Delta G_{fT}^\circ(Q) - a \Delta G_{fT}^\circ(A) - b \Delta G_{fT}^\circ(B)$$

$$\Delta n = r + q - a - b$$

$$\Delta G_T^\circ = \Delta H_T^\circ - T \Delta S_T^\circ$$

$$\Delta H_T^\circ = r \Delta H_{fT}^\circ(R) + q \Delta H_{fT}^\circ(Q) - a \Delta H_{fT}^\circ(A) - b \Delta H_{fT}^\circ(B)$$

$$\Delta S_T^\circ = r S_T^\circ(R) + q S_T^\circ(Q) - a S_T^\circ(A) - b S_T^\circ(B).$$

$$\left[\frac{\partial(\Delta H^\circ)}{\partial T}\right]_P = \Delta C_P^\circ \qquad \left[\frac{\partial(\Delta S^\circ)}{\partial T}\right]_P = \frac{\Delta C_P^\circ}{T}$$

$$\Delta C_P^\circ = r C_P^\circ(R) + q C_P^\circ(Q) - a C_P^\circ(A) - b C_P^\circ(B)$$

$$\Delta H_T^\circ = \Delta H_{T_0}^\circ + \int_{T_0}^{T} (\Delta C_P^\circ) \, dT;$$

$$\Delta S_T^\circ = \Delta S_{T_0}^\circ + \int_{T_0}^{T} \left(\frac{\Delta C_P^\circ}{T}\right) dT.$$

Basic Thermochemical Quantities: ΔH_f°; S°; C_{pT}°

From 0 - 1500°K. C_{pT}° can be represented by

3 parameter equation: $C_{pT}^\circ = a + \dfrac{b}{C+T}$

2. Additivity Approach to Physical and Chemical Properties.

   Raoult's Law for Molecules:

   Consider the balanced stoichiometric, disproportionation reaction:

   $$RNR + SNS \rightleftarrows 2RNS$$

   If Φ represents a molecular property then for the reaction:

   $\Delta\Phi \rightarrow 0$ (or $\Delta\Phi_{stat}$) as diameter N >> diameter R, S

3. Law of Additivity of Atomic Properties. Symmetry corrections to entropy. Optical isomerism.

   $\Phi_{molecular}\, [A_a B_b C_c \ldots] = a\Phi_A + b\Phi_B + \ldots$

   Data Base → Table of Atomic Properties

4. Law of Additivity of Bond Properties.

   $\Phi_{molecular}\, [A_a B_b C_c \ldots] = n_{AB} \Phi(A-B) + N_{AC} \Phi(A-C) + \ldots$

   Data Base → Table of Bond Properties

5. Law of Additivity of Group Properties.
   Definition of Group

   Corrections for non-bonded interactions—gauche correction—cis and ortho interactions. Applications to ring compounds. Ring strain. Application to free radicals

   $\Phi_{molecular}\, [A_a B_b C_c \ldots] = \Sigma\Phi$ (groups)

   A group is a polyvalent atom + its near neighbors

   Data Base → Table of Group Properties

$$C_p^\circ(i\text{-butane}) = 3[C-(C)(H)_3] + [C-(C)_3(H)]$$
$$= 18.57 + 4.54 = 23.1$$
$$C_p^\circ(\text{obs}) = 23.1 \text{ cal/mole-}^\circ K$$

$$S^\circ(\text{pentene-2}) = [C-(C_d)(H)_3] + 2[C_d-(C)(H)] + [C-(C_d)(C)(H)_2]$$
$$+ [C-(C)(H)_3] - 2R \ln 3 \text{ (symmetry of CH}_3\text{)}$$
$$= 30.41 + 16.0 + 9.8 + 30.41 - 4.32$$
$$= 82.3$$
$$S^\circ(\text{pentene-2})(\text{obs})_{\text{trans}} = 82.0 \text{ cal/mole-}^\circ K$$

The law of group additivity is probably as far as it is reasonable and actually feasible to carry the scheme of molecular additivity. The reason is that steric interactions between non-bonded neighbors, ring strain and non-bonded electrostatic interactions still remain as significant corrections to additivity properties and may transcend any available data base. One example is the methyl-methyl repulsion found in the compound 2,2,4,4, tetra-methyl pentane. This is an interaction between the hydrogen atoms in these methyl groups, separated by 5 intervening carbon atoms. Such strains like ring strains can only be dealt with from structural considerations involving the entire molecular, not from any additivity rule.

6. Delocalized Electron Systems.
   Treatment of aromatics and polynuclear aromatics. (PNA)

Comparison of Calculated and Experimental $\Delta H_f^\circ{}_{298}$ (kcal mol^{-1})

Compound	ΔH_f°(expt)	$\Delta H_f^\circ{}_{298}$(calcd) − $\Delta H_f^\circ{}_{298}$(expt)		
		GA	Laidler-type bond additivity	Resonance theory
Benzene	19.8	0.0	0.0	+1.6
Naphthalene	35.8	+0.2	0.7	−1.0
Anthracene	54.4	−2.2	−1.1	−2.6
Phenanthrene	50.0	+0.5	+0.4	−0.5
Tetracene	67.8	+0.7	+2.4	+2.4
Benz[a]anthracene	66.0	+0.2	+0.7	−1.5
Chrysene	62.8	+1.2	+0.5	+1.2
Triphenylene	61.9	−0.1	−2.0	+2.2
Pyrene	51.6	+3.6	+5.0	+1.8
Perylene	73.7	−6.7	−7.0	−3.0
Benz[c]phenanthrene	69.6	−4.2	−6.3	−10.7
Average		1.8[1.5]	2.4[2.0]	2.6[1.8]

Defining 4 groups occurring in PNA we can derive values for thermochemical properties. The above Table shows the agreement found for group additivity (GA) compared with alternate methods.

7. Deviations From Group Additivity.
Very polar compounds.

Effect of Polarity and Electronegativity on Differences in Heat of Formation Between Hydrogen (HX) and Methyl Derivative (CH$_3$X)

X	ΔH_f°(HX)	ΔH_f°(CH$_3$X)	$\Delta(\Delta H_f^\circ)$
F	−64.8	−55 ± 2	−9.8 ± 2
OH	−57.8	−48.0	−9.8
O(SO$_3$)CH$_3$	−170.5	−164	−6.5
O(CO)CH$_3$	−103.8	−98	−5.8
NH$_2$	−11.0	−5.5	−5.5
OCH$_3$	−48.0	−44.0	−4.0
ONO$_2$	−32.1	−28.6	−3.5
ONO	−18.3	−15.6	−2.7
Cl	−22.0	−19.6	−2.4
O$_2$H	−32.6	−31.3	−1.3
NH(CH$_3$)	−5.5	−4.5	−1.0
Br	−8.7	−9.5	0.8
SH	−4.8	−5.4	0.6
N(CH$_3$)$_2$	−4.5	−5.9	1.4
CH$_3$	−17.9	−20.2	2.3

X	ΔH_f°(HX)	ΔH_f°(CH$_3$X)	$\Delta(\Delta H_f^\circ)$
I	6.3	3.3	3.0
SCH$_3$	−5.4	−8.9	3.5
C$_2$H$_5$	−20.2	−24.8	4.6
S$_2$H	3.8	−1.0	4.8
n-C$_3$H$_7$	−24.8	−30.2	5.4
C$_2$H$_3$	12.5	4.9	7.6
C$_6$H$_5$	19.8	12.0	7.8
NO	23.8	16	8
COCH$_3$	−39.7	−51.7	12.0
CN	32.3	[19 ± 2]	13 ± 2
COOH	−90.5	−103.8	13.3
CF$_3$	−167	−178	11
SiH$_3$	8	−4	12
SnH$_3$	39	[28 ± 3]	11 ± 3

The above tables show that deviations from bond additivity are not random but systematic and monotomic with electronegativity differences of the elements in the bond (C-X vs. H-X).

The following tables in section 8 show that a scheme of bond-bond corrections to bond additivity will not account for ΔH_f^0 and that a non-linear scheme is needed.

8. Data Base Requirements for Group Additivity Table and for Bond Additivity.

ΔH_f^0 in the series $CH_n(CH_3)_{4-n}$

Compound	$-\Delta H_f^0$	$\Delta(-\Delta H_f^0)$	$\Delta^2(-\Delta H_f^0)$
CH_4	17.9		
		2.3 ± 0.2	
$CH_3(CH_3)$	20.2		2.3 ± 0.5
		4.6 ± 0.35	
$CH_2(CH_3)_2$	24.8		2.8 ± 0.7
		7.4 ± 0.5	
$CH(CH_3)_3$	32.2		0.9 ± 1.0
		8.3 ± 0.65	
$C(CH_3)_4$	40.5		

ΔH_f^0 in the series CH_nCl_{4+n}

Compound	$-\Delta H_f^0$	$\Delta(-\Delta H_f^0)$	$\Delta^2(\Delta H_f^0)$
CH_4	17.9		
		1.7 ± 1	
CH_3Cl	19.6		1.5 ± 1.4
		3.2 ± 1	
CH_2Cl_2	22.8		−1.8 ± 1.4
		1.4 ± 1	
$CHCl_3$	24.2		−2.7 ± 1.2
		−1.3 ± 0.5	
CCl_4	22.9		

ΔH_f^0 in the series CH_nF_{4-n}

Compound	$-\Delta H_f^0$	$\Delta(-\Delta H_f^0)$	$\Delta^2(-\Delta H_f^0)$
CH_4	17.9		
		39 ± 2	
CH_3F	56.8 ± 2		12 ± 4
		51 ± 2.3	
CH_2F_2	108 ± 1		7 ± 3
		58 ± 1.4	
CHF_3	166 ± 1		−1 ± 2.5
		57 ± 1.4	
CF_4	223 ± 1		

Recommended Reading-Ref. 1 - Chap. 1,2; Ref. 4; Ref. 9; Ref. 10

SECTION II THE ELECTROSTATIC BOND

INTRODUCTION

Valence forces and Van der Waals forces are short range in nature. They are significant out to distances of the order of 3-4Å. The reason that the laws of additivity of ΔH_f° work as well as they do is largely because of this short range. But one consequence of this is that we should expect to see deviations in molecules of large polarity or in ions and this is indeed the case. In the resent section we shall explore these exceptions in detail and explore a method for dealing with them.

1. Local nature of chemical forces, bond lengths, bond strengths, stretching and bending frequencies.

2. Systematic nature of deviations from bond additivity.

	$-\Delta H_{f298}^\circ$ [kcal/mol]	$\Delta(-\Delta H_{f298}^\circ)$	$\Delta^2(\Delta H_{f298}^\circ)$
CH_4	17.9		
CH_3F	56.8 ± 2	39 ± 2	
CH_2F_2	108 ± 1	51 ± 2.3	12 ± 4.1
CHF_3	166 ± 1	58 ± 1.4	7 ± 3.0
CF_4	233 ± 1	57 ± 1.4	−1 ± 2.5

	$-\Delta H_{f298}^\circ$ [kcal/mol]	$\Delta(-\Delta H_{f298}^\circ)$	$\Delta^2(-\Delta H_{f298}^\circ)$
CH_4	17.9		
CH_3OH	48.0	30.1	
$CH_2(OH)_2$	93 ± 1	45 ± 1	15 ± 1.5

	$-\Delta H_{f298}^\circ$ [kcal/mol]	$\Delta(-\Delta H_{f298}^\circ)$	$\Delta^2(-\Delta H_{f298}^\circ)$
H_2O	57.9		
$HO(OH)$	32.6	−25.3	
$O(OH)_2$	13.6 ± 1	−19.0 ± 1	6.3 ± 1

3. Allen's scheme for bond-bond interactions.

It is important to realize that group additivity includes all near-neighbor interactions and is thus equivalent to any scheme such as Allen's for dealing with bond-bond, or next-nearest neighbor interactions. Compounds such as CF_3CH_3; CCl_3CH_3; $(CN)_3C-CH_3$ and related species show large deviations from group additivity. These deviations cannot be treated by any bond-bond interaction scheme.

4. Principle of alternating polarity.

Compounds in which elements of alternate electronegativity are bonded together are the most stable. A simple system of formal charges associated with chemical bonds can be used to illustrate this. One consequence of this is that disproportionation between symmetrical species is usually endothermic.

$$(+2a) X \underset{A\,(-a)}{\overset{A\,(-a)}{\diagup\!\!\!\diagdown}} 2\theta \;+\; (+2b) X \underset{B\,(-b)}{\overset{B\,(-b)}{\diagup\!\!\!\diagdown}} 2\theta \;\rightleftarrows\; 2 X \underset{B\,(-b)}{\overset{A\,(-a)}{\diagup\!\!\!\diagdown}} 2\theta \quad (+a+b)$$

Electrostatic Energy in a Simple Disproportionation

$$\Delta E_{el} = 2E_{el}(XAB) - E_{el}(XA) - E_{el}(XB_2)$$

$$= \left(\frac{-2a^2}{r_{XA}} - \frac{2b^2}{r_{XB}} - \frac{2ab}{r_{XA}} - \frac{2ab}{r_{XB}} + \frac{2ab}{r_{AB}} \right)$$

$$+ \left(\frac{4a^2}{r_{XA}} - \frac{a^2}{r_{AA}} \right) + \left(\frac{4b^2}{r_{XB}} - \frac{b^2}{r_{BB}} \right)$$

$$= \frac{2a^2}{r_{XA}} + \frac{2b^2}{r_{XB}} - 2ab \left[\frac{1}{r_{XA}} + \frac{1}{r_{XB}} \right] - \left[\frac{a^2}{r_{AA}} + \frac{b^2}{r_{BB}} - \frac{2ab}{r_{AB}} \right]$$

$$\Delta E_{el}(r_{XA} \cong r_{XB}) \approx \frac{2(a-b)^2}{r_{XA}} - \frac{(a-b)^2}{r_{AA}} > 0$$

since $r_{AA} = 2 r_{XA} \sin \theta$.

Some heats of disproportionation ΔH_r^0 of symmetrical species:

$A_2X + B_2X \rightleftharpoons 2ABX$

Reaction	ΔH_r^0 [kcal/mol]	Ref.
$H_2O + Me_2O \rightarrow 2 MeOH$	6.0	[15]
$H_2O + F_2O \rightarrow 2 HOF$	5±3	[1]
$H_2O + Cl_2O \rightarrow 2 HOCl$	1±1	[1, 16]
$H_2S + Me_2S \rightarrow 2 MeHS$	2.8	[5]
$H_2S + (HS)_2S \rightarrow 2 HSSH$	4.4	[5]
$H_2NH + Me_2NH \rightleftharpoons 2 MeNH_2$	4.5	[5]
$Me_2NMe + H_2NMe \rightleftharpoons 2 HNMe_2$	2.4	[5]
$H_2CH_2 + Me_2CH_2 \rightleftharpoons 2 MeCH_3$	2.5	[5]
$H_2CF_2 + F_2CF_2 \rightleftharpoons 2 HCF_3$	0±2	[5]
$H_2CH_2 + F_2CH_2 \rightleftharpoons 2 H_3CF$	12±2	[5]
$H_2CH_2 + Cl_2CH_2 \rightleftharpoons 2 ClCH_3$	1.5±1.4	[5]
$Cl_2CH_2 + Cl_2CCl_2 \rightleftharpoons 2 HCCl_3$	−2.7±1	[5]
$H_2CH_2 + I_2CH_2 \rightleftharpoons 2 ICH_3$	−4.4	[1]
$H_2C=CH_2 + Me_2C=CH_2 \rightleftharpoons 2 HMeC=CH_2$	1.3	[5]
$H_2C=CF_2 + F_2C=CF_2 \rightarrow 2 HFC=CF_2$	4.5±4	[5]
$H_2C=CCl_2 + Cl_2C=CCl_2 \rightarrow 2 HClC=CCl_2$	−2±4	[5]
$H_2CO + Me_2CO \rightarrow 2 HMeCO$	−1.7±1.5	[1]
$Cl_2CO + Me_2CO \rightarrow 2 ClMeCO$	−13.5	[1]
$Me_2Hg + Cl_2Hg \rightarrow 2 MeHgCl$	−12±3	[1, 15]

5. Electrostatic measure of the perturbation of bonding electrons.

Since group additivity includes all short range interactions out to about 3A, we must ascribe longer range interactions to coulombic or dipole effects. The electrostatic model of the chemical bond is an attempt to do this in terms of formal changes and local dipoles. It derives historically from the nineteenth century idea of electronegativity, later used by Pauling to describe deviations of bond lengths from simple additivity.

6. Electrostatic model of the paraffin hydrocarbons. Formalism of the model. Stability of branched isomers.

It is possible to reproduce the ΔH_f^o for all alkanes using a formal charge model in the C-H bond and a single CH_2 bond energy term, in all two parameters.

This scheme can be extended to olefins, acetylene compounds, aromatics and free radicals.

7. Application to olefins, aromatics and acetylenes.

Bond energies and partial charges used for reproducing the heats of formation $\Delta H_{f\,298}$ of hydrocarbons and alkyl radicals at 298 K. — For olefins and aromatics, a 1.0 kcal/mol repulsion is assigned to *cis*- or *ortho*-alkyl groups and 0.3 kcal/mol repulsion to geminal alkyl groups in olefins. For large, highly-branched alkanes 0.7 kcal/mol is assigned to *gauche* interactions and 1.5 kcal/mol to 1,5 methyl interaction.

Bond	Energy [b] [kcal/mol]	Partial charge [c] [10^{-10} esu]
C—C	0.25	0
H—H	0	0
C—H	−1.13	0.278 (H)
C_d—H	−3.3	0.32 (H)
C_B—H	−2.2	0.32 (H)
C_t—H	−5.2	0.36 (H)
C*—H	11.9	0.12 (H)
C_d—C	−5.5	0.12 (C)
C_t—C	−8.0	0.16 (C)
C_B—C	−2.0	0.12 (C)
C*—C	14.0	0.04 (C*)
C=C	0.25	0
C_B—C_B	6.5	0
C_d=C_d	34.0	0
C_t≡C_t	67.7	0

8. Application to Free Radicals

Comparison between Electrostatic Energy of the
Free Radical and Parent Molecule (E_{el} in kcal mol^{-1})

Radical	Parent molecule	E_{el} (molecule)	E_{el} (radical)	ΔE_{el} (molecule − radical)
Methyl	Methane	−12.55	−1.41	−11.14
Ethyl	Ethane	−13.57	−8.12	−5.45
n-Propyl	Propane	−16.44	−10.73	−5.71
Isopropyl	Propane	−16.44	−15.55	−0.89
sec-Butyl	n-Butane	−19.36	−18.24	−1.12
tert-Butyl	Isobutane	−21.17	−23.68	+2.51
Neopentyl	Neopentane	−27.73	−21.56	−6.17

The electrostatic model suggests that the large bond strength in CH_4 is a consequence of the large loss in electrostatic energy on removing a H atom from CH_4.

9. "Distorted" geometries in alkanes, cyclics and polycylics.

10. Dipole moments in isobutane and propane.

$\mu_{CH_3} = 2\mu_{CH} \cos 53.1° = 1.20\, \mu_{CH}$

$\mu_{CH_2} = 3\mu_{CH} \cos 68.8° = 1.90\, \mu_{CH}$

$\mu_{C_2H_6} = \mu_{CH_3} - 2\mu_{CH_3} \cos 56.2° = 0$

$\mu_{ind}\, (CH_3)_2 = 0.151\, D$

$\mu_{ind}\, (CH_3)_2 = 0.062\, D$

$\mu_{net}\, (C_3H_8) = 0.089\, D$ (calc.)
         $= 0.083\, D$ (obs.)

Intrinsic and induced dipole moments for CH_3 and CH_2 groups in propane calculated from the observed geometrical structure of the molecule.

11. Model of halogen compounds. Comparisons of MeX and tBuX. Lone pair contributions to electrostatics. Current status of halogenated methanes. Paradox of C_2F_6 and CH_3CF_3.

(Recommended Reading: Ref. 5,6,7; Ref. 3; Ref. 2)

SECTION III THERMOCHEMICAL KINETICS

1. Notion of elementary reactions. Mechanism of chemical reactions.

Chemical reactions seldom occur as one concerted chemical event. Instead they are composed of a series of what are termed "elementary steps". These series or sequences needed to describe a chemical reaction are called the "mechanism" of the reaction. Thus the simple, spontaneous chemical reaction:

$$Cl_2 + H_2 \rightarrow 2HCl + 44 \text{ kcal}$$

in practise occurs by an atom—catalyzed chain:

(initiation) $\rightarrow Cl_2 + M \rightleftarrows 2Cl$ (termination)

(chain) $\begin{cases} Cl + H_2 \rightleftarrows HCl + H \\ H + Cl_2 \rightleftarrows HCl + Cl \end{cases}$

The goal of chemical kinetics is to be able to predict the rates at which these elementary steps occur. With this information it is then possible to deduce the important steps which may participate in the reaction mechanism and then finally to deduce a mathematical expression for the overall, global reaction to product products. This latter is called the "rate law". For many systems such as oxidation of methane the mechanism may become so complex that it involves 140 elementary steps and defy any simple effort to represent it in simple, closed form. In such cases modern computing techniques allow us to "model" the reaction if the rate constants are known and obtain a quantitative statement of the rate of change of all species' concentrations.

In the present section we shall examine the transition state theory of elementary reactions and empirical methods which have been developed for estimating rate constants.

2. Free radical processes: Abstraction reactions. Addition reactions

$$CH_3 + CH_3\text{-}CH_3 \rightleftarrows CH_4 + \dot{C}H_2\text{-}CH_3 \quad \text{(metathesis)}$$

$$CH_3 + CH_2\text{=}CH\text{-}CH_3 \rightleftarrows CH_3\text{-}CH_2\text{-}\dot{C}HCH_3 \quad \text{(addition)}$$

3. Molecular processes: Concerted reactions, Fission reactions. Complex processes. Isomerization reactions.

Simple fission:
$C_2H_5I \rightarrow C_2H_5 + I$
$CH_3\text{—}CH_3 \rightarrow 2CH_3$

Complex fission:

$$CH_3\text{—}\underset{I}{CH}\text{—}CH_3 \rightarrow CH_3\text{—}CH\text{=}CH_2 + HI$$

$$CH_3\text{—}\underset{H}{CH}\text{—}\underset{NO_2}{CH_2} \rightarrow CH_3CH\text{=}CH_2 + HNO_2$$

$$CH_3C\overset{O}{\underset{OC_2H_5}{\diagup}} \longrightarrow CH_3\text{—}C\overset{OH}{\underset{O}{\diagup}} + C_2H_4$$

$$\underset{CH_2\text{—}CH_2}{\overset{CH_2\text{—}CH_2}{|\quad\quad|}} \rightarrow 2C_2H_4$$

Isomerization:

$$\underset{CH_3}{\overset{H}{\diagdown}}C\text{=}C\underset{CH_3}{\overset{H}{\diagup}} \text{ (cis)} \longrightarrow \underset{H}{\overset{CH_3}{\diagdown}}C\text{=}C\underset{CH_3}{\overset{H}{\diagup}} \text{ (trans)}$$

$$\underset{CH_2\text{—}\!\!\!\!-CH_2}{\overset{CH_2}{\diagup\diagdown}} \rightarrow CH_3\text{—}CH\text{=}CH_2$$

$$\overset{CH_2}{\|}\underset{CH}{\diagdown}\overset{CH_2}{\diagup}\underset{\underset{CH_3}{|}}{CH}\overset{CH}{\diagdown}\overset{\|}{CH_2} \longrightarrow \overset{CH_2}{\|}\underset{CH}{\diagdown}\overset{CH_2}{\diagup}\underset{}{CH_2}\overset{CH_2}{\diagdown}\underset{\underset{CH_3}{|}}{\overset{CH}{\|}}CH$$

4. Equilibrium Kinetics. Transition State Theory, Thermochemical Model of the Transition State.

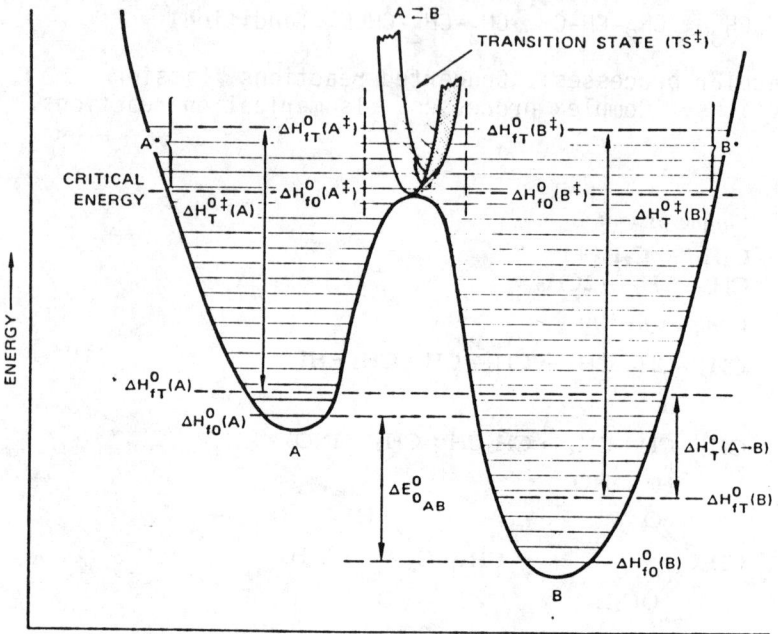

Arrhenius Parameters for Gas Phase: Unimolecular Reactions

Transition state theory is an equilibrium theory of chemical reactions. It assumes that reactants are in equilibrium with transition states, namely those species having the energy and conformation corresponding to the potential energy maximum. In the above isomerization diagram:

$$(A^{\ddagger}) = K^{\ddagger}(A)$$

and the rate of isomerization is given by:

$$\frac{-d(A)}{dt} = \nu^{\ddagger}(A^{\ddagger}) = \nu^{\ddagger}K^{\ddagger}$$

5. Loose and Tight Transition States. Entropy of Activation.

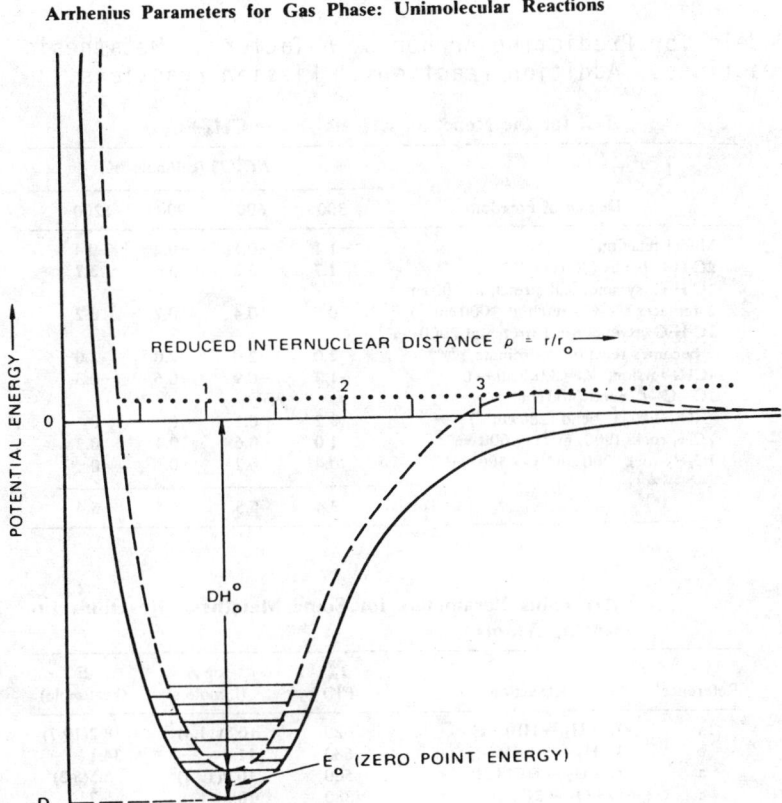

Arrhenius Parameters for Gas Phase: Unimolecular Reactions

Rules for tight transition states (metathesis)

$$\dot{C} + A-B \rightleftarrows [C \cdot A \cdot B]^\ddagger \rightleftarrows C-A + \dot{B}$$

In TS bonds are lengthened by 0.3Å
       find constants are diminished by factor of 2
       hence frequencies decrease by factor of 0.7

    In simple bond fissions, the rotational maximum occurs at distances of about $3r_o$ where r_o is the normal bond length. These are called loose transition states.

The following tables illustrate the methods of calculation of Arrhenius A-factors and the change in A-factor with temperature:

$A = A'T^n$ where $n = (\Delta C_v^{\ddagger}/R)$

6. Models for Predicting Arrhenius A-factors. Metathesis reactions. Addition reactions. Fission reactions.

$\Delta C_v^{\ddagger}$ for the Reaction $CH_3 + C_2H_6 \rightarrow CH_4 + C_2H_5$

Degree of Freedom	$\Delta C_v^{\ddagger}(T)$ (cal/mole-°K)			
	300	600	900	1200
Model reaction[a]	−1.1	−0.1	−0.2	−0.4
2C·H·C bends (700 cm^{-1})[b]	1.7	3.2	3.6	3.7
1C·H·C (symmetrical stretch at 700 cm^{-1} replaces C—C stretch at 1000 cm^{-1})	0.5	0.4	0.2	0.2
1C·H·C (asymmetrical stretch at 2000 cm^{-1} becomes reaction coordinate $\nu^{\ddagger}$)[b,c]	2.0	2.0	2.0	2.0
1CH$_3$ torsion, $V_0 = 3.0$ kcal $\rightarrow$ 0	−1.3	−0.9	−0.6	−0.3
1C—C—C bend (400 cm^{-1}) $\rightarrow$ (C·H)·C—C bend (280 cm^{-1})	0.2	0.1	0	0
2CH$_3$ rocks (900 cm^{-1}) $\rightarrow$ 600 cm^{-1}	1.0	0.6	0.4	0.2
1C$_2$H$_5$ rock (700 cm^{-1}) $\rightarrow$ 500 cm^{-1}	0.4	0.2	0.1	0
Totals	3.4	5.5	5.5	5.4

Arrhenius Parameters for Some Metathesis Reactions Involving Atoms

Reference[1]	Reaction	T_m° (°K)	log A (l./mole-s)	E (kcal/mole)
a	$Br + H_2 \rightarrow HBr + H$	620	10.8(11.4)	18.2(19.7)
b	$I + H_2 \rightarrow HI + H$	680	11.4	34.1
a	$Cl + H_2 \rightarrow HCl + H$	500	10.9(10.7)[g]	5.5(5.3)[g]
c	$O + O_3 \rightarrow 2O_2$	380	10.5	5.7
a	$Br + C_2H_6 \rightarrow HBr + C_2H_5$	500	10.8	13.5
a	$Cl + C_2H_6 \rightarrow HCl + C_2H_5$	500	11.0	1.0
d	$I + CH_4 \rightarrow HI + CH_3$	630	11.7	33.5
e	$H + C_2H_6 \rightarrow H_2 + C_2H_5$	300–1100	11.1	9.7
f	$H + D_2 \rightarrow HD + D$	900	10.7	9.4
d	$I + CH_3I \rightarrow I_2 + CH_3$	600	11.4	20.5
g	$O + NO_2 \rightarrow O_2 + NO$	330	10.3	1.0
h	$O + H_2 \rightarrow OH + H$	600	10.50(11.34)	10.2(13.7)
i	$H + HOH \rightarrow H_2 + OH$	(300–2500)	10.9	20.5
j	$H + N_2O \rightarrow HO + N_2$	(450–1487)	10.7	13
k	$N + O_2 \rightarrow NO + O$	300	9.3	6.3
k	$N + NO \rightarrow N_2 + O$	(300–5000)	10.2	0
l	$K + Br_2 \rightarrow KBr + Br$	(beam)	11.2	−0.4
k	$O + C_2H_6 \rightarrow HO + C_2H_5$	(300–650)	10.4	6.4
i	$O + OH \rightarrow O_2 + H$	300	10.3	0
m	$F + H_2 \rightarrow HF + F$	300	11.1	1.7
m	$Na + CCl_4 \rightarrow NaCl + CCl_3$	550	10.3	3.5

1. Bibliographic sources:
   a. G. C. Fettis and J. H. Knox, *Progress in Reaction Kinetics*, McMillan, New York, 1964, Vol. 2, p. 26. Also see p. 17 for Cl + H$_2$. Values in parentheses are "best" estimates by F and K.

Arrhenius Parameters for Some Metathesis Reactions not Involving Atoms

Reference	Reaction	T_m (°K)	log A (l./mole-s)	E (kcal/mole)
a	$CH_3 + \overset{*}{C}H_4 \rightarrow CH_4 + \overset{*}{C}H_3$	500	8.8	14.6
b	$CH_3 + C_2H_6 \rightarrow CH_4 + \cdot C_2H_5$	420	8.5	10.8
	$CH_3 + C(CH_3)_4 \rightarrow CH_4 + \cdot CH_2C(CH_3)_3$	420	8.5	10.4
	$CH_3 + benzene \rightarrow CH_4 + phenyl$	450	7.5^c	9.6^c
d	$C_2H_5 + C_2H_5COEt \rightarrow C_2H_6 + \dot{C}_2H_4COEt$	450	8.0	7.8
e	$2C_2H_5 \rightarrow C_2H_4 + C_2H_6$	450	9.6(8.3)^f	0
g	$C_2H_6 + C_2H_4 \rightarrow 2C_2H_5$	450	11.3(10.4)^h	60.0
i	$2C_2H_4 \rightarrow C_2H_5 + C_2H_3$	1300	11.1(10.4)^j	62
k	$2NO_2 \rightarrow (sym)\,NO_3 + NO$	800	9.7	23.6
l	$CH_3 + CCl_4 \rightarrow CH_3Cl + CCl_3$	400	8.6	9.1
m	$CF_3 + CCl_4 \rightarrow CF_3Cl + CCl_3$	450	8.5	10.4
n	$CF_3 + CHD_3 \rightarrow CF_3H + CD_3$	420	8.1	10.5
	$\rightarrow CF_3D + CHD_2$		8.5	12.7
o	$C_6H_5 + CH_4 \rightarrow C_6H_6 + CH_3$	450	8.6	11.1
p	$CF_3 + CH_3Br \rightarrow CF_3Br + CH_3$	420	7.5	8.1
	$+ CH_3Cl \rightarrow CF_3Cl + CH_3$	420	—	>17.0
	$+ CH_3I \rightarrow CF_3I + CH_3$	420	6.8	3.3

a. F. S. Dainton, K. J. Irvin, and F. Wilkinson, *Trans. Faraday Soc.*, **55**, 929 (1959); F. S. Dainton and D. E. McElcheran, *Trans. Faraday Soc.*, **51**, 657 (1955).
b. A. F. Trotman-Dickinson and E. W. R. Steacie, *J. Chem. Phys.*, **19**, 329 (1951).
c. It is doubtful if these are the correct parameters of this reaction.
d. M. H. J. Wijnen and E. W. R. Steacie, *J. Chem. Phys.*, **20**, 205 (1952).
e. A. Shepp and K. O. Kutschke, *J. Chem. Phys.*, **26**, 1020 (1957).
f. R. Hiatt and S. W. Benson, *J. Amer. Chem. Soc.*, **94**, 25 (1972); **94**, 6886 (1972). These values in parentheses are calculated from a combination of rate data on recombination together with thermochemical estimates of the differences in $S°$ and $\Delta H_f°$ of the alkyl radicals.
g Calculated from reference e and thermochemical data.
h Revised estimates, see reference i.
i. Estimates by S. W. Benson and G. R. Haugen, *J. Phys. Chem.*, **71**, 1735, (1935) (1967), from data on hydrogenation of C_2H_4.
j. Revised estimates, see reference f.
k. P. G. Ashmore and M. G. Burnett, *Trans. Faraday Soc.*, **58**, 253 (1962).
l. J. Currie, H. Sidebottom, and J. Tedder, *Int. J. Chem. Kinet.*, **6**, 481 (1974).
m. W. G. Alcock and E. Whittle, *Trans. Faraday Soc.*, **62**, 139, 664 (1966).
n. T. E. Sharp and H. S. Johnston, *J. Chem. Phys.*, **37**, 1541 (1962).
o. F. J. Duncan and A. F. Trotman-Dickson, *J. Chem. Soc.*, 4672 (1962).
p. W. G. Alcock and E. Whittle, *Trans. Faraday Soc.*, **61**, 244 (1965).

Estimation of the A-Factor for the Reaction $CF_3 + BrCH_3 \rightarrow CF_3Br + CH_3$

Degree of Freedom	$\Delta S^{\ddagger}_{300}$	Contributions to $\Delta C^{\ddagger}_v$	
		300°K	600°K
Reference reaction ($TS = CF_3SCH_3$)	−41.2	0.3	5.0
Spin	1.4	—	—
Symmetry (no change)	0	—	—
Translation	1.0	—	—
External rotation (increase in moments)	1.5	—	—
Two reduced barriers for internal rotation ($V_0 = 2$ kcal $\rightarrow 0$)	1.8	−2.1	−2.1
Reaction Coordinate C—S stretch ($600 cm^{-1}) \rightarrow \nu^{\ddagger}$	−0.5	1.0	0.3
Four rocking modes (CF$_3$ and CH$_3$) decrease to 70% (400 cm$^{-1} \rightarrow$ 280) (900 cm$^{-1} \rightarrow$ 630)	1.8	1.3	1.0
C—S—C bend (400 cm$^{-1} \rightarrow$ 280)	0.6	0.2	0.1
Totals	−33.6	0.7	4.3
		$\langle \Delta C^{\ddagger}_v \rangle_{600} = 2.5$	

$A_{300} = 10^{7.7}$ l./mole-a (note bent TS)
$A_{600} = 10^{8.1}$ l./mole-s

Recommended Reading: Ref. 1 - Chapter 3, 4
Ref. B

BIBLIOGRAPHY

Most of the material to be presented has been published. The best sources are as follows:

1. S. W. Benson, Thermochemical Kinetics, 2nd Ed., John Wiley and Sons, Inc., New York, 1976.

2. Predicting Chemical Reactivity, S. W. Benson, Chem. Tech. 10, 121, 1980.

3. Electrostatics, The Chemical Bond and Molecular Stability, S. W. Benson, Angewandte Chem. 17, 1978, Int. Ed. (English).

4. S. E. Stein, D. M. Golden and S. W. Benson, Predictive Scheme for Thermochemical Properties of Polycyclic Aromatic Hydrocarbons, J. Phys. Chem. 81, 314 1977.

5. Electrostatics and the Chemical Bond, I. Saturated Hydrocarbons, Menachem Luria and S. W. Benson, J. Amer. Chem. Soc., 97, 704, 1975.

6. Electrostatics and the Chemical Bond, II. Unsaturated Hydrocarbons, Menachem Luria and S. W. Benson, J. Amer. Chem. Soc., 97, 3337, 1975.

7. Electrostatics and the Chemical Bond, III. Free Radicals, Menachem Luria and S. W. Benson, J. Amer. Chem. Soc., 97, 3342, 1975.

8. Methods for the Estimation of Rate Parameters of Elementary Processes, David M. Golden and S. W. Benson, Chap. 2, Physical Chemistry, Vol. VII (edited by Eyring, Henderson and Jost), Academic Press, New York, 1975.

9. Additivity Rules for the Estimation of Thermochemical Properties, F. R. Cruickshank, D. M. Golden, G. R. Haugen, H. E. O'Neal, A. S. Rodgers, R. Shaw, R. Walsh and S. W. Benson, Chem. Rev. 69, 279 1969.

10. Entropies and Heat Capacities of Free Radicals, H. E. O'Neal and S. W. Benson, Int. J. Chem. Kinet., 1, 221, 1969.

BARRIERS TO INTERNAL ROTATION IN INORGANIC SPECIES

Stanley Abramowitz

National Bureau of Standards, Washington, D.C. 20234
U.S.A.

ABSTRACT

Barriers to internal rotation have been determined using infrared and Raman spectroscopy for some inorganic species. The methods used for the determination of these barriers will be described for a high barrier (BCl_2SH), a medium barrier (PF_5, AsF_5 and VF_5) and a low or zero energy barrier ($B(CH_3)_3$).

INTRODUCTION

The determination of the energy barriers to internal rotation and the vibrational energy levels associated with these barriers are important for the computation of thermodynamic functions. This paper will illustrate the experimental determination of barriers to internal rotation for the species BCl_2SH (D), XF_5 where X = P, As or V, and $B(CH_3)_3$. These represent a high, medium and low (zero) barrier respectively. The techniques of vibrational spectroscopy were utilized to determine the vibrational energy levels necessary for the determination of the barriers. In the case of $B(CH_3)_3$ a comparison of the third law entropy with an experimentally determined entropy was made.

BCl_2SH

The BCl_2SH molecule and its deuterium analog have a planar C_s symmetry. This symmetry has 7a' and 2a" modes (1). (It should be noted that another C_s structure is possible with a σ_v plane bisecting the ClBCl angle. This structure has 6a' and 3a" vibrations.) The vibrational assignment for the BCl_2SH and BCl_2SD

M. A. V. Ribeiro da Silva (ed.),
Thermochemistry and Its Applications to Chemical and Biochemical Systems, 789–802.

Table 1. Vibrational Assignment BCl_2 SH and BCl_2 SD

Vibrational Assignment		Approximate Description	BCl_2SH	BCl_2SD
	ν_1	SH (D) str.	2602	1863
	ν_2	BCl_2 asymm. str.	994	999
	ν_3	BS str.	900	900
A'	ν_4	SH(D) bend	779	599
	ν_5	BCl_2 sym str.	461.8	459
	ν_6	ClBCl bend	255	255
	ν_7	ClBS bend	230	227
A"	ν_8	Cl o.o.p.	474	475
	ν_9	SH(D) torsion*	386	283

*The 0-1 frequency is given.

Hot band transitions are given in Table 2.

Table 2. Observed vs. Calculated ν_9 Torsional Frequencies (cm^{-1})

	Transition	$V = (V_2/2)(1-\cos2\phi)$				$V = A(z^4 + Bz^2)$		
		Obs'd.	Calc'd.[1]	Δ		Obs'd.	Calc'd.[1]	Δ
BCl$_2$SH F^2 = 9.581 cm^{-1}		$\bar{s}$ = 430.204	V_2 = 4121.78 cm^{-1}			$\bar{B}$ = -21.896	V_2 = 3574.67 cm^{-1}	
	0 → 1	386.1	387.6	-1.5		386.1	386.1	0.0
	1 → 2	376.6	377.4	-0.8		376.6	377.1	0.5
	2 → 3	367.9	355.7	+1.1		367.9	367.4	+0.5
	3 → 4	356.9	355.7	+1.2		356.9	356.9	0.0
BCl$_2$SD F^2 = 4.790 cm^{-1}		$\bar{s}$ = 903.091	V_2 = 4325.81 cm^{-1}			$\bar{B}$ = -26.45	V_2 = 3459.36 cm^{-1}	
	0 → 1	282.9	282.9	0.0		282.9	282.9	+0.1
	1 → 2	277.8	277.9	-0.1		277.8	278.0	-0.2
	2 → 3	272.9	272.8	+0.1		272.9	272.9	0.0
	3 → 4	267.5	267.5	0.0		267.5	267.5	0.0

[1] 1 cm^{-1} = 2.859 cal/mol = 11.963 J/mol.
[2] Computed from assumed molecular dimensions (ref. 1).

species is given in Table 1. (Where B and/or Cl isotope effects have been observed the frequency given is for the $^{11}B^{35}Cl_2SH$ (D) species.) This assignment was arrived at through interpretation of the infrared and Raman spectra and the boron, chlorine and hydrogen vibrational isotopic shifts. The vibrational assignment for BCl_2SH and BCl_2SD was used to calculate the force field for these species. The computed force field successfully predicts the isotopic splittings for these species (1). The off diagonal force constants were taken from BCl_3 (2). Other off-diagonal force constants were used only if they improved the overall frequency fit. The symmetry coordinates and force field and the assumed geometric structure for the BCl_2SH (D) species can be found in reference 1.

The SH (D) torsional mode shows a hot band structure in the infrared spectrum with 0-1, 1-2, 2-3, and 3-4 modes observed for both species. Spectra observed at temperatures higher than ambient verified the assignments. These modes could be fit with either of two potentials

$$V = A(z^4 - Bz^2), \text{ where } z = 2\phi$$

or,

$$V = (V_2/2)(1-\cos 2\phi)$$

This is because the barrier is quite high and the first potential is simply $(1-\cos n\phi)$ with a truncated power series substituted for $\cos n\phi$.

The $A(z^4 - Bz^2)$ potential which is most useful for ring puckering modes does not require a knowledge of the molecular dimensions. This is an advantage for BCl_2SH (D) since experimentally determined bond lengths and angles are not available. The $V_2/2(1-\cos 2\phi)$ potential which requires a knowledge of the molecular dimensions is a better model for the molecular motion involved in the SH(D) torsional mode, ν_7.

For this potential the moments of inertia are computed from assumed moleculer dimensions. Table 2 presents a summary of the calculated and observed vibrational frequencies for the torsional vibrational modes about the B-S bond. As one can see the barrier is quite high. The more physically meaningful potential, $V=(V_2/2)(1-\cos 2\phi)$ yields a barrier of about 4200 cm^{-1} for these species.

XF_5 where X=P, As and V

All of these species have been shown to be trigonal bipyramids having D_{3h} structure. The F atom interchange in these species can be accomplished through the motions along the E' ν_7 coordinate which can lead to axial equatorial interchange through a C_{4v} intermediate (3,4). It was therefore very interesting to have obtained the Raman gas phase spectra of these species in order to directly observe the hot band structure. This vibrational hot band structure must be interpreted using a doubly degenerate harmonic oscillator function (3,4).

The Hamiltonian for axial-equatorial interchange is given by

$$H = (\hbar^2/m)^{1/2} H_D$$

where $H_D = [1/2(Px^2 + Py^2) + 1/2(x^2+y^2) - B(x^3-3xy^2) + C(x^2+y^2)^2]$

where B and C contain a term which is a scale factor involving the reduced mass involved in the ν_7 vibration and the harmonic force constant describing the ν_7 vibration. One can produce a table of dimensionless eigenvalues for this potential (4), where the barrier for F atom interchange is $-(1/32)B^2$. The important feature of this potential is that it requires (n + 1) vibrational levels where n is the vibrational quantum number.

The advantages of working with the above reduced Hamiltonian are (1) one need not calculate an explicit reduced mass and (2) there is one less potential constant to vary than in the explicit form. Thus a fit of the reduced Hamiltonian H_D to observed energy levels will yield B and C, the scale factor and the barrier to axial - equatorial interchange.

The observed and calculated ν_7 transitions for PF_5, AsF_5, and VF_5 are given in Tables 3, 4, and 5 respectively. The barriers decrease monotonically from about 1050 to 475 cm^{-1}. More features are observed in VF_5 then AsF_5 and more in AsF_5 then PF_5 because of the Boltzmann factor. The vibrational frequency decreases from 175 to 133 to 108 cm^{-1} in going from PF_5 to AsF_5 and VF_5. Therefore at room temperature the largest number of vibrational levels are populated in VF_5; with less in AsF_5 and even less in PF_5. Further features of the dynamics of axial - equatorial fluorine atom interchange can be found in the references cited in (4).

$B(CH_3)_3$

The third law entropy of $B(CH_3)_3$ computed from the available vibrational assignment and inertial parameters assuming free rotation of the CH_3 groups is not in agreement with the experimentally measured entropy (6,7). (The difference is such that any barrier to rotation would increase the discrepancy). Since the Raman gas phase spectrum for this species had not been previously observed a reinvestigation of the vibrational spectrum was initiated in order to better characterize the vibrational assignment. Boron trimethyl has a planar BC_3 skeleton. Conventional point group theory would indicate an average D_{3h} symmetry for freely rotating CH_3 groups with 20 normal modes. However, the observed infrared and Raman gas phase spectra showed fewer features. The observed spectra indicated that the G_{324} symmetry group based upon the molecular symmetry (MS) group theory first proposed by Longuet-Higgins (8,9) was necessary. Since the discussion of the vibrational structure in terms of

Table 3. Observed and Calculated ν_7 Transitions for PF_5

Transitions	Symmetry	Calculated (cm^{-1}) B= 0.07	Calculated (cm^{-1}) B = 0.075	Observed[a] (cm^{-1}) Raman
0 → 1	$A_1' \to E'$	175.0	175.0	175
1 → 2	$E_1' \to A_1'$	161.8	159.8	159
1 → 2	$E' \to E'$	177.9	178.0	178
2 → 3	$A_1' \to E'$	164.8	163.5	164
2 → 3	$A_1' \to A_1'$	192.7	194.0	194
2 → 3	$A_1' \to A_1'$	198.9	199.4	b
2 → 3	$E' \to E'$	148.8	145.3	b
2 → 3	$A_1' \to A_1'$	177.5	175.7	c
2 → 3	$E' \to A_1'$	182.9	184.3	180
Barrier height (cm^{-1})	---	1139	995	---

[a] The additional peaks at 168 cm^{-1} and 187 cm^{-1} probably correspond to OP and RS maximum, respectively.

[b] Not observed.

[c] This feature is probably obscured by 0 → 1 transition.

Table 4. Observed and Calculated ν_7 Transitions for AsF_5

Transition	Symmetry	Calculated (cm^{-1}) B = 0.07	Calculated (cm^{-1}) B = 0.075	Observed (cm^{-1}) Raman
0 → 1	A_1' → E'	132.8	132.8	133
1 → 2	E' → A_1'	122.8	121.2	121
1 → 2	E' → E'	135.0	135.1	136
2 → 3	A_1' → E'	125.1	124.1	125
2 → 3	A_1' → A_1'	146.2	147.2	148
2 → 3	A_1' → A_1'	151.0	153.7	157
2 → 3	E' → E'	112.9	110.2	113
2 → 3	E' → A_1'	134.7	133.3	a
2 → 3	E' → A_1'	138.8	139.9	141
3 → 4	A_1' → A_1'	117.6	116.2	118
3 → 4	E' → E'	126.7	125.3	129
3 → 4	E' → E'	161.7	165.6	166?
3 → 4	A_1' → A_1'	96.5	93.1	b
3 → 4	A_1' → E'	105.4	102.2	b
3 → 4	A_1' → E'	140.5	142.5	143
3 → 4	A_1' → A_1'	91.7	86.6	b
3 → 4	A_1' → E'	100.7	95.7	b
3 → 4	A_1' → E'	135.8	135.9	c
Barrier height (cm^{-1})		---	864	755	---

^{a} Obscured by 0 → 1, ^{b} not observed, ^{c} obscured by 1 → 2 (E' → E').

Table 5. Observed and Calculated ν_7 Transitions for VF_5

Transition	Symmetry	Calculated (cm^{-1}) B = 0.08	Calculated (cm^{-1}) B = 0.09	Observed (cm^{-1}) Infrared[a]	Raman[b]
0 → 1	A' → E'	107.4	197.3	---	108
1 → 2	E' → A$_1'$	96.8	94.0	91.9	93
1 → 2	E' → E'	109.4	109.4	109.4	108
2 → 3	A$_1'$ → E'	99.3	97.0	98.1	100
2 → 3 R	A$_1'$ → A$_1'$	119.5	119.5	c	120
2 → 3 R	A$_1'$ → A$_1'$	126.6	131.6	c	126,120
2 → 3	E' → E'	86.7	81.6	85.8	5,83
2 → 3	A$_1'$ → A$_1'$	107.0	104.2	103.6	---
2 → 3	E' → A$_1'$	114.0	116.2	---	114
Barrier height (cm^{-1})		---	539	428	---	---

[a] Reference 5.

[b] There are additional Raman features which can be attributed to 3 → 4 transitions. Limitations in our calculations did not enable us to determine these transitions.

[c] Only Raman active.

the G_{324} group and its relationship to D_{3h} has not been previously given a brief discussion of the vibrational analysis follows. Finally a program developed to compute the thermodynamic functions for anharmonic oscillators was utilized. Somewhat better agreement between the computed third law entropy and that experimentally determined is obtained.

The vibrational modes of $B(CH_3)_3$ are shown in Table 6. For a D_{3h} symmetry with freely rotating methyl groups one expects 20 normal modes. The A_1', E' and species are Raman active while the A_2'' and E' species are infrared active. If one considers the G_{324} molecular symmetry group there are 15 normal modes of which the A_1, I, G and E_2 are Raman active while the A_3, I, G and E_2 are infrared active. Table 5 shows a correlation diagram between D_{3h} and G_{324} symmetries. It should be noted that G_{324} requires 8 Raman active modes and nine infrared active modes while D_{3h} symmetry requires 13 Raman active modes and 11 infrared active modes. The significant feature of the MS group is the relative simplicity of the expected vibrational spectrum. The observed infrared and Raman vibrational spectra confirm the MS symmetry group as the better representation of this molecule's symmetry. A survey Raman spectrum of $B(CH_3)_3$ is shown in Figure 1. Excitation

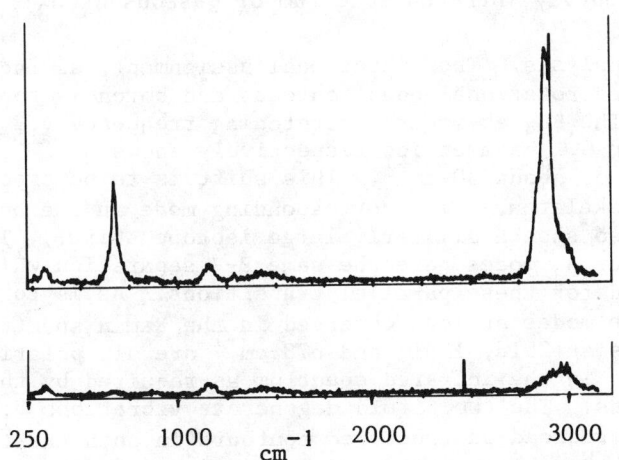

Figure 1. Survey Raman spectrum of $B(CH_3)$ upper spectrum depolarized lower spectrum polarized.

was provided by an Argon ion laser operating at 514.5 nm at a power of 3.5 watts. The survey spectrum was run at a resolution of about 3 cm^{-1}. The polarized and depolarized spectra are shown in the lower and upper frames respectively. Higher resolution spectra of selected regions were obtained for the vibrational analysis. A survey infrared spectrum is shown in Figure 2. Higher resolution spectra were utilized for the

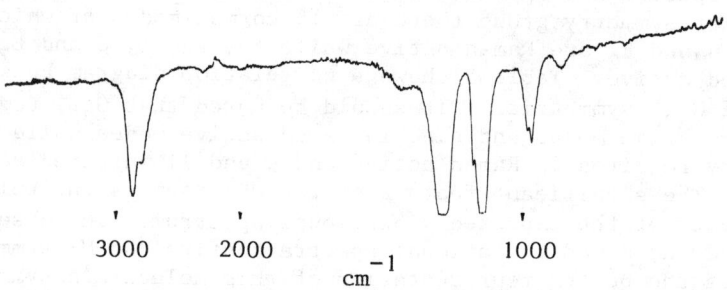

Figure 2. Survey infrared spectrum of gaseous $B(CH_3)$.

vibrational analysis. The vibrational assignment was facilitated by the observed rotational band contours and boron isotopic splittings. The BC_3 asymmetric stretching frequency ν_{15} for D_{3h} or ν_{14} for G_{324} symmetries respectively shows a $^{10}B-^{11}B$ isotope shift of about 30 cm^{-1}. This shift is to be expected for a planar BC_3 skeleton. (The corresponding mode in the boron trihalides also show a similarly large isotope shift.) Three out of the four A_3 modes have the same P-R separation which is to be expected for these parallel transitions. As is to be expected these modes are not observed in the Raman spectrum. The three A_1 modes at 2916, 1291, and 678 cm^{-1} are all polarized and are absent in the infrared spectrum as required by the selection rules. The two 6 fold degenerate vibrations ν_{11} and ν_{12} have rather broad unstructured contours in both the infrared and Raman spectra. In summary the infrared and Raman spectra are consistent with a G_{324} molecular symmetry assignment. There are not enough vibrational bands observed to justify a D_{3h} assignment for the $B(CH_3)_3$ molecule. A summary of the vibrational assignment is given in Table 7.

A calculation of the third law entropy of $B(CH_3)_3$ can be made using the rigid rotor harmonic oscillator approximation and by assuming free rotation of the three methyl groups about the B-C bonds. The vibrational assignment given in Table 7 is used together with the bond distances and angles determined using

Table 6. Correlation Table for D_{3h}- G_{324} and for $B(CH_3)_3$

Point Group D_{3h}				Molecular Symmetry Group G_{324}			
Species	Vibration	Activity	Mode	Vibration	Species	Activity	
A_1'	1	R(p)	CH_3 stretch	1	A_1	R(p)	
	2		CH_3 deformation	2			
	3		BC_3 stretch	3			
A_2'	4	IA	CH_3 stretch	4	A_4	IA	
	5		CH_3 deformation	5			
	6		CH_3 rock	6			
A_2''	7	IR	CH_3 stretch	7	A_3	IR	
	8		CH_3 deformation	8			
	9		CH_3 rock	9			
	10		BC_3 deformation	10			
E'	11	IR,R	CH_3 stretch	11	I	IR,R	
	12		CH_3 stretch				
	13		CH_3 deformation	12			
	14		CH_3 deformation				
	15		BC_3 stretch				
	16		CH_3 rock	13	G	IR,R	
	17		BC_3 deformation				
E"	18	IR,R	CH_3 stretch	14			
	19		CH_3 deformation	15	E_2	IR,R	
	20		CH_3 rock				

Table 7. Vibrational Assignment of $B(CH_3)_3$ According to G_{324} MS Group

Species	Vibration	IR	Raman	Inactive (Estimate)
A_1	1		2916	
	2		1291	
	3		678	
A_2	4			2915
	5			1400
	6			900
A_3	7	2914		
	8	1303		
	9	975		
	10	326		
I	11	2984	2985	
	12	1450	1445	
G	13	867	868	
E_2	14	1184[a]	1162	
		1154		
	15	310	314	

a ^{10}B, ^{11}B isotopic splitting.

electron diffraction techniques by Bartell and Carroll (10). An entropy of 67.83 cal/K at 199.92 K is computed assuming free rotation of the CH_3 groups. This compares with the measured value of 68.29 ± 0.10 cal/K (1). The fact that the measured entropy is greater than that computed using spectroscopic data indicates essentially free rotation of the methyl groups. Any significant barrier to the rotation of the methyl groups results in a lowering of the calculated entropy and therefore an increase in the discrepancy between the measured and computed entropy.

A computer program has been written to compute the effect of anharmonicity on the vibrational contribution to the entropy. In this computation all vibrational states were assumed to have a 1% anharmonicity (other values of the anharmonicity can be chosen for each vibration). This computer code obtains a state sum by counting those levels whose partition coefficient exceeds a chosen value, in this case 10^{-8}. (A choice of 10^{-11} does not effect the computed entropy.) An increase of 0.03 e.u. at 199.92K is obtained in this manner. The projected difference increases for higher temperatures. Additional entropy is to be expected by taking proper account of the rotational distortion constants due to non-rigidity of the molecule.

This program will be especially useful in computing estimated errors in entropy caused by vibrational anharmonicity. It will therefore be useful for comparing third law (statistically calculated) entropies with those derived from thermodynamic measurements. By identifying expected differences the evaluation of thermodynamic data particularly for high temperature species should be expedited.

ACKNOWLEDGMENT

This research was supported in part by the Air Force Office of Scientific Research.

REFERENCES

1. Kirklin, D.R., Ritter, J.J., and Abramowitz, S.: 1977, J. Mol. Spectroscopy, 67, 322.
2. Levin, I.W., and Abramowitz, S.: 1965, J. Chem., 43, 4213.
3. Bernstein, L.S., Kim, J.J., Pitzer, K.S., Abramowitz, S., and Levin, I.W.: 1975, J. Chem. Phys., 62, 3671.
4. Bernstein, L.S., Abramowitz, S., and Levin, I.W.: 1976, 64, 3228.
5. Holmes, R.R. Sr., Deiters, R.M., and Hora, C.J.: 1974, J. Chem. Soc. Commun., 175.

6. Furukawa, G.T. and Park, R.P.: 1954, National Bureau of Standard Report 3644.
7. Woodward, L.A., Hall, T.R., Dixon, R.N.: 1959, Spectrochim. Acta 15, 249.
8. Longuet-Higgins, H.C.: 1963, Molecular Phys., 6, 445.
9. Bunker, P.R.: 1979, Molecular Symmetry and Molecular Structure, Academic Press, New York.
10. Bartell, L.S. and Carroll, B.L.: 1965, J. Chem. Phys., 42, 3076.

CRITICAL EVALUATION OF THERMODYNAMIC DATA - A RESEARCH ACTIVITY

S. Abramowitz, D.D. Wagman, V.B. Parker, and D. Garvin

National Bureau of Standards, Washington, D.C. 20234
U.S.A.

ABSTRACT

The principles underlying the critical evaluation of thermodynamic data are described. The role of modern computer technology in both data evaluation and the creation of data bases is described.

INTRODUCTION

This is an appropriate time to describe the process involved in the evaluation thermochemical data since NBS has just completed the omnibus thermochemical tables entitled "Selected Values of Chemical Thermodynamic Properties," which have been issued in eight volumes over the past sixteen years as National Bureau of Standards Technical Note 270 (1). These tables cover the compounds of all the chemical elements except the transuranics. These tables contain two and a half times as many entries as their predecessor, NBS Circular 500 which was issued in 1952 (2). This new work covers 14,300 inorganic and low molecular weight organic (generally those containing not more than two carbon atoms) substances and is based on 60,000 references. A page from the final volume, NBS Technical Note 270-8, is shown in Figure 1. Those items in boxes show the expansion since 1952. However it must be noted that every number is new. This work is a complete reevaluation of the thermochemical data from the original papers and presents a self-consistent set of data for use by scientists and engineers.

Selected Values of Chemical Thermodynamic Properties — Series I
National Bureau of Standards — Technical Note 270-8

Table 99 — Enthalpy and Gibbs Energy of Formation; Entropy and Heat Capacity
Sodium

Substance Formula and Description	State	Formula weight	0 K $\Delta Hf°_0$ kcal/mol	298.15 K (25°C) $\Delta Hf°$ kcal/mol	$\Delta Gf°$ kcal/mol	$H°_{298}-H°_0$	$S°$ cal/deg mol	$C°_p$
Na_2F_2	g	83.9764	−197.03	−198.2	−197.8	4.02	71.5	17.9
$NaHF_2$	c	61.9946	−218.62	−219.95	−203.68	3.334	21.73	17.93
from HF_2	a			−212.73	−200.77		36.2	
in 400 H_2O				−214.1				
NaH_2F_3	c	82.0010		−295.2				
NaCl	c	58.4428	−98.168	−98.268	−91.815	2.536	17.24	12.07
	g		−41.9	−42.22	−47.00	2.298	54.90	8.55
	a			−97.34	−93.965		27.6	−21.5
in 9 H_2O				−97.820				
in 10 H_2O				−97.809				
in 12 H_2O		58.4428		−97.770				
in 15 H_2O				−97.707				
in 20 H_2O				−97.614				
in 25 H_2O				−97.547				
in 30 H_2O				−97.496				

Figure 1. Sample entries from the table of compounds of sodium from NBS Tech. Note 270-8. The highlighted items are new material, showing expansion from circular 500.

DISCUSSION

This paper will describe important features of the process by which these tables have been constructed. Emphasis will be on recent applications of automation because they will become very important in future work, both for the evaluation of thermochemical data and for the creation of the data bases.

In order to proceed in an orderly fashion the following definitions will be helpful.

Thermophysical properties - Thermodynamic functions of a single unchanging substance as a function of temperature and pressure and composition in non-reacting mixtures.

Thermochemical properties - Thermodynamic functions that record changes as the chemical identity of the material changes.

"Chemical thermodynamic properties" is an all inclusive term which includes both thermochemical and thermophysical properties as they apply to the thermodynamics of chemical reactions. Obviously the engineer needs both of these types of properties in order to have the data appropriate to the process of interest at the temperature and pressure of interest. Alternatively the data evaluator needs thermophysical data in order to bring experimental reaction data to a common temperature and pressure so that they can be compared and the best values determined.

Reactions

Changes that occur in a typical chemical reaction are expressed in terms of formation properties. These properties, $\Delta_f H$, $\Delta_f G$, $\Delta_f S$ and C_p, which are commonly tabulated, can be algebraically added to give the changes in the properties of interest. As an example consider the reaction

$$Rb_2O(c) + H_2O(l) \rightarrow 2RbOH(aq)$$
$$\Delta H = -338 \pm 3 \text{ kJ mol}^{-1}$$
$$\Delta H = 2 \cdot \Delta H_f(RbOH) - \Delta H_f(Rb_2O) - \Delta H_f(H_2O)$$

or $\quad y \pm u = -338 \pm 3 = 2x_3 - x_2 - x_1$

Networks

In order to get from the process data to the formation properties one considers networks of data. An example of a thermochemical network for compounds of barium is shown in Figure 2.

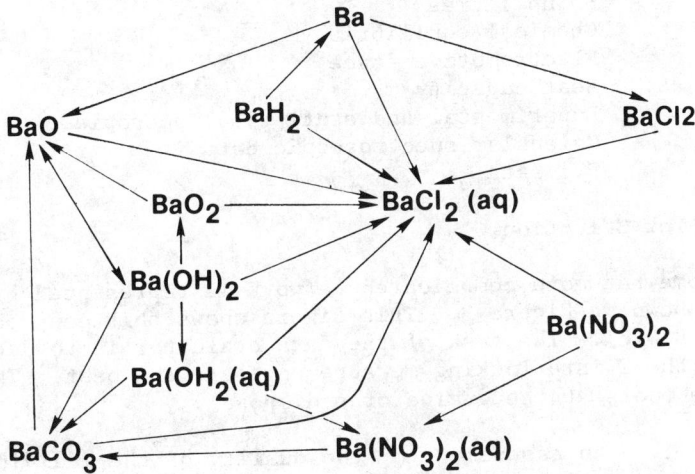

Figure 2. A thermochemical cycle showing measurement paths from elemental barium to its compounds.

As one can see Ba(c), the defined starting point, can be linked to $BaCO_3$ or $Ba(NO_3)_2$ even though measurements of direct chemical reactions involving the reaction of Ba(c) with the necessary elements to yield $BaCO_3$ or $Ba(NO_3)_2$ do not exist. In these networks the arrows indicate the reaction or process studied. A double headed arrow indicates that measurements have been made on both forward and reverse reactions. For example the enthalpy of hydrogenation of Ba(c) to form BaH_2 has been measured as well as enthalpy of the process

$$BaH_2(c) \rightarrow Ba(c) + H_2(g)$$

These networks exist in Gibbs energy, enthalpy and entropy space. They are further constrained by the thermodynamic relationship

$$\Delta G = \Delta H - T\Delta S$$

It has to be emphasized that all pathways for obtaining a value of a thermochemical property must be considered when one is developing a recommended value.

Measurement Base for Thermochemistry

The types of experimental data which are used to obtain these chemical thermodynamic formation data are

    H: Experimental heats of reaction, fusion, transition, vaporization, sublimation, solution and dilution.
    G: Vapor pressures
           Solubilities
           Chemical equilibria
           Electromotive force
    S: Heat capacity
           Experimental and statistical entropies
           Molecular spectroscopic data
           PVT data

Criteria for Selection

A somewhat more complicated network is represented by the Be network shown in Figure 3. This figure shows thirteen routes from Be(c) to BeO(c). The task of the data evaluator is to decide which of these interlocking measurements are the best. There are three tools at the evaluator's disposal.

1. An assessment of the quality of the individual's measurements.
2. A determination of how well they agree.
3. A comparison with similar processes (estimates or comparisons with theory).

A typical network for compounds of one chemical element is shown in figure 4. These networks get quite large. Figure 4 summarizes the core of the measurement on the compounds of lithium. Only those compounds for which properties have been determined by at least two paths are included. The one-shot cases are more than those represented in the figure. The total

CRITICAL EVALUATION OF THERMODYNAMIC DATA

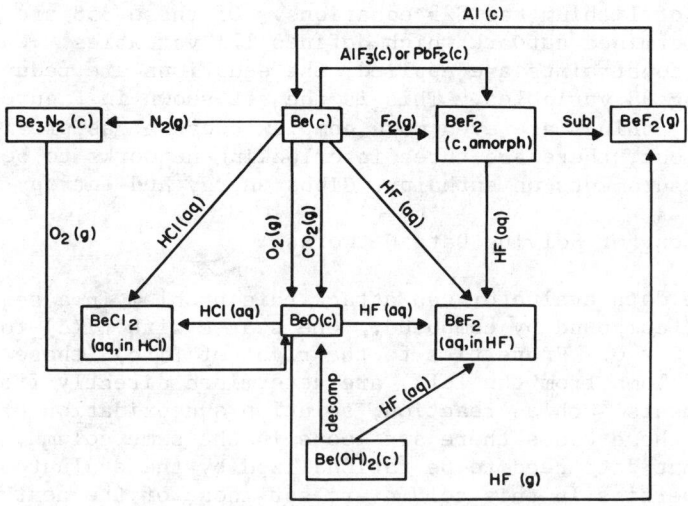

Figure 3. A thermochemical cycle showing measurement path leading from elemental beryllium to beryllium oxide.

Li Thermochemical Network

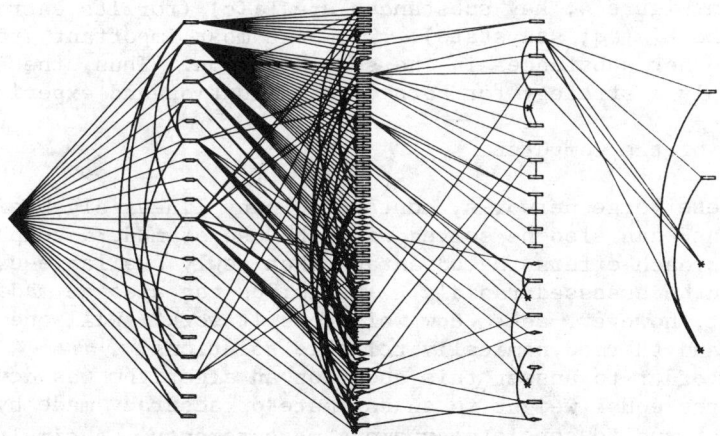

Figure 4. The Lithium Thermochemical Network. Each box is a compound and each line a measurement. The common starting point at the left is Li(c) and the top two compounds in the next column are Li(g) and Li$^+$ (aq., std. state). Rectangles are enthalpies, diamonds are Gibb's energies and hexagons are entropies.

network of lithium has 825 equations. Of these 358 are in an
over-determined network which defines 171 variables. When thermodynamic constraints are applied, the equations are reduced to 280
involving 93 variables. This is the set shown in Figure 4.
Actually, things are even more complex because, as previously
pointed out, there are three interlocking networks to be considered,
with measurements on enthalpy, Gibbs energy and entropy planes.

Techniques for Solving Data Networks

The data evaluator can attack this problem in a sequential
manner: compound by compound. One starts with Li(c) for which
$\Delta_f H = \Delta_f G = 0$. Properties to the right of Li(c) (those in the
second column from the left) are determined directly from
measurements such as reaction, solution and oxidation of Li(c).
Even in these cases there are loops in the same column. This
means that data need to be rationalized by the evaluator. Once
the properties in this column are set those of the next are
considered and determined. The properties of the complete
network are set in this way.

The graphical representation of the network has another
significant advantage. It very clearly indicates which substances
are key ones. They are the substances upon whose properties
those of many others depend. Their properties should be determined
as precisely and accurately as possible. For the lithium network
shown in Figure 4, key substances are Li(c) (for its entropy),
Li(g) and Li^+ (aq; std state). The next most important group would
be the other substances in the second column. Thus, these diagrams
can supply a strategy for both data evaluation and experiment.

The Use of the Computer

These large networks, containing many linear algebraic
equations, can also be solved with the use of modern computers.
This approach offers the advantage that newly available experimental
data can be assessed rapidly. Revisions can also be made quickly.
One must, however, ask: How well does it work? Will one get
stable and thermodynamically reliable solutions?

In order to answer this question an algorithm was developed
that gives equal weight to an estimate of accuracy made by the
evaluator and to consistency among measurements. A simultaneous
solution (made by the computer) and a sequential solution (using
the conventional stepwise approach of the evaluator described
above) were then compared. The residuals for each data item were
calculated and plotted in Figures 5 and 6. The residuals represented
as histograms in the two figures are quite similar.

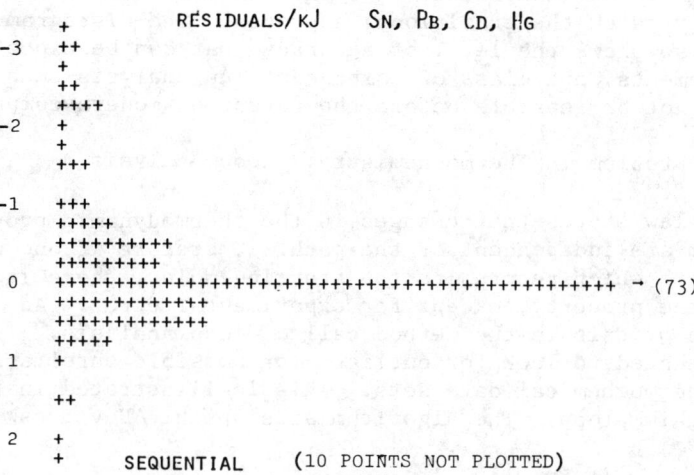

Figure 5. Residuals of sequential solution for Sn, Pb, Cd, Hg compounds.

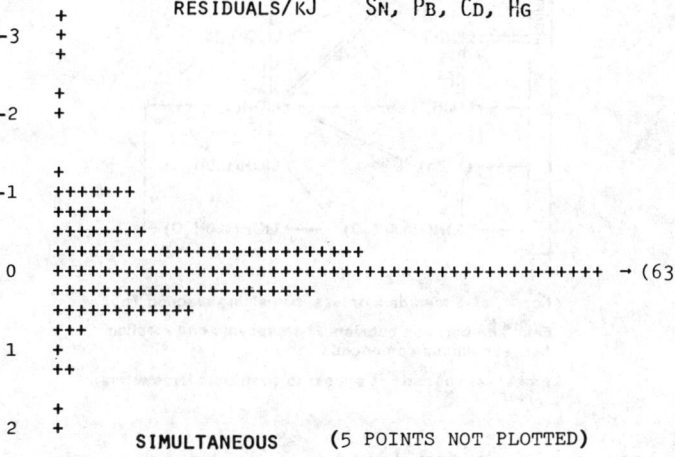

Figure 6. Residuals of simultaneous solution for Sn, Pb, Cd, Hg compounds.

This gives one confidence that the computer solution can successfully match the traditional approach. The histogram analysis also shows the level of accuracy that can be expected for measurements on a class of compounds. An analysis such as this would not be feasible before the advent of modern computers.

The Survey Problem in Thermochemistry: Loop Analysis

Hess' law states that changes in the thermodynamic properties of a system are independent of the path. Therefore if one chooses a closed path which returns to the starting point, there is no change in the property, except for experimental error. An application of this is the method called "loop analysis." It has been implemented to look for outliers (or possible unreliable data) in thermochemical data sets. This is illustrated in Figure 7 for a lithium loop. The algorithm adds up the ΔH values (or

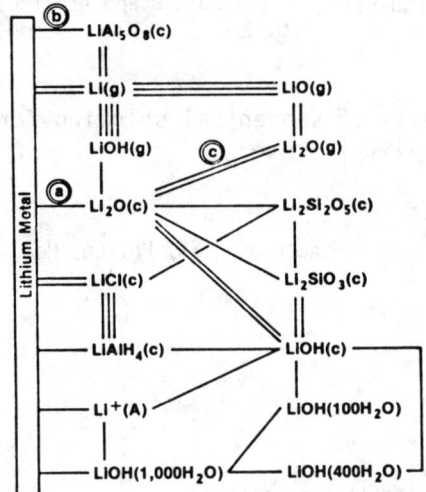

Loops, of 5 members or less, containing reaction (a).
Each line between substances represents one reaction between lithium compounds.
Processes (b) and (c) appear to contribute large errors.

Figure 7. A loop analysis for lithium and its compounds of 5 members or less containing reaction a. Each line represents one reaction between lithium compounds. Process b and c appear to contribute large errors.

any other thermodynamic state function) around all loops that include a particular measurement. It tabulates the errors of closure and considers those that exceed experimental errors.

Figure 7 shows such a loop involving all loops with less than 6
measurements that include reaction a. There are 15 substances,
40 measurements and 86 loops. The residuals for these loops are
shown in the histogram (Fig. 8). The dash marked ones are

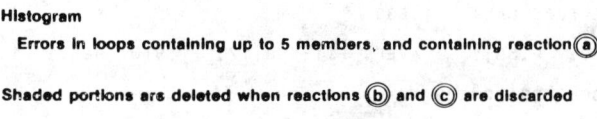

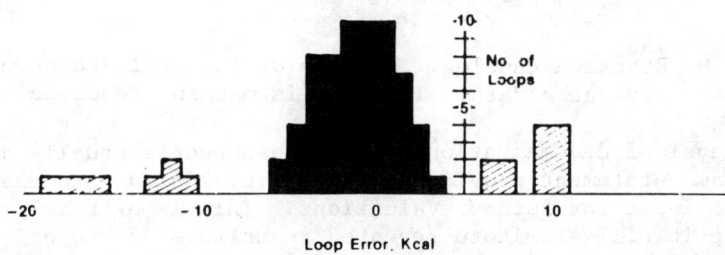

Figure 8. A histograph of the errors in loops containing up to
5 members and reaction a (Figure 7). Shaded portions
are deleted when b and c are discarded.

outliers. They represent a greater closure error than would
be expected. Subsequent study indicated that all of these outliers
are traceable to two measurements which, after investigation by
the evaluator, were discarded. The histogram also shows that all
the other measurements are probably valid since they are
concentrated around zero.

Documentation and Archives

The use of modern computers has also simplified the
documentation of results and made it feasible to give users more
complete information on what has been done by the experimentalist
and the evaluator. The input used for the solution of the network
(thermochemistry in machine-readable form) can also be used to
show the user what the measurement was, how accurate it was and
how well this datum was fit by the selected values. Figure 9
shows the documentation for the process

$$Th_3N_4(c) \rightarrow 3\ Th(c) + 2N_2(g)$$

```
                    NETWORK SOLUTION OUTPUT

REACTION:  TH'3N4(C) = 3 TH'N(C) + 0.5 N2(GS')

PROPERTY MEASURED:  H

OBSERVED +- UNCERT.: 37.0 +- 4.5;       WEIGHT:  0.4

RESIDUAL (OBS-CAL):  1.833

STD. DEV.: 2.59;      AVER. FIT: 3.17;    STD. RES.:  1.00

REFERENCE:  66ARO/AUS

COMMENT:  AUTHORS' 2ND-LAW YIELDS 36.35 AT APPROX. 1923 K
          ADJ'D BY 0.65 USING ESTD. CP(TH3N4)= 39.33+
          0.00624T-533000/T2.
```

Figure 9. Reaction catalog. Example of the stylized summary of evaluated data. The form is machine readable.

This level of documentation is more than people usually ask for. This same statement provides a file of evaluated data that can be used as input for future evaluations. This is extremely important because it will eliminate repeated reanalysis of the original paper. The archive is then prospective as well as retrospective.

Cooperation in Thermochemistry

Future evaluations probably will be organized as an international effort involving many laboratories. This is shown schematically in Figure 10. The projects will be coordinated by

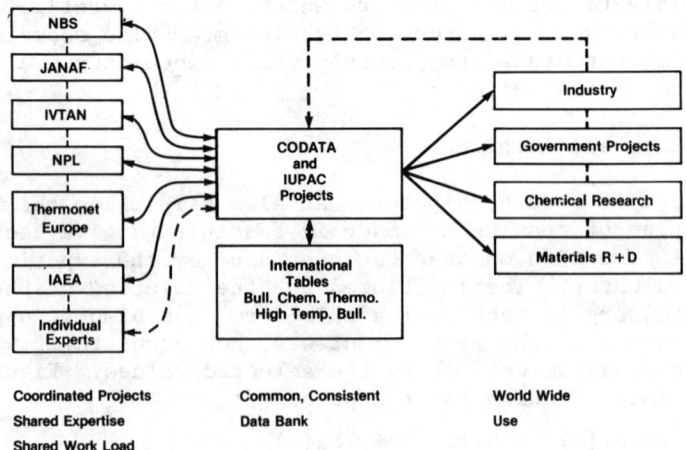

Figure 10. International cooperation in data evaluation.

CODATA, IUPAC and other organizations and the results published in the form of data banks and tables.

The Essential Ingredient

After extolling the use of modern computer methodology in chemical thermodynamic data evaluation one needs to address the issue of the role of the thermochemical data evaluator. The computer has the ability of looking at large arrays of interlocking thermochemical data. It can raise well defined questions. The data evaluator must make the decisions. The evaluator oversees the computer solution, prepares the input for the reaction catalog and loop analysis which are subject to automatic computer verification. Figure 11 shows the process of thermochemical data evaluation. The analyst represented by a diamond is appropriately placed at the center.

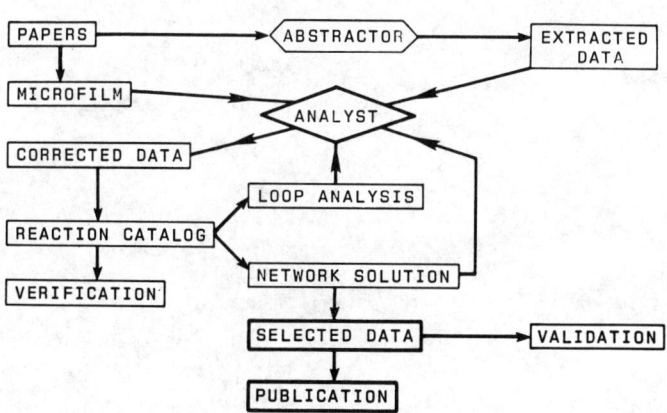

Figure 11. Thermochemical data evaluation. A summary of the process.

REFERENCES

1. Wagman, D.D., Evans, W.H., Parker, V.B., Halow, I., Bailey, S.M., and Schumm, R.H., "Selected Values of Chemical Thermodynamic Properties", U.S. Nat. Bur. Standard Tech. Note 270-3 (1968); idem, Tech. Note 270-4 (1969); Wagman, D.D., Evans, W.H., Parker, V.B., Halow, I., Bailey, S.M., Schumm, R.H., and Churney, K.L., Tech. Note 270-5 (1971); Parker, V.B., Wagman, D.D., and Evans, W.H., Tech. Note 270-6 (1971); Schumm, R.H., Wagman, D.D., Bailey, S.M., Evans, W.H., and

Parker, V.B., Tech. Note 270-7 (1973); Wagman, D.D., Evans, W.H., Parker, V.B., Schumm, R.H., and Nuttall, R.L., Tech. Note. 270-8 (1981). U.S. Govt. Printing Office, Washington, D.C.

2. Rossini, F.D., Wagman, D.D., Evans, W.H., Levine, S., and Jaffe, I., "Selected Values of Chemical Thermodynamic Properties", NBS Circular 500 (1952), pp. 1268.

THERMOCHEMISTRY TODAY AND ITS ROLE IN THE IMMEDIATE
FUTURE

H.A. Skinner

Chemistry Department,
University of Manchester,
Manchester, M13 9PL.

Although the primary objective of the thermochemist is
continuously to add new information to the growing data bank
on thermodynamic properties of chemical compounds (and of
mixtures of compounds), there is nothing haphazard in his
choice of subjects for investigation. These invariably reflect
current chemical interests, many of which arise from the
problems of the day in chemical industry, in medical science,
and in military matters; others reflect theoretical advances
in understanding, in particular of the nature of chemical bonding.
Accordingly, we have experienced a continually changing scene in
Thermochemistry where effort has concentrated in different areas
with the passage of time. The present era is one of <u>accelerating</u>
change - catalysed by advances in scientific instrumentation -
and the chosen title for this NATO Advanced Study Institute accepts
that we recognize this, and prepare ourselves for the broader
canvas about to unfold.

My own involvement with Thermochemistry stretches over
nearly 40 years, since the time of the Second World War. I have
seen some significant changes happen, and for those of you just
beginning your careers as Thermochemists, it may be helpful to
be guided back along the road as I saw it. Thermochemistry was
already a well-established subject, closely allied with
calorimetry, long before I ventured therein; indeed, within the
IUPAC organisation, the Commission on Thermodynamics and
Thermochemistry was one of the earliest to be constituted. The
subject underwent a renaissance in the late twenties, largely
due to the advent of the automobile, and the growth of the
petroleum industry, particularly in the United States. The prime
requirement then was to isolate pure hydrocarbons, to determine

their heats of formation with high precision, and to establish
good values for entropies and free energies of formation over a
range of temperature. The outcome of this enormous effort was
(and continues to be) the A.P.I. Tables of 'Thermodynamic
Properties of Hydrocarbons', which most of us now take for granted
and use without question, in much the same way as we accept
present values for the atomic weights of the elements. To high-
light the activities, we mention:-

1930 - 1950.

Combustion bomb-calorimetry Combustion flame-calorimetry	ROSSINI, PROSEN (NBS, Washington)
Heats of hydrogenation	KISTIAKOWSKY
Low-temperature C_p values Entropies	GIAUQUE
Bond Energies Resonance Energies	PAULING
Barriers to Free Rotation Statistical Thermodynamics	PITZER
Molecular Orbital Theory	COULSON MULLIKEN
Bond-Dissociation Energies (Kinetics; Mass-Spectroscopy)	POLANYI, SZWARC STEVENSON

There was a hiatus on paper during the War years. But not
in reality; the development of rocket technology, of atomic
energy, of synthetic polymers, of radar, and of scientific
instrumentation occurred in this period of closed communication,
and its effect on physical science was all the more pronounced
once the aftermath of war subsided.

1950 - 1960.

Combustion bomb-calorimetry enjoyed a period of rapid growth,
following the development of more reliable techniques for the study
of organic compounds of S,F,Cl,Br,B,Si,P, and certain metals. The
conduction microcalorimeter became established as a viable
instrument, for studies over a wide range of temperature.
Critical assessment of thermodynamic data became a skilled
profession (the NBS Circular 500 appeared in 1952; the JANAF

Tables were in preparation). The Experimental Thermodynamics Conference was founded in the UK, and the Bulletin of Chemical Thermodynamics made its debut at this time. The first IUPAC Conference on Thermodynamics was held in Fritzens-Wattens (Austria) in 1959. Specifically, we may mention:-

Rotating-Bomb Calorimetry
(Bartlesville, Lund)
    WADDINGTON
    McCULLOUGH
    SUNNER
    GOOD

Microcalorimetry
(Marseilles)
    CALVET

Heats of atomization, including carbon
(Knudsen Cell/Mass Spec.)
    CHUPKA
    INGHRAM

Fluorine Bomb-Calorimetry
    HUBBARD

Ligand-Field Theory
    GRIFFITH
    ORGEL

1960 - 1970.

The applications of calorimetry increased in the sixties, mainly due to improvements in commerically available instrumentation, coupled with early signs of computerization. The need for data-computation and compilation on an international basis led to the founding of the CODATA office; the Journal of Chemical Thermodynamics made its appearance in 1969. The 'Calorimetry Conference' became established as a National event (following the examples of the USA and the UK) in the Soviet Union and in Japan.

Specifically, we may mention:-

Precision Calorimetry

(LKB)   SUNNER, WADSÖ
(Tronac)   CHRISTIANSEN, IZATT

Quartz-Thermometry
  (Hewlett-Packard)

Fluorine flame calorimetry
    ARMSTRONG
    (NBS)

Biochemical Calorimetry	STURTEVANT
	BRANDTS
Liquid Mixtures	STOKES
	McGLASHAN
Metal Oxides	HOLLEY
	(Los Alamos)
Flame Calorimetry (ethers, alkyl chlorides)	PILCHER
Molecular Mechanics	
Force field	SCHLEYER
	ALLINGER
Mindo	DEWAR
Extended Hückel	POPLE
a priori	

1970 - 1980.

The computer and the microprocessor came into prominence in calorimetry (as in most physico-chemical equipment) during this decade. The desk-computer became commonplace; efficient calorimeters and microcalorimeters of excellent quality were to be found in chemical, biochemical, metallurgical and medical laboratories world-wide. In particular, we may mention developments in:-

Titration Calorimetry	CHRISTENSON (Provo)
Flow Calorimetry	WADSÖ
	GILL
	PICKER
High Temperature Calorimetry	MARGRAVE
	KLEPPA
Pulse Calorimetry	CEZAIRLIYAN (NBS)
Mini-Bomb Calorimetry	MÅNSSON (Lund)
	SUNNER
'Drop' Microcalorimetry	SKINNER (Manchester)
	CONNOR

There was a noticable fall in the number of bomb-combustion studies, accompanied by the termination of this particular activity in some laboratories - (including that of the NPL in London). On the other hand, there was an upsurge in solution calorimetric and differential scanning studies, particularly in relation to biochemical thermodynamics, and to binary liquid mixtures. Calorimetry played a major role in studies on organometallic compounds (mainly of the transition-metals) : for these novel compounds, the high-temperature Calvet microcalorimeter in the 'drop' mode, and semi-micro solution-calorimeters proved especially advantageous.

Critical evaluation of thermodynamic data continued with the revision of Circular 500 (now completed) and the similar compilation 'Thermal Constants of Chemical Compounds' (Academy of Sciences, Moscow; V.P. Glyshko, V.A. Medvedev, L.V. Gurvich and others), both of which appeared in sectionalized parts as work progressed. The CATCH Tables (computer analysed thermochemical data) on certain elements and their compounds, and on organic and organometallic compounds were issued (J.B. Pedley and the Sussex University/NPL team); 'Selected Data on Mixtures' (H.V. Kehiaian, B.J. Zwolinski) began a continuing series in 1973.

The journals 'Thermochimica Acta', 'High Temperature Chemistry', and the 'International J. Chemical Kinetics', and a considerably expanded 'Bulletin of Chemical Thermodynamics' emerged, reflecting the broadening of chemical thermodynamic activities during the decade.

Perhaps most significant was the development of techniques other than calorimetry enabling the study of gaseous ions and free-radicals, and leading to enthalpies and free-energies of formation of these species. We may mention:-

Photoionization Spectroscopy	INGRAM
Photo dissociation Spectroscopy	CHUPKA
	BERKOWITZ
Ion-Cyclotron Resonance	KEBARLE
Spectroscopy	McIVER
	BEAUCHAMP
V.L.P.P. Kinetics	BENSON

1980 - 1990.

The present decade has encountered an atmosphere of depression, scarred by cut-backs in Universities, National

Laboratories, and Industry in general. In view of expansion during the previous two decades, a deceleration was perhaps to be expected; but research activities in thermochemistry in its broader sense will not necessarily undergo a serious decline. Certain activities will diminish - notably the more traditional ones involving combustion-bomb calorimetry and low-temperature heat capacity calorimeters. Others, now very active, should continue at a steady pace - (in effect, the field of organometallic thermochemistry is only beginning to be explored, and the same applies to studies in biothermodynamics). Some activities may even increase in volume, and undoubtedly there will be new adventures, as yet in embryo. Looking into the future haze, I could foresee:-

<u>Diminishing</u> → Combustion calorimetry
Low temperature C_p measurements

<u>Keeping Steady</u> → Biochemical studies
Organometallic thermochemistry

<u>Changing emphasis</u>   Data compilation
↓
On-line information

Binary mixtures
↓
Ternary (and higher) mixtures.

<u>Rising steadily</u> → P.I.M.S. ; I.C.R.S. (PIPECO, + ?)
Microcalorimetric applications

<u>On the horizon</u> - for calorimetric study in various ways:-

PHOTOCHEMISTRY   -   Light-induced reactions

ELECTROCHEMISTRY   -   Solid-state cells at elevated temperatures

SURFACE ADSORPTION   -   Catalytic activity on surfaces

THERMOKINETICS   -   Solid-state reactions
                           Natural growth processes.

HIGH PRESSURE   -   Various.
SYSTEMS

   Thermochemists and calorimetricists should now be prepared to spread their technical skills more widely. The Keywords are

Liaison in Research
and
Service to Industry

The calorimeter is no longer just a specialist research instrument. It compares more closely with the spectroscope, and has many different forms and uses, both as an analytical and as a diagnostic tool. Its applications extend beyond the traditional ones of measurement of ΔH and C_p. Liaison in research implies a more extensive usage of the modern calorimeter - in medical research, in hospitals, in pharmacy departments and the pharmaceutical industry, in agricultural research institutes. Already we have seen the profitable results of liaison between biochemists and calorimetricists, which began in earnest less than 20 years ago. Similar liaison with other disciplines is already evident, and must surely become increasingly important.

Service to industry has tended to be left to National Research Institutes, and not sought after seriously by University research groups. The intense specialization of modern research needs the help of specialist research groups, often to be found only in Universities.

We have covered some current aspects of Thermochemistry in this Advanced Institute. We have had the chance to broaden perspectives, and strengthen links with colleagues from abroad. We are indeed indebted to the NATO Organization for the generous support given, and to our Portuguese hosts for the work willingly undertaken to make this meeting a successful and memorable event.

LIST OF PARTICIPANTS

Abraham, M. H.: Dep. Chemistry, University of Surrey, Guildford
             GU2 5XH, England
Abramowitz, S.: National Bureau of Standards, Chemical Thermody-
             namics Division, Washington, D.C. 20234, USA
Bastos, M. M.: Dep. Chemistry, Faculty of Sciences, University of
             Oporto, 4000 Porto, Portugal
Benson, S. W.: Dep. Chemistry, University of Southern California,
             Los Angeles, USA
Bickerton, J.: Dep. Chemistry, The University of Manchester,
             Manchester M13 9PL, England
Carson, A. S.: Dep. Physical Chemistry, University of Leeds,
             Leeds LS2 9JT, England
Carson, E. M.: Dep. Physical Chemistry, University of Leeds,
             Leeds LS2 9JT, England
Chatgilaloglu, C.: Division of Chemistry, National Research
             Council Canada, Ottawa, Canada K1A OR6
Christensen, J. J.: Dep. Chemical Engineering Science, Brigham
             Young University, Provo-Utah 84602, USA
Connor, J. A.: The Chemical Laboratory, University of Kent,
             Canterbury - Kent CT2 7NH, England
Costa, M. R. G.: LNETI, Poço do Bispo - Lisboa, Portugal
Date, R.: Dep. Chemistry, University of Surrey, Guildford
             GU2 5XH, England
Dias, A. R.: Centro de Química Estrutural, I.S.T.,
             1096 Lisboa Codex, Portugal
Erbil, M.: Çukorova Universitesi, Temel Bilimler Fakultesi,
             P.K. 171, Adana, Turkey
Ferrão, M. L. C. C. H.: Dep. of Chemistry, Faculty of Sciences,
             University of Oporto, 4000 Porto, Portugal
Gather, B.: Universitat-GH-Siegen, FB 8/Anorg Chemie,
             5900 Siegen 21, Federal Republic of Germany
Girauld, H.: Dep. Chemistry, University of Southampton,
             Southampton, England
Gonçalves, R.: Faculty of Sciences, Lisboa, Portugal
Hastie, J. W.: National Bureau of Standards, High Temperature
             Processes Group, Washington, DC 20234, USA
Irving, R. J.: Home Economics Department, University of Surrey,
             Guildford, Surrey GU2 5XH, England

Klump, H.: Institut Für Physikalische Chemie, Lehrstuhl II,
    D-7800 Freiburg, Federal Republic of Germany
De Kruif, C. G.: Dep. General Chemistry, State University of
    Utrecht, Utrecht 2506, Netherlands
De Maria, G.: Institute of Physical Chemistry, University of
    Rome, 00185-Rome, Italy
Mason, R.: Chemistry: Molecular Sciences, University of Warwick,
    Coventry CV4 7AL, England
Mateo, P. L.: Dep. Physical Chemistry, Faculty of Sciences,
    Granada, Spain
Monte, M. J. S.: Dep. Chemistry, Faculty of Sciences, University
    of Oporto, 4000 Porto, Portugal
Mortimer, C. T.: Dep. Chemistry, University of Keele, Keele,
    Staffordshire, ST5 5BG, England
Palheiros, I. M. B.: Dep. Chemistry, Faculty of Sciences, University of Oporto, 4000 Porto, Portugal
Palumbo, M.: Istituto di Chimica Organica dell'Universita,
    35100 Padova, Italia
Paoletti, P.: Institute of General Chemistry, University of
    Florence, 50132 Firenze, Italia
Parody-Morreale, A.: Dep. Quimica Fisica, Facultad de Ciencias,
    Granada, Spain
Paz-Andrade, M. I.: Dep. Fisica Fundamental, Facultad de Fisica,
    Santiago de Compostela, Spain
Piedade, M. E. R. M.: Centro de Química Estrutural, I.S.T.
    1096 Lisboa Codex, Portugal
Pilcher, G.: Dep. Chemistry, The University of Manchester,
    Manchester M13 9PL, England
Reis, A. M. M. V.: Dep. Chemistry, Faculty of Sciences, University of Oporto, 4000 Porto, Portugal
Rialdi, G.: Centro di Studi Chimico-Fisici di Macromolecole Sintetiche e Naturali, Istituto di Chimica Industriale - Universita, 16132 Genova, Italia
Ribeiro da Silva, M. A. V.: Dep. Chemistry, Faculty of Sciences,
    University of Oporto, 4000 Porto, Portugal
Ribeiro da Silva, M. D. M. C.: Dep. Chemistry, Faculty of Sciences, University of Oporto, 4000 Porto, Portugal
Salema, M. M.: LNETI, Sacavem, Portugal
Schulz, R. A.: Dep. Chemistry, University of Surrey, Guildford
    GU2 5XH, England
Silva, A. F. S.: Dep. Chemistry, Faculty of Sciences, University
    of Oporto, 4000 Porto, Portugal
Simões, J. A. M.: Centro de Química Estrutural, I.S.T.,
    1096 Lisboa Codex, Portugal
Skinner, H. A.: Dep. Chemistry, The University of Manchester,
    Manchester M13 9PL, England
Soccorsi, L.: Istituto Superiore di Sanità, Laboratorio degli
    Alimenti, 00161 Roma, Italia
Somsen, G.: Dep. Chemistry, Free University of Amsterdam,
    1081 hv Amsterdam, Netherlands

LIST OF PARTICIPANTS

Sottomayor, M. J. F.: Dep. Chemistry, Faculty of Sciences, University of Oporto, 4000 Porto, Portugal
Staveley, L. A. K.: Inorganic Chemistry Laboratory, University of Oxford, Oxford OX1 3QR, England
Tachoire, H.: Laboratoire de Thermochimie, Université de Provence F-13331 Marseille, France
Taylor, J. R.: Johnson Matthey Research Centre, Sonning Common, Nr. Reading-Berkshire, England
Teixeira, C.: Centro de Química Estrutural, I.S.T., 1096 Lisboa Codex, Portugal
Tieje, J.: Dep. Chemistry, University of Southampton, Southampton, England
Wadsö, I.: Thermochemistry Chemical Center, University of Lund, P.O.B. 740, S-220 07 Lund 7 - Sweden
Westrum, E. F.: Dep. Chemistry, The University of Michingan, Ann Arbor, Michingan 48109, USA

SUBJECT INDEX

ab-initio methods, 641
Acetamide, 433
Acetic acid, 151, 320
Acetylacetone, 319, 320
    enol percentage, 323
    enthalpy of formation, 322
    enthalpy of vaporization 323
Acetylacetone complexes, 317
    enthalpies of combustion, 327
    enthalpies of formation, 327
    enthalpies of sublimation, 328, 329
Acetylenes, 778
Acid proteases, 553, 560
Acidic site
    nature of, 639
Acidity, 628
    absolute, 633
Activation energy, 565
Activation enthalpy, 565
Activity, 157, 164, 212, 222, 225
    data, 250
    measurements, 163
    thermodynamic, 235
Activity coefficient, 214, 279, 280, 283, 399, 400, 518
    corrections, 281
    ionic, 281
Adamantane, 154
Additivity
    bond, 774, 775
    group, 774
    Laws of, 775
    relations, 769
    rule, 772

Adenosine diphosphate, 25
    binding to bovine liver glutamate, 25
Adhesion, 456
Adiabatic system, 543
Aerobic processes, 454
L-Alanine, 151
Alcohols, 324, 331, 426
    secondary, 428
    solvation of, 425
    tertiary, 428
Algorithm of Biltonen and Freire, 542
Aliphatic alkanols
    dissociation of hydrogen in, 331
Alkali halide, 170, 228, 405
Alkali hydrogen
    difluorides, 685
    azides, 685
Alkali metal species, 208
Alkali nitrites, 667
Alkanes, 779
    branched-chain, 574
    bromo-, 584
    chloro-, 582
    enthalpies of formation, 573
    fluoro-, 586
    iodo-, 585
    polychloro-, 583
N-Alkanols
    enthalpies of formation, 575
    branched-chain, 576
Alkoxyacetates, 324
Alkyl cations, 594, 595
Alkyl manganese carbonyls, 372

Alkyl radicals, 592, 593
Alkyl thiols, 580
Allen scheme, 569, 570, 599, 776
    parameter values, 572
Almost-symmetrical-top molecules, 677
Al_2O_3, 235
Aluminosilicate minerals, 725
Amides, 426
    Solvation of, 425
Amines, 426, 579
    cyclo-, 431
    di-, 431
    mono-, 431
    primary, 431
    secondary, 431
    solvation of, 425
    terteary, 431
    tri-, 431
D-Amino acid oxidase, 563
Ammonium ions, 684
Ammonium salts, 100
Ampoule breaking, 280
Anabolic reactions, 532
Analysis, differential thermal, 725
Anemic patients, 449
Angular overlap model, 289, 298, 300
Anilinium
    salt, 265
    ion, 268
Animal cells,
    blood cells, 441
    cultured tissue cells, 441
    fat cells, 441
    liver cells, 441
    macrophages, 441
    tumor cells, 441
Animals, small, large, 411
Anomalous stability
    sequence, 270
Antibiotic activity
    testing of, 445
A.P.I. Tables, 816
Apolipoprotein A-1, 562
Appearance Energies, 609
Aprotic solvents, 406
Archives, 811, 812

Arene, 372, 385
    carbonyls, 372
    displacement reactions, 389
Arenechromium tricarbonyls, 374
Arenediazonium salt, 265
Aromatic compounds, 591, 778
    adsorption by zeolite, 21, 22
Aromatic hydrocarbons, 590
    condensed polynuclear, 680
Aromaticity, 317
Arrhenius
    A-factors, 784, 786
    parameters, 783, 784, 785
Arrhenius plot, 563
    non linear, 564
Artefacts, 451
AsF_5, 789, 792
Ash, 62, 67
Aspergillus Awamori, 561
Associated compounds, 325
Astrophysics, 178
Atom-atom potential, 143
Automated calorimeters, 757
Axial-equatorial interchange,
    fluorine atom, 793
Azonia-spiro-alkane bromides, 416
Azulene, 680

Bacteria, 441, 536
Bacterial cell
    response to cytotoxic agents, 22
Bacterial growing, 532
Barriers
    to free rotation, 816
    to internal rotation, 789
Baseline heat capacity, 656, 663
Baseline stability, 543
Basicity, 628
Batch
    Calorimetry, 275, 277
    Experiments, 278
    Method, 283
BCl_3 skeleton, 798

SUBJECT INDEX

$B(CH_3)_3$, 789
    Raman spectrum, 797
    third law entropy, 793
    vibrational assignment, 800
    vibrational modes of, 797
BCl_2SD vibrational assignment, 790, 792
BCl_2SH vibrational assignment, 790, 792
Beam
    molecular, 183, 186
    supersonic, 187
Bending frequences, 775
Benzoic acid, 149, 154, 320
Benzophenone, 149, 154
Benzoquinone, 151
Benzothiazole complexes, 56
Benzoxazole complexes, 56
Benzoylacetone, 319, 320
    enthalpy of combustion, 321
    enthalpy of formation, 322
    enthalpy of sublimation, 325
Bicycloalkanes, 679
Bimetallic molecules, 169
Binding
    of Con A to saccharines, 525
    of ion, 525
Binding energies metal-ligand, 642
Binding reactions in biochemistry, 511
Bio-calorimetry, 440
Biochemical reactions, 37
Biochemistry
    binding reactions, 511
Biomembranes, 562
Bixbyite sesquioxides, 713
Boltzmann factor, 793
Bond
    additivity, 775
    angles, 798
    dissociation energies, 816
    distances, 798
    energies, 816
    energy scheme, 322
    length, 333, 775
    metal-oxygen, 326, 334
    strength, 775, 779
Bond-bond interactions, 569, 571
Bond-contributions, 569, 571

Bond dissociation energies, 289
Bond dissociation enthalpy, 358
    heterolytic, 317
    homolytic, 317, 331
    in metal-β-diketonates, 332
Bond energy schemes, 363
Bond enthalpy, 127
    contribution, 359, 389
    dissociation, 331, 332
    heterolytic, 317
    homolytic, 317
    metal-oxygen, 317, 332
Bond strengths, 289
Bonding electrons
    perturbation of, 778
Bourdon spoon gauge, 128
Bovine liver glutamate binding to adenosine diphosphate, 25
Bracketing, 615
Brewer's rule, 165
Butanoic acid, 151
Butylacetamide, 433
n-Butylamine, 430
t-Butylamine, 430
n-Butylmethylamine, 430
Butyramide, 433

Calibration, 452
    process, 44
Calorific value, 61
Calorimeter
    Arnett solution, 11, 12
    automated, 757
    capacity bomb, 61
    commercially available, 10, 11, 12
    control, 747
    C.R.M.T., 11, 12
    differential scanning, 724
    drop heat capacity, 36
    flow, 11, 12, 36
    flow heat capacity, 35
    gram-size bomb, 67
    kilogram-size, 62
    large scale, 61

L.K.B., 11, 12
oxygen flow, 61
Parr solution, 11, 13
Picker, 35
pressure-scanning, 43
thermopile heat conduction, 442
Tronac, 11, 13
Vessel, 73, 75
Calorimetric, 511
  enthalpies, 542, 546
Calorimetry, 461
  A.C., 33
  biochemical, 818
  bomb, 62
  combustion bomb, 816
  combustion flame, 816
  drop, 725
  flame, 818
  flow, 818
  high temperature, 818
  low-temperature, 653
  mini-bomb, 818
  pulse, 818
  solution and reaction, 326, 327
  static bomb, 325
  titration, 818
Calvet microcalorimeter, 139
CaO, 235
Carbides, 174
  non stoichiometric, 174
Carbonyls, 371, 384
Carboranes, 677
Carboxylic acids, 150, 151
  lower, 324
C-A-S-H system minerals, 722, 737
Catabolic reactions, 532
CATCH tables, 819
Cell-cell interaction, 536
Cell multiplication, 532
Cell orifices, 159
Cellular thermochemistry, 531
Ceramic science, 171
Chalcogenide
  family, 171
  minerals, 722
Charge transfer
  complexes, 682

compounds, 683
Chatelier effect, 204
Chelate effect, 343
Chemical equilibria, 806
Chemical forces, 775
Chemical industry, 815
Chemical thermodynamic properties, 804
Chupka-De Maria rule, 175
Circumstellar atmospheres, 178
Clapeyron equation, 127
Clausing factor, 159
Clausius-Clapeyron equation, 166
Clinical studies, 450
Clusters, 186
Coal ash, 250
Codata, 813, 817
Coil transition, 521
Cold denaturation, 552
Collection efficiency, 138
Combustion
  bomb, 62
  bomb-calorimetry, 816
  calorimetry, 32
  flame-calorimetry, 816
  kilogram, 67
Combustor, 61, 62, 64, 71, 72, 73, 74, 75
  25 gram, 63
  2.5 kilogram, 69
  kilogram-size, 61
Complexes
  charge-transfer, 682
Compressibility
  isothermal, 178
Computer technology, 803
Concanavalin A, 523, 536
Condensation, 127
  coefficient, 131
Condensed polynuclear aromatic hydrocarbons, 680
Conduction calorimetry, 77
Conductors, superionic, 655
Configurational entropy gain, 666
Conformational change, 512
Cooperative length, 497
Co-operativity, 546
  intramolecular, 546, 547

structural domains, 566
substructures, 557
unit of the transition, 547
Coordination bonds
strengths of, 328
Correlation diagram, 797
Correction to ΔH^{298}
Cu(alaninate)$_2 \cdot$2NH$_3$(c), 55
CuSO$_4 \cdot$5H$_2$O(c), 54
Ni(glycinate)$_2 \cdot$2H$_2$O(c), 55
Cosmology, 178
Coulombic effects, 778
Cp$\left[$Lu(OH)$_3\right]$, 710

Cp$\left[$Y(OH)$_3\right]$, 709

Critical evaluation, 803, 819
Critical point, 177
18-Crown-6, 280
18-Crown-6-complexes, 265, 268, 270, 271
Crown ethers, 275, 280
complexes, 269
CsCl, 228
Cryptand 222, 286
Cryptand ligand, 285
Crystal field
effects, 727
splittings, 700
theory, 317
Crystalline
electric field levels, 686
metal-β-diketonates, 330
metal oxides, 334
Cu(alaninate)$_2 \cdot$2NH$_3$(c), 55
CuSO$_4 \cdot$5H$_2$O(c), 54
Cyclics, 779
Cycloalkenes, 589
Cytochrome b$_5$, 563

Data
acquisition, 747, 750
bases, 803
evaluater, 804
evaluation, 803
processing, 78
thermochemical, 803
thermodynamic, 803

Debye-Hückel
coefficients, 396
expressions, 281, 282, 396
Debye T^3 law, 658
Debye theta approximation, 704
Decomposition studies, 32
Deconvolution, 78, 82, 86, 90, 111, 117, 122, 542
calorimetric curves, 558
techniques, 81
Degree of
association, 325
dissociation, 399
freedom, 786
Delocalization energy, 322, 331
Denaturation
bacteriorhodopsin, 562
reversible, 553
temperature, 548
Deposits
metamorphosed sulfide, 739
Dialkyl ethers, 577
Dialkyl sulphides, 581
Diamines, 579
Diatomic molecules
homonuclear, 168
heteronuclear, 169
Dibenzoylmethane, 319
enthalpy of formation, 322
enthalpy of sublimation, 325

Dicarbides, 175
metal, 174
species, 176
β-Dicarbonyl compounds, 318
1,2-Dichloroethane, 400, 402
Dielectric
constant, 398, 406
profile, 406
Dienes, 589
Diethylamine, 430
Diethyldithiocarbamate complexes, 311
Differential scanning calorimetry, 292, 328, 541, 543
Differential thermal analysis, 725
Diisobutyrylmethane, 319
enol percentage, 323

enthalpy of formation, 322
enthalpy of vaporization, 323
Diketonate ring, 317
β-Diketones
  enthalpies of enolization, 321
  enthalpies of formation, 321, 322
  enthalpies of sublimation, 323
  enthalpies of vaporization, 323, 324, 325
  keto-enol equilibrium, 322
  practical applications, 320
Dimer species, 196
Dimerization, 151
N,N-Dimethylacetamide, 433
N,N-Dimethylformamide, 433
1,1-Dimethylurea, 433
1,3-Dimethylurea, 433
Di-i-propylamine, 430
Di-n-butylamine, 430
Di-n-propylamine, 430
Dionate complexes, 309
Dioxides, 176
Dipivaloylmethane, 319, 320
  enol percentage, 323
  enthalpy of formation, 322
  enthalpy of vaporization, 323
Dipole
  effects, 778
  moments, 779
Disorder
  third-law, 726
Disordered layers, 407
Disproportionation, 776, 777
Dissociated ions, 402
Dissociation enthalpies
  metal-oxygen, 333
  of enolic hydrogen, 331
  of hydrogen in
    acetone, 331
    alkanols, 331
    phenol, 331
Distorted geometries, 779
Dithiols, 580
DNA, 471, 472
Z-DNA structures, 492
Documentation, 811, 812
  of results, 811

Dolomite, 211
  vaporization, 210
Domain, 558
  structural, 553
Donor-acceptor crystals, 151
Double helix, 475, 491
"Drop" calorimetry, 725
D_{3h} structure, 792
DSC, 32, 47, 724
  Perkin-Elmer DSC-2, 48
  Seteram DSC 111, 48
DTA, 47
Dynamic correction, 443

Ebulliometric method, 129
Effusion
  Cell, 145, 154
    Knudsen, 154
    Langmuir, 154
  torque, electromagnetic compensation, 145
  torsion and mass-loss, 143
Einstein
  functions, 704
  heat-capacity functions, 699
Electrolytes
  enthalpies of solution, 393
Electromotive forces, 306
Electron
  affinities, 606, 607
  diffraction, 801
  impact, 194
  impact fragmentation, 228
  pair bond description, 384
Electronegativity, 773
Electronic
  density in metal, 331
  energy level, 698
Electrostatic
  Bond, 775
  energy, 777, 779
  equations, 406
  interactions, 772
  model, 778, 779
  theories, 405
Elementary reactions, 780
Empirical methods, 780
Energy

SUBJECT INDEX

atomization, 167, 569
bond, 167, 174, 175
cohesive, 167
dissociation, 167, 171, 174
states, macroscopic, 557
Enol
  enthalpy of formation, 322
  percentage, 323
  structure, 318
Enthalpic effect of hydrophobic hydration, 419, 429
Enthalpimeter, 11, 12
Enthalpy function
  avertage excess, 557
Enthalpy of
  ampoule breaking, 393, 394
  atomization, 359
  combustion, 61, 327
  dilution, 280, 282, 283, 393, 394, 396, 397
  disruption, 362
  dissociation, 326, 330
  enolization, 321, 322
  evaporation, 394
  formation, 325, 327, 328, 735
    of n-alkanes, 573
    of n-alkanols, 575
    of β-diketones, 321
    of gaseous metals, 360
    of metal-β-diketonates, 326
    of metal-tropolonates, 328
  fusion, 51
  halogenation, 357
  hydroborination, 357
  hydrolysis, 356
  mixing
    determination of, 112
    parcial molar, 103
  redistribution reaction, 357
  solution, 393, 395, 411, 413, 425
    of $BuNH_3Br$ in DMF/water, 414
    of CsBr in DMF/water, 413
    of electrolytes, 393
    of NH_4Br in DMF/water, 413, 414
    of tetraalkylammonium bromides, 415
  sublimation, 166, 326, 328, 329, 330, 371

  of metal complexes, 301
  transfer, 285, 393, 402, 405, 411, 433
    of alkalibromides, 412
    of tetraalkylammonium bromides, 412
    standard, 286
  vaporization, 51, 322
Entropy
  activation, 783
  effects, 634
  experimental, 806
  saturated, 178
  standard calorimetric, 653
  statistical, 806
  third law, 676
  transfer, 405
Enzyme assays, 441
Equilibrium constants, 275, 276, 277, 279, 283
Errors, 451
Erythrocytes, 448
Equilibria phase, 734
Equilibrium
  Calculation, 258
  Constant, 257, 258
  kinetics, 782
Esteres, 324
$Et_3(EtOH)N^+$, 417

$Et_3(EtOH)NBr$, 417

$(EtOH)_4NBr$, 417

Ethylurea, 433
$EuCl_3$, 708

$Eu(OH)_3$, 706

Evaluation
  data, 812
  future, 812
Evaporation, 127
  coefficient, 134, 154
Excess enthalpy, 116
  measurement of, 117
Excess heat capacity
  curve, 546
  function, 559

833

Expansion, isobaric thermal, 178

FA, 619
Facial configuration, 349
Fermi level, 686, 688
Ferromagnetic materials, 732
Fiber optics, 37
Filtering
   digital, 90
   digital inverse, 99
   electronic, 93
   electronic inverse, 117
   electronic or digital, 86
   inverse, 86, 90, 103, 114, 116, 122, 124
   techniques, 120
   two step, 90
First order transitions, 696
Fission products, 173
Flame calorimetry, 68
   calibrants, 68
Flow calorimeter, 61, 67
   25 gram, 64
   2.5 kilogram, 74
   oxygen, 61
Flow experiments, 62
Flowing afterglow, 619
Fluorescence, 563
   lifetime, 565
Fluorine
   bomb calorimetry, 817
   flame calorimetry, 817
Fluorocarbons, 586
Force field, 792
Formamide, 433
Fourier
   space, 80
   transform, 82, 84, 86, 120, 123, 124
Franck-Condon, 228
   electron impact, 194
Free
   energy functions, 127, 167
   molecular flow, 132
   radicals, 569, 771, 779
   rotation, 801
Fugacity, 163

G_{324} group, 797

Gas dynamic, 188
Gas phase ions, 601
Gaseous species
   dissociation energy, 157
   high temperature, 157
Geochemical
   modeling, 730
   processes, 731
Geochemistry, 178
Geoplanetological, 720
Glasses, 183
Global control parameters, 751
Globular molecule, 675
Granulocyte, 534
Grønvold and Westrum scheme, 701, 703, 704
Group
   definition of, 771
   law of additivity, 772
   scheme, 321

ΔH
   calculation of, 256
   calorimetric determination of, 265
   evaluation of, 253
Halobacterium halobium, 562
Halogen-substituted pyridine complexes, 57
Harmonic
   analysis, 78, 82, 87, 99, 103, 111
   oscillator, double degenerate, 792
Heat capacity, 514, 806
   measurements, 32
   of an electron, conventions for estimating, 630
Heat conduction flow calorimetry, 117
Heats
   metal-ligand, 253
Heats of
   atomization, 817
   combustion, 67
   dilution, 806

SUBJECT INDEX

fusion, 806
hydrogenation, 816
ionization of water, 23
reaction, 806
solution, 806
sublimation, 806
transition, 806
vaporization, 806
Heating value, 61
Helix, 521
Heptane-3,5-diones
methyl-substituted, 321
Heterogeneous fuels, 61
Heterolytic bond dissociation
energy, 628
Heteronuclear species, 157
Hexafluoroacetylacetone, 319, 320
enol percentage, 323
enthalpy of formation, 321, 322
enthalpy of vaporization, 323
n-Hexylamine, 430
$Hg(CN)_2$, 265
High pressure, 183
sampling, 188
High temperature, 183, 235
mass spectrometry, 228
Homoleptic alkyls, 384
Homolytic bond, dissociation
enthalpies of metal-
-oxygen, 317, 330
Homonuclear species, 157
Hoogsteen, 560
Hospitals, 821
Hot band structure, 792
Hot corrosion, 184, 209, 235
Hot wire, 130
Hückel, extended, 818
Human neutrophils, 536
Hybridized band, 686
Hydration model, clathrate-like, 418, 429
Hydrocarbons, condensed polynu-
clear aromatic, 680
Hydrogen azides alkali, 685
Hydrogen bonded, 151
Hydrogen bonds, 407, 550
Energy in the enol, 322
internal, 555

intramolecular, 325
Hydrogen difluorides, alkali, 685
Hydrophobic, 286
contributions, 551
cores, 551, 558
effects, 414
enthalpic effect, 411, 429
hydration, 411, 425
interactions, 558
Hydrophilic, 286
Hydroquinone, 151
Hydroxo complexes, 341
Hydroxylic solvents, 406

Immobilized enzyme, 39
Immunoglobulin, 528
Immunology, 528
Impulse response, 122, 124
Incoloy 825, 78
Inertial parameters, 793
Infrared spectroscopy, 789, 792
Inhibited pepsin, 554
Injection techniques, 43
Inner complexes, 317
Inorganic species, 789
Interatomic distances, 331
Intermediate states, 551, 552
Intermetallic compounds, 685
Intermolecular
elimination, 388
forces, 128, 328
Internal
motility, 560
rotation, energy barriers, 789
Intramolecular
elimination, 388
hydrogen bonding, 325, 331
Inverse filtering, 78
Invertion temperature points, 177
Ion
association, 281, 393, 394, 398, 399
chestering, 646
-cyclotron resonance, 603, 618, 819

-molecule equilibrium measurements, 615
molecule reactions, 601, 627
-pair, association constant, 399
-size parameter, 281
solvation, nature of, 646
Ionic crystals, orientation disorder in, 684
Ionic strength, 516
Ionization
  cross section, 138, 163, 198, 201
  potentials, 606
Ionizing electron, 162
Iron disulfides, marcasite/pyrite, 728
$IrX(CO)A_2L$, 58
t-$IrX(CO)(PPh_3)_2$, 59
$IrX(CO)(PPh_3)_2L$, 59
Iso-anionic series, 711
Isobutyrylpivaloylmethane, 319
  enol percentage, 323
  enthalpy of formation, 322
  enthalpy of vaporization, 323
Isomerization
  diagram, 782
  rate, 782
  reactions, 781
Isoperibol calorimetry, 2
Isoteniscope, 128
  Technique, 325, 328, 329
Isothermal calorimetry, 2
  titration, 7
Isotope exchange, 135
Isotopic
  distribution, 161
  splittings, 792
IUPAC, 813

JANAF tables, 816, 817
Joule calibration, 120

logK
  calorimetric determination of, 265

evaluation of, 253
K values
  approximate calculation of, 258
  calculation of, 257
  evaluation of, 260
K_p, 127
KCl, 220
Keto structure, 318
Ketones, 324
  aliphatic, 325
Key substances, 808
Kirchoff equation, 654
Kirchoff's law, 550
Knudsen
  cell, 159, 162
  cell mass spectrometry, 157, 158, 163
  cell method, 325
  cell ion source, 161
  effusion, 158, 194, 202
    cells, 160, 196
    equation, 162
    mass spectrometry, 236
    technique, 329
    torsional, 135
  equation, 161
  number, 230
K_2O, 235
  activities, 239
KOH, 223
KOH(ℓ) vaporization, 239

Lactic acid, 532
Laidler
  scheme, 363
  parameters for estimation of ΔH_f^o, 364
Langmuir effusion, 183
Lantanide
  chloride complexes, 270, 271
  pnictides, 701
  sesqueoxides, 713
  sesquisulfides, 712
  pnictides, 701
Laplace space, 86
Lattice
  dynamics, 657

heat capacity, 731
  contribution, 702
Laws of additivity of
  atomic properties, 771
  bond properties, 771
  group properties, 771
Lead tetraphenyl, 134
Leukocytes, 534
Life sciences, 461
Ligand field theory, 817
Light spot follower, 136
Limiting law, 396
Lindemann's relationship, 705
Lipids, 483
Lipoproteins
  human high density, 562
Liquid
  crystal system, 674
  mixtures, 818
Ln_2O_3

Living systems, 41
Logaritmic relationship, 385
Loop, 808
  analysis, 810
Low temperature
  adiabatic calorimetry cryostat, 676
  calorimetry, 653
  cp values, 816
  entropies, 816
  molecules, 170
Lunar samples, 178
Lysozyme, 547

Mach number, 230
Macrocalorimeters, 31
Macrocyclic effect, 345
  in crown ether complexes, 269
  thermodynamic origin, 269, 345

Macrophages, 532
Magnetic
  circular dichroism, 169
  ordering, 662
  Verwey transition, 727
Manometers, 128
Martensitic type, 103

Mass
  discrimination, 190
  effects, 709
  -loss effusion, 143
  spectrometer, 138
  spectrometric Technique, 159, 174
  spectrometry, high pressure pulsed source, 616
  -to-charge ratio, 161
  transport, 133
Materials transport, 184
Matrix isolation technique, 169
McLeod gauge, 130
Mean
  bond dissociation
    energy, 359
    enthalpy, 359, 360, 368
  free path, 162
  velocity, 161
Mechanism, 780
Medical
  research, 821
  science, 815
Melting, specific enthalpy, 555
Membrane
  proteins, 562
  purple, 562
Mercury diphenyl, 134
Merril-Sanford bands, 178
Metabolic burst, 534
Metal
  acetylacetonates, enthalpies of formation, 317
  bellows, 129
  cyclopendadienyl derivatives, 379
  cyclopentadienyls, mean bond dissociation enthalpies, 368
  β-diketonates, 317, 318
    bond dissociation enthalpies, 326, 331, 332
    combustion of, 326
    enthalpies of formation, 325, 327, 328
    enthalpies of sublimation, 326, 328, 329, 330
  -formyl complexes, 387

-ligand
  bond strengths, 289
  heats, 253
  mean bond dissociation energies, 289
-metal
  bond enthalpies, 362
  multiple bonds, 377
  quadruple bonds, 377
  oxides, 818
-oxygen bond enthalpies in metal-β-diketonates, 317, 331
-polyamine complexes, 339
-to-metal bonds, 390
tropolonates
  enthalpies of formation, 328
Metals, enthalpies of fusion of, 51
Metamorphosed sulfide deposits, 739
pp'-Methane diphenyl isocyanate, 153
Methanol, 407, 696
Methylene group
  increment in vaporization, 325
N-Methyl substituted pyridine complexes, 57
N-Methyl substitution, 434
N-Methylformamide, 433
2-Methylpropanol, 427
Methylurea, 433
Methyltropolone, 320
  enthalpy of formation, 323
MHD, 215
Microbial growth curves, 444
Microcalorimeters, 31
  batch, 37
  batch, titration unit, 37
  Calvet, 11, 12
  Calvet high temperature, 370
  LKB systems, 11, 12
  Picker, 11, 13
  stopped-flow, 41
  Tian-Calvet, 50
Microcalorimetry, 817
  drop, 818
  historical background, 439

Microphase, 118
Mindo, 818
Mineral
  and rock systems, 719
  systems, 719
Minerals, 183
  aluminosilicate, 725
  C-A-S-H system, 722
  chalcogenide, 722
  miscellaneous, 722
  oxide, 722
  phases, 238
  silicate, 722
  transition-element -bearing, 730
Miscellaneous minerals, 722
Mixing
  enthalpies of, 240
  entropies of, 240
Mixtures of water and
  DMA, 423
  DMF, 412, 423, 426, 427
  DMSO, 422
Modeling, geochemical, 730
Molecular
  beam, 183, 186
  donor-acceptor compounds, 682
  evolution, 562
  ions, 231
  mechanics, 818
  orbital theory, 816
  species, gaseous, 168
  symmetry group theory, 793
  spectroscopy, 157
Molecule-ions, 569
Molecules, 569
  almost-symmetrical-top, 677
  planar, 680
Moments of inertia, 792
Mononucleotides, 514
Monoxides, 175
MS_2 compounds, 712
MS_2 symetry group, 797
Multiple rotating-cells, 163
Mycoplasma, 441

NaCl, 192, 220
  vaporization, 192

SUBJECT INDEX

mass spectral fragmentation, 193
Na_2CO_3, 214

Naphthalene, 149, 681
  -pyromellitic dianhydride, 683
  -tetracyanobenzene, 683
  -tetracyanoethylene, 683
Na_2SO_4, 199, 203

Naturally occurring samples, 723
NBS circular, 803, 816
Needs, thermodynamic data, 740
Neopentane, 673
Nd_2O_3, 700

Networks, 805
  Be, 806
  Li thermochemical, 807
  techniques for solving data, 808
Neutrophils, human, 536
NH_4Br, 696

Ni(glycinate)$_2 \cdot 2H_2O$(c), 55

Nonaqueous solvents, 402
Non-polar contacts/residue, 555
Nuclear
  fuels, 177
  oxides, 173
Nucleation, 648
Nucleic acids, 489
Nucleosomes, 480

Octahedral complexes, 298, 300
Optical spectroscopy, 169
Olefines, 586, 588, 778
  protonated, 596
Olefinic ligands, 388
Oligonucleotides, 522
Orbital methods
  atomic, 157
  molecular, 157
Orbital overlap in tetrahedral complexes, 308

Order-disorder transformation, 726
Organic salts, 688
Organomethallic
  Chemistry, 642
  Compounds
    of transition metal, 367
    main group complexes, 353
Organs from animals, 441
Orientational disorder, 662
  in crystals, 672
  in ionic crystals, 684
Orifices, 145
Oxide minerals, 722
Oxides
  nuclear, 173
  refractory, 172
  ternary, 173
Parallel transitions, 798
Partial
  molal heat capacity, 421
  specific heat capacity of protein, 547
  specific volumes of protein, 547
Partion functions, 127
Partition function, 167, 177, 178, 557
Pellets, 62
  kilogram-size, 68
Pentane, 673
Pepsinogen, 553
Pepstatin, 558
Perfusion experiments, 39
Perhydroazepine, 430
Petroleum industry, 815
PF_5, 789, 792

Pharmacy departments, 821
Phase
  diagram calculation, 734
  equilibria, experiments, 734, 736
  transitions, 696
Phonon dispersion, 704
Phorbol-myristate-acetate, 536
Phospholipids bilayers, 547
Photochemical reactions, 37
Photodetachement spectroscopy
  ICR, 607

Laser, 607
Photoelectron spectroscopy, 606
Physiological conditions, 44
Pizoelectric voltage, 112
Pivaloylpropionylmethane, 319
  enol percentage, 323
  enthalpy of formation, 322
  enthalpy of vaporization, 323
Plastic crystal transition, 677, 679
Plastically-cristalline state, 673
Polar compounds, 773
Polarisability of alkyl groups, 637
Polarity, 773
  alternating, 776
Polycyclics, 779
Poly-L-glutamate, 520
Poly-L-lysine, 521
Poly-L-ornithine, 521
Polynuclear aromatic hydrocarbons, condensed, 680
Polynucleotides, 547
Polyoma virus, 504
Polypeptides, 547
Porcine pepsin, 553
Positional disorder, 656
Positive ion current, 138
Potassium, partial pressure, 216
Potential curve, 228
Power measurements in living systems, 41
$PrCl_3$, 709

Precision calorimetry, 817
Pressure multiplier, 128
Probe, 190
Processes
  complex, 781
  molecular, 781
1-Propanol, 399
Properties
  chemical thermodynamics, 804
  thermochemical, 804
  thermophysical, 804
Protein
  concentration, 548
  simples globular, 550
  solutions, 541

  stability, 548, 550
Proton affinity, 628
  absolute, 629
  scales, reference standard, 631
Proton transfer, 628
  structural effects, 637
Protonated complexes, 341
$t-PtX(CH_3)A_2$, 59

$PtX(CH_3)A_2L$, 59

Pulsed trapped ion cell, 605
Pyrene
  pyromellitic dianhydride, 683

  tetracyanobenzene, 683
Pyridine Complexes
  halogen substituted, 293
  methyl substituted, 293
Pyridine derivatives, 592
Pyridine and methylcyanide complexes, 372
N-(2-pyridyl)acetamide complexes, 302
Piromellitic dianhydride
  naphthalene, 683
  pyrene, 683
Pyrrolidine, 430
2-Pyrrolidine complexes, 297

$Q_{c,p}$ values, 255

Quadrupole gas analyser, 149
Quantum-mechanical tunnelling, 657
Quartz-thermometry, 817
Quenching process, dynamic, 565

Raoult's Law, for molecules, 771
Rate
  constant, 780
  law, 780
Reaction

SUBJECT INDEX

abstraction, 781
addition, 781, 784
calorimetric method, 328
catalog, 812
concerted, 781
elementary, 780
fission, 781, 784
isodesmic, 641
isomerization, 781
mechanism of chemical, 780
measurement of rate, 39
metathesis, 784, 785
notation of elementary, 780
redistribution, 389
unimolecular, 783
Real stability, 164
Reduced mass, 793
Refractory
compounds, 165
material, 157
oxides, 172
Relaxation processes, 565
Residual
entropies, 733
moisture, 65
Resonance energies, 316
Resting state, 531
Restricted rotation, 660
Rigid rotor harmonic oscillator approximation, 798
Ring puckering modes, 792
$RMn(CO)_5$, 372

T-RNA, 523
Rock systems, 719
Rotating Bomb Calorimetry, 817
applications, 355
Rotation band contours, 798
Run control parameters, 751

Salicylaldoxime, 52
Salts with organic anions, 688
Samples
naturally occuring, 723
synthesized, 723
Satured entropy, 178
Schottky
contributions, 727
effect, 658
heat-capacity contribution, 733
heat-capacity functions, 699
Second-Law method, 127, 166
Second-order transitions, 696
Sedimentation, 455
Seed germination, 441
Selected ion flow tube, 619
Sequential manner, 808
ΔSi, calorimetric determination, 265
SIFT, 619
Significant structure theory, 177
Silicate, 223, 726
minerals, 722
systems, 235
Single-ion, ΔH_t^o values, 403
SiO_2, 235
Slag, 183, 219, 221
channel, 215
coal, 215, 235
Slightly soluble compounds, solution of, 37
Soda-lime-silica glass, 210
Sodium
acetate, 688
chloride, 393, 397
ethanoate, 688
Sofa conformation, 348
Soil, microbial activity, 446
Solar system, 178
Solid
bodies, 178
phases, 243
waste, 61
Solid-solid
transformation, 103
transition, 109, 110, 122
Solubilities, 806
Solubility, water vapour, 224
Solution
sequential, 808
simultaneous, 808
Solution calorimeters
Commercially available, 10, 11, 12
isothermal-jacketed, 326

operating characteristics, 12
Solution calorimetry, 294
Solvation, 406, 644
   alcohols, 425
   ammides, 425
   ammines, 425
   enthalpy, 405
   parameters, 406
   ureas, 425
Solvent effects, 645
Solvents
   aprotic, 406
   hydroxylic, 406
   methanol, 407
   nonaqueous, 402
   water, 406, 407
Specific heats, 52
Spectroscopic
   data, 167
   entropy, 677
Spectroscopy
   photo dissociation, 819
   photoionization, 819
Spiral, 128
Stability
   chemical, 157, 168, 170
   physical, 157
Stabilization energy, 298
Stacking enthalpy, 475
Standard
   calorimetric entropy, 653
   enthalpies of reaction, 275, 277
Static-bomb
   combustion calorimetry, 367
   for Cd, Al, Si, Ge, Pb compounds, 354
   for P, As, Sb, Bi compounds, 354
   for Sn compounds, 354
   for Zn, Hg, B, Sn compounds, 353
Statistical thermodynamics, 157, 816
Steric interactions, 772
Stimulation, 531
Stochastic processes, 565
Strain, 772
   ring, 771
Structural transitions, 489

Succionitrile, 675
Sublimation
   bulb technique, 329
   enthalpies of, 323, 326, 328
Sulfide deposits, metamorphosed, 739
Superambient temperature adiabatic calorimeter, thermostat, 747
Supercoiling, 503
Superhelicity, 489
Superionic conductors, 655
Surface diffusion, 133
SV 40, 503
Symmetrical ions, 684
Synthesized samples, 723
Systems
   C-A-S-H, 737, 738
   mineral and rock, 719

Tautomeric equilibrium, 318
$Tb(OH)_3$, 708

Techniques, thermochemical, 725
Tensimetric determinations, 158
Test
   experiments, 452
   process, 44
Tetraalkylammonium bromides, 411
   enthalpies of transfer, 412
   enthalpies of solution, 415
Tetraaza-cycloalkanes, 345
Tetracyanobenzeno
   naphtalene, 683
   pyrene, 683
Tetracyanoethylene
   naphtalene, 683
Tetracyanoquinodimethane, 151
Tetracarbides, 176
Tetrahedral complex
   molecular orbital description, 304
   molecular orbital diagram, 305, 307
Tetramethylurea, 433
1,4,8,11-Tetramethyl-1,4,8,11--Tetraazacyclotetradecane, 343

Tetrathiofulvalene, 151
Theophylline, 153
Theoretical
   calculations, 641
   chemistry, 157
Thermistors, 130, 149
Thermochemical
   Bond strengths, 358
   data, 803, 813
   model, 782
   properties, 804
   techniques, 725
Thermochemistry
   cellular, 531
   of β-diketones, 317
   of free radicals, 569
   of metal-β-diketonates, 317
   of molecule-ions, 569
   of molecules, 569
Thermochemistry and kinetics, 769, 780
Thermodynamic
   data, 803
   data needs, 740
   equilibrium, 203, 207, 225
   functions, 127, 789
   measurements, 801
   of crystals, 671
   stability, 157
Thermoelastic process, 103
Thermogenesis, 77, 78
Thermograms, 3, 254, 531
   continuous isoperibol titration, 6
   typical, 3
   typical calorimetric run, 254
Thermokinetic studies, 81
Thermolysin binding to
   1,10-phenanthroline, 27
   Zn^{2+}, 27
Thermophysical
   properties, 804
   techniques, 724
Thermotitrators, Sanda, 11, 13
Thiourea, 265
   complexes, 295
Third Law
   disorder, 726
   entropy, 676, 789, 793
   method, 127, 167

Threshold
   measurements, 609
   photoion photoelectron coincidence, 612
Time constant, 79, 92, 443
Tin tetraphenyl, 134
Tissues, 441
Titration calorimeter, 39
   main components, 3, 4
   scheme for isothermal, 8, 9
Titration calorimetry, 1, 2, 253, 257, 275, 283
   analytical determinations, 2
   applications of, 17
      -to adsorption of compounds, 21
      -to airborne matter, 19
      -to bacterial cell response, 22
      -to binding adenosine diphosphate, 25
      -to binding 1,10-phenanthroline, 27
      -to binding Zn^{2+}, 27
      -to measure heat of ionization of water, 23
      -to redox reactions, 18
   calculation of K, 14
   calculation of ΔH, 14
   determination of
      Fe(II) in acid solution, 18
      ΔG, 2
      ΔH, 2
      nitrite, 20
      reducing agents, 20
      S(IV), 20
      sulfate, 20
   identification of species, 2
   isoperibol, 4
   isothermal data output, 10
   reaction vessel, 5
   Techniques of, 1
   theory of, 1
Torker Technique, 135
Torsion
   pendulum, 149

vacuum microbalances, 134
Torsinal
  constant, 136
  frequences, observed/calculated, 791
  mode, 792
  vibrational modes, B-S bond, 792
TPIPECO, 612
Trans stilbene, 149, 154
Transfer
  RNAs, 498
  function, 79, 80, 98, 120
Transform, 80
Transformations, order-disorder, 726
Transition
  element-bearing minerals, 730
  enthalpy, 475, 560
  first-order, 696
  metal complexes, 56
  non-denaturational conformational, 563
  phase, 696
  phase, schottky contributions, 695, 698
  plastic-crystal, 677, 679
  second-order, 696
  state
    loose, 783
    theory, 780, 782
    tight, 783
  temperature, 560
  λ, 696
Transpiration, 183, 187, 212, 221
  mass spectrometry, 183
  methods, 129
Transport phenomena, 173
Triethylamine, 430
Trifluoroacetylacetone, 319, 320
  enol percentage, 323
  enthalpy of formation, 321, 322
  enthalpy of vaporization, 323

Tri-n-propylamine, 430
Trio-interactions, 569, 571
Triple helix, 499
Tropolone, 320

enthalpy of formation, 323
enthalpy of sublimation, 325

Tummor cells, 450
Tritium gas, 134
Twin-cell assembly, 163
Twin-Knudsen cell source, 164
Two-dimensional solids, 662
Two-state process, 542, 544

Unfolding, heat capacity change, 566
Univalent ion, 406
Uranium intermetallic compounds, 685
Urea, 154, 426, 433
  methylsubstituted-, 435
  solution of, 425

Valente forces, 775
Vaporization calorimetry, 32
Vapor pressure, 127
Viscosity, 132
Vibrating quartz fibre, 129
Vacuum
  evaporation, 154
  microbalance, 145
  sublimation drop Technique, 329
Van't Hoff
  enthalpies, 542, 563
  equation, 166, 545
Vaporization
  by dissociation, 171
  congruent, 206
  glass, 213
  methods, 183
  of carbides, 174
  of oxides, 171
  studies, 157
Wadsö calorimeter, 323
Vapour pressure, 195, 235, 806
  equation, 149
  measurements, 325
  models, 235
Venusian troposphere, 739

Verwey transition, magnetite, 727

Vibration cooling, 229
Vibration contribution, 551
   anharmonicity, 801
Vibration relaxation, 231
Vibrational spectroscopy
   analysis, 798
   assignement, 789, 790, 798
   bands, 798
   energy levels, 789
   isotopic shifts, 792
Viroids, 476, 498
Viscosity gauge, 325
Vitreous system, 674
VF_5, 789, 792
V.L.P.P. kinetics, 819
Volumetric scheme, 705

Water, 406, 407
   content, 402
   heat of ionization, 23
$W(CO)_5(CH_3CN)$, 57

Whole body calorimetry, 441
Wobble pair, 494
Work function, 178

Yeast, 441